# Computer Modeling Applications for Environmental Engineers

## Second Edition

Isam Mohammed Abdel-Magid
Mohammed Isam Mohammed Abdel-Magid

**CRC Press**
Taylor & Francis Group
Boca Raton  London  New York

CRC Press is an imprint of the
Taylor & Francis Group, an **informa** business

CRC Press
Taylor & Francis Group
6000 Broken Sound Parkway NW, Suite 300
Boca Raton, FL 33487-2742

First issued in paperback 2019

ISBN-13: 978-1-4987-7654-7 (hbk)
ISBN-13: 978-0-367-88966-1 (pbk)

---

**Library of Congress Cataloging-in-Publication Data**

---

Names: ʿAbd al-Majīd, ʿIṣām Muḥammad, author. | Abdel-Magid, Mohammed
Isam Mohammed, author.
Title: Computer modeling applications for environmental engineers / Isam
Mohammed Abdel-Magid Ahmed and Mohammed Isam Mohammed Abdel-Magid.
Other titles: Modeling methods for environmental engineers.
Description: Second edition. | Boca Raton : Taylor & Francis, CRC Press,
2017. | Revised edition of: Modeling methods for environmental engineers /
Isam Mohammed Abdel-Magid, Abdel-Wahid Hago Mohammed, Donald R. Rowe.
1997. | Includes bibliographical references.
Identifiers: LCCN 2016054370 | ISBN 9781498776547 (print : alk. paper)
Subjects: LCSH: Pollution control equipment--Design and construction--Data
processing. | Pollution control equipment--Computer simulation. | Sanitary
engineering--Data processing. | Environmental management--Data processing.
| Sanitary engineering--Computer simulation.
Classification: LCC TD192 .A25 2017 | DDC 628.0285--dc23
LC record available at https://lccn.loc.gov/2016054370

---

**Visit the Taylor & Francis Web site at**
**http://www.taylorandfrancis.com**

**and the CRC Press Web site at**
**http://www.crcpress.com**

# Computer Modeling Applications
# for Environmental Engineers

# Dedication

---

**Dedications to the Second Edition**

*To their souls, Mother Umsalama Al-Tahir Ahmed, Father Mohammed Abdel-Magid Ahmed, and Wife Eng. Layla Salih Mahmoud*

*I. M. Abdel-Magid*

*To mom, till we meet again*

*M. I. M. Abdel-Magid*

**Dedications to the First Edition**

*To all my former teachers and professors*

*I. M. Abdel-Magid*

*To all knowledge seekers*

*A. W. Hago*

*To Amjad and Nasra*

*D. R. Rowe*

# Contents

# List of Figures

# List of Tables

# List of Computer Programs

# Appendix Contents

# Preface

Books and text material in the field of environmental, public health, and sanitary engineering are numerous for addressing the principles, fundamentals, methods, models, concepts, designs, remedial activities, and science-related ideas. Nonetheless, rare are the books that deal with practical problems, problem solving in environmental engineering, and fundamental utilization of computer programming perceptions and applications.

The broad objectives of this book are to facilitate teaching, improve learning, shoulder training, and assist research in the application of computer programming based on well-established models, equations, formulae, and procedures used in the environmental engineering domain. Visual Basic .NET was selected and used as the programming language of choice throughout this book for its simplicity, comprehensive usage, and easiness. It is a language that was traditionally considered a model language of teaching purposes, not to undermine its wide range of applicability; it has been introduced into scripting, application programming (VBA), databases (VB and SQL), and web scripting (VBScript), to mention but a few uses.

Section I of first edition has been rewritten from scratch to reflect the advancements in computer modeling and software programming that took place over the past two decades. The first edition was written in Quick BASIC 4.5, a widely used and popular model programming language of the time. As BASIC faded away and gave the stage to its descendants Visual Basic and Visual Basic .NET, programming models and methods have changed vastly, new programming concepts emerged, and graphical user interface (GUI) systems took over command-line environments. Object-oriented programming (OOP) became the *de facto* programming model, and a whole set of new programming languages has erupted to exploit these concepts.

In addition, severe competition emerged between *Closed Source* and *Open Source* camps. Companies fighting to protect their royalties and copyrights by defending the *Closed Source* regime and closing arms with the emerging and ever-growing *Open Source* movement, which has novel concepts such as the freedom of source code and software *copyleft*ing (a pun on the word *copyright* to reflect the exact opposite meaning, that is, freedom of information). These concepts are changing the face of contemporary programming, more details on which can be found in Section I of the first edition.

Lastly, there are more operating systems (OSs) in active use today than when the first edition of the book was published. Microsoft™ and Apple® were once the only horses in the OS arena (not to undermine UNIX, but it was majorly a teaching OS used by universities and big institutions, not by regular users on their home portable computers); nowadays, other systems such as Linux, Solaris, Android, and iOS are actively contributing to the OS industry. As such, programmers are being faced with the problems of *system*

*compatibility* (ensuring software runs on different OSs) and *architecture compatibility* (ensuring software runs on different platforms such as x86 and x86_64).

In the face of all these intermingled points, the authors were faced with the question of which programming language to use in writing the examples of this text:

1. Using a cross-platform language such as Java would seem logical. This ensures all the examples will work on the widest possible range of target machines without recompilation. In fact, this would have been the best choice had the intent been commercial software production, not teaching. Java is an OOP language by nature; and the performance hit of using the intermediate Java Virtual Machine (JVM) is negligible in most cases. This is all good, but using Java means the need to discuss OOP in depth, having a few sections on the JVM, the garbage collector, Java packages and built-in classes, and so on. This will shift the focus of the text from modeling computer applications for solving environmental engineering problems to a text on Java and OOP.

2. Using a simple procedural language such as C is perfect in a teaching setting like this, as it will remove all the headache of OOP, virtual machines, intermediate compilers, garbage collectors, and other things that are not primal to the teaching process. The reader will be able to focus on the specific modeling techniques under discussion. In addition, pure C is text based by nature, which is an important feature when programming console programs (such as Linux terminal programs or MS-DOS prompt-like programs). However, unfortunately, console programs have fallen out of favor to GUI programs, and most users (except power users and system administrators) feel uncomfortable when faced with the black screen and its blinking cursor. As such, using C alone is defective. One would usually combine C with a GUI toolkit; the most popular combination is using C++ with Qt (pronounced "cute"; stands for the Q toolkit). However, again, this will make it obligatory to include a whole chapter on Qt, its history, structure, object classes, and its uses, not to mention the mandatory chapter on core C programming. This will undoubtedly unnecessarily inflate the book out of scope.

3. Use of a traditional teaching language such as Pascal or Basic. This ensures that the focus of the text will remain on problem solving and application modeling. The chapter on programming itself will be supplementary, and habituated programmers can skip it and start directly on the text proper. This is the solution we opted to use throughout this text.

That being said, there are important points and topics pertaining to computer programming in general, and to modeling engineering applications specifically, which need to be discussed, or to the least, pointed to. These topics have been handled in Section I of the first edition.

The more specified objectives are proposed to accomplish the main learning outcomes for students enrolled in environmental engineering courses, which are as follows:

1. To help the reader in the design of treatment and disposal facilities
2. To introduce the basic concepts of computer design related to environmental engineering
3. To aid in teaching of environmental engineering unit operations and processes design
4. To improve the understanding by students and researchers in environmental engineering field
5. To improve the understanding of relevant physical, chemical, and biological processes, and their mutual relationships within various sanitation components
6. To promote environmental protection, improvement, and sustainability, enhancing the overall quality of life
7. To demonstrate effective problem-solving abilities
8. To utilize the current technology of waste collection, handling, treatment, final disposal, and management
9. To apply appropriate regulations and guidelines in safeguarding environmental conditions
10. To be able to design and apply necessary procedures and precautions to establish sound environmental engineering systems

Recently, the Department of Environmental Engineering of the College of Engineering at the Imam Abdulrahman Bin Faisal University, Kingdom of Saudi Arabia (KSA), sought the Accreditation Board for Engineering and Technology Accreditation Commission (http://www.abet.org/). As such, student outcomes were adhered to as per requirements. The program must have documented student outcomes that prepare graduates to attain the program educational objectives. Student outcomes are outcomes (a) through (k) plus any additional outcomes that may be articulated by the program are as follows:

a. An ability to apply the knowledge of mathematics, science, and engineering
b. An ability to design and conduct experiments, as well as to analyze and interpret data
c. An ability to design a system, component, or process to meet the desired needs within realistic constraints such as economic, environmental, social, political, ethical, health and safety, manufacturability, and sustainability

d. An ability to function on multidisciplinary teams
e. An ability to identify, formulate, and solve engineering problems
f. An understanding of professional and ethical responsibility
g. An ability to communicate effectively
h. The broad education necessary to understand the impact of engineering solutions in a global, economic, environmental, and societal context
i. A recognition of the need for, and an ability to engage in, lifelong learning
j. A knowledge of contemporary issues
k. An ability to use the techniques, skills, and modern engineering tools necessary for engineering practice

*Computer-Aided Modeling in Environmental Engineering* is chiefly intended as a supplement and a complementary guide to basic principles of environmental engineering. Nonetheless, it can be sourced as a stand-alone problem-solving text in the field. The book targets university students taking first-degree courses in environmental, construction, civil, mechanical, biomedical, and chemical engineering or related fields. The manuscript is estimated to have valuable benefits to postgraduate students and professional sanitary and environmental engineers. Equally, it is anticipated that the book will stimulate problem-solving learning and accelerate self-teaching. By writing such a script, it is hoped that the included worked examples, set problems, and compiled computer programs will warrant that the booklet is a treasured support to student-centered learning. To accomplish such goals, great attention was paid to offer solutions to selected problems in a well-defined, clear, and discrete layout exercising step-by-step procedures and explanation of the related solution employing crucial procedures, methods, approaches, equations, data, figures, and calculations.

Chapter 6 has been added to the format of the book to address solid waste engineering and management. The first part deals with municipal solid waste classification, quantities, and properties stressing on introductory items, and physical and chemical properties of solid waste. The second part deals with solid waste collection, processing, material separation, and financing facilities. Treatment methods tackle incorporated sanitary landfill, biochemical processes, and combustion and energy recovery. This chapter highlights financing solid waste facilities. Likewise, this part of the book incorporates homework problems shouldering both discussion and special mathematical types of practical questions to be encountered. Computer programs were written for every stage discussed in this chapter.

The authors hope this work will contribute toward the achievement of United Nations Sustainable Development

Goals (UN SDGs[*]) numbers 1 (no poverty), 3 (good health and decent work and economic growth), 4 (quality education), 6 (clean water and sanitation), 8 (decent work and economic growth), 9 (industry, innovation, and infrastructure), 11 (make cities inclusive, safe, resilient, and sustainable, i.e., sustainable cities and communities), 12 (Reasonable consumption and production), 13 (climate action), 14 (life under water), 15 (life on land), and 17 (partnership for the goals).

Users of this book will be able to access the contents of Appendix 9 (*A User's Manual and Programs Screenshots*), the source codes and executables of all example programs, along with the solutions manual on the downloads tab of the book's CRC Press site: https://www.crcpress.com/Computer-Modeling-Applications-for-Environmental-Engineers-Second-Edition/Ahmed-Abdel-Magid-Abdel-Magid/p/book/9781498776547.

**Prof. Dr. CEng. Isam Mohammed Abdel-Magid Sr.**

**Dr. Mohammed Isam Mohammed Abdel-Magid Jr.**
*Dammam, Doha, 2017*

---

[*] UN SDGs include (1) end poverty in all its forms everywhere, that is, no poverty; (2) end hunger, achieve food security and improved nutrition, and promote sustainable agriculture, that is, zero hunger; (3) ensure healthy lives and promote well-being for all at all ages, that is, good health and welfare; (4) ensure inclusive and quality education for all and promote lifelong learning, that is, quality education; (5) gender equality; (6) ensure access to water and sanitation for all, that is, clean water and sanitation; (7) affordable and clean energy; (8) promote inclusive and sustainable economic growth, employment, and decent work for all, that is, decent work and economic growth; (9) industry, innovation, and infrastructure; (10) reduced inequalities; (11) make cities inclusive, safe, resilient, and sustainable, that is, sustainable cities and communities; (12) responsible consumption and production; (13) climate action; (14) life below water; (15) life on land; (16) peace, justice, and strong institutions; (17) partnership for the goals. (Source: United Nations, 2015, http://www.un.org/sustainabledevelopment/sustainable-development-goals/#.)

# Preface and Acknowledgment to First Edition

Environmental engineering is the branch of engineering dealing with activities in the areas of air, water, wastewater, solid waste, radiation, and hazardous materials. Many textbooks have been published in all of these areas; however, most of them lack the fundamental utilization and application of computer programming. Computer programming is now of paramount importance not only for the students but also for the practicing engineers. This book presents computer programs in the air, water, wastewater, and solid waste fields that can be used to evaluate and assist in the design of environmental control systems. This book will be of value to environmental engineers, sanitary engineers, civil engineers, town planners, industrial engineers, agricultural engineers, treatment plant designers, consultation firms, and researchers. The BASIC programming language was selected for its simplicity and used throughout this book.

The broad objectives of this book are to facilitate improving teaching, learning, and research in the application of computer programming based on well-established models, equations, formulae, and procedures used in the environmental engineering field.

The more specified objectives are as follows:

1. To help the reader in the design of treatment and disposal facilities
2. To introduce the basic concepts of computer design
3. To aid in teaching of environmental engineering unit operations and processes design
4. To improve the understanding by students and researchers in the environmental engineering field

This book contains many illustrative figures, schematic diagrams, solved practical examples, design problems, and tutorial homework problems. A wide range and spectrum of environmental engineering subjects are covered. Many computer programs are included, and they cover both the environmental design principles and the structural design parameters using different materials. This book is written in a clear, concise manner in order to enhance understandability. SI units are used throughout this book.

The format of the book is divided mainly into four sections. Section I deals with an introductory material that contains functions and layout of the computer, programming and modeling concepts, and the basic outlines for the BASIC language, which is discussed in Chapter 1.

Section II covers water sources and resources, and is divided into three chapters. Chapter 2 deals with the characteristics of water and wastewater, supported by mathematical equations in order to help programming, water resources and sources, and groundwater flow. Emphasis is laid on water abstraction from wells with the consequent drawdown relationships. Chapter 3 discusses water treatment unit operations/

processes and water storage and distribution. It outlines the design concepts adopted in treatment plant unit operations (e.g., screens, sedimentation and flotation, and aeration) and treatment unit processes (e.g., coagulation and flocculation, filtration, desalinization, and water stabilization). Each treatment process is briefly introduced and emphasis is given to the design of the units using appropriate equations, assumptions, and design tools. Chapter 4 deals with water storage and distribution. It highlights the design of water storage tanks (communa and, dwelling) and the methods used for both urban and rural water distribution.

Section III deals with wastewater engineering, measurement, collection, treatment units, and disposal. It is divided into two main chapters. Chapter 5 focuses on wastewater collection systems for rural areas and urban fringes. Sewerage system design is introduced within the context of this chapter. Chapter 5 also discusses the selected methods for wastewater disposal for rural Inhabitants. Chapter 6 is introduced to aid the design of wastewater treatment and disposal units. More stress is placed on in-depth design considerations and parameters for selected systems universally adopted for both rural and urban regions. This chapter emphasizes preliminary, primary, tertiary, and sludge treatment methods and the disposal of treated wastewater to rivers, lakes, or oceans.

Section IV deals with air pollution control technology. It presents a brief overview of the air pollution field followed by the fundamental concepts that are necessary for calculations dealing with air pollution control. It also includes computer programs based on mathematical equations and models related to these fundamental concepts. The most commonly used air pollution control devices for gaseous and particulate pollutants are presented (e.g., absorption, adsorption, and combustion for gaseous contaminants and settling chambers, cyclones, electrostatic precipitators, venturi scrubbers, and baghouse filters for controlling particulate emissions). Each control device or technique is accompanied by a computer program that can aid in the design or evaluation of air pollution control equipment. This section concludes with mathematical models that can be used for the determination of effective stack heights (plume rise) as well as dispersion models that can help estimate the concentrations of air pollutants dispersed in the atmosphere. Computer programs for both stack height calculations and dispersion models are included. All computer programs in the book are included on diskettes.

The authors gratefully acknowledge the support, patience, encouragement, and understanding of their families while they were engaged in the enjoyable task of writing this book.

CRC Press/Lewis editors are acknowledged for their forbearance and patience during all stages of the manuscript preparation, review, editing, and publishing. Particular thanks and appreciation go to Andrea Demby, Jan Boyle, and Victoria Held for their contributions and outstanding efforts to bring this book into existence.

The authors are indebted to a number of institutions, organizations, and individuals for permission to reproduce and draw published material, and these are acknowledged in the references. Thanks are also extended to the reviewers of the manuscript for their helpful comments and numerous constructive suggestions.

The authors extend their thanks to all who contributed in different ways toward the production of this book. In the air pollution control section, special acknowledgment goes to Asa Raymond for typing, reviewing, and developing the computer programs. Also a special thanks goes to Gary Whittle for his significant contribution in locating pertinent literature in the air pollution control field. Special thanks are also extended to Majid Soud Rashid Al-Khanjri and Taiseer Marhoon Al-Riyami for redrawing and tracing Figures 7.16, 7.17, 7.32, and 7.33.

The authors trust that the book will serve the needs of the profession and provide help and stimulation for candidates to develop their own computer models. They invite and welcome any comments, suggestions, improvements, and constructive criticism from users of the text. Although diligent efforts have been made by the authors, reviewers, editors, and publisher to produce an error-free book, nonetheless, this may not have been achieved and some errors may have escaped the scrutiny of all. The authors would appreciate if any omissions or errors found be brought to their attention so that corrections can be made, which will enhance the usefulness of this publication.

**Isam Mohammed Abdel-Magid**

**Abdel-Wahid Hago Mohammed**

**Donald R. Rowe**

ISBN 1-56670-172-4
© 1997 by CRC Press, Inc.

# Authors

**Isam Mohammed Abdel-Magid**, BSc, PDH, DSE, PhD, FSES, CSEC, MSECS, is the professor of water resources and environmental engineering. He received his BSc in civil engineering from the University of Khartoum, Khartoum, Sudan; diploma in hydrology from the University of Padova, Padova, Italy; MSc in sanitary engineering from Delft University of Technology, Delft, the Netherlands; and PhD in public health engineering from the University of Strathclyde, Glasgow, in 1977, 1978, 1979, and 1982, respectively.

Professor Abdel-Magid authored or coauthored many papers, publications, scientific text and reference books, technical reports, and lecture notes in the areas of water supply; wastewater disposal, reuse, and reclamation; solid waste disposal; water resources and industrial wastes; and slow sand filtration in both English and Arabic. He has participated in several workshops, symposia, seminars, and conferences. He has edited and coedited many conference proceedings and college bulletins.

The Sudan Engineering Society awarded him the prize for the Best Project in Civil Engineering in 1977. In 1986, the Khartoum University Press awarded him Honourly Scarf for Enrichment of Knowledge. Ministry of Irrigation (MoI) awarded him the prize for the Second Best Performance in fourth-year civil engineering. The Sudanese Press Council of Ministry of Information and Culture (MoIC) awarded him the best book of the year. ALECSO awarded him the prize for a book in engineering.

Professor Abdel-Magid has taught environmental engineering subjects and supervised research projects for graduate and postgraduate students at the University of Khartoum, United Arab Emirates University (Al Ain, the United Arab Emirates), Sultan Qaboos University (Muscat, Oman), Sudan University for Science and Technology (Khartoum, Sudan), Juba University (Juba, Sudan), Industrial Research and Consultancy Centre (Khartoum, Sudan), Sudan Academy for Sciences (Khartoum, Sudan), King Faisal University (Dammam, KSA), University of Dammam and Imam Abdulrahman Bin Faisal University (Dammam, KSA). Professor Abdel-Magid acted as an external examiner for different institutions and a founder of some institutions, centers, and refereed journals.

The current address of Prof. Abdel-Magid is as follows: Chair Development and Training Unit of Postgraduate Studies Deanship, Head Proofreading and Revision Department of the Center of Scientific Publications and member of its Council, and Professor of Environmental Engineering and Water Resources at Environmental Engineering Department of the College of Engineering, Imam Abdulrahman Bin Faisal University, Box 1982, Dammam 31451, KSA. Fax: +96638584331, phone: +966530310018, +96633331686, e-mail: iahmed@uod.edu.sa and isam.abdelmagid@gmail.com. Web sites: http://www/sites.google.com/site/isamabdelmagid/; http://www.isam-abdelmagid.net/. Twitter: twitter.com/IsamAbdelmagid. Facebook: https://www.facebook.com/isam.m.abdelmagid. Researchgate: https://www.researchgate.net/profile/Isam_Abdel-Magid. Google scholar: https://www.facebook.com/isam.m.abdelmagid. LinkedIn: https://www.linkedin.com/nhome/?trk=Isam%20Abdel-Magid. Amazon: https://author-central.amazon.com/author/isamabdelmagid.

**Mohammed Isam Mohammed Abdel-Magid,** MBBS, BLS, ALS, MRCP(UK), PgDip in Diabetes, is a graduate of the College of Medicine, the University of Khartoum, Khartoum, Sudan, in 2008. He completed basic training with the Ministry of Health, Khartoum, Sudan, then worked as a physician in the department of internal medicine, Ribat University Hospital, Khartoum, Sudan, and the Ministry of Health, Sultanate of Oman. He is currently working with the Health Assistance Medical Services, Doha, Qatar.

He completed his higher training with the membership of the Royal Colleges of Physicians of the United Kingdom and accomplished postgraduate diploma in diabetes from the University of South Wales, Wales.

He tutored in problem-based learning teaching sessions in the department of Internal Medicine, Sudan International University, Sudan.

He is a registered practicing physician with the Sudan Medical Council, the Health Authority of Abu-Dhabi, and the Saudi Commission of Health Specialties. He is a full member of the Society for Acute Medicine (Edinburgh, UK), the European Society for Emergency Medicine (Brussels, Belgium), and the European Respiratory Society (H.Q. Lausanne, Switzerland).

He is a peer reviewer of the *Science Journal of Medicine & Clinical Trial* and the *Pan-African Journal of Medical Sciences.*

He is a qualified Linux system administrator, as he obtained the Linux Foundation Certified SysAdmin qualification in 2015. He is an experienced computer programmer with more than 15 years of programming, with a special focus on C/C++ and Visual Basic. He is the maintainer of several Fedora GNU/Linux packages and is an active contributor to the GNU Project (with his GnuDOS and Fontopia packages). He is a member of the Free Software Foundation (Boston, MA). He has designed and packaged several fonts, all of which can be accessed and downloaded from the Open Font Library (http://fontlibrary.org) or from the author's Web site (http://sites.google.com/site/mohammedisam2000).

# Authors of First Edition

**Isam Mohammed Abdel-Magid Ahmed** received his BSc degree in civil engineering from the University of Khartoum, Khartoum Sudan; diploma in hydrology from the University of Padova, Padova, Italy; MSc in sanitary engineering from Delft University of Technology, Delft, the Netherlands; and PhD in public health engineering from the University of Strathclyde, Glasgow, U.K. in 1977, 1978, 1979, and 1982, respectively.

Abdel-Magid has authored more than 28 publications, 10 scientific textbooks, and numerous technical reports, and lecture notes in the areas dealing with water supply and resources; wastewater treatment, disposal, reclamation, and reuse; industrial wastes; and slow sand filtration, in both English and Arabic. He has participated in several workshops, symposia, seminars, and conferences. He has edited and coedited many conference proceedings and college bulletins.

The Sudan Engineering Society awarded him the prize for the Best Project in civil engineering in 1977. In 1986, the Khartoum University Press awarded him Honourly Scarf and Badge for Enrichment of Knowledge.

Abdel-Magid has taught at the University of Khartoum; United Arab Emirates University, Al Ain, Abu Dhabi, the United Arab Emirates; Sultan Qaboos University, Muscat, Oman; and Omdurman Islamic University, Omdurman, Sudan.

**Abdel-Wahid Hago Mohammed** received his BSc in civil engineering from the University of Khartoum, Khartoum, Sudan in 1976. He received his PhD degree from Glasgow University, Scotland in 1982. Hago has authored more than 13 publications, numerous reports, lecture notes, and more than 95 translated articles. He headed the Design Office of Newtech Industrial and Engineering Group, Sudan for 3 years, where he contributed in the design of buildings and factories in different parts of that country. In 1986, he acted as a Senior Analyst for Associated Consultants and Partners, Khartoum, Sudan for the design of the ElObied town water distribution system.

Hago has participated in several conferences, and was a member of the organizing committee of the First National Conference on Technology of Buildings organized by the Sudan Engineering Society in 1984. Hago is a member of the Sudan Engineering Society. He has taught both undergraduate and postgraduate courses at the University of Khartoum and Sultan Qaboos University, Muscat, Oman. He has supervised and cosupervised many MSc and PhD students.

**Donald R. Rowe**, PhD* is president of D. R. Rowe Engineering Services, Inc., Bowling Green, Kentucky.

He received his MS (1962) and PhD (1965) from the University of Texas at Austin, and his BSc degree (1948) in civil engineering from the University of Saskatchewan, Saskatoon, Canada.

From 1964 to 1969 he was associate professor, Department of Civil Engineering, Tulane University, New Orleans, Louisiana; from 1969 to 1982, he was associate professor, Department of Engineering Technology, Western Kentucky University, Bowling Green, Kentucky; and from 1982 to 1988 he was professor, Department of Civil Engineering, King Saud University, Saudi Arabia. From 1971 to 1993, he was vice president of the Larox Research Corporation, Kentucky and from 1990 to the present, he has been president of his own company.

Rowe has authored or coauthored more than 70 reports and publications in the areas of air pollution, solid waste, water treatment, and wastewater reclamation and reuse. He also coauthored *Environmental Engineering* (McGraw-Hill Publishers) and the *Handbook of Wastewater Reclamation and Reuse* (CRC/Lewis Publishers).

Rowe holds, with coinventors, more than 10 patents on catalytic conversion removal processes of air contaminants in the United States, Canada, Great Britain, and Japan.

In 1980, he was recipient of a Fulbright–Hays Award to Ege University in Izmir, Turkey. Rowe also carried our research on wastewater reclamation and reuse at King Saud University, Riyadh, Saudi Arabia.

From 1984 to the present, he has served on the Peer Review Evaluation Committee for research projects funded by King Abdulaziz City for Science and Technology (KACST), Riyadh, Saudi Arabia.

---

* Professor Donald Richard Rowe, 82, died on May 21, 2009, at his residence. The Late Donald R. Rowe owned and operated D. R. Rowe Engineering Services, Inc., Bowling Green, Kentucky, until his passing. This Vantage, Saskatchewan, Canada native was born in October 12, 1926. He moved from Saskatchewan to Austin, Texas, in the 1950s and married Nora Alice Crittenden in 1954. His passions in life were traveling and his farm situated on Cape Breton Island in Nova Scotia. He was son of the late William Hazen Rowe and Donalda McIntyre.

# List of Symbols and Abbreviations

| Symbol | Description | Unit |
|---|---|---|
| $A$ | Area | $m^2$ |
| AI | Aggressiveness index | |
| Alk | Total alkalinity | mg/L $CaCO_3$ |
| $A_i$ | Area of heat exchanger number $i$ | $m^2$ |
| $A_n$ | Area of an individual unit | $m^2$ |
| $A_t$ | Total area | $m^2$ |
| $A$ | Constant, coefficient, factor, fraction | |
| $a'$ | Effective sulfide flux coefficient | m/h |
| $a_a$ | Percentage open area for a clogged screen | % |
| $a_c$ | Cell constant | |
| $a_o$ | Percentage open area for a clean screen | % |
| | | |
| $B$ | Width, thickness | m |
| BOD | Biochemical oxygen demand | mg/L |
| $BOD_e$ | Effective biochemical oxygen demand | mg/L |
| $BOD_{load}$ | BOD load | kg/$m^3$ |
| $BOD_s$ | BOD of standard sewage | gBOD$_s$/c/day |
| $BOD_t$ | BOD that has been exerted in time interval $t$ | mg/L |
| $BOD_5$ | 5-Day BOD of wastewater | mg/L |
| $b'$ | Constant | |
| $b$ | Slope of the straight line of $t/V$ versus $V$ | s/$m^6$ |
| | | |
| $C$ | Concentration | mg/L |
| $C$ | Idle resistance factor | Dimensionless |
| $C$ | Efficiency factor | |
| $C$ | Total amount percolating within the top layer of soil | mm/year |
| $C$ | Capacity | tons/h |
| CA | Sludge age | day |
| Cd | Constant or shape factor | Dimensionless |
| CON | Conductance | Ohm$^{-1}$ |
| $C'$ | Solubility of oxygen at barometric pressure $P$ and given temperature | mg/L |
| $C_D$ | Newton's drag coefficient | Dimensionless |
| $C_e$ | Concentration in effluent | mg/L, g/$m^3$ |
| $C_f$ | Friction coefficient, coefficient of DeChezy | $m^{0.5}$/s |
| $C_m$ | Concentration of the pollutant in mixture of river water and wastewater discharge, or concentration of pollutant in the river downstream point of discharge | mg/L |
| $C_o$ | Concentration at time $t = 0$ (initial) | mg/L |
| $C_p$ | Carbon concentration in the mixture prior to composting, as the percentage of total wet mass of mixture | % |
| $C_s$ | Saturation concentration | mg/L |
| $C_s$ | Carbon concentration in the sludge, as the percentage of total wet sludge mass | % |
| $C_r$ | Concentration of the same pollutant in river upstream discharge point | mg/L |
| $C_r$ | Carbon concentration in the refuse, as the percentage of total wet refuse mass | % |

| Symbol | Description | Unit |
|---|---|---|
| $C_v$ | Volumetric concentration of particles (volume of particles divided by the total volume of the suspension) | Dimensionless |
| $C_w$ | Concentration of pollutant in wastewater | kg/$m^3$ |
| $C$ | Coefficient, constant | |
| $C$ | Velocity of the sound in a given medium | m/s |
| ch | Chlorinity | g/kg |
| $c_g$ | Gas concentration in gas phase | g/$m^3$ |
| | | |
| $D$ | Cumulative passing weight fraction | % |
| $D$ | Diameter | m |
| $D$ | Disintegrating organic material | |
| $D$ | Cylinder diameter | m |
| $D_{60}$ | Particle (sieve) size where 60% of the particles are smaller than that size | |
| $D_{10}$ | Particle (sieve) size where 10% of the particles are smaller than that size | |
| Diff | Molecular diffusion coefficient | $m^3$/s |
| $D_m$ | Mean hydraulic depth | ft |
| $DO_c$ | Critical oxygen deficit | mg/L |
| $D_I$ | Dielectric constant | |
| $DO_o$ | Initial Oxygen deficit at the point of waste discharge, at time $t = 0$ | mg/L |
| $DO_t$ | Oxygen deficit at time $t$ | mg/L |
| DR | Dilution rate | Dimensionless |
| DWF | Dry weather flow | L/day |
| $d$ | Depth, diameter | m |
| $d$ | Particle size | |
| $d'$ and $n$ | Characteristic of the material studied | |
| $d_o$ | Maximum particle size | |
| $dP$ | Differential change in pressure | Pa |
| du/dy | Velocity gradient (rate of angular deformation, rate of shear) | s$^{-1}$ |
| $dV$ | Differential change in volume | $m^3$ |
| $d_{max}$ | Maximum depth | m |
| $d_p$ | The distance between blades or the scroll pitch | m |
| | | |
| $E$ | Coefficient of Eddy diffusion, or turbulent mixing | $m^2$/s |
| $E$ | Evapotranspiration | mm/year |
| EC | Electrical conductivity | Ohm/m |
| Eff | Efficiency | % |
| EM | Electrophoretic mobility | m/s/V/m |
| $E_b$ | Bulk modulus | N/$m^2$ |
| $E_v$ | Rate of evaporation | L/day |
| $E'$ | Activation energy | calories |
| $E_e$ | Electrochemical equivalent | g/Coulumb |
| $e_e$ | Expanded bed porosity | Dimensionless |
| $e$ | Porosity | Dimensionless |

*(Continued)*

| Symbol | Description | Unit |
|---|---|---|
| $F$ | Collection frequency, number of collections per week | |
| $F$ | Speed factor | Dimensionless |
| $F$ | Volume of reduction | |
| Far | Faraday's constant | Ampere*h |
| F/M | Food-to-microorganisms ratio | day$^{-1}$ |
| Fr | Froude number | Dimensionless |
| $F_t$ | Shear force | N |
| $F_1$, $F_2$ | Recirculation factor | Dimensionless |
| $f$ | Factor, function, coefficient | |
| $f$ | Frequency | cycle/s (Hertz) |
| $G$ | Food waste, percentage by weight of total MSW, on a dry basis | % |
| $G$ | Velocity gradient | s$^{-1}$ |
| $g$ | Gravitational acceleration | m/s$^2$ |
| $H$ | Lift | ft |
| $H$ | Total energy (total head, energy head) | m |
| Hard | Hardness | mg equivalent CaCO$_3$/L |
| HCa | Calcium hardness | mg/L CaCO$_3$ |
| HE$_i$ | Heat exchanged in effect number $i$ | J |
| H$_o$ | Initial resistance for a clean screen | m |
| H$_s$ | Resistance for a clogged screen | m of water |
| [H$^+$] | Concentration of hydrogen ions | mol/L |
| $h$ | Height, depth, thickness | m |
| $h_f$ | Friction head | m |
| $h_l$ | Head loss | m |
| $h_t$ | Height of proposed tank | m |
| $I$ | Mean rainfall intensity for a duration equal to the time of concentration | mm/h |
| $I_r$ | Average infiltration into the sewer owing to poor joints or pervious material | L/day |
| $i$ | Annual interest rate | fraction |
| $i$ | Current | A |
| $i$ | Current density | A/cm$^2$ |
| $j$ | Slope, gradient | m/m |
| $K$ | Constant, factor, coefficient | |
| $K$ | Coefficient of permeability | |
| $K$ | Hydraulic conductivity of the drainage layer | ft/day |
| $K$ | Intrinsic permeability or specific permeability | |
| $K_b$ | Bunsen absorption coefficient | g/J |
| $K_D$ | Distribution coefficient | |
| $K_s$ | Half-velocity constant | mg/L |
| $k$ | Coefficient, constant | |
| $k$ | Constant | Dimensionless |
| $k$ | Exponent | Dimensionless |
| $k$ | Landfill gas emission constant | Time$^{-1}$ |
| $k_H$ | Henry's constant | g/m$^3$*Pa (g/J) |
| $k_l$ | Loss coefficient | |
| $k_n$ | Kinetic coefficient | Dimensionless |

| Symbol | Description | Unit |
|---|---|---|
| $k_p$ | Removal rate constant for waste stabilization pond | day$^{-1}$ |
| $k_t$ | Decay constant for the particular decay reaction | day$^{-1}$ |
| $k'''$ | Reaeration constant | day$^{-1}$ |
| $k'$ | First-order reaction rate constant | day$^{-1}$ |
| $k_1$ | Bacterial die-away rate | day$^{-1}$ |
| $L$ | Length of conveyor belt | ft |
| $L$ | Length of cylinder | m |
| $L_b$ | BOD$_5^{20}$ of benthal deposit | g/kg volatile matter |
| $L_e$ | Effluent BOD | mg/L |
| $L_i$ | Influent BOD | mg/L |
| $L_m$ | Maximum daily benthal oxygen demand | g/m$^2$ |
| $L_o$ | Methane generation potential | volume/mass of waste |
| $L_o$ | BOD remaining at time $t = 0$ (total or ultimate first stage BOD initially present) | mg/L |
| $L_t$ | Amount of first-stage BOD remaining in the sample at time $t$ | mg/L |
| $L_{EP},d$ | Daily personal noise exposure level | dB |
| $L_{EP},w$ | Weekly average of the daily personal noise exposure | dB(A) |
| $(L_{EP},d)_i$ | Daily personal noise exposure of day $i$ | dB(A) |
| $l$ | Length, depth | m |
| $l_e$ | Expanded bed depth | m |
| $l_e$ | Equivalent pipe length | m |
| $M$ | Moisture content, percentage (on a wet basis) | % |
| $M$ | Mass of the material | lb, tons, kg |
| MCRT | Mean cell residence time | day |
| MLVSS | Mixed liquor suspended solids | mg/L |
| MW | Molecular weight of the gas | g |
| $M_a$ | Moisture in solids as in shredded and screened refuse | % |
| $M_d$ | Moisture content, percentage (on a dry basis) | % |
| $M_i$ | Mass of wet waste, placed at time $i$ | |
| $M_P$ | Moisture in mixed pile (heap) ready to begin process of composting | % |
| $m$ | Mass | kg |
| $N$ | Number | Dimensionless |
| $N$ | Number of conveyor leads or number of blades wrapped around conveyor hub | |
| $N$ | Number of sound sources | |
| $N$ | Number of collection vehicles needed | |
| $N$ | Number of working days taken into account during the week | |
| ND | Nondecomposable material | |
| $N_e$ | Effluent bacterial number, number of bacteria | /100 mL |
| $N_i$ | Influent bacterial number, number of bacteria | /100 mL |
| $N_o$ | Number of viable microorganisms of one type at time $t = 0$ | /100 mL |

| Symbol | Description | Unit |
|---|---|---|
| $n$ | Number | Dimensionless |
| $n$ | Number of installments | |
| $n$ | Number of years | |
| $N$ | Roughness factor, Manning and Kutter factor | $m^{1/6}$ |
| $N$ | Sound intensity level | dB |
| $N$ | Total time periods of waste placement | |
| OL | Organic loading rate of the sewage effluent | g/L |
| $P$ | Distance between collection pipes | ft |
| $P$ | Level of sound intensity to be measured | W |
| $P$ | Number of people served by the sewer | Dimensionless |
| $P$ | Paper, percentage by weight of total MSW, on a dry basis | |
| $P$ | Precipitation | mm/year |
| $P$ | Pressure | mmHg, N/m$^2$ (Pa) |
| $P$ | Pressure of sound wave | Pa |
| $P$ | Pulley friction horsepower | |
| $PA(t)$ | Time-varying value of A-weighted instantaneous sound pressure | Pa |
| $P_{ref}$ | Some reference sound pressure | Pa |
| PE | Population equivalent | Dimensionless |
| POW | Power | J/s, W |
| $P_g$ | Partial pressure of the respective gas in the gas phase | Pa, atm |
| $P_{osm}$ | Osmotic pressure | atm |
| pOH | Negative logarithm of the hydroxyl ion concentration | |
| $p_w$ | Pressure of saturated water vapor | mmHg, N/m$^2$ |
| $P_o$ | A reference level, usually the intensity of a note of the same frequency at the threshold of audibility = intensity of the least audible sound | W |
| $P_{ref}$ | Some reference sound pressure | |
| $P_{x1}$ | Purity of the first output stream in terms of $x$ | % |
| pAlk | Negative logarithm of total alkalinity | Equivalent CaCO$_3$/L |
| pCa$^{2+}$ | Negative logarithm of calcium ion | mol/L |
| Pf | Peaking factor | Dimensionless |
| pH | Negative logarithm of the hydrogen ion concentration | |
| pH$_a$ | Actual [measured] pH of the water | |
| pH$_s$ | Saturation pH | |
| $Q$ | Rate of flow, fire demand | m$^3$/s |
| $Q$ | Delivery of refuse | m$^3$/min |
| $Q_a$ | Average flow | m$^3$/s |
| $Q_I$ | Total inflow volume during a specified period | m$^3$/s |
| $Q_{max}$ | Maximum flow rate | m$^3$/s |
| $Q_{min}$ | Minimum wastewater flow | m$^3$/day |
| $Q_O$ | Total outflow volume during a specified period | m$^3$/s |
| $Q_p$ | Peak flow | m$^3$/s |
| $Q_R$ | Recirculated flow | m$^3$/s |
| $Q_T$ | Total gas emission rate from a landfill | volume/time |
| OR$_{x,y}$ | Overall recovery for components $x$ and $y$ | |
| $Q_w$ | Waste sludge flow | m$^3$/day |
| $q^+$ | Charge on the colloid | Coulomb |
| $q$ | Flow that can be applied per unit area | L/m$^2$*day |
| $q$ | Vertical inflow (infiltration) | ft/day |
| $R$ | Plastics, percentage by weight of total MSW, on a dry basis | |
| $R$ | Rotational speed of screw | rpm |
| $R$ | Universal gas constant | J/K*mol, L*atm/ mol*K, J/kg*K |
| Re | Reynolds number | Dimensionless |
| RES | Resistance of a conductor | W |
| RES$_s$ | Specific resistance of the suspension | W*cm |
| RI | Ryzner index | |
| $R_m$ | Resistance of filter medium | m$^{-1}$ |
| $R_{mr}$ | Rate of corrosion | |
| $R_s$ | Specific resistance | W$^{-1}$ |
| $R_T$ | Total removal | % |
| $R_u$ | Recycling ratio | Dimensionless |
| $R_{x1}$ | Recovery of component $x$ in the first output stream (1) | |
| $r$ | Coefficient of runoff (can be estimated for different types of soil) | |
| $r$ | Radius | m |
| $r_D$ | Rate of deoxygenation | |
| $r_H$ | Hydraulic radius | m |
| $r_r$ | Rate of reaeration | |
| $r_s$ | Specific resistance of sludge cake | m/kg |
| $r_s'$ | Constant | |
| $S$ | Salinity correction term | |
| $S$ | Storage in the soil or solid waste | mm/year |
| $S$ | Speed of belt | ft/min |
| $S$ | Total number of customers serviced | |
| SA | Sludge age | days |
| SDI | Sludge density index | g/mL |
| SLR | Sludge loading rate | day$^{-1}$ |
| SRT | Solids retention time | m |
| SS | Suspended solids | mg/L |
| SVI | Sludge volume index | mL/g |
| SWL | sound power level | dB |
| SPL | Sound pressure level | dB |
| SPLN | Net sound pressure level | dB |
| SLP$_A$ | Sound level at distance $D_A$ from the noise source | m |
| SLP$_B$ | Sound level at distance $D_B$ from the noise source | m |
| SPL$_i$ | Sound pressure level from source $i$ | dB |
| $S_c$ | Specific capacity of well | m$^3$/day*m |
| $S_u$ | Sulfide concentration | mg/L*h |
| $S_{udis}$ | Dissolved sulfide concentration in wastewater | mg/L |

(Continued)

| Symbol | Description | Unit | Symbol | Description | Unit |
|---|---|---|---|---|---|
| $s$ | Distance of separation between cylinders (discs) | m | $v_{sco}$ | Scour velocity | m/s |
| sa | Salinity | g/kg | $v_w$ | The difference in the rotational velocity between the bowl and the conveyor | rad/s |
| s.g. | Specific gravity of settling particle | Dimensionless | $v_z$ | Updraft velocity | m/s |
| $s^*$ | Growth-limiting substrate concentration in solution | mg/L | | | |
| | | | $W$ | Loading rate | kg/day |
| $T$ | Capacity | tons/h | $W$ | Number of workdays per week | |
| $T$ | Temperature | °C, °F, °K | $W$ | Organic loading of the trickling filter | g BOD/day |
| TDS | Total dissolved solids | mg/L | $W$ | Power of sound wave | W |
| TOR | Net torque input | dyne/cm | $W$ | Water, percentage by weight, on a dry basis | % |
| TR | Transmissibility of aquifer | m³/day*m | $W$ | Weight of SW to be buried | |
| $T_c$ | Temperature correction factor | Dimensionless | $Wt$ | Weight | kg |
| $T_w$ | Average trade waste discharge | L/day | $W_w$ | Initial (wet) weight of sample | |
| $T_e$ | Duration of the person's personal exposure to noise | | $W_d$ | Final (dry) weight of the sample | |
| | | | $W_o$ | Some reference sound power | W (pW) |
| $T_o$ | Length of the working day | h | $w_p$ | Wetted perimeter | m |
| $t$ | Time, detention time | days | | | |
| $t_c$ | Critical time | days | $X$ | Amount borrowed, amount available every year | |
| $t_{½}$ | Half-life of the particular nuclide | s, days | | | |
| $t_{25}$ | Time required for water surface to fall 25 mm | s | $X$ | Number of customers a single truck can service per day | |
| | | | $X_T$ | Total removal | % |
| $t_i$ | Age of the $i$th section of waste | Time | $X_s$ | Total mass of sludge, wet tons per day | |
| | | | $X_r$ | Total mass of refuse, wet tons per day | |
| U | Velocity at free plate surface | LT⁻¹ | $x_1$ | First component emerging of the first output stream (1) | Mass/time |
| UC | Uniformity coefficient | | | | |
| $u_a$ | Average velocity | m/s | $x_o$ | $x$ Component entering to the binary separator | Mass/time |
| $u_i$ | Heat transfer coefficient for heat exchanger number $i$ | J | | | |
| | | | $x_g$ | Mole fraction of gas | Dimensionless |
| $V$ | Volume | m³ | $x_c$ | Critical distance | m |
| $V$ | Volume of refuse between each pitch | m³ | $x$ | Distance, drawdown | m |
| $V$ | Volume of sanitary landfill area | | $x$ | Percentage of compressed SW volume | % |
| $Vr$ | Volume of a layer of coverage required | | $x_a$ | Mass of solids | wet tons |
| VO | Volatile organics | | $x_s$ | Mass of sludge or other source of water | tons |
| $V_o$ | Original size (initial) | | | | |
| $V_c$ | Volume after compaction | | $Y$ | Distance from a fixed plate | m |
| VOL | Volumetric organic loading rate | kg/m³*s | $Y$ | Installment cost, amount that has to be invested | |
| VS | Concentration of volatile solids | kg/m³ | | | |
| $V_A$ | Volume of material (A) | | $Y_{max}$ | Maximum saturated depth over the liner | ft |
| $V_B$ | Volume of material (B) | | $Y_t$ | Ultimate gas yield | m³ gas/kg VS added |
| $V_g$ | Volumetric gas production rate (specific yield) | m³ gas/m³ digester/day | $y$ | Distance, depth, width | m |
| $V_s$ | Settled volume of sludge in a 1000 mL graduated cylinder in 30 minutes | mL/L, % | $Z$ | Depth, elevation | m |
| | | | ZP | Zeta potential | V |
| $v$ | Velocity | m/s | | | |
| $v$ | Speed of cylinders | rpm | $\alpha$ | Constant | |
| $v'$ | Hindered settling velocity | m/s | $\alpha$ | Inclination of liner from horizontal | ° |
| $v_a$ | Approach, actual velocity | m/s | $\beta$ | Beta function for a centrifuge | |
| $v_c$ | Centerline velocity | m/s | $\sigma$ | Surface tension | N/m |
| $v_f$ | Filtration velocity | m/s | $\phi$ | Angle | ° |
| $v_H$ | Velocity of horizontal water movement | m/s | $\phi$ | Particle shape factor | Dimensionless |
| $v_r$ | Rotational velocity of the bowl | rad/s | $\phi_{sw}$ | Hydrogen sulfide flux to pipe wall | g/m²*h |
| $v_s$ | Settling rate, displacement velocity, loading rate, overflow rate | m/s | $\gamma$ | Specific weight | N/m³ |
| | | | $\gamma$ | Specific weight of water | |
| $v_{sc}$ | Self-cleansing velocity | m/s | $\tau$ | Shear stress | N/m² |

| Symbol | Description | Unit |
|---|---|---|
| $\Sigma$ | Parameter related to the characteristics of centrifuge | |
| $\Sigma F$ | Sum of all forces acting on fluid contained between two cross sections | N |
| $\rho$ | Density | $kg/m^3$ |
| $\rho$ | Average density of solid waste and garbage | |
| $\rho$ | Density of the material | $g/cm^3$ |
| $\rho_o$ | Original apparent density | |
| $\rho_c$ | Apparent density after compaction | |
| $\rho_c = \rho_{A+B}$ | Bulk density of the mixture of material (A) and material (B) | |
| $\rho_A$ | Bulk density of material (A) | |
| $\rho_B$ | Bulk density of material (B) | |
| $\delta$ | Thickness of laminar layer | m |
| $\Psi$ | Rate of deposition | $m^2/s$ |
| $\nu$ | Kinematic viscosity | $m^2/s$ |
| $\zeta$ | Eddy viscosity | $N*s/m^2$ |

| Symbol | Description | Unit |
|---|---|---|
| $\mu$ | Dynamic (absolute) viscosity | $N*s/m^2$ |
| $\mu$ | Dynamic viscosity of water | |
| $\xi$ | Ionic strength | |
| $\theta$ | Temperature coefficient | Dimensionless |
| $\varepsilon$ | Roughness of the pipe wall | m |
| $\lambda_o$ | Filtration coefficient | $m^{-1}$ |
| $\lambda$ | wave length | m |
| $\omega$ | Impeller rotations | Number of rotations/min |
| $\chi$ | Constant | |
| $\mu_s$ | Growth rate of microorganism | $day^{-1}$ |
| $(\mu_s)_{max}$ | Maximum specific growth rate of microorganisms | $day^{-1}$ |
| $\upsilon$ | Specific volume of the fluid | $m^3/kg$ |
| $\Delta T$ | Change in temperature | °C |
| $\Delta S$ | Change in storage volume during specified interval | |
| $\Delta P$ | Pressure difference | Pa |

# List of Acronyms

| Abbreviation | Meaning |
|---|---|
| ALGOL | algorithmic language |
| ALU | arithmetic and logic unit |
| AMD | Advanced Micro Devices, Inc. |
| BASIC | beginner's all-purpose symbolic instruction code |
| BIOS | basic input/output system |
| BMP | bitmapped picture |
| CD | compact disc |
| CIL | common intermediate language |
| CLI | command line interface |
| COBOL | common business-oriented language |
| CPU | central processing unit |
| CRT | cathode ray tube |
| DVD | digital versatile disc, or digital video disc |
| ENIAC | electronic numerical integrator and computer |
| FORTRAN | formula translation |
| FOSS | free and open source software |
| FSF | free software foundation |
| GNU | GNU's not Unix |
| GPL | GNU general public license |
| GUI | graphical user interface |
| IBM | International Business Machines Corporation |
| IDE | integrated developer environment |
| IoT | internet of things |
| JIT | just-in-time translator |
| JPEG | joint of picture experts group |
| JVM | Java virtual machine |
| LAN | local area network |

| Abbreviation | Meaning |
|---|---|
| LCD | liquid crystal display |
| LED | light emitting diodes |
| MMIO | memory mapped input/output |
| MS-DOS | microsoft disk operating system |
| MSDN | microsoft developer network |
| OCR | optical character recognition |
| OOP | object-oriented programming |
| OS | operating system |
| OSI | open source initiative |
| PC | personal computer |
| PDF | portable document format |
| PNG | portable network graphics |
| PROLOG | programming logic |
| QWERTY | The first six letters appearing from the left-hand side of a keyboard |
| QWERTZ | same as QWERTY, German version |
| RAM | random access memory |
| ROM | read-only memory |
| SQL | structured query language |
| TUI | text-user interface |
| USB | universal serial bus |
| VB.NET | Visual Basic.NET |
| VBA | VB for applications |
| VS | Visual Studio |
| WINE | Windows Emulator |

# 1 Programming Concepts

This chapter presents an introduction to the basic functions, layout, and components of a computer workstation. Computer programming languages are briefly outlined and summarized, delineating their history, types, and uses, with an emphasis being placed on Visual Basic.NET (VB.NET) language as is the language of choice to this book.

The management of water resources, wastewater collection, treatment, and disposal, solid waste collection and ultimate recycling, reuse and disposal as well as air pollution and noise abatement and control involves complex processes. This is due to many interacting parameters, some of which are difficult in presenting a straightforward mathematical model, equation, or formula. Mathematical modeling techniques can be used in predicting the quality and sequence of relationships that can help management in an effort to solve potential environmental problems. A model can be regarded as an assembly of concepts in the form of one or more mathematical equations that approximate the behavior of a natural system or phenomenon (Ji 2008). Models can be divided into simulation, optimization, and computer-aided design models (Chapra 2008). *Simulation models* address the presentation of a mathematical model that simulates a specific situation, with the formulation of a mathematical relationship and solution through a structured and valid process. *Optimization models* use mathematical techniques to achieve a reasonable solution from a range of possibilities. *Computer-aided design models* help in the preparation of design drawings and computation of quantities.

Modeling methods rely on computers for their validation, testing, speediness, and efficient use. The basic concepts of mathematical modeling and formulation of a well-written operative computer program form the central theme of this book. The aim is to train individuals to write, develop, validate, and use their own environmental engineering control programs and simulation models.

In many of the examples tackled in this book, the best *teaching* method was adhered to. This is not always the best *working* or the best method with regards to its real-world *performance*. Sometimes the teaching method is longer, slower, or even more complex than what a production-level model should be, but the authors opted to make all examples, presented throughout the manuscript, as readable and understandable as possible. As such the reader is free to edit any selected examples to be more performance efficient if so he or she desires. A live example is editing controls'

properties in the form _ load() function of each form. In reality, the programmer (including oneself for that matter) will edit these properties in the design view. This will not only reduce code clutter, but it also helps separate design from code and reduces the work that the program has to do when loading each new instance of the form to memory upon execution. In this book, however, everything is placed inside code blocks to reduce the complexity of the text. As such, the reader is able to follow easily the code and understand exactly what it does without really looking into the form design. Suffice it to know that there are five labels, five textboxes, and one button in the form. Every other detail about these 11 controls is found in one place inside the form _ load() function. Again, this design choice is not suitable for production-level software as it is solely intended for teaching models like the ones presented in this script. Furthermore, error checking has been dropped in some examples to keep the programs as direct, explicit, clear, and simple as possible.

## 1.1 HOW IT ALL STARTED

Long before computers came to life, machines have been developed to perform certain tasks. These ranged from ancient and simple devices like the Greek Abacus, which was used for counting (around 50 BC), to the Pascaline (a mechanical calculator), which was named after its inventor, *Blaise Pascal*, in the seventeenth century, to help his father do his tax work. It functioned mainly to add numbers. The first real calculator that could do the four basic calculations (addition, subtraction, multiplication, and division) was developed by the Austrian *Antonius Braun* (1727). Later on, in 1822, *Charles Babbage* (1791–1871), known as the father of computers, designed a machine that would do complex calculations and print the results. This was a revolutionary concept in his time. Although he never actually built the machine due to the limitations present in the technology at that time, which was not precise enough to match his designs, the design and functions he set are present in nearly all today's computers (see Figure 1.1). During World War II, German submarines were receiving encrypted messages from the bases that were coded using a coding machine called the *Enigma*. Around that time, in the U.S. Army, a machine for computing tables for ballistic trajectories, named electronic numerical integrator and computer, was being developed (Murdocca and Heuring 1999).

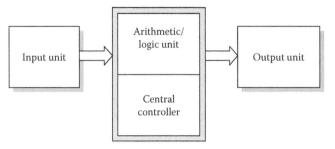

**FIGURE 1.1**   Components of a computing machine.

## 1.2   COMPONENTS OF A COMPUTER

Essentially, a computer is a machine that can perform complex, long, and repetitive sequences of operations at very high speeds (Welsh and Elder 1988). It captures data in the form of instructions or electronic signals, processes it, and then supplies the results in the form of information or electronic signals to control some other device or process (Schneider 2013). Theoretically speaking, conventional computers have five major components (as designed by von Neumann and others, often referred to as *the von Neumann model*) (see Figure 1.2) (Murdocca and Heuring 1999):

1. An input unit
2. An arithmetic and logic unit (ALU)
3. An output unit
4. A memory unit
5. A control unit

A typical personal computer (PC) includes four hardware components (in analogy to the von Neumann model): a processor (central processing unit [CPU], which contains both the control unit and the ALU), input devices (a keyboard and a mouse), output devices (video display devices such as a monitor or a projector, printers, and other peripherals), and storage media (memory, hard disks, optical media, and flash memory).

The main components of a PC system can be detailed as follows (see Figures 1.2 through 1.6):

1. *The CPU*: This component represents the brain or the control center of the computer. It carries out the sequence of operations specified by a program, and

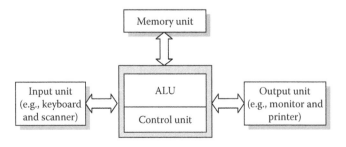

**FIGURE 1.2**   Components of a computer (von Neumann model).

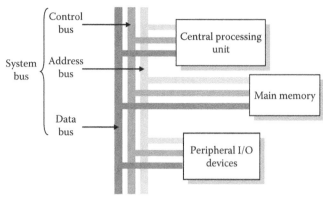

**FIGURE 1.3**   The system bus. I/O, input/output.

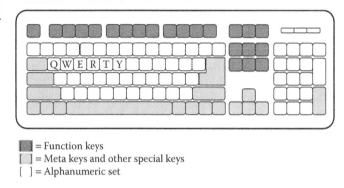

■ = Function keys
[ ] = Meta keys and other special keys
[ ] = Alphanumeric set

**FIGURE 1.4**   QWERTY keyboard layout.

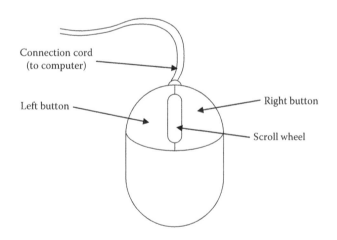

**FIGURE 1.5**   Structure of a computer mouse.

it controls all computer activities. The functions of the CPU can be divided as follows (Hager and Wellein 2010):

a. The control unit controls electronic signals passing through the processor, and it monitors the operation of the whole system. It also controls the flow of code and data through the processor, on system buses, to main memory and other destinations (such as hard disks, Universal Serial Bus [USB] disks, and output devices such as printers).

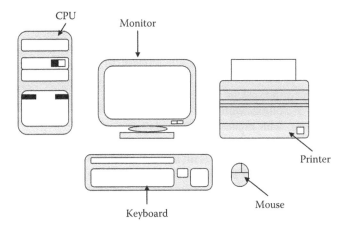

CPU

Monitor

Printer

Keyboard

Mouse

**FIGURE 1.6**   Components of a PC system.

b. The ALU executes arithmetic and logical operations. These are of paramount importance, as most of the commands performed by the processor involve data moving or operations on data. As the processor's golden currency is time, this must be handled as efficiently as possible. As such, modern processors apply very complex logical operations to determine the best path to take, how to handle data, and what operation to do next, and when to do it.

As the CPU is the maestro of the orchestra, it must have a method of contacting the other components under its control (see Figure 1.2). It negotiates tasks with other components of the system using a common channel known as the system bus (see Figure 1.3), which has three components (Murdocca and Heuring 1999):

- *The control bus*: It carries information about the operation being requested, for example, read from memory and write to disk.
- *The address bus*: It carries the address of the destination (or source) of the data, be it in the main memory or saved on the disk.
- *The data bus*: It carries the data being read/written.

2. *Input devices*: There are several programmable input devices such as the keyboard, digital pen, joystick, mouse, microphone, and document scanner. These input devices function much the same as human sensory organs in that they provide input from different sources into the computer:

a. The keyboard is by far the primary and mostly used input device for entering commands and data. It transfers characters to a special "keyboard buffer" in memory in response to each key event. Key events include key press, key release, and "sticky" or persistent key presses. Each key event corresponds to a special and well-defined "scan code" (Brouwer 2009), which is a special byte number indicating the key that generated

the event and the event itself (press, release, repeat). Most key events generate one to three scan codes. In addition to the scan codes, most keyboards generate special control codes that are used to convey messages to the host computer (e.g., the presence of key input) or to respond to commands from the host computer. A typical PC keyboard has 101 keys (some Windows-specific keyboards have 104). The keys generate scan codes as described previously. The operating system (OS) does the job of transforming each key event to its respective key mapping (e.g., the user pressed the key with "A" printed on it. This might result in "a" printed on the screen, or "A" if the user pressed Shift simultaneously, or has turned Caps Lock on beforehand, or maybe it prints a whole different letter in another language if the user is using international language mapping). The keys on a typical keyboard can be divided into: (1) alphabetical letters and numbers (also called the alphanumeric set), (2) special symbols and shapes, and (3) function and control keys. As each scan code is 1 byte in length, the keyboard driver can generate 256 different scan codes (a byte is 8 bits, thus $2^8$ different values can be represented in a single byte). This is not an awful lot, but remembering that each key event can generate one to three different scan codes means that all key events (press, release, repeat) can be generated by using multiple scan codes for individual key events. There are different ways of arranging the keys on a keyboard, which is termed the keyboard "layout" (see Figure 1.4). The most commonly used layout is the QWERTY layout, which corresponds to the first six letters appearing from the left hand side. This layout was adopted from early typewriters. Other layouts exist, such as QWERTZ (a German layout) and others.

b. The mouse is a positioning device that is used to navigate graphical user interfaces (GUIs) by controlling a cursor (commonly shaped as an arrow with changing sizes, shapes, and colors). It eases user interaction with the system by providing a means of working with files, operating on the disk file system, launching programs, and other functions, by the user clicking on one (or more) of its two to three buttons. Older mice used to have three buttons by default. More current mice dropped the middle button (but can usually be emulated by pressing both buttons concurrently) and substituted it for the scrolling wheel (see Figure 1.5). Although the mouse is commonly associated with GUI environments, a lot of contemporary OSs provide some (albeit sometimes rudimentary) mouse support in text-based environments (text–user interface; also called command-line interface). For example, most Linux distributions provide mouse support by default on their virtual terminals.

c. The digital pen is a type of input device that can be used to convert analog data (like user handwriting or hand drawing) to digital data that can be accepted, processed, and stored by a computer system. A popular use is in sending handwritten articles (by authors or journalists, for example) to the computer, after which it can be run through a special software known as optical character recognition (OCR) software, which can analyze the digital data, perform complex calculations on it, and ultimately convert the raw pixel data it received into distinct language characters, thus recovering the original piece of handwritten article. Text can then be edited by a word processor to add visual effects and can thereafter be printed or emailed.

d. Document scanners are used to scan pages and save them in an image format (such as JPEG, PNG, BMP, and others) or in the portable document format (PDF). The data can then be handled as is or processed through the same type of OCR software described previously to retrieve the data contained in it.

3. *Output devices* use different means to display the output of calculations, the results of different computations, whole documents, spreadsheets, or even audio and video.

a. *The monitor* (*visual display, display screen*): This output unit permits the display of input information as it is entered and allows the display of the output produced by computer software. Many types of monitors are found on the market, with different vendors, sizes, and capabilities. Current monitors are usually light-emitting diode displays, which use an array of light-emitting diodes to present the pixels on screen. Some other monitors are using the inferior liquid crystal display technique, which takes advantage of the light-modulating capabilities of liquid crystals. The old cathode ray tube displays are rendered extinct by these new technologies. The contents of the display are governed by data stored in a special area of the memory known as the video memory. The video memory is typically built into the monitor itself to allow for fast transfer of images from the memory to the screen proper. The OS usually either writes directly to the video memory or channels the graphics (or text) it wants output on the screen to a special area of the main memory, known as memory mapped input/output (MMIO), which is connected to the main video memory. Writing to this special MMIO area is akin to writing to the main video memory, and changes in one are directly reflected on the other. This technique is not exclusively used by video memory. In fact, most (though not all) modern I/O devices use some sort of MMIO in their communication with the system. Some vendors provide special capabilities with their video cards to enable programs with higher visual needs to be run seamlessly, a feature that is very important in video game programming (Hill and Pilkgton 1990).

b. *Printers and plotters*: A printer allows the generation of a permanent record or a hard copy of the output. Printers can be broadly classified according to the following:

– Users connected to the printer:
  – Personal printer: It is connected to one host computer. It is usually office or home based. It is easy to install, operate, and maintain, and is suitable for personal or small business use.
  – Shared or multiuser printer: It is connected to a network (usually local area network) of users. Typically, it is found in corporate offices and is shared and used simultaneously by different users at the same time. This type must be durable and highly maintained to withstand the workload expected.

– Physical versus virtual: Physical printers are real-world printers like the one sitting on top of one's own desk. A virtual printer is a piece of software that acts and behaves as if it was a real printer. From the application's point of view, the virtual printer is a printer, and all data that were destined to be printed will be sent to the virtual printer as such. A typical example is the indispensable "generic" PDF printer, which is installed by default on almost all modern OSs.

– Dimensions of the printed document: Typical, everyday use printers are two dimensional. With the advent of three-dimensional printers (3D), a new era of printing has emerged, and it may not take long for 3D printing to be integrated into our daily life activities and routine.

– Mechanisms of character formation:
  – Dot matrix printers form individual characters by using a series of dots, one line at a time. Although it is considered legacy technology, dot matrix printers are still used in special situations where low cost, low quality, or large volume printing is required, for example, in cash registers.

- Solid font mechanisms produce fully formed characters via the formation of sharp images. They are to be preferred to dot matrix printers when letter-quality output is needed.
  - Method of transfer of data, or print, to page:
    - Impact printers form characters by striking the type against an inked ribbon that presses the image onto the paper. This is obsolete nowadays.
    - Nonimpact printers do not involve physical contact but instead transfer the image to the paper using ink spray, heat, xerography, or laser. They are usually quieter and faster than impact printers. Most modern printers are nonimpact printers.

A plotter allows the sequential drawing of objects rather than using the raster-scan principle adopted for printers. The plotter produces professional quality drawings on different paper sizes and allows the production of final, multiple-colored drawings.

4. *Data storage media*: Storage of programs and data can be made on (fixed) internal hard disks, optical media (including compact discs [CDs] and DVDs), memory sticks (USB or flash drives), and external hard disk stores. Memory is considered storage medium, albeit temporary, as all the information stored therein is lost whenever the electrical current is lost; hence, it is often called "volatile memory."

Generally, storage media are used to store information, either permanently (like CDs and hard disks) or temporarily (like memory). They may be divided as follows:

- *Floppy disks* (*diskettes*): These portable media are made of flexible plastics that can be magnetized. Data are stored on the disks as magnetized spots or concentric tracks on its magnetic coating by the head of the disk drive. They are largely obsolete and are rarely used today as the other technologies have superseded them in terms of efficiency, reliability, storage size, and cost effectiveness.
- *Hard disks* (*Winchester disks*): These are made of inflexible metal and are installed and fixed in the computer. They are more reliable with larger storage capacities and faster access times than the floppies. Each computer system usually contains at least one hard disk, which is fixed inside the computer case. This serves as the primary storage medium for the system. Otherwise, disks can be used as a large medium for transferring data between systems or as backup storage for data recovery in case of system failure.

- CDs are made of polycarbonate plastic with a thin layer of aluminum (the reflective surface). Data are stored on the surface of the disc as a series of microscopic indentations; this is read later using a laser beam. These discs can be formatted to store media (e.g., music CDs), data (e.g., software CDs), and video (e.g., movie CDs).
- The primary memory is volatile, that is, when the computer is turned off, the information is lost (also known as the main store, random access memory [RAM], or read/write memory). It is used by the processor as a temporary information storage device. The information can be an input to the system, or it may be generated from internal calculations. The RAM is also used to store and retrieve computer programs during program execution. The basic measurement unit of computer memory is denoted as a byte (corresponds to 8 bits, a bit being the smallest unit of information handled by a computer and either 1 or 0 in the binary number system). Bytes are grouped into kilobytes (1 kB = 1024 or $2^{10}$ bytes), megabytes (1 MB = 1024 kB), giga bytes (1 GB = 1024 MB), and tera bytes (1 TB = 1024 GB). The immutable memory is called read-only memory (ROM) and contains the basic input/output system (BIOS). The BIOS controls the major I/O devices in the system. ROM is not a volatile memory, it is built into the computer, and this information is retained even when the computer is off.

Figure 1.6 shows the main components of a computer. The inside of the computer contains the main board, power supply, adapter cards, disk drives, processor, RAM, ROM, math coprocessor, and support chips. For effective, trouble-free operation, routine corrective and preventive maintenance should be performed regularly for the computer. Computer failures can be caused by excessive heat, dust buildup, noise interference, powerline problems, corrosion, and the presence of magnetic flux (Andrews 2013).

## 1.3 TYPES OF PROGRAMS

A program is an orderly collection of instructions directed to the computer, instructing it to perform specific tasks. Programs collectively available to the computer to be performed are termed software. This is in contrast with the hardware, which is the solid component of a computer system, described in Section 1.2. Software is needed to direct the computer to perform the required tasks. Hardware without software is just a collection of assembled materia. Likewise, software has no purpose to do if there is no h/ ware to control. Software can be divided into the follc categories:

### 1.3.1 OS (Supervisor Programs)

This is a group of programs that efficiently control the operation of the computer. It is usually the first program to load into memory, immediately after the BIOS finishes inspecting the hardware, checks memory for errors, does some household work, and then looks for a suitable OS to run. The BIOS looks into its list of bootable media (editable through BIOS setup on all modern computers and laptops). For example, if the first bootable medium is the CD ROM, the BIOS looks into the controller to see if there is a CD in it. If there is, it inspects the disk to see if it is bootable. If so, it looks for the OS executable and loads it into a special address of the main memory and handles it control of the system. If the CD is not bootable (or no media is inserted in the drive), the BIOS checks the next bootable medium, typically one of the hard disks, and if it finds an OS installed there, it loads it and passes control to it. The part of the OS that forms the core of the system and is loaded by the BIOS is called the OS *kernel*. The kernel permanently resides in the computer memory throughout the system operation. It continuously supervises and coordinates other programs running on the machine and monitors device activities. Many OSs have been, and will continue to be, developed, such as Mac O/S by Apple, Microsoft Windows by Microsoft Corporation, and Unix and its successor Linux by Bell Laboratories.

Linux was originally developed by Linus Torvalds (started as a hobby project back in 1991). Since then, Linux became a pillar of its own. It expanded vigorously, and with many contributions from the GNU project (http://www.gnu.org), GNU/Linux has become an important and respected player in today's OS arena. There are many systems built around the Linux kernel, which are called "distributions," or shortly "distros." Such distros include Fedora Linux, SuSE Linux, and Gentoo Linux.

### 1.3.2 Utility Programs (User Programs)

These programs are designed by manufacturers or software companies to perform specific tasks. Examples of utility programs include text editors, debugging aids, disk formatters, file copiers, data sorters, and data file mergers. The major difference between user and supervisor programs is that user programs run in user mode, which means they are not granted direct control of the system except through special requests to the kernel. For example, when the user opens a document in his/her word processor, the user program (the word processor) asks the OS (the kernel) to do the operation of opening and reading the document. The process of shielding the user programs and preventing them from doing such important actions on the system is an important concept. Consider what happens if that same word processor had frozen because of a bug in the program logic, or even worse, due to malware or viral infection. If the word processor was given liberty while handling the disk and the files on it, a disaster could have happened when the program is frozen: When a program goes down, it will take the whole system down with it. This was the case in the old Microsoft Disk OS (MS-DOS) era,

when system crash was common, and every user knew the life-saving Ctrl-Alt-Del combination to reboot the system and start anew. In modern OSs however, user programs must ask the kernel to do almost every task that involves interaction with the computer system or its peripherals. As such, when a program freezes for whatever reason, the system remains intact, and the user can simply click the "kill this program" button to get rid of the offending software, leaving the rest of the system out of harms' way.

## 1.4 PROGRAMMING AND LANGUAGES

As computers became more and more complex, so did the functions performed by a computer. Old-fashioned computers were general purpose, whereas most of today's microprocessors serve specific goals, such as driving a user's smartphone, smart watch, or iPad, or even controlling a smart TV or a smart fridge; the interconnection of all these objects together has led to the term "Internet of Things." With the universal explosion of information technology that resulted in virtually each person having at least a personal laptop, a desktop PC at work, and a smartphone in his or her pocket, focus has shifted from the older PC era, when the old-fashioned desktop PC was the center of the programming world, and every effort was made to suit the programming process to ensure compatibility with the specific system vendor's specifications and needs.

Complexities in programs and their goals have led to the development of new programming languages, some of which are specific, sometimes highly specific to the extent that one programming language might be associated with just programming into another program (scripting). As such, there are different ways of looking at the programming languages and grouping them into different categories:

- *Object-oriented versus procedural programming*: Almost every programming language in today's world is object oriented (OOP). In this paradigm, everything is an object: you, your car, your dog, and every other thing in your world. Every object belongs to a class of objects, which defines in general terms what are the common features held by objects of its type. Every class has *members* that define its characteristics (such as color, model, and number of doors a car has) and *methods* that define how the outer world interacts with it (gas pedal to accelerate the car, user does not need to know how it actually works internally). As such, every object contains two things: data (*members* or *fields*) and code (*methods*) that works on this data. This programming model is different from the old way in which programs were written, which is known as procedural programming. In procedural programming, the program is written as one clump of code, which is read and executed by the processor sequentially until the commands are finished. This method of programming is considered retrospective in most today's programming aspects, but it still has its important applications in selected cases. The user is directed to the references where more information about OOP is provided (Buyya et al. 2009).

- *Event-driven environments*: In an OOP world, everything is event driven; nothing happens on its own. For example, the word processor program will continue staring back at the user unless he or she strikes a key, clicks a mouse button, or touches a touch screen, instructing the program to do some action. As such, every event (key press, mouse click, touch on a touch screen) has an event handler (special part of the program's code that determines how to respond to certain events in specific ways). Contrast this with procedural programming where the only event needed was the user invoking the program's executable from the command line, after which the program will continue running from the top down in a predetermined and well-known path. Each and every time the program is run, it will run in the exact same way. It is worth mentioning that reference concerns the output (or results), and not focuses on the *steps* taken to reach these results. In an event-driven environment, how the program runs exactly depends on the events generated by the user, which are different every time unless the user generates the exact set of events every time he or she runs the program.

- *High-, intermediate-, and low-level programming* (see Figure 1.7) (Dandamudi 2005): As more basic functionality is hidden from the user (termed *encapsulation* [Buyya et al. 2009] under the OOP umbrella), as more the language becomes high level. High-level programmers usually program desktop application software; they do not need to know exactly how the OS or the underlying hardware works. Examples of high-level languages are VB, (Visual) C and C++, Java, structured query language (SQL), and many more. On the contrary, OS programmers need to have a detailed knowledge of how the monitor paints pixels to display images, how the keyboard sends key strokes in the form of scan codes, how the hard disk retrieves a sector from the disk to read part of a file, and so on.

In essence, these programmers need to know how the system behaves under the hood. As such, they need a language that does not have all the ornaments of a high-level language but is actually more close to the understanding of the underlying processor. This is known as the intermediate language (or the *assembly* language, which differs slightly according to the type of the underlying processor architecture). The lowest level of functionality is the bare bones, actual machine language, which is not human readable, and it translates to the letter what the processor does, usually expressed in the form of hexadecimal (letters and numbers) words.

- *General- and special-purpose languages*: Languages used to program desktop software, games, general web applications, and servers are general-purpose languages (most of the commonly used programming languages fall into this category, including VB, C/C++, Java, among others). Special-purpose languages are tailored toward more specific tasks, for example, database handling, program scripting, or programming logic, and artificial intelligence (e.g., SQL for database handling, Visual Basic for Applications, JavaScript for web scripting, and PROLOG [Programming in Logic] for artificial intelligence programs).

- *Compiled and translated languages*: Traditionally, every programming language was compiled. Compilation means transforming a program's source code (usually written in a high-level language code) into a machine-specific low-level (binary) code. The compiler is a program that understands both the high-level and machine languages for the specific platform, something akin to the human language translators in international meetings, and as such, each language has its own compiler, being specific for each and every processor. This is not a good scheme for certain conditions, for example, web scripting. As the web page needs to be fast loading and small in size, having a precompiled program is not always the good answer.

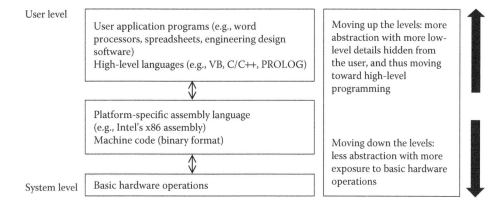

**FIGURE 1.7** Different levels of abstraction in the programming model.

Some languages were developed for this very purpose, being resident as a source code, which will be executed by a mediocre program (the translator) at execution time (e.g., JavaScript for web scripting). This scheme involves another step in program's life: writing program source, distributing it, and letting another program (the translator) read it and tell the processor what to do. Some languages took it even further, as there arose the problem of system compatibility. A program written for MS Windows will not work on Unix or Linux. A program written for Intel's x86 architecture will not work on Motorola AMD processor, at least not natively and easily. Therefore, the compatibility problem has two facets: *OS* and *architecture* compatibility. Some languages such as Java solved this problem by inventing the Java virtual machine (JVM). Java source code is compiled into an executable containing intermediate Java binary code (similar, but not identical to, machine code). This executable needs a small program (the JVM) to run it. The JVM is like a processor inside a processor; it reads the binary code and tells the real processor what to do. Thus, programmers need not worry about distributing their original source code (a headache for many corporations and programmers), and users only need to install a small-sized virtual machine that is specific to their systems (e.g., JVM for x86 under Linux, JVM for AMD under Windows). Every program written in Java will work on the fly (provided the JVM is installed on the target system), with the price of some slowness of execution compared to a native machine-coded program.

Writing one's own programs has many benefits over buying and using packaged programs. These advantages include the following:

- Solve specific problems, as the new program will be written specifically to answer the question at hand
- Speed, performance, and interpretation of calculations, compared to manual interpretation of data. Computers are better than humans in doing complex calculations, provided the correct input is given
- Enhancing graphical work, especially with the advent of new hardware that is specially tailored toward graphical design and the continuous improvements in graphical design and modeling software
- Formulating cost-effective design and operation of products. Certainly, a company that invents its own software solutions manages to save more money than the one that relies on presold software
- Computer-aided design and manufacturing is a continuously growing market
- Performing preliminary testing enables preemption of bugs and early correction of problems before deploying software solutions to users

- Controlling manufacturing and assembly operations
- Control and automation of processes
- Effective business and management operations
- Communicating technical results in a reliable and effective way
- Educational and training programs. Virtual tutors are becoming a common practice in many educational institutions and corporate training schedules

Writing one's own software can have some drawbacks and problems also as follows:

- Reinventing the wheel, which is the most fatal mistake a programmer might do. If there is some problem that needs solving, one might better do some search around, looking at some programming forums, running through a specialized website (like the Microsoft Developer Network—http://msdn. microsoft.com), or even asking Google or Bing: someone might have had the answer in a more effective and elegant way, so there is no need to waste time and brain power.
- Not knowing where to start. Sitting there with a void mind and an empty source file, staring at the blinking cursor with no notion of what code to write. Therefore, good programming design and preemptive thinking are paramount before grasping the keyboard and starting on writing code.
- Not knowing which language to use. Having a look at the examples of programming languages and their uses as discussed previously helps make this decision.
- Working single handedly. Usually as programs grow bigger, team work becomes more efficient and effective. Sometimes assigning team members with specific tasks help speed up the process and make it more effective and reliable, such as having one software programmer, one debugger, and one testing person.

Once a program has been written, it needs to be verified for its accuracy through the use of specific data. This results in a cycle of checking and correcting the program to obtain required accuracy and specifications. This process is called debugging, and it is carried out by running the program on a computer with suitable test data to eliminate (almost) all compile time, runtime, and logical errors. Figure 1.8 shows the steps to be followed in order to obtain a *virtually* error-free program (we say virtually because it is impossible to write a 100% error-free program. Each and every program—except the most trivial ones—will fail at some point in some way under the wrong—or right?—conditions). The process of debugging is simply a trial to reduce these errors to a minimum and to ensure that the program runs as expected under most *normal* conditions.

High-level programming languages are machine independent; after all, this is why they were developed in the first

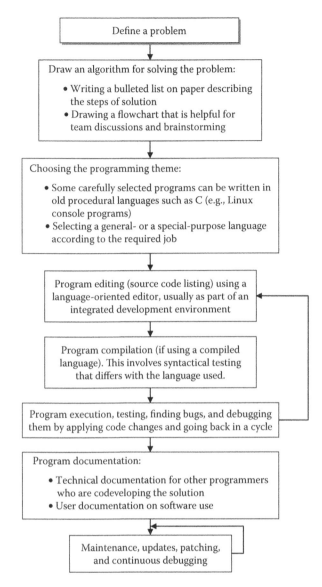

**FIGURE 1.8** Steps of programming.

place. The features considered essential to the success of a program include simplicity of the algorithm or technique, program clarity, and efficiency in executing the program. Some of the most commonly used high-level languages are described briefly as follows:

1. Java (Schildt 2007), first released in 1995 by Sun Microsystems, is a general-purpose, OOP language. It became widely popular and is used in almost all electronic and digital devices of today's use. From the desktop programming point of view, Java was intended to be a cross-platform language, that is, a program written in Java language under one platform (e.g., Windows on an Intel processor) will be able to run without change in the source code to fit another platform (e.g., Linux on AMD processor). Java does this by implementing the JVM and by compiling the source code into an intermediate Java byte code that

is run by the JVM. Java is derived from C/C++, but it was made more stable, secure, and memory efficient. One of the advancements introduced in Java is the garbage collector, which is a small program (part of the Java machine) that scans the program's memory periodically to look for variables and data that are no longer used, which are then deleted by the garbage collector, and their memory space is revoked for other uses.

2. FORTRAN (Formula Translation) uses a notation that facilitates writing of mathematical formulas. The language is useful especially when dealing with large amounts of data such as in the case of business-oriented work. The language is excellent for sophisticated computations and is fast and precise with relatively powerful I/O capabilities. FORTRAN programs execute quite rapidly because the language is almost a compiled one. In program writing, a number of commands are used throughout the different stages of the program to its completion. Each instruction is written as a single line and, in some systems, aligned in specific columns to constitute the source program. The compiler translates the program for the computer to yield the object program (Hager and Wellein 2010, Chivers and Sleightholme 2015). Many changes and various versions have emerged since the language first appeared in the IBM labs in the 1950s (FORTRAN, FORTRAN 77, FORTRAN 95, FORTRAN 2008), and the language is becoming more user-friendly and object oriented.

3. VB.NET (Petroutsos 2002) (the successor to the older VB, which in turn is coming all the way from BASIC, the Beginner's All-purpose Symbolic Instruction Code) is widely used as it is one of the easiest programming languages to learn and provide an excellent platform for novice programmers. It is an OOP language. It relies on the .NET framework for its functionality (discussed later). It is usually programmed using Visual Studio (VS) integrated development environment (IDE) or the free Visual Basic Express Edition supplied by Microsoft (downloadable from their web-site: http://www.visualstudio.com/en-us/downloads/download-visual-studio-vs.aspx). It is also used in most educational settings, university, and other programming courses. VB.NET is interactive as all .NET languages (the *IntelliSense* feature of VS allows the IDE to evaluate each statement on the fly and diagnose any potential errors and then provides possible completion or correction options to the user). VB.NET syntax is much easier than many other languages, and it is much simpler to use.

4. C++ (pronounced "see plus plus"), appeared as a successor to old C language. It started with adding object-oriented features to C and was originally named "C with classes", starting at Bell Labs in 1979; it then

was renamed C++ in 1983. It is a general-purpose, object-oriented, high-level language, although most programmers consider C as an intermediate language (as assembly language) because of many of the low-level and memory-manipulating features. C/C++ is widely used in programming software, especially software needing fine-tuning with the underlying hardware as OSs. Most programs are a hybrid with intervening assembly code inserted in the middle of C/C++ code to do some basic hardware functions or specific system calls.

5. Pascal is a structured language used to perform large-scale programming. The language uses a restricted set of design structures and other organizing principles. The language rules are made up of syntax and semantics. The former rules define how the words of the language can be put together to form sentences; the latter ascribes meaning and significance to these combinations of words (Welsh and Elder 1988). The language is excellent for large-scale and complicated programs. Although Pascal per se is not largely used today, its successor, Delphi, is widely used as a general-purpose OOP language.

6. Other programming languages include a wide variety of languages developed for specific applications. The Assembly language is the one used primarily by system programmers to address the hardware. It is rarely used for scientific applications, although it can be, but with great difficulty and effort. It is excellent for direct negotiation with the hardware, but it is limited by its complexity and lack of the programming structs that are present in high-level languages. Another drawback is that it is platform dependent, that is, each platform (or processor family) has its own instruction set, which dictates what assembly instructions can be performed on this specific platform. Currently, high-level languages such as C and its extension C++ are satisfying the needs of most programmers, even those requiring interaction with the hardware, and are being used for almost all purposes such as scientific, educational, system programming, database, equipment control, games, and many other applications. However, special languages such as SQL are intended for database programming only. Other languages include ALGOL, ADA, COBOL, MODULA, and many others.

## 1.5  THE OPEN-SOURCE MOVEMENT

In the late 1970s and the early 1980s, computers have been becoming a commonplace that people started to obtain their own PCs. At that time, computer world was dominated by big names such as IBM, Microsoft, and Apple. Some people, specifically a visionary man named Richard Stallman, have felt that something was wrong. Big companies were selling their software to customers for a price that included copyright royalties and customer support fees, which is all right. The problem was that these proprietary programs were *closed source*, meaning that the customer did not know what the program was *really* doing, he or she has been promised a specific behavior, and he or she got that, but he or she did not know how he or she got there. In other words, what steps did the program take to arrive at the result it displayed? Was the algorithm undertaken by the program the best one in this case? Was there another way of solving the problem that the original programmer did not use or did not know existed? Was the program doing any side work (like spying on the user's system determining what programs he or she is using and sending a report to the mother company, a technique commonly used by many online companies nowadays to tailor their advertisements to the user's browsing history)? All these questions gave rise to the novel idea of Dr. Stallman: software ought to be "free" (synonymous to "open" in this context).

Up to the time of this writing, there is a tribal war in the open-source community regarding whether code should be *open* or *free*. This is reflected by the fact that there are two different governing bodies in the open-source world: the Free Software Foundation (FSF) (Free Software Foundation 2016), founded by Richard Stallman, and the Open Source Initiative (Open Source Initiative 2016). From the authors' point of view, the difference between the two is pure semantics. Whether one uses the term "open-source" or "free-source" programming is entirely up to ones' liking. They are essentially one and the same, and troops in both camps are fighting for the exact same cause. In fact, it is common to hear the term FOSS, referring to *free- and open-source software*.

The basic idea is that the user has the right to know not only *what* the program does but also *how* it does it. This is why most open-source code is distributed under one of the GNU licenses (like the GNU General Public License GPL [Free Software Foundation 2007]. Large programs and software, sometimes whole OSs like the Linux kernel, are distributed under the GPL. This means the user is provided not only with the program executable but also promised the source code. Sometimes even the executable is dropped, and the software is distributed as source code to save space and avoid malware infection of executable images. The software is usually protected with the GPL (or a similar) license that copyrights the intellectual property of the author. Nonetheless, instead of building a wall around the code royalty, it grants the user the permission to use, reuse, modify, enhance, and distribute the code, as long as the original author and his or her license are not removed, and in turn, the user is granting the same permissions to the other users he or she is redistributing the code to.

Some authorities are criticizing the movement out of misconception of the word "free." The word free means—in this context—freedom of information, not freedom of price. This is depicted by the GNU organization's motto: "think of free as in free speech, not free beer." The GNU and the FSF outline four freedoms for the user of a software in their scheme (Free Software Foundation 2015):

- The freedom to run the program in any way the user wishes, including using the software for any purpose (called freedom 0)
- The freedom to study how the program works and modify it to the user needs. This is achieved by providing the source code for each and every program under one of the GNU licenses (called freedom 1)
- The freedom to redistribute the software (called freedom 2), including the source code itself and any modified code thereof (see the next point)
- The freedom to improve and modify the program and release the improved version publicly, including the modified source code (called freedom 3)

The open-source movement has become an important paradigm embraced by more and more programmers and institutions every day. As mentioned previously, whole systems have contributed to the advancement of the movement; of note is GNU/Linux, the Apache web server, among other software.

An important idiom here is that the sharing spirit of the open-source movement is not limited to any specific language as many authors believe. In fact, code written in "proprietary" programming languages such as VS family of languages can be distributed under the GPL or another open-source license.

The choice of which language to use in programming these text's examples was a debatable topic. The authors support the open-source movement and believe that it had contributed much to the advancement of software technology, and it will continue to do for the foreseeable future. This means the choice should be writing the code in C/C++ as it is the *de facto* language(s) for writing Unix/Linux software (the biggest contender under the open/free-source paradigm as discussed previously). Another option was writing in Java as it is the best cross-platform language, which means the authors' code will be guaranteed to reach the largest number of audiences. The point against the first choice was that C is a little uncomfortable for beginner programmers, and C++ is much more so, although the authors are avid C programmers by heart. As for Java, it did not deviate much from the C/C++ line of complexity; in addition, it would have involved another level of indirection (the intermediate language) that would have needed another discussion.

Finally, VB.NET has been selected for the models incorporated in this book for the following reasons:

- MS Windows is one of the most (if not *the* most) popular OSs in use today. Market shares of MS Windows are by far outgrowing all other competitive OSs collectively (Stat Counter 2015, W3Schools 2016). Chances are, when buying a new PC or laptop, the user will face this OS first and foremost (although other platforms are being dominated by other OSs, such as the Android system which is dominating on smartphones and tablets).
- The .NET framework is becoming very popular as an effective programming platform in computer industry that it will be counterproductive not to learn and understand how to implement engineering models using this platform. This is in part driven by the fact that .NET is a Microsoft technology modeled for Windows, a Microsoft OS.
- VB language is the *de facto* language of programming under MS Windows family of OSs. VB.NET came from the good old BASIC (the Beginner's All-purpose Symbolic Instruction Code), which was very appealing to novice programmers as it was known to be an easy-to-learn language, although it had modest-to-moderate capabilities compared with powerful languages of the time like C and Pascal. One of its important aspects was that VB had many layers of abstraction (hiding details) that the programmer did not need to know about the low-level workings of the system, he or she just needed to focus on programming the solution at hand. This is why BASIC, its child VB, and its grandchild VB.NET are all considered the easiest programming languages to learn, especially for beginners and those with no prior programming experience.
- Simplicity of the language in structure and content. Compared to code written in C/C++ or Java, VB is much clearer and easier to comprehend.
- Ease of learning the language and its interactive use.
- Flexibility in program alteration, updating, and patching.
- Universal language, popular, interactive, and common in use.
- Relatively standard language with many versions simple to run on different computer architectures with little or no modification, provided it is run under MS Windows or a compatible OS (like Linux with Windows Emulator WINE) installed, for example).
- Relevance to engineering applications and scientific computations being a general-purpose language, which is widely used in programming many mathematical and engineering models.
- Availability of the language software on PCs (either as part of VS or as the free Visual Basic Express Edition from Microsoft).
- Immediate detection of errors (using the *IntelliSense* feature of the VS common language IDE, a feature shared by all VS languages).
- Production of acceptable quality graphics and ease of embedding video/sound and other object files.
- The language requires no hardware knowledge, and it shields the user from the OS of the computer (Kiong 2012).

Computer programs usually are formed of subprograms (subroutines and functions) that are easily developed and tested separately. Generally, a subroutine functions to serve some specific goal and then returns control to the main program.

A function serves similar goals with the difference that it is required to return a result to the main program. Most of the current OOP use an event-driven approach. In this model, every task done by the program is triggered by an event (key press or mouse click or moving a joystick), and the program responds by executing a specific subroutine (called the event handler). This modular design approach results in short and highly focused programs.

## 1.6 ABOUT THE .NET FRAMEWORK

The newer version of the well-known VB language, which is known as VB.NET, does most of its work by relying on the .NET (pronounced "dot net") framework (shorthanded "Fx"), as does all the other members of the .NET family of programming languages (C#—pronounced "see sharp," Visual C++, J++, and others). What the framework essentially is, in simple terms, a large group of functions. These functions serve to do virtually whatever a programmer wants to do. It consists of classes grouped into different namespaces, with each class containing related functions that work on certain targets. For example, the Drawing class holds the functionality of drawing lines, rectangles, circles, coloring, and much more (Petroutsos 2002).

As such, Windows programming has come into a new paradigm, the OOP space. The user is not bound to learning specific languages to do complex tasks that cannot be done in other languages (VB used to be the *sandboxed* language, that is, it protected the user from knowing the lower level functionality of the OS). But now this has changed, as with the .NET framework, every program that is written in a .NET compatible language will be compiled to an intermediate language, known as the common intermediate language (CIL). One can think of this as something similar to assembly language, but being specific to the .NET framework (like Java byte code). All .NET programs translate to the CIL, which will finally be compiled into the final binary executable to be run later by the machine.

At run time, a special program, part of the .NET Fx, known as the *just-in-time translator*, or, as affectionately called between programmers, the *Jitter*, loads the program to be run into memory, reads in the program instruction by instruction, and directs the processor to execute each instruction in turn. This is akin to the JVM described previously, which was developed for making Java platform independent.

As such, .NET is being ambitious in trying to be platform independent in its own way. Any OS that can run the .NET Fx or a similar platform can—theoretically at least—run any CIL-compiled program. For example, Windows native programs can be run under Linux by using an emulator (like the WINE, http://www.winehq.org). Similarly, .NET programs can be programmed, tested, and debugged under Linux (and other Unix-like systems) by using a cross-platform IDE such as MonoDevelop (http://www.monodevelop.com). A major drawback that breaks the flow of this ambitious reach is the issue of copyright, as the .NET Fx and all the .NET compatible languages are registered trademarks of Microsoft.

## 1.7 WHAT IS VS.NET?

Traditionally, writing a working program used to involve many steps (see Figure 1.8), starting with entering the raw program code into some sort of a text editor (usually part of a programming IDE), checking the code for errors, running the program, and using a compiler to convert the program from a source code into a machine language. Most of these steps are essential and so are retained in today's programming, with some modifications into how they are done.

In the .NET programming world, one will eventually revert to the VS. VS is an IDE for developing Windows and Web applications (Petroutsos 2002). Many languages use VS, including VB, Visual C++, and others. What VS does is that it groups all the necessary tools used by a programmer and put it under one umbrella. VB programmers, C++ programmers, and other .NET compatible languages use the VS to program their code. Some parts of the programming process are not affected by which language is used, such as the design of a form or the contents of a database, and these parts are provided for every language's use in the common IDE of VS.NET.

In this text, we are using VB.NET (as the programming language) and VS (as the IDE).

Indeed, VS is a dynamic piece of software that changes very frequently, but changes are in terms of patches, bug fixes, and security enhancements. VS itself may be regarded as a fancy (editor + compiler) front end to make programmer's life easier. We could have chosen any other IDE front end, like MonoDevelop or Eclipse, the choice would not change anything in the source code listings of the book. Those tools are just the front ends; the user is welcome to use his or her preferred IDE if he or she does not like using VS. Though, in practice, most Windows programmers would use VS as it is the IDE developed by the same company that devised the programming languages like VB. This is like choosing between MS Word and another word processing program to edit our manuscript. It does not make any difference as long as the final work is a DOCX or DOC file.

However, VB itself is a stern and solid programming language. The core of the language changed very little through the years. This makes the programming listings of the book the same whether we were using VB 10, VB 2013, or VB 2015, or even a future version of VB. We are not using any language extensions, and we are not doing any funny business with templates, delegates, abstract classes, and other mind-boggling topics, but we are just using the core language features that are mostly the same since the original Windows 3.1 version of VB. In fact, the original manuscript we wrote over a year ago was compiled using VS 2013, and this one is done using VS 2015. We changed exactly zero lines of code in the transition. We just changed the screenshots as the window headers looked different under Windows 7 and Windows 8, that's all!

As such, all in all, although the VS will surely change in the future, VB will not (or at least not so dramatically), and so our programming modules should hold water for years to come.

As an additional security measure, we recompiled and ran all the source programs using VS versions 2010, 2013, and 2015, Professional and Enterprise editions. All programs ran smoothly and efficiently under Windows XP, 7, and 8. Moreover, we ran a couple of the executables on a Fedora Linux box (with WINE), and they ran like a charm; we edited them using MonoDevelop IDE on Linux, which is a world of difference, compared to VS on Windows, and it worked also. As such, no matter what the editor is, be it VS or something else, the user will get the programs up and running smoothly enough, provided that VB is the language used, and the .NET framework is installed on his or her system.

In summary, the version of VS used in writing this book's programs is VS 2015 Professional Edition, but the programs should compile and run on other versions (e.g., VS 2010 or 2013, Enterprise, Basic, or Express). Most of the examples have been tested and verified to be working under Linux (Fedora Linux 23) with WINE installed as an emulator (see Section 1.6).

For a step-by-step guide to using VS.NET for new users, refer to the user's guide at the end of this book (see Appendices).

## 1.8 FUNDAMENTALS OF THE VB.NET LANGUAGE

In VB.NET, the instructions can be entered in two ways (Gottfried 2002, Petroutsos 2002, Loffelmann and Purohit 2011, Kiong 2012, Schneider 2013):

1. Directly entering program instructions into the editor (usually an IDE such as VS.NET)
2. Loading previously coded program from a saved file or project

During the stage of program preparation and execution, errors (bugs) may occur. These errors can be grouped as follows:

- Syntax errors (coding mistakes, diagnostic errors) are mistakes made in the usage of the high-level language (e.g., misspelling of a key word or giving a function an inadequate number of parameters). These are easy to spot as most modern text editors provide syntax highlighting to help see erroneous words and missing braces. Furthermore, the compiler will spot such errors and will report them to the programmer.
- Runtime errors develop during program execution and are usually unforeseen during program coding and compiling, for example, accessing a memory location that is inaccessible or referencing an object

that has not been initialized. These are harder to foresee as they are not visible until running time.
- Logical errors are due to faulty program logic. They are like incorrect order of expressions or incorrect ordering of statements. Usually these types of errors are skipped by the diagnostics of the compiler, as the compiler checks the syntax of the language and makes sure that references are correct and language usage is plausible, but it has no idea about the intention of the programmer when he or she wrote the code. As such, it cannot judge if the logic of the program is correct or not. This is a hard type of error to spot, because sometimes the logical error is not apparent until there is an abnormal result of a given calculation or an abnormal response to a specific task.

Debugging and testing should be carried out to produce a reliably operating program. After debugging the program, the next phases are documentation, storage, maintenance, and updating. Documentation signifies addition of language descriptions (may be English or any other Latin-letter based) to ease implementation of the program by the user. Storage refers to saving the program in a suitable media, whereas maintenance concerns the upgrading of the program and posting patches to fix errors and security issues.

VB.NET is a simple, reliable language. The instructions resemble elementary algebraic formulas and use plain English words (those reserved by the language are known as the language's "keywords"). Use of the language covers various disciplines such as science, engineering, business, mathematics, finance, database management, commercial, and computer games, to name but a few.

The original BASIC language was developed by John Kemeny and Thomas Kurtz at Dartmouth College, Hanover, New Hampshire, in the early 1960s. It was running under the command line DOS. It then underwent extensive work and modification, and reemerged as VB. This then underwent the evolution to become the current VB.NET version. Throughout this book, VB.NET 2015, of Microsoft, is adopted for writing programs and models. There is an extensive help, both locally and online (from Microsoft—the Microsoft Developer Network and from other programming forums).

The simplest VB.NET program structure is a single Windows form, which has a *design interface* (the part holding the textboxes and buttons, and having everything to do with what the user sees) and a *code part* (the processes and calculations that happen behind the scenes), as depicted in Figure 1.9. The code is composed of a sequence of statements that are grouped into subroutines, each one being executed whenever a specific action happens (e.g., when a user clicks a button to do some calculations). Sections 1.8.1 through 1.8.17 summarize the important statements and functions used in VB.NET programming and are used in the models in this book. They do not cover all the commands in VB.NET; such a list will be exhaustive and lengthy. The interested reader

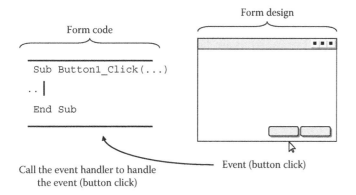

**FIGURE 1.9** Form code and design.

is directed to refer to the Microsoft documentation or other publications dealing with this language (see References).

### 1.8.1 VARIABLES AND CONSTANTS

Data are expressed in VB directly as raw data or constants and indirectly as variables. Constants refer to values unaltered during program execution, that is, the value that was hard coded during design time is preserved during execution time. Some constants are predefined as part of the .NET Fx, for example, Pi is predefined as the constant `System.Math.PI`. Variables signify symbolic names to which a changing (during execution) value is assigned. They are names that refer to a certain place in memory ready to store values of a specific type (see Figure 1.10). The programmer usually specifies which type the variable is during the writing of the code. Generally the types of variables are as follows:

- Numbers can be positive or negative and can be expressed as integer quantities (whole numbers without fraction part) or floating point (real) quantities (which include fractions). Integer quantities are written as numbers without a decimal point and can be declared as short, integer, or long, with a capacity to store 2, 4, and 8 bytes of memory, respectively (according to the platform, but these are the commonest used quantities). This means short integers can have values in the range of $-32,768$ to $+32,767$ (i.e., $-2^{15}$ to $2^{15} - 1$) for signed shorts and 0 to $+65,535$ ($2^{16} - 1$) for unsigned shorts (similarly, signed integers range from $-2^{31}$ to $2^{31} - 1$ and signed longs range from $-2^{63}$ to $2^{63} - 1$). In writing numbers, the following points need to be considered:

  - A number can be preceded by a (+, optional) or a (−) sign.
  - A number can be written in a scientific format as $x$E$y$, where $x$ is the number and $y$ is the exponent. The exponent can be positive or negative but is devoid of a decimal point. For example, 3E2 means 3 times 100.
  - Letters and symbols such as dollar signs and commas are not allowed in numeric constants.
  - Real numbers can be expressed in single or double precision. A single-precision number is a real number stored with seven significant digits plus a decimal point; a double-precision number is stored with 16 digits plus a decimal point. Both numbers are floating point numbers, that is, the decimal point floats in its position in the number depending on the value of the number. Usually, single-precision floating points are stored in 4 bytes, and double precisions are stored in 8 bytes of memory.

- Strings, which consist of a sequence of characters, are used to represent nonnumeric information. The value is written between two double quotations.
- Characters—a variable of this type stores only one character, represented as a two-byte Unicode char.

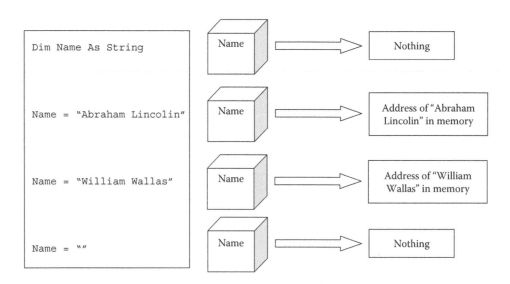

**FIGURE 1.10** Changing the value of a variable.

- Booleans—a variable of this type accepts only one of the two values: True or False, commonly used in logical comparisons to direct program flow through loops, jumps, and the conditional (IF...ELSE) statements.
- Others, less used types, like the Date and Object variables.

Examples of variable and constant declaration are as follows:

```
REM Declaring Gravitational Acceleration as a
    Constant
Const g as Double = 9.81
REM Declare some variables
Dim simpleInteger as Integer = 30
Dim myName as String = "Abdel-Magid"
Dim falseLogic as Boolean = False
```

### 1.8.2 OPERATORS

Operators are used to indicate arithmetic operations such as addition, subtraction, multiplication, division, and exponentiation as follows:

| Operator | Function |
| --- | --- |
| + | Addition |
| − | Subtraction |
| * | Multiplication |
| / | Division |
| ^ | Exponentiation |
| Mod | Modulus (remainder of integral division) |

Order of execution gives priority to exponentiations to be followed by multiplication and division, with addition and subtraction to be carried out last. The hierarchy of operation can be altered through the use of pairs of parentheses at proper places in an equation or formula. If two operands of the same priority exist in one formula, the order is left to right in execution, unless parentheses are present. For example:

$$X = 2 + 3 * 4$$

results in the value of 14 stored in X, whereas

$$X = (2 + 3) * 4$$

results in the value of 20.

### 1.8.3 THE REM STATEMENT (REMARK OR THE SINGLE QUOTE, " ' ")

The REMark statement adds descriptive comments anywhere into a VB.NET program. The compiler ignores statements beginning with REM or a single quote. Comments are an important—and seldom overlooked—practice in computer programming. Not only are comments important for the users and other programmers looking into one's code, but they are of the utmost importance for the programmer himself or herself. Consider pulling out a several month-old software to find a bug or apply an update. More often than none, you will stop at one function staring at the screen and thinking "what on earth was I thinking when I wrote this piece of code?" This is not only common, but it is *standard*. So comments are important to keep the programmer sane several months down the road when he or she wants to revise the code, and he or she *will*.

### 1.8.4 THE END STATEMENT

The End statement is used to indicate the end of a subroutine or a function, like End Sub or End Function, or as an isolated End statement to dictate the end of execution, although this last one is not needed, as an End is implied whenever a subroutine or a function comes to its last command. Sometimes, the programmer needs to force the closure of a loop without executing the remaining rounds; in this case, the word Exit is used. A piece of code is as follows:

```
Dim J as Integer = 3
For I = 1 to 10
        If I > J Then Exit For
Next I
```

It will result in four iterations of the loop, with I being 1, 2, 3, then 4, at which point Exit For will cause the loop to break.

### 1.8.5 THE GoTo STATEMENT (TRANSFER OF CONTROL AND UNCONDITIONAL BRANCHING)

The GoTo statement is used to transfer the program to some other point of execution, changing the sequence of commands. It can be combined with IF clauses to bind the branching decisions to different situations, also known as conditional branching. The GoTo statement needs a label to jump to. A label is just a name given to a specific point in the program, usually as a string followed by a colon. For example:

```
...
If X >= 0 Then GoTo PositiveX
GoTo NegativeX
PositiveX:
  ...
NegativeX:
  ...
```

### 1.8.6 INTRINSIC FUNCTIONS (LIBRARY OR BUILT-IN FUNCTIONS)

Table 1.1 outlines some of the mathematical functions that are commonly used in a scientific VB program. Most of the mathematical functions are grouped in the namespace System. Math. A namespace is a name used to define a group of classes that are related to each other in certain ways. The namespace can contain class definitions, constants (such as "Pi" in the Math namespace), and functions (such as the mathematical functions described next). Some of the popularly used namespaces in VB.NET programming are System.Math (for mathematical functions), Microsoft.VisualBasic. FileIO (for file reading, writing, and other operations), and

**TABLE 1.1**

**Some of the Mathematical Functions Used in VB (Part of the `System.Math` Namespace)**

| Function | Description |
|---|---|
| Abs(x) | Determines the absolute value of the function $x$; $y = |x|$ |
| Sqrt(x) | Determines the square root of $x$, $x > 0$; $y = \sqrt{x}$ |
| Log(x) | Determines the natural logarithm of $x$, $x > 0$; $y = \log x$ |
| Log10(x) | Determines the logarithm to base 10 |
| Exp(x) | Determines the exponential of $x$; $y = e^x$ |
| Sin(x) | Determines the sine of $x$ ($x$ in radians); $y = \sin(x)$ |
| Cos(x) | Determines the cosine of $x$ ($x$ in radians); $y = \cos(x)$ |
| Tan(x) | Determines the tangent of $x$ ($x$ in radians); $y = \tan(x)$ |
| Atan(x) | Determines the arctangent of $x$ ($x$ in radians); $y = \text{atan}(x)$ |

**TABLE 1.2**

**Relational (Boolean or Logical) Operators (Example Column Uses $A = 1$ and $B = 0$)**

| Operator | Relationship | Example Usage | Example's Result |
|---|---|---|---|
| = | Equal | $A = B$ | False |
| <> | Not equal | $A <> B$ | True |
| < | Less than | $A < B$ | False |
| > | Greater than | $A > B$ | True |
| <= | Less than or equal to | $A <= B$ | False |
| >= | Greater than or equal to | $A >= B$ | True |
| Not | Negates the following expression | Not $A$ | False |

`Microsoft.Windows.`*Forms* (general Windows forms functions). If a namespace is required to be incorporated into the program, it will need to be *imported* or inserted at the top of the code, before the `Public Class Form` statement (essentially, before any of its components are used), such as

```
Imports System.Math
Imports Microsoft.VisualBasic.FileIO

Public Class Form1
...
```

However, if only a few functions or declarations of the namespace are required and importing the whole namespace seems like a lot, then accessing a specific member of the namespace can be done by prefixing the namespace before the desired member and separate them by a dot, such as

```
REM Define Radius
Dim Radius As Double = 6.0
REM Calculate Area
Dim Area As Double = Math.PI * Math.Pow(Radius, 2)
```

### 1.8.7 The `IF...THEN` Construct (Conditional Transfer)

The `IF...THEN` statement construct is used to perform a conditional branching operation. The condition occurring between the `IF` and the `THEN` relates an expression (variables, formulas, constants) to another through one of the operators shown in Table 1.2. The result of the operation should be a Boolean expression evaluating to True or False. It can be used in an `IF...THEN...ELSE` block as follows:

```
IF {condition1} THEN
      (What to do in case the condition is TRUE)
ELSEIF {condition2} THEN
      (What else to do in the case second
      condition is TRUE)
ELSE
      (What to do in case all the conditions
      are FALSE)
END IF
```

If the operator is omitted, the condition is assumed to be TRUE, for example:

```
IF Me.Visible = TRUE THEN ...
is syntactically equal to
IF Me.Visible THEN ...
```

### 1.8.8 The `CASE` Construct

The `IF...ELSE` construct is very useful and often used, but it can become tricky in case the conditions are numerous, or in case one needs to nest many `IF` conditions inside each other. The `CASE` construct allows the selection of one path from a group of alternative paths. It is syntactically equivalent to a long list of `IF...ELSEIF...ELSE` statements in a more elegant and structured way. It has the following structure:

```
SELECT CASE x
CASE a
      What to do
CASE b
      What else
...
CASE ELSE
      Default case if none is applicable
END SELECT
```

### 1.8.9 The `FOR...TO...NEXT` Block (Building a Loop)

The `FOR...TO` statement specifies the number of executions of a loop in the program. The statement includes a running variable whose value changes each time the loop is executed. The `NEXT` statement is used to end the loop. For example:

```
FOR X = 1 TO 12 STEP 2
      Debug.Print(3*X)
NEXT
```

will print (in the Debug window):

```
3
9
15
21
27
33
```

If the index variable is intended to be incremented by 1 each time the loop is run, the STEP clause can be omitted, such as

```
FOR I = 1 TO 12
      Debug.Print(3*I)
NEXT
```

This will print the multiplication table of 3 in the Debug window.

### 1.8.10  DO...LOOP

The DO..LOOP construct is a control flow statement that repeats a block of statements while a condition is true or until a condition becomes true. This is handy if the number of loops is not known beforehand, in which case a FOR...NEXT loop will be impossible to use. For example, looping while waiting for user input until the user gives a certain response means one needs to loop until some condition (user response equals expected response) becomes true. This is known as the Until loop. However, one can loop as long as the user is giving a specific response, which is known as the While loop. A While loop can be written as follows:

```
Do
      ...
Loop While X = 0
```

The same loop can be written as a Do...Until loop:

```
Do
      ...
Loop Until X <> 0
```

Notice that the previous loops will be executed at least once, even if the condition is false (first loop) or true (second loop). If this is not the desired behavior, the condition can be put on top of the loop as follows:

```
Do While X = 0
      ...
Loop
```

In this case, the loop will be executed *only* if $X = 0$; otherwise, it will *never* be entered.

### 1.8.11  THE STRING CLASS

The STRING class is a class that contains string processing functions, such as string concatenation, returning substring of a string, changing the string's capitalization case, and getting the length of a string. This is an important and handy class when dealing with strings, a common and everyday practice.

### 1.8.12  THE FORMAT FAMILY OF FUNCTIONS

The FORMAT family of functions is used to format a string for output in a certain way. Some of the styles used are

predefined in the function. It takes a string input and a style, then applies the style to the input string, and gives the result as a string object. One function that is helpful when dealing with number is the FormatNumber() function, which is used as

```
FormatNumber(X,3)
```

The first operand is the number to format, and the second is the number of decimal places to put. The return value is the number formatted as a string with the requested decimal digits.

### 1.8.13  THE DRAWING FUNCTIONS

Some of the drawing functions with relevance to engineering applications are:

1. DrawLine() is a graphics function that draws a line on the graphics screen. The coordinates of the ends of the line and the drawing color must be provided as parameters to the function.
2. DrawRectangle() is a graphics function that draws a rectangle (a box if the sides are of equal length) on the graphics screen. The coordinates must be provided, along with the color of the border.
3. DrawEllipse() and DrawArc() are used to draw circles, ellipses, and arcs, provided the *X/Y* coordinates, and the width and height are given as parameters. The functions also need drawing color to be given with the parameters.
4. DrawString() draws a string text on the graphics screen, using the Font selected, with the specified color.

These drawing functions need a drawing board to draw on; this board is known as a "Graphics object." A *Graphics object* is associated with controls that have a drawing surface, such as a form (the window on which the other controls appear) or a *PictureBox* (an object that is used to display pictures). The *Graphics object* to draw on must be first created and then used to draw on. A simple example goes as follows:

```
Dim g As Graphics
g = PictureBox1.CreateGraphics
REM Parameters are: Color, X, Y, Width, Height
g.DrawLine(Pens.Black, 0, 0, 10, 10)
g.DrawString("Hello", New Font(FontFamily.
  GenericMonospace, 8), Brushes.Blue, 50, 50)
```

### 1.8.14  THE DIM STATEMENT

The DIM (dimension) statement is used to declare variables. If used on arrays (which represent a collection of variables of the same type and size, accessed all under one name, useful for representing a list or a table), it allows specifying the maximum number of elements (size of array). Beware that arrays are zero indexed (i.e., indexing starts at 0, not 1!), so the

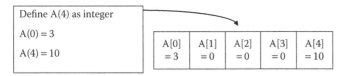

**FIGURE 1.11**  Declaring an array.

array's dimension(s) is usually one size more than the given dimension(s), for example, the declaration

```
DIM x(3) As Integer
```

Declares a one-dimensional array of integers, with four elements in the first dimension. Notice there are four elements, not three, as indexing goes as 0, 1, 2, 3 and not as 1, 2, 3. The individual elements within an array are called subscripted variables. The declared variable can be assigned a type on declaration (called variable initialization) or later when assigned a value (Figure 1.11).

Arrays can be multidimensional (arrays of arrays):

```
Dim xy(3, 5) As Double
```

which declares an array, xy, which has two dimensions: three rows by five columns (or vice versa). If you declare a size-less array such as

```
Dim noSize() as String
```

you will need to specify its size later with the ReDim command:

```
ReDim noSize(10)
```

### 1.8.15  Functions and Subroutines

As programs become more complex and bigger in size, the need to fragment a program into smaller chunks that can be handled separately is imperative. A section of a program that handles a specific function is called a *subroutine*, or simply a Sub. A subroutine that performs some calculations and returns a result to the caller is called a Function. Hence, the difference between subroutines and functions is that the latter returns a result to the caller, whereas the former does not. This indeed is such a subtle difference that some languages such as C and most of its descendents ignore it completely.

In an event-driven environment where events trigger actions, the special subroutine that handles a specific event is known as the *event handler*. Each and every object has a set of events associated with it, for example, each action such as window resize, maximize, and minimize triggers a specific event that calls on its specific event handler. To start coding an event handler to the default event associated with an object, double-click that object in the Design window to open the Code window and point the cursor at that event handler.

For example, when designing, if a programmer double-clicks on a button, the IDE automatically creates an empty stub subroutine that functions as an event handler for the button_click event, because the *click* event is the default event for a button object.

The CALL statement is used to transfer control to the subroutine, and the End Sub and Return statements are used to cause control to transfer back to the calling part of the program or to the statement following the point of reference. The function call is similar, except it ought to return a value to the caller.

An example of a subroutine definition is as follows:

```
Sub doSomeAction()
     ...
End Sub
```

And an example of a function is as follows:

```
Function addTwoNumbers(ByVal x as Double,
ByVal y as Double) As Double
     Return x+y
End Function
```

Passing parameters to a function or a subroutine can be done in two ways:

1. Passing by value (written as ByVal), which means the function is handed a copy of the original variable. Any changes applied by the function to this copy do not affect the original variable in the caller.
2. Passing by reference (written as ByRef), which means the function is handed a reference to the variable, that is, a pointer that tells it where the variable resides in memory. This means any changes applied by the function on the variable are visible and seen by the caller.

### 1.8.16  The MSGBOX and INPUTBOX Functions

These are used to provide a predefined dialog box to the user, containing a message with response options (e.g., a message box with the button YES/NO or OK/CANCEL) or providing a means of input for the user (INPUTBOX). They are used frequently to convey messages to the user and to gather simple input from the user during program execution. The parameters passed to the MSGBOX are the message to the user (the prompt), the title (optional), and the buttons to display and/or the icon to show (a big "I" for information, a red cross for error, etc.).

### 1.8.17  Objects and Their Properties

Everything is an object, and objects appear in programming as controls. Some of the most used controls are as follows:

- *Windows form*: A window that holds other controls inside it
- *TextBox*: A text entry field that is commonly used to gather data input from user

- *Label*: A control to show some text, like a message to the user, commonly used with textboxes to show the user what data to enter in a textbox
- *CheckBox*: Used to give the user the option of enabling/disabling certain features
- *RadioButton*: Used in groups; only one RadioButton can be checked at any time in a RadioButtons group
- *ListBox*: Used to provide multiple options to the user to select from or to display many data as a list
- *ComboBox*: Similar to a ListBox, except the items are shown in a drop-down list
- *PictureBox*: Originally used to display images, but can be used as a drawing board to provide custom drawings to the user. This is done using functions described earlier as `DrawLine()` and `DrawRectangle()`

Every control has properties that define how it looks and behaves. Some of the common properties are as follows (see Table 1.3):

- *Text*: What is displayed on the control or on the title bar of the form.
- *Visible*: Shows/hides the control or form.
- *Size (height, width)*: Sets the coordinate size of the control (in pixels or other measurement units).
- *Location (x, y)*: Sets the coordinate location of the control on the screen (for forms) or inside a form (for controls).
- *Items*: Members of a ListBox or a ComboBox contain the items that are displayed on the control.
- *Checked*: Boolean value that tells if the checkbox is checked or cleared.

To simplify the programming process, all programs in this book were written in a uniform way, with most of the calculations contained in a function called `calculateResults()`. In some of the examples, an additional *showResults()* function is responsible for showing the results on the form. Most of the other assignments (giving names to labels, buttons, textboxes, etc) were written inside the code (usually the `Form _ Load()` subroutine) to make the process of following the code easy. The default names of the controls were unchanged. This allows the reader easily to follow between the code and the design (e.g., *TextBox1, Label2* and *Button3* were left with their default names). Usually, in real-world programming, giving names to controls, setting the text of a form, resizing a button, and other simple assignments will be done during design time using the Properties window. In fact, the reader is encouraged to modify the programming examples in this book by reading the assignments given in the `Form _ Load()` sub of each program, reverting to the design mode, and changing the properties as indicated. For example, if an assignment is given *Me.Text = "Program 2.1,"* go to the design, select the form, then select *Text* in the Properties window, and change its value to *"Program 2.1"*, then delete the statement from the code, and so on.

Apart from using statements for entering data in a program, data may also be introduced using files. A data file is a collection of records maintained on a disk or other storage media. Data files can easily be read and updated by a VB.NET program. Files can be grouped as sequential and direct access files. Sequential files are those files in which individual items are entered and accessed sequentially, one after another, usually in the form of bytes or characters. Direct access (random access) files consist of individual data items that are not arranged in any particular pattern or order. The latter files retrieve data more quickly and more efficiently, and individual records can easily be modified. Table 1.4 presents a summary of common VB.NET statements, functions, and commands.

## 1.9 PROGRAMS ON THE ACCOMPANYING CD/ROM

The programs on the CD accompanying this book are all in source form with their executables. Each set of programs carries the name of the chapter in the book to which they pertain. For example, all the programs of Chapter 2 are under the folder "Projects\CHAP2\".

If a project's folder is opened in Windows Explorer (like "Projects\CHAP2\CHAP2.1\"), one will find the following files and folders (this list is not comprehensive):

- *CHAP2.1.sln*: This is the solution file (a solution contains one or more projects). It contains all information VS needs to open your solution. When opening a project later after being closed, one will double-click this file to reopen the project in VS.
- *CHAP2.1.suo*: This is the solution's user options. This is used internally by VS.

### TABLE 1.3
### Some of the Common Control Properties

| Property | Purpose | Example |
|---|---|---|
| Text | What is displayed on the control? | `TextBox1. Text = "hello" Form1.Text = "Example Program"` |
| Visible | Shows/hides the control | `Me.Visible = True TextBox1. Visible = False` |
| Checked | Gets or sets whether the checkbox or radio button is checked | `CheckBox1. Checked = True` |
| Size (height, width) | Sets the size of the control | `TextBox1.Height = 20 TextBox1.Width = 100` |
| Multiline | Determines if a textbox supports multiple lines of output | `textBox1. Multiline = True` |

**TABLE 1.4**

**Summary of VB.NET Statements, Functions, and Commands**

| Statement/Function/Command | Purpose | Example |
|---|---|---|
| Abs | Gives absolute value | Y = ABS(x) |
| Atan | Returns arctangent | Y = ATAN(x) |
| Call | Calls a subroutine | CALL doSomething() |
| DrawEllipse, DrawArc | Draws a circle, an ellipse, or an arc | g.DrawEllipse(Pens.Black,10,10,40,40) |
| DrawLine | Draws lines | g.DrawLine(Pens.Black, 0, 0, 10, 10) |
| DrawRectangle | Draws a rectangle or a box | g.DrawRectangle(Pens.Black, 0, 0, 20, 40) |
| DrawString | Draws a text string, given the brush color, the *X–Y* coordinates of where to start drawing, the string to be written, and the font to be used | g.DrawString("Hello", New Font(FontFamily.GenericMonospace, 8), Brushes.Blue, 50, 50) |
| FileOpen | Opens a file, given file number (ID), file name, and open mode | FileOpen(5, "Hello.txt", OpenMode. Input) |
| FileClose | Closes a file | FileClose(5) |
| Cos | Determines cosine function | Y = COS(x) |
| Dim | Defines variables or arrays | DIM x as Integer |
| | | DIM y(5,200) as Float |
| End | Ends a function or subroutine | END SUB |
| | | END FUNCTION |
| Exp | Determines the exponential function (E to exponent *x*) | Y = EXP(x) |
| For...Next | Defines the beginning and end of a FOR...TO loop | FOR m = 1 to 39 NEXT m |
| GoTo | Transfers program to a remote statement | GOTO someLabel |
| If...Then | Conditional run | IF y < 8 THEN |
| | | ELSE |
| | | END IF |
| Log | Returns the natural logarithm | Y = LOG(x) |
| Log10 | Returns base 10 logarithm | Y = LOG10(x) |
| REM | Inserts comments and remarks in the program | REM Program listing |
| Sin | Determines the sine function | Y = SIN(x) |
| Sqrt | Determines the square root of a value | Y = SQRT(x) |
| Atan | Determines the arctangent | Y = ATAN(x) |
| Val | Changes a string to a numerical value | Y = VAL(x) |

- *CHAP2.1* (*folder*): This contains the files and resources of the project. Inside this folder, the following files and subfolders are present:
  a. *CHAP2.1.vbproj*: The project's file. If there is more than one project in the solution, each one will have its own project file describing the respective projects.
  b. *CHAP2.1.vbproj.user*: User's options, similar to the solution's user options, except this one is for the individual project.
  c. *Form1.vb*: Code for Form1 controls (button and form event handlers and user-defined functions and subroutines). What is present in this file is what is found in the program listings in the text.
  d. *Form1.Designer.vb*: Code generated by the designer to describe the layout of the form and the different controls it contain.
  e. *Form1.resx*: Resources associated with the form, such as images and icons.
  f. *Bin* (*folder*): This folder contains two folders— *Debug* and *Release*. According to the build mode selected when building the project in VS, the project executable files will be under one (or both) of these directories. Inside either folder, there will be some files; the file named *CHAP2.1.exe* is the program executable proper, which can be invoked by double-clicking it to tell Windows to run the executable.

In order to use the accompanying CD/ROM, the following system hardware and software specifications are proposed:

- A PC or a laptop
- A CD/ROM drive (or a CD emulator if using an ISO copy of the CD)
- A hard disk drive with at least 70 MB free space (for the code and EXE files)
- A math coprocessor chip recommended for enhancing performance (usually incorporated into modern Intel processors)
- A Windows OS version XP, 2003, 7 or higher

- An installed .NET framework package version 4.x or later (see the User's Manual on how to download and install the latest version)
- To edit, compile, and test the source files, VS is required (see the User's Manual on how to download and install it)

## 1.10 HOMEWORK PROBLEMS IN PROGRAMMING CONCEPTS

1. What are the hardware components of a computer system?
2. Describe briefly the von Neumann model and relate it to Babbage's model structure.
3. What storage medium is most commonly used with microcomputers?
4. Compare between RAM and ROM.
5. What is the "brain" of a computer? What component functions much the same as the human sensory organs?
6. Distinguish between bit and byte.
7. What are the various types of software?
8. Assess the three basic types of mathematical models.
9. Appraise the four types of special software available to direct the computer to perform required tasks.
10. Grade at least five computer languages other than VB.
11. Justify the difference between high-level programming language and machine code.
12. Briefly describe the difference between the three levels of programming languages.
13. Choose three of the special-purpose languages and their uses.
14. What does the term "debugging" mean?
15. Revise the different types of errors or program bugs.
16. Value the term "high-level language."
17. What are the steps involved in designing a computer program?
18. What is an algorithm?
19. What are the arithmetic operators and what are their meanings? Give an example of each.
20. What is the order a computer follows when executing an arithmetic expression? How to change this order?
21. What is a keyword? Give five examples of VB keywords and their uses.
22. What is the purpose of the keyword REM?
23. What is the purpose of the End statement?
24. What is a constant? Give an example of a statement that declares a constant.
25. What is an integer? Give an example.
26. What is the range of signed and unsigned short integers?
27. Differentiate between supervisor and user programs.
28. What is a loop?
29. What is an infinite loop?
30. Predict what happens when the computer executes an IF...THEN statement.
31. What is a character string?
32. How is a character variable formed?
33. Assess the difference between a function and a subroutine.
34. Write a short essay about free- and open-source software.

## REFERENCES

Andrews, J. 2013. *PC Troubleshooting Pocket Guide.* Boston, MA: Course Technology.

Brouwer, A. 2009. *Keyboard Scancodes.* http://www.win.tue.nl/~aeb/linux/kbd/scancodes-1.html (accessed February 10, 2016).

Buyya, R., Selvi, S. T., and Chu, X. 2009. *Object Oriented Programming with Java: Essentials and Applications.* New Delhi, India: Tata McGraw Hill Education Private Limited.

Chapra, S. C. 2008. *Surface Water-Quality Modeling.* Long Grove, IL: Waveland Press.

Chivers, I., and Sleightholme, J. 2015. *Introduction to Programming with Fortran: With Coverage of Fortran 90, 95, 2003, 2008 and 77,* 3rd ed. Cham, Switzerland: Springer.

Dandamudi, S. P. 2005. *Guide to Assembly Language Programming in Linux.* New York: Springer Science.

Free Software Foundation. 2007. *The GNU General Public License,* Version 3. http://www.gnu.org/licenses/gpl.html (accessed February 10, 2016).

Free Software Foundation. 2015. *The GNU Operating System.* http://www.gnu.org (accessed February 9, 2016).

Front Page – Free Software Foundation 2016 – working together for free software. http://www.fsf.org (accessed February 1, 2016).

Gottfried, B. S. 2002. *Schaum's Outline of Theory and Problems of Programming with Visual Basic.* Schaum's Outline Series. New York: McGraw-Hill.

Hager, G., and Wellein, G. 2010. *Introduction to High Performance Computing for Scientists and Engineers.* Boca Raton, FL: CRC Press.

Hill, A. E., and Pilkgton, R. D. 1990. *A Complete AutoCAD Databook.* Englewood Cliffs, NJ: Prentice-Hall.

Ji, Z. G. 2008. *Hydrodynamics and Water Quality: Modeling Rivers, Lakes, and Estuaries.* Hoboken, NJ: Wiley-Interscience.

Kiong, L. V. 2012. *Visual Basic 2010 Made Easy.* North Charleston, SC: CreateSpace Independent Publishing Platform.

Loffelmann, K., and Purohit, S. 2011. *Microsoft Visual Basic 2010 Developer's Handbook.* Developer Reference Series. Ottawa, Canada: Microsoft Press.

Murdocca, M. J., and Heuring, V. P. 1999. *Principles of Computer Architecture.* Upper Saddle River, NJ: Prentice Hall.

Open Source Initiative. 2016. http://www.opensource.org (accessed February 2, 2016).

Petroutsos, E. 2002. *Mastering Visual Basic.NET.* Alameda, CA: SYBEX Inc.

Schildt, H. 2007. *Java: A Complete Reference,* 7th ed. Osborne, KS: McGraw-Hill.

Schneider, D. I. 2013. *An Introduction to Programming Using Visual Basic 2010 (with Visual Studio 2012 Express Edition DVD),* 9th ed. Upper Saddle River, NJ: Pearson.

Stat Counter. 2015. *Top 8 Operating Systems in Asia from Jan to Aug 2015.* http://gs.statcounter.com/#all-os-as-monthly-201501-201508 (accessed February 12, 2016).

Welsh, J., and Elder, J. 1988. *Introduction to Pascal,* 3rd ed. Englewood Cliffs, NJ: Prentice-Hall.

W3Schools.com. 2016. *OS Platform Statistics.* http://www.w3schools.com/browsers/browsers_os.asp (accessed February 12, 2016).

# 2 Computer Modeling Applications for Water and Wastewater Properties

## 2.1 INTRODUCTION

This chapter presents water and wastewater characteristics that can be expressed by mathematical equations or models and then formulated into workable computer programs. Also included in this chapter are computer programs dealing with water resources or sources, usage, and groundwater flow.

The importance of water for all ecosystems is well known, and the availability of water resources has governed the location and evolution of civilizations, their progress, development, and the well-being of man's socioeconomic situation (or condition). The importance of water is manifested in the following (Abdel-Magid 1986; Abdel-Magid and ElHassan 1986; Grigg 2012; Howe et al. 2012; Lin 2014; London and General Water Purifying Company 2012; McGhee and Steel 1991; Nemerow et al. 2009; Peavy et al. 1985; Ricketts et al. 2003; Rowe and Abdel-Magid 1995; Spellman 2013; Vesilind and Peirce 1997; Viessman et al. 2015):

1. Existence and evolution of the three kingdoms of plant, animal, and protista
2. Dissolution of organic and inorganic materials
3. Initiation and implementation of agricultural and industrial development projects
4. Initiation, execution, and operation of recreational activities such as boating, site-seeing areas, fishing, swimming, and golf courses
5. Transportation of raw materials and navigational activities
6. Conveyance and final disposal of wastes

## 2.2 PROPERTIES OF WATER AND WASTEWATER

The characteristics of water or wastewater are important for various purposes such as the following (Rowe and Abdel-Magid 1995):

* Selecting treatment facilities
* Evaluating operation of existing units
* Evaluating relevance of methods and policies
* Estimating past, present, and future degrees of pollution
* Managing discharge, recreation, and reuse
* Adopting guidelines, regulations, bylaws, and standards
* Employing appropriate surveillance and monitoring systems
* Introducing or updating operational, maintenance, and training schemes and programs
* Managing and controlling environmental quality

Classification of water and wastewater characteristics involves physical, chemical, biological, and radiological parameters.

## 2.3 PHYSICAL PROPERTIES

Physical characteristics are those parameters that are governed by forces of a physical nature. The most important physical parameters include temperature, turbidity, taste, odor, color, conductivity, salinity, solids content, viscosity, and moisture content. Those parameters that are most easily represented by a mathematical equation or formula are presented in this chapter, as well as the computer programs associated with that equation, formula, or model.

### 2.3.1 TEMPERATURE

Water and wastewater may experience variations in temperature due to climatic influences, hot discharges, and industrial discharges. An increase in temperature can pose certain problems such as the following:

* Affecting the performance of treatment units
* Reducing the concentration of dissolved oxygen (DO)
* Accelerating the rate of chemical and biochemical reactions
* Reducing the solubility of gaseous substances
* Increasing the rate of corrosion of materials
* Increasing the toxicity of dissolved substances
* Increasing undesirable growths
* Increasing problems of taste and odor

Equations 2.1 through 2.4 can be used to convert between Centigrade (C), Fahrenheit (F), Kelvin (K), and Rankine (R) temperatures.

$$°C = \frac{5}{9}(°F - 32) \tag{2.1}$$

$$°K = °C + 273.15 \tag{2.2}$$

$$°F = \frac{9}{5}°C + 32 \tag{2.3}$$

$$R = °F + 459.67 \tag{2.4}$$

**Example 2.1**

1. Write a short computer program that enables converting temperatures between Fahrenheit, Centigrade, Kelvin, and Rankine temperatures.
2. Convert 20°F, 160°F, and −40°F to Celsius and Kelvin reading. Convert 30°C, 75°C, and −10°C to Fahrenheit and Rankine degrees.

**Solution**

1. For solution to Example 2.1 (1), see the listing of Program 2.1.

2. Solution to Example 2.1 (2):
   a. Given: $T = 20°F$, $160°F$, and $−40°F$; $T = 30°C$, $75°C$, and $−10°C$.
   b. Determine $°C_{20} = (5/9)*(°F−32) = (5/9)*(20−32) = −6.7$. Similarly, $°C_{160} = 71.1$ and $°C_{−40} = −40$.
   c. Find $K−_{6.7} = °C + 273.15 = −6.7 + 273.15 = 266.45$. Similarly, $K_{71.1} = 344.25$ and $K_{−40} = 233.15$.
   d. Find $°F_{30} = (9/5)*C + 32 = (9/5)*30 + 32 = 86$. Similarly, $°F_{75} = 167$ and $°F_{−10} = 14$.
      * Compute $°R_{86} = 86 + 459.67 = 545.67$. Similarly, $°R_{167} = 626.67$ and $°R_{14} = 473.67$.

---

### LISTING OF PROGRAM 2.1 (CHAP2.1\FORM1.VB): CONVERTING TEMPERATURE TO DIFFERENT SCALES

```
'*********************************************************************
'Program 2.1
'Converts temperature to different scales
'*********************************************************************

Public Class Form1
    Dim ttl(3) As String            'The members of the list of selections
    Dim f, c, k, r As Double        'Variables used in calculations
    '****************************************************************
    'Code in Form_Load sub will be executed when the form is loading,
    'i.e., first thing after program loads into the memory
    '****************************************************************

    Private Sub Form1_Load(ByVal sender As System.Object,
ByVal e As System.EventArgs) Handles MyBase.Load
        ttl(0) = "Fahrenheit"
        ttl(1) = "Centigrade"
        ttl(2) = "Kelvin"
        ttl(3) = "Rankine"
        Me.FormBorderStyle = Windows.Forms.FormBorderStyle.FixedSingle
        Me.MaximizeBox = False
        r = f = c = k = 0           'Initialize all variables to '0'
        Me.MaximizeBox = False
'Text shown on top
        Me.Text = "Program 2.1:Converts temperature to different scales"
'Label control on left upper part of the form
        Label1.Text = "Convert temperature from:"
        ListBox1.Items.Clear()      'Initialize
        For i = 0 To ttl.Length - 1
            ListBox1.Items.Add(ttl(i))      'Add members to the list
        Next

        'The labels on left and right lower part of the form will be invisible initially,
        'together with the Textbox and the NumericUpDown control.
        Label2.Visible = False
        TextBox1.Visible = False
        Label4.Text = "Decimal Places:"
        NumericUpDown1.Maximum = 10
        NumericUpDown1.Minimum = 0
```

```
          NumericUpDown1.Value = 2
          Label4.Visible = False
          NumericUpDown1.Visible = False
          'Set sizes for label at upper right side of the form.
          'AutoSize is set to false to allow user resizing.
          Label3.AutoSize = False
          Label3.Height = 95
          Label3.Width = 150
          viewResults()
      End Sub

      '******************************************************************************
      'Code in SelectedIndexChanged member of the ListBox will be executed
      'when the user clicks on a member of the listbox and changes selection.
      'The following code will recalculate the results and show them when ever
      'the user changes his selection.
      '******************************************************************************

      Private Sub ListBox1_SelectedIndexChanged(ByVal sender As System.Object,
  ByVal e As System.EventArgs) Handles ListBox1.SelectedIndexChanged
          'Make sure the selected unit is a valid one, else exit the sub safely.
          If ListBox1.SelectedIndex < 0 Then
              Label2.Visible = False
              TextBox1.Visible = False
              Label4.Visible = False
              NumericUpDown1.Visible = False
              Exit Sub
          End If
          Label2.Text = "Enter temperature in " + ListBox1.SelectedItem.ToString
          Label2.Visible = True
          TextBox1.Visible = True
          Label4.Visible = True
          NumericUpDown1.Visible = True
          calculateResults()
          viewResults()
      End Sub

      '******************************************************************************
      'Code in TextChanged member of the TextBox will be executed
      'when the user changes the text.
      'The following code will recalculate the results and show them when ever
      'the user changes the input.
      '******************************************************************************

      Private Sub TextBox1_TextChanged(ByVal sender As System.Object,
  ByVal e As System.EventArgs) Handles TextBox1.TextChanged
          calculateResults()
          viewResults()
      End Sub

      Function FnCen(ByVal f)                    'Function to change Fahrenheit to Centigrade

          Return 5 / 9 * (f - 32)
      End Function

      Function FnFah(ByVal c)                    'Function to change Centigrade to Fahrenheit
          Return 9 / 5 * c + 32
      End Function
```

```
Function FnKel(ByVal c)                  'Function to change Centigrade to Kelvin
    Return c + 273.15
End Function

Function FnRan(ByVal f)                   'Function to change Fahrenheit to Rankine
    Return f + 459.67
End Function

'*************************************************************************************
'The following function is used to format the output of the calculations
'it uses the value selected in the NumericUpDown control as a guide to
'the number of decimal places in output, then returns 'n' as a string
'formatted according to the set number of decimals.
'*************************************************************************************

Function formatN(ByVal n As Double) As String
    'Use the function as FormatNumber(NumberToFormat, NumberOfDecimals).
    Return FormatNumber(n, NumericUpDown1.Value).ToString
End Function

'*************************************************************************************
'The following code calculates the results of changing the degrees.
'It first checks for valid input in the TextBox, then calculates based
'on the selected item in the ListBox.
'*************************************************************************************

Sub calculateResults()
    Dim t = Val(TextBox1.Text)
    If Not IsNumeric(t) Or t = 0 Then
        r = f = c = k = 0
        Exit Sub
    End If

    Select Case ListBox1.SelectedIndex
        Case 0
            f = Val(TextBox1.Text)
            c = FnCen(f)
            k = FnKel(c)
            r = FnRan(f)
        Case 1
            c = Val(TextBox1.Text)
            f = FnFah(c)
            k = FnKel(c)
            r = FnRan(f)
        Case 2
            k = Val(TextBox1.Text)
            c = k - 273.15
            f = FnFah(c)
            r = FnRan(f)
        Case 3
            r = Val(TextBox1.Text)
            f = r - 459.67
            c = FnCen(f)
            k = FnKel(c)
        Case Else
            'Safeguard mechanism. If any error in the selection,
'Zero all the variables.
            r = f = c = k = 0
    End Select
End Sub
```

```
'*******************************************************************************
'Show the calculations done in calculateResults() in the Label3 control.
'*******************************************************************************
Sub viewResults()
    Label3.Text = "C=" + formatN(c) + Chr(10) + Chr(13) _
                + "F=" + formatN(f) + Chr(10) + Chr(13) _
                + "K=" + formatN(k) + Chr(10) + Chr(13) _
                + "R=" + formatN(r)
End Sub


'*******************************************************************************
'Whenever a different number of decimals is selected, re-show the results.
'*******************************************************************************

    Private Sub NumericUpDown1_ValueChanged(ByVal sender As System.Object,
ByVal e As System.EventArgs) Handles NumericUpDown1.ValueChanged
        viewResults()
    End Sub
End Class
```

### 2.3.2 Conductivity

Conductivity denotes the ability of an aqueous solution to carry an electric current. This ability is a function of concentration, mobility, valence and relative concentration of ions, and temperature. Generally, solutions of most inorganic acids, bases, and salts are relatively good conductors. The conductivity may also be defined as "the electrical conductance of a conductor of unit length and unit cross-sectional area" (Rowe and Abdel-Magid 1995).

The relationship between conductivity and concentration of dissolved solids in a solution may be expressed as follows:

$$EC = \frac{TDS}{a} \qquad (2.5)$$

where:

EC is the electrical conductivity, micromho/cm or microsiemens/cm

TDS is the total dissolved solids, mg/L

$a$ is the constant

### Example 2.2

1. Write a computer program that enables the computation of the electrical conductivity of a sample of known concentration of total dissolved solids ($TDS_s$), given the total dissolved solids and electrical conductivity of another sample as $TDS_1$ and $EC_1$, respectively.
2. The TDS of a certain sample is recorded as 890 mg/L, and its EC amounted to 1025 mmhos/cm. Determine the electrical conductivity of another sample that has a TDS of 1450 mg/L.

#### Solution

1. For solution to Example 2.2 (1), see the listing of Program 2.2.
2. Solution to Example 2.2 (2):
   a. Find the constant $a$ for the sample as $a = TDS/EC$ by $a = 890/1025 = 0.87$.
   b. Determine the electrical conductivity for the second sample by $EC_2 = TDS_2/a = 1450/0.87 = 1667$ mmohs/cm.

---

**LISTING OF PROGRAM 2.2   (CHAP2.2\FORM1.VB): ELECTRIC CONDUCTIVITY FOR TDS VALUE**

```
'*******************************************************************************
'Program 2.2: Conductivity
'Computes electric conductivity of a sample of known TDS concentration
'*******************************************************************************

Public Class Form1
    Dim TDS1, EC1, TDS2, EC2 As Double 'Variables used in calculations
    '*******************************************************************************
    'Code in Form_Load sub will be executed when the form is loading,
    'i.e., first thing after program loads into the memory
    '*******************************************************************************
```

```vb
    Private Sub Form1_Load(ByVal sender As System.Object,
ByVal e As System.EventArgs) Handles MyBase.Load
        Me.MaximizeBox = False
        Me.Text = "Electric conductivity of a sample of known TDS concentration"
        Label1.Text = "Enter total Dissolved Solids (TDS) concentration(Sample1) mg/L:"
        Label2.Text = "Enter electrical conductivity of Sample1    umhos/cm:"
        Label3.Text = "Enter total Dissolved Solids (TDS) concentration(Sample2) mg/L:"
        Label4.Text = "Electrical conductivity for Sample2:"
        Label5.Text = "umhos/cm"
        Label6.Text = "Decimal Places:"
        NumericUpDown1.Maximum = 10
        NumericUpDown1.Minimum = 0
        NumericUpDown1.Value = 2
        Me.FormBorderStyle = Windows.Forms.FormBorderStyle.FixedSingle
        Me.MaximizeBox = False
    End Sub

    '*************************************************************************************
    'The following code calculates the result of EC2 depending on the input
    'values in textboxes 1-3.
    'The Val(x) function is used to convert from String to Numerical.
    'The FormatNumber is used to format the output to specific decimals.
    '*************************************************************************************

    Sub calculateResults()
        Dim a As Double
        TDS1 = Val(TextBox1.Text)
        EC1 = Val(TextBox2.Text)
        TDS2 = Val(TextBox3.Text)
        a = TDS1 / EC1
        EC2 = TDS2 / a
        'Use the function as FormatNumber(NumberToFormat, NumberOfDecimals).
        TextBox4.Text = FormatNumber(EC2, NumericUpDown1.Value).ToString
    End Sub

    '*************************************************************************************
    'Whenever a different number of decimals is selected, re-show the results.
    '*************************************************************************************

    Private Sub NumericUpDown1_ValueChanged(ByVal sender As System.Object,
ByVal e As System.EventArgs) Handles NumericUpDown1.ValueChanged
        calculateResults()
    End Sub

    Private Sub TextBox1_TextChanged(ByVal sender As System.Object,
ByVal e As System.EventArgs) Handles TextBox1.TextChanged
        calculateResults()
    End Sub

    Private Sub TextBox2_TextChanged(ByVal sender As System.Object,
ByVal e As System.EventArgs) Handles TextBox2.TextChanged
        calculateResults()
    End Sub

    Private Sub TextBox3_TextChanged(ByVal sender As System.Object,
ByVal e As System.EventArgs) Handles TextBox3.TextChanged
        calculateResults()
    End Sub
End Class
```

### 2.3.3 SALINITY

Salinity may be described as the total solids in water after all carbonates have been converted to oxides, all bromide and iodide have been replaced by chloride, and all organic matter has been oxidized (APHA 2012). The chloride content of natural waters may be augmented from geological formations containing chlorides, or through salt water intrusion, or from agricultural, industrial, and domestic discharges (Rowe and Abdel-Magid 1995). Salinity as a function of chlorinity is presented as follows (APHA 2012):

$$sa = 0.03 + 1.805\,ch \qquad (2.6)$$

where:
sa is the salinity, g/kg
ch is the chlorinity, g/kg

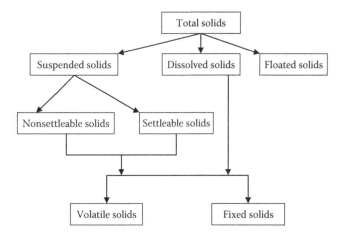

**FIGURE 2.1**  Types of solids.

### 2.3.4 SOLIDS CONTENT

The term *solids* designates matter that is suspended or dissolved in water and wastewater. Generally, solids are grouped as dissolved, suspended, volatile, fixed, and settleable solids. Most of the procedures used to quantify solids are gravimetric tests involving measurement of the mass of residues as a function of volume (see Figure 2.1).

**Example 2.3**

1. Write a short computer program that enables the computation of total, volatile, and fixed solids of a certain sample. The following data are given: mass of crucible dish ($m_1$) in grams, volume of sample ($V$) in milliliters, constant mass of dish plus solids dried at 104°C ($m_2$), and weight of cool dish (after being placed in 550°C furnace) ($m_3$).
2. Given for a certain sample that $m_1 = 35.6248$ g, $m_2 = 35.6452$ g, $m_3 = 35.6319$ g, and $V = 150$ mL, find the total, volatile, and fixed solids of the sample.

**Solution**

1. See the listing of Program 2.3.
2. Solution for Example 2.3 (2):
   a. Given: $m_1 = 35.6248$ g, $m_2 = 35.6452$ g, $m_3 = 35.6319$ g, $V = 150$ mL.
   b. Evaluate the mass of total solids as follows:
      $m_{solids}$ = (mass of dish + solids) – mass of dish
      = $m_2 - m_1$ = 35.6452 – 35.6248 = 0.0204 g = 20.4 mg.
   c. Find the concentration of total solids:
      = mass of solids (mg)/volume of sample = $m_{solids}/V$
      = 20.4 * 1000/150 = 136 mg/L.
   d. Determine the mass of volatile solids as follows:
      $m_{volatile}$ = (mass of dish + volatile solids) – mass of dish = $m_3 - m_1$ = 35.6319 – 35.6248 = 0.00071 g.
   e. Determine the concentration of volatile solids as follows: $m_{volatile}/V$ = 7.1 * 1000/150 = 47 mg/L.
   f. Compute the concentration of fixed solids as follows:
      $m_{fixed}$ = content of total solids – content of volatile solids
      = $m_{solids} - m_{volatile}$ = 136 – 47 = 89 mg/L.

---

**LISTING OF PROGRAM 2.3    (CHAP2.3\FORM1.VB): SOLIDS CONTENT**

```
'*************************************************************************************
'Program 2.3: Salinity
'Computes solid contents
'*************************************************************************************
Public Class Form1
    Dim m1, m2, m3, volume As Double
    Const T1 = 104        'Centigrade
    '*********************************************************************************
    'Code in Form_Load sub will be executed when the form is loading,
    'i.e., first thing after program loads into the memory
    '*********************************************************************************

Private Sub Form1_Load(ByVal sender As System.Object,
ByVal e As System.EventArgs) Handles MyBase.Load
        Me.FormBorderStyle = Windows.Forms.FormBorderStyle.FixedSingle
        Me.MaximizeBox = False
```

```vb
              Me.Text = "Program 2.3: Determines the solids content"
              Label1.Text = "Enter weight of crucible m1 (gm):"
              Label2.Text = "Enter volume of sample V (mL):"
              Label3.Text = "Enter weight of dish+solids at 104 C (mg):"
              Label4.Text = "Enter weight of dish+ solids after heating to 550 C (mg):"
              Label5.Text = "Concentration of total solids (mg/L)="
              Label6.Text = "Concentration of volatile solids (mg/L)="
              Label7.Text = "Concentration of fixed solids (mg/L)="
              Label8.Text = "Decimal Places:"
              NumericUpDown1.Maximum = 10
              NumericUpDown1.Minimum = 0
              NumericUpDown1.Value = 2
       End Sub

       Sub calculateResults()
              m1 = Val(TextBox1.Text)
              volume = Val(TextBox2.Text)
              m2 = Val(TextBox3.Text)
              m3 = Val(TextBox4.Text)
              Dim msds, csds, mvols, cvols, cfixs As Double
              msds = m2 - m1                          'solids
              csds = msds * 10 ^ 6 / volume           'concentration of solids
              mvols = m3 - m1                         'volatiles
              cvols = mvols * 10 ^ 6 / volume
              cfixs = csds - cvols
              TextBox5.Text = FormatNumber(csds, NumericUpDown1.Value).ToString
              TextBox6.Text = FormatNumber(cvols, NumericUpDown1.Value).ToString
              TextBox7.Text = FormatNumber(cfixs, NumericUpDown1.Value).ToString
       End Sub

       '**************************************************************************************
       'Whenever a different number of decimals is selected, re-show the results.
       '**************************************************************************************

       Private Sub NumericUpDown1_ValueChanged(ByVal sender As System.Object,
ByVal e As System.EventArgs) Handles NumericUpDown1.ValueChanged
              calculateResults()
       End Sub

       Private Sub TextBox1_TextChanged(ByVal sender As System.Object,
ByVal e As System.EventArgs) Handles TextBox1.TextChanged
              calculateResults()
       End Sub

       Private Sub TextBox2_TextChanged(ByVal sender As System.Object,
ByVal e As System.EventArgs) Handles TextBox2.TextChanged
              calculateResults()
       End Sub

       Private Sub TextBox3_TextChanged(ByVal sender As System.Object,
ByVal e As System.EventArgs) Handles TextBox3.TextChanged
              calculateResults()
       End Sub

       Private Sub TextBox4_TextChanged(ByVal sender As System.Object,
ByVal e As System.EventArgs) Handles TextBox4.TextChanged
              calculateResults()
       End Sub
End Class
```

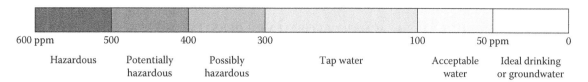

**FIGURE 2.2** Total dissolved solids as related to water quality.

### 2.3.5 TOTAL DISSOLVED SOLIDS

TDS is the total amount of mobile charged ions, including minerals, salts, or metals dissolved in a given volume of water, expressed in units of mg per unit volume of water (mg/L), also referred to as parts per million (ppm). It is directly related to the purity of water and the quality of water purification systems and affects everything that consumes, lives in, or uses water, whether organic or inorganic, whether for better or for worse (see Figure 2.2).

### 2.3.6 DENSITY, SPECIFIC VOLUME, SPECIFIC WEIGHT, AND SPECIFIC GRAVITY

Density is defined as the mass of a substance per unit volume. It is temperature dependent. The following equation may be used to find the density:

$$\rho = \frac{m}{V} \qquad (2.7)$$

where:
  $\rho$ is the density of the fluid, kg/m$^3$
  $m$ is the mass, kg
  $V$ is the volume, m$^3$

The specific volume is the volume per unit mass, that is, it is the reciprocal of the density as shown by the following equation:

$$\upsilon = \frac{1}{\rho} \qquad (2.8)$$

where:
  $\upsilon$ is the specific volume of the fluid, m$^3$/kg
  $\rho$ is the density of the fluid, kg/m$^3$

The specific weight is the weight of unit volume as shown in the following equation:

$$\gamma = \frac{mg}{V} = \rho g \qquad (2.9)$$

where:
  $\gamma$ is the specific weight, N/m$^3$
  $m$ is the mass, kg
  $g$ is the gravitational acceleration, m/s$^2$
  $V$ is the volume, m$^3$
  $\rho$ is the density of the fluid, kg/m$^3$

The specific gravity is the ratio of the density of fluid to the density of water at some specified temperature. The following equation indicates the specific gravity concept:

$$s.g. = \frac{\rho}{\rho_{\text{water at } 4°C}} \qquad (2.10)$$

### Example 2.4

1. Write a computer program that allows the computation of density, specific volume, specific weight, and specific gravity of a particular fluid, given the temperature, mass, and volume of the fluid.
2. The specific weight of water at ordinary temperature and pressure conditions is 9.806 kN/m$^3$. The specific gravity of mercury is 13.55. Find
   a. Density of mercury.
   b. Specific weight of mercury.
   c. Density of water.

### Solution

1. For solution to Example 2.4 (1), see the listing of Program 2.4.
2. Solution to Example 2.4 (2):
   a. Given: $\gamma = 9.806$ kN/m$^3$, $s.g._{\cdot Hg} = 13.55$.
   b. Determine the density of water as follows: $\rho = \gamma_{\text{water}}/g = 9.806$ kN/m$^3$/9.81 m/s$^2$ = 1 Mg/m$^3$ = 1 g/cm$^3$.
   c. Find the specific weight of mercury as follows: $\gamma_{Hg} = s.g._{\cdot Hg} * \gamma_{\text{water}} = 13.55 * 9.806 = 133$ kN/m$^3$.
   d. Compute the density of mercury as follows: $\rho_{Hg} = s.g._{\cdot Hg} * \rho_{\text{water}} = 13.55 * 1 = 13.55$ Mg/m$^3$.

### LISTING OF PROGRAM 2.4 (CHAP2.4\FORM1.VB): DENSITY AND SPECIFIC GRAVITY

```vb
'*********************************************************************************
'Program 2.4: Density
'Computes Density and specific gravity
'*********************************************************************************

Public Class Form1
    Const gamma = 1000              'density of water KN/m3
    Const g = 9.801                 'acceleration due to gravity
    Dim t(3) As String
    Dim m, Ro, V, sv, sw, sg As Double      'Variables used in calculations
    '*********************************************************************************
    'Code in Form_Load sub will be executed when the form is loading,
    'i.e., first thing after program loads into the memory
    '*********************************************************************************

    Private Sub Form1_Load(ByVal sender As System.Object,
ByVal e As System.EventArgs) Handles MyBase.Load
        t(0) = "Compute density"
        t(1) = "Compute specific volume"
        t(2) = "Compute specific weight"
        t(3) = "Compute specific gravity"
        ListBox1.Items.Clear()          'Initialize
        For i = 0 To t.Length - 1
            ListBox1.Items.Add(t(i))        'Add members to the list
        Next
        Me.FormBorderStyle = Windows.Forms.FormBorderStyle.FixedSingle
        Me.MaximizeBox = False
        Me.Text = "Program 2.4:Computes Density and specific gravity"
        Label5.Text = "Decimal Places:"
        NumericUpDown1.Maximum = 10
        NumericUpDown1.Minimum = 0
        NumericUpDown1.Value = 2
        Label1.Text = "What to compute:"
        Label2.Visible = False
        Label3.Visible = False
        Label4.Visible = False
        Label5.Visible = False
        TextBox1.Visible = False
        TextBox2.Visible = False
        TextBox3.Visible = False
        NumericUpDown1.Visible = False
    End Sub

    '*********************************************************************************
    'The following code shows/hides the labels and textboxes
    'according to the selected item from the list. It also sets
    'the text displayed in each label.
    '*********************************************************************************

    Private Sub ListBox1_SelectedIndexChanged(ByVal sender As System.Object,
ByVal e As System.EventArgs) Handles ListBox1.SelectedIndexChanged
        If ListBox1.SelectedIndex < 0 Then Exit Sub
        Select Case ListBox1.SelectedIndex
            Case 0
                Label2.Text = "Enter the mass in Kg:"
                Label3.Text = "Enter the volume in m3:"
                Label4.Text = "The density 'Ro' (Kg/m3):"
```

```
            Case 1
                Label2.Text = "Enter the the density 'Ro' (Kg/m3):"
                Label4.Text = "The Specific volume (m3/Kg):"
            Case 2
                Label2.Text = "Enter the the density 'Ro' (Kg/m3):"
                Label4.Text = "Specific wight (N/m2):"
            Case 3
                Label2.Text = "Enter the the density 'Ro' (Kg/m3):"
                Label4.Text = "Specific gravity:"
        End Select

        TextBox1.Text = ""
        TextBox2.Text = ""
        TextBox3.Text = ""
        Label2.Visible = True
        Label4.Visible = True
        Label5.Visible = True
        TextBox1.Visible = True
        TextBox3.Visible = True
        NumericUpDown1.Visible = True
        If ListBox1.SelectedIndex = 0 Then
            Label3.Visible = True
            TextBox2.Visible = True
        Else
            Label3.Visible = False
            TextBox2.Visible = False
        End If
    End Sub

    '********************************************************************************
    'The following code calculates the results according to the
    'selected item from the list box and displays the result
    'in TextBox3.
    '********************************************************************************

    Sub calculateResults()
        If ListBox1.SelectedIndex < 0 Then Exit Sub
        Select ListBox1.SelectedIndex
            Case 0
                m = Val(TextBox1.Text)
                v = Val(TextBox2.Text)
                Ro = m / v        'the density
                TextBox3.Text = formatN(Ro)
            Case 1
                Ro = Val(TextBox1.Text)
                sv = 1 / Ro       'specific volume
                TextBox3.Text = formatN(sv)
            Case 2
                Ro = Val(TextBox1.Text)
                sw = Ro * g  'specific weight
                TextBox3.Text = formatN(sw)
            Case 3
                Ro = Val(TextBox1.Text)
                sg = Ro / gamma  'specific gravity
                TextBox3.Text = formatN(sg)
        End Select
    End Sub
```

```
      '***********************************************************************
      'Whenever a different number of decimals is selected, re-show the results.
      '***********************************************************************
      Private Sub NumericUpDown1_ValueChanged(ByVal sender As System.Object,
ByVal e As System.EventArgs) Handles NumericUpDown1.ValueChanged
          calculateResults()
      End Sub


      Private Sub TextBox1_TextChanged(ByVal sender As System.Object,
ByVal e As System.EventArgs) Handles TextBox1.TextChanged
          calculateResults()
      End Sub


      Private Sub TextBox2_TextChanged(ByVal sender As System.Object,
ByVal e As System.EventArgs) Handles TextBox2.TextChanged
          calculateResults()
      End Sub


      '***********************************************************************
      'The following function is used to format the output of the calculations
      'it uses the value selected in the NumericUpDown control as a guide to
      'the number of decimal places in output, then returns 'n' as a string
      'formatted according to the set number of decimals.
      '***********************************************************************
      Function formatN(ByVal n As Double) As String
          'Use the function as FormatNumber(NumberToFormat, NumberOfDecimals).
          Return FormatNumber(n, NumericUpDown1.Value).ToString
      End Function
End Class
```

### 2.3.7 Viscosity (Rheological Properties)

Viscosity of a fluid is a measure of its resistance to shear or angular deformation (Finnemore and Franzini 2001). Viscosity is the property that relates the applied forces to the rates of deformation of the fluid (see Figure 2.3). It may be found from Newton's law as presented in the following equation:.

$$\tau = \frac{F_\tau}{A} = -\mu \frac{du}{dy} \qquad (2.11)$$

where:

$\tau$ is the shear stress, N/m$^2$ (see Figure 2.4)

$F_\tau$ is the shear force acting on a very wide plate free to move, N

$A$ is the area, m$^2$

$\mu$ is the dynamic (absolute) viscosity, N*s/m$^2$

$du/dy$ is the velocity gradient (rate of angular deformation, rate of shear), s$^{-1}$

Coefficient of dynamic viscosity, $m$, may be described as "The shear force per unit area required to drag one layer of fluid with unit velocity past another layer through unit distance in the fluid" (Rowe and Abdel-Magid 1995). Equation 2.12 gives the relationship between viscosity and density:

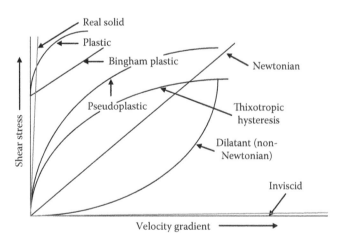

**FIGURE 2.3** Different types of fluids.

$$\mu = \rho\nu \qquad (2.12)$$

where:

$\mu$ is the dynamic (absolute) viscosity, N*s/m$^2$

$\nu$ is the kinematic viscosity, m$^2$/s (usually defined as the ratio of dynamic viscosity to mass density)

$\rho$ is the density, kg/m$^3$

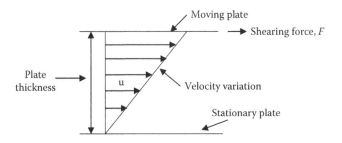

**FIGURE 2.4** Viscosity concept.

For many liquids, the stress–velocity gradient relationship is not a simple ratio. Such liquids are termed non-Newtonian, for example, sewage sludge. Figure 2.3 illustrates the variation of shear stress with velocity gradient for different types of fluids. The factors that influence viscosity include temperature, rate of variation of shearing stress, flow conditions, and characteristics of the fluid.

**Example 2.5**

1. Write a short computer program to determine the value of the kinematic viscosity of a fluid, given its dynamic viscosity, density, and temperature.
2. For a liquid with a dynamic viscosity of $4.5 * 10^{-3}$ Pa*s and a density of 912 kg/m$^3$, determine its kinematic viscosity.
3. Use the program developed in (1) to verify data presented in (2).

**Solution**

1. For solution to Example 2.5 (1), see the listing of Program 2.5.
2. Solution to Example 2.5 (2):
   a. Given: $\mu = 4.5 * 10^{-3}$ Pa*s, $\rho = 912$ kg/m$^3$.
   b. Determine the kinematic viscosity as follows: $\nu = \mu/\rho = 4.5 * 10^{-3}$ Pa*s/912 kg/m$^3 = *10^{-6}$ m$^2$/s.

---

**LISTING OF PROGRAM 2.5   (CHAP2.5\FORM1.VB): KINEMATIC VISCOSITY FROM DYNAMIC VISCOSITY**

```
'*************************************************************************************
'Program 2.5: Viscosity
'Computes kinematic viscosity from dynamic viscosity
'*************************************************************************************

Public Class Form1
    Dim mu, rho, FnKinem      'Variables used in calculations
    '*********************************************************************************
    'Code in Form_Load sub will be executed when the form is loading,
    'i.e., first thing after program loads into the memory
    '*********************************************************************************

    Private Sub Form1_Load(ByVal sender As System.Object,
       ByVal e As System.EventArgs) Handles MyBase.Load
        Me.FormBorderStyle = Windows.Forms.FormBorderStyle.FixedSingle
        Me.MaximizeBox = False
        Me.Text = "Computes kinematic viscosity from dynamic viscosity"
        Label1.Text = "Enter dynamic viscosity u  in  Ns/m2:"
        Label2.Text = "Enter density of liquid Ro in  Kg/m3:"
        Label3.Text = "The Kinematic viscosity equals:"
        Label4.Text = "m2/s"
        Label5.Text = "Decimal Places:"
        NumericUpDown1.Maximum = 10
        NumericUpDown1.Minimum = 0
        NumericUpDown1.Value = 2
    End Sub


    '*********************************************************************************
    'Code in TextChanged member of the TextBox will be executed
    'when the user changes the text.
    'The following code will recalculate the results and show them when ever
    'the user changes the input.
    '*********************************************************************************
```

```
    Private Sub TextBox1_TextChanged(ByVal sender As System.Object,
      ByVal e As System.EventArgs) Handles TextBox1.TextChanged
        calculateResults()
    End Sub

    Private Sub TextBox2_TextChanged(ByVal sender As System.Object,
      ByVal e As System.EventArgs) Handles TextBox2.TextChanged
        calculateResults()
    End Sub

    '******************************************************************************
    'The following code calculates the result according to the inputs in
    'the text boxes. It then shows result in TextBox3.
    '******************************************************************************

    Sub calculateResults()
        mu = Val(TextBox1.Text)
        rho = Val(TextBox2.Text)
        FnKinem = mu / rho
        TextBox3.Text = FormatNumber(FnKinem, NumericUpDown1.Value).ToString
    End Sub

    Private Sub NumericUpDown1_ValueChanged(ByVal sender As System.Object,
      ByVal e As System.EventArgs) Handles NumericUpDown1.ValueChanged
        calculateResults()
    End Sub
End Class
```

### 2.3.8 Surface Tension and Capillary Rise

Surface tension is a property of a liquid that permits the attraction between molecules to form an imaginary film that is able to resist tensile forces at the interface between two immiscible liquids or at the interface between a liquid and a gas (Munson et al. 2012). The capillary rise in a small vertical open tube of circular cross section dipping into a pool of liquid may be obtained (Green and Perry 2007) from the following equation (see Figure 2.5):

$$h = \frac{4\sigma\cos\phi}{g\,D(\rho_l - \rho_g)}$$  (2.13)

where:

$h$ is the column height by which liquid rose along the capillary tube, m

$\sigma$ is the surface tension, N/m

$\phi$ is the angle of contact subtended by the heavier fluid and tube (= 0° for most organic liquids and water against glass, provided the glass is wet with a film of the liquid [Green and Perry 2007]; = 130° for mercury against glass [Munson et al. 2012])

$\rho_l$ is the density of liquid, kg/m³

$\rho_g$ is the density of gas (or light liquid), kg/m³

$g$ is the local gravitational acceleration, m/s²

$D$ is the inside diameter of capillary tube, m

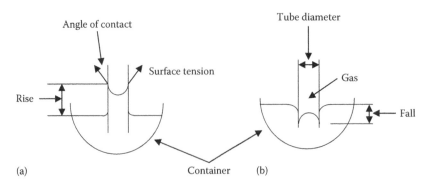

**FIGURE 2.5** Surface tension: (a) capillary rise (wetting liquid); (b) capillary depression (nonwetting liquid).

## Example 2.6

1. Write a computer program to enable the determination of the rise of water along capillary tubes for a temperature range between 0°C and 100°C and for tubes with diameters ranging between 10 and 60 mm.
2. Compute the diameter of a clean glass tube required so that the rise of water at 30°C in the tube due to capillary action does not exceed 2 mm.

## Solution

1. For solution to Example 2.6 (1), see the listing of Program 2.6 (fluid properties' calculators may also be used such as: http://www.mhtl.uwaterloo.ca/old/onlinetools/airprop/airprop.html).
2. Solution to Example 2.6 (2):
   a. Given: $h = 2 * 10^{-3}$ m, $T = 30°C$.
   b. For a temperature of 30°C and from Appendix A1, $\sigma = 0.07118$ N/m and $\gamma = 9.765$ kN/m$^3$.
   c. Compute the diameter as follows: $d = 4*s*\cos\phi/\gamma*h = [4 * 0.07118*\cos 0]/[9.765 * 10^3 * 2*10^{-3}] = 14.6$ mm.

---

### LISTING OF PROGRAM 2.6  (CHAP2.6\FORM1.VB): CAPILLARY HEIGHTS IN SURFACE TENSION

```
'*********************************************************************************
'Program 2.6: Tension
'Computes capillary heights in surface tension problems
'*********************************************************************************

Imports System.Math
Public Class Form1
    'Arrays used to store temperature values
    Dim temp(40), sigma(40), density(40) As Double
    Dim ttl(2) As String
    Dim diam, T, Rog As Double                      'Variables used in calculations
    Dim sigmaData, densityData As String            'Raw termperature data
    'constants
    Const pi = 3.142857
    Const g = 9.81           'acceleration due to gravit
    Const phiw = 0           'angle of contact water & glass
    Const phim = 130         'angle of contact mercury & glass
    Const Hgd = 13.55

    Private Sub Form1_Load(ByVal sender As System.Object,
ByVal e As System.EventArgs) Handles MyBase.Load
        ttl(0) = "Determine capillary height h"
        ttl(1) = "Determine the diameter of the glass tube for a specific h"
        ttl(2) = "Determine surface tension"
        'Temperature-Surface tension table for water- in dyne/cm
        sigmaData = "74.92,74.78,74.64,74.50,74.36,74.22,74.07,73.93,73.78,73.64,"
        sigmaData = sigmaData + "73.49,73.34,73.19,73.05,72.90,72.75,72.59,72.44,72.28,72.13,"
        sigmaData = sigmaData + "71.97,71.82,71.66,71.5,71.35,71.18,71.02,70.86,70.7,70.53,"
        sigmaData = sigmaData + "70.38,70.21,70.05,69.88"
        'Temperature-Density table- kg/m3
        densityData = "999.965,999.941,999.902,999,849,999.781,999.7,999.605,999.498,"
        densityData = densityData + "999.377,999.244,999.099,998.943,998.774,998.595,998.405,998.203,"
        densityData = densityData + "997.992,997.77,997.538,997.296,997.044,996.783,996.512,996.232,"
        densityData = densityData + "995.944,995.646,995.34,995.025,994.702,994.374,994.03,993.68,"
        densityData = densityData + "993.33,992.296"
        'Parse the above data, storing it from String format into Array format..
        readData(sigmaData, densityData)
        Me.FormBorderStyle = Windows.Forms.FormBorderStyle.FixedSingle
        Me.MaximizeBox = False
        Me.Text = "Program 2.6:Computes capillary heights in surface tension problems"
```

```
            'Add the options to the list box
            ListBox1.Items.Clear()
            For i = 0 To ttl.Length - 1
                ListBox1.Items.Add(ttl(i))
            Next
            'Options used when calculating the capillary height
            RadioButton1.Text = "Water tube"
            RadioButton2.Text = "Mercurial tube"
            RadioButton1.Checked = True
            RadioButton1.Visible = False
            RadioButton2.Visible = False
            'Hide all labels and text boxes at the start
            Label2.Visible = False
            Label3.Visible = False
            Label4.Visible = False
            Label5.Visible = False
            Label6.Visible = False
            TextBox1.Visible = False
            TextBox2.Visible = False
            TextBox3.Visible = False
            TextBox4.Visible = False
            TextBox5.Visible = False
            Label7.Text = "Decimal Places:"
            NumericUpDown1.Maximum = 15
            NumericUpDown1.Minimum = 0
            NumericUpDown1.Value = 8
            Label7.Visible = False
            NumericUpDown1.Visible = False
        End Sub


        '*******************************************************************************************
        'This function reads the data stored in two strings,
        'the sigmaData and densityData strings. It uses call-back
        'to itself to add elements to the sigma and destiny arrays,
        'chopping the first value stored in each string every time
        'it loops, until the last value in the string is reached.
        '*******************************************************************************************


        Sub readData(ByVal sigmaData As String, ByVal densityData As String)
            Static count = 5         'Declared static so one instance of it is used
            temp(count) = count
            If sigmaData.IndexOf(",") < 0 Then  'if last value in the string is reached..
                sigma(count) = Val(sigmaData)
                sigma(count) = sigma(count) / 1000        'convert to N/m
                density(count) = Val(densityData)
            Else                     'if there are still values in the string..
                'Read from inside-out: first determine the index of the first ','
        'in the string,
                'then take the substring from the beginning to just before the ',' and then
                'convert this string to numerical value.. Simple!!
                sigma(count) = Val(sigmaData.Substring(0, sigmaData.IndexOf(",")))
                sigma(count) = sigma(count) / 1000        'convert to N/m
                density(count) = Val(densityData.Substring(0, densityData.IndexOf(",")))
                'Remove the first value, which is already stroed in the array,
        'from the string,
                'using the index of the ',' removing everything before the comma (equals the
                'first value)..
                sigmaData = sigmaData.Remove(0, sigmaData.IndexOf(",") + 1)
                densityData = densityData.Remove(0, densityData.IndexOf(",") + 1)
                count += 1        'increase the count by one
```

```vbnet
        'call-back to the function, passing the new strings
            readData(sigmaData, densityData)
        End If
    End Sub

    Private Sub ListBox1_SelectedIndexChanged(ByVal sender As System.Object,
ByVal e As System.EventArgs) Handles ListBox1.SelectedIndexChanged
        If ListBox1.SelectedIndex < 0 Then
            Label7.Visible = False
            NumericUpDown1.Visible = False
            Exit Sub
        End If

        Select Case ListBox1.SelectedIndex
            Case 0
                'Determine capillary height h
                Label2.Text = "Enter diameter of tube in (mm) (between 10 to 60mm):"
                Label3.Text = "Enter temperature t (C) (5<t<38):"
                Label4.Text = "Enter the density of the gas (kg/m3):"
                Label6.Text = "The column height h (m):"
            Case 1
                'Determine the diameter of the glass tube for a specific h
                Label2.Text = "Enter the density of the liquid or gas, kg/m3:"
                Label3.Text = "Enter surface tension coefficient, N/m:"
                Label4.Text = "Enter angle of contact in degrees:"
                Label5.Text = "Enter the exent of the liquid rise in the tube, m:"
                Label6.Text = "the required diameter of the tube (m):"
            Case 2
                'Determine sigma
                Label2.Text = "Enter the density of the liquid or gas, kg/m3:"
                Label3.Text = "Enter the diameter of the capillary tube,m:"
                Label4.Text = "Enter angle of contact in degrees:"
                Label5.Text = "Enter the exent of the liquid rise in the tube, m:"
                Label6.Text = "the surface tension coefficient (N/m)="
        End Select

        'Make the labels and text boxes visible after the user makes the selection

        Label2.Visible = True
        Label3.Visible = True
        Label4.Visible = True
        Label6.Visible = True
        Label7.Visible = True
        NumericUpDown1.Visible = True
        TextBox1.Text = ""
        TextBox2.Text = ""
        TextBox3.Text = ""
        TextBox4.Text = ""
        TextBox5.Text = ""
        TextBox1.Visible = True
        TextBox2.Visible = True
        TextBox3.Visible = True
        TextBox5.Visible = True

        If ListBox1.SelectedIndex = 0 Then
            Label5.Visible = False
            TextBox4.Visible = False
            RadioButton1.Visible = True
            RadioButton2.Visible = True
        Else
            Label5.Visible = True
            TextBox4.Visible = True
```

```
                RadioButton1.Visible = False
                RadioButton2.Visible = False
            End If
        End Sub

        '**************************************************************************************
        'Code in TextChanged member of the TextBox will be executed
        'when the user changes the text.
        'The following code will recalculate the results and show them when ever
        'the user changes the input.
        '**************************************************************************************

        Private Sub TextBox1_TextChanged(ByVal sender As System.Object,
    ByVal e As System.EventArgs) Handles TextBox1.TextChanged
            calculateResults()
        End Sub

        Private Sub TextBox2_TextChanged(ByVal sender As System.Object,
    ByVal e As System.EventArgs) Handles TextBox2.TextChanged
            calculateResults()
        End Sub

        Private Sub TextBox3_TextChanged(ByVal sender As System.Object,
    ByVal e As System.EventArgs) Handles TextBox3.TextChanged
            calculateResults()
        End Sub

        Private Sub TextBox4_TextChanged(ByVal sender As System.Object,
    ByVal e As System.EventArgs) Handles TextBox4.TextChanged
            calculateResults()
        End Sub

        Private Sub NumericUpDown1_ValueChanged(ByVal sender As System.Object,
    ByVal e As System.EventArgs) Handles NumericUpDown1.ValueChanged
            calculateResults()
        End Sub

        '**************************************************************************************
        'The following code calculates the result according to the inputs in
        'the text boxes. It then shows result in TextBox5.
        '**************************************************************************************

        Sub calculateResults()
            If ListBox1.SelectedIndex < 0 Then
            Exit Sub
            Dim s, Roww, h, Ro, sig, phi, d As Double
            Select ListBox1.SelectedIndex
                Case 0
                    diam = Val(TextBox1.Text)
                    T = Val(TextBox2.Text)
                    Rog = Val(TextBox3.Text)
                    If diam < 10 Or diam > 60 Then
                        TextBox5.Text = "The Diameter should be between 10 and 60 mm."
                        Exit Sub
                    End If
                    If T < 5 Or T > 38 Then
                        TextBox5.Text = "The Temperature should be between 5 and 38
    degrees."
                        Exit Sub
                    End If
                    'get the value of sigma and density of water corresponding to temp t
                    For i = 5 To 38
```

```
                        If T = temp(i) Then
                            s = sigma(i)
                            Roww = density(i)
                            Exit For
                        End If
                    Next i
                    'compute height h
                    If RadioButton1.Checked Then     'Use water tube
                        h = 4 * s * Cos(phiw * pi / 180) / (g * diam * (Roww - Rog))
                    Else                             'Use mercurial tube
                        h = 4 * s * Cos(phim * pi / 180) / (g * diam * (Roww - Rog))
                    End If
                    TextBox5.Text = formatN(h)
                Case 1
                    Ro = Val(TextBox1.Text)
                    sig = Val(TextBox2.Text)
                    phi = Val(TextBox3.Text)
                    h = Val(TextBox4.Text)
                    phi = phi * pi / 180     'in radians
                    d = 4 * sig * Cos(phi) / (Ro * g * h)
                    TextBox5.Text = formatN(d)
                Case 2
                    Ro = Val(TextBox1.Text)
                    d = Val(TextBox2.Text)
                    phi = Val(TextBox3.Text)
                    h = Val(TextBox4.Text)
                    phi = phi * pi / 180     'in radians
                    sig = h * Ro * g * d / (4 * Cos(phi))
                    TextBox5.Text = formatN(sig)
            End Select
        End Sub

        '*****************************************************************************************
        'The following function is used to format the output of the calculations
        'it uses the value selected in the NumericUpDown control as a guide to
        'the number of decimal places in output, then returns 'n' as a string
        'formatted according to the set number of decimals.
        '*****************************************************************************************

        Function formatN(ByVal n As Double) As String
            'Use the function as FormatNumber(NumberToFormat, NumberOfDecimals).
            Return FormatNumber(n, NumericUpDown1.Value).ToString
        End Function

        Private Sub RadioButton1_CheckedChanged(ByVal sender As System.Object,
ByVal e As System.EventArgs) Handles RadioButton1.CheckedChanged
            calculateResults()
        End Sub

        Private Sub RadioButton2_CheckedChanged(ByVal sender As System.Object,
ByVal e As System.EventArgs) Handles RadioButton2.CheckedChanged
            calculateResults()
        End Sub
End Class
```

## 2.3.9 Bulk Modulus (Bulk Modulus of Elasticity)

Bulk modulus is a property that is used to evaluate the degree of compressibility. The following equation gives the bulk modulus:

$$E_b = -\frac{dP}{dV/V} = -\frac{dP}{\rho/d\rho} \qquad (2.14)$$

where:

$E_b$ is the bulk modulus, N/m$^2$

$dP$ is the differential change in pressure, Pa

$dV$ is the differential change in volume, m$^3$

$V$ is the volume, m$^3$

$\rho$ is the density of fluid, kg/m$^3$

The negative sign shows that an increase in pressure produces a reduction in volume. Large values of bulk modulus indicate that the fluid is relatively incompressible (i.e., a large pressure change is needed to create a small change in volume) (Abdel-Magid and Abdel-Magid 2015).

## Example 2.7

1. Write a computer program to compute the bulk modulus, given any differential change in pressure needed to create any differential change in volume of a specified volume.

2. A liquid compressed in a cylinder has a volume of 2 L at 1000 kPa and a volume of 1805 cm$^3$ at 2000 kPa. What is its bulk modulus of elasticity?

**Solution**

1. For solution to Example 2.7 (1), see the listing of Program 2.7.
2. Solution to Example 2.7 (2):
   a. Given: $V_1 = 2000$ mL, $V_2 = 1805$ ml, $P_1 = 1000$ kPa, $P_2 = 2000$ kPa.
   b. Determine the bulk modulus of the liquid as follows: $E_b = -dP/(dV/V) = -(2000-1000)$ kPa/$[(1805 - 2000)/2000] = 10.3$ MPa.

---

### LISTING OF PROGRAM 2.7   (CHAP2.7\FORM1.VB): BULK MODULUS

```
'********************************************************************************
'Program 2.7: Bulk
'Computes Bulk modulus given dV and dP
'********************************************************************************

Public Class Form1
    Dim v1, v2, p1, p2, Eb As Double
    Private Sub Form1_Load(ByVal sender As System.Object,
ByVal e As System.EventArgs) Handles MyBase.Load
        Me.FormBorderStyle = Windows.Forms.FormBorderStyle.FixedSingle
        Me.MaximizeBox = False
        Me.Text = "Program 2.7: Computes Bulk modulus given dV and dP"
        Label1.text = "Enter initial volume V1 in mL:"
        Label2.text = "Enter final volume V2 in mL:"
        Label3.text = "Enter initial pressure P1 in kPa:"
        Label4.text = "Enter final pressure P2 in kPa:"
        Label5.Text = "The bulk modulus Eb in kPa:"
        Label6.Text = "Decimal Places:"
        NumericUpDown1.Maximum = 10
        NumericUpDown1.Minimum = 0
        NumericUpDown1.Value = 2
    End Sub


    '********************************************************************************

    'Code in TextChanged member of the TextBox will be executed
    'when the user changes the text.
    'The following code will recalculate the results and show them when ever
    'the user changes the input.
    '********************************************************************************

    Private Sub TextBox1_TextChanged(ByVal sender As System.Object,
ByVal e As System.EventArgs) Handles TextBox1.TextChanged
        calculateResults()
    End Sub

    Private Sub TextBox2_TextChanged(ByVal sender As System.Object,
ByVal e As System.EventArgs) Handles TextBox2.TextChanged
        calculateResults()
    End Sub

    Private Sub TextBox3_TextChanged(ByVal sender As System.Object,
ByVal e As System.EventArgs) Handles TextBox3.TextChanged
        calculateResults()
    End Sub

    Private Sub TextBox4_TextChanged(ByVal sender As System.Object,
```

```
ByVal e As System.EventArgs) Handles TextBox4.TextChanged
        calculateResults()
    End Sub

    Private Sub NumericUpDown1_ValueChanged(ByVal sender As System.Object,
ByVal e As System.EventArgs) Handles NumericUpDown1.ValueChanged
        calculateResults()
    End Sub

    '***************************************************************************
    'The following code calculates the result according to the inputs in
    'the text boxes. It then shows result in TextBox5.
    '***************************************************************************

    Sub calculateResults()
        v1 = Val(TextBox1.Text)
        v2 = Val(TextBox2.Text)
        p1 = Val(TextBox3.Text)
        p2 = Val(TextBox4.Text)
        Eb = -(p2 - p1) / ((v2 - v1) / v1)
        TextBox5.Text = FormatNumber(Eb, numericupdown1.value).ToString
    End Sub
End Class
```

## 2.4 CHEMICAL CHARACTERISTICS

Chemical characteristics of significance in the field of water and wastewater include pH, alkalinity, acidity, hardness, DO, dissolved gases, chloride content, nitrogen (ammonia-nitrogen, nitrite, nitrate), nutrients, protein content, oil and grease, carbohydrates, phenols, detergents, toxic metals, biochemical oxygen demand (BOD), and chemical oxygen demand (COD). The chemical characteristics that are supported by mathematical equations, formulae, or models are presented here along with computer programs associated with the equation, formula, or model.

### 2.4.1 Hydrogen Ion Concentration

The hydrogen ion concentration, pH, is a measure of the acidity or alkalinity of a solution. The pH influences the treatment methods and quality of water supply or wastewater discharge. The following equation gives a numerical expression for the determination of the pH:

$$pH = -Log[H^+(aq)] = Log\frac{1}{[H^+(aq)]} \qquad (2.15)$$

where:

pH is the negative logarithm of the hydrogen ion concentration

$[H^+ (aq)]$ is the concentration of hydrogen (hydroxonium ion, $H_3O^+$) ions, mole/L

pH ranges from a value of 0–14. Solutions with pH of 7 are regarded as neutral, whereas solutions with pH < 7 constitute acids, and those above pH of 7 are alkalis. A strongly acidic solution can, in theory, have a pH < 0, and a strongly alkaline solution can have a pH > 14 (Davis 2011).

At a temperature of 298 K, the following equation is valid:

$$pH + pOH = 14 \qquad (2.16)$$

where:

pOH is the negative logarithm of the hydroxyl ion concentration

pH is the negative logarithm of the hydrogen ion concentration

### Example 2.8

1. Write a short computer program to determine the pH, given one of the following:
   a. Hydrogen ion concentration
   b. Hydroxyl ion concentration or its numerical value
   c. The molar concentration of hydrogen ions
2. Determine which of the following solutions is more acidic or alkaline: A solution with a pH of 7.2 or another containing $1.95 * 10^{-9}$ g of H+ per liter?
3. Calculate the pH and H+ concentration in grams of H+ per liter of a solution containing $1 * 10^{-8.3}$ mole of OH− per liter.
4. Use the program developed in 1(a) to verify computations conducted on 1(b) and 1(c).

### Solution

1. For solution to Example 2.8 (1), see the listing of Program 2.8.
2. Solution to Example 2.8 (1):
   a. Given: pH of the first solution = 7.2, H+ and pH of the second solution = $1.95 * 10^{-9}$ M.

b. Find pH of the second solution as follows:
   pH = −Log [H⁺] = Log 1.95 * 10⁻⁹ = 8.7.

   c. Since the pH of the second solution is lower, it is more alkaline than the first solution.

3. The dissociation constant for water is $[H^+][OH^-] = 10^{-14}$. Find the concentration of hydrogen ions as follows: $[H^+] = 1 * 10^{-14}/1*10^{-8.3} = 10^{-5.7}$ M. pH = −Log [H⁺] = −Log $1 * 10^{-5.7}$ = 5.7. Hydrogen ion concentration = $[H^+] = 10^{-5.7}*MW = 10^{-5.7} * 1.008 = 2.01 * 10^{-6}$ g H⁺/L.

---

### LISTING OF PROGRAM 2.8    (CHAP2.8\FORM1.VB): HYDROGEN ION CONCENTRATION ASCERTAINING

```
'*****************************************************************************
'Program 2.8: Ph
'Computes Hydrogen ion concentration
'*****************************************************************************
Imports System.Math

Public Class Form1
    Dim TTL(2) As String
    Dim H, OH, POH As Double          'Variables used in calculations
    '*************************************************************************
    'Code in Form_Load sub will be executed when the form is loading,
    'i.e., first thing after program loads into the memory
    '*************************************************************************

    Private Sub Form1_Load(ByVal sender As System.Object,
ByVal e As System.EventArgs) Handles MyBase.Load
        Me.MaximizeBox = False
        Me.Text = "Program 2.9: Computes Hydrogyn ion concentration"
        TTL(0) = "Given Concentration of Hydrogen ions H+ in mg/L or molar"
        TTL(1) = "Given Hydroxyl ion concentration in mg/L or molar"
        TTL(2) = "Given negative logarithm of hydroxyl ion concentration"
        Label1.Text = "Select an option:"
        ListBox1.Items.Clear()
        For i = 0 To TTL.Length - 1
            ListBox1.Items.Add(TTL(i))
        Next
        Label2.Text = "Select an option from list above"
        Label4.Text = "Decimal Places:"
        NumericUpDown1.Maximum = 10
        NumericUpDown1.Minimum = 0
        NumericUpDown1.Value = 2
    End Sub

    Private Sub ListBox1_SelectedIndexChanged(ByVal sender As System.Object,
ByVal e As System.EventArgs) Handles ListBox1.SelectedIndexChanged
        If ListBox1.SelectedIndex < 0 Then Exit Sub
        Select Case ListBox1.SelectedIndex
            Case 0
                Label2.Text = "Enter concentration of Hydrogen ion H+ in mg/L or molar:"
            Case 1
                Label2.Text = "Enter the Hydroxyl ion concentration in mg/L or molar:"
            Case 2
                Label2.Text = "Enter the negative logarithm of hydroxyl ion
concentration:"
        End Select
    End Sub

    Sub calculateResults()
        Dim PH As Double
        Select Case ListBox1.SelectedIndex
            Case 0
                H = Val(TextBox1.Text)
                PH = FnLg10(H)
```

```
                Case 1
                     OH = Val(TextBox1.Text)
                     PH = -FnLg10(10 ^ -14 / OH)
                Case 2
                     POH = Val(TextBox1.Text)
                     PH = 14 - POH
           End Select
           Label3.Text = "PH of the solution="
           TextBox2.Text = FormatNumber(PH, NumericUpDown1.Value).ToString + " (" +
checkPH(PH) + ")"
     End Sub


     '*****************************************************************************
     'Whenever a different number of decimals is selected, re-show the results.
     '*****************************************************************************

     Private Sub NumericUpDown1_ValueChanged(ByVal sender As System.Object,
ByVal e As System.EventArgs) Handles NumericUpDown1.ValueChanged
           calculateResults()
     End Sub

     Private Sub TextBox1_TextChanged(ByVal sender As System.Object,
ByVal e As System.EventArgs) Handles TextBox1.TextChanged
           calculateResults()
     End Sub

     Function FnLg10(ByVal x) As Double
           Return -0.4342944818# * LOG(x)     'this is log base 10 of x
     End Function

     Function checkPH(ByVal PH) As String
           If PH < 7 Then Return "Acidic"
           If PH = 7 Then Return "Neutral"
           If PH > 7 Then Return "Alkaline"
           Return ""
     End Function
End Class
```

### 2.4.2 Hardness

Hardness prevents the formation of a soap lather and is generally associated with divalent metallic cations of calcium, $Ca^{2+}$; magnesium, $Mg^{2+}$; strontium, $Sr^{2+}$; ferrous ion, $Fe^{2+}$; and manganous ions, $Mn^{2+}$, as presented in Table 2.1. ($Ca^{2+}$ and $Mg^{2+}$ are the major ions generally associated with hardness). Hardness may be computed by using Equation 2.17 (APHA 2012).

Lenntech water hardness calculator (Lenntech 2016), or Calculator 8 water hardness calculator (Groundwater Software 2016), or Casio water hardness calculator (Casio 2016), and so on may be used to compute its value taking into account the limits set by them for indication of the degree of water hardness.

$$Hard = 2.497\,Ca^{2+} + 4.118\,Mg^{2+} \qquad (2.17)$$

where:

Hard is the hardness, milliequivalent $CaCO_3/L$
$Ca^{2+}$ is the concentration of calcium ion, mg/L
$Mg^{2+}$ is the concentration of magnesium ion, mg/L

**TABLE 2.1**
**Principal Cations and Anions Associated with Hardness**

| Cations | Anions |
| --- | --- |
| $Ca^{2+}$ | $HCO_3^-$ |
| $Fe^{2+}$ | $NO_3^-$ |
| $Mg^{2+}$ | $SO_4^{--}$ |
| $Mn^{2+}$ | $Cl^-$ |
| $Sr^{2+}$ | $SiO_3$ |

*Source:* Rowe, D. R. and Abdel-Magid, I. M., *Handbook of Wastewater Reclamation and Reuse*, CRC Press/Lewis Publishers, Boca Raton, FL, 1995. With permission.

- When alkalinity < total hardness, carbonate hardness = alkalinity (mg/L).
- When alkalinity > total hardness, carbonate hardness = total hardness (mg/L) (Rowe and Abdel-Magid 1995).

*Advantages and disadvantages of hardness in water*:
Advantages:

1. Hard water aids in the growth of teeth and bones.
2. Hard water reduces lead oxide (PbO) toxicity from pipelines made of lead.

Disadvantages:

1. Hard waters are thought to be associated with cardiovascular disease
2. Financial losses due to consumption of soap
3. Scaling of hot water systems, boilers, domestic appliances, fittings, utensils, bath tubs, sinks, and dishwashers
4. Staining of clothes and household utensils
5. Hard water residues can remain in the pores of skin
6. Increased laxative effect of hard waters containing magnesium sulfates

Table 2.2 offers a classification of waters according to their degree of hardness.

### Example 2.9

1. Write a computer program that will determine the concentration of anions and cations in milliequivalents per liter and milligrams per liter calcium carbonate for the range of ions to be found in hard water. Design the program so that the total, carbonate, and noncarbonate hardness; experimental error; and the degree of water hardness can be calculated as presented in Table 2.2.

**TABLE 2.2**

**Degrees of Hardness**

| mg/L as $CaCO_3$ | Degrees of Hardness |
|---|---|
| 0–75 | Soft |
| 75–150 | Moderately soft |
| 150–175 | Moderately hard |
| 175–300 | Hard |
| ≥300 | Very hard |

*Source:* Rowe, D. R. and Abdel-Magid, I. M., *Handbook of Wastewater Reclamation and Reuse*, CRC Press/Lewis Publishers, Boca Raton, FL, 1995. With permission.

2. Laboratory analysis of a water sample yielded the following data:

| Cation (mg/L) | Anion (mg/L) |
|---|---|
| $Na^+ = 17.25$ | $HCO_3^- = 91.5$ |
| $Ca^{2+} = 19$ | $SO_4^- = 19.2$ |
| $Mg^{2+} = 12.2$ | $Cl^- = 28.4$ |
| $Sr^{2+} = 3.1$ | |

a. Express the concentrations of ions in milliequivalents per liter.
b. Comment about the experimental error, assuming that an error of 10% is acceptable.
c. Draw a bar graph of the sample.
d. Determine the total, carbonate, and noncarbonate hardness of the sample.

3. Use program developed in (1) to verify results obtained in (2).

### Solution

1. For solution to Example 2.9 (1), see the listing of Program 2.9.
2. Solution to Example 2.9 (2):
   a. Given: concentration of ions.
      i. Determine the concentration of given species in units of mEq/L by dividing the given concentration (mg/L) by the equivalent weight of each substance, that is, $C_{meq} = C_{mg/L}$/equivalent weight (EW).
      ii. Convert the concentrations (mEq/L) to units of mg/L as $CaCO_3$. This is achieved by multiplying the concentrations (mEq/L) by the EW of $CaCO_3$. The EW of calcium carbonate = molecular weight of calcium carbonate (MW)/valency (Z), that is, EW of $CaCO_3$ = MW/Z = (40 + 12 + 16 * 3)/2 = 50. Tabulate results as in the table shown below:

| | Concentration | | | |
|---|---|---|---|---|
| Constituent | EW (mg/mEq) | C (mg/L) | c = C/EW (mEq/L) | c*50 (mg/L $CaCO_3$) |
| **Cations** | | | | |
| $Ca^{2+}$ | 20 | 19 | 0.95 | 47.5 |
| $Mg^{2+}$ | 12.2 | 12.2 | 1.00 | 50 |
| $Sr^{2+}$ | 43.8 | 3.1 | 0.07 | 3.5 |
| $Na^+$ | 23 | 17.25 | 0.75 | 37.5 |
| | | | 2.77 | |
| **Anions** | | | | |
| $HCO_3^-$ | 61 | 91.5 | 1.5 | 75 |
| $SO_4^-$ | 48 | 19.2 | 0.4 | 20 |
| $Cl^-$ | 35.5 | 28.4 | 0.8 | 40 |
| | | | 2.70 | |

Compute experimental error as follows: Percentage of experimental error = (cations − anions)*100/cations = (anions − cations)*100/anions = [(2.77 − 2.7)/2.77]*100 = 3%. Because the computed experimental error is < 10%, the analysis can be considered acceptable.

b. Plot the bar graph as indicated in the following diagram:

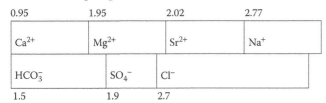

c. Determine the hardness as follows: total hardness = $Ca^{2+}$ + $Mg^{2+}$ + $Sr^{2+}$ = 0.95 + 1.0 + 0.07 = 2.02 mEq/L = 2.02 * 50 = 101 mg/L as $CaCO_3$. Carbonate hardness = $HCO_3^-$ = 1.5 mEq./L = 75 mg/L $CaCO_3$. Noncarbonate hardness = total hardness − carbonate hardness = 101 − 75 = 26 mg/L $CaCO_3$.

---

### LISTING OF PROGRAM 2.9   (CHAP2.9\FORM1.VB): HARDNESS ESTIMATION

```
'********************************************************************************
'Program 2.9: Hardness
'Program to compute hardness
'********************************************************************************

Imports System.Drawing
Public Class Form1
    Dim EW(12) As Double, SYMBOL(12) As String, CONCENTRATION(12), OPT, COL2(12),
COL3(12)
    Dim SCATION = 0
    Dim SANION = 0
    '****************************************************************************
    'Code in Form_Load sub will be executed when the form is loading,
    'i.e., first thing after program loads into the memory
    '****************************************************************************

    Private Sub Form1_Load(ByVal sender As System.Object,
ByVal e As System.EventArgs) Handles MyBase.Load
        defConstants()
        Me.FormBorderStyle = Windows.Forms.FormBorderStyle.FixedSingle
        Me.MaximizeBox = False
        Me.Text = "Program 2.10: Computes hardness"
        Label1.Text = "Select concentration:"
        ListBox1.Items.Clear()
        ListBox1.Items.Add("Concentration in mg/Litre")
        ListBox1.Items.Add("Concentration in mequivalent/Litre")
        Dim I As Integer = 1
        For Each c As Control In Me.Controls
            If c.Name.StartsWith("Label") Then
                If Not c.Name.Equals("Label1") Then
                    c.Text = "Concentration of " + SYMBOL(I) + ":"
                    I += 1
                End If
            End If
        Next
    End Sub

    Sub defConstants()
        EW(1) = 20.0485
        EW(2) = 12.1525
        EW(3) = 27.9235
        EW(4) = 27.4654
        EW(5) = 43.81
        EW(6) = 22.98977
        EW(7) = 30.0246
        EW(8) = 61.019
        EW(9) = 48.033
        EW(10) = 35.453
        EW(11) = 62.0049
```

```
            EW(12) = 38.0419
            SYMBOL(12) = "Ca2+"
            SYMBOL(11) = "Mg2+"
            SYMBOL(10) = "Fe2+"
            SYMBOL(9) = "Mn2+"
            SYMBOL(8) = "Sr2+"
            SYMBOL(7) = "Na+"
            SYMBOL(6) = "CO3-"
            SYMBOL(5) = "HCO3"
            SYMBOL(4) = "SO4-"
            SYMBOL(3) = "CL-"
            SYMBOL(2) = "NO3-"
            SYMBOL(1) = "SiO3-"
    End Sub

    Private Sub ListBox1_SelectedIndexChanged(ByVal sender As System.Object,
ByVal e As System.EventArgs) Handles ListBox1.SelectedIndexChanged
        `calculateResults()
        `drawResults()
    End Sub

    Sub calculateResults()
        SCATION = 0
        SANION = 0
        Dim i As Integer
        For Each c As Control In Me.Controls
            If c.Name.StartsWith("TextBox") Then
                i = Val(c.Name.Remove(0, 7))
                CONCENTRATION(i) = Val(c.Text)
                If ListBox1.SelectedIndex = 0 Then
                COL2(i) = CONCENTRATION(i) / EW(i) 'C/EW
                If ListBox1.SelectedIndex = 1 Then COL2(i) = CONCENTRATION(i)
                If i <= 6 Then SCATION = SCATION + COL2(i) 'Sum cat.
                If i > 6 Then SANION = SANION +
                COL2(i) 'Sum anions
                COL3(i) = COL2(i) * 50
            End If
        Next
        Dim EERROR = (SCATION - SANION) * 100 / SCATION
        DataGridView1.Columns.Clear()
        ListBox2.Items.Clear()
        If ListBox1.SelectedIndex = 0 Then
            DataGridView1.Columns.Add("ION", "CATION/ANION")
            DataGridView1.Columns.Add("CONCMGL", "CONC.(MG/L)")
            DataGridView1.Columns.Add("CONCML", "CONC.(MEQUIV/L)")
            DataGridView1.Columns.Add("CONCMGL", "CONC.(MG/L) AS CACO3")
        Else
            DataGridView1.Columns.Add("ION", "CATION/ANION")
            DataGridView1.Columns.Add("CONCML", "CONC.(MEQUIV/L)")
            DataGridView1.Columns.Add("CONCMGL", "CONC.(MG/L) AS CACO3")
        End If
        For i = 1 To 12
            If i = 7 Then
            DataGridView1.Rows.Add("TOTAL CATIONS: ", fN(SCATION))
            End If
            If ListBox1.SelectedIndex = 0 Then
            DataGridView1.Rows.Add(SYMBOL(i), fN(CONCENTRATION(i)), fN(COL2(i)),
fN(COL3(i)))
            Else
            DataGridView1.Rows.Add(SYMBOL(i), fN(COL2(i)), fN(COL3(i)))
            End If
```

```
            Next
            DataGridView1.Rows.Add("TOTAL ANIONS: ", fN(SANION))
            DataGridView1.Rows.Add("TOTAL ERROR: ", fN(EERROR) + "%")
            Dim TOTHARD1 = COL2(1) + COL2(2)              'Total hardness as Ca2+ Mg2+
            Dim TOTHARD2 = COL3(1) + COL3(2)
            ListBox2.Items.Add("Total Hardness =" + fN(TOTHARD1) + " mEq/L=" _
                              + fN(TOTHARD2) + " mg/L as CaCO3")
            Dim TOTHARD3 = COL2(7) + COL2(8)              'Carbonate hardness as CO3-+ HCO3-
            Dim TOTHARD4 = COL3(7) + COL3(8)
            ListBox2.Items.Add("Carbonate Hardness =" + fN(TOTHARD3) + " mEq/L=" _
                              + fN(TOTHARD4) + " mg/L as CaCO3-")
            Dim NONCARB = TOTHARD1 - TOTHARD3
            Dim NONCARB2 = TOTHARD2 - TOTHARD4
            If NONCARB < 0 Then NONCARB = 0
            If NONCARB2 < 0 Then NONCARB2 = 0
            ListBox2.Items.Add("Noncarbonate Hardness =" + fN(NONCARB) + " mEq/L=" _
                              + fN(NONCARB2) + " mg/L as CaCO3-")
            'Determine the degree of hardness
            Dim chk = TOTHARD2, d$
            If chk > 0 And chk <= 75 Then d$ = "Soft"
            If chk > 75 And chk <= 150 Then d$ = "Moderately Soft"
            If chk > 150 And chk <= 175 Then d$ = "Moderately Hard"
            If chk > 175 And chk <= 300 Then d$ = "Hard"
            If chk > 300 Then d$ = "Very Hard"
            ListBox2.Items.Add("The degree of hardness is " + d$)
        End Sub


        Function fN(ByVal N As Double) As String
    'Formats a number into two decimal points
            Return FormatNumber(N, 2).ToString
        End Function


        '****** Plot the results into bar graph
        Sub drawResults()
            On Error Resume Next
            Dim P(12)
            Dim SUMCAT As Double = 0
            Dim SUMAN As Double = 0
            For I = 1 To 12
                If I <= 6 Then SUMCAT = SUMCAT + COL2(I)
                If I > 6 Then SUMAN = SUMAN + COL2(I)
                P(I) = COL2(I)
            Next I
            'SCALE VALUES
            Dim sum1 As Double = 0
            For I = 1 To 6
                P(I) = Int(P(I) * (PictureBox1.Width - 10) / SUMCAT)
                sum1 = sum1 + P(I)
            Next
            For I = 7 To 12
                P(I) = Int(P(I) * sum1 / SUMAN)
            Next
            'Draw the bar graph. Each ion is represented by a RectangleShape shape,
            'which is then filled with a preselected brush color.
            Dim lastX As Double = 5
            Dim font = New Font("Microsoft Sans Serif", 14)
            Dim g = PictureBox1.CreateGraphics
            Dim XSUM As Double = 0
            Dim brush(16) As Brush
            brush(1) = Brushes.AliceBlue
            brush(2) = Brushes.Aqua
```

```
            brush(3)  = Brushes.Azure
            brush(4)  = Brushes.Beige
            brush(5)  = Brushes.BlanchedAlmond
            brush(6)  = Brushes.BlueViolet
            brush(7)  = Brushes.BurlyWood
            brush(8)  = Brushes.Chocolate
            brush(9)  = Brushes.Cornsilk
            brush(10) = Brushes.Cyan
            brush(11) = Brushes.DarkGreen
            brush(12) = Brushes.DeepPink
            brush(13) = Brushes.Fuchsia
            brush(14) = Brushes.Green
            brush(15) = Brushes.Lavender
            brush(16) = Brushes.LightBlue
            g.FillRectangle(Brushes.White, 0, 0, PictureBox1.Width-1, PictureBox1.Height-1)
            For I = 1 To 6
                g.DrawRectangle(Pens.Black, CInt(lastX),
CInt((PictureBox1.Height / 2) - 30), CInt(P(I)), 30)
                g.FillRectangle(brush(I), New Rectangle(CInt(lastX) + 1,
CInt((PictureBox1.Height / 2) - 28), CInt(P(I)) - 1, 28))
                If P(I) > 0 Then
                    g.DrawString(SYMBOL(I), font, Brushes.Black, New Point(lastX,
(PictureBox1.Height / 2) - 30))
                    XSUM = XSUM + COL2(I)
                    g.DrawString(FormatNumber(XSUM, 2), font, Brushes.Black, _
                            New Point(lastX, (PictureBox1.Height / 2) -
30 - font.GetHeight))
                End If
                lastX += P(I)
            Next
            lastX = 5
            XSUM = 0
            For I = 7 To 12
                g.DrawRectangle(Pens.Black, CInt(lastX), CInt((PictureBox1.Height / 2) + 1),
CInt(P(I)), 30)
                g.FillRectangle(brush(I), New Rectangle(CInt(lastX) + 1,
CInt((PictureBox1.Height / 2) + 2), CInt(P(I)) - 1, 28))
                If P(I) > 0 Then
                    g.DrawString(SYMBOL(I), font, Brushes.Black, New Point(lastX,
(PictureBox1.Height / 2) + 1))
                    XSUM = XSUM + COL2(I)
                    g.DrawString(FormatNumber(XSUM, 2), font, Brushes.Black, _
                            New Point(lastX, (PictureBox1.Height / 2) + 31))
                End If
                lastX += P(I)
            Next
        End Sub

    Private Sub Button1_Click(ByVal sender As System.Object,
ByVal e As System.EventArgs) Handles Button1.Click
            If ListBox1.SelectedIndex < 0 Then Exit Sub
            calculateResults()
            drawResults()
        End Sub
End Class
```

## 2.5 BIOLOGICAL PROPERTIES

The biological properties of water and wastewater are of notable importance. Of great interest and essential consideration are growth, decay, turn over time, biological half-life, retention time, composition and its change, and active ingredients. Biological properties allow identification of biological concepts, reactions, and effects on tissues. Of significance in this area are DO and biological oxygen demand.

### 2.5.1 DISSOLVED OXYGEN

DO is of paramount importance in aerobic metabolic reactions. The saturation concentration of oxygen may be determined as follows:

$$C_s = K_D C_g \tag{2.18}$$

where:

$C_s$ is the saturation concentration, g/m$^3$
$K_D$ is the distribution coefficient
$C_g$ is the gas concentration in gas phase, g/m$^3$, which may be calculated from the following equation:

$$C_g = \frac{P_g MW}{RT} \tag{2.19}$$

where $P_g$ is the partial pressure of the respective gas in the gas phase, Pa, which may be calculated from the following equation:

$$P_g = x_g k_H \tag{2.20}$$

where $x_g$ is the mole fraction of gas, which may be calculated from the following equation:

$$x_g = \frac{N_g}{N_g + N_w} \tag{2.21}$$

where:

$N_g$ is the moles of gas
$N_w$ is the moles of water
$k_H$ is the Henry's constant
MW is the molecular weight of the gas
$R$ is the universal gas constant = 8314.3 J/kg*K
$T$ is the absolute temperature, K

Equation 2.22 relates oxygen solubility concentration to pressure:

$$C' = \frac{C_s(P - p_w)}{(760 - p_w)} \tag{2.22}$$

where:

$C'$ is the solubility of oxygen at barometric pressure $P$ and given temperature, mg/L
$C_s$ is the saturation concentration at given temperature, mg/L
$P$ is the barometric pressure, mmHg
$p_w$ is the pressure of saturated water vapor at the temperature of the water, mmHg

Oxygen is slightly soluble in water. The actual content of oxygen that can be found in a solution is influenced by different interacting parameters such as solubility of the gas, partial pressure of the gas in the gas phase, temperature, and degree of water purity (e.g., level of salinity, concentration of suspended solids, etc.). Drinking water saturated with oxygen has a pleasant taste, whereas water devoid of DO has an insipid taste.

### Example 2.10

1. Write a computer program to evaluate the saturation concentration of oxygen dissolved in water for any value of pressure and at any temperature and chloride concentration in water.
2. Calculate the concentration of DO in a water sample (with zero salinity) having a DO saturation concentration of 9.2 mg/L at a temperature of 20°C for an atmospheric gauge pressure of 695 mmHg.
3. Use the computer program developed in (a) to verify computations achieved in (b).

### Solution

1. For solution to Example 2.10 (1), see the listing of Program 2.10.
2. Solution to Example 2.10 (1):
   * Given: $T = 20°C$, $C' = 9.2$ mg/L, $P = 695$ mmHg, Salinity = 0.
      * From Appendix A2 (Viessman et al. 2015; Whipple and Whipple 1911), for $T = 20°C$, find the vapor pressure as $p_w = 17.535$ mmHg.
      * Find the oxygen saturation concentration as follows:
   $C_s = C'*(760 - p_w)/(P - p_w) = 9.2*(760 - 17.535)/(695 - 17.535) = 11.07$ mg/L.

**LISTING OF PROGRAM 2.10    (CHAP2.10\FORM1.VB): DISSOLVED OXYGEN CONCENTRATION**

```vb
'********************************************************************************
'Program 2.10: DO
'Computes Dissolved oxygen concentration
'********************************************************************************

Public Class Form1
    Dim T(31), Pwi(31) As Double
    Dim pTable As String
    '********************************************************************************
    'Code in Form_Load sub will be executed when the form is loading,
    'i.e., first thing after program loads into the memory
    '********************************************************************************

    Private Sub Form1_Load(ByVal sender As System.Object,
ByVal e As System.EventArgs) Handles MyBase.Load
        pTable = "05, 05, 05, 06, 06, 07, 07, 08, 08, 09, 09, 10, 11, 11, 12, 13,"
        pTable += " 14, 15, 16, 17, 18, 19, 20,"
        pTable += " 21, 22, 24, 25, 27, 28, 30, 32"
        Me.FormBorderStyle = Windows.Forms.FormBorderStyle.FixedSingle
        Me.MaximizeBox = False
        Me.Text = "Program 2.10: Computes dissolved oxygen concentration"
        'Table of temperature-pressure relationship
        For i = 0 To 30
            T(i) = i
            Pwi(i) = Val(pTable.Substring(i * 4, 2))
        Next i
        Label1.Text = "Enter temperature in the range 0 to 30 C          ="
        Label2.Text = "Enter gauge pressure in mm Hg                     ="
        Label3.Text = "Enter oxygen solubility concentration in mg/L     ="
        Label4.Text = "Oxygen saturation concentration at the given temp. (mg/L)="
        Label5.Text = "Salinity assumed=0"
        Label6.Text = "Decimal places:"
        NumericUpDown1.Maximum = 10
        NumericUpDown1.Minimum = 0
        NumericUpDown1.Value = 2
    End Sub

    '********************************************************************************
    'Whenever a different number of decimals is selected, re-show the results.
    '********************************************************************************

    Private Sub NumericUpDown1_ValueChanged(ByVal sender As System.Object,
ByVal e As System.EventArgs) Handles NumericUpDown1.ValueChanged
        calculateResults()
    End Sub

    Private Sub TextBox1_TextChanged(ByVal sender As System.Object,
ByVal e As System.EventArgs) Handles TextBox1.TextChanged
        calculateResults()
    End Sub

    Private Sub TextBox2_TextChanged(ByVal sender As System.Object,
ByVal e As System.EventArgs) Handles TextBox2.TextChanged
        calculateResults()
    End Sub

    Private Sub TextBox3_TextChanged(ByVal sender As System.Object,
ByVal e As System.EventArgs) Handles TextBox3.TextChanged
        calculateResults()
    End Sub
```

```
    Sub calculateResults()
        Dim T1, P, C As Double
        T1 = Val(TextBox1.Text)
        P = Val(TextBox2.Text)
        C = Val(TextBox3.Text)
        'Error checking
        If T1 < 0 Or T1 > 30 Then
            MsgBox("Temp should be between 0 and 30",
                    vbOKOnly Or vbCritical)
            Exit Sub
        End If
        'Calculations
        'Find Pw corresponding to temperature T
        Dim Pw As Double
        For i = 0 To 30
            If T1 <= T(i) Then
                Pw = Pwi(i)
                Exit For
            End If
        Next
        Dim Ps = 760 'mmHg
        Dim Cx = C * (Ps - Pw) / (P - Pw)
        TextBox4.Text = FormatNumber(Cx, NumericUpDown1.Value).ToString
    End Sub
End Class
```

## 2.5.2 Biochemical Oxygen Demand

The BOD test measures the relative amount of oxygen that is needed to biologically stabilize organic matter present in a sample. The advantages of the test include estimation of the size of treatment units, evaluation of treatment efficiency, and estimation of the relative amount of oxygen required for oxidation of organic pollutants (Rowe and Abdel-Magid 1995).

The BOD reaction is assumed to follow a first-order reaction, and it may be expressed as follows:

$$\frac{L_t}{L_0} = e^{-k't} = 10^{-k'_1 t} \qquad (2.23)$$

where:

$L_t$ is the amount of first-stage BOD remaining in the sample at time $t$, mg/L

$L_0$ is the initial remaining BOD, total, or ultimate first-stage BOD present at zero time

$k'$ is the rate constant (to base e), per day

$k'_1$ is the rate constant (to base 10), per day $(=0.4343*k')$

$t$ is the time, days

Equation 2.24 gives the value of BOD exerted at any time $t$ and temperature $T$ (see Figure 2.6):

$$BOD_t^T = L_0 - L_t = L_0(1 - 10^{-k'_1 t}) \qquad (2.24)$$

The usual standard for reporting BOD is based on the 5-day BOD at a temperature of 20°C ($BOD_5^{20}$). Oxidation of the organic matter is from 60% to 70% complete within the 5-day test period. The 20°C temperature represents an average value for slow-moving streams in temperate climates and is easily duplicated in the laboratory (Metcalf and Eddy 2013). Table 2.3 gives the wastewater strength in terms of $BOD_5$ or COD.

The COD test is employed to measure the amount of oxygen required to oxidize organic matter by using a strong chemical oxidizing agent in an acidic medium. The standard reagent used in the determination is a boiling mixture of concentrated sulfuric acid and potassium dichromate ($K_2Cr_2O_7$),

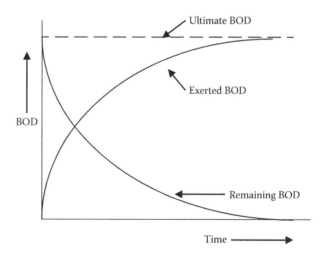

**FIGURE 2.6** Variation of BOD with time.

## TABLE 2.3
### Sewage Strength in Terms of BOD and COD

| Strength | BOD$_5$ (mg/L) | COD (mg/L) |
|---|---|---|
| Weak | <200 | <400 |
| Medium | 200–350 | 400–700 |
| Strong | 351–500 | 701–1000 |
| Very strong | >750 | >1500 |

*Source:* Rowe, D. R. and Abdel-Magid, I. M., *Handbook of Wastewater Reclamation and Reuse*, CRC Press/Lewis Publishers, Boca Raton, FL, 1995. With permission.

together with silver sulfate as a catalyst. Generally, the COD of a wastewater is greater than its 5-day BOD. This is for the following reasons:

1. More compounds are likely to be oxidized chemically than can be oxidized biologically.
2. The 5-day BOD does not equal the ultimate BOD.

The advantages of the COD test include the following:

1. The test can be conducted in a short period of time (2–3 h). This is more convenient than the 5-day period required for determination of the BOD.
2. The test can be used to aid in design, operation, and control of treatment works (if related with BOD).

The Thomas method can be used to determine values for $k'$ and $L_0$ from a series of BOD measurements. This method is based on the similarity of two series functions (expansions for $1 - e^{-k'*t}$ and $k'*t*(1 + k'*t/6)^{-3}$). Thomas developed the approximate linearized formula presented in the following equation:

$$\sqrt[3]{\frac{t}{\text{BOD}}} = \sqrt[3]{k'L_0} + \frac{t \cdot \sqrt[3]{(k')^2}}{6\sqrt[3]{L_0}}$$ (2.25)

where:
$t$ is the time, s
BOD is the BOD exerted in time $t$, mg/L
$k'$ is the reaction rate constant (to base e), day$^{-1}$
$L_0$ is the ultimate BOD, mg/L

Equation 2.25 can be put in the form of a straight line as indicated in the following equation:

$$y = a x + b$$ (2.26)

where:
$y$ is the $(t/\text{BOD})^{1/3}$
$b$ is the intercept of the line:

$$b = (k' L_0)^{-1/3}$$ (2.27)

$a$ is the slope of the straight line:

$$a = \frac{\sqrt[3]{k}^2}{6\sqrt[3]{L_0}}$$ (2.28)

The reaction rate constant and ultimate BOD may therefore be determined as shown in the following equations:

$$L_0 = \frac{1}{k' b^3}$$ (2.29)

$$k' = \frac{6a}{b}$$ (2.30)

### Example 2.11

1. Write a computer program to determine the BOD of a wastewater at any time $t$, rate constant $k$, and temperature $T$. Design the program so that $t$ or $k$ or $L_o$ can be determined, given the other two parameters. The program should show the strength of the sewage in accordance with Table 2.3.
2. Find the amount of the 5-day BOD compared to the ultimate BOD for a sewage sample, which has a rate constant ($k_1'$) value of 0.103/day.
3. A wastewater sample has a 5-day BOD of 324 mg/L and an ultimate BOD of 405 mg/L. Find the rate of biological oxidation for the wastewater.
4. The 3-day BOD of a sample is 255 mg/L at a temperature of 20°C. Determine its 5-day BOD, assuming that the rate constant is 0.22/day (to base 10).

### Solution

1. For solution to Example 2.11 (1), see the listing for Program 2.11.
2. Solution to Example 2.11 (2):
   a. Given: $k_1 = 0.103$/day.
   b. Find the ratio of 5-day BOD to ultimate BOD for the sample as follows: $\text{BOD}^{20}{}_5/L_0 = (1 - 10^{-k_1*t}) = 1 - 10^{-0.103 * 5} = 0.69$
   c. This shows that the 5-day BOD is ~69% of the ultimate BOD.
3. Solution to Example 2.11 (3):
   a. Given: $\text{BOD}^{20}{}_5 = 324$ mg/L, $L_0 = 405$.
   b. Determine the reaction rate constant as follows: $324 = 405(1 - 10^{-5k_1})$. This yields a value for $k'_1$ equal to 0.14/day.
4. Solution to Example 2.11 (4):
   a. Given: $\text{BOD}_3 = 255$ mg/L and $k_1' = 0.22$/day.
   b. Determine the ultimate BOD as follows: $L_0 = \text{BOD}/(1 - 10^{-k_1*t}) = 255/(1 - 10^{-0.22 * 3}) = 326.4$ mg/L.
   c. Find the 5-day BOD as follows: $\text{BOD}_5 = 3264*(1 - 10^{-0.22 * 5}) = 300$ mg/L.

## LISTING OF PROGRAM 2.11   (CHAP2.11\FORM1.VB): BOD COMPUTATIONS

```vb
'********************************************************************************
'Program 2.11: BOD-1
'Estimates the BOD for wastewater for a given rate k, temperature T
' and at any time t
'********************************************************************************
Imports System.Math
Public Class Form1
    'Variables used in calculations
    Dim k1, kd, t, BOD5, BOD, Lo As Double
    '********************************************************************************
    'Code in Form_Load sub will be executed when the form is loading,
    'i.e., first thing after program loads into the memory
    '********************************************************************************

    Private Sub Form1_Load(ByVal sender As System.Object,
ByVal e As System.EventArgs) Handles MyBase.Load
        Me.MaximizeBox = False
        Me.Text = "Program 2.11: Estimates the BOD for wastewater"
        Dim ttl(3) As String
        ttl(0) = "Determine BOD at any time t"
        ttl(1) = "Determine the reaction rate constant k1 to base 10"
        ttl(2) = "Determine the reaction rate constant k1 to base e"
        ttl(3) = "Determine the ultimate BOD (Lo)"
        For i = 0 To 3
            ListBox1.Items.Add(ttl(i))
        Next
        Label1.Text = "Select an option:"
        Label2.Visible = False
        Label3.Visible = False
        Label4.Visible = False
        Label5.Visible = False
        Label6.Visible = False
        Label7.Visible = False
        Label8.Visible = False
        Label9.Visible = False
        TextBox1.Visible = False
        TextBox2.Visible = False
        TextBox3.Visible = False
        TextBox4.Visible = False
        TextBox5.Visible = False
        NumericUpDown1.Visible = False
        NumericUpDown1.Maximum = 10
        NumericUpDown1.Minimum = 0
        NumericUpDown1.Value = 2
    End Sub

    Private Sub ListBox1_SelectedIndexChanged(ByVal sender As System.Object,
ByVal e As System.EventArgs) Handles ListBox1.SelectedIndexChanged
        If ListBox1.SelectedIndex < 0 Then Exit Sub
        Select Case ListBox1.SelectedIndex
            Case 0
                'Determine BOD at any time t
                Label2.Text = "Enter rate constant k1 (/day) to base 10:"
                Label3.Text = "Enter the time t (days), t:"
                Label4.Text = "Enter the 5-day BOD in mg/L:"
                Label5.Text = "The value of BOD at" + t.ToString + " days="
                Label6.Text = " of the ultimate BOD"
                Label7.Text = "The value of BOD at " + t.ToString + " days(mg/L)="
            Case 1
                'Determine k1 to base 10
```

```
            Label2.Text = "Enter the value of the ultimate BOD mg/L:"
            Label3.Text = "Enter the value of the 5-day BOD mg/L:"
            Label5.Text = "Rate at which water is being oxidized k1(to base 10)="
            Label6.Text = " /day"
        Case 2
            'Determine k1 to base e
            Label2.Text = "Enter the value of the ultimate BOD mg/L:"
            Label3.Text = "Enter the value of the 5-day BOD mg/L:"
            Label5.Text = "Rate at which water is being oxidized k1(to base e)="
            Label6.Text = " /day"
        Case 3
            'Determine the ultimate BOD (Lo)
            Label2.Text = "Enter rate constant k1 (/day):"
            Label3.Text = "Enter the time t (days), t:"
            Label4.Text = "Enter the BOD at that time, BOD, mg/L:"
            Label5.Text = "The ultimate BOD ="
            Label6.Text = " mg/day"
    End Select
    TextBox1.Visible = True
    TextBox2.Visible = True
    TextBox4.Visible = True
    Label2.Visible = True
    Label3.Visible = True
    Label5.Visible = True
    Label6.Visible = True
    Label9.Visible = True
    NumericUpDown1.Visible = True
    If ListBox1.SelectedIndex = 0 Or ListBox1.SelectedIndex = 3 Then
        'First and fourth selections have three inputs, so show the third textbox
        TextBox3.Visible = True
        Label4.Visible = True
        TextBox5.Visible = True
        Label7.Visible = True
    Else
        'The other options have two inputs, so hide the third textbox
        TextBox3.Visible = False
        Label4.Visible = False
        TextBox5.Visible = False
        Label7.Visible = False
    End If
    calculateResults()
End Sub


Sub calculateResults()
    If ListBox1.SelectedIndex < 0 Then Exit Sub
    Select Case ListBox1.SelectedIndex
        Case 0
            'Determine BOD at any time t
            k1 = Val(TextBox1.Text)
            t = Val(TextBox2.Text)
            BOD5 = Val(TextBox3.Text)
            'First order reaction rate
            Dim BODRATIO = (1 - 10 ^ (-k1 * t))
            Label5.Text = "The value of BOD at " + t.ToString + " days="
            TextBox4.Text = FormatNumber(BODRATIO, NumericUpDown1.Value).ToString
            Label6.Text = " of the ultimate BOD"
            Dim B = BODRATIO * BOD5
            Label7.Text = "The value of BOD at " + t.ToString + " days(mg/L)="
            TextBox5.Text = FormatNumber(B, NumericUpDown1.Value).ToString
            'Classify sewage strength
            Dim c As String
```

```
                    If B <= 200 Then
                        c = "Weak"
                    ElseIf B > 200 And B <= 350 Then
                        c = "Medium"
                    ElseIf B > 350 And B <= 500 Then
                        c = "Strong"
                    ElseIf B > 750 Then
                        c = "Very Strong"
                    Else
                        c = "Unknown"
                    End If
                    Label8.Text = "The strength of sewage is " + c
                Case 1
                    'Determine k1 to base 10
                    Lo = Val(TextBox1.Text)
                    BOD5 = Val(TextBox2.Text)
                    t = 5
                    k1 = -Log(1 - (BOD5 / Lo)) / (t * Log(10))
                    TextBox4.Text = FormatNumber(k1, NumericUpDown1.Value).ToString
                Case 2
                    'Determine k1 to base e
                    Lo = Val(TextBox1.Text)
                    BOD5 = Val(TextBox2.Text)
                    t = 5
                    k1 = -Log(1 - (BOD5 / Lo)) / (t * Log(10))
                    kd = k1 / 0.4343
                    TextBox4.Text = FormatNumber(kd, NumericUpDown1.Value).ToString
                Case 3
                    'Determine the ultimate BOD (Lo)
                    k1 = Val(TextBox1.Text)
                    t = Val(TextBox2.Text)
                    BOD = Val(TextBox3.Text)
                    Lo = BOD / (1 - 10 ^ (-k1 * t))
                    TextBox5.Text = FormatNumber(Lo, NumericUpDown1.Value).ToString
            End Select
            If ListBox1.SelectedIndex = 0 Then Label8.Visible = True _
            Else Label8.Visible = False
        End Sub

    '****************************************************************************************
    'Whenever a different number of decimals is selected, re-show the results.
    '****************************************************************************************

    Private Sub NumericUpDown1_ValueChanged(ByVal sender As System.Object,
ByVal e As System.EventArgs) Handles NumericUpDown1.ValueChanged
        calculateResults()
    End Sub

    Private Sub TextBox1_TextChanged(ByVal sender As System.Object,
ByVal e As System.EventArgs) Handles TextBox1.TextChanged
        calculateResults()
    End Sub

    Private Sub TextBox2_TextChanged(ByVal sender As System.Object,
ByVal e As System.EventArgs) Handles TextBox2.TextChanged
        calculateResults()
    End Sub

    Private Sub TextBox3_TextChanged(ByVal sender As System.Object,
ByVal e As System.EventArgs) Handles TextBox3.TextChanged
        calculateResults()
    End Sub
End Class
```

## Example 2.12

1. Write a computer program for the estimation of the ultimate BOD and the rate constant by using the Thomas method for the following data.
2. A BOD test for a wastewater sample, at a temperature of 20°C, gave the following results:

| $t$ (days) | 0.5 | 1 | 1.5 | 2 | 2.5 | 3 | 3.5 | 4 | 4.5 | 5 |
|---|---|---|---|---|---|---|---|---|---|---|
| BOD (mg/L) | 63 | 112 | 150 | 180 | 202 | 220 | 234 | 241 | 248 | 254 |

Using the Thomas method, find
  a. The reaction rate constant.
  b. The ultimate first-stage BOD.
  c. The value of $k_1'$ (to base 10).

## Solution

1. For solution to Example 2.12 (1), see the listing for Program 2.12.
2. Solution to Example 2.12 (2):
   a. Use the Thomas method to construct the following table:

| $t$ (days) | BOD (mg/L) | $(t/\text{BOD})^{1/3}$ |
|---|---|---|
| 0.5 | 63 | 0.007937 |
| 1.0 | 91 | 0.008929 |
| 1.5 | 20 | 0.01 |
| 2.0 | 142 | 0.011111 |
| 2.5 | 158 | 0.012376 |
| 3.0 | 169 | 0.013636 |
| 3.5 | 177 | 0.014957 |
| 4.0 | 184 | 0.016598 |
| 4.5 | 187 | 0.018145 |
| 5.0 | 190 | 0.019685 |

   b. Plot a graph of $(t/\text{BOD})^{1/3}$ versus $t$.
   c. From the constructed straight line, estimate the slope and intercept as follows: $a = \text{slope} = 0.015729 = k'^{2/3}/6L_0^{1/3}$ (1); $b = \text{intercept} = 0.191745 = (k'*L_0)^{-1/3}$ (2).
   d. Use Equations (1) and (2), find: $k' = 0.492$/day (to base e) and $L_0 = 288.5$ mg/L.
   e. Determine the reaction rate constant to base 10 as follows: $k_1' = 0.4343*k' = 0.4343 * 0.492185 = 0.214$/day.

---

### LISTING OF PROGRAM 2.12   (CHAP2.12\FORM1.VB): BOD DETERMINATION ACCORDING TO THOMAS

```
'********************************************************************************************
'Program 2.12: BOD-2
'Thomas method for determining BOD
'********************************************************************************************

Public Class Form1
    'Variables used in calculations
    Dim T(), BOD(), tbod() As Double
    Dim a, b As Double
    Dim k1, kd, Lo As Double
    '****************************************
    'get_max(): Gets the largest number of an
    '           array, given a reference to
    '           the array and member count.
    '****************************************

    Private Function get_max(ByRef array() As Double, ByVal count As Integer) As Double
        Dim i As Integer
        Dim max As Double = array(0)
        For i = 1 To count - 1
            If max < array(i) Then
                max = array(i)
            End If
        Next
        Return max
    End Function

    '**********************************************
    'Calc's a straight line from the scattered
    'point data. The algorithm used is simple:
    '(1) Divide data into two sets
    '(2) Find a mid-point in each set
    '(3) Find line equation from these two points
    '(4) Find the first and last points in this line
    '(5) Get the line equation.
    '**********************************************
```

```
    Private Sub get_straight_line()
        Dim count As Integer = DataGridView1.Rows.Count - 1
        Dim max_T As Double = get_max(T, count)
        Dim max_BOD As Double = get_max(tbod, count)
        Dim mid_count As Integer = count / 2
        Dim sum1X, sum1Y, sum2X, sum2Y As Double
        Dim i As Integer
        sum1X = 0 : sum2Y = 0
        sum1Y = 0 : sum2Y = 0
        For i = 0 To mid_count - 1
            sum1X += T(i)
            sum1Y += tbod(i)
        Next
        sum1X /= mid_count
        sum1Y /= mid_count
        For i = mid_count To count - 1
            sum2X += T(i)
            sum2Y += tbod(i)
        Next
        sum2X /= (count - mid_count)
        sum2Y /= (count - mid_count)
        Dim m As Double
        m = (sum2Y - sum1Y) / (sum2X - sum1X)
        'find straight line equation:
        'y = a + bx
        b = -(m * sum1X) + sum1Y
        a = m
    End Sub

    '**********************************************************************************
    'Code in Form_Load sub will be executed when the form is loading,
    'i.e., first thing after program loads into the memory
    '**********************************************************************************
    Private Sub Form1_Load(ByVal sender As System.Object,
      ByVal e As System.EventArgs) Handles MyBase.Load
        Me.MaximizeBox = False
        Me.FormBorderStyle = Windows.Forms.FormBorderStyle.FixedSingle
        Me.Text = "Program 2.12: Thomas method for determining BOD"
        Label1.Text = "a (slope):"
        Label2.Text = "b (intercept):"
        Label3.Text = "k' (per day, to base E):"
        Label4.Text = "k' (per day, to base 10):"
        Label5.Text = "Lo (mg/L):"
        Label6.Text = "Decimal points:"
        NumericUpDown1.Maximum = 10
        NumericUpDown1.Minimum = 0
        NumericUpDown1.Value = 2
        DataGridView1.Rows.Clear()
        DataGridView1.Columns.Clear()
        DataGridView1.Columns.Add("tCol", "t (days)")
        DataGridView1.Columns.Add("bodCol", "BOD (mg/L)")
    End Sub

    Sub calculateResults()
        If DataGridView1.ColumnCount <> 3 Then
            DataGridView1.Columns.Add("tbodCol", "(t/BOD)^(1/3)")
        End If
        'prepare the third column
        Dim i As Integer
        Dim count As Integer = DataGridView1.Rows.Count - 1
        If count = 0 Then Exit Sub
        ReDim T(count), BOD(count), tbod(count)
```

```
                For i = 0 To count - 1
                    T(i) = Val(DataGridView1.Rows(i).Cells("tCol").Value)
                    BOD(i) = Val(DataGridView1.Rows(i).Cells("bodCol").Value)
                    tbod(i) = Math.Pow((T(i) / BOD(i)), (1 / 3))
                    ''tbod(i) = (T(i) / BOD(i))
                    DataGridView1.Rows(i).Cells("tbodCol").Value = tbod(i).ToString
                Next
                'use our data to mathematically estimate
                'a straight line equation.
                get_straight_line()
                'then calculate...
                'Reaction rate constant (to base e)
                kd = 6 * a / b
                'Ultimate first stage BOD
                Lo = 1 / (b ^ 3 * kd)
                'Reaction rate constant (to base 10)
                k1 = 0.4343 * kd
            End Sub


            Sub showResults()
                'Output results
                TextBox1.Text = FormatNumber(a, NumericUpDown1.Value)
                TextBox2.Text = FormatNumber(b, NumericUpDown1.Value)
                TextBox3.Text = FormatNumber(kd, NumericUpDown1.Value)
                TextBox4.Text = FormatNumber(k1, NumericUpDown1.Value)
                TextBox5.Text = FormatNumber(Lo, NumericUpDown1.Value)
            End Sub

            Private Sub NumericUpDown1_ValueChanged(ByVal sender As System.Object,
                ByVal e As System.EventArgs) Handles NumericUpDown1.ValueChanged
                showResults()
            End Sub

            Private Sub Button1_Click(ByVal sender As System.Object,
                ByVal e As System.EventArgs) Handles Button1.Click
                calculateResults()
                showResults()
            End Sub
        End Class
```

## 2.6  RADIOACTIVITY

Radioactivity is a property of unstable atoms. It arises from the spontaneous breaking up of certain heavy atoms into other atoms, which themselves might be radioactive. This phenomenon continues producing a transformation series. This disintegration results in three kinds of radioactive emissions known as alpha particles, beta particles, and gamma rays (Figure 2.7).

Each radioactive substance is characterized by its half-life. The half-life period is the time required for half the atoms in any given sample of the substance to decay. The half-life periods are different for each element. The unit of radioactivity is the Curie, which is the number of disintegrations occurring per second in 1 g of pure radium (Rowe and Abdel-Magid 1995).

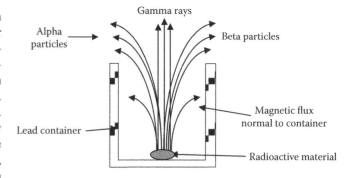

**FIGURE 2.7**  Radioactive decay.

The rate of decay of any nuclide is directly proportional to the number of atoms present and follows a first-order reaction as given by the following equation:

$$Ln \frac{N_t}{N_0} = -k_t \qquad (2.31)$$

where:

$N_0$ is the number of atoms at time $t_0$
$N_t$ is the number of atoms found at time $t$
$k_t$ is the decay constant

Equation 2.32 gives the relationship between decay time and half-life:

$$k_t = \frac{0.6932}{t_{1/2}} \qquad (2.32)$$

where $t_{1/2}$ is the half-life of a particular nuclide.

## Example 2.13

1. Write a computer program for the computation of the number of atoms present after time $t$ for any fraction of the remaining mass and design the program so as to determine the time needed by the following substances to lose any fraction of their initial mass: polonium, $Po^{218}$ (3.05 min); bromine, $Br^{78}$ (6.4 min); lead, $Pb^{214}$ (26.8 min); sodium, $Na^{24}$ (15 h); fluorine, $F^{131}$ (8 days); phosphorous, $P^{32}$ (14.3 days); thorium, $Th^{234}$ (24.1 days); cobalt, $Co^{60}$ (5.3 years); hydrogen, $H^3$

(12.3 years); cesium, $Cs^{137}$ (30 years); radium, $Ra^{226}$ (1600 years); carbon, $C^{14}$ (5730 years); strontium, $Sr^{90}$ (28.1 years); uranium, $U^{234}$ (2.48 * $10^5$ years); potassium, $K^{40}$ (1.28 * $10^9$ years); and uranium, $U^{238}$ (4.51 * $10^9$ years). (*Note*: Number within parentheses indicates the half-life period of the radioactive substance).

2. Radioactive radium has a half-life of 1600 years. Find the mass remaining unchanged of 50 units of this radioactive material after 3200, 4800, and 6400 years. Compute the time needed for the substance to lose 10% of its mass.

**Solution**

1. For solution to Example 2.13 (1), see the listing of Program 2.13.
2. Solution to Example 2.13 (2):
   a. Determine the decay constant as follows: $k_t = 0.693/t_{1/2} = 0.693/1600 = 4.33 * 10^{-4}$ year$^{-1}$.
   b. Find the mass remaining after time $t$ as follows: $N = N_0*e^{-k*t}$ for $t = 1600$, $N = 50*e^{-0.000433 * 1600} = 25$; for $t = 3200$, $N = 12.5$; for $t = 4800$, $N = 6.25$; for $t = 6400$, $N = 3.13$.
      i. Plot a graph of mass remaining versus time, and find from the graph for a remaining mass of [$=50 - (10 * 50/100)$] 45, the required time = 243 years, or weight lost = 10 * 50/100 = 5; weight remaining = 50 − 5 = 45.
      ii. Or use Ln 45/50 = −4.33 * $10^{-4}*t$ to find $t = 243$ years.

---

**LISTING OF PROGRAM 2.13 (CHAP2.13\FORM1.VB): RADIOACTIVITY RATE OF DECAY**

```
'***************************************************************************************
'Program 2.13: radioactivity
'Computes radioactivity
'***************************************************************************************

Imports System.Math
Public Class Form1
    Dim elem(16), sym(16) As String
    Dim t_half(16) As Double
    Dim th, kt, no, p As Double
    Dim tmp1, tmp2, tmp3 As String
    'convert days to years
    Const F1 As Double = 1 / 365.25
    'convert hours to years
    Const F2 As Double = F1 / 24
    'convert minutes to years
    Const F3 As Double = F2 / 60
    '********************************
    'load calculation tables
    '********************************

    Sub load_tables()
        sym(0) = "Br 78"
        sym(1) = "C 14"
```

```
        sym(2) = "Co 60"
        sym(3) = "Cs 137"
        sym(4) = "F 131"
        sym(5) = "H 3"
        sym(6) = "Pb 124"
        sym(7) = "K 40"
        sym(8) = "Na 24"
        sym(9) = "P 32"
        sym(10) = "Po 218"
        sym(11) = "Ra 226"
        sym(12) = "Sr 90"
        sym(13) = "Th 234"
        sym(14) = "U 238"
        sym(15) = "U 234"
        elem(0) = "Bromine"
        elem(1) = "Carbon"
        elem(2) = "Cobalt"
        elem(3) = "Cesium"
        elem(4) = "Flourine"
        elem(5) = "Hydrogen"
        elem(6) = "Lead"
        elem(7) = "Potassium"
        elem(8) = "Sodium"
        elem(9) = "Phosphorus"
        elem(10) = "Polonium"
        elem(11) = "Radium"
        elem(12) = "Strontium"
        elem(13) = "Thorium"
        elem(14) = "Uranium"
        elem(15) = "Uranium"
        'convert all half times to years
        t_half(0) = 6.4 * F3
        t_half(1) = 5730
        t_half(2) = 5.3
        t_half(3) = 30
        t_half(4) = 8 * F1
        t_half(5) = 12.3
        t_half(6) = 26.8 * F3
        t_half(7) = 1280000000.0
        t_half(8) = 15 * F2
        t_half(9) = 14.3 * F1
        t_half(10) = 3.05 * F3
        t_half(11) = 1600
        t_half(12) = 28.1
        t_half(13) = 24.1 * F1
        t_half(14) = 4510000000.0
        t_half(15) = 248000.0
    End Sub


    '*********************************************************************************
    'Code in Form_Load sub will be executed when the form is loading,
    'i.e., first thing after program loads into the memory
    '*********************************************************************************

    Private Sub Form1_Load(ByVal sender As System.Object,
ByVal e As System.EventArgs) Handles MyBase.Load
        Me.MaximizeBox = False
        Me.FormBorderStyle = Windows.Forms.FormBorderStyle.FixedSingle
        Me.Text = "Program 2.8: Radioactivity"
        'load the tables with default values
        load_tables()
```

```
            Label1.Text = "Select material from below:"
            ListBox1.Items.Clear()
            For i = 0 To 15
                ListBox1.Items.Add(sym(i))
            Next
            ListBox1.Items.Add("Other material..")
            Label2.Text = "Initial mass of material="
            Label3.Text = "What period of time for decay in years:"
            GroupBox1.Text = " Decay: "
            label4.text = "Mass remaining after time period t is:"
            Label5.Text = "Time to loose that percentage (years):"
            GroupBox2.Text = " Mass Loss: "
            Label6.Text = "What percentage of mass to lose:"
            Label7.Visible = False
            TextBox3.Enabled = False
            TextBox4.Enabled = False
        End Sub

    Private Sub ListBox1_SelectedIndexChanged(ByVal sender As System.Object,
ByVal e As System.EventArgs) Handles ListBox1.SelectedIndexChanged
            If ListBox1.SelectedIndex < 0 Then Exit Sub
            Select Case ListBox1.SelectedIndex
                Case 16
                    Dim sym = InputBox("Enter the symbol of the material:",
"Other material")
                    th = Val(InputBox("Enter the half life period:", "Other material"))
                    Label7.Text = "Using material (" + sym + "), t1/2 of " + th.ToString
                    Label7.Visible = True
                Case Else
                    th = t_half(ListBox1.SelectedIndex)
                    Label7.Visible = False
            End Select
            'Decay constant
            'th is the half-life period
            kt = 0.693 / th
    End Sub

    Private Sub TextBox1_TextChanged(ByVal sender As System.Object,
ByVal e As System.EventArgs) Handles TextBox1.TextChanged
            no = Val(TextBox1.Text)
    End Sub

    Private Sub TextBox2_TextChanged(ByVal sender As System.Object,
ByVal e As System.EventArgs) Handles TextBox2.TextChanged
            Dim t = Val(TextBox2.Text)
            'Mass remaining after any time t for an initial mass no
            Dim n = no * EXP(-kt * t)
            textbox3.text = n.tostring
    End Sub

    Private Sub TextBox5_TextChanged(ByVal sender As System.Object,
ByVal e As System.EventArgs) Handles TextBox5.TextChanged
            p = Val(TextBox5.Text)
            'to find the time to loose a percentage p
            Dim n1 = (100 - p) / 100 * no
            Dim t1 = Log(no / n1) / kt
            TextBox4.Text = t1.ToString
    End Sub
End Class
```

## 2.7 HOMEWORK PROBLEMS IN WATER AND WASTEWATER PROPERTIES

### 2.7.1 DISCUSSION PROBLEMS

1. Why is water vital for human survival?
2. What benefits are to be gained by acquiring knowledge about the characteristics of water and wastewater?
3. Outline the main physical quality parameters relevant in water quality management.
4. What are the main effects of an increase in the temperature of water or wastewater?
5. Define conductivity and indicate the units of measurements.
6. Indicate the different methods that contribute to increase water salinity.
7. Describe a method for measuring the solids content of a water sample.
8. How can you relate mathematically density, specific volume, specific weight, and specific gravity?
9. What are some of the problems associated with viscous fluids?
10. Give one example for each of the following: dilatant, thixotropic, ideal liquid, ideal solid, non-Newtonian, Newtonian, plastic, and pseudoplastic fluid.
11. Illustrate the differences between kinematic and absolute viscosity coefficients.
12. What is surface tension?
13. What are the practical benefits of surface tension?
14. Define bulk modulus.
15. Describe the differences between alpha, beta, and gamma radiation.
16. Outline the main chemical characteristics relating to water quality.
17. Outline the factors that affect the pH of water and wastewater.
18. How can you differentiate between an alkaline and an acidic solution?
19. Define hardness of water. Note the main classes of hardness, and discuss cons and pros.
20. Indicate how to differentiate between a hard and a soft water sample.
21. Discuss the factors that influence the solubility of a gas in water.
22. What is BOD?
23. What are the advantages of the BOD test?
24. Potable palatable water that is supplied by a certain municipality may contain chemicals which were initially not in the source (raw) water. In the following table, give three examples of such chemicals. Indicate the way they entered the water. Outline their potential health impacts and effects. Suggest methods for avoiding such ailments.

**Table (1) Chemicals not in source (raw) water**

| Chemical | First Chemical | Second Chemical | Third Chemical |
|---|---|---|---|
| Example | | | |
| How chemical entered water? | | | |
| Potential health impacts and effects | | | |
| Suggested methods for getting rid of harmful chemical | | | |

25. Describe the hydraulic cycle.
26. Outline the preferable water source characteristics.
27. Rearrange group I with the corresponding relative ones of group II in the area allocated for the answer shown in the following table.

| Group I | Rearranged Group II | Group II |
|---|---|---|
| Pollution | | Hand washing |
| True solutions removal | | Water loss |
| Pathogen removal | | Quality related |
| Colloidal suspensions removal | | Anopheles mosquitoes |
| Physical treatment | | Fecal related |
| Preliminary treatment | | Water based |
| Leakage | | Quantity related |
| Sanitation barrier | | Water site related |
| House fly | | African sleeping sickness |
| Scistosomiasis | | Blue babies syndrome |
| Malaria | | Chlorination |
| Eye disease | | Tooth decay |
| Water contact diseases | | Grit removal |
| Water-washed diseases | | Gas transfer |
| Sanitation-related diseases | | Chemical coagulation |
| Water-borne diseases | | Sedimentation |
| Water insect vector | | Scistosoma |
| Trypanosomiasis | | Trachoma |
| Methemoglobinemia | | *Musca domestica* |
| Fluoride | | Improper storage |
| | | Desalination |

28. Indicate whether the following sentences are true (T) or false (F):
    a. Water project components, structures, and equipment are assumed to serve the project for its design period. ( )
    b. Reasons for water treatment include removal or reduction of SS or floating solids, colloids, dissolved salts, oil, color, taste, and odor. ( )
    c. Choice of type of needed water treatment requires completing surveys on source, operation, and

presence of skilled workers, funding, and costs of construction maintenance. ( )

d. Water-based diseases are caused by aquatic organisms that spend part of their life cycle in the water and another part as a parasite affecting humans or animals. ( )

e. Hardness is related to scale deposition (soap curd) and taste, and it hampers soap lather formation. ( )

f. Although *E. coli* is considered to be the best indicator of fecal contamination, but the disclosure of temperature-tolerant coliform bacteria is considered an acceptable alternative. ( )

g. Standard methods for the examination of water and wastewater provide hundreds of laboratory tests of water quality and properties. These methods represent the best current practice of water and wastewater analysis. ( )

h. Langelier Index is an expression of the tendency for deposition (noncorrosiveness) or dissolution of calcium carbonate scale in a pipe. ( )

29. Complete missing titles and fill in blanks by using the following words and phrases (health, corrosive, lead, lethal dose, guidelines, poverty, disease causing)

a. An objective of treatment is removal of bacteria, viruses, pathogens, and other ........................................... organisms that hamper health of the individual.

b. ........................... is a state of complete physical, mental, and social well-being and not merely the absence of disease or infirmity.

c. Unsafe and bad water quality is the basis of causes of .......................

d. ......................... may accumulate in bones (skeletal changes) and is associated with kidney damage, constipation, loss of appetite, anemia, abdominal pain, paralysis, poisoning, metabolic interference, and central and peripheral nervous system toxicity.

e. Factors affecting the toxic effects of chemicals include chemical characteristics, ..............................., average daily intake, ability of chemical to reach and accumulate in the tissues, and routes of exposure to the chemical

f. ....................................................... envisage drinking water free from pathogenic microbes, or types and quantities of toxins that affect the public health in the short or long term.

g. A water supply that exhibits ........................... characteristics may cause health, aesthetic, and economic problems in distribution pipelines and residential plumbing systems.

## 2.7.2 SPECIFIC MATHEMATICAL PROBLEMS

1. Write a short computer program to illustrate the effect of angle of contact on surface tension.

2. Write a short computer program to enable determination of the temperature when the readings on a Celsius and a Fahrenheit thermometer coincide.

3. Write a simple computer program to find the weight of a liquid, given its volume and specific gravity.

4. Write a short computer program that allows for the conversion of viscosity (dynamic and kinematic) between the SI system of units and the British system.

5. Write a short computer program that allows the determination of the molecular weight and equivalent weight of the materials found in the periodic table. Test the program by determining the molecular weights of the following compounds: barium acetate ($Ba(C_2H_3O_2)_2$), nitrous oxide ($N_2O$), ammonium aluminum sulfate ($(NH_4)_2Al_2(SO_4)_4$), strontium sulfate ($SrSO_4$) and copper sulfate, (Blue Vitriol) ($CuSO_4 \cdot 5H_2O$).

6. Write a computer program to determine the normality ($N$) and molarity ($M$) of a solution given that $N = (wt/EW)/V$ and $M = (wt/MW)/V$, where wt is the weight, EW is the equivalent weight, MW is the molecular weight, and $V$ is the volume. Test your program by finding the molarity of a solution that contains 2.5 g of sulfuric acid, in 4 L of solution.

7. Write a computer program to determine the percentage of elements found in a compound composed, say, of nine substances: carbon (C), hydrogen (H), nitrogen (N), and oxygen (O). Test your program by determining the percent carbon in fructose $CH_2OH(CHOH)_3COCH_2OH$ and the percent zinc in the compound zinc dimethyl-dithiocarbamate $Zn[S_2CN(CH_3)_2]_2$.

8. Write a computer program that computes the pH of a solution. Design the program so as to indicate whether the solution is acidic or alkaline. The program is to differentiate between the acidity of different solutions, given $[H^+]$ or $[OH^-]$ ion concentration.

9. Write a computer program that enables estimating the concentration of a cation or an anion from data of cations involving calcium, magnesium, sodium, strontium, and iron, and anions containing chloride, sulfate, nitrate, carbonate, and bicarbonate, given the total hardness and percent experimental error.

10. Assuming an experimental error of 3%, use the program developed in problem 9 to find the missing chloride concentration for the following data of a sample of water (all values are in mg/L):

| Ca$^{++}$ | 35 | Cl$^-$ | 18 |
|-----------|----|--------|----|
| Mg$^{++}$ | 40 | HCO$_3^-$ | 122 |
| Sr$^{+++}$ | 9 | SO$_4^-$ | 34 |
| Fe$^{+++}$ | 23 | NO$_3^-$ | 22 |
|           |    | SiO$_3^-$ | 14 |

11. Write a computer program that enables the computation of the BOD of organic compounds (containing hydrogen, carbon, nitrogen, and oxygen) with a known concentration (in mg/L). Assume reaction products to be carbon dioxide, water, or nitrogen oxide according to reactants entering chemical reaction. Check the validity of your program by determining the total BOD of the following compounds:

    a. 0.1 molar solution of ether [$(CH_3CH_2)_2O$]

    b. 5 g of glycerol [$CH_2OH \cdot CHOH \cdot CH_2OH$]

    c. 20 mg/L uric acid [$C_5H_4O_3N_4$]

## REFERENCES

Abdel-Magid, I. M. 1986. *Water Treatment and Sanitary Engineering*. Khartoum, Sudan: Khartoum University Press (Arabic). doi:10.13140/RG.2.1.1962.0883.

Abdel-Magid, I. M., and ElHassan, B. M. 1986. *Water Supply in the Sudan*. Khartoum, Sudan: Khartoum University Press, Sudan National Council for Research (Arabic). doi:10.13140/RG.2.1.2371.6882.

Abdel-Magid, I. M., and Abdel-Magid, M. I. M. 2015. *Problem Solving in Environmental Engineering*, 2nd ed. Dammam, Saudi Arabia: CreateSpace Independent Publishing Platform.

APHA. 2012. *Standard Methods for the Examination of Water and Wastewater*, 22nd ed. Washington, DC: American Water Works Association.

Casio Computer Co. 2016. *Water Hardness Calculator*. http://keisan.casio.com (accessed on February 20, 2016).

Davis, W. M. 2011. *Physical Chemistry: A Modern Introduction*, 2nd ed. Boca Raton, FL: CRC Press.

Finnermore, E., Franzini, J. 2001. Fluid Mechanics With Engineering Applications, 10th ed. New York: McGraw-Hill Education.

Green, D. W., and Perry, R. H. 2007. *Perry's Chemical Engineers Handbook*, 8th ed. New York: McGraw-Hill Professional.

Grigg, N. 2012. *Water, Wastewater, and Storm water Infrastructure Management*, 2nd ed. Boca Raton, FL: CRC Press.

Groundwater Software. 2016. *Calculator 8 - Water Hardness Calculator*. http://www.groundwatersoftware.com (accessed on February 18, 2016).

Howe, K. J., Hand, D. W., Crittenden, J. C., Trussell, R. R., and Tchobanoglous, G. 2012. *Principles of Water Treatment*, 1st ed. Hoboken, NJ: Wiley.

Lenntech. 2016. *Water Hardness Calculator*. http://www.lenntech.com/ro/water-hardness.htm#ixzz2iQsTNzz5 (accessed on February 17, 2016).

Lin, S. D. 2014. *Water and Wastewater Calculations Manual*, 3rd ed. New York: McGraw-Hill Education.

London and General Water Purifying Company. 2012. *Water: Its Impurities and Purification*. Nabu Press.

McGhee, T. J., and Steel, E. W. 1991. *Water Supply and Sewerage*, 6th ed. New York: McGraw-Hill.

Metcalf and Eddy, Inc., Tchobanoglous, G., Stensel, H. D., Tsuchihashi R. and Burton, F. 2013. *Wastewater Engineering: Treatment and Resource Recovery*, 5th ed. New York: McGraw-Hill Science/Engineering/Math.

Munson, B. R., Rothmayer, A. P., Okiishi, T. H., and Huebsch, W. W. 2012. *Fundamentals of Fluid Mechanics*. New York: Wiley.

Nemerow, N. L., Agardy, F. J., and Salvato, J. A. 2009. *Environmental Engineering*, 6th ed. New York: Wiley.

Peavy, H. S., Rowe, D. R., and Tchobanoglous, G. 1985. *Environmental Engineering*. New York: McGraw-Hill Book Co.

Ricketts, J., Loftin, M., and Merritt, F. S. 2003. *Standard Handbook for Civil Engineers*. New York: McGraw-Hill Professional.

Rowe, D. R., and Abdel-Magid, I. M. 1995. *Wastewater Reclamation and Reuse*. New York: Lewis Publishers.

Spellman, F. R. 2013. *Handbook of Water and Wastewater Treatment Plant Operations*, 3rd ed. Boca Raton, FL: CRC Press.

Vesilind, P. A., and Peirce, J. J. 1997. *Environmental Pollution and Control*, 4th ed. London: Butterworth-Heinemann.

Viessman, W., Perez, E. M., Chadik, P. A., and Hammer, M. J. 2015. *Water Supply and Pollution Control*. Upper Saddle River, NJ: Pearson Publishing.

Whipple, G. C., and Whipple, M. C. 1911. Solubility of oxygen in sea water. *J. Am. Chem. Soc.* 33:362.

# 3 Computer Modeling Applications for Water Resources, Usage, Groundwater, and Water Storage and Distribution

## 3.1 INTRODUCTION

This chapter emphasized on conventional and nonconventional water sources (reclaimed and renovated supplies) and source selection. Water footprints were stressed upon within the chapter. Demographic and population growth estimates were stressed upon due to their direct effects on consumption design rates and for design and operation of water supply units and wastewater processes. Likewise, the chapter looked into water requirements for extinguishing fires and preventing them from flaring into uncontrollable firestorm. Also, the chapter expanded on groundwater flow, aquifer characteristics, well yield and groundwater recharge. Groundwater recharge is stressed upon to overcome aquifer water reduction, augment usage of surface runoff, and conserve and increase yield of wealth. Various types of water storage reservoirs were dealt with together with computations for storage volume and hydraulic and structural design of storage tanks. Use of mass curve techniques signified a study of storage effects on stream flow conditions. Water distribution analysis and design received attention for appropriate design merits for urban sectors and rural fringes using suitable design techniques and procedures. The chapter introduced relevant computer programs modeled for solving specific problems in water resources, water storage and distribution within areas and domains of importance.

## 3.2 WATER SOURCES AND FOOTPRINTS

### 3.2.1 INTRODUCTION

The sources for a water supply can be grouped into three major categories: rainwater (precipitation), surface water (runoff), and groundwater. Surface water includes both rainwater and groundwater where the water table lies above the ground surface. Surface waters include water in rivers, streams, lakes, lagoons, ponds, and oceans. Groundwater is contained in the soil or rocks below the water table (Smith 2005).

The movement of water on the earth follows a complex cyclic pattern from the sea to the atmosphere and then by precipitation to the earth, where it collects in rivers, streams, and lakes, and then runs back to the sea. The cycle may be short-circuited at several stages; there is no uniformity in the time a cycle takes; and the intensity and frequency of the cycle depend on geographical and climatic conditions (Gupta 2007) (see Figure 3.1).

### 3.2.2 WATER FOOTPRINT

The *water footprint* of a country is defined as "the volume of water needed for the production of goods and services consumed by the inhabitants of the country". The global water footprint is 1240 $m^3$/cap/year (Chapagain and Hoekstra 2004).

The water footprint of a country can be calculated with either the top-down approach or the bottom-up approach.

In the top-down approach, the water footprint is being calculated as the sum of water use in the country plus gross virtual water import (entering) into the country minus gross virtual water export (leaving the country), as shown in the following equation:

$$\text{Water footprint} = \Sigma(\text{water use in country})$$
$$+ (\text{gross virtual water import into country}) \quad (3.1)$$
$$- (\text{gross virtual water export})$$

The bottom-up approach aggregates the individual water footprints of the inhabitants of a country to get the total water footprint of a country. It considers the sum of all goods and services consumed multiplied with their respective virtual water content, where the virtual water content of a good will vary as a function of place and conditions of production. Individual water footprints are calculated by multiplying all consumed goods and services with their respective virtual water content as presented in the following equation (Chapagain and Hoekstra 2004, Zhang et al. 2013, National Geographic 2016, Water Footprint Network 2016):

$$\text{Water footprints} = (\text{all consumed goods and services})$$
$$\times (\text{respective virtual water content}) \quad (3.2)$$

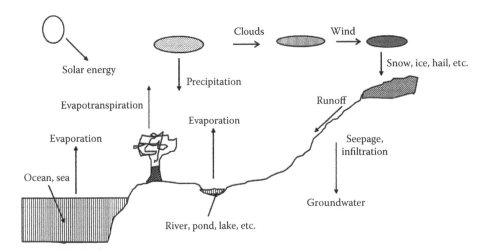

**FIGURE 3.1**  Schematic diagram of the hydrological cycle.

The national water footprint may be determined from the website of water footprint network, and the global average water footprint (for comparison) may be taken as 1385 m³/year per capita (Water Footprint Network 2016). To find out one's ecological footprint, the global footprint network computations may be used to discover one's biggest areas of resource consumption and to learn how to effectuate changes (Global Footprint Network 2016).

### 3.2.3  SOURCE SELECTION

The process of selecting a water source must consider the following items:

1. Is the source free from toxic or undesirable chemicals?
2. Does the source contain:

a.  *Domestic sewage*: It may contain disease-causing agents or undesirable substances that pose a health hazard.
b.  *Industrial waste*: It may contain materials that may render the water unsuitable for the intended use.
c.  *Agricultural waste*: It may contain residues of herbicides, insecticides, pesticides, and nutrients.

Figure 3.2 gives an outline that can be followed during the process of source selection.

Water use and consumption can be grouped into several broad categories: domestic, industrial, commercial and public, agricultural, and recreational. Unaccounted for water consumption in general makes up approximately 10%–15% of the total water consumption. The total consumption is the sum of all the

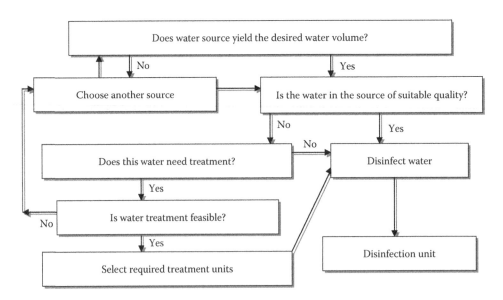

**FIGURE 3.2**  Source selection.

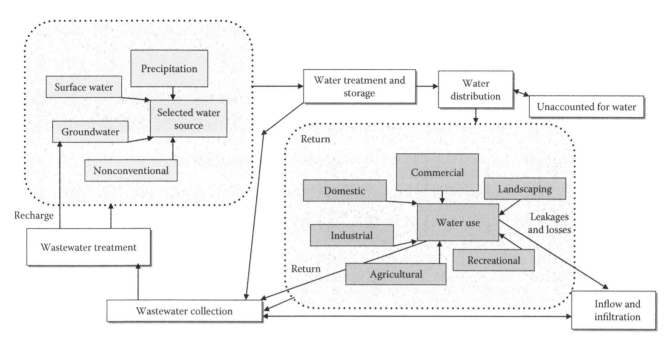

**FIGURE 3.3**   Water use cycle.

aforementioned uses and losses. Water use and consumption data are generally expressed in liters per capita per day, (L/c/d).

There is a wide range of uses for water and properly treated and reclaimed wastewater such as in power plant and industrial cooling water, groundwater recharge, aquaculture (fish farming), silviculture, landscape irrigation, recreational purposes, firefighting, industrial processes, stock watering, stream flow augmentation, and agricultural irrigation. Figure 3.3 shows an overlap of the water use cycle.

## 3.3   POPULATION GROWTH AND CONSUMPTION DESIGN RATES

Population estimates are required for design and operation of water supply and wastewater units operations and processes. The most important of these estimates used are estimates for short periods (within 1–10 years) and estimates for a long term (within 10–50 years and more). Many methods are used to estimate population such as arithmetic progression and graphical methods to estimate future population. Census data in previous years are used to assess the prospects for the current and foreseeable circumstances, without taking into account unusual variables such as migration of people due to new industries; or the migration of population to the discovery of oil, gold, and other metals of value; or escape, displacement, and refugees due to the ravages of war or as a result of disasters and calamities; or changes in industry and industrial growth; or factors of births and deaths; or military and government activities and other considerations.

Methods of calculating short-period estimates include computational mathematical, geometrical growth, interim increase method, decreasing rate of increase, and graphical extension method.

Mathematical methods for calculating long-period estimates include graphical comparison with rate of growth of large similar cities and choice of a mathematical method such as the logical curve and its satisfaction to observational data. These methods of estimates are not to be adopted for a long period for the possibility of impact of unexpected factors on outcomes and estimates.

1. In the *arithmetic progression method* for population estimates for a short period, a steady increase of growth is added in it for periods. The number of population can be found as follows:

$$Pn = P + n * i \qquad (3.3)$$

where:

$Pn$ is the number of population after $n$ years or a decade (10 years)

$P$ is the number of present population

$i$ is the increase in the number of the population annually or every decade

2. The *moderate rate of growth* for population estimates for arithmetic growth (to be used for large established cities) is calculated as presented in Equation 3.4

$$P_m = P_e + \left(P_l - P_e\right)\left(\frac{t_m - t_e}{t_l - t_e}\right) \qquad (3.4)$$

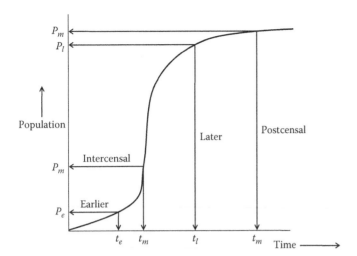

**FIGURE 3.4**   Moderate rate of growth method.

where:

$P_m$ is the population of the desired area for time $t_m$

$P_e$ is the population of earlier census at time $t_e$

$P_l$ is the population of later census for time $t_l$

Estimate of the number of the population censuses by arithmetic growth for postcensal estimates is given as follows

$$P_m = P_l + \left(P_l - P_e\right)\left(\frac{t_m - t_l}{t_l - t_e}\right)$$   (3.5)

3. Geometric estimates progression method gives intercensal population geometric estimates as depicted in Equation 3.6, whereas Equation 3.7 offers the postcensal population estimates:

$$\text{Log } P_m = \text{Log } P_e + \left(\text{Log } P_l - \text{Log } P_e\right)\left(\frac{t_m - t_e}{t_l - t_e}\right)$$   (3.6)

$$\text{Log } P_m = \text{Log } P_l + \left(\text{Log } P_l - \text{Log } P_e\right)\left(\frac{t_m - t_l}{t_l - t_e}\right)$$   (3.7)

**Example 3.1**

1. Write a computer program that enables the computation of population growth in a certain area.

$$P_m = P_l + \left(P_l - P_e\right)\frac{t_m - t_l}{t_l - t_e}$$

$$P_m = P_l + \left(P_l - P_e\right)\left(\frac{t_m - t_l}{t_l - t_e}\right)$$

$$\text{Log } P_m = \text{Log } P_e + \left(\text{Log } P_l - \text{Log } P_e\right)\left(\frac{t_m - t_e}{t_l - t_e}\right)$$

$$\text{Log } P_m = \text{Log } P_l + \left(\text{Log } P_l - \text{Log } P_e\right)\left(\frac{t_m - t_l}{t_l - t_e}\right)$$

2. A city recorded a population of 135,000 in its earlier decennial census and 1,95,000 in its later one. Assuming a census date of April 1, estimate the midyear (July 1) populations for the following:
   a. For the fifth intercensal year
   b. For the ninth postcensal year by geometric increase

**Solution**

1. For solution to Example 3.1 (1), see the listing of Program 3.1.
2. Solution to Example 3.1 (2):
   a. Given: $P_l = 195,000$ and $P_e = 135,000$
   b. Intercensal geometric growth estimates after the fifth year

$$t_m - t_e = 5.25 \text{ years, then } t_l - t_e = 10 \text{ years}$$

$$\frac{t_m - t_e}{t_l - t_e} = \frac{5.25}{10} = 0.525$$

$$\text{Log } P_m = \text{Log } P_e + \left(\text{Log } P_l - \text{Log } P_e\right)\frac{t_m - t_e}{t_l - t_e}$$

$$= \text{Log } 135,000 + \left(\text{Log } 195,000 - \text{Log } 135,000\right) * 0.525$$

$$P_m = 163,749$$

   c. Postcensal geometric growth estimates for the ninth year

$$t_m - t_l = 9.25 \text{ year, then: } t_l - t_e = 10 \text{ years}$$

$$\frac{t_m - t_l}{t_l - t_e} = \frac{9.25}{10} = 0.925$$

$$\text{Log } P_m = \text{Log } P_l + \left(\text{Log } P_l - \text{Log } P_e\right)\frac{t_m - t_l}{t_l - t_e}$$

$$= \text{Log } 195,000 + \left(\text{Log } 195,000 - \text{Log } 135,000\right) * 0.925$$

$$P_m = 274,005$$

**LISTING OF PROGRAM 3.1   (CHAP3.1\FORM1.VB): POPULATION ESTIMATES**

```vb
'******************************************************************************
'Program 3.1: Population
'Computes Population Growth Estimates
'******************************************************************************

Imports System.Math
Public Class Form1
    '**************************************************************************
    'Code in Form_Load sub will be executed when the form is loading,
    'i.e., first thing after program loads into the memory
    '**************************************************************************

 Private Sub Form1_Load(ByVal sender As System.Object,
ByVal e As System.EventArgs) Handles MyBase.Load
    Me.Text = "Example 3.1: Population Growth Estimates"
    Me.FormBorderStyle = Windows.Forms.FormBorderStyle.FixedSingle
    Me.MaximizeBox = False
    Label1.Text = "(tm-te) in years:"
    Label2.Text = "(t1-te) in years:"
    Label3.Text = "P1 (population at time t1):"
    Label4.Text = "Pe (population at time te):"
    Label5.Text = "Calculation:"
    Label6.Text = "Pm (population at time tm):"
    Button1.Text = "&Calculate"
    ComboBox1.Items.Clear()
    ComboBox1.Items.Add("Intercensal population growth")
    ComboBox1.Items.Add("Postcensal population growth")
    ComboBox1.SelectedIndex = 0
 End Sub

 Private Sub Button1_Click(ByVal sender As System.Object,
ByVal e As System.EventArgs) Handles Button1.Click
    '*** Sanity check
    If ComboBox1.SelectedIndex = -1 Then
       MsgBox("Please select a calculation from the list.",
          vbOKOnly Or vbExclamation)
       Exit Sub
    End If
    Dim tmte, t1te As Double
    Dim Pm, P1, Pe As Double
    '*** Gather inputs
    tmte = Val(TextBox1.Text)
    t1te = Val(TextBox2.Text)
    P1 = Val(TextBox3.Text)
    Pe = Val(TextBox4.Text)
    '*** Calculate
    Dim logPm As Double
    If ComboBox1.SelectedIndex = 0 Then
       logPm = Log(Pe) + ((Log(P1) - Log(Pe)) * (tmte / t1te))
    Else
       logPm = Log(P1) + ((Log(P1) - Log(Pe)) * (tmte / t1te))
    End If
    Pm = Pow(Math.E, logPm)
    TextBox5.Text = FormatNumber(Pm, 0)
 End Sub
End Class
```

## 3.4   FIRE DEMAND

Water is required to extinguish fires and to prevent them from spreading into uncontrollable firestorm. Usually, the quantity of water to be used for firefighting is small. Nevertheless, the rate of application is high. For a community of 20,000 or less (IPHE 1984), the following equation may be used to estimate the fire demand:

$$Q = 3860\sqrt{P} \cdot \left(1 - 0.01\sqrt{P}\right) \qquad (3.8)$$

where:

$Q$ is the fire demand, L/min
$P$ is the population, in thousands

The U.S. Insurance Service Office advocates the use of the formula presented in the following equation to estimate the fire flow required for a given floor area:

$$Q = 3.7 C_f \sqrt{A} \qquad (3.9)$$

where:

$Q$ is the required fire flow, L/s
$C_f$ is the coefficient related to the type of construction
$A$ is the total floor, area including all stories in the building, but excluding its basement, m$^2$

The values of the coefficient $C_f$ may be taken as presented in Table 3.1.

Regardless of the computed value, the fire flow need not exceed 500 L/s for wood frame or ordinary construction, or 380 L/s for noncombustible or fire-resistive constructions. For normal one-storey construction of any type, the fire flow may not exceed 378 L/s. The minimum fire flow may not be less than 32 L/s, and the maximum for all purposes for a single fire is not to exceed 760 L/s. Extra flow may be needed to protect adjacent buildings. For groups of single-family residences and small two-family residences not exceeding two stories in height, the fire flow outlined in Table 3.2 may be used to evaluate the required fire demand (McGhee and Steel 1991).

### TABLE 3.2
### Residential Fire Flows

| Distance between Adjacent Units (m) | Required Fire Flow (L/min) |
| --- | --- |
| >30.5 | 1890 |
| 9.5–30.5 | 2835–3780 |
| 3.4–9.2 | 3780–5670 |
| ≤3 | 5670–7660[a] |

*Source:*   McGhee, T. J. and Steel, E. W., *Water Supply and Sewerage*, 6th ed., New York: McGraw-Hill, 1991.

*Note:*   [a] For continuous construction, use 9450 L/min.

The maximum flow required for an individual fire is 45.4 m$^3$/min for all purposes. In large communities, the possibility of concurrent fires should also be considered. In residential districts, the required fire flow ranges from a minimum of 1.9 m$^3$/min to a maximum of 9.5 m$^3$/min (McGhee and Steel 1991). Hydrant spacing is dictated by the required fire flow because the capacity of a single hydrant is limited. Table 3.3 presents the proposed distances for fire hydrants. The required duration for a fire flow is indicated in Table 3.4.

### TABLE 3.3
### Fire Hydrant Spacing

| Hydrant | Distance (m) | Location |
| --- | --- | --- |
| Ordinary | 60–150 | At street intersections where streams can be taken in any direction |
| High-value districts | As close as 30 | Additional hydrants may be necessary in the middle of long blocks |
| Low-risk areas | 100–150 | In towns |

*Source:*   McGhee, T. J. and Steel, E. W., *Water Supply and Sewerage*, 6th ed., New York: McGraw-Hill, 1991.

### TABLE 3.1
### Values of the Coefficient $C_f$

| Description | $C_f$ |
| --- | --- |
| Wood frame construction | 1.5 |
| Ordinary construction | 1.0 |
| Noncombustible construction | 0.8 |
| Fire-resistive construction | 0.6 |

*Sources:*   Viessman, W. et al., *Water Supply and Pollution Control*, Pearson Pub, 2015; McGhee, T. J. and Steel, E. W., *Water Supply and Sewerage*, 6th ed., New York: McGraw-Hill, 1991.

### TABLE 3.4
### Fire Flow Duration

| Required Fire Flow (L/min) | Duration (h) |
| --- | --- |
| <3780 | 4 |
| 3780–4725 | 5 |
| 4725–5670 | 6 |
| 5670–6615 | 7 |
| 6615–7560 | 8 |
| 7560–8505 | 9 |
| >8505 | 10 |

*Source:*   McGhee, T. J. and Steel, E. W., *Water Supply and Sewerage*, 6th ed., New York: McGraw-Hill, 1991.

## Example 3.2

1. Write a computer program that enables the computation of fire demand for communities of different numbers of populations taking in to consideration the type of construction and restrictions outlined in Section 3.4.
2. Find the fire demand for a town having a population of 0.9 million.
3. Estimate the fire demand for a single-storey building made of a wood frame construction, given that the total floor area amounts to 10,000 $m^2$.

## Solution

1. For solution to Example 3.2 (1), see the listing of Program 3.2.

2. Solution to Example 3.2 (2):
   a. Given: $P = 900,000 = 900$ in thousands.
   b. Determine the fire demand as follows: $Q = 3860(P)^{1/2}$ $[(1-0.01(P^{1/2})] = 3860 * 900^{1/2}*(1-0.01 * 900^{1/2}) = 81$ $m^3$/min.
3. Solution to Example 3.2 (3):
   a. Given: $A = 10000$ $m^2$.
   b. Determine the fire flow as follows: $Q = 3.7 * C_f* (A)^{0.5}$.
   c. From Table 2.4 for a wood frame construction, the coefficient, $C_f = 1.5$.
   d. Compute the fire demand as follows: $Q = 3.7 * 1.5*(10000)^{0.5} = 555$ L/s.
   e. Since this is a wood frame single-storey building, the maximum fire demand may not exceed 380 L/s.

---

### LISTING OF PROGRAM 3.2 (CHAP3.2\FORM1.VB): FIRE DEMAND APPROXIMATION

```
'********************************************************************************
'Program 3.2: Fire
'Computes Fire demand for communities of different population
'********************************************************************************

Public Class Form1
  Dim ttl(3), Cw(4), limit(4)
  Dim opt1
  'maximum fire flow for all purposes
  Const global_max = 760
  'minimum fire flow for all purposes
  Const global_min = 32
  '********************************************************************************
  'Code in Form_Load sub will be executed when the form is loading,
  'i.e., first thing after program loads into the memory
  '********************************************************************************

  Private Sub Form1_Load(ByVal sender As System.Object,
ByVal e As System.EventArgs) Handles MyBase.Load
    Me.MaximizeBox = False
    Me.FormBorderStyle = Windows.Forms.FormBorderStyle.FixedSingle
    Me.Text = "Program 3.2: Computes Fire demand for communities"
    limit(1) = 500
    limit(2) = 500
    limit(3) = 300
    limit(4) = 300
    Cw(1) = 1.5
    Cw(2) = 1
    Cw(3) = 0.8
    Cw(4) = 0.6
    RadioButton1.Text = "Estimate fire demand for a building"
    RadioButton2.Text = "Estimate fire demand for a whole town"
    RadioButton1.Checked = True
    RadioButton3.Text = "Wood frame construction"
    RadioButton4.Text = "Ordinary construction"
    RadioButton5.Text = "Non conmbustible construction"
    RadioButton6.Text = "Fire-resistive construction"
    RadioButton3.Checked = True
    opt1 = 1
```

```vb
      Label1.Text = "Enter total floor area in sq.m:"
      GroupBox1.Text = " Select construction: "
      GroupBox2.Text = " Output: "
      TextBox1.Text = ""
      TextBox2.Text = ""
  End Sub

  Sub calculateResults()
    If RadioButton1.Checked Then
       Dim A, Q As Double
       A = Val(TextBox1.Text)
       Q = 3.7 * Cw(opt1) * A ^ 0.5
       If Q < global_min Then
          TextBox2.Text = global_min.ToString + " (original result is " +
Q.ToString + ")"
       ElseIf Q > limit(opt1) Then
          TextBox2.Text = limit(opt1).ToString + " (original result is " +
Q.ToString + ")"
       Else
          TextBox2.Text = Q.ToString
       End If
    Else
       Dim pop, Q As Double
       pop = Val(TextBox1.Text)
       pop = (pop / 1000) ^ 0.5
       Q = 3860 * pop * (1 - 0.01 * pop)
       TextBox2.Text = Q.ToString
    End If
  End Sub

  Private Sub RadioButton3_CheckedChanged(ByVal sender As System.Object,
ByVal e As System.EventArgs) Handles RadioButton3.CheckedChanged
    If RadioButton3.Checked Then opt1 = 1
    calculateResults()
  End Sub

  Private Sub RadioButton4_CheckedChanged(ByVal sender As System.Object,
ByVal e As System.EventArgs) Handles RadioButton4.CheckedChanged
    If RadioButton4.Checked Then opt1 = 2
    calculateResults()
  End Sub

  Private Sub RadioButton5_CheckedChanged(ByVal sender As System.Object,
ByVal e As System.EventArgs) Handles RadioButton5.CheckedChanged
    If RadioButton5.Checked Then opt1 = 3
    calculateResults()
  End Sub

  Private Sub RadioButton6_CheckedChanged(ByVal sender As System.Object,
ByVal e As System.EventArgs) Handles RadioButton6.CheckedChanged
    If RadioButton6.Checked Then opt1 = 4
    calculateResults()
  End Sub

  Private Sub RadioButton1_CheckedChanged(ByVal sender As System.Object,
ByVal e As System.EventArgs) Handles RadioButton1.CheckedChanged
    showHideGroups()
    TextBox1.Text = ""
    TextBox2.Text = ""
  End Sub
```

```
    Private Sub RadioButton2_CheckedChanged(ByVal sender As System.Object,
 ByVal e As System.EventArgs) Handles RadioButton2.CheckedChanged
        showHideGroups()
        TextBox1.Text = ""
        TextBox2.Text = ""
    End Sub


    Sub showHideGroups()
      If RadioButton1.Checked Then
          'View both groupboxes, and put the second one below the first
          GroupBox1.Visible = True
          GroupBox2.Location = New Point(GroupBox2.Location.X, 171)
          Label1.Text = "Enter total floor area in sq.m:"
          Label2.Text = "Total fire demand for this building (L/sec):"
      Else
          'Hide the first groupbox and put the second groupbox in its place
          GroupBox1.Visible = False
          GroupBox2.Location = New Point(GroupBox2.Location.X, GroupBox1.Location.Y)
          Label1.Text = "Enter population of town:"
          Label2.Text = "FIRE DEMAND FOR THIS TOWN (L/min):"
      End If
    End Sub


    Private Sub TextBox1_TextChanged(ByVal sender As System.Object,
 ByVal e As System.EventArgs) Handles TextBox1.TextChanged
        calculateResults()
    End Sub
End Class
```

## 3.5  GROUNDWATER FLOW

Groundwater aquifers are depleted by pumping, natural discharge to surface water systems, and to a minor extent evaporation and evapotranspiration. The balance between the depletion rate and the recharge of an aquifer determines the average annual water table level (Viessman and Lewis 2002). The principal method for depletion of an aquifer is the pumping of water. The main factors that influence groundwater flow are the following:

- Liquid characteristics such as density and viscosity
- Media, through which the liquid flows, such as porosity and permeability
- Existing boundary conditions

Movement of groundwater starts from levels of higher energy to levels of lower energy. Basically, its energy is due to elevation and pressure. The velocity heads are neglected because flow is essentially laminar (the velocity of flow in laminar conditions reaches approximately 1 cm/s, and the Reynolds number achieves values ranging between 1 to 10) (Valsaraj and Melvin 2009).

From the continuity equation, the groundwater flow for a particular aquifer may be determined from the following equation:

$$Q = vA = kjA \tag{3.10}$$

where:

$Q$ is the groundwater flow, m$^3$/s

$v$ is the average velocity of flow in voids of water bearing material, m/s

$$v = v_a e \tag{3.11}$$

$v_a$ is the actual velocity of flow, m/s

$e$ is the porosity, dimensionless (see Table 3.5)

$k$ is the coefficient of permeability (or conductivity), m/s (see Table 3.6)

$j$ is the hydraulic gradient, dimensionless

$A$ is the cross-sectional area perpendicular to flow direction, m$^2$

The area may be computed from the following equation:

$$A = Bh \tag{3.12}$$

where:

$B$ is the width of aquifer, m

$h$ is the saturated thickness of aquifer, m

## TABLE 3.5
### Typical Total Porosities for Selected Materials

| Material | Porosity (%) |
| --- | --- |
| Unaltered granite and gneiss | 0–2 |
| Quartzite | 0–1 |
| Shales, slates, mica-schists | 0–10 |
| Chalk | 5–40 |
| Sandstones | 5–40 |
| Volcanic tuff | 30–40 |
| Gravels | 25–40 |
| Sands | 15–48 |
| Silt | 35–50 |
| Clays | 40–70 |
| Fractured basalt | 5–50 |
| Karst limestone | 5–50 |
| Limestone, dolomite | 0–20 |

*Source:* Nielsen, D. M. ed., *Practical Handbook of Groundwater Monitoring*, Lewis Publishers, Chelsea, 1991. With permission.

## TABLE 3.6
### Typical Hydraulic Conductivities for Selected Materials

| Geologic Material | Range of K (m/s) |
| --- | --- |
| Coarse gravel | $10^{-1}$–$10^{-2}$ |
| Sands and gravels | $10^{-2}$–$10^{-5}$ |
| Fine sands, silts, loess | $10^{-5}$–$10^{-9}$ |
| Clay, shale, glacial till | $10^{-5}$–$10^{-13}$ |
| Dolomite limestone | $10^{-3}$–$10^{-5}$ |
| Weathered chalk | $10^{-3}$–$10^{-5}$ |
| Unweathered chalk | $10^{-6}$–$10^{-9}$ |
| Limestone | $10^{-3}$–$10^{-9}$ |
| Sandstone | $10^{-4}$–$10^{-10}$ |
| Unweathered granite, gneiss, compact basalt | $10^{-7}$–$10^{-13}$ |

*Source:* Nielsen, D. M. ed., *Practical Handbook of Groundwater Monitoring*, Lewis Publishers, Chelsea, 1991. With permission.

As such, the groundwater flow may be determined as follows:

$$Q = kjBh \qquad (3.13)$$

The transmissibility, or hydraulic conductivity, TR, of a certain aquifer may be defined as the capacity of a unit prism of aquifer to yield water. The transmissibility may also be defined as the rate at which water of prevailing kinematic viscosity is transmitted through a unit width of an aquifer under a unit hydraulic gradient. Equation 3.14 may be used for determination of transmissibility of an aquifer:

$$TR = kh \qquad (3.14)$$

where:
    TR is the transmissibility of aquifer, $m^3/d*m$
    $k$ is the permeability coefficient, m/d
    $h$ is the saturated thickness of aquifer, m

Using the concept of transmissibility, the groundwater flow may be determined as presented in Equation 3.15:

$$Q = TR \; jB \qquad (3.15)$$

where:
    $Q$ is the groundwater flow, $m^3/s$
    TR is the transmissibility of aquifer, $m^3/day*m$
    $j$ is the hydraulic gradient, dimensionless
    $B$ is the width of aquifer, m

### Example 3.3

1. Write a computer program to determine the actual velocity of flow in an aquifer, quantity of groundwater flow, and transmissibility, given porosity, permeability, hydraulic gradient, saturated thickness, and width of aquifer.
2. An aquifer has a porosity of 20% and an average grain size of 1.5 mm. Experimental studies reveal that a tracer needs 540 min to move between two observation wells, 20 m apart. Find the permeability of the aquifer to yield a difference in water surface elevation of 80 cm and a water temperature of 15°C.

### Solution

1. For solution to Example 3.3 (1), see the listing of Program 3.3.
2. Solution to Example 3.3 (2):
   a. Given: $e = 0.2$, $d = 1.5 * 10^{-3}$ m, $l = 20$ m, $t = 540$ min $= 9$ h, and $h = 0.8$ m.
   b. For a temperature of $T = 15°C$, find from Appendix A1 the coefficient of viscosity as follows: $\mu = 1.1447 * 10^{-3}$ N*s/m² and density $\rho = 999.099$ kg/m³.
   c. Find the velocity of the traveling tracer by using velocity = distance/time; thus, $va = l/t = 20/9 = 2.22$ m/h.
   d. Determine the seepage velocity: $v = k * j = k * h/l = k * 0.8/20 = 0.04*k$.
   e. Find the actual velocity as follows: $v = va* e = 2.22*(20/100) = 0.444$ m/h $= 1.235 * 10^{-4}$ m/s.
   f. Compute the coefficient of permeability as follows: $k = 0.444/0.04 = 11.1$ m/h $= 3.08 * 10^{-3}$ m/s.
   g. Determine the value of the Reynolds number as follows: Re $= \rho*v*d/\mu = 999.1 * 1.235 * 10^{-4} * 1.5 * 10^{-3}/1.1447 * 10^{-3} = 0.162$ (since this value is less than 1, computations are justifiable).

**LISTING OF PROGRAM 3.3   (CHAP3.3\FORM1.VB): FLOW VELOCITY WITHIN AQUIFERS**

```vb
'*******************************************************************************
'Program 3.3: Aquifers
'Computes Actual velocity of flow in aquifers
'*******************************************************************************
Public Class Form1
  Dim const$(4), ttl$(4), Cw(4)
'*******************************************************************************
'Code in Form_Load sub will be executed when the form is loading,
'i.e., first thing after program loads into the memory
'*******************************************************************************
  Private Sub Form1_Load(ByVal sender As System.Object,
ByVal e As System.EventArgs) Handles MyBase.Load
    Me.MaximizeBox = False
    Me.FormBorderStyle = Windows.Forms.FormBorderStyle.FixedSingle
    Me.Text = "Program 3.3: Computes Actual velocity of flow in aquifers"
    ListBox1.Items.Clear()
    ListBox1.Items.Add("Determine actual velocity of flow")
    ListBox1.Items.Add("Determine flow rate")
    ListBox1.Items.Add("Determine transmissibility")
    Label1.Text = "Select an option:"
    Label1.Visible = True
    Label2.Visible = False
    Label3.Visible = False
    Label4.Visible = False
    Label5.Visible = False
    Label6.Visible = False
    Label7.Visible = False
    TextBox1.Visible = False
    TextBox2.Visible = False
    TextBox3.Visible = False
    TextBox4.Visible = False
    TextBox5.Visible = False
    TextBox6.Visible = False
  End Sub

  Private Sub ListBox1_SelectedIndexChanged(ByVal sender As System.Object,
ByVal e As System.EventArgs) Handles ListBox1.SelectedIndexChanged
    If ListBox1.SelectedIndex < 0 Then Exit Sub
    Label2.Visible = True
    Label3.Visible = True
    Label4.Visible = True
    TextBox1.Visible = True
    TextBox2.Visible = True
    TextBox3.Visible = True
    Select Case ListBox1.SelectedIndex
      Case 0
        Label2.Text = "Enter the porosity of the material (as a ratio):"
        Label3.Text = "Enter the permeability of the material in m3/day.m2:"
        Label4.Text = "Enter the hydraulic gradient:"
        Label5.Text = "Enter the starturated thickness of the aquifer m:"
        Label6.Text = "Enter the width of the aquifer in m:"
        Label7.Text = "Actual velocity of flow (m/s):"
        Label7.Visible = True
        Label6.Visible = True
        Label5.Visible = True
```

```vbnet
            TextBox4.Visible = True
            TextBox5.Visible = True
            TextBox6.Visible = True
        Case 1
          'Determine flow rate
          Label2.Text = "Enter transmissibilty (m2/day):"
          Label3.Text = "Enter hydraulic gradient:"
          Label4.Text = "Enter width of aquifer (m):"
          Label5.Text = "The flow rate (m3/day):"
          Label5.Visible = True
          Label6.Visible = False
          Label7.Visible = False
          TextBox4.Visible = True
          TextBox5.Visible = False
          TextBox6.Visible = False
        Case 2
          'Determine transmissibility
          Label2.Text = "Enter the permeability (m/day):"
          Label3.Text = "Enter saturated thickness of aquifer (m):"
          Label4.Text = "The transmissibilty (m2/day):"
          Label5.Visible = False
          Label6.Visible = False
          Label7.Visible = False
          TextBox4.Visible = False
          TextBox5.Visible = False
          TextBox6.Visible = False
        End Select
        resetTextBoxes()
    End Sub


    Private Sub TextBox1_TextChanged(ByVal sender As System.Object,
ByVal e As System.EventArgs) Handles TextBox1.TextChanged
      calculateresults()
    End Sub


    Private Sub TextBox2_TextChanged(ByVal sender As System.Object,
ByVal e As System.EventArgs) Handles TextBox2.TextChanged
      calculateResults()
    End Sub


    Private Sub TextBox3_TextChanged(ByVal sender As System.Object,
ByVal e As System.EventArgs) Handles TextBox3.TextChanged
      If ListBox1.SelectedIndex = 2 Then Exit Sub
      calculateResults()
    End Sub


    Private Sub TextBox4_TextChanged(ByVal sender As System.Object,
ByVal e As System.EventArgs) Handles TextBox4.TextChanged
      If ListBox1.SelectedIndex >= 1 Then Exit Sub
      calculateResults()
    End Sub


    Private Sub TextBox5_TextChanged(ByVal sender As System.Object,
ByVal e As System.EventArgs) Handles TextBox5.TextChanged
      calculateResults()
    End Sub


    'This Subroutine resets all textboxes into Null value.
    Sub resetTextBoxes()
      TextBox1.Text = ""
      TextBox2.Text = ""
```

```
    TextBox3.Text = ""
    TextBox4.Text = ""
    TextBox5.Text = ""
    TextBox6.Text = ""
  End Sub

  Sub calculateResults()
  Dim por, per, TR, h, Q, velocity, j, B, k
  Select Case ListBox1.SelectedIndex
    Case 0
      por = Val(TextBox1.Text)
      per = Val(TextBox2.Text)
      j = Val(TextBox3.Text)
      h = Val(TextBox4.Text)
      B = Val(TextBox5.Text)
      per = per / (24 * 3600) 'permeability in m/s
      TR = per * h            'Transmissibility
      'Actual equations are:
      'Q = TR * j * B
      'and v = Q / (B * h)
      Q = TR * j               'Groundwater flow
      velocity = Q / h         'velocity of flow
      TextBox6.Text = velocity.ToString
    Case 1
      'Determine flow rate
      TR = Val(TextBox1.Text)
      j = Val(TextBox2.Text)
      B = Val(TextBox3.Text)
      Q = TR * j * B
     TextBox4.Text = Q.ToString
    Case 2
      'Determine transmissibility
      k = Val(TextBox1.Text)
      h = Val(TextBox2.Text)
      TR = k * h    'transmissibility
      TextBox3.Text = TR.ToString
    End Select
  End Sub
End Class
```

Field experiments were proposed to find the permeability coefficient of an aquifer by drilling observation wells in the cone of depression of a well and noting the corresponding drawdown in observation wells (McGhee and Steel 1991). By applying the continuity equation and Darcy's law, the steady-state discharge from a well with radial flow in a steady confined aquifer may be predicted by the following equation (see Figures 3.5 and 3.6):

$$k = \frac{Q\mathrm{Ln}\left(\dfrac{r_2}{r_1}\right)}{2\pi h(x_1 - x_2)} \tag{3.16}$$

where:

$k$ is the coefficient of permeability, m/s

$Q$ is the steady-state discharge from the well (pumping rate from supply well), m³/s

$x_1$, $x_2$ are the drawdowns at observation wells, m

$r_1$, $r_2$ are the distances of observation wells from the pumped well (radial direction), m

$h$ is the depth of aquifer (average thickness of bed at $r_1$ and $r_2$ for water table conditions), m

Equation 3.16 is the Thiem formula or equilibrium equation. The equation may be used to estimate the permeability of the aquifer for measurements around a pumping well.

The specific capacity of a well is defined as the output of the well as divided by the drawdown. This definition is presented mathematically in the following equation:

$$S_c = \frac{Q}{x} \tag{3.17}$$

where:

$S_c$ is the specific capacity of well, m³/day*m

$Q$ is the discharge (output) from well, m³/s

$x$ is the drawdown, m

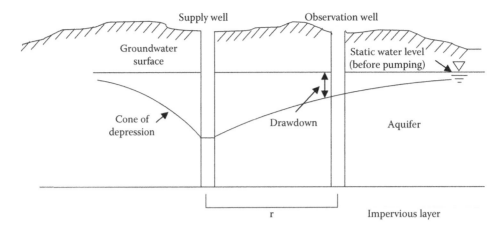

**FIGURE 3.5** A supply and an observation well.

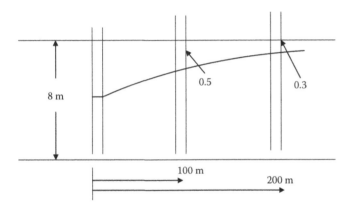

**FIGURE 3.6** Solution to Example 3.4.

## Example 3.4

1. Write a computer program to find the draw-down, coefficient of permeability, transmissibility, or specific capacity for a steady confined aquifer with known characteristics and pattern of flow between two observation wells.
2. Pumping in an artisan well is conducted at the rate of 0.04 m³/s. At observation wells 100 m and 200 m away, the drawdowns noted are 0.5 and 0.3 m, respectively. The average thickness of the aquifer at the observation wells is 8 m. Compute the coefficient of permeability of the aquifer.
3. A 30-cm well penetrates 25 m below the static water table. After a long period of pumping at a rate of 2500 L/min, the draw down in wells 20 and 60 m from the pumped well was 1.9 and

0.6 m, respectively. Determine the transmissibility of the aquifer.
4. A well after prolonged pumping produces 800 m³/h. For a drawdown from the static level of 75 cm, estimate the specific capacity.

## Solution

1. For solution to Example 3.4 (1), see the listing of Program 3.4.
2. Solution to Example 3.4 (2):
    a. $Q = 0.04$ m³/s (see Figure 3.6)
    b. Given: $Q = 0.04$ m³/s, $r_1 = 100$ m, $r_2 = 200$ m, $x_1 = 0.5$ m, $x_2 = 0.3$ m, and $h = 8$ m.
    c. Compute the permeability coefficient as follows: $k = (0.04)\text{Ln}(200/100)/[2\pi*8(0.5 - 0.3)] = 2.8 * 10^{-3}$ m/s.
3. Solution to Example 3.4 (3):
    a. Given: $h = 25$ m, $Q = 2500$ L/min, $r_1 = 20$, $r_2 = 60$ m, $x_1 = 1.9$ m, and $x_2 = 0.6$ m.
    b. Use Thiem equation to determine the permeability coefficient as follows: $k = Q*\text{ln}(r_2/r_1)/[(h_1)^2-(h_o)^2]$.
    c. Determine $h_1 = h - x_2 = 25 - 0.6 = 24.4$ m and $h_o = h - x_1 = 25 - 1.9 = 23.1$ m.
    d. Find the permeability as follows: $k = [(2500/1000 * 60) * \text{Ln} (60/20)]/[(24.4)^2 - (23.1)^2] = 7.41 * 10^{-4}$ m/s $= 64$ m/day.
    e. Compute the value of the transmissibility of the aquifer as follows: TR $= k * h = 64 * 25 = 1.6 * 10^6$ Lpd/m.
4. Solution to Example 3.4 (4):
    a. Given: $Q = 800$ m³/h and $x = 0.75$ m.
    b. Determine the specific capacity of the well as follows: $S_c = Q/x = 800/75 = 10.7$ m³/h/cm.

**LISTING OF PROGRAM 3.4    (CHAP3.4\FORM1.VB): DEPRESSIONS IN OBSERVATION WELLS**

```vb
'*********************************************************************************
'Program 3.4: Aquifers-2
'Computes Properties of aquifers
'*********************************************************************************
Imports System.Math
Public Class Form1
   Const pi = 3.142857
   '*********************************************************************************
   'Code in Form_Load sub will be executed when the form is loading,
   'i.e., first thing after program loads into the memory
   '*********************************************************************************
   Private Sub Form1_Load(ByVal sender As System.Object,
ByVal e As System.EventArgs) Handles MyBase.Load
      Me.MaximizeBox = False
      Me.FormBorderStyle = Windows.Forms.FormBorderStyle.FixedSingle
      Me.Text = "Program 3.4: Computes Properties of aquifers"
      ListBox1.Items.Clear()
      listbox1.items.add("To determine permeability")
      listbox1.items.add("To determine transmissibility")
      listbox1.items.add("To determine the specific capacity")
      Label1.Text = "Select an option:"
      Label2.Visible = False
      Label3.Visible = False
      Label4.Visible = False
      Label5.Visible = False
      Label6.Visible = False
      Label7.Visible = False
      Label8.Visible = False
      Label9.Visible = False
      Label10.Visible = False
      TextBox1.Visible = False
      TextBox2.Visible = False
      TextBox3.Visible = False
      TextBox4.Visible = False
      TextBox5.Visible = False
      TextBox6.Visible = False
      TextBox7.Visible = False
      TextBox8.Visible = False
      TextBox9.Visible = False
      Button1.Visible = False
   End Sub

   Sub clear_text()
      TextBox1.Text = ""
      TextBox2.Text = ""
      TextBox3.Text = ""
      TextBox4.Text = ""
      TextBox5.Text = ""
      TextBox6.Text = ""
      TextBox7.Text = ""
      TextBox8.Text = ""
      TextBox9.Text = ""
   End Sub
```

```vbnet
   Private Sub ListBox1_SelectedIndexChanged(ByVal sender As System.Object,
ByVal e As System.EventArgs) Handles ListBox1.SelectedIndexChanged
    If ListBox1.SelectedIndex < 0 Then Exit Sub
    TextBox1.Visible = True
    TextBox2.Visible = True
    TextBox3.Visible = True
    Label2.Visible = True
    Label3.Visible = True
    Label4.Visible = True
    Button1.Visible = True
    Select Case ListBox1.SelectedIndex
      Case 0
        Label2.Text = "Enter distance to the first test well r1 (m):"
        Label3.Text = "Enter distance to the second test well r2 (m):"
        Label4.Text = "Enter the drawdown at the first test well x1 (m):"
        Label5.Text = "Enter the drawdown at the second test well x2 (m):"
        Label6.Text = "Enter the average depth of the aquifer h (m):"
        Label7.Text = "Enter the pumping flow rate from the well m3/s:"
        Label8.Text = "Coefficient of permeability K (m/s):"
        TextBox4.Visible = True
        TextBox5.Visible = True
        TextBox6.Visible = True
        TextBox7.Visible = True
        TextBox8.Visible = False
        TextBox9.Visible = False
        Label5.Visible = True
        Label6.Visible = True
        Label7.Visible = True
        Label8.Visible = True
        Label9.Visible = False
        Label10.Visible = False
        TextBox3.Enabled = True
        TextBox7.Enabled = False
      Case 1
        Label2.Text = "Enter distance to the first test well r1 (m):"
        Label3.Text = "Enter distance to the second test well r2 (m):"
        Label4.Text = "Enter the draw down at the first test well x1 (m):"
        Label5.Text = "Enter the draw down at the second test well x2 (m):"
        Label6.Text = "Enter the average depth of the aquifer h (m):"
        Label7.Text = "Enter pumping flow rate from the well Q (L/min):"
        Label8.Text = "Coefficient of permeability K (m/s):"
        Label9.Text = "Coefficient of permeability K1 (m/day):"
        Label10.Text = "Transmissibility of the aquifer (m3/day*m):"
        TextBox4.Visible = True
        TextBox5.Visible = True
        TextBox6.Visible = True
        TextBox7.Visible = True
        TextBox8.Visible = True
        TextBox9.Visible = True
        Label5.Visible = True
        Label6.Visible = True
        Label7.Visible = True
        Label8.Visible = True
        Label9.Visible = True
        Label10.Visible = True
        TextBox3.Enabled = True
        TextBox7.Enabled = False
        TextBox8.Enabled = False
        TextBox9.Enabled = False
      Case 2
        Label2.Text = "Enter flow rate from the well Q (m3/h):"
        Label3.Text = "Enter drawdown from static level X (m):"
```

```vbnet
        Label4.Text = "The specic capacity Sc (m3/h/m):"
        TextBox4.Visible = False
        TextBox5.Visible = False
        TextBox6.Visible = False
        TextBox7.Visible = False
        TextBox8.Visible = False
        TextBox9.Visible = False
        Label5.Visible = False
        Label6.Visible = False
        Label7.Visible = False
        Label8.Visible = False
        Label9.Visible = False
        Label10.Visible = False
        TextBox3.Enabled = False
    End Select
    clear_text()
  End Sub

  Sub calculateResults()
    Dim X, K, r1, r2, x1, x2, h, Q As Double
    Select Case ListBox1.SelectedIndex
      Case 0
        r1 = Val(TextBox1.Text)
        r2 = Val(TextBox2.Text)
        x1 = Val(TextBox3.Text)
        x2 = Val(TextBox4.Text)
        h = Val(TextBox5.Text)
        Q = Val(TextBox6.Text)
        X = r2 / r1
        If r1 > r2 Then X = r1 / r2
        K = ABS(Q * LOG(X) / (2 * pi * h * (x1 - x2)))
        TextBox7.Text = K.ToString
      Case 1
        r1 = Val(TextBox1.Text)
        r2 = Val(TextBox2.Text)
        x1 = Val(TextBox3.Text)
        x2 = Val(TextBox4.Text)
        h = Val(TextBox5.Text)
        Q = Val(TextBox6.Text) / 1000 * 60
        X = r2 / r1
        'If r1 > r2 Then X = r1 / r2
        Dim h1 = h - x2
        Dim ho = h - x1
        K = Q * Log(X) / (Pow(h1, 2) - Pow(ho, 2))
        Dim K1 = K * 3600 * 24 'in m/day
        TextBox7.Text = K.ToString
        TextBox8.Text = K1.ToString
        Dim TR = K1 * h
        TextBox9.Text = TR.ToString
      Case 2
        Q = Val(TextBox1.Text)
        X = Val(TextBox2.Text)
        Dim Sc = Q / X
        TextBox3.Text = Sc.ToString
    End Select
  End Sub

  Private Sub Button1_Click(ByVal sender As System.Object,
ByVal e As System.EventArgs) Handles Button1.Click
    calculateResults()
  End Sub
End Class
```

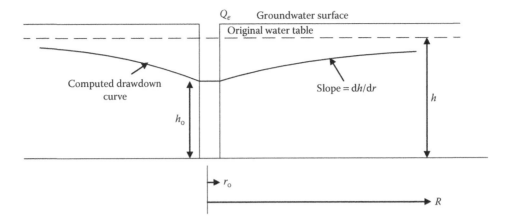

**FIGURE 3.7**  Steady unconfined flow.

For a steady unconfined flow, the discharge of a well may be estimated as presented in the following equation (see Figure 3.7):

$$Q_o = \frac{\pi k [(h_1)^2 - (h_o)^2]}{\text{Ln}(R/r_o)} \qquad (3.18)$$

where:
$Q_o$ is the steady-state discharge from the well, $m^3/s$
$k$ is the coefficient of permeability, m/s
$h_1$ is the depth to original water table when $r = R$, m
$h_o$ is the depth to original water table when $r = r_o$, m
$R$ is the radius of influence, m
$r_o$ is the radius of well, m

## Example 3.5

1. Write a computer program to find the radius of zero drawdown, coefficient of permeability, and drawdown in a pumped well for a steady unconfined aquifer with known characteristics and pattern of flow between two observation wells.
2. A well of diameter 0.3 m contains water to a depth of 50 m before pumping commences. After completion of pumping, the drawdown in a well 20 m away is found to be 5 m, whereas the drawdown in another well 40 m further away reaches 3 m. For a pumping rate of 2500 L/min, determine
   a. The radius of zero drawdown.
   b. The coefficient of permeability.
   c. The drawdown in the pumped well.

### Solution

1. For solution to Example 3.5 (1), see the listing of Program 3.5.

2. Solution to Example 3.5 (2):
   a. Given: $h = 50$ m, $r_1 = 20$m, $x_1 = 5$ m, $r_2 = 40$ m, $x_2 = 3$ m, and $Q = 2500$ L/min.
      i. Find $h_1 = h - x_1 = 50 - 5 = 45$ m and $h_2 = h - x_2 = 50 - 3 = 47$ m.
      ii. Use Equation 2.43 for both observation wells:

$$\left( \pi k \frac{h^2 - h_1^2}{\text{Ln}(R/r)} \right)_{\text{First well}} = \left( \pi k \frac{h^2 - h_1^2}{\text{Ln}(R/r)} \right)_{\text{Second well}}$$

      iii. By substituting given values into the previous equation, then

$$\frac{50^2 - 45^2}{\text{Ln}(R/20)} = \frac{50^2 - 47^2}{\text{Ln}(R/40)}$$

   This yields $R = 119.7$ m.
   b. Find the permeability coefficient by using the data of one of the wells. For $h = 50$ m, $h_o = 45$ m, $r = 20$ m, $R = 119.7$ m, $Q = 2500 * 10^{-3} * 60 * 24 = 3600$ $m^3/day$,

$$k = \frac{3600\,\text{Ln}(119.7/20)}{\pi(50^2 - 45^2)} = 4.32 \text{ m/day}$$

      i. Depth of the water in the pumped well may be found as follows:

$$h_1^2 = h^2 - Q\frac{\text{Ln}(R/r)}{\pi k} = 50^2 - 3600\frac{\text{Ln}(119.7/0.15)}{4.32\pi}$$

   This yields $h = 27$ m.
   c. Determine the drawdown at the well as follows: $x = h - h_1 = 50 - 27 = 23$ m.

**LISTING OF PROGRAM 3.5    (CHAP3.5\FORM1.VB): STEADY
UNCONFINED AQUIFER DRAWDOWN COMPUTATIONS**

```vb
'*************************************************************************************
'Program 3.5: Aquifers-3
'Computes drawdown from a steady unconfined aquifer
'with known characteristics
'*************************************************************************************
Imports System.Math
Public Class Form1
   Const pi = 22.0! / 7.0!
   '*********************************************************************************
   'Code in Form_Load sub will be executed when the form is loading,
   'i.e., first thing after program loads into the memory
   '*********************************************************************************
   Private Sub Form1_Load(ByVal sender As System.Object,
ByVal e As System.EventArgs) Handles MyBase.Load
      Me.MaximizeBox = False
      Me.FormBorderStyle = Windows.Forms.FormBorderStyle.FixedSingle
      Me.Text = "Program 3.5: Computes draw down from unconfined aquifers"
      Label1.Text = "Enter distance to the first test well r1 (m):"
      Label2.Text = "Enter distance to the 2nd test well r2 (m):"
      Label3.Text = "Enter the draw down at the first test well x1 (m):"
      Label4.Text = "Enter the draw down at the 2nd test well x2 (m):"
      Label5.Text = "Enter the average depth of the aquifer h (m):"
      Label6.Text = "Enter the pumping flow rate from the well (m3/s):"
      Label7.Text = "Enter the diameter of the well (m):"
      Label8.Text = "Radius of zero draw down R (m):"
      Label9.Text = "Coefficient of permeability K (m/s):"
      Label10.Text = "The draw down at the well X (m):"
   End Sub


   Private Sub Button1_Click(ByVal sender As System.Object,
ByVal e As System.EventArgs) Handles Button1.Click
      calculateResults()
   End Sub


   Sub calculateResults()
    Dim r1, r2, x1, x2, h, Q, diam
    Dim h1, h2, c, R, K
    r1 = Val(TextBox1.Text)
    r2 = Val(TextBox2.Text)
    x1 = Val(TextBox3.Text)
    x2 = Val(TextBox4.Text)
    h = Val(TextBox5.Text)
    Q = Val(TextBox6.Text)
    diam = Val(TextBox7.Text)
    h1 = h - x1
    h2 = h - x2
    c = (h ^ 2 - h1 ^ 2) / (h ^ 2 - h2 ^ 2)
    R = EXP(1 / (c - 1) * (c * LOG(r2) - LOG(r1)))
    TextBox8.Text = R.ToString
    Dim Qday = Q * 60 * 24 / 1000
    K = Qday * Log(R / r1) / (pi * (h ^ 2 - h1 ^ 2))
    TextBox9.Text = K.ToString
    Dim radius = diam / 2
    Dim ho = (h ^ 2 - Qday * Log(R / radius) / (pi * K)) ^ 0.5
    Dim x = h - ho
    TextBox10.Text = x.ToString
   End Sub
End Class
```

Groundwater recharge is useful for a number of reasons such as overdraft reduction, surface runoff conservation, and increasing yield of groundwater sources. Groundwater recharge can be placed in two main categories: incidental recharge (e.g., surplus irrigation water) or intentional (deliberate) recharge (e.g., municipal waste, surface water). Artificial or intentional recharge may be defined as augmenting the natural replenishment of groundwater storage by some man-made processes such as surface spreading of water or water well injection (Rowe and Abdel-Magid 1995, Todd and Mays 2004). Groundwater recharge by reclaimed water has been practiced to augment the groundwater volume and increase its development. The factors that influence this practice include the location of natural recharge areas, geological formations, soil structure, hydrological conditions, the quantity of water withdrawn, and the degree of wastewater treatment (Rowe and Abdel-Magid 1995).

## 3.6   WATER STORAGE

The first section of this chapter deals with the various types of water storage reservoirs. The mass flow curve that can be used to evaluate the volume of water needed in storage to meet the uniform user's demand is included here. Also, accompanying this is a computer program that enables determination of the required volume of water in storage, provided the necessary data are available. The next section deals with the water distribution system and includes patterns and configurations of pipeline networks and fundamental hydraulic principles for both laminar and turbulent flows. Computer programs for both laminar (Poiseuille's equation) and turbulent flow (Darcy's equation) are presented. The hydraulic design and analysis of flow in pipes in the distribution system—such as the equivalent-velocity-head method, the equivalent-pipe-length method, the pipes-in-parallel method, the Hardy Cross method, and the analysis of pipe networks by the finite element method—are covered. In all of these cases, computer programs are provided, and each computer program incorporates all the elements involved in the design and evaluation of flow of water in pipes in network in water distribution systems. The last section of this chapter deals with the structural design of water storage tanks, circular or rectangular (deep or shallow). Computer programs for the design of both steel and reinforced concrete water tanks are included.

### 3.6.1   Water Storage

Water storage is needed before and after treatment in order to maintain a reliable, uniform, and constant water supply; to balance out fluctuations in flow; and to furnish large volumes of water during emergencies such as fires. Fire insurance rates generally depend upon water storage capacity available. Water storage in the distribution system is also used to equalize operating pressures as well as to equalize pumping rates.

The fluctuations in water demand depend upon many variables, such as the following:

- Time of day
- Day of the week
- Season of the year
- Weather conditions
- Living standards in the community
- Industrial demands
- Fires
- Main breaks

Benefits of water storage reservoirs include flood control, hydroelectric power, irrigation, water supply, navigation, sailing, preservation of aquatic life, fire protection, emergency needs, recreation, and pollution abatement (Henry and Heinke 1996, Smith and Scott 2002, Davis and Cornwell 2006, Viessman et al. 2015). Reservoirs can be grouped as follows:

1. *Storage (conservation, impounding, direct-supply) reservoirs*: These reservoirs store water in excess of demand, from a natural source in periods of high flow, to be used during periods of dry weather or low flows. Water is stored for periods ranging from a few days to several months or even longer (Smith and Scott 2002).

2. *Distribution reservoirs*: These reservoirs can be either elevated (see Figure 3.8) or ground (reserve water supply under pressure). They store water to provide for varying demands of the community over a period of a day or several days. Water stored in distribution reservoirs is often supplied at a steady rate from a storage reservoir. The distribution reservoirs (elevated storage tanks and towers) provide equalizing storage, provide emergency storage (fire-fighting, power blackouts, pump station failure), and reduce necessary size and capacity of pipes and treatment facilities (Nathanson 2007). Generally, the volume of water needed to balance or equalize peak hourly flows is approximately 20% of average daily demand in the service area (Nathanson 2007). When the height of the reservoir exceeds its diameter, it is referred to as stand pipe. Stand pipes are basically tall cylindrical tanks, the upper portion of which constitutes useful storage to produce the necessary lead; the lower portion serves to support the structure (Henry and Heinke 1996). A water tower (elevated tank) is a service reservoir or tank (>15 m in height) raised above ground level (Henry and Heinke 1996, Smith and Scott 2002).

3. *Plant storage reservoirs (clear wells)*: Clear wells are important for the storage of filtered water and to provide operational storage to average out high and low demands. Clear wells with sufficient capacity can prevent the need for varying filtration rates and can prevent frequent on/off cycling of water pumps.

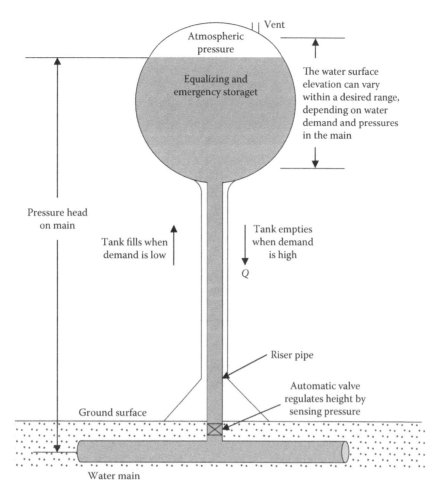

**FIGURE 3.8**    An elevated water storage tank.

The clean well also provides storage for filtered water before it is introduced into the distribution system.

The storage volume depends on the water demand as well as the purpose of storage. The amount of storage that needs to be provided can be determined by using the following equation (Viessman et al. 2015):

$$\Delta S = Q_1 - Q_0 \qquad (3.19)$$

where:

$\Delta S$ is the change in storage volume during a specified time interval

$Q_1$ is the total inflow volume during a specified time interval

$Q_0$ is the total outflow volume during a specified time interval

The storage volume required above the minimum operating conditions can be determined by

- Analytical techniques, which are numerical analyses of historical flow records, especially during periods of low flow.

- The graphical method (Rippl mass curve) which evaluates the cumulative deficiency between outflow and inflow and selects the maximum value as the required storage (Viessman et al. 2015).

### 3.6.2    MASS CURVE (RIPPL DIAGRAM AND S-CURVE)

A flow-mass curve may be defined as a graph of cumulative values of a hydrologic quantity (e.g., runoff or other flows) plotted as ordinates versus time on the abscissa. Use of the mass curve technique includes a study of the storage effects on the stream flow conditions. The ordinate of any point in the Rippl diagram indicates the total amount of water flowing past a given station during a certain time. Equation 3.20 mathematically represents the flow-mass curve:

$$v = \int_{t_1}^{t_2} Q_t dt \cong \sum_{t_1}^{t_2} Q_t \Delta t \qquad (3.20)$$

where:

$v$ is the volume of runoff, m³

$Q$ is the discharge as a function of time $t$, m³/s

$t$ is the time, s

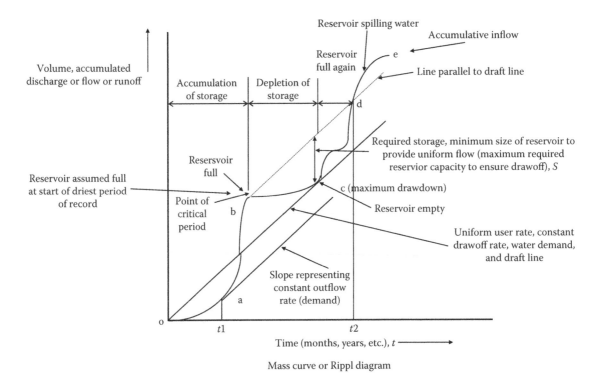

Mass curve or Rippl diagram

**FIGURE 3.9** Mass diagram. See text for details.

The capacity of a reservoir needed to maintain a uniform flow in a stream may be estimated by drawing, on the mass curve, a draft line tangent to a point at the beginning of the critical period. The slope of the draft line signifies the uniform regulated discharge. The storage capacity, to meet the uniform demand, is represented by the maximum ordinate between the draft line and the curve. Other ordinates, between the draft line and the mass curve, represent the amount of draft on the reservoir corresponding to other times. Where the draft line intercepts the curve once more, the reservoir is again full, and supply to the reservoir exceeds demand, resulting in overflow (spillage) conditions. The maximum storage required to meet demand also should consider the local rate of evaporation.

The mass diagram provides the following information regarding the condition of the reservoir (see Figure 3.9) (Gupta 2007):

1. From a to b, inflow rate exceeds use rate, and reservoir is full and overflowing.
2. At b, inflow rate is equal to use rate, and reservoir is full but not overflowing.
3. From b to c, use rate or demand exceeds inflow rate, and the amount of depletion (drawdown) is increasing.
4. At c, use rate is equal to inflow rate, and drawdown is at a maximum.
5. From c to d, inflow rate exceeds use rate, and drawdown is decreasing.
6. At d, the reservoir is full.
7. From d to e, conditions are the same as from a to b.

8. The greatest vertical distance(s) between bd and bcd occurs at c. This is the storage required to maintain a use rate during the low flow period from b to c. The largest value, such as s for the entire period of record, is the minimum size of reservoir that would provide this uniform user rate or demand.

## Example 3.6

1. Write a computer program that enables determination of the storage requirement to meet a uniform user water draw off (with no losses from the reservoir). The reservoir is to supply a constant flow rate of $Q$ (in m³/s) and uses the water flowing from a particular catchment area. The monthly stream flow records are given in total cubic meters.
2. A reservoir is to supply a constant flow rate of 3.1 m³/s using the amounts of water flowing from a particular catchment area. The monthly stream flow records, in one total cubic meter, are as follows:

| Month | J | F | M | A | M | J | J | A | S | O | N | D |
|---|---|---|---|---|---|---|---|---|---|---|---|---|
| m³ of water (×10⁶) | 12.8 | 6 | 1.2 | 4.5 | 9 | 18.5 | 46.9 | 25.1 | 15 | 10.5 | 7.5 | 7 |

   Assuming uniform water draw off and no losses, determine the storage requirement to meet the aforementioned uniform draw off.
3. Use the program developed in (1) to verify the computations in (2).

**Solution**

1. For solution to Example 3.6 (1), see the listing of Program 3.6.
2. Solution to Example 3.6 (2):
   a. Given: stream flow records.
   b. Determine the cumulative total flow as tabulated in the following table.
   c. Draw the mass curve for the data given by drawing the cumulative total flow versus months (see Figure 3.10).

d. Determine the annual constant draw off (for the month of December) = 3.1 $m^3/s*3600$ s/h*24 h/d*365 d/year = 97.8 * $10^6$ $m^3$.
e. Plot the uniform draw off as a straight line from the origin to the point of 97.8 $Mm^3$ on the mass curve. Draw a parallel line from the point where reservoir is full.
f. From the mass curve, determine the storage-minimum reservoir required to give demand = $14 \times 10^6$ $m^3$.

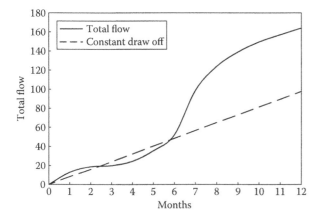

**FIGURE 3.10**    Mass diagram for Example 3.6.

| Month | Monthly Flow | Total Flow |
|-------|--------------|------------|
| 1 | 12.8 | 12.8 |
| 2 | 6 | 18.8 |
| 3 | 1.2 | 20 |
| 4 | 4.5 | 24.5 |
| 5 | 9 | 33.5 |
| 6 | 18.5 | 52 |
| 7 | 46.9 | 98.9 |
| 8 | 25.1 | 124 |
| 9 | 15 | 139 |
| 10 | 10.5 | 149.5 |
| 11 | 7.5 | 157 |
| 12 | 7 | 164 |

## LISTING OF PROGRAM 3.6    (CHAP3.6\FORM1.VB): RESERVOIR CAPACITY

```
'******************************************************************************
'Example 3.6: Reservoir capacity
'******************************************************************************
Public Class Form1
  Dim g As Graphics
  Dim mult_factor As Integer
  Dim vAcc() As Double
  Dim month_str() As String = {"Jan", "Feb", "Mar", "Apr", "May", "Jun", "Jul", "Aug",
"Sep", "Oct", "Nov", "Dec"}
  Dim ys(12), yd(12), ysd(12), x(12), diff(12) As Double
  Dim supply, max As Double
  Dim Y1, P, Maxd, Qt, q As Double
  Dim mon As Integer
  Private Sub Form1_Load(ByVal sender As System.Object, ByVal e As System.EventArgs)
Handles MyBase.Load
    Label1.Text = "Constant flow rate, Q (m3/s):"
    Label2.Text = ""
    Label3.Text = ""
    Button1.Text = "Calculate"
    Me.Text = "Example 3.6: Reservoir capacity"
    Me.FormBorderStyle = Windows.Forms.FormBorderStyle.FixedSingle
    DataGridView1.Columns.Clear()
    DataGridView1.Rows.Clear()
    DataGridView1.Columns.Add("monthCol", "Month")
    DataGridView1.Columns.Add("volCol", "Water flow *10^6 (m3)")
    DataGridView1.Columns("monthCol").ReadOnly = True
    DataGridView1.AllowUserToAddRows = False
    DataGridView1.AllowUserToDeleteRows = False
    DataGridView1.Rows.Add(12)
    DataGridView1.Rows(0).Cells("monthCol").Value = "January"
```

```
    DataGridView1.Rows(1).Cells("monthCol").Value = "February"
    DataGridView1.Rows(2).Cells("monthCol").Value = "March"
    DataGridView1.Rows(3).Cells("monthCol").Value = "April"
    DataGridView1.Rows(4).Cells("monthCol").Value = "May"
    DataGridView1.Rows(5).Cells("monthCol").Value = "June"
    DataGridView1.Rows(6).Cells("monthCol").Value = "July"
    DataGridView1.Rows(7).Cells("monthCol").Value = "August"
    DataGridView1.Rows(8).Cells("monthCol").Value = "September"
    DataGridView1.Rows(9).Cells("monthCol").Value = "October"
    DataGridView1.Rows(10).Cells("monthCol").Value = "November"
    DataGridView1.Rows(11).Cells("monthCol").Value = "December"
  End Sub

  Private Sub Button1_Click(ByVal sender As System.Object, ByVal e As System.EventArgs)
Handles Button1.Click
    Dim count As Integer = 12
    Dim flow As Double = Val(TextBox1.Text)
    Dim i As Integer
    ReDim vAcc(count)
    Dim v As Double = 0
    For i = 0 To count - 1
      vAcc(i) = v + Val(DataGridView1.Rows(i).Cells("volCol").Value)
      v = vAcc(i)
    Next
    Dim flowDec As Double
    flowDec = flow * 3600 * 24 * 365
    Label2.Text = "Annual constant draw off (m3): " + FormatNumber(flowDec, 2)
    Dim m As Double = flowDec / 12.0 'slope of demand curve
    m /= 1000000
    Dim last As Integer = 0
    For i = 1 To 12
      x(i) = i
      ys(i) = vAcc(i - 1)
      yd(i) = m * i 'demand
      diff(i) = ys(i) - yd(i)
    Next i
    max = 0
    mon = 0
    For i = 1 To 12
      If diff(i) < 0 Then GoTo jump
      If diff(i) > max Then
        max = diff(i)
        mon = i
      End If
    Next i
jump:
    If mon = 0 Then
      MsgBox("Couldn't find maximal flow!",
        vbOKOnly Or vbCritical)
      Exit Sub
    End If
    Label3.Text = "Slope max at: " + month_str(mon - 1) + " = " + FormatNumber(max, 2)
    'parallel line to demand curve
    For i = 0 To 12
      ysd(i) = yd(i) + max
      diff(i) = ys(i) - ysd(i)
    Next i
    Y1 = mon
```

```
      For i = mon + 2 To 12
        If Math.Sign(diff(i)) <> Math.Sign(diff(i - 1)) Then GoTo jump2
      Next i
jump2:
    P = i - 1
    Q = i
    Maxd = 0
    For i = mon To P
      If Math.Abs(diff(i)) > Maxd Then Maxd = Math.Abs(diff(i))
    Next i
    drawResults()
  End Sub

  Sub drawResults()
    Dim Maxy, Miny, Y, X As Double
    Dim bmp As Bitmap = New Bitmap(PictureBox1.Width, PictureBox1.Height)
    g = Graphics.FromImage(bmp)
    g.Clear(Color.White)
    Maxy = ys(12)
    If yd(12) > Maxy Then Maxy = yd(12)
    Miny = -Maxy / 15
    'sX and sY are variables used to scale the points to fit in the PictureBox
    'each point with be multiplied by these factors to scale it to fit.
    Dim sX, sY, pY As Double
    sX = PictureBox1.Width / 12
    sY = PictureBox1.Height / Maxy
    'drawing starts at zero point (0,0) which is the upper-left corner of PictureBox
    'all values on the y-scale will be substracted from this to make the figure upside down
    'i.e., start at the lower-left corner instead.
    pY = PictureBox1.Height
    'black rectangle as a border to the figure
    g.DrawRectangle(Pens.Black, 0, 0, CInt(12 * sX) - 1, CInt(Maxy * sY) - 1)
    Dim lastPoint As Point
    lastPoint = New Point(0, 0)
    For X = 0 To 12
      Y = ys(X)
      g.DrawLine(Pens.Red, lastPoint, New Point((X * sX), pY - (Y * sY)))
      lastPoint = New Point(X * sX, pY - Y * sY)
    Next X
    lastPoint = New Point(0, 0)
    For x = 0 To 12
      Y = yd(x)
      g.DrawLine(Pens.Black, lastPoint, New Point((X * sX), pY - (Y * sY)))
      lastPoint = New Point(X * sX, pY - Y * sY)
    Next x
    lastPoint = New Point(0, 0)
    For x = 0 To 12
      Y = ysd(x)
      g.DrawLine(Pens.Blue, lastPoint, New Point((X * sX), pY - (Y * sY)))
      lastPoint = New Point(X * sX, pY - Y * sY)
    Next x
    Label3.Text += vbCrLf + "Capacity = " + FormatNumber(Maxd, 2) + " m3"
    PictureBox1.Image = Image.FromHbitmap(bmp.GetHbitmap)
    g.Dispose()
    bmp.Dispose()
  End Sub
End Class
```

## 3.7 WATER DISTRIBUTION

### 3.7.1 INTRODUCTION

Systems used for distribution of water for public water supplies are networks of pipes within networks of streets. The distribution system is designed to serve each individual property, household, commercial establishment, public building, or industrial factory, and so on. Flow type and character depend on interrelated factors that include street plan, topography of land, location of supply works, and service storage. The analysis of a water distribution system includes quantifying flows and head losses in the network of pipes and finding the resulting residual pressures.

### 3.7.2 PATTERNS OF PIPELINES IN WATER NETWORKS

Basically, there are two patterns of pipelines (Fair et al. 1966, Henry and Heinke 1996, Smith and Scott 2002, Davis and Cornwell 2006, Nathanson 2007, Abdel-Magid and Abdel-Magid 2015, Viessman et al. 2015):

1. *Branching pattern with dead ends*: This system resembles the branching of a tree. The flow of the water continues in the same direction, and single pipe supplies water to a particular area. The advantages of the branching pattern include the following:
   a. Simplicity of water distribution
   b. Simplicity in design of pipe network
   c. Usage of economical pipe dimension
   However, the disadvantages include the following:
   a. Accumulation of sediments (due to stagnation of dead ends)
   b. Tastes and odors (due to absence of regular flushing)
   c. Water shortage (during breakdowns, repair, and maintenance)
   d. Occurrence of insufficient water pressure (especially during extension of services)
2. *Grid pattern*: This system serves an area with a high demand for water using a large diameter pipe loop around the area. In this plan, all the pipes are interconnected together with no dead ends. Water can reach any point from more than one direction. The advantages of this system include the following:
   a. Free movement of water in more than one direction
   b. Less possibility of stagnation (comparable to the branching system)
   c. Continuous water flow (no disconnection during repair or breakdown)
   d. Little adverse effect on supply when there are large variations in consumption
   The disadvantages of this system include the following:
   a. Complicated calculations when designing pipes
   b. More pipes and fittings required

### 3.7.3 VISCOUS FLOW IN CLOSED CONDUITS

A closed conduit may be defined as a pipe or a duct through which fluid flows while completely filling its cross section. Since the flowing fluid has no free surface, it can be either a liquid or a gas, its pressure may be above or below atmospheric pressure, and this pressure may vary from one cross section to another along its length (Douglas et al. 2006). A closed conduit is commonly called a pipe if it has a round cross section or a duct if it is not round, and they are designed to withstand a considerable pressure difference across their walls without undue distortion of their shape (Munson et al. 2012). Fluid flow in a pipe may be laminar or transitional or turbulent (see Table 3.7). The criterion used to determine whether the flow is laminar or turbulent is the Reynolds number presented in the following equation:

$$\text{Re} = \rho v D / \mu \qquad (3.21)$$

where:
Re is the Reynolds number, dimensionless
$\rho$ is the density of fluid, kg/m$^3$
v is the average velocity in the pipe, m/s
D is the pipe diameter, m
$\mu$ is the dynamic viscosity of fluid, N*s/m$^2$ or kg/m*s

### 3.7.3.1 Fundamental Equations for an Incompressible Flow

For steady, fully developed, incompressible flow (neglecting gravitational effects), the relationship of pressure difference and shear for an element of fluid in a horizontal pipe can be determined by using the following equation (see Figure 3.11):

$$\Delta P = \frac{4 * l * \tau_w}{D} \qquad (3.22)$$

where:
$\Delta P$ is the pressure difference, Pa
l is the length of circular cylindrical element, m
$\tau_w$ is the maximum shear at pipe wall (wall shear stress), N/m$^2$
D is the pipe diameter, m

Equation 4.4 indicates that a small shear stress can produce a large pressure difference if the pipe is relatively long (l/D >> 1).

---

**TABLE 3.7**
**Turbulent and Laminar Flows**

| Criterion | Value of Reynolds Number |
|---|---|
| Laminar | <2100 |
| Transitional | 2100 > Re > 4000 |
| Turbulent | >4000 |

---

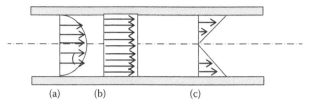

Keys:

(a) Laminar flow velocity profile

(b) Inviscid flow velocity profile

(c) Shear stress distribution (laminar or turbulent)

**FIGURE 3.11** Pipe flow velocity profile and shear stress distribution.

### 3.7.4 Flow in Pipes

#### 3.7.4.1 Laminar Flow (Hagen–Poiseuille Flow)

For laminar flow of a Newtonian fluid within a horizontal pipe, the velocity and discharge may be obtained from the Poiseuille's law presented in the following equations:

$$v = \Delta P * \frac{D^2}{32\mu * l} \qquad (3.23)$$

where:

v is the average velocity, m/s

$\Delta P$ is the pressure difference, Pa

$D$ is the diameter of pipe, m

$l$ is the length, m

$\mu$ is the dynamic viscosity, N*s/m$^2$

$$Q = \pi * \frac{D^4}{l} * \frac{\Delta P}{128\mu} \qquad (3.24)$$

where $Q$ is the flow rate through the pipe, m³/s

To account for a pressure drop in nonhorizontal pipes, the Darcy–Weisbach equation may be used as presented in the following equation:

$$h_l = f * \left(\frac{1}{D}\right) * \left(\frac{V^2}{2g}\right) \qquad (3.25)$$

where:

$f$ is the friction factor (Darcy friction factor $= 64\mu/(v*D*\rho)$

$\quad = 64/\text{Re}$ for laminar flow)

$g$ is the gravitational acceleration, m/s$^2$

Since $\Delta P = \gamma * h_l$ and from Equation 3.25,

$$h_l = 32\mu * l * \frac{v}{(\gamma * D^2)} = f * \left(\frac{1}{D}\right) * \frac{V^2}{2g} \qquad (3.26)$$

where $h_l$ is the head loss (drop in hydraulic grade line), m*N/N

#### Example 3.7

1. Write a computer program to determine the viscosity of a fluid that flows in a pipe of diameter $d$ (m) and of length $l$ (m) at a rate of flow of $Q$ (m³/s), with the head loss in the pipe given as $h_f$ (m). Let the program indicate whether the flow is laminar or turbulent by checking its Reynolds number.
2. Given the values of $d = 5$ mm, $l = 20$ cm, $Q = 0.95$ L/min, and $h_f = 0.8$ m, determine v, V, and Re.
3. Use the program developed in (1) to check the computations in (2).

#### Solution

1. For solution to Example 3.7 (1), see the listing of Program 3.7.
2. Solution to Example 3.7 (2):
   a. Given: $d = 5$ mm, $l = 20$ cm, $Q = 0.95$ L/min, and $h_f = 0.8$ m.
   b. Find $Q = 0.95 * 10^{-3}$ m³ / 15 * 60 s = 1.06 * $10^{-6}$ m³/s.
   c. Determine the kinematic viscosity as: $v = (\pi*D^4*g*h_l)/(128*l*Q) = (\pi*(5*10^{-3})^4 * 9.81 * 0.8)/(128 * 20 * 10^{-2} * 1.06 * 10^{-6}) = 5.7 * 10^{-4}$ m²/s.
   d. Compute the velocity of flow as follows: $v = Q/A = 1.06 * 10^{-6}/(\pi*(5 * 10^{-3})^2/4) = 0.054$ m/s.
   e. Determine the Reynolds number as: Re = $V*d/v = 0.054 * 5 * 10^{-3}/5.7 * 10^{-4} = 0.5$.
   f. Since the Reynolds number is less than 2100, the flow is laminar.

---

**LISTING OF PROGRAM 3.7  (CHAP3.7\FORM1.VB): VISCOUS FLOW IN PIPES**

```
'*****************************************************************************
'Program 3.7: Viscous flow in pipes
'*****************************************************************************
Public Class Form1
    Const Pi = 3.1415962#
    Const g = 9.81
    Dim d, L, Q, hf As Double
    Dim Nu, A, V, Re As Double
```

```
    Private Sub Form1_Load(ByVal sender As System.Object, ByVal e As System.EventArgs)
Handles MyBase.Load
        Me.Text = "Program 3.7: Viscous flow in pipes"
        Me.FormBorderStyle = Windows.Forms.FormBorderStyle.FixedSingle
        Me.MaximizeBox = False
        Label1.Text = "Diameter of pipe, D (mm):"
        Label2.Text = "Length of pipe, L (m):"
        Label3.Text = "Flow rate, Q (L/min):"
        Label4.Text = "Head loss, hf (m):"
        button1.text = "&Calculate"
        Label5.Text = ""
        Label6.Text = ""
    End Sub

    Sub calculateResults()
        d = Val(TextBox1.Text)
        L = Val(textbox2.text)
        Q = Val(textbox3.text)
        hf = Val(textbox4.text)
        d = d / 1000 'convert to metre
        'convert to m3/s
        Q = (Q / 1000) / (15 * 60)
        Nu = (Pi * (d ^ 4) * g * hf) / (128 * L * Q)
        A = Pi * (d ^ 2) / 4
        V = Q / A
        Re = V * d / nu
        Label5.Text = "Kinematic viscosity (m2/s) = " + FormatNumber(Nu, 5)
        Label6.Text = "Reynolds number = " + FormatNumber(Re, 2)
        If Re < 2100 Then
          label6.text += " LAMINAR"
        Else : label6.text += " TURBULENT"
        End If
    End Sub

    Private Sub Button1_Click(ByVal sender As System.Object, ByVal e As System.EventArgs)
Handles Button1.Click
        calculateResults()
    End Sub
End Class
```

### 3.7.4.2 Turbulent Flow

Turbulent pipe flow actually is more likely to occur than laminar flow in a practical situation (White 2010; Munson et al. 2012). The main differences between laminar and turbulent flows occur due to variables such as velocity, pressure, shear stress, and temperature. Shear stress for turbulent flow can be determined by using the following equation:

$$\tau = \xi\left(\frac{du_a}{dy}\right) \tag{3.27}$$

where:
   $\xi$ is the Eddy viscosity, $N \cdot s/m^2$
   $u_a$ is the average velocity, m/s

The velocity profile may be obtained by the empirical power law velocity profile as presented in the following equation:

$$\frac{u_a}{v_c} = \left[1 - \left(\frac{r}{R}\right)\right]^{1/b} \tag{3.28}$$

where:
   $b$ is the constant (function of Reynolds number)
   $v_c$ is the centerline velocity, m/s

The one-seventh power law velocity profile ($b = 7$) is often used as a reasonable approximation for many practical flows (White 2010; Munson et al. 2012).

Using the technique of dimensional analysis of pipe flow, the experimental data and semiempirical formulas used in analysis of turbulent flow can be grouped in a dimensionless form. For steady, incompressible turbulent flow in a horizontal round pipe, the pressure drop and head loss can be determined by using the following equations:

$$\Delta P_f = f * \left(\frac{l}{D}\right) * \left(\frac{\rho * v^2}{2}\right) \qquad (3.29)$$

where:

$f$ is the friction factor = $\Phi(Re, \varepsilon/D)$ = $64/Re$ for laminar flow (independent of $\varepsilon/D$) = $\phi(\varepsilon/D)$ for completely (wholly) turbulent flow

$\varepsilon$ is the roughness of the pipe wall, m

$D$ is the pipe diameter, m

$l$ is the length of pipe, m

v is the average velocity, m/s

$$h_l = f * \left(\frac{1}{D}\right) * \left(\frac{v^2}{2g}\right) \qquad (3.30)$$

where $h_l$ is the head loss, m.

It is difficult to estimate the dependence of the friction factor on Reynolds number and relative roughness. Experimental findings have correlated data in terms of relative roughness of commercially available (new and clean) pipe materials and plotting this information in a diagram termed the Moody chart. The Moody chart is valid for steady, fully developed, incompressible pipe flows. The Colebrook formula gives the roughness for the entire nonlaminar range of the Moody chart, as indicated in the following equation (White 2010; Munson et al. 2012):

$$\frac{1}{\sqrt{f}} = -2\log\left[\frac{(\varepsilon/D)}{3.7} + \frac{2.51}{\left(Re * \sqrt{f}\right)}\right] \qquad (3.31)$$

## Example 3.8

1. Write a computer program that enables the computation of head loss due to friction and the power needed to maintain flow $Q$ (m³/s) through a circular horizontal pipe of diameter $d$ (m) and length $l$ (m). Use different pipe materials (different roughness coefficients) and different temperatures (different viscosities) in the program.

2. Water flows at the rate of 0.08 L/s through a horizontal, circular 500 m long pipe of diameter 6 cm at a temperature of 20°C. Assuming an absolute roughness of 0.08 mm for the pipe, determine the head loss due to friction using both Poiseuille and Darcy equations. Estimate the power required to maintain the flow in the pipe.

3. Verify your computations in (2) by using the program developed in (1).

### Solution

1. For solution to Example 3.8 (1), see the listing of Program 3.8.

2. Solution to Example 3.8 (2):

   a. Find the viscosity corresponding to the given temperature from Appendix A1 as follows: $\mu$ = $1.0087 * 10^{-3}$ N*s/m², $\rho$ = 998.203 kg/m³, and $\gamma$ = 9792 N/m³.

   b. Determine the Reynolds number as follows: Re = $\rho *V*D/\mu$ = $\rho *Q*D/\mu*A$ = 998.203 * $0.08 * 10^{-3}$ * 6 * $10^{-2}/1.0087$ * $10^{-3}*[\pi*(6 * 10^{-2})^2/4]$ = 1680. Therefore, the flow is laminar since Re < 2000.

   c. For laminar flow, find the loss of head due to friction by using Poiseuille's equation as follows:

   $$h_f = 128 *\mu * l * \frac{Q}{\pi} *d^4 *\gamma$$

   $$= 128*1.0087*10^{-3} *500$$

   $$*0.08*\frac{10^{-3}}{\pi}*\left(6*10^{-2}\right)^4 *9792$$

   $$= 0.013\,\text{m of water}$$

   d. Determine the head loss using Darcy's equation ($f$ = 64/Re) as follows: $h_f$ = $f*l*v^2/2*D*g$ = $512*l*Q^2/D^5*g*Re*\pi^2$ = 512 * 500*(0.08 * $10^{-3})^2/(6 * 10^{-2})^5$ * 9.81 * 1680*$\pi^2$ = 0.013 m.

---

### LISTING OF PROGRAM 3.8    (CHAP3.8\FORM1.VB): HEAD LOSS DUE TO FRICTION IN PIPES

```
'*****************************************************************************
'Program 3.8: Head loss due to friction in pipes
'*****************************************************************************
Imports System.Math
Public Class Form1
   Const g = 9.81
   Dim D, L, Q, Mu, Rho, roughness As Double
   Dim T As Integer
   Dim gamma, A, V, Re As Double
   Dim f, hf, hf1, hf2, EPSILON As Double
   Dim temp_table() As Integer =
      {5, 6, 7, 8, 9, 10, 11, 12, 13, 14, 15, 16, 17, 18,
      19, 20, 21, 22, 23, 24, 25, 26, 27, 28, 29, 30,
      31, 32, 33, 34, 35, 36, 37, 38
      }
```

```
    Dim viscosity() As Double =
      {1.5188, 1.4726, 1.4288, 1.3872, 1.3476, 1.3097,
      1.2735, 1.239, 1.2061, 1.1748, 1.1447, 1.1156,
      1.0875, 1.0603, 1.034, 1.0087, 0.9843, 0.9608,
      0.938, 0.9161, 0.8949, 0.8746, 0.8551, 0.8363,
      0.8181, 0.8004, 0.7834, 0.767, 0.7511, 0.7357,
      0.7208, 0.7064, 0.6925, 0.6791
      }
    Dim density() As Double =
      {0.999956, 0.999941, 0.999902, 0.999849, 0.999781,
      0.9997, 0.999605, 0.999498, 0.999377, 0.999244,
      0.999099, 0.998943, 0.998774, 0.998595, 0.998405,
      0.998203, 0.997992, 0.99777, 0.997538, 0.997296,
      0.997044, 0.996783, 0.996512, 0.996232, 0.995944,
      0.995646, 0.99534, 0.995025, 0.994702, 0.994371,
      0.99403, 0.99368, 0.99333, 0.99296
      }
'****************************************************************************************
'Find viscosity as a function of temp (Table A1).
'****************************************************************************************
  Function find_viscosity(ByVal T As Integer)
      For i = 0 To temp_table.Length - 1
        If temp_table(i) = T Then Return viscosity(i)
      Next
      Return 0
  End Function
'****************************************************************************************
'Find density as a function of temp (Table A1).
'****************************************************************************************
  Function find_density(ByVal T As Integer)
      For i = 0 To temp_table.Length - 1
        If temp_table(i) = T Then Return density(i)
      Next
      Return 0
  End Function

  Private Sub Form1_Load(ByVal sender As System.Object, ByVal e As System.EventArgs)
Handles MyBase.Load
      Me.Text = "Program 3.8: Head loss due to friction in pipes"
      Me.FormBorderStyle = Windows.Forms.FormBorderStyle.FixedSingle
      Me.MaximizeBox = False
      Label1.Text = "Diameter of pipe, D (m):"
      Label2.Text = "Length of pipe, L (m):"
      Label3.Text = "Flow rate, Q (m3/s):"
      Label4.Text = "Temperature (C):"
      Label5.Text = "Absolute Roughness (mm):"
      Label6.Text = ""
      Label7.Text = ""
      Label8.Text = "Decimals:"
      Button1.Text = "&Calculate"
      NumericUpDown1.Value = 3
      NumericUpDown1.Maximum = 10
      NumericUpDown1.Minimum = 0
  End Sub

  Sub calculateResults()
      D = Val(textbox1.text)
      L = Val(textbox2.text)
      Q = Val(TextBox3.Text) / 1000
      T = Val(TextBox4.Text)
      roughness = Val(TextBox5.Text)
```

```vb
      If T < 5 Or T > 38 Then
        MsgBox("Please enter temperature between 5 and 38 inclusive.",
        vbOKOnly Or vbInformation)
        Exit Sub
      End If
      Mu = find_viscosity(T) / 1000
      Rho = find_density(T) * 1000 'kg/m3
      gamma = Rho * g
      A = Pi * (D ^ 2) / 4
      V = Q / A
      Re = Rho * V * D / Mu
      Dim decimals As Integer = NumericUpDown1.Value
      Label6.Text = "Reynolds number = " + FormatNumber(Re, decimals)
      If Re < 2000 Then
        label6.text += " LAMINAR"
        hf1 = 128 * Mu * L * Q / (Pi * (D ^ 4) * gamma) 'Poiseuille eqn.
        f = 16 / Re
        hf2 = 4 * f * L * V ^ 2 / (2 * D * g) 'Darcy eqn.
        Label7.Text = "hf = " + FormatNumber(hf1, decimals) + " POISEUILLE hf = " +
FormatNumber(hf2, decimals) + " DARCY"
      Else
        label6.text += " TURBULENT"
        f = getf(E, D, Re)
        hf = f * L / D * V ^ 2 / (2 * g)
        Label7.Text = "hf = " + FormatNumber(hf, decimals)
      End If
   End Sub

   Function getf(ByVal e, ByVal D, ByVal Re)
      Dim RHS, LHS As Double
      For i = 0.000001 To 0.001 Step 0.000001
        RHS = f ^ -0.5
        LHS = -2 * log10(e / (D * 3.7) + 2.51 / (Re * f ^ 0.5))
        If ABS(RHS - LHS) < EPSILON Then
          Return i
        End If
      Next i
      Return Nothing
   End Function

   Private Sub Button1_Click(ByVal sender As System.Object, ByVal e As System.EventArgs)
Handles Button1.Click
      calculateResults()
   End Sub

   Private Sub NumericUpDown1_ValueChanged(ByVal sender As System.Object, ByVal e As
System.EventArgs) Handles NumericUpDown1.ValueChanged
      If Re = 0 Then Exit Sub
      Dim decimals As Integer = NumericUpDown1.Value
      Label6.Text = "Reynolds number = " + FormatNumber(Re, decimals)
      If Re < 2000 Then
        Label6.Text += " LAMINAR"
        Label7.Text = "hf = " + FormatNumber(hf1, decimals) + " POISEUILLE hf = " +
FormatNumber(hf2, decimals) + " DARCY"
      Else
        Label6.Text += " TURBULENT"
        Label7.Text = "hf = " + FormatNumber(hf, decimals)
      End If
   End Sub
End Class
```

### 3.7.4.3 Minor Losses

Losses (pressure drops) are either major or minor in nature. Major losses are associated with friction in the straight portions of the pipes, whereas minor losses are due to losses from valves, bends, and tees. Minor head losses may be determined by specifying the loss coefficient as shown in the following equation:

$$h_l = k_1 * \frac{v^2}{2g} \qquad (3.32)$$

where $k_1$ is the loss coefficient (depends on the geometry of the component and fluid properties) = $\phi$(geometry, Re).

### 3.7.5 Pipes in Series

#### 3.7.5.1 Equivalent-Velocity-Head Method

This method concerns a pipe composed of sections of different diameters. In pipes in series, the same fluid flows through all the pipes, and the head losses are cumulative (see Figure 3.12). As such, continuity and energy equations yield the following equations:

$$Q = Q_1 = Q_2 = Q_3 = \cdots = Q_N \qquad (3.33)$$

$$h_l = h_{l1} + h_{l2} + h_{l3} + \cdots + h_{li} = \sum_{i=1}^{N} h_{li} \qquad (3.34)$$

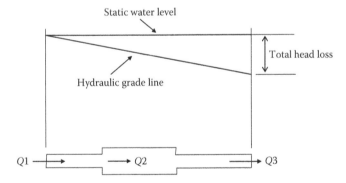

**FIGURE 3.12** Pipes in series.

where:

$h_{li}$ is the contribution of head loss from $i$th section, m
$N$ is the number of pipes

$$h_l = \left(\frac{f_1 * l_1 v_1^2}{2g * D_1}\right) + \left(\frac{f_2 * l_2 v_2^2}{2g * D_2}\right) + \cdots + \left(\frac{f_i * l_i v_i^2}{2g * D_i}\right) + \frac{k_1 * v^2}{2g} \qquad (3.35)$$

where $D_i$ is the diameter of $i$th pipe, m.

### Example 3.9

1. Write a computer program that enables the determination of flow from a large tank to which two pipes are connected in series using the equivalent-velocity-head method. The two pipes have the following characteristics: their diameters are $D_1$ mm and $D_2$ mm, their lengths are $l_1$ m and $l_2$ m, and their friction factors are $f_1$ and $f_2$, respectively. The head loss is equal to $h_l$.
2. Two pipes of diameters 200 and 150 mm and lengths 250 and 150 m are connected in series from a large tank. For a head loss of 8 m and given friction factors of 0.018 and 0.21, respectively, determine the rate of flow from the tank to the second pipe using the equivalent-velocity-head method.

### Solution

1. For solution to Example 3.9 (1), see the listing of Program 3.9.
2. Solution to Example 3.9 (2):
   a. Given: $d_1 = 200 * 10^{-3}$ m, $l_1 = 250$ m, $f_1 = 0.018$ m, $d_2 = 150 * 10^{-3}$ m, $l_2 = 150$ m, $h_f = 8$ m, $f_2 = 0.021$ m.
   b. Use continuity equation as follows: $v_2^2 = v_1^2*(D_1/D_2)^4 = v_1^2*(200/150)^4 = 3.16* v_1^2$.
   c. Use energy equation for the two pipes as follows: $hf = (f_1*l_1*v_1^2/2g*D_1) + (f_2*l_2*v_2^2/2g*D_2) = (0.018 * 250*v_1^2/2 * 9.81 * 200 * 10^{-3}) + (0.21 * 150* v_2^2/2 * 9.81 * 150 * 10^{-3}) = 1.15* v_1^2 + 1.07* v_2^2 = (1.15 + 1.07 * 3.16) v_1^2 = 4.53* v_1^2$. Thus, $v_1 = 1.32$ m/s.
   d. Determine the flow rate as follows: $Q = v*A = 1.32*\pi*(200 * 10^{-3})^2/4 = 0.041$ m³/s.

### LISTING OF PROGRAM 3.9   (CHAP3.9\FORM1.VB): EQUIVALENT-VELOCITY-HEAD METHOD

```
'******************************************************************************
'Program 3.9: Equivalent-Velocity-Head Method
'******************************************************************************
Public Class Form1
  Const Pi = 3.1415962#
  Const g = 9.81
  Dim N As Integer
  Dim d(), l(), f() As Double
  Dim sum, R, V1, Q, H, V As Double
```

```
    Private Sub Form1_Load(ByVal sender As System.Object, ByVal e As System.EventArgs)
Handles MyBase.Load
        Me.Text = "Program 3.9: Equivalent-Velocity-Head Method"
        Me.FormBorderStyle = Windows.Forms.FormBorderStyle.FixedSingle
        Me.MaximizeBox = False
        Label1.Text = "Enter how many pipes connected in series:"
        DataGridView1.Columns.Clear()
        DataGridView1.Rows.Clear()
        DataGridView1.Columns.Add("colD", "Diameter")
        DataGridView1.Columns.Add("colL", "Length")
        DataGridView1.Columns.Add("colF", "Friction factor")
        DataGridView1.AllowUserToAddRows = False
        Button1.Text = "&Calculate"
        TextBox1.Text = "2"
        Label2.Text = ""
        Label3.Text = "Decimals:"
        NumericUpDown1.Maximum = 10
        NumericUpDown1.Minimum = 0
        NumericUpDown1.Value = 2
    End Sub


    Sub calculateResults(ByVal calcHead As Boolean)
        N = CInt(Val(TextBox1.Text))
        Dim decimals As Integer = NumericUpDown1.Value
        For i = 1 To N
            d(i) = Val(DataGridView1.Rows(i - 1).Cells(0).Value)
            l(i) = Val(DataGridView1.Rows(i - 1).Cells(1).Value)
            f(i) = Val(DataGridView1.Rows(i - 1).Cells(2).Value)
        Next
        If calcHead Then
            Dim H = Val(InputBox("Enter head loss H (m):", "Enter H", "0"))
            sum = f(1) * l(1) / (2 * g * d(1))
            For i = 2 To N
                R = (d(1) / d(i)) ^ 4
                sum = sum + f(i) * l(i) * R / (2 * g * d(i))
            Next i
            V1 = (H / sum) ^ 0.5 'velocity in pipe 1
            Q = V1 * Pi / 4 * d(1) ^ 2 'discharge
            Label2.Text = "The flow rate for this head loss = " + FormatNumber(Q, decimals) +
" m3/s"
        Else 'compute head loss in pipes
            Q = Val(InputBox("Enter discharge Q(m3/s):", "Enter Q", "0"))
            H = 0
            For i = 1 To N
                V = 4 * Q / (Pi * d(i) ^ 2)
                H = H + f(i) * l(i) * v ^ 2 / (2 * g * d(i))
            Next i
            Label2.Text = "Total head loss (m): " + FormatNumber(H, decimals)
        End If
    End Sub


    Private Sub TextBox1_TextChanged(ByVal sender As System.Object, ByVal e As
System.EventArgs) Handles TextBox1.TextChanged
        If N > Val(TextBox1.Text) Then
            'delete some rows
            For i = N To Val(TextBox1.Text) + 1 Step -1
                DataGridView1.Rows.RemoveAt(i - 1)
            Next
```

```
        Else
            'add some rows
            For i = N + 1 To Val(TextBox1.Text)
              DataGridView1.Rows.Add()
            Next
        End If
        N = Val(TextBox1.Text)
        ReDim d(N)
        ReDim l(N)
        ReDim f(N)
    End Sub

    Private Sub Button1_Click(ByVal sender As System.Object, ByVal e As System.EventArgs)
Handles Button1.Click
        Dim i = MsgBox("You want to compute discharge for a given head loss?", vbYesNo,
"Prompt")
        If i = vbYes Then
            calculateResults(True)
        Else
            calculateResults(False)
        End If
    End Sub
End Class
```

### 3.7.5.2 Equivalent-Length Method

In this method, pipes are replaced by equivalent lengths of a selected pipe size. Usually the selected pipe size is one that figures most prominently in the system. An equivalent length may be defined as length, $l_e$, of pipe of a certain diameter, $D_e$, which for the same flow will give the same head loss as the pipe of length $l$ and diameter $D$ under consideration (Currie 2012). Thus, pipe friction and continuity equations yield the following equation:

$$l_e = l * \left(\frac{f * v^2}{2g * D}\right) / \left(\frac{f_e * v_e^2}{2g * D_e}\right) = l * \left(\frac{f}{f_e}\right) * \left(\frac{D_e}{D}\right)^E \quad (3.36)$$

where:
  $l_e$ is the equivalent length for pipe to be replaced, m
  $l$ is the length of pipe to be replaced, m
  $f$ is the friction factor of selected pipe, m
  v is the velocity of flow through selected pipe, m/s
  $g$ is the gravitational acceleration, m/s$^2$
  $D$ is the diameter of pipe to be replaced, m
  $f_e$ is the friction factor of selected pipe, m
  $v_e$ is the velocity of flow through equivalent pipe, m/s
  $D_e$ is the diameter of selected pipe

The equivalent-length method is of value when there are minor losses, such as bends, which are expressed in terms of equivalent lengths of pipe.

### Example 3.10

1. Write a computer program to solve Example 3.9 using the equivalent-length method.
2. Solve Example 3.9 using the equivalent-length method.

### Solution

1. For solution to Example 3.10 (1), see the listing of Program 3.10.
2. Solution to Example 3.10 (2):
   a. Given: $d_1 = 200 * 10^{-3}$ m, $l_1 = 250$ m, $f_1 = 0.018$ m, $d_2 = 150 * 10^{-3}$ m, $l_2 = 150$ m, $h_f = 8$ m, $f_2 = 0.021$ m.
   b. Choose the 200-mm diameter pipe as the standard pipe. The equivalent length for the second pipe may be found as follows: $l_e = l * (f/f_e) * (D_e/D)^5 = 150*(0.021/0.18)*(200/150)^5 = 737$ m of 200 mm diameter.
   c. Determine the total effective length as 250 + 737 = 987 m of 200 mm pipe.
   d. Use the energy equation for the equivalent pipe as follows: $h_f = (f_1*l_e \, v_1^2/2g* D_1) = 0.018 * 987* v_1^2/2 * 9.81*(200/1000) = 8$, which yields $v_1 = 1.33$ m/s.
   e. Determine the flow rate as follows: $Q = v * A = 1.33*(\pi*(200 * 10^{-3})^2/4) = 0.042$ m$^3$/s.

### LISTING OF PROGRAM 3.10    (CHAP3.10\FORM1.VB): EQUIVALENT-LENGTH METHOD

```vb
'*********************************************************************************
'Program 3.10: Equivalent-Length Method
'*********************************************************************************
Public Class Form1
  Const g = 9.81
  Dim N, EQP As Integer
  Dim d(), l(), f(), MAXD As Double
  Dim sum, R, V1, Q, H, V As Double
  Private Sub Form1_Load(ByVal sender As System.Object, ByVal e As System.EventArgs)
Handles MyBase.Load
    Me.Text = "Program 3.10: Equivalent-Length Method"
    Me.FormBorderStyle = Windows.Forms.FormBorderStyle.FixedSingle
    Me.MaximizeBox = False
    Label1.Text = "Enter how many pipes connected in series:"
    DataGridView1.Columns.Clear()
    DataGridView1.Rows.Clear()
    DataGridView1.Columns.Add("colD", "Diameter")
    DataGridView1.Columns.Add("colL", "Length")
    DataGridView1.Columns.Add("colF", "Friction factor")
    DataGridView1.AllowUserToAddRows = False
    Button1.Text = "&Calculate"
    TextBox1.Text = "2"
    Label2.Text = ""
    Label3.Text = "Decimals:"
    NumericUpDown1.Value = 2
    NumericUpDown1.Maximum = 10
    NumericUpDown1.Minimum = 0
  End Sub

  Sub calculateResults(ByVal calcHead As Boolean)
    N = CInt(Val(TextBox1.Text))
    Dim decimals As Integer = NumericUpDown1.Value
    For i = 1 To N
      d(i) = Val(DataGridView1.Rows(i - 1).Cells(0).Value)
      l(i) = Val(DataGridView1.Rows(i - 1).Cells(1).Value)
      f(i) = Val(DataGridView1.Rows(i - 1).Cells(2).Value)
      If MAXD < d(i) Then
          MAXD = d(i)
          EQP = i
      End If
    Next
    'determine the equivalent length
    sum = 0
    For i = 1 To N
      sum = sum + l(i) * f(i) / f(EQP) * (MAXD / d(i)) ^ 5
    Next i
    If calcHead Then
      Dim H = Val(InputBox("Enter head loss H (m):", "Enter H", "0"))
      V1 = (2 * g * MAXD * H / (sum * f(EQP))) ^ 0.5 'velocity in pipe
      Q = V1 * Math.Pi / 4 * d(1) ^ 2 'discharge
      Label2.Text = "The flow rate for this head loss (m3/s): " + FormatNumber(Q,
decimals)
    Else 'compute head loss in pipes
      Q = Val(InputBox("Enter discharge Q(m3/s):", "Enter Q", "0"))
```

```
            V = 4 * Q / (Math.Pi * MAXD ^ 2)
            H = f(EQP) * sum * V ^ 2 / (2 * g * MAXD)
            Label2.Text = "Total head loss (m): " + FormatNumber(H, decimals)
        End If
    End Sub

    Private Sub TextBox1_TextChanged(ByVal sender As System.Object, ByVal e As
System.EventArgs) Handles TextBox1.TextChanged
        If N > Val(TextBox1.Text) Then
            'delete some rows
            For i = N To Val(TextBox1.Text) + 1 Step -1
                DataGridView1.Rows.RemoveAt(i - 1)
            Next
        Else
            'add some rows
            For i = N + 1 To Val(TextBox1.Text)
                DataGridView1.Rows.Add()
            Next
        End If
        N = Val(TextBox1.Text)
        ReDim d(N)
        ReDim l(N)
        ReDim f(N)
    End Sub

    Private Sub Button1_Click(ByVal sender As System.Object, ByVal e As System.EventArgs)
Handles Button1.Click
        Dim i = MsgBox("You want to compute discharge for a given head loss?", vbYesNo,
"Prompt")
        If i = vbYes Then
            calculateResults(True)
        Else
            calculateResults(False)
        End If
    End Sub
End Class
```

### 3.7.6 Pipes in Parallel (Pipe Network)

A pipe network is formed by a group of pipes that are interconnected to allow the flow of a fluid from a particular inlet to a certain outlet through different directions. A loop is defined as a string of connected pipes. Loops may be classified as follows (Abdel-Magid et al. 1996):

1. *Closed loop*: Last pipe in the closed loop is connected to the first pipe in it.
2. *Pseudo loop*: It contains a series of pipes that are not closed, but whose starting and ending nodes have fixed hydraulic grade line elevations, for example, a loop connecting two reservoirs.
3. *Open loop*: It is a loop that does not end at its starting node or a pseudo loop that does not have fixed-head end nodes.

Generally, the number of loops is equal to the number of pipes minus the number of nodes minus 1 as shown in the following equation:

$$Nloops = Npipes - Nnodes - 1 \tag{3.37}$$

In parallel pipes, head losses are the same in any of the lines, whereas discharges are cumulative. Equations 3.38 and 3.39 can be formulated from Figure 4.6 (size of pipes, fluid properties, and roughness are assumed to be known):

$$Q = Q_1 + Q_2 + Q_3 + \cdots + Q_i = \sum_{i=1}^{N} Q_i \tag{3.38}$$

where:

$Q$ is the total flow through a node, m/s$^3$
$Q_i$ is the flow through pipe number $i$, m/s$^3$
$N$ is the number of pipes

$$h_l = h_{l1} + h_{l2} + h_{l3} + \cdots + h_{li} = \sum_{i=1}^{N} h_{li} \tag{3.39}$$

where:

$h_l$ is the total head loss, m
$h_{li}$ is the head loss at pipe number $i$, m

The cases that can be met are

1. Unknown discharge for known elevation of hydraulic grade line.
2. Unknown direction and quantity of flow and head loss in each line for known discharge.

Analysis of a network of pipes by Bernoulli's and continuity equations would result in formulation of a large volume of simultaneous equations that are tedious to solve. Instead of this lengthy procedure, the method of successive approximations may be employed. This method assumes values for flow in each pipe or head losses at each node, and it checks that the chosen values satisfy the following requirements (Douglas et al. 2006):

1. The head loss between any two junctions is the same for all routes between these junctions.
2. The inflow to each junction equals the outflow from that junction.

Assumptions are corrected by successive approximations until they satisfy the requirement within the required degree of accuracy (Douglas et al. 2006). The procedure used in the corrections is the Hardy Cross (relaxation) method. This method considers the network as a combination of simple loops (circuits). Each loop in the network is balanced to maintain compatible flow conditions in the system.

The procedure adopted for individual loops is summarized as follows:

1. Continuity equation is satisfied at all junctions. This indicates that at every junction, the total quantity of water entering is equal to the algebraic sum of water leaving, including any water removed from or added to the system at that junction, that is,

$$\sum_{i=1}^{N} Q_i = 0 \tag{3.40}$$

where:
$Q_i$ is the flow at node $i$, m$^3$/s
$N$ is the number of nodes

Equation 3.40 is referred to as Kirchoff's node law.

2. Law of conservation of energy is valid. Energy loss is the same for all routes through which water passes. This indicates that around any closed loop of pipes, the algebraic sum of the energy losses must be zero. Furthermore, the sum of the pipe head losses (in a series of pipes connecting two constant head sources) must equal the difference in head between the two sources, that is, in every loop, the algebraic sum of head losses through any chosen path is zero.

$$\Sigma(h)_{\text{loop}} = 0 \tag{3.41}$$

Equation 3.41 is termed Kirchoff's loop law.

For solving a network problem, the following methods can be used:

1. Balancing heads by correcting assumed flows
2. Balancing flows by correcting assumed heads

The Hardy Cross method analyzes the network by using the former concept of assuming initial flows. The imbalance in the energy equations is then determined, and flows in each loop are corrected accordingly. Corrections are carried out until convergence is achieved when the largest correction is less than some tolerance level. Energy losses may be found using any one of the standard energy loss equations, neglecting similar minor losses and kinetic energy. All energy loss equations can be expressed as indicated in the following equation:

$$h = k * Q^n \tag{3.42}$$

where:
$h$ is the head loss, m
$k$ is the resistance coefficient, numerical constant (a function of pipe geometry, diameter, length, material and age of conduit, and fluid viscosity)
$Q$ is the volume rate of flow in the pipe, m$^3$/s
$n$ is the constant exponent for all pipes. If the Darcy–Weisbach equation is used ($h = f * l * v^2/2D*g$), $n = 2$; if Manning's formula is used ($v = r_H^{2/3}*j^{1/2}/n$), $n = 2$ (turbulent flow). If the Hazen–Williams formula is used ($v = 0.894*c*(r_H)^{0.63}*j^{0.54}$), $n = 1.85$; $c$ is the Hazen–Williams roughness factor (depends on the material and age of the conduit). Generally, $v = \text{constant}*D^x(h_f/l)^y$, where $x$ and $y$ are constants depending upon the equation used.

The relationship of the discharges and corrections may be determined from the following equation:

$$Q_2 = Q_1 + \Delta Q_1 \tag{3.43}$$

where:
$Q_2$ is the second set of assumed values of discharge
$Q_1$ is the initial set of assumed values of discharge
$\Delta Q_1$ is the first-order correction to the discharge

The sum of the energy losses around a loop based on the first assumption is as indicated in the following equation:

$$h_1 = \Sigma k * Q_1^n \tag{3.44}$$

The sum of energy losses when the first correction is applied yields the following equation:

$$h_2 = \sum \left[ k \left( Q_1 + \Delta Q_1 \right)^n \right] \tag{3.45}$$

Equation 3.46 is obtained by expanding Equation 3.45 into a series and neglecting small values:

$$h_2 = \sum \left[ k \left( Q_1^n + n * Q_1^{(n-1)} * \Delta Q_1 \right) \right] \qquad (3.46)$$

where $h_2$ is the algebraic sum of the losses around a loop (=0).

Putting Equation 3.46 equal to zero and solving for $Q$ gives the following equation:

$$\Delta Q_1 = \frac{-\sum h}{\left( n \sum (h/Q) \right)} \qquad (3.47)$$

The negative sign illustrates that the positive (clockwise) discharges ought to be reduced, and the negative (anticlockwise, Tawaf-wise) discharges ought to be increased. The process is iterative and must be continued until the desired degree of accuracy is achieved (Douglas et al. 2006).

In summary, the Hardy Cross method involves the following steps:

1. Geometric configuration of the network is established.
2. Appropriate flows are assumed in each pipe (continuity must be satisfied at each junction, clockwise flows are positive, and they produce positive head losses).
3. For each elementary loop in the system:
   a. Sign convention is established.
   b. Head loss in each pipe is estimated.
   c. Algebraic sum of the head loss around the loop is evaluated.
   d. Correction factor is determined for each loop.
   e. Flow adjustment is conducted for the loop.
4. The aforementioned step is applied to each circuit in the network, with flow correction established to

each pipe, and the procedure is repeated to fulfill the required accuracy. It is to be noted that a common element in two different loops receives two different flow rate corrections.

5. For large complex networks, the analysis needs to be conducted by using a computer program.

In modeling a water supply network, the necessary data needed include system maps or pipe and valve grid maps; geometric data, such as pipe diameters, lengths of pipes, materials, pipe junction elevations, pump characteristic curves, valve types, and sets; operational data, such as total water production, consumption and system losses, control valve set options, and reservoir levels.

### Example 3.11

1. Write a computer program to compute the flow through each pipe of the network of the three pipes (A, B, and C) in parallel as shown in Figures 3.13 and 3.14, given the rate of flow

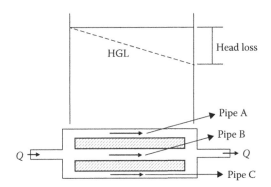

**FIGURE 3.13** Pipes in parallel.

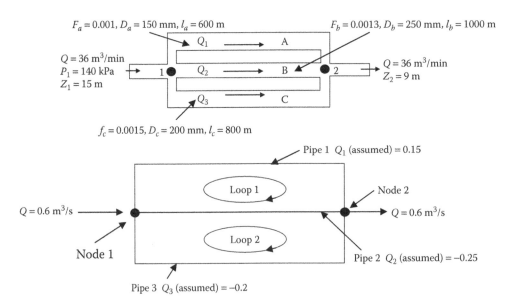

$F_a = 0.001$, $D_a = 150$ mm, $l_a = 600$ m        $F_b = 0.0013$, $D_b = 250$ mm, $l_b = 1000$ m

$Q = 36$ m³/min
$P_1 = 140$ kPa
$Z_1 = 15$ m

$Q = 36$ m³/min
$Z_2 = 9$ m

$f_c = 0.0015$, $D_c = 200$ mm, $l_c = 800$ m

Pipe 1 $Q_1$ (assumed) = 0.15

$Q = 0.6$ m³/s     Loop 1     Node 2

$Q = 0.6$ m³/s

Node 1     Loop 2

Pipe 2 $Q_2$ (assumed) = −0.25

Pipe 3 $Q_3$ (assumed) = −0.2

**FIGURE 3.14** Example 3.11.

to system $Q$, water temperature $T$, pipe diameters $D_a$, $D_b$, and $D_c$; friction factors $f_a$, $f_b$, and $f_c$; lengths $l_a$, $l_b$, and $l_c$; pressure at point 1, with point 1 at a level of $z_1$; and point 2 at elevation of $z_2$. Let the program determine the pressure at point 2. Use Darcy–Weisbach equation.

2. Solve for flow through each pipe of the network discussed in (1), given values of $D_a$ = 150 mm, $D_b$ = 250 mm, $D_c$ = 200 mm, $f_a$ = 0.001, $f_b$ = 0.0013, $f_c$ = 0.0015, $l_a$ = 600 m, $l_b$ = 1000 m, $l_c$ = 800 m, $P_1$ = 140 kPa, $z_1$ = 15 m, $z_2$ = 9 m, $Q$ = 36 m³/min, and temperature = 20°C.

3. Use the program written in (1) to verify computations in (2).

**Solution**

1. For solution to Example 3.11 (1), see the listing of Program 3.11.
2. Solution to Example 3.11 (2):
   a. Given: $D_a$ = 150 mm, $D_b$ = 250 mm, $D_c$ = 200 mm, $f_a$ = 0.001, $f_b$ = 0.0013, $f_c$ = 0.0015, $l_a$ = 600 m, $l_b$ = 1000 m, $l_c$ = 800 m, $P_1$ = 140 kPa, $z_1$ = 15 m, and $z_2$ = 9 m, $Q$ = 36 m³/min.
   b. Use Darcy–Weisbach equation for the flow as follows: $h = f*l*v^2/2D*g = 8*f*l*Q^2/(\pi^2*D^5*g) = k*Q^2$.

c. Assume flow passing through each pipe (long, narrow pipe offers more resistance to flow) and take clockwise flow (anti-Tawaf-wise) to be positive.

d. Find the values of $k$ $(=8*f*l/(\pi^2*D^5*g))$, $h$, and $h/Q$ for each pipe in a loop.

e. Compute the correction factor for each loop, $\Delta Q = -\Sigma h/\{2\Sigma(h/Q)\}$, and make adjustments of flow as shown in Table 3.8.

f. Repeat previous steps, taking into consideration the correction factor for each pipe in a loop and especially treating the common element 1B2 as shown in Table 3.8.

g. Adjustments are carried on until a desirable conversion is achieved.

h. Values of flow in pipes are $Q_{1A2}$ = 0.123 m³/s, $Q_{1B2}$ = 0.229 m³/s, and $Q_{1C2}$ = 0.178 m³/s.

i. Pressure at point 2 can be determined by using Bernoulli's equation between points 1 and 2 as follows: $(P_1/\rho) + (v_1^2/2g) + z_1 = (P_2/\rho) + (v_2^2/2g) + z_2 = h_{f1B2}$.

j. The head loss through pipe 1B2 can be found as follows: $h_{f1B2} = k*Q^2 = 109.9*(0.299)^2 = 9.8$ m.

k. For the same pipe: $v_1 = v_2$; therefore, $P_2 = P_1 + \rho (z_2 - z_1) - \rho h_{f1B2} = 140 * 10^3 + 998.2* (15 - 9) - 998.2 * 9.8 = 136.2$ kPa.

**TABLE 3.8**

**First Iteration**

| Pipe | F | l | d | k | Q | h | h/Q | ΔQ |
|---|---|---|---|---|---|---|---|---|
| [1A2] | 0.001 | 600 | 0.15 | 652.3289506 | 0.15 | 14.67740139 | 97.84934259 | −0.031152405 |
| [1B2] | 0.0013 | 1000 | 0.25 | 109.9043816 | −0.25 | −6.86902385 | +27.4760954 | |
| Result | | | | | | 7.808377539 | 125.325438 | |
| [1B2] | 0.0013 | 1000 | 0.25 | 109.9043816 | 0.25 | 6.86902385 | +27.4760954 | 0.030845955 |
| [1C2] | 0.0015 | 800 | 0.2 | 309.6014355 | −0.2 | −12.38405742 | +61.92028711 | |
| Result | | | | | | −5.515033572 | 89.39638251 | |

**Second Iteration**

| Pipe | F | l | d | k | Q | h | h/Q | ΔQ |
|---|---|---|---|---|---|---|---|---|
| [1A2] | 0.001 | 600 | 0.15 | 652.3289506 | 0.118847595.15 | 9.21398393 | 77.52772709 | 0.006637748 |
| [1B2] | 0.0013 | 1000 | 0.25 | 109.9043816 | −0.31199836 | −10.69841963 | +34.28998678 | |
| Result | | | | | | −1.484435698 | 111.8177139 | |
| [1B2] | 0.0013 | 1000 | 0.25 | 109.9043816 | 0.3119836 | 10.69841963 | 34.28998678 | −0.010614809 |
| [1C2] | 0.0015 | 800 | 0.2 | 309.6014355 | −1.69154045 | −8.858654037 | +52.37033518 | |
| Result | | | | | | 1.839765591 | 86.66032196 | |

**Third Iteration**

| Pipe | F | l | d | k | Q | h | h/Q | ΔQ |
|---|---|---|---|---|---|---|---|---|
| [1A2] | 0.001 | 600 | 0.15 | 652.3289506 | 0.125485344 | 10.27194444 | 81.85772253 | −0.00316841 |
| [1B2] | 0.0013 | 1000 | 0.25 | 109.9043816 | −0.294745802 | −9.547952797 | +32.39385509 | |
| Result | | | | | | 0.723991643 | 114.2515776 | |
| [1B2] | 0.0013 | 1000 | 0.25 | 109.9043816 | 0.294745802 | 9.547952797 | 32.39385509 | 0.002597301 |
| [1C2] | 0.0015 | 800 | 0.2 | 309.6014355 | −0.179768854 | −10.00534037 | +55.65669539 | |
| Result | | | | | | −0.457387574 | 88.05055048 | |

*(Continued)*

**TABLE 3.8 (*Continued*)**

**Fourth Iteration**

| Pipe | F | l | d | k | Q | H | h/Q | ΔQ |
|------|---|---|---|---|---|---|-----|-----|
| [1A2] | 0.001 | 600 | 0.15 | 652.3289506 | 0.122316934 | 9.759775429 | 79.79087707 | 0.00073294 |
| [1B2] | 0.0013 | 1000 | 0.25 | 109.9043816 | −0.300511513 | −9.9251536 | +33.02753198 | |
| Result | | | | | | −0.165378172 | 112.818409 | |
| [1B2] | 0.0013 | 1000 | 0.25 | 109.9043816 | 0.300511513 | 9.9251536 | 33.02853198 | −0.001176825 |
| [1C2] | 0.0015 | 800 | 0.2 | 309.6014355 | −0.177171553 | −9.718314548 | +54.85256726 | |
| Result | | | | | | 0.206839053 | 87.88009924 | |

**Fifth Iteration**

| Pipe | F | l | d | k | Q | h | h/Q | ΔQ |
|------|---|---|---|---|---|---|-----|-----|
| [1A2] | 0.001 | 600 | 0.15 | 652.3289506 | 0.123049873 | 9.877089656 | 80.26899483 | −0.000343475 |
| [1B2] | 0.0013 | 1000 | 0.25 | 109.9043816 | −0.298601748 | −9.799404796 | +32.81764044 | |
| Result | | | | | | 0.07768486 | 113.0866353 | |
| [1B2] | 0.0013 | 1000 | 0.25 | 109.9043816 | 0.298601748 | 9.799404796 | 32.81764044 | 0.000275132 |
| [1C2] | 0.0015 | 800 | 0.2 | 309.6014355 | −0.178348379 | −9.847847101 | +55.21691407 | |
| Result | | | | | | −0.048442305 | 88.03455451 | |

**Sixth Iteration**

| Pipe | F | l | d | k | Q | h | h/Q | ΔQ |
|------|---|---|---|---|---|---|-----|-----|
| [1A2] | 0.001 | 600 | 0.15 | 652.3289506 | 0.122706398 | 9.822025831 | 80.04493615 | 7.9799E-05 |
| [1B2] | 0.0013 | 1000 | 0.25 | 109.9043816 | −0.299220355 | −9.840049319 | +32.8856281 | |
| Result | | | | | | 0.018023487 | 112.9305642 | |
| [1B2] | 0.0013 | 1000 | 0.25 | 109.9043816 | 0.299220355 | 9.840049319 | 32.8856281 | −0.000128172 |
| [1C2] | 0.0015 | 800 | 0.2 | 309.6014355 | −0.178073246 | −9.81748662 | +55.1317327 | |
| Result | | | | | | 0.022562698 | 88.0173608 | |

---

### LISTING OF PROGRAM 3.11    (CHAP3.11\FORM1.VB): HARDY CROSS METHOD

```vb
'*********************************************************************************
'Program 3.11: Hardy Cross method
'*********************************************************************************
Imports System.Math
Public Class Form1
    Const g = 9.81
    Dim Npipes As Integer = 3
    Dim Nnodes As Integer = 2
    Dim Nloops = Npipes - (Nnodes - 1)
    Dim iter As Integer = 1
    Dim d(Npipes), f(Npipes), l(Npipes), k(Npipes), Q(Npipes) As Double
    Dim h(Npipes), hQ(Npipes) As Double
    Dim Qtotal, P1, P2, z1, z2 As Double
    Dim deltaQ(Nloops) As Double
    Dim T As Integer
    Dim dTotal As Double = 0
    Dim Rho_table() As Double =
    {0.999956, 0.999941, 0.999902, 0.999849,
    0.999781, 0.9997, 0.999605, 0.999498,
    0.999377, 0.999244, 0.999099, 0.998943,
    0.998774, 0.998595, 0.998405, 0.998203,
    0.997992, 0.99777, 0.997538, 0.997296,
    0.997044, 0.996783, 0.996512, 0.996232,
    0.995944, 0.995646, 0.99534, 0.995025,
    0.994702, 0.994371, 0.99403, 0.99368,
    0.99333, 0.99296
    }
```

```
Dim temp_table() As Double =
{5, 6, 7, 8, 9, 10, 11, 12, 13, 14, 15, 16,
17, 18, 19, 20, 21, 22, 23, 24, 25, 26, 27,
28, 29, 30, 31, 32, 33, 34, 35, 36, 37, 38
}

Private Sub Form1_Load(ByVal sender As System.Object, ByVal e As System.EventArgs)
Handles MyBase.Load
    Me.Text = "Example 3.11: Hardy Cross Method"
    Me.MaximizeBox = False
    Me.FormBorderStyle = Windows.Forms.FormBorderStyle.FixedSingle
    Button1.Text = "&Calculate"
    Button1.Enabled = True
    Button2.Text = "&Next >>"
    Button2.Enabled = False
    DataGridView1.Columns.Clear()
    DataGridView1.Columns.Add("DCol", "Pipe diameter, mm")
    DataGridView1.Columns.Add("FCol", "Friction factor")
    DataGridView1.Columns.Add("LCol", "Pipe length, m")
    DataGridView1.Rows.Clear()
    DataGridView1.Rows.Add(3)
    DataGridView1.AllowUserToAddRows = False
    Label1.Text = "Pipe data:"
    Label2.Text = "Flow rate, Q (m3/min):"
    Label3.Text = "Temperature, T (C):"
    Label4.Text = "Pressure P1 (kPa):"
    Label5.Text = "Point 1 level, z1 (m):"
    Label6.Text = "Point 2 level, z2 (m):"
    Label7.Text = "Pressure P2 (kPa):"
    GroupBox1.Text = " Output: "
End Sub


Private Sub Button1_Click(ByVal sender As System.Object, ByVal e As System.EventArgs)
Handles Button1.Click
    Qtotal = Val(TextBox1.Text) / 60
    T = Val(TextBox2.Text)
    P1 = Val(TextBox3.Text) * 1000
    z1 = Val(TextBox4.Text)
    z2 = Val(TextBox5.Text)
    dTotal = 0
    For i = 0 To Npipes - 1
      d(i) = Val(DataGridView1.Rows(i).Cells("DCol").Value) / 1000
      f(i) = Val(DataGridView1.Rows(i).Cells("FCol").Value)
      l(i) = Val(DataGridView1.Rows(i).Cells("LCol").Value)
      k(i) = 8 * f(i) * l(i) / ((PI ^ 2) * (d(i) ^ 5) * g)
      dTotal += d(i)
    Next
    For i = 0 To Npipes - 1
      'estimate Q and h for each pipe
      Q(i) = Qtotal * (d(i) / dTotal)
      h(i) = k(i) * (Q(i) ^ 2)
      hQ(i) = h(i) / Q(i)
    Next
    iter = 1
    calc_deltaQ()
    calc_P2()
    showOutput()
    Button2.Enabled = True
End Sub


Sub showOutput()
```

```
'*********************************************************************************
'Output calculations
'*********************************************************************************
Dim j As Integer = 0
If DataGridView1.Columns.Count < 8 Then
    DataGridView1.Columns.Add("kCol", "k")
    DataGridView1.Columns.Add("qCol", "Q")
    DataGridView1.Columns.Add("hCol", "h")
    DataGridView1.Columns.Add("hQCol", "h/Q")
    DataGridView1.Columns.Add("deltaQCol", "dQ")
End If
For i = 0 To Npipes - 1
    DataGridView1.Rows(i).Cells("kCol").Value = k(i)
    DataGridView1.Rows(i).Cells("qCol").Value = Q(i)
    DataGridView1.Rows(i).Cells("hCol").Value = h(i)
    DataGridView1.Rows(i).Cells("hQCol").Value = hQ(i)
Next
j = 0
For i = 0 To Nloops - 1
    DataGridView1.Rows(j).Cells("deltaQCol").Value = deltaQ(i)
    j += 2
Next
Label1.Text = digit_to_str(iter) + " Iteration"
TextBox6.Text = FormatNumber(P2, 2)
End Sub

Sub calc_deltaQ()
    'Calculate correction factor (delta Q) for each pipe in a loop
    Dim j As Integer = 0
    Dim hTotal As Double = 0
    Dim hQTotal As Double = 0
    For i = 0 To Nloops - 1
        hTotal = h(j) - h(j + 1)
        hQTotal = hQ(j) + hQ(j + 1)
        j += 1
        deltaQ(i) = -hTotal / (2 * hQTotal)
    Next
End Sub

Sub calc_P2()
    Dim Rho As Double = -1
    For i = 0 To temp_table.Length - 1
        If T <= temp_table(i) Then
            Rho = Rho_table(i)
            Exit For
        End If
    Next
    'Not found?
    If Rho = -1 Then
        MsgBox("Please enter a temperature between 5 and 38.",
            vbOKOnly Or vbInformation)
        Exit Sub
    End If
    Rho *= 1000
```

```vbnet
    P2 = P1 + (Rho * Abs(z2 - z1)) - (Rho * h(1))
    P2 /= 1000
  End Sub

  Function digit_to_str(ByVal n As Integer) As String
    Select Case n
      Case 1 : Return "First"
      Case 2 : Return "Second"
      Case 3 : Return "Third"
      Case 4 : Return "Fourth"
      Case 5 : Return "Fifth"
      Case 6 : Return "Sixth"
      Case 7 : Return "Seventh"
      Case 8 : Return "Eighth"
      Case 9 : Return "Ninth"
      Case 10 : Return "Tenth"
      Case 11 : Return "Eleventh"
      Case 12 : Return "Twelfth"
      Case 13 : Return "Thirteenth"
      Case 14 : Return "Fourteenth"
      Case 15 : Return "Fifteenth"
      Case 16 : Return "Sixteenth"
      Case 17 : Return "Seventeenth"
      Case 18 : Return "Eighteenth"
      Case 19 : Return "Nineteenth"
      Case Else
        If n Mod 10 = 1 Then
            Return n.ToString + "st"
        ElseIf n Mod 10 = 2 Then
            Return n.ToString + "nd"
        ElseIf n Mod 10 = 3 Then
            Return n.ToString + "rd"
        Else
            Return n.ToString + "th"
        End If
    End Select
  End Function


  Private Sub Button2_Click(ByVal sender As System.Object, ByVal e As System.EventArgs)
Handles Button2.Click
    Dim dQ As Double = deltaQ(0) - deltaQ(1)
    Q(0) += deltaQ(0)
    Q(1) -= dQ
    Q(2) = Qtotal - (Q(0) + Q(1))
    For i = 0 To Npipes - 1
      'calc h for each pipe
      h(i) = k(i) * (Q(i) ^ 2)
      hQ(i) = h(i) / Q(i)
    Next
    iter += 1
    calc_deltaQ()
    calc_P2()
    showOutput()
  End Sub
End Class
```

### 3.7.6.1 Disadvantages of the Hardy Cross Method

1. A lot of time and tedious work is exhausted in assuming initial flows.
2. Limitations in usage for large flows result in convergence problems.
3. Direction of flow is assumed incorrectly occasionally.
4. There are complications in the method for complex systems such as reservoirs, interior pumps, and valves (McCormich and Bellamy 1968; Shamir and Howard 1968; Liu 1969; Epp and Fowler 1970; Wood and Charles 1972; Collins and Johnson 1975; Abdel-Magid 1987; Abdel-Magid et al. 1991).

### 3.7.6.2 Analysis of Pipe Networks by the Finite Element Method

The finite element method for the analysis of water distribution networks may be used with the aid of personal computers, which make tedious, iterative calculations more amenable to quick solutions (Lam and Wolla 1972; Wood and Charles 1972; Gientke 1974; Collins and Johnson 1975; Abdel-Magid 1987; Abdel-Magid et al. 1991). Application of the finite element method to a structural problem demands the subdivision of the structure into a number of discrete elements. Each of these elements must satisfy the following three conditions (Collins and Johnson 1975; Abdel-Magid 1987; Abdel-Magid et al. 1991):

1. Equilibrium of forces
2. Compatibility of strains
3. Force–displacement relationship specified by the geometric and elastic properties of the discrete element

An equivalent set of conditions for a pipe network exists; hence, the ability to draw the analogy (Abdel-Magid 1987; Abdel-Magid et al. 1991):

1. The algebraic sum of the flows at any point or node must be zero.
2. The value of the piezometric head at a joint or node is the same for all pipes connected to that joint.
3. The flow–head relationship (such as Darcy–Weisbach or Hazen–Williams) must be satisfied for each element or pipe.

For a direct application of the finite element method involving a matrix solution, a linear relationship is required to define the element or pipe (see Figure 3.15). Hence, there is a relationship of the form indicated in the following equation:

$$q = c * h \qquad (3.48)$$

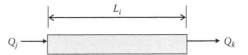

**FIGURE 3.15** Pipe element. (From Abdel-Mageid et al. *Water International J,* 16, 96–101, 1991, With permission.)

where:
$q$ is the flow, $m^3/s$
$h$ is the head loss, m
$c$ is the hydraulic properties of the pipe (to be assumed)

The solution technique can be subdivided into three steps (Abdel-Magid 1987; Abdel-Magid et al. 1991):

1. An initial value of the pipe coefficient, $c$, is selected for each pipe and is then combined to yield the system matrix coefficient ($C$). The system matrix is then solved for the value of the piezometric head at each joint.
2. The individual pipe flows, $q$, are computed by means of Equation 3.48 using the difference between the determined piezometric heads. These flows are then substituted in the Darcy–Weisbach equation to calculate the pipe head losses. If the pipe head losses obtained from the Darcy–Weisbach equation correspond to those obtained from the matrix solution, then the unique solution, satisfying both the Darcy–Weisbach equation and the linear Equation 3.48, has been found.
3. If there is a difference between the values of head loss calculated by the two methods, the values of $c$ are changed to cause the problem to converge to a solution.

In analysis of a pipe network, Figure 3.16 is used (Abdel-Magid 1987; Abdel-Magid et al. 1991) to show the application of the finite element method. Nodes and pipes are numbered for identification purposes. As a sign convention, any external

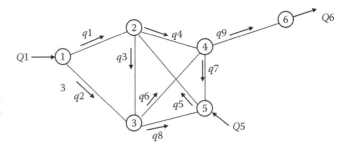

**FIGURE 3.16** Pipe network. (From Abdel-Mageid et al. *Water International J.* 16, 96–101, 1991, With permission.)

flow at a joint will be positive when fluid is input and negative when actual consumption occurs. Applying the equilibrium flow criteria at each node then gives

$$Q_1 = q_1 + q_2$$

$$Q_3 = q_2 + q_3 + q_6 + q_8 \qquad (3.49)$$

$$Q_4 = q_4 + q_6 + q_7 + q_9$$

However, the pipe flows in individual elements are given by

$$q_1 = c_1(H_1 - H_2)$$

$$q_2 = c_2(H_1 - H_3)$$

$$q_3 = c_3(H_2 - H_3)$$

$$q_4 = c_4(H_2 - H_4)$$

$$q_5 = c_5(H_5 - H_2) \qquad (3.50)$$

$$q_6 = c_6(H_3 - H_4)$$

$$q_7 = c_7(H_4 - H_5)$$

$$q_8 = c_8(H_3 - H_5)$$

$$q_9 = c_9(H_4 - H_6)$$

Applying the sign convention on the flow in the element, it will be noted that although the flow at one end, $j$, of the element will be input (i.e., positive), it is an output (i.e., negative) at the other end, $k$, of the element. It follows that if the flow in element $m$, whose ends are nodes $j$ and $k$, is given by $q_m = c_m(H_j - H_k)$ at end $j$, the same flow when considering end $k$ will be $q_m = c_m(H_k - H_j)$. Thus, substituting Equation 3.50 in Equation 3.49, observing the sign convention and writing in matrix form, we get the following equation:

Equation 3.51 may be put in a compact form as shown in the following equation:

$$Q = C * H \qquad (3.52)$$

where:
   $Q$ is the network consumption vector
   $C$ is the network characteristics matrix
   $H$ is the network head vector

The input data in the finite element procedure consist of the following:

1. The number of nodes, elements, known consumption nodes, known head nodes, and fluid properties
2. The element number, its diameter, its length, and the number of its first and second node connectivities
3. The node number and its known head or consumption

The Darcy–Weisbach relationship may be used in the analysis of head loss, as presented in the following equation:

$$h_f = \left(\frac{f * L}{2g * D * A^2}\right)q^2 = C_1 * q^2 \qquad (3.53)$$

where:
   $h_f$ is the head loss, m
   $f$ is the friction factor
   $L$ is the length of pipe, m
   $D$ is the diameter of pipe, m
   $A$ is the cross-sectional area of pipe, m²
   $g$ is the acceleration of gravity, m/s²
   $q$ is the flow discharge, m³/s

The friction factor may be determined from the following equation (Wood and Charles 1972, Wood 1980, Abdel-Magid 1987, Abdel-Magid et al. 1991):

$$f = 0.094K^{0.225} + 0.53K + 88.0K^{0.44}Re^{-a} \qquad (3.54)$$

where:
   $f$ is the friction factor
   Re is the Reynolds number

$$\begin{bmatrix} Q \\ Q \\ Q \\ Q \\ Q \\ Q \end{bmatrix} = \begin{bmatrix} (C+C_2) & -C_1 & -C_2 & 0 & 0 & 0 \\ -C_1 & (C_1+C_3+C_4+C_5) & -C_3 & -C_4 & -C_5 & 0 \\ -C_2 & -C_3 & (C_3+C_3+C_6+C_8) & -C_6 & -C_8 & 0 \\ 0 & -C_4 & -C_6 & (4+C_6+C_7+C_8) & -C_7 & -C_9 \\ 0 & -C_5 & -C_8 & -C_7 & (C_5+C_7+C_8) & 0 \\ 0 & 0 & 0 & -C_9 & 0 & C_9 \end{bmatrix} = \begin{bmatrix} H_1 \\ H_2 \\ H_3 \\ H_4 \\ H_5 \\ H_6 \end{bmatrix} \qquad (3.51)$$

$$K = \frac{\varepsilon}{D} \qquad (3.55)$$

where $\varepsilon$ = roughness coefficient

$a$ is the constant to be found from the following equation:

$$a = 1.62K^{0.134} \qquad (3.56)$$

The initial value of pipe coefficient, $C_0$, was chosen to correspond to Re = 200,000 in each pipe, a typical value for practical problems (Collins and Johnson 1975). Accordingly, and since Re = $qD/(Av)$, the initial flow in the pipe becomes

$$q_0 = 200,000v * A/D \qquad (3.57)$$

where:

$A$ is the area of the pipe, $m^2$

$D$ is the diameter, m

$v$ is the kinematic viscosity of the fluid, $m^2/s$

The value of head loss, $h_0$, corresponding to the flow, $q_0$, can be calculated from Equation 3.53. The pipe coefficient is then found from Equation 3.48 as indicated in the following equation:

$$c_0 = \frac{q_0}{h_0} \qquad (3.58)$$

This initial value of the pipe coefficient, $c_0$, for each pipe was then combined, according to the geometry of the network, to obtain the initial network characteristic matrix $[C_1]_0$. Using a standard finite element procedure, the matrix assembly process to form the global matrix can be summarized. Starting with a zero matrix, the following operations are performed for each element:

1. Add coefficient—$C_1$ to position $(k, k)$ and $(j, j)$.
2. Add coefficient—$C_1$ to position $(k, j)$ and $(j, k)$, according to the connectivity table. Once all elements are considered, the $[C]$ matrix is assembled.

Before the total system of equations can be solved, it is necessary to introduce proper boundary conditions for at least some of the nodes of the network. The two possible types of boundary conditions involve specifying either head or consumption for any given node. The introduction of the boundary conditions for prescribed heads can be implemented by performing the following steps:

1. Add the contribution of the prescribed unknown $H_j$ to the vector of nodal consumption $[Q]$.
2. Zero the $j$th column and $j$th row of the matrix $[C]$.
3. Make the $j$th coefficient of the vector nodal consumption equal to $H_j$.

A boundary condition of the second type, where the discharge rather than the head is prescribed, is handled by simply placing the value of the prescribed discharge or consumption in the proper position in the vector of nodal consumption. The solution of the simultaneous algebraic equations can be done using the Gauss elimination technique (Burden and Faires 2010). The solution of the system of equations could provide the values of the previously unknown nodal heads. With these, discharges could be computed for every element using Equation 3.48.

During the checking procedure, the flow, $q_c$, for each pipe calculated via Equation 3.48, and the matrix solution may be used to determine the head loss, $h_c$, from the Darcy–Weisbach equation (Collins and Johnson 1975; Abdel-Magid 1987; Abdel-Magid et al. 1991). The first step in the development of the program is to obtain the correction for the $c$ value for each pipe by assuming that the point $(h_c, q_c)$ is the unique solution. Thus, the correct linear relationship is defined by the straight line joining this point to the origin and is defined by the following equation:

$$h = \left(\frac{q_c}{h_c}\right)q \qquad (3.59)$$

The new value of $c$ is then set equal to $q_c/h_c$. When all pipe coefficients are corrected in a similar way, the flow distribution obviously is altered, and this method can be overcorrected when the matrix is resolved. To dampen this overcorrection effect, an averaging technique is introduced. The corrected value of $c$ is taken to be the mean of the $c$ value defined by Equation 3.59 and the value of $c$ used to obtain the matrix solution (Abdel-Magid 1987; Abdel-Magid et al. 1991).

### Example 3.12

Write a computer program to predict flows in a network that consists of $E$ number of elements and $N$ number of nodes, given input and withdrawals at nodes, pipe characteristics for each pipe in the system, and water level in tank at a fixed node.

## LISTING OF PROGRAM 3.12    (A) LISTING OF FORM1 (CHAP3.12\ FORM1.VB): FEM ANALYSIS OF PIPE NETWORKS

```vb
'********************************************************************************
'Program 3.12: FEM Analysis of Pipe Networks
'********************************************************************************
Imports System.Math
Public Class Form1
  Dim loadFromFile As Boolean
  Dim N
  Dim ICHECKER, EE, WW, DIF As Double
  Private Sub Form1_Load(ByVal sender As System.Object, ByVal e As System.EventArgs)
Handles MyBase.Load
    loadFromFile = False
    Me.Text = "Program 3.12: FEM Analysis of Pipe Networks"
    Me.Width = 322
    TextBox6.Multiline = True
    TextBox6.Height = 291
    TextBox6.ScrollBars = ScrollBars.Vertical
    TextBox6.Font = New Font(FontFamily.GenericMonospace, 10)
    Button1.Text = "Browse"
    Label1.Text = "Open data file:"
    Label2.Text = ""
    GroupBox1.Text = " Load data from file: "
    GroupBox2.Text = " OR, Enter new data: "
    Button2.Text = "Browse"
    Label3.Text = "Select a file to save data.."
    Label4.Text = ""
    Label5.Text = "Enter number of elements:"
    Label6.Text = "Enter number of nodes:"
    Label10.Text = "Are you using metric units?:"
    Label7.Text = "Enter number of nodes with known head:"
    Label8.Text = "Enter number of nodes with known flow:"
    Label9.Text = "Enter roughness (enter zero if unknown):"
    RadioButton1.Text = "Yes"
    RadioButton2.Text = "No"
    RadioButton1.Checked = True
    Button3.Text = "Calculate"
  End Sub


  Sub calculateResults()
    N = NELEM
    If NNODE > NELEM Then N = NNODE
    If loadFromFile Then
        ReDim DIAM(N), XLEN(N), AK(N), LNODS(N, 2), HEAD(N), H(N), Q(N), C(N), F(N)
        ReDim D(N), NODEH(N), Z(N, N), DELH(N), QO(N), CO(N), R(N), DH(N)
    End If
    For I = 1 To NNODE
       H(I) = 0.0!
       Q(I) = 0.0!
    Next I
```

```
        If loadFromFile Then
            For I = 1 To NELEM
               Input(1, DIAM(I))
               Input(1, XLEN(I))
               Input(1, LNODS(I, 1))
               Input(1, LNODS(I, 2))
            Next I
            For I = 1 To NPFLOW
               Input(1, NODEQ)
               Input(1, Q(NODEQ))
            Next I
            For I = 1 To NVFIX
               Input(1, NODEH(I))
               Input(1, HEAD(NODEH(I)))
            Next I
            FileClose(1)
        End If
        'CONSTANT VALUES
        If UCase$(met$) = "Y" Then unit$ = "Metric" Else unit$ = "fps"
        Re = 200000.0! 'ASSUMED INITIAL REYNOLDS NUMBER
        EPSILON = 10 ^ -3 'ITERATION CONTROL FACTOR
        If ROUGHNESS = 0.0! Then
           ROUGHNESS = 0.004 'IN FEET
           If unit$ = "Metric" Then ROUGHNESS = 0.0013 'IN METRES
        End If
        PI = 22 / 7.0!
        viscous = 1.059 * 10 ^ -5
        G = 32.2 'GRAVITATIONAL ACCELERATION
        CNVRT = 1.0!
        If unit$ = "Metric" Then
           viscous = 0.000000976
           G = 9.807
           CNVRT = 3600.0! * 24 'CONVERT TO M3/DAY
        End If
        '*********************************************************************************
        '*********************************************************************************
        For I = 1 To NNODE
            QO(I) = Q(I)
        Next I
        'CALCULATE PIPE COEFFICIENT C(I)
        For I = 1 To NELEM
           AK(I) = ROUGHNESS / DIAM(I)
           FF = -1.62 * AK(I) ^ 0.134
           BK = 88.0! * AK(I) ^ 0.44 * Re ^ FF
           F(I) = 0.094 * AK(I) ^ 0.225 + 0.53 * AK(I) + BK
           Q1 = Re * viscous * PI / 4.0! * DIAM(I)
           CT = 8.0! * F(I) * XLEN(I) / (PI ^ 2 * G * DIAM(I) ^ 5)
           H1 = CT * Q1 ^ 2
           CO(I) = Q1 / H1
           C(I) = CO(I)
        Next I
        'Assembly of global matrix z(i,j)
        ITER = 0
Start:
        Dim J
        ITER = ITER + 1
        For I = 1 To NELEM
            Q(I) = QO(I) / CNVRT
```

```
        For J = 1 To NNODE
            Z(I, J) = 0.0!
        Next J
    Next I
    For I = 1 To NELEM
        L1 = LNODS(I, 1)
        L2 = LNODS(I, 2)
        Z(L1, L1) = Z(L1, L1) + C(I)
        Z(L2, L2) = Z(L2, L2) + C(I)
        Z(L1, L2) = Z(L1, L2) - C(I)
        Z(L2, L1) = Z(L2, L1) - C(I)
    Next I
        'IMPOSE BOUNDARY CONDITIONS
        For INODE = 1 To NVFIX
        J = NODEH(INODE)
        Z(J, J) = Z(J, J) * 100000000.0
        Q(J) = HEAD(J) * Z(J, J)
    Next INODE
    'SOLVE THE EQUATIONS BY GAUSSIAN ELIMINATION
    Call GAUSS(NNODE, Z, Q, H)
    'COMPUTE HEADLOSSES AND DISCHARGES
    ICHECKER = 0
    For I = 1 To NELEM
        L1 = LNODS(I, 1)
        L2 = LNODS(I, 2)
        DELH(I) = Abs(H(L1) - H(L2))
        D(I) = C(I) * DELH(I)
        R(I) = 4.0! * D(I) / (PI * viscous * DIAM(I))
        EE = -1.62 * AK(I) ^ 0.134
        WW = 88.0! * AK(I) ^ 0.44 * R(I) ^ EE
        F(I) = 0.094 * AK(I) ^ 0.225 + 0.53 * AK(I) + WW
        DH(I) = XLEN(I) * D(I) ^ 2 / DIAM(I) ^ 5
        DH(I) = DH(I) * 8.0! * F(I) / (G * PI ^ 2)
        DIF = Abs(DH(I) - DELH(I))
        If DIF > EPSILON Then
            ICHECKER = ICHECKER + 1
            C(I) = 0.5 * (C(I) + (D(I) / DH(I)))
        End If
    Next I
    If ICHECKER > 0 Then GoTo start
    'RESULTS
    Dim s As String
    s = "Element No. Discharge HEAD LOSS"
    For I = 1 To NELEM
        s += vbCrLf + Format(I, "####   ")
        s += Format((D(I) * CNVRT), "####.#####   ")
        s += Format(DELH(I), "####.#####")
    Next
    s += vbCrLf + "Node Number Head"
    For I = 1 To NNODE
        s += vbCrLf + Format(I, "####   ")
        s += Format(H(I), "####.#####")
    Next
    s += vbCrLf + "TOTAL ITERATIONS = " + ITER.ToString
    TextBox6.Text = s
    Me.Width = 691
End Sub
```

```vb
  Sub GAUSS(ByVal NNODE, ByVal Z(,), ByVal Q(), ByVal H())
  '*********************************************************************************
  'SOLVES A SYSTEM OF LINEAR EQUATIONS BY GAUSSIAN ELIMINATION
  '*********************************************************************************

  Dim ZZ, COF As Double
  'FORWARD ELIMINATION
  For I = 1 To NNODE - 1
    ZZ = Z(I, I)
    Q(I) = Q(I) / ZZ
    For J = 1 To NNODE
      Z(I, J) = Z(I, J) / ZZ
    Next J
    For K = I + 1 To NNODE
      COF = -Z(K, I)
      Q(K) = Q(K) + COF * Q(I)
      For J = 1 To NNODE
        Z(K, J) = Z(K, J) + COF * Z(I, J)
      Next J
    Next K
  Next I
  'BACK SUBSTITUTION
  H(NNODE) = Q(NNODE) / Z(NNODE, NNODE)
  For I = NNODE - 1 To 1 Step -1
    H(I) = Q(I)
    For J = I + 1 To NNODE
      H(I) = H(I) - Z(I, J) * H(J)
    Next J
  Next I
  End Sub


  Private Sub Button1_Click(ByVal sender As System.Object, ByVal e As System.EventArgs)
Handles Button1.Click
    Dim f = OpenFileDialog1.ShowDialog
    If f = DialogResult.Cancel Then Exit Sub
    Label2.Text = OpenFileDialog1.FileName
    FileOpen(1, OpenFileDialog1.FileName, OpenMode.Input)
    Input(1, NELEM)
    Input(1, NNODE)
    Input(1, met$)
    met$ = met$.Trim()
    Input(1, NVFIX)
    Input(1, NPFLOW)
    Input(1, ROUGHNESS)
    loadFromFile = True
  End Sub


  Private Sub Button2_Click(ByVal sender As System.Object, ByVal e As System.EventArgs)
Handles Button2.Click
    Dim f = SaveFileDialog1.ShowDialog
    If f = DialogResult.Cancel Then Exit Sub
    Label4.Text = SaveFileDialog1.FileName
    saveFileName = SaveFileDialog1.FileName
    loadFromFile = False
  End Sub


  Private Sub Button3_Click(ByVal sender As System.Object, ByVal e As System.EventArgs)
Handles Button3.Click
    If Not loadFromFile Then
      NELEM = Val(TextBox1.Text)
```

```
        NNODE = Val(TextBox2.Text)
        NVFIX = Val(TextBox3.Text)
        NPFLOW = Val(TextBox4.Text)
        ROUGHNESS = Val(TextBox5.Text)
        If RadioButton1.Checked Then met$ = "Y" Else met$ = "N"
        Dim f2 As New Form2
        f2.ShowDialog()
      End If
      calculateResults()
    End Sub
End Class
```

## (B) LISTING OF FORM2 (CHAP3.12\FORM2.VB)

```
Public Class Form2
  Private Sub Form2_Load(ByVal sender As System.Object, ByVal e As System.EventArgs)
Handles MyBase.Load
      TabPage1.Text = "Pipe data"
      GroupBox1.Text = ""
      DataGridView1.Columns.Clear()
      DataGridView1.Columns.Add("DCol", "Diameter")
      DataGridView1.Columns.Add("LCol", "Length")
      DataGridView1.Columns.Add("E1Col", "End 1")
      DataGridView1.Columns.Add("E2Col", "End 2")
      DataGridView1.Rows.Add(CInt(NELEM))
      DataGridView1.AllowUserToAddRows = False
      TabPage2.Text = "Ext. Node flow"
      GroupBox2.Text = ""
      DataGridView2.Columns.Clear()
      DataGridView2.Columns.Add("NCol", "Node No.")
      DataGridView2.Columns.Add("ECol", "Ext. flow")
      DataGridView2.Rows.Add(CInt(NPFLOW))
      DataGridView2.AllowUserToAddRows = False
      TabPage3.Text = "Node head"
      GroupBox3.Text = ""
      DataGridView3.Columns.Clear()
      DataGridView3.Columns.Add("NCol", "Node No.")
      DataGridView3.Columns.Add("HCol", "Known head")
      DataGridView3.Rows.Add(CInt(NVFIX))
      DataGridView3.AllowUserToAddRows = False
      Button3.Text = "&Save data"
  End Sub

  Sub saveData()
    Dim N = NELEM
    ReDim DIAM(N), XLEN(N), AK(N), LNODS(N, 2), HEAD(N), H(N), Q(N), C(N), F(N)
    ReDim D(N), NODEH(N), Z(N, N), DELH(N), QO(N), CO(N), R(N), DH(N)
    FileOpen(2, saveFileName, OpenMode.Output)
    Print(2, NELEM, NNODE, Chr(34), met$, Chr(34))
    Print(2, NVFIX, NPFLOW, ROUGHNESS)
    For I = 1 To NELEM
      DIAM(I) = Val(DataGridView1.Rows(I - 1).Cells("DCol").Value)
      XLEN(I) = Val(DataGridView1.Rows(I - 1).Cells("LCol").Value)
      LNODS(I, 1) = Val(DataGridView1.Rows(I - 1).Cells("E1Col").Value)
      LNODS(I, 2) = Val(DataGridView1.Rows(I - 1).Cells("E2Col").Value)
      Print(2, DIAM(I), XLEN(I), LNODS(I, 1), LNODS(I, 2))
    Next
    'INPUT EXTERNAL NODAL FLOWS
    For I = 1 To NPFLOW
```

```
            nodeq = Val(DataGridView2.Rows(I - 1).Cells("NCol").Value)
            Q(nodeq) = Val(DataGridView2.Rows(I - 1).Cells("ECol").Value)
            Print(2, NODEQ, Q(NODEQ))
        Next
        'INPUT KNOWN HEAD AT NODES
        For I = 1 To NVFIX
            nodeH(I) = Val(DataGridView3.Rows(I - 1).Cells("NCol").Value)
            head(nodeh(I)) = Val(DataGridView3.Rows(I - 1).Cells("HCol").Value)
            Print(2, NODEH(I), HEAD(NODEH(I)))
        Next
        FileClose(2)
    End Sub

    Private Sub Button3_Click(ByVal sender As System.Object, ByVal e As System.EventArgs)
Handles Button3.Click
        saveData()
        Me.Hide()
    End Sub
End Class
```

### (C) LISTING OF MODULE1 (CHAP3.12\MODULE1.VB)

```
Module Module1
    Public NELEM, NNODE, NVFIX, NPFLOW, ROUGHNESS As Double
    Public ITER, CNVRT, L1, L2, FF, BK, Re, Q1, viscous, CT, H1 As Double
    Public DIAM(), XLEN(), AK(), LNODS(,), HEAD(), H(), Q(), C(), F()
    Public D(), NODEH(), Z(,), DELH(), QO(), CO(), R(), DH()
    Public EPSILON, NODEQ As Double
    Public saveFileName As String
    Public PI, G As Double
    Public met$, unit$
End Module
```

### 3.7.6.3 Advantages of the Finite Element Method

The major advantages of the finite element method over the Hardy Cross method include the following (Abdel-Magid 1987, Abdel-Magid et al. 1991):

1. The speed of convergence and the apparent lack of convergence problems
2. No need for an initial guess of flow distribution
3. Flexibility in applying boundary conditions
4. Ease of modifying and extending network without interrupting the whole system
5. Absence of artificial loops
6. The choice of flow–head loss relationships
7. Ease of input data
8. The ability to account for temperature effects
9. The unlimited network size (depending only on computer storage capacity)

The effect of pumps, boosters, pressure-reducing valves, nonreturn valves, and so on can easily be incorporated in the program if their actual head–flow relationships are known.

## 3.8 STRUCTURAL DESIGN OF STORAGE TANKS

The structural design procedure involves two steps:

1. *Analysis of forces*: Finding the distribution of the vertical and horizontal bending moments and the axial forces in the tank
2. *Design of sections*: Proportioning the sections and determining the necessary amounts of steel required

### 3.8.1 ANALYSIS OF FORCES

#### 3.8.1.1 Circular Tanks

Assuming that the thickness of the walls of the tank is small in comparison with the overall dimensions of the tank, the tank can be analyzed as a cylindrical shell. Considering a tank with a uniform wall thickness under lateral fluid pressure, the governing equilibrium equation in terms of the lateral deflection of the tank $w$ is given by the following differential equation (Reddy 2006):

$$\frac{d^4w}{dx^4} + 4\beta^4 w = \frac{-y(H-x)}{D} \qquad (3.60)$$

where:

$w$ is the lateral deflection of the tank
$x$ is the distance along the height of the tank measured from its base
$H$ is the height of the tank
$\gamma$ is the density of liquid, $KN/m^3$
$\beta$ is a constant given by the following equation:

$$\beta^4 = \frac{3\left(1-v^2\right)}{a^2 t^2} \qquad (3.61)$$

where:

$v$ is the Poisson's ratio of the material of the tank
$a$ is the radius of the tank
$t$ is the thickness of the walls of the tank
$D$ is the flexural rigidity of the tank, given by the following equation:

$$D = \frac{Et^3}{\left[12\left(1-v^2\right)\right]} \qquad (3.62)$$

where $E$ is the Young's modulus of the material of the tank, $KN/m^2$.

For a tank with its walls fixed to the base, a solution to Equation 3.62 is given by the following equation:

$$w = e^{-\beta x}\left(C_1\cos\beta x + C_2\sin\beta x\right) - \frac{\left(H-x\right)a^2}{Et} \qquad (3.63)$$

The constants $C_1$ and $C_2$ are obtained by applying the boundary conditions and are given by the following equations:

$$C_1 = \frac{\gamma a^2 H}{\left(Et\right)} \qquad (3.64)$$

$$C_2 = \gamma a^2 (H - 1/\beta) / (Et) \qquad (3.65)$$

Using the solution given by Equation 3.64, together with the strain displacement relationship and Hook's law, the bending moments in the walls of the tank can be found as shown in Equation 3.66:

$$M_x = -\frac{Dd^2w}{dx^2}$$

$$= \gamma aHt\left[-\varsigma + \left(1 - \frac{1}{(\beta H)}\theta\right)\right] \bigg/ \left(12\left(1-v^2\right)\right)^{0.5} \qquad (3.66)$$

where $M_x$ is the vertical moment on the wall per unit width, KNm.

$$\varsigma = e^{''x}\sin\beta x \qquad (3.67)$$

$$\theta = e^{-x}\cos\beta x \qquad (3.68)$$

The horizontal moment $M_y$ on the walls per unit width is given by the following equation:

$$M_y = vM_x \qquad (3.69)$$

where $v$ is the Poisson's ratio of the material of the tank.

The axial thrust on the walls of the tank per unit width $N_y$, which acts in the horizontal direction, is given by the following equation:

$$N_y = \frac{-Etw}{a} \qquad (3.70)$$

The maximum shear force at the foot of the walls will be given by the following equation:

$$Q_0 = -aHt\left(2\beta - \frac{1}{H}\right)\bigg/\left[12\left(1-v^2\right)\right] \qquad (3.71)$$

## Example 3.13

1. Write a computer program to compute the bending moments, axial thrust, and shearing forces in a circular tank of height $H$, radius $R$, and wall thickness $t$. The tank retains a fluid with density $\gamma$.
2. A cylindrical tank with radius of 9 m and height of 8 m has a wall thickness of 0.35 m and retains a liquid with a density of 9.99 $KN/m^3$. Assuming Poisson's ratio of 0.25, compute the bending moments, axial thrust, and shearing forces at midheight and at the base of the walls.

### Solution

1. For solution to Example 3.13 (1), see the listing of Program 3.13.
2. Using the data with Program 3.13, the following results are obtained:
   a. At a depth of 4 m: $M_x = 4.17$ kNm/m, $M_y = -1.04$ kNm/m, $N = 390.41$ kN/m, $Q = 5.39$ KN/m.
   b. At a depth of 8 m: $M_x = 62.2$ kNm/m, $M_y = 15.5$ kNm/m, $N = 0$, $Q = 100.15$ KN/m.

### LISTING OF PROGRAM 3.13    (CHAP3.13\FORM1.VB): ANALYSIS OF CIRCULAR TANKS

```vb
'******************************************************************************
'Program 3.13: Analysis of Circular Tanks
'******************************************************************************
Imports System.Math
Public Class Form1
   Dim H, t, D, gamma, Mu, y As Double
   Private Sub Form1_Load(ByVal sender As System.Object, ByVal e As System.EventArgs)
Handles MyBase.Load
      Me.Text = "Program 3.13: Analysis of Circular Tanks"
      Me.FormBorderStyle = Windows.Forms.FormBorderStyle.FixedSingle
      Me.MaximizeBox = False
      Label1.Text = "Enter the height of tank, H (m):"
      Label2.Text = "Enter the thickness of wall, t (m):"
      Label3.Text = "Enter the diameter of the tank, Dt (m):"
      Label4.Text = "Enter the density of the liquid, (KN/m3):"
      Label5.Text = "Enter Poisson's ratio of the material of tank:"
      Label6.Text = "Enter depth at which forces are required, x(m):"
      Button1.Text = "Load another section"
      Button2.Text = "Calculate"
      TextBox7.Multiline = True
      TextBox7.Height = 144
      TextBox7.ScrollBars = ScrollBars.Vertical
   End Sub

   Sub calculateResults()
      Dim decimals As Integer = 2
      H = Val(TextBox1.Text)
      t = Val(TextBox2.Text)
      D = Val(TextBox3.Text)
      gamma = Val(TextBox4.Text)
      Mu = Val(TextBox5.Text)
      y = Val(TextBox6.Text)
      Dim X, R, b, bh, c0, c1, c2, N, M, V As Double
      X = H - y
      R = D / 2
      b = (3 * (1 - (Mu ^ 2))) / ((R ^ 2) * (t ^ 2))
      b = (b ^ 0.25)
      bh = b * H
      c0 = (12 * (1 - Mu ^ 2)) ^ 0.5
      Dim theta As Double = Exp(-X) * Cos(b * X)
      N = gamma * R * H * (1 - X / H - Exp(-b * X) * Cos(b * X) - (1 - 1 / bh) * Exp(-b * X) *
Sin(b * X))
      M = gamma * R * H * t * (-Exp(-b * X) * Sin(b * X) + (1 - 1 / bh) * Exp(-b * X) *
Cos(b * X)) / c0
      c1 = 1 / bh
      c2 = -(2 - (1 / bh))
      V = -gamma * R * H * t / c0 * (b * Exp(-b * X)) * (c1 * Sin(b * X) + c2 * Cos(b *
X))
      Dim _My As Double = Mu * M
      Dim s As String
      s = "At a depth = " + y.ToString + " m from the top of the tank:" + vbCrLf
      s += "Vertical moment Mx = " + FormatNumber(M, decimals) + " KNm/m" + vbCrLf
      s += "Horizontal moment My = " + FormatNumber(_My, decimals) + " KNm/m" + vbCrLf
      s += "Axial ring force N = " + FormatNumber(N, decimals) + " KN/m" + vbCrLf
      s += "Shearing force V = " + FormatNumber(V, decimals) + " KN/m"
      TextBox7.Text = s
   End Sub
```

```
    Private Sub Button2_Click(ByVal sender As System.Object, ByVal e As System.EventArgs)
Handles Button2.Click
        calculateResults()
    End Sub

    Private Sub Button1_Click(ByVal sender As System.Object, ByVal e As System.EventArgs)
Handles Button1.Click
        TextBox6.Text = ""
        TextBox7.Text = ""
        TextBox6.Focus()
    End Sub
End Class
```

### 3.8.1.2  Rectangular Tanks

Rectangular tanks can be divided into two types: deep and shallow tanks.

*Deep tanks*: Here, the height of the walls $H$ is greater than twice the lateral dimensions, that is, $H > 2L_x$ and $H > 2L_y$, where $L_x$ and $L_y$ are the plan length and width of the tank, respectively. Such tanks will resist the applied forces primarily horizontally. The moments in the horizontal strips of the walls can be found using the equation of three moments as shown in Equations 3.72 through 3.74.

At the center of the longer side $L_x$:

$$M_{x1} = \frac{\left(0.5a^2 + a - 1\right)PL_y^2}{12} \tag{3.72}$$

At the center of the shorter side $L_y$:

$$M_{x2} = \frac{\left(0.5 + 5a - a^2\right)PL_y^2}{12} \tag{3.73}$$

where:
$M_{x1}$ and $M_{x2}$ are the horizontal moments at the center of the long and short sides of the tank, respectively
$a$ is the side ratio = $L_x/L_y$
$P$ is the fluid pressure at any depth $z$ from the top of the tank = $\gamma z$
$\gamma$ is the density of the retained material
$z$ is the depth at which the moments and forces are calculated

The horizontal moment at the junction between the walls of the tank $M_s$ will be opposite in direction to both $M_{x1}$ and $M_{x2}$ and will be given by the following equation:

$$M_s = \left(1 - a - a^2\right)PL_y^2 \tag{3.74}$$

The moment on the vertical strips of the walls of the tank $M_y$ per unit width of the walls can be calculated from the following equations:

$$M_{y\,max}^{-} = \frac{-\gamma HL_y^2}{24} \tag{3.75}$$

$$M_{y\,max}^{+} = \frac{+\gamma HL_y^2}{12} \tag{3.76}$$

The axial force in the wall acts horizontally and is equal to the reaction on the other wall. For the side with length $L_x$, the axial force is as given by the following equation:

$$N_{x1} = \frac{PL_y}{2} \tag{3.77}$$

And that on the side with length $L_y$ is given by the following equation:

$$N_{x2} = \frac{PL_x}{2} \tag{3.78}$$

where $P$ is the pressure of the fluid at the depth under consideration.

*Shallow tanks*: Here, the depth $H < 2L_x$ and $H < 2L_y$. In such a case, the tank walls act as two-dimensional two-way plates. The analysis of the walls in this case requires the solution of the equilibrium equation for plates (Reddy 2006):

$$\frac{\partial^4 w}{\partial x^4} + \frac{\partial^4 w}{\partial x^2 \partial^2} + \frac{\partial^4 w}{\partial y^4} = \frac{-q}{D} \tag{3.79}$$

where:
$w$ is the lateral deflection of the tank walls
$q$ is the lateral pressure on the walls
$D$ is the flexural rigidity of the walls of the tank

The usual procedure is to assume each plate to be completely fixed to its neighboring plate, whereas its top can be free or simply supported. The moments are then found by solving Equation 3.79. Various methods exist for the solution, which includes the approximate elasticity methods described by Timoshinko and Krieger (Timoshinko and Krieger 1989), or the numerical approximate methods of the finite differences (Reddy 2006), or the finite elements (Arya 2002). The moments found in this way at the edges of the plates will not be in equilibrium and can be distributed using the stiffness of the mating walls in the normal way

(Obrien and Dixon 2012). Details of the method of solution for Equation 3.79 using the finite difference method are given in Chapter 5. The method is suitable for both shallow and deep tanks (see Program 5.5).

## Example 3.14

1. Write a computer program to analyze a rectangular tank, given its plan dimensions $L_x$ and $L_y$, the wall thickness $t$, its depth $H$, and the density of the retained material $\gamma$.
2. Use the program developed in (1) to analyze a rectangular tank of plan dimensions 3 m × 2 m with a depth of 8 m and a wall thickness of 0.2 m,

at a depth of 4 m and at a depth of 8 m. The material retained has a density of 9.81 KN/m³. Take Poisson's ratio = 0.25.

### Solution

1. For solution to Example 3.14, see the listing of Program 3.14.
2. Using the following data with the program: $H = 8$ m, $t = 0.2$ m, $L_x = 3$ m, $L_y = 2$ m, $\gamma = 9.81$ KN/m³, and Poisson's ratio = 0.25, we get the following results: at a depth of 4 m: $M_x = 21.255$ KNm/m, $M_y = -3.27$ KNm/m, $M_c = -35.97$ KNm/m, $N_x = 39.24$, $N_y = 58.86$ NK/m; at a depth of 8 m, $M_x = 42.51$ KNm/m, $M_y = -6.54$ KNm/m, $M_c = -71.48$ KNm/m $N_x = 78.48$ KN/m, $N_y = 117.72$ KN/m.

---

**LISTING OF PROGRAM 3.14    (CHAP3.14\FORM1.VB): ANALYSIS OF RECTANGULAR TANKS**

```vb
'*********************************************************************************
'Program 3.14: Analysis of Rectangular Tanks
'*********************************************************************************

Imports System.Math
Public Class Form1
  Dim H, t, Lx, Ly, gama, mu, y As Double

  Private Sub Form1_Load(ByVal sender As System.Object, ByVal e As System.EventArgs)
Handles MyBase.Load
      Me.Text = "Program 3.14: Analysis of Rectangular Tanks"
      Me.FormBorderStyle = Windows.Forms.FormBorderStyle.FixedSingle
      Me.MaximizeBox = False
      Label1.Text = "Enter the height of tank, H (m):"
      Label2.Text = "Enter the thickness of wall, t (m):"
      Label3.Text = "Enter the length of the tank, Lx (m):"
      Label4.Text = "Enter the width of the tank, Ly (m):"
      Label5.Text = "Enter the density of the liquid (KN/m3):"
      Label6.Text = "Enter Poisson's ratio of the material of tank:"
      Label7.Text = "Enter depth at which forces are required, x (m):"
      Button1.Text = "Load another section"
      Button2.Text = "Calculate"
      TextBox7.Multiline = True
      TextBox7.Height = 144
      TextBox7.ScrollBars = ScrollBars.Vertical
  End Sub


  Sub calculateResults()
    H = Val(textbox1.text)
    t = Val(textbox2.text)
    Lx = Val(TextBox3.Text)
    Ly = Val(TextBox4.Text)
    gama = Val(TextBox5.Text)
    mu = Val(TextBox6.Text)
    y = Val(TextBox8.Text)
    Dim a, P, Mx1, Mx2, Mc, Nx, Ny As Double
    a = Lx / Ly
    P = gama * y
    Mx1 = (0.5 * a ^ 2 + a - 1) * P * Ly ^ 2 / 12
    Mx2 = (0.5 + a - a ^ 2) * P * Ly ^ 2 / 12
    Mc = (1 - a - a ^ 2) * P * Ly ^ 2 / 12
    Nx = P * Ly / 2
```

```
        Ny = P * Lx / 2
        Dim s As String
        s = "At a depth = " + y.ToString + " m from the top of the tank:" + vbCrLf
        s += "Horizontal moment at the center of the long side Mx = " + Mx1.ToString + "
KNm/m" + vbCrLf
        s += "Horizontal moment at the center of the short side My = " + Mx2.ToString + "
KNm/m" + vbCrLf
        s += "Horizontal moment at the joints of the sides Mc = " + Mc.ToString + " KNm/m" +
vbCrLf
        s += "Horizontal thrust on the long sides Nx = " + Nx.ToString + " KN/m" + vbCrLf
        s += "Horizontal thrust on the short sides Ny = " + Ny.ToString + " KN/m"
        TextBox7.Text = s
    End Sub

    Private Sub Button2_Click(ByVal sender As System.Object, ByVal e As System.EventArgs)
Handles Button2.Click
        calculateResults()
    End Sub

    Private Sub Button1_Click(ByVal sender As System.Object, ByVal e As System.EventArgs)
Handles Button1.Click
        TextBox8.Text = ""
        TextBox7.Text = ""
        TextBox8.Focus()
    End Sub
End Class
```

### 3.8.2   Design of the Section

For a section subjected to a bending moment $M$, the amount of steel needed can be computed from the following equation:

$$Ast = \frac{M}{(0.87 f_y Z)} \qquad (3.80)$$

where:

Ast is the area of steel needed, $mm^2$

$M$ is the applied bending moment, Nmm

$f_y$ is the yield strength of the steel, $N/mm^2$

$Z$ is the level arm, mm

For a certain moment $M$, the lever arm $Z$ can be obtained from the following equation:

$$Z = d\left[0.5 + \sqrt{(0.25 - K / 0.9)}\right] \qquad (3.81)$$

where $d$ is the effective depth of the section (mm), that is, total depth minus the cover to steel.

$$K = M / (f_{cu} * b * d_2)$$

where:

$M$ is the bending moment, N*mm

$b$ is the breadth of the section, mm

$d$ is the effective depth, mm

$f_{cu}$ is the concrete strength, $N/mm^2$

If the section is subjected to an axial tensile force $N$ in addition to the bending moment $M$ (as is the case with the horizontal strips in tanks), the total area of steel can be obtained from the following equation:

$$Ast = \frac{N}{fst} + \left[M - N * (d - 0.5 * t)\right] / (fst * Z) \qquad (3.82)$$

where:

$N$ is the axial tensile force on the section, N

$fst$ is the design steel stress, $N/mm^2$

$M$ is the bending moment on the section, Nmm

$d$ is the effective depth of the section, mm

$t$ is the thickness of the section, mm

$Z$ is the lever arm given by Equation 3.81, mm

To cover the effects of thermal and shrinkage, a check for cracks can be made. However, most codes of practice recommend that the area of the steel in any section must not be less than $Ast_{min}$, given by the following equation:

$$Ast_{min} = F * b * t \qquad (3.83)$$

where:

$b$ is the breadth of section, mm

$t$ is the thickness of walls, mm

$F$ = 0.0024 for mild steel reinforcement (i.e., grade 250); = 0.0015 for high-yield steel reinforcement (i.e., grade 460)

## Example 3.15

1. Write a computer program to design the rein-forcement for a section of a concrete tank for a certain moment $M$ and axial tension $N$, given the wall thickness $t$, the concrete strength (grade of concrete) $f_{cu}$, and the steel yield strength $f_y$.
2. Use the program developed in (1) to design a concrete tank section subject to a vertical bending moment $M_x = 4.17$ KNm/m, horizontal bending moment of 1.04 KNm/m, an axial (ring) tension of 390.41 KN/m, and a shearing force of 5.39 KN/m. Take $f_{cu} = 30$ N/mm², $f_y = 250$ N/mm²,

and a wall thickness of 0.2 m, assuming severe exposure conditions.

### Solution

1. For solution to Example 3.15 (1), see the listing of Program 3.15.
2. Input for the program will be in the following form: $f_{cu} = 30$, $f_y = 250$, $M_h = 1.04$, $M_v = 4.17$, $N = 390$, $V = 5.39$, $t = 0.2$, severe exposure conditions. The following results will be obtained: cover to steel $= 40$ mm, horizontal steel $= 1739.1$ mm², and vertical steel $= 300$ mm².

---

### LISTING OF PROGRAM 3.15 (CHAP3.15\FORM1.VB): STRUCTURAL DESIGN OF TANKS

```
'****************************************************************************************
'Program 3.15: Structural design of tanks
'****************************************************************************************

Imports System.Math
Public Class Form1
    Public d(3), A(4), spac(4), NB(4)
    Public CV, fcu, fy, Mh, Mv, N, V, t As Double
    Public Astv, Ra, Vc, fu, Stress As Double

    Private Sub Form1_Load(ByVal sender As System.Object, ByVal e As System.EventArgs)
Handles MyBase.Load
        Me.Text = "Program 3.15: Structural design of tanks"
        Label1.Text = "Enter grade of concrete, fcu (N/mm2):"
        Label2.Text = "Enter the steel yield stress, fy (N/mm2):"
        Label3.Text = "Enter the horizontal moment, Mh (KNm/m):"
        Label4.Text = "Enter the vertical moment, Mv (KNm/m):"
        Label5.Text = "Enter the ring tension, N (KN/m):"
        Label6.Text = "Enter the maximum shear, V (KN/m):"
        Label7.Text = "Enter the thickness of the wall, t (m):"
        Label8.Text = "Select exposure conditions:"
        Button1.Text = "&Calculate"
        TextBox8.Multiline = True
        TextBox8.Height = 114
        TextBox8.ScrollBars = ScrollBars.Vertical
        ComboBox1.Items.Clear()
        ComboBox1.Items.Add("Mild")
        ComboBox1.Items.Add("Moderate")
        ComboBox1.Items.Add("Severe")
        ComboBox1.Items.Add("Very severe")
        ComboBox1.Items.Add("Extreme")
        ComboBox1.SelectedIndex = 0
    End Sub


    Sub calculateResults()
        fcu = Val(TextBox1.Text)
        fy = Val(TextBox2.Text)
        Mh = Val(TextBox3.Text)
        Mv = Val(TextBox4.Text)
        N = Val(TextBox5.Text)
        V = Val(TextBox6.Text)
        t = Val(TextBox7.Text)
```

```
      If t < 0.2 Then
        MsgBox("Please enter 't' value more than 0.2", vbOKOnly + vbInformation, "Prompt")
        Exit Sub
      End If
      Dim K, Z, Ast, Ast2, Br, Astmin, Fst, maxMu, d As Double
      'determine the cover to steel reinforcement
      Dim index = ComboBox1.SelectedIndex
      If index = 0 Then
        If fcu <= 30 Then CV = 25
        If fcu >= 35 Then CV = 20
      ElseIf index = 1 Then
        If fcu <= 35 Then CV = 35
        If fcu = 40 Then CV = 30
        If fcu = 45 Then CV = 25
        If fcu = 50 Then CV = 20
      ElseIf index = 2 Then
        If fcu <= 40 Then CV = 40
        If fcu = 45 Then CV = 30
        If fcu = 50 Then CV = 25
      ElseIf index = 3 Then
        If fcu <= 40 Then CV = 50
        If fcu = 45 Then CV = 40
        If fcu = 50 Then CV = 30
      ElseIf index = 4 Then
        If fcu <= 45 Then CV = 60
        If fcu = 50 Then CV = 50
      Else
        MsgBox("Please select exposure condition.",
          vbOKOnly Or vbInformation)
        Exit Sub
      End If
      Br = 1000 '1m width
      t = t * 1000 'convert to mm
      d = t - CV - 10 'effective depth, assuming 20mm bars
      Astmin = 0.0015 * br * t 'minimum area of steel for grade 250 steel
      If fy > 250 Then Astmin = 0.0024 * br * t 'for grade 460 steel
      Mh = Abs(Mh) * 10 ^ 6 'convert to KNmm
      Mv = Abs(Mv) * 10 ^ 6 'convert to KNmm
      N = Abs(N) * 10 ^ 3 'convert to KN
      V = Abs(V) * 10 ^ 3 'convert to KN
      Fst = 0.87 * fy 'design steel stress
RECYCLE:
      'Horizontal steel
      maxMu = 0.156 * fcu * br * d ^ 2
      Do
        d = d + 100
        t = t + 100
      Loop While Mh > maxMu
      K = Mh / (fcu * br * d ^ 2)
      Z = d * (0.5 + (0.25 - K / 0.9) ^ 0.5)
      If z > 0.94 * d Then z = 0.94 * d
      Ast = N / fst
      Ast2 = (Mh - N * (d - t / 2)) / (fst * z)
      If Ast2 > 0 Then Ast = Ast + Ast2
      If Ast < Astmin Then Ast = Astmin
      'Vertical steel
      Do
        d = d + 100
        t = t + 100
      Loop While Mv > maxMu
      K = Mv / (fcu * br * d ^ 2)
```

```
    Z = d * (0.5 + (0.25 - K / 0.9) ^ 0.5)
    If z > 0.94 * d Then z = 0.94 * d
    Astv = Mv / (fst * z)
    If Astv < Astmin Then Astv = Astmin
    'Check for shear stress
    Ra = 100 * Ast / (Br * d)
    If d > 400 Then d = 400
    Vc = 0.79 * Ra ^ (1 / 3) * (400 / d) ^ 0.25 / 1.25
    If fu > 40 Then fu = 40
    If fu > 25 Then Vc = Vc * (fu / 25) ^ (1 / 3)
    Stress = V / (br * d)
    If stress > Vc Then
       d = d * V / Vc * 1.1
       t = d + CV + 10
       GoTo RECYCLE
    End If
    Dim Mmax, Asm, Dr, Ec, Es, Alpha, R, X, Fsa, Fs As Double
    'check for cracking at serviceability limit state
    N = N / 1.5 'service ring tension
    Mmax = Mh
    Asm = Ast
    Dr = 1
    If Mmax < Mv Then
       Mmax = Mv
       Asm = Astv
       Dr = 2
    End If
    Mmax = Mmax / 1.5 'service maximum moment
    Ec = 4500 * fcu ^ 0.5 'Young's modulus of concrete
    Ec = 0.5 * Ec 'service Youngs modulus
    Es = 2 * 10 ^ 5 'Youngs modulus for steel
    Alpha = Es / Ec 'modular ratio
    R = Alpha * Asm / (Br * d) 'steel ratio
    X = d * R * ((1 + 2 / R) ^ 0.5 - 1)
    Z = d - X / 3 'level arm
    Fs = Mmax / (Asm * z) 'service stress in steel
    Fsa = 0.58 * fy 'allowable service stress
    If fs > fsa Then
       'cracking is excessive, so increase area of steel
       Dim Fact = Fs / Fsa
       If dr = 1 Then Ast = Ast * fact Else Astv = Astv * fact
    End If
    Dim s As String
    Dim decimals As Integer = 2
    s = "The cover to reinforcement (mm) = " + FormatNumber(CV, decimals) + vbCrLf
    s += "Area of horizontal steel (sq.mm/m): " + FormatNumber(Ast, decimals) + vbCrLf
    s += "Area of vertical steel (sq.mm/m): " + FormatNumber(Astv, decimals) + vbCrLf
    s += "Shear stress (N/mm2): " + FormatNumber(Stress, decimals) + vbCrLf
    s += "Service steel stress (N/mm3): " + FormatNumber(Fs, decimals) + " O.K.: No
Cracking" + vbCrLf
    s += "Final depth of wall required (mm): " + FormatNumber(t, decimals) + vbCrLf
    TextBox8.Text = s
  End Sub

  Private Sub Button1_Click(ByVal sender As System.Object, ByVal e As System.EventArgs)
Handles Button1.Click
    calculateResults()
  End Sub
End Class
```

## 3.9 HOMEWORK PROBLEMS IN WATER RESOURCES, WATER STORAGE, AND WATER DISTRIBUTION

### 3.9.1 DISCUSSION PROBLEMS

1. What are the main categories into which water usage may be grouped?
2. What are the factors that influence fire demand?
3. Define the following terms: transmissibility coefficient, confined aquifer, unconfined aquifer, drawdown, and radius of influence.
4. Compare between confined, unconfined, and perched aquifers in accordance with items outlined in the following table. Illustrate your answer with suitable sketches.

| Item | Unconfined | Confined |
|---|---|---|
| Water movement | | |
| Possibility of contamination | | |
| Pressure | | |
| Depth | | |
| Recharge and replenishment | | |
| Groundwater age | | |
| Typical examples | | |

5. Define the term "water foot print." Indicate how to measure its value for your country.
6. How do you estimate future population in your locality with the available data?
7. In the following table, give two examples of vital standards related to water and its use, describe the most essential influencing parameters and responsible authority within a certain municipality (Dammam University, B.Sc. exam 2013).

### Table of Examples of Standards Related to Water

| Standard | Affecting Factors | Responsible Body |
|---|---|---|
| | | |

8. In the following table, outline the importance of each term as related to water supply from an engineering prospective (Dammam University, B.Sc. exam 2013).

### Table of Water Supply Engineering Issues

| Item | Significance |
|---|---|
| Hand as a germ farm | |
| Concentration of desalination plants along the Arabian Gulf | |
| Water awareness message highlighted in a municipality water storage tank | |
| Hand washing | |
| KSA hydrogeological map | |

*(Continued)*

### Table of Water Supply Engineering Issues

| Item | Significance |
|---|---|
| Pipe material in KSA | |
| WHO drinking water guidelines | |
| Personal distance | |
| Wash hand basin water reuse | |
| Sampling techniques | |
| Water leakage rate | |
| Water footprints | |

9. Compare between ground water and surface water from different points of view as per activity parameters shown in the following table:

| Activity | Groundwater | Surface Water |
|---|---|---|
| Natural recharge | | |
| Availability | | |
| Quality | | |
| Production cost | | |
| Storage | | |
| Properties | | |

10. Why is water storage needed?
11. Outline the classification used for water storage reservoirs.
12. What are the differences between an elevated water storage tower and a stand pipe?
13. What parameters govern the size of storage reservoirs?
14. Define the following terms: mass curve and water supply network.
15. Show how to estimate the required reservoir storage capacity in order to maintain uniform user flow.
16. Rate accuracy of a rating (mass, $S$-curve, Rippll diagram) curve and its representation of the relationship between measured water level (at a particular measuring station) and the discharge passing.
17. What are the main types of pipeline patterns? Indicate the advantages and disadvantages of each system.
18. What is the difference between a pipe and a duct?
19. Differentiate between laminar, transitional, and turbulent flows.
20. Define the following numbers: Reynolds number, Froude number, and Weber number. Show that each number is a dimensionless quantity. Show examples and the use of each of these numbers.
21. Differentiate between steady-state and non-steady-state flow.
22. What are the differences between compressible and incompressible flows?
23. Define the following: shear wall, eddy viscosity, dimensional analysis, and network loop.

24. What are the advantages of the Moody chart? Indicate the range of the validity of the Moody chart.

25. What is the value of the equivalent-velocity-head method in analysis of pipe flow? Indicate the differences between this method and the method of equivalent length.

26. Outline the classification of loops in pipe networks. What are the differences between them?

27. What are the disadvantages of analyzing networks by the Bernoulli and continuity equations?

28. Define the method of successive approximations and finite element method. Indicate their main differences.

29. Indicate the assumptions made in the method of successive approximations. Show how to correct these assumptions.

30. Briefly describe the Hardy Cross method of network analysis. Indicate its disadvantages and limitations.

31. What are the required data needed for appropriate modeling of a water supply network?

32. What are the steps involved in the structural design of a water storage tank?

33. How can you structurally differentiate between a cylindrical shell, a deep rectangular tank, and a shallow rectangular tank?

34. Indicate the assumptions made in the structural design of a water storage tank.

35. Briefly describe the following methods for determining moments: approximate elasticity method, numerical approximate methods of finite differences, and finite elements method.

36. Why is it important to check a structural design for cracks?

## 3.9.2 Specific Mathematical Problems

1. A city recorded a population of 245,000 in its earlier decennial census and 315,000 in its later one. Write a computer program that enables the computation of the midyear (1 Muharram) populations (Dammam University, B.Sc. exam 2013).
   a. For the fifth intercensal year.
   b. For the ninth postcensal year by both arithmetic and geometric increases. Assume a census date of 1 Rabi-2.
   c. Comment on your results.
   d. Select your design engineering population. State your reasons for the selection.
   e. Determine the amount of water to be supplied to the city if the average water consumption amounts to 450 L/c/day.

2. Write a computer program that enables to find the future fire demand for a town having a population of 18,000 inhabitants with a population growth rate of 2% for a design period of 15 years (Dammam University, B.Sc. exam 2013).

3. a. Using the following equation:

$$S_1 - S_2 = \frac{Q_o}{2\pi kH} \, \mathrm{Ln}\!\left(\frac{r_2}{r_1}\right), \quad S_o = \frac{Q_o}{2\pi kH} \, \mathrm{Ln}\!\left(\frac{R}{r_o}\right),$$

$$Q_o = \frac{\pi k \left(H^2 - h_o^2\right)}{\mathrm{Ln}\left(R/r_o\right)}$$

write a computer program to calculate transmissivity ($T$) of an aquifer, hydraulic conductivity ($K$) of an aquifer material, and radius of influence ($R$) of a well.

b. A 250-mm-diameter well that fully penetrates a confined aquifer of thickness 20 m is pumped at a constant rate of 1875 m³/day. The steady-state drawdown in the well is 9 m, and the drawdown in a piezometer 120 m from the well is 1.8 m, ignoring well losses (Dammam University, B.Sc. exam 2015):
   i. Calculate the transmissivity ($T$) of the aquifer.
   ii. Find the hydraulic conductivity ($K$) of the aquifer material.
   iii. Compute the radius of influence ($R$) of the well.
   iv. Repeat the calculation for an unconfined aquifer with a saturated thickness of 25 m.
   v. What would be the calculated values of $T$, $K$, and $R$ in (a, b, and c) if the well is 80% efficient?

4. a. Write a computer program to determine for a given aquifer radius of zero drawdown, coefficient of permeability, drawdown in the pumped well, transmissibility of the aquifer, and suitability of well diameter to its yield.

$$Q = \frac{\pi k \left(H^2 - h^2\right)}{\mathrm{Ln}\left(R/r\right)}$$

b. A well is 30 cm diameter and penetrates 50 m below the static water table. After 36 h of pumping at 4.0 m³/min, the water level on a test well 200 m distance is lowered by 1.2m, and in a well 40 m away, the drawdown is 2.7 m (Dammam University, B.Sc. exam 2013).
   i. Determine the radius of zero drawdown.
   ii. Find the coefficient of permeability.
   iii. Compute the drawdown in the pumped well.
   iv. What is the transmissibility of the aquifer?

c. Using expected well yield estimates as presented in the following table, comment about the yield of the well as related to its diameter, and suggest a more suitable well diameter. Explain and validate your answer.

**Table of Expected Well Yield**

| Well Diameter (cm) | Expected Well Yield (m³/day) |
|---|---|
| 15 | <500 |
| 20 | 400–1,000 |
| 25 | 800–2,000 |
| 30 | 2,000–3,500 |
| 35 | 3,000–5,000 |
| 40 | 4,500–7,000 |
| 50 | 6,500–10,000 |
| 60 | 8,500–17,000 |

5. a. Write a computer program to plot a mass curve for given river discharges. Determine the minimum capacity of a reservoir if the entire annual inflow is to be drawn off at a uniform rate (with no flow going into waste over the spillway) and to compute the amount of water, which must be initially stored to maintain the uniform draw-off.

   b. The following table is a record of the mean monthly discharges of a river in a dry year (Dammam University, B.Sc. exam 2015).

   i. Plot the mass curve for the river discharges given.

   ii. Determine the minimum capacity of a reservoir if the entire annual inflow is to be drawn off at a uniform rate (with no flow going into waste over the spillway).

   iii. Compute the amount of water that must be initially stored to maintain the uniform draw-off.

| Month | Mean Flow (cumec) |
|---|---|
| January | 42 |
| February | 88 |
| March | 79 |
| April | 69 |
| May | 35 |
| June | 38 |
| July | 80 |
| August | 62 |
| September | 87 |
| October | 79 |
| November | 53 |
| December | 38 |

6. Write a computer program to determine the pressure difference in a horizontal circular pipe, given its length, $l$ (m); diameter, $D$ (m); and the maximum shear that can develop at pipe wall, $\tau_w$ (N/m²). Check the validity of your program, given $D = 300$ mm and $\tau_w = 0.4$ kPa per unit length.

7. Write a short program to estimate minor losses in a piping system, $h_1$ (m), given the loss coefficient, $k_1$, and the velocity of flow, v (m/s). Use the relationship:

$h_1 = k_1 * v^2/2g$. Check the validity of your program for the following data: v = 15 m/s, $k_1 = 0.5$.

8. Write a computer program that enables the determination of the pressure drop in a pipe using Poiseuille's law ($\Delta P = 32\mu l v/D^2$). Let the program also find the pressure drop through the Darcy–Weisbach equation ($\Delta P = f l \rho v^2/2D$) with the friction factor found either for laminar flow ($f = 64/Re$) or for turbulent flow, $f$, determined from Colebrook's equation: $1/(f)^{1/2} = -2\log[((\varepsilon/D)/3.7) + (2.51/(Re*(f)^{1/2}))]$ and given the roughness of pipe wall, $\varepsilon$ (m); the pipe diameter, $D$ (m); the velocity of flow through pipe, v (m/s); and fluid properties (density, $\rho$ [kg/m³], dynamic viscosity, $\mu$ [N*s/m²], and kinematic viscosity, v [m²/s]). Let the program determine whether the flow is laminar, transition, or turbulent. Check the validity of your program for a gas flowing through a tube at v = 10 m/s, $D = 5$ mm, $\varepsilon = 0.002$ m, $l = 0.2$ m, $\rho = 1.33$ kg/m³, $\mu = 2 * 10^{-5}$ N*s/m², and v = $1.5 * 10^{-5}$ m²/s.

## REFERENCES

Abdel-Magid, H. 1987. *Analysis and Design of Networks Using Finite Element Method*. MSc thesis, Department of Civil Engineering, University of Khartoum, Khartoum, Sudan.

Abdel-Magid, H., Hago, A., and Abdel-Magid, I.M. 1991. Analysis of pipe networks by the finite element method. *J. Water Int.* 16(2):96–101.

Abdel-Magid, I. M. 1986. *Water Treatment and Sanitary Engineering*. Khartoum, Sudan: Khartoum University Press (Arabic). doi:10.13140/RG.2.1.1962.0883.

Abdel-Magid, I. M. and Abdel-Magid, M. I. M. 2015. *Problem Solving in Environmental Engineering*, 2nd ed. Dammam, KSA: CreateSpace Independent Publishing Platform.

Abdel-Magid, I. M. and ElHassan, B. M. 1986. *Water supply in the Sudan*. Khartoum, Sudan: Khartoum University Press, Sudan National Council for Research (Arabic). doi:10.13140/RG.2.1.2371.6882.

Abdel-Magid, I. M., Mohammed, A. H., and Rowe, D. R. 1996. *Modeling Methods for Environmental Engineers*. Boca Raton, FL: CRC Press.

Arya, C. 2002. *Design of Structural Elements: Concrete, Steelwork, Masonry and Timber Designs to British Standards and Eurocodes*, 2nd ed. New York: CRC Press.

Burden, R. L. and Faires, J. D. 2010. *Numerical Analysis*, 9th ed. Boston, MA: Cengage Learning.

Chapagain, A. K. and Hoekstra, A.Y. 2004. *Water Footprints of Nations, Volume 1: Main Report*. Value of Water: Research Report Series No. 16. Delft, the Netherlands: UNESCO-IHE Institute for Water Education.

Collins, A.G. and Johnson R. L. 1975. Finite element method for water distribution networks. *J. Am. Waterworks Assoc.* 67(7):385–389.

Currie, I.G. 2012. *Fundamental Mechanics of Fluids*, 4th ed. Boca Raton, FL : CRC Press.

Davis, M. L. and Cornwell, D. A. 2006. *Introduction to Environmental Engineering*, 4th ed. Boston, MA: McGraw-Hill Science/Engineering/Math.

Douglas, J. F., Gasiorek, J. M., Swaffield, J.A., and Jack, L. 2006. *Fluid Mechanics*, 5th ed. New York: Prentice Hall.

Epp, R. and Fowler, A. G. 1970. Efficient code for steady state flows in networks. *Proc. Am. Soc. Civil Engineers, J. Hydraulics Div.* 96(HY1):43–56.

Fair, G. M., Geyer, J. C., and Okum, D. A. 1966. *Water and Wastewater Engineering,* Vols I and II. New York: John Wiley & Sons.

Gientke, F. J. 1974. Finite element solution for flow in noncircular conduits. *Proc. Am. Soc. Civil Engineers. J. Hydraulics Div.* 100(HY3):425–442.

Global Footprint Network. 2016. *Footprint Calculator.* http://www.footprintnetwork.org/resources/footprint-calculator/ (accessed on May 20, 2016).

Grigg, N. 2012. *Water, Wastewater, and Storm Water Infrastructure Management,* 2nd ed. Boca Raton, FL: CRC Press.

Gupta, R. S. 2007. *Hydrology and Hydraulic Systems,* 3rd ed. Long Grove, IL: Waveland Press Inc.

Henry, J. G. and Heinke, G. W. 1996. *Environmental Science and Engineering,* 2nd ed. Englewood Cliffs, NJ: Prentice Hall.

Howe, K. J., Hand, D. W., Crittenden, J. C., Trussell, R. R., and Tchobanoglous, G. 2012. *Principles of Water Treatment,* 1st ed. Hoboken, NJ: Wiley.

IPHE. 1984. *The Public Health Engineering Data Book 1983/84.* Edited by Bartlett, R. London: Sterling Publishers.

Lam, C.F. and Wolla, M. L. 1972. Computer analysis of water distribution systems, Part I. formulation of equations. *Proc. Am. Soc. Civil Engineers, J. Hydraulics Div.* 98 (HY2):335–344.

Lin, S. D. 2014. *Water and Wastewater Calculations Manual,* 3rd ed. New York: McGraw-Hill Education.

Liu, K. T. 1969. *The Numerical Analysis of Water Supply Networks by a Digital Computer.* Proceedings of the IAHR 13th Congress, Kyoto, Japan, August 31–September 5. pp. 35–42.

London and General Water Purifying Company. 2012. *Water: Its Impurities and Purification.* London, UK: Nabu Press.

McCormich, M. and Bellamy, C. J. 1968. A computer program for the analysis of networks of pipes and pumps. *J. Inst. Eng.* 38(3):51–58.

McGhee, T. J. and Steel, E. W. 1991. *Water Supply and Sewerage,* 6th ed. New York: McGraw-Hill.

Munson, B. R., Rothmayer, A. P., Okiishi, T. H., and Huebsch, W. W. 2012. *Fundamentals of Fluid Mechanics,* 7th ed. Hoboken, NJ: Wiley.

Nathanson, J. A. 2007. *Basic Environmental Technology: Water Supply, Waste Management & Pollution Control,* 5th ed. New Delhi, India: Prentice Hall.

National Geographic. 2016. *Water Footprint Calculator.* http://environment.nationalgeographic.com/environment/freshwater/change-the-course/water-footprint-calculator/ (accessed on May 21, 2016).

Nemerow, N. L., Agardy, F. J., and Salvato, J. A. 2009. *Environmental Engineering,* 6th ed. New York: Wiley and Sons.

Nielsen, D. M. (Ed.) 1991. *Practical Handbook of Ground-Water Monitoring.* Boca Raton, FL: CRC Press.

Obrien, E. and Dixon, A. 2012. *Reinforced and Prestressed Concrete Design to EC2: The Complete Process,* 2nd ed. Boca Raton, FL: CRC Press.

Peavy, H. S., Rowe, D. R., and Tchobanoglous, G. 1985. *Environmental Engineering.* New York: McGraw-Hill Book Co.

Reddy, J. N. 2006. *Theory and Analysis of Elastic Plates and Shells,* 2nd ed. (Series in Systems and Control). Boca Raton, FL: CRC Press.

Ricketts, J., Loftin, M., and Merritt, F. S. 2003. *Standard Handbook for Civil Engineers.* New York: McGraw-Hill Professional.

Rowe, D. R. and Abdel-Magid, I. M. 1995. *Wastewater Reclamation and Reuse.* New York: Lewis Pub.

Shamir, U. and Howard, C. D. 1968. Water distribution systems analysis. *Proc. Am. Soc. Civil Engineers, J. Hydraulics Div.* 94 (HY1):219–234.

Smith, P. G. 2005. *Dictionary of Water and Waste Management,* 2nd ed. Burlington, MA: Butterworth-Heinemann.

Smith, P. G. and Scott, J. S. 2002. *Dictionary of Water and Waste Management.* London: International Water Association.

Spellman, F. R. 2013. *Handbook of Water and Wastewater Treatment Plant Operations,* 3rd ed. Boca Raton, FL: CRC Press.

Timoshenko, S.P. and Woinowsky-Krieger, S. 1989. *Theory of Plates and Shells,* 28th ed. McGraw-Hill international editions.

Todd, D. K. and Mays, L. W. 2004. *Groundwater Hydrology,* 3rd ed. New York: John Wiely and Sons.

Valsaraj, K. T. and Melvin, E. M. 2009. *Elements of Environmental Engineering: Thermodynamics and Kinetics,* 3rd ed. Boca Raton, FL: CRC Press.

Vesilind, P. A. and Peirce, J. J. 1997. *Environmental Pollution and Control,* 4th ed. London: Butterworth-Heinemann.

Viessman, W. and Lewis, G. L. 2002. *Introduction to Hydrology,* 5th ed. New Delhi, India: Prentice Hall.

Viessman, W., Perez, E. M., Chadik, P. A., and Hammer, M. J. 2015. *Water Supply and Pollution Control.* Upper Saddle River, NJ: Pearson.

Water Footprint Network (WFN). 2016. *Water Footprints.* http://www.waterfootprint.org/en/water-footprint/ (accessed on April 21, 2016).

White, F. 2010. *Fluid Mechanics with Student DVD,* 7th ed. New York: McGraw-Hill Science/Engineering/Math.

Wood, D. J. 1980. Slurry flow in pipe networks. *Proc. Am. Soc. Civil Engineers, J. Hydraulics Div.* 102 (HY1):57–70.

Wood, D. J. and Charles, C. O. 1972. Hydraulic network analysis using linear theory. *Proc. Am. Soc. Civil Engineers, J. Hydraulics Div.* 98 (HY7):1157–1170.

Zhang, G. P., Hoekstra, A. Y., and Mathews, R. E. 2013. Water Footprint Assessment (WFA) for better water governance and sustainable development. *Water Resources and Industry,* 1–2: 1–6.

# 4 Computer Modeling Applications for Water Treatment

This chapter outlines the concepts used in the design of treatment facilities for both unit operations and processes. Unit operations are physical in nature, whereas unit processes involve chemical and biological principles. Each treatment process, whether physical, chemical, or biological or a combination thereof, is briefly described with an emphasis being placed on the design of the units, applying well-established formulae, equations, and models that are then used to develop computer programs that can aid in the various design procedures.

## 4.1 WATER TREATMENT SYSTEMS

Generally, water treatment processes are broadly classified as follows:

1. *Physical treatment, unit operations*: Generally, these are the simplest forms of water treatment. Factors that govern treatment here are physical in nature and include particle size, specific gravity, and viscosity. Examples of such treatment units include screening, sedimentation, flotation, aeration, and filtration.
2. *Chemical treatment, unit processes*: These are capable of changing the nature of the pollutants and existing contaminants to by-products that are no longer objectionable. Examples include coagulation, color and odor removal, and disinfection.
3. *Biological treatment, unit processes*: These are capable of removing organic matter and soluble and colloidal particles in an engineered system under controlled environmental conditions. An example of such a process is biological filtration.

## 4.2 AIMS OF WATER TREATMENT

Municipal water works are required to produce a water supply that is hygienically safe (potable), aesthetically attractive (palatable), and economically satisfactory for its intended uses. Water treatment unit operations and unit processes are both used to provide a water supply that is free of the following:

- Suspended or floating substances (e.g., leaves, debris, soil particles)
- Colloidal substances (e.g., clay, silt, microorganisms)
- Dissolved solids (e.g., inorganic salts)
- Dissolved gaseous substances (e.g., carbon dioxide, hydrogen sulfide)
- Immiscible liquids, such as oils and greases
- Substances that cause corrosion or encrustation
- Odor/taste/color-producing substances

- Pathogenic organisms (bacteria, viruses, amoebas, worms, fly nymphs, Cyclops, etc.)
- Compounds (organic or inorganic) that have an adverse effect, acute or long term, on human health

Some of the first information needed in order to design an economically feasible water treatment facility includes the source of the water supply, the quality and quantity of water available, and the population to be served, as well as the commercial and industrial development in that community so that present demands and further trends can be evaluated. Design capacity must also include social and economic conditions within the community. Estimates must be made as to the design period, design population, design flow, peak flows, design area, and design hydrology.

## 4.3 SCREENING

Screening devices are installed to facilitate the removal of coarse suspended and floating matter. Presence of coarse solids may damage or interfere with the operation of the pumps and other mechanical equipment. Solids can also obstruct valves and other devices in the plant. Generally, these materials consist of leaves, rags, sticks, vegetable matter, broken stones, tree branches, boards, and other large objects that are relatively inoffensive. The coarse solids are usually removed by simple screening devices.

The minimum velocity, through a screen, that is needed to prevent the deposition of sand and suspended matter is related to size and density of the suspended impurities and to flow velocity in the channel from which water or wastewater emerges. The approach velocity of the water in the channel upstream should not be less than 0.3–0.5 m/s. This velocity is recommended to prevent settling out of suspended matter. The passing velocity of the water through the openings between the bars should not exceed 0.7–1 m/s to prevent fluffy and soft matter from being forced through the openings in the screen.

With the blockage of screen openings, resistance to flow rapidly increases. Flow resistance increases can be evaluated according to the following equation (Huisman 1977):

$$H_s = \left(\frac{a_0}{a_a}\right)^2 * H_0 \tag{4.1}$$

where:

$H_s$ is the resistance for a clogged screen, m of water
$H_0$ is the initial resistance for a clean screen, m
$a_a$ is the percentage open area for a clogged screen
$a_0$ is the percentage open area for a clean screen

The development of this head loss will produce a considerable load on the bar screen, which necessitates an adequate structural design to meet the developing hydraulic load. Frequent cleaning of the screen openings helps to limit the effects of this hydraulic load. Cleaning of openings keeps the maximum value of the resistance below a practical value of 0.5 m. In general, two screens should be provided so that one screen can be shut down, whereas the other is being cleaned or repaired.

## 4.4 SEDIMENTATION

### 4.4.1 INTRODUCTION

Sedimentation may be defined as the gravitational settling of suspended particles that are heavier than the surrounding fluid. The water to be clarified by sedimentation is held in a tank for a considerable period of time. For tanks with large cross-sectional areas, flow velocities are small. This condition induces a state of virtual quiescence (Huisman 1977). As such, heavier particles (with a mass density greater than that of the surrounding fluid) start settling under gravitational forces (a process referred to as sedimentation; see Figure 4.1). Lighter particles (with a mass density less than that of the surrounding fluid) tend to move vertically upward (a process denoted as flotation; see Figure 4.1). Accordingly, suspended particles are retained either in the scum layer (top layer) at the water surface or in the sludge layer at the tank floor (bottom layer). Removal of settled and floatable particles clarifies the raw water. With the same capacity and tank volume, long, narrow, and shallow basins have greatest potential for solids removal.

The aims of the sedimentation process include the following:

- Removal of grit (grit chamber)
- Reduction of content of particulate solids, both settleable and floatable (primary settling)
- Reduction of $BOD_5$ (primary clarifier)
- Removal of chemical floc (chemical coagulation)
- Removal of biological floc (activated sludge system)
- Separation of solids from mixed liquor (activated sludge)

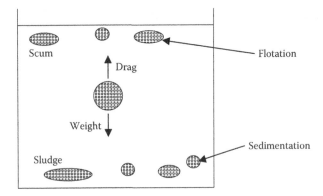

**FIGURE 4.1** Sedimentation and flotation.

- Production of a concentrated return sludge flow to sustain biological treatment (secondary clarifiers)
- Increase of solids concentration (sludge thickeners and dewatering units)
- Provision of recirculated water flow (high-rate trickling filtration)

Sedimentation can be improved by adding coagulant aids, such as lime, soda ash, NaOH, $H_2SO_4$, clay, bentonite, powdered stone, or polyelectrolytes. Flotation can be facilitated by bubbling air or chlorine gas into the liquid at the bottom of the sedimentation tank.

Sedimentation tanks may be constructed as rectangular, circular, or square in plan. The water in the tank may be at rest or continuously flowing, either vertically or horizontally. Figure 4.2 illustrates the two types of sedimentation tanks. Square or circular basins may be preferred, for one or all of the following reasons:

- They make better use of the land area allocated for the project (multiple rectangular tanks require less area than multiple circular tanks).
- They offer savings in cost.
- They use smaller amounts of raw construction materials.
- They allow use of more durable materials (e.g., pre-stressed concrete).

Best results can be obtained with vertical flow basins, which have large depths and inlet structures that spread the incoming water equally over the entire tank plan area (Huisman et al 1986, Wang and Hung 2009, Hammer and Hammer 2011, Riffat 2012).

### 4.4.2 TYPES OF SETTLING PHENOMENON

The settling phenomenon is governed by the particle shapes, sizes, and characteristics. Particle settling is often classified as follows:

*Class I: Discrete particle settling.* Under discrete particle settling, nonflocculating conditions develop, and particles maintain their identity throughout the settling process with a constant velocity, that is, particles experience no change in their size, shape, or density. General features of discrete settling can be summarized as follows (Viessman et al. 2008):

- Particles settle as individual entities.
- No significant interaction occurs between adjacent particles.
- Settling is unhindered by the basin walls.

This class of particle settling can be represented by the settling of sand grains in water treatment or the removal of grit in grit chambers in the field of wastewater treatment.

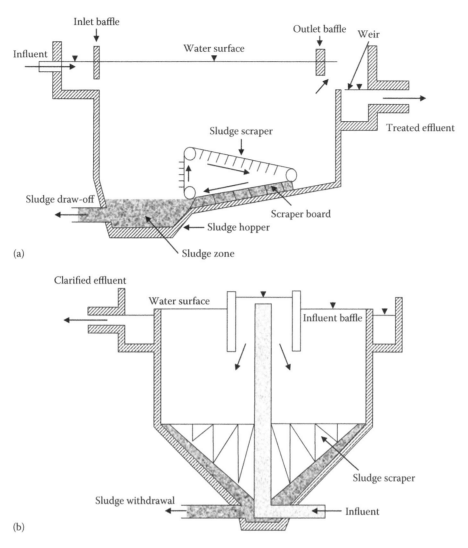

**FIGURE 4.2** Sedimentation tanks: (a) schematic diagram of a rectangular horizontal sedimentation tank; (b) schematic diagram of a circular sedimentation tank.

*Class II: Flocculent settling.* Under flocculent settling conditions, particles flocculate during sedimentation, that is, they experience a change in their size, weight, and shape. The main features of this class of particle settling include the following:

- Particles coalesce together or flocculate.
- Particles are hindered by their neighbors.
- Due to the close proximity of particles, the change in their size is mainly due to particle contact.
- Particles increase in mass with depth and settling time.
- Particles settle rapidly.

Examples of this class of settling occur in alum or iron coagulation (Davis and Cornwell 2006).

*Class III: Hindered (zone) settling.* Hindered settling occurs when high-density particles interact, and individual particles are so close together that the displacement of water by

the settling of one affects the relative settling velocities of its neighbors (Steel and McGhee 1991). With the cross-sectional area of a settling tank being fixed, the water displaced by the settling particle must flow in the opposite direction to that of the moving particles. The essential features of this kind of settling include the following:

- Settling characteristics of adjacent particles are influenced by existence of interparticle forces.
- Moving particles have similar settling velocities.
- Settlement occurs for a mass of particles that are able to settle as one unit.
- Particles remain in fixed places relative to each other.

Examples of this kind of settling may be encountered in lime-softening tanks, sludge thickeners, and activated sludge sedimentation (Davis and Cornwell 2006). Compared to the stationary walls of the sedimentation tank, the settling rate under hindered settling conditions is reduced from a value of v to a value of v' as given by Equation 4.2 (Huisman 1977):

$$\frac{v'}{v} = 1 - \left(f * C_v\right)^{\frac{2}{3}} \qquad (4.2)$$

where:
  v' is the hindered settling velocity, m/s
  v is the initial settling velocity of the particle, free settling velocity, m/s
  $C_v$ is the volumetric concentration of particles (volume of particles divided by the total volume of the suspension) = $V_{particle}/V_{suspension}$
  f is the coefficient that varies with the type of particles

An estimate of the extent to which settling is hindered can also be calculated from Equation 4.3, which is valid for Reynolds numbers less than 0.2, which is generally the situation with hindered settling (Steel and McGhee 1991):

$$\frac{v'}{v} = \left(1 - C_v\right)^{4.65} \qquad (4.3)$$

where:
  v' is the hindered settling velocity, m/s
  v is the free settling velocity of the particle, m/s
  $C_v$ is the volume of particles divided by the total volume of the suspension

*Class IV: Compression settling.* In this class, the concentration of particles forms a distinct structure. Compression, under the weight of the particles, governs the settling process. This kind of settling occurs at the bottom of deep secondary settling tanks and takes place under sludge settling conditions (Metcalf and Eddy 2013). Usually, more than one type of settling occurs in a sedimentation basin.

### 4.4.3 CLASS I SETTLING

When a particle is placed in a liquid of a density lesser than its own, its settling velocity will accelerate until a limiting (terminal) velocity is reached. Beyond this velocity, the settling forces and the submerged weight of the particle are in balance with the frictional drag forces. This yields turbulent flow conditions. The relationship for this settling velocity is presented in the following equation:

$$v = \sqrt{3.3 * g * d\left(s.g. - 1\right)} \qquad (4.4)$$

where:
  v is the settling (terminal) velocity of particle, m/s
  g is the gravitational acceleration, m/s$^2$

d is the diameter of spherical particle, m
s.g. is the specific gravity of the particle

The settling velocity of a spherical particle under laminar flow conditions can be expressed by Stoke's law, which is presented in the following equation:

$$v = \frac{[g * d^2 (s.g. - 1)]}{(18 * v)} \qquad (4.5)$$

where:
  v is the settling (terminal) velocity of particle, m/s
  g is the gravitational acceleration, m/s$^2$
  d is the diameter of spherical particle, m
  s.g. is the specific gravity of the particle = $\rho_s/\rho$
  $\rho$ is the density of fluid, kg/m$^3$
  $\rho_s$ is the particle density, kg/m$^3$
  v is the kinematic velocity, m$^2$/s

$$v = \frac{\mu}{\rho} \qquad (4.6)$$

where $\mu$ is the dynamic viscosity (absolute), N*s/m$^2$

### Example 4.1

1. Write a computer program to determine the settling velocity for spherical discrete particle settling under different flow conditions (laminar or turbulent). Provide a program to check the Reynolds number for the computed overflow rate (at different temperatures).
2. Particles are allowed to settle in a settling column test. The particles have a specific gravity of 1.3 and an average diameter of 0.1 mm. Using Stoke's law, determine the settling velocity of the particles in water (the temperature of the water is 20°C).
3. Use the program developed in (1) to verify the computations in (2).

### Solution

1. For solution to Example 4.1 (1), see the listing of Program 4.1.
2. Solution to Example 4.1 (2):
   a. Given: s.g. = 1.3, d = 0.1 * 10$^{-3}$ m.
   b. Find the viscosity coefficient from Appendix A1 that corresponds to a temperature of T = 20°C as follows: $\mu$ = 1.0087 * 10$^{-3}$ N*s/m$^2$ and $\rho$ = 998.2 kg/m$^3$.
   c. Use Stoke's law to find the settling velocity of the particles, $v = \rho * g * d^2 (s.g. - 1)/(18 * \mu)$.
   d. Substitute the given values in Stoke's equation to find the settling velocity as follows: v = 998.2 * 9.81(0.1 * 10$^{-3}$)$^2$(1.3 − 1)/18 * 1.0087 * 10$^{-3}$ = 1.62 * 10$^{-3}$ m/s = 1.62 mm/s.

**LISTING OF PROGRAM 4.1 (CHAP4.1\FORM1.VB): SETTLING VELOCITY FOR SPHERICAL DISCRETE PARTICLES**

```vb
'**********************************************************************************
'Program 4.1: Settling
'computes settling velocity for spherical
'discrete particles
'**********************************************************************************
Public Class Form1
  Dim temp(34) As Integer
  Dim density(34), viscosity(34) As Double
  Const G = 9.81 'gravity constant
  Dim Velocity, sg, d, Ro, Mu As Double
  Dim T As Integer
  '**********************************************************************************
  'Fill the tables with default values,
  'taken from Appendix A1.
  '**********************************************************************************
  Private Sub fill_tables()
    temp(0)  = 5
    temp(1)  = 6
    temp(2)  = 7
    temp(3)  = 8
    temp(4)  = 9
    temp(5)  = 10
    temp(6)  = 11
    temp(7)  = 12
    temp(8)  = 13
    temp(9)  = 14
    temp(10) = 15
    temp(11) = 16
    temp(12) = 17
    temp(13) = 18
    temp(14) = 19
    temp(15) = 20
    temp(16) = 21
    temp(17) = 22
    temp(18) = 23
    temp(19) = 24
    temp(20) = 25
    temp(21) = 26
    temp(22) = 27
    temp(23) = 28
    temp(24) = 29
    temp(25) = 30
    temp(26) = 31
    temp(27) = 32
    temp(28) = 33
    temp(29) = 34
    temp(30) = 35
    temp(31) = 36
    temp(32) = 37
    temp(33) = 38
    density(0) = 999.956
    density(1) = 999.941
    density(2) = 999.902
    density(3) = 999.849
    density(4) = 999.781
    density(5) = 999.7
    density(6) = 999.605
    density(7) = 999.498
```

```
    density(8) = 999.377
    density(9) = 999.244
    density(10) = 999.099
    density(11) = 998.943
    density(12) = 998.774
    density(13) = 998.595
    density(14) = 998.405
    density(15) = 998.203
    density(16) = 997.992
    density(17) = 997.77
    density(18) = 997.538
    density(19) = 997.296
    density(20) = 997.044
    density(21) = 996.783
    density(22) = 996.512
    density(23) = 996.232
    density(24) = 995.944
    density(25) = 995.646
    density(26) = 995.34
    density(27) = 995.025
    density(28) = 994.702
    density(29) = 994.371
    density(30) = 994.03
    density(31) = 993.68
    density(32) = 993.33
    density(33) = 992.96
    viscosity(0) = 1.5188
    viscosity(1) = 1.4726
    viscosity(2) = 1.4288
    viscosity(3) = 1.3872
    viscosity(4) = 1.3476
    viscosity(5) = 1.3097
    viscosity(6) = 1.2735
    viscosity(7) = 1.239
    viscosity(8) = 1.2061
    viscosity(9) = 1.1748
    viscosity(10) = 1.1447
    viscosity(11) = 1.1156
    viscosity(12) = 1.0875
    viscosity(13) = 1.0603
    viscosity(14) = 1.034
    viscosity(15) = 1.0087
    viscosity(16) = 0.9843
    viscosity(17) = 0.9608
    viscosity(18) = 0.938
    viscosity(19) = 0.9161
    viscosity(20) = 0.8949
    viscosity(21) = 0.8746
    viscosity(22) = 0.8551
    viscosity(23) = 0.8363
    viscosity(24) = 0.8181
    viscosity(25) = 0.8004
    viscosity(26) = 0.7834
    viscosity(27) = 0.767
    viscosity(28) = 0.7511
    viscosity(29) = 0.7357
    viscosity(30) = 0.7208
    viscosity(31) = 0.7064
    viscosity(32) = 0.6925
    viscosity(33) = 0.6791
End Sub
```

```
   Private Sub Form1_Load(ByVal sender As System.Object, ByVal e As System.EventArgs)
Handles MyBase.Load
     'read tables
     fill_tables()
     RadioButton1.Text = "Liquid is water"
     RadioButton2.Text = "Any other fluid"
     RadioButton1.Checked = True
     Me.MaximizeBox = False
     Me.FormBorderStyle = Windows.Forms.FormBorderStyle.FixedSingle
     Me.Text = "Program 4.1: Computes settling velocity for discrete particles"
     Label5.Text = "The settling velocity (m/s)="
     GroupBox1.Text = " Output: "
     button1.text = "&Calculate"
     Label8.Text = "Decimal Places:"
     NumericUpDown1.Maximum = 10
     NumericUpDown1.Minimum = 0
     NumericUpDown1.Value = 2
   End Sub

   Private Sub RadioButton1_CheckedChanged(ByVal sender As System.Object, ByVal e As
System.EventArgs) Handles RadioButton1.CheckedChanged
     If RadioButton1.Checked Then
        Label1.Text = "Enter the temperature (Centigrade):"
        Label2.Text = "Specific gravity of the particles:"
        Label3.Text = "Diameter of particles d (m):"
        Label4.Visible = False
        TextBox4.Visible = False
     Else
        Label1.Text = "Density of liquid (kg/m3):"
        Label2.Text = "Specific gravity of the particles:"
        Label3.Text = "Diameter of particles d (m):"
        Label4.Text = "Enter viscosity of fluid (Pa.s):"
        Label4.Visible = True
        TextBox4.Visible = True
     End If
   End Sub

   Sub calculateResults()
     If RadioButton1.Checked Then
        T = Val(TextBox1.Text)
        If T = 0 Then Exit Sub
        sg = Val(TextBox2.Text)
        d = Val(TextBox3.Text)
        'determine the density and the viscosity
        Dim kr As Integer = -1
        For i = 0 To 33
          If T = temp(i) Then
             kr = i
             Exit For
          End If
          If T > temp(i) And T < temp(i + 1) Then
             kr = i
             Exit For
          End If
        Next i
        'not found?
        If kr = -1 Then
          Exit Sub
        End If
        Ro = density(kr)
        Mu = viscosity(kr) / 1000
```

```
        Else
            Ro = Val(TextBox1.Text)
            sg = Val(TextBox2.Text)
            d = Val(TextBox3.Text)
            Mu = Val(TextBox4.Text)
        End If
        'compute settling velocity (Stoke's law)
        Velocity = Ro * G * (d ^ 2) * (sg - 1) / (18 * Mu)
    End Sub

    Private Sub showResults()
        TextBox5.Text = FormatNumber(Velocity, NumericUpDown1.Value).ToString
    End Sub

    Private Sub Button1_Click(ByVal sender As System.Object, ByVal e As System.EventArgs)
Handles Button1.Click
        If RadioButton1.Checked Then
            '** Check the validity of temp as we will use it
            '** to index our tables later in calculateResults()
            T = Val(TextBox1.Text)
            If T < 5 Or T > 38 Then
                MsgBox("Please enter a valid temp. between 5 & 38",
                    vbOKOnly Or vbInformation)
                Exit Sub
            End If
        End If
        calculateResults()
        showResults()
    End Sub

    '***************************************************************************************
    'Whenever a different number of decimals is selected, re-show the results.
    '***************************************************************************************

    Private Sub NumericUpDown1_ValueChanged(ByVal sender As System.Object, ByVal e As
System.EventArgs) Handles NumericUpDown1.ValueChanged
        showResults()
    End Sub
End Class
```

### 4.4.4 SETTLING CHARACTERISTICS

Laboratory experiments are often conducted to evaluate the settling phenomenon of suspended particles, as well as to provide the information needed for the design of sedimentation tanks. The tests are conducted in a settling column apparatus (Figure 4.3), which consists of a cylindrical container having a uniform surface area and tapping points (ports) at preset level. The cylinder is of a height equal to that of the proposed tank. The temperature of the container can be kept constant by using a water bath. The test procedure may be summarized as follows:

1. The initial suspended solids (SS) concentration, $C_0$, is determined.
2. The settling column is filled with a well-mixed suspension of the sample to be tested.

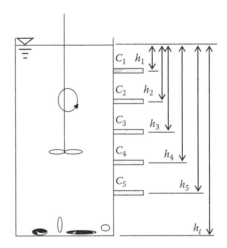

**FIGURE 4.3** Settling column test.

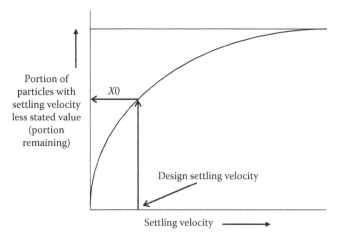

**FIGURE 4.4** Cumulative frequency distribution plot.

3. Samples are taken at various depths ($h_1$, $h_2$, ... etc.) and analyzed for solid concentrations, $c_1$, $c_2$, and so on. Thus, all particles with a settling velocity less than the design settling velocity ($v_{so}$) will not be removed during the settling time. Particles with a settling velocity greater than $v_{so}$ will have settled past the sampling point. As such, the ratio of removal for particles with $v < v_{so}$ (i.e., portion of particles $x_1$ that have settling velocity less than $v_1$) is given by $x_1 = c_1/c_0$. The entire range of settling velocities present in the system can be obtained by making use, for example, of a sieve analysis plus a hydrometer (wet sieve analysis), or a Coulter counter, or any other appropriate device or a technique according to standard procedures.
4. The above procedure is repeated at various time intervals ($t_1$, $t_2$, etc.).
5. Data are then plotted to give the cumulative settling characteristics curve for the suspension as presented in Figure 4.4. The overall removal efficiency of the sedimentation tank (degree of removability of particles) can be found though the relationship indicated in the following equation:

$$X_T = 1 - X_0 + \left\{ \frac{\int_0^{X_0} v*dx}{v_{so}} \right\} \quad (4.7)$$

The settling velocity for a discrete particle that enters the sludge zone at the end of the sedimentation tank can be determined from the following equation:

$$v_{so} = \frac{h}{t} = \frac{h}{(V/Q)} = h*\frac{Q}{V} = \frac{Q}{(V/h)} = \frac{Q}{A} \quad (4.8)$$

where:
$v_{so}$ is the settling velocity of the discrete particle, m/s
$h$ is the depth of tank, m
$t$ is the detention time, s (see Equation 4.9)

$V$ is the volume of tank, $m^3$
$Q$ is the flow rate entering the tank, $m^3/s$
$A$ is the area perpendicular to direction of flow, $m^2$

$$t = \frac{V}{Q} \quad (4.9)$$

where:
$V$ is the volume of tank, $m^3$
$Q$ is the flow rate entering the tank, $m^3/s$

Equation 4.8 illustrates that for a discrete particle, the solids removal is independent of the depth of the sedimentation tank. To account for the less than optimum conditions encountered in practice, the design settling velocity ($v_{so}$) obtained from the column test is often multiplied by a factor of 0.65–0.85 (Metcalf and Eddy 2013), and the computed detention times are multiplied by a factor of 1.25–1.5. Efficiently designed and operated, primary clarifiers remove 50%–65% of the total SS and 25%–40% of the $BOD_5$ (Riffat 2012).

**Example 4.2**

1. Write a computer program to compute the removal efficiency and concentration of SS in effluent, given the relevant column test results.
2. A wastewater treatment plant incorporates four sedimentation tanks for treating a wastewater flow of 1200 $m^3/h$. Each of the circular sedimentation tanks has a diameter of 21 m. The settling column test conducted for a sample of the wastewater indicated the data tabulated below. Determine the SS removal efficiency of the sedimentation units.

| Sampling Depth (m) | Concentration of SS Removed (mg/L) | | | | |
|---|---|---|---|---|---|
| | Sampling Time (h) | | | | |
| | 0 | 1 | 2 | 3 | 4 |
| 0.2 | 190 | 101 | 166 | 183 | 186 |
| 0.5 | 190 | 57 | 85 | 114 | 124 |
| 1.0 | 190 | 53 | 58 | 70 | 86 |
| 1.5 | 190 | 50 | 54 | 60 | 66 |
| 2.0 | 190 | 44 | 54 | 56 | 57 |
| 2.5 | 190 | 40 | 50 | 51 | 56 |

**Solution**

1. For solution to Example 4.2 (1), see the listing of Program 4.2.
2. Solution to Example 4.2 (2):
   a. Given: $Q = 1200$ $m^3/h$, $D = 21$ m, number of sedimentation tanks = 4, SS concentrations at different depths for different times.

b.  Determine the settling velocity as follows: settling velocity = depth/time = h/t.

c.  Find the percentage SS remaining, $C_r$, in the sample for different time intervals and sampling ports indicated as follows: $C_r = 100 -$ SS concentration removed $= 100 - (C/C_0)$.

d.  Determine the percentage SS with settling velocity less than that stated as follows: $X = C_r/C_0$, as shown in the tabular form.

e.  Plot the percentage SS with settling velocity less than that stated versus settling velocity to obtain the cumulative frequency distribution curve for the settling particles.

f.  Find the flow rate introduced to each tank = total flow/number of tanks = (1200/(60 * 60))/4 = 0.083 m³/s.

g.  Compute the surface area of each tank: $A = \pi (21)^2/4 = 346.4$ m².

h.  Determine the surface loading rate for the settling particles: $v_{so} = Q/A = 0.083/346.4 = 2.4 * 10^{-4}$ m/s = 0.24 mm/s.

i.  From the plotted graph and for a design surface loading of 0.24 mm/s, the value of $X_0 = 72\%$.

j.  Determine the removal efficiency of the tank by using the following equation:

$$X_T = 1 - X_0 + \frac{\left\{ \int_0^{X_0} v * dx \right\}}{v_{so}}$$

k.  The integral part of the previous equation $\left\{ \int v * dx \right\}$ may be determined by using a planimeter or by making use of Simpson's rule, by manual means, or by any other appropriate method. In this case, the integral part = $0.76 * 10^{-3}$.

l.  Find the removal efficiency as follows: $X_T = 100 - 72 + (0.76 * 10^{-3}/2.4 * 10^{-4}) = 31\%$.

3.  Use the computer program presented in (1) to check the results obtained in (2).

| Depth (mm) | Time (s) | Velocity (mm/s) | SS Remaining (mg/L) | SS with $v$ Less Than Stated (%) |
|---|---|---|---|---|
| 200 | 3600 | 0.056 | 89 | 46.8 |
| 200 | 7200 | 0.028 | 24 | 12.6 |
| 200 | 10,800 | 0.019 | 7 | 3.7 |
| 200 | 14,400 | 0.014 | 4 | 2.1 |
| 500 | 3600 | 0.139 | 133 | 70 |
| 500 | 7200 | 0.069 | 105 | 55.3 |
| 500 | 10,800 | 0.046 | 76 | 40 |
| 500 | 14,400 | 0.035 | 66 | 34.7 |
| 1000 | 3600 | 0.278 | 137 | 72.1 |
| 1000 | 7200 | 0.139 | 132 | 69.5 |
| 1000 | 10,800 | 0.093 | 120 | 63.2 |
| 1000 | 14,400 | 0.069 | 104 | 54.7 |
| 1500 | 3600 | 0.417 | 140 | 73.7 |
| 1500 | 7200 | 0.208 | 136 | 71.6 |
| 1500 | 10,800 | 0.139 | 130 | 68.4 |
| 1500 | 14,400 | 0.104 | 124 | 65.3 |
| 2000 | 3600 | 0.556 | 146 | 76.8 |
| 2000 | 7200 | 0.278 | 136 | 71.6 |
| 2000 | 10,800 | 0.185 | 134 | 70.5 |
| 2000 | 14,400 | 0.139 | 133 | 70 |
| 2500 | 3600 | 0.694 | 150 | 78.9 |
| 2500 | 7200 | 0.347 | 140 | 73.7 |
| 2500 | 10,800 | 0.231 | 139 | 73.2 |
| 2500 | 14,400 | 0.174 | 134 | 705 |

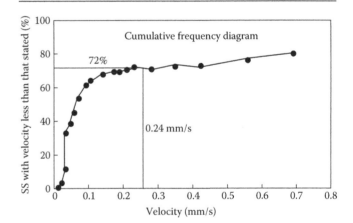

---

### LISTING OF PROGRAM 4.2   (A) LISTING OF FORM1 (CHAP4.2\FORM1.VB): SUSPENDED SOLIDS IN A SETTLING TANK EFFLUENT

```
'*****************************************************************************
'Program 4.2: Removal
'computes percentage of suspended solids in a settling
'tank effluent
'*****************************************************************************

Public Class Form1
    Dim timeS, depthS, ssStr As String
    Dim Numdepth, Numsampletime, k As Integer
    Dim f2 As Form2
```

```
  Private Sub Form1_Load(ByVal sender As System.Object, ByVal e As System.EventArgs)
Handles MyBase.Load
    RadioButton1.Text = "Input new data"
    RadioButton2.Text = "Use the built-in data"
    RadioButton2.Checked = True
    Me.MaximizeBox = False
    Me.FormBorderStyle = Windows.Forms.FormBorderStyle.FixedSingle
    Me.Text = "Program 4.2: Computes percentage of suspended solids in a settling tank
effluent"
    GroupBox1.Text = " Inputs: "
    Label1.Text = "Number of sampling times:"
    Label4.Text = "Enter sampling times (h):"
    Label2.Text = "Number of sampling depths:"
    Label3.Text = "Enter depths (m):"
    Label5.Text = "Enter concentration of SS removed (mg/L) for each depth and time"
    Button1.Text = "Calculate"
    DataGridView1.AllowUserToAddRows = False
    DataGridView2.AllowUserToAddRows = False
    DataGridView3.AllowUserToAddRows = False
  End Sub

  Sub loadDefaults()
    'Data of example 3.2
    timeS = "0, 1, 2, 3, 4,"
    depthS = "0.2, 190, 101, 166, 183, 186,"
    ssStr = "0.5, 190, 57, 85, 114, 124,"
    ssStr += "1, 190, 53, 58, 70, 86,"
    ssStr += "1.5, 190, 50, 54, 60, 66,"
    ssStr += "2, 190, 44, 54, 56, 57,"
    ssStr += "2.5, 190, 40, 50, 51, 56,"
    Numdepth = 6
    Numsampletime = 5
    Dim last = 0
    For i = 1 To Numsampletime
      time(i) = Val(timeS.Substring(last, timeS.IndexOf(",", last) - last))
      last = Val(timeS.IndexOf(",", last)) + 1
      time(i) = 3600 * time(i) 'time converted to seconds
    Next
    last = 0
    For i = 1 To Numdepth
      depth(i) = Val(depthS.Substring(last, depthS.IndexOf(",", last) - last))
      last = Val(depthS.IndexOf(",", last)) + 1
      depth(i) = depth(i) * 1000 'convert depth to mm
    Next
    last = 0
    For i = 1 To Numdepth
      For j = 1 To Numsampletime
        SS(i, j) = Val(ssStr.Substring(last, ssStr.IndexOf(",", last) - last))
        last = Val(ssStr.IndexOf(",", last)) + 1
      Next
    Next
    'put the defaults in the DataGridViews
    TextBox1.Text = Numsampletime.ToString
    TextBox2.Text = Numdepth.ToString
    DataGridView1.Columns.Clear()
    DataGridView1.Columns.Add("timeCol", "time")
    DataGridView2.Columns.Clear()
    DataGridView2.Columns.Add("depthCol", "depth")
    DataGridView3.Columns.Clear()
```

```
    For i = 1 To Numsampletime
       DataGridView1.Rows.Add(time(i) / 3600)
    Next
    For i = 1 To Numdepth
       DataGridView3.Columns.Add("depth" + i.ToString, depth(i).ToString)
    Next
    For j = 1 To Numsampletime
       DataGridView3.Rows.Add()
       DataGridView3.Rows(j - 1).HeaderCell.Value = time(j)
    Next
    For i = 1 To Numdepth
       DataGridView2.Rows.Add(depth(i))
       DataGridView2.Rows(i - 1).HeaderCell.Value = time(i)
       For j = 1 To Numsampletime
          DataGridView3.Rows(j - 1).Cells(i - 1).Value = SS(i, j)
       Next
    Next
 End Sub

 Sub calculateResults()
    If RadioButton1.Checked Then
       Numdepth = Val(TextBox2.Text)
       Numsampletime = Val(TextBox1.Text)
       'read in sampling times
       For i = 1 To Numsampletime
          time(i) = Val(DataGridView1.Rows(i - 1).Cells(0).Value)
          time(i) = 3600 * time(i) 'time converted to seconds
       Next
       'obtain depth and SS concentration
       For i = 1 To Numdepth
          depth(i) = Val(DataGridView2.Rows(i - 1).Cells(0).Value)
          'Convert from m to mm
          depth(i) *= 1000
          For j = 1 To Numsampletime
             SS(i, j) = Val(DataGridView3.Rows(j - 1).Cells(i - 1).Value)
          Next
       Next
    End If
    'Prepare output form
    Try
       f2.Label1.Text = "Output results:"
    Catch ex As Exception
       f2 = New Form2
       f2.Label1.Text = "Output results:"
    End Try
    f2.DataGridView1.Columns.Clear()
    f2.DataGridView1.Columns.Add("depthCol", "Depth (mm)")
    f2.DataGridView1.Columns.Add("timeCol", "Time (s)")
    f2.DataGridView1.Columns.Add("velCol", "Velocity (mm/s)")
    f2.DataGridView1.Columns.Add("ssrCol", "Suspended Solids remaining (mg/L)")
    f2.DataGridView1.Columns.Add("sspCol", "Suspended solids with v less than stated
 (%)")
    'Computations
    k = 0
    For i = 1 To Numdepth
       For j = 2 To Numsampletime
          k = k + 1
          VELOCITY(k) = depth(i) / time(j)
          ssr(k) = SS(i, 1) - SS(i, j)
          SSP(k) = ssr(k) / SS(i, 1) * 100
```

```
            f2.DataGridView1.Rows.Add(FormatNumber(depth(i), 2), FormatNumber(time(j), 2), _
                        FormatNumber(VELOCITY(k), 2), FormatNumber(ssr(k), 2), _
FormatNumber(SSP(k), 2))
        Next
    Next
        COUNT = k
        f2.ShowDialog()
    End Sub


    Private Sub RadioButton2_CheckedChanged(ByVal sender As System.Object, ByVal e As
System.EventArgs) Handles RadioButton2.CheckedChanged
        If RadioButton2.Checked Then
            loadDefaults()
        End If
    End Sub


    Private Sub Button1_Click(ByVal sender As System.Object, ByVal e As System.EventArgs)
Handles Button1.Click
        calculateResults()
    End Sub


    Private Sub RadioButton1_CheckedChanged(ByVal sender As System.Object, ByVal e As
System.EventArgs) Handles RadioButton1.CheckedChanged
        If RadioButton1.Checked Then
            DataGridView1.Rows.Clear()
            DataGridView2.Rows.Clear()
            DataGridView3.Rows.Clear()
            DataGridView1.Rows.Add(Numsampletime)
            DataGridView2.Rows.Add(Numdepth)
            DataGridView3.Rows.Add(Numsampletime)
        End If
    End Sub


    Private Sub TextBox1_TextChanged(ByVal sender As System.Object, ByVal e As System.
EventArgs) Handles TextBox1.TextChanged
        Dim i As Integer
        i = CInt(Val(TextBox1.Text))
        If i > Numsampletime Then
            'add some rows
            DataGridView1.Rows.Add(i - Numsampletime)
        ElseIf i < Numsampletime Then
            'delete some rows
            For j = Numsampletime To i + 1 Step -1
              DataGridView1.Rows.RemoveAt(j - 1)
            Next
        End If
        Numsampletime = i
    End Sub


    Private Sub TextBox2_TextChanged(ByVal sender As System.Object, ByVal e As System.
EventArgs) Handles TextBox2.TextChanged
        Dim i As Integer
        i = CInt(Val(TextBox2.Text))
        If i > Numdepth Then
            'add some rows
            DataGridView2.Rows.Add(i - Numdepth)
        ElseIf i < Numdepth Then
            'delete some rows
            For j = Numdepth To i + 1 Step -1
              DataGridView2.Rows.RemoveAt(j - 1)
```

```
        Next
      End If
      Numdepth = i
  End Sub
End Class
```

## (B) LISTING OF FORM2 (CHAP4.2\FORM2.VB): SETTLING TANK EFFICIENCY

```
Public Class Form2
    Dim Q, D, Q1, a, Vs, IND, sum, RAV, Xr As Double
    Dim numTanks As Integer
    Dim Kount, Notav As Integer

    Private Sub Form2_Load(ByVal sender As System.Object, ByVal e As System.EventArgs)
Handles MyBase.Load
        Me.MaximizeBox = False
        Me.FormBorderStyle = Windows.Forms.FormBorderStyle.FixedSingle
        Me.Text = "Calculate"
        Label2.Text = "Flow Rate Q (m3/min):"
        Label3.Text = "Number of tanks:"
        Label4.Text = "Diameter of the tank (m):"
        Button1.Text = "&Calculate"
        Button2.Text = "Close"
        Me.CancelButton = Button2
        Me.AcceptButton = Button2
        TextBox4.Multiline = True
        TextBox4.Text = ""
        TextBox4.Height = 85
    End Sub

    Sub calculateResults()
        Q = Val(TextBox1.Text)
        numtanks = Val(TextBox2.Text)
        D = Val(TextBox3.Text)
        'flow in m3/sec
        Q = Q / (60 * 60)
        'flow per tank
        Q1 = Q / numTanks
        'surface are of a tank
        a = pi * (D ^ 2) / 4
        'surface loading rate for settling particles mm/s
        Vs = Q1 * 1000 / a
        'search for an estimate for SS corresponding to Vs
        Call integrate(COUNT, Vs)
    End Sub

    Sub integrate(ByVal N, ByVal Vs)
        '*******************************************************************************
        Dim X(N), Y(N), X1(N), X2(N), Y2(N)
        Dim last As Integer
        Dim I As Integer
        Dim SS As Double
        For I = 1 To N
            X(I) = VELOCITY(I)
            Y(I) = SSP(I)
        Next I
        'sort out equations based on x values
        For I = 1 To N : X1(I) = X(I) : Next I
```

```
        For I = 1 To N
          Call MIN(X1, N, IND)
          X2(I) = X(IND)
          Y2(I) = Y(IND)
          X1(IND) = 20000000000.0
        Next I
          For I = 1 To N
          X(I) = X2(I)
          Y(I) = Y2(I)
        Next I
          'remove repeated equations
          kount = 1
          X2(1) = X(1)
          For I = 2 To N
            If X(I) <> X(I - 1) Then
                Kount = kount + 1
                X2(kount) = X(I)
                Y2(kount) = Y(I)
            End If
          Next I
          'Search for SS corresponding to velocity Vs
          Notav = 0
          For I = 1 To Kount
            If Vs = X2(I) Then
                SS = Y2(I)
                GoTo jmp
            End If
            If Vs < X2(I) Then
                If I = 1 Then SS = Y2(1)
                If I = Kount Then SS = Y2(Kount)
                If I > 1 And I < Kount Then SS = 0.5 * (Y2(I) + Y2(I + 1))
                GoTo jmp
            End If
          Next I
          notav = 1
jmp:    'CONTINUE
          last = I
          If Notav = 1 Then
            TextBox4.Text = "Velocity out of range, No solution found!!!!"
            Exit Sub
          End If
          TextBox4.Text = "Corresponding SS% TO Vs = " + FormatNumber(Vs, 4) + " Is SS% = "
+ FormatNumber(SS, 4)
          'FIND THE INTEGRATION OF VDX BY TRAPEZOIDAL RULE
          sum = 0.5 * Y2(1) * X2(I)
          For I = 1 To last - 1
            'INTEGRATION OF VDX
            sum += 0.5 * (X2(I) + (X2(I + 1)) * (Y2(I + 1) - Y2(I)))
          Next I
          sum /= 10000
          TextBox4.Text += vbCrLf + "The integral part vdx = " + FormatNumber(sum, 4)
          RAV = sum / (Vs / 1000)
          'REMOVAL EFFICIENCY
          Xr = 100 - SS + RAV
          TextBox4.Text += vbCrLf + "Removal efficiency of the tank = " +
FormatNumber(Xr, 2) + " %"
    End Sub

    Sub MIN(ByRef X(), ByRef N, ByRef IND)
        Dim XMIN = 10000000000.0
        For I = 1 To N
```

```
          If XMIN > X(I) Then
              XMIN = X(I)
          IND = I
          End If
       Next I
   End Sub


   Private Sub Button1_Click(ByVal sender As System.Object, ByVal e As System.EventArgs)
Handles Button1.Click
       calculateResults()
   End Sub


   Private Sub Button2_Click(ByVal sender As System.Object, ByVal e As System.EventArgs)
Handles Button2.Click
       Me.Hide()
       End Sub
       End Class
```

### (C) LISTING OF MODULE1 (CHAP4.2\MODULE1.VB)

```
Module Module1
    Public depth(20), SS(20, 20), time(20), VELOCITY(50), ssr(50), SSP(50)
    Public COUNT As Integer
    Public Const pi = 3.142857
End Module
```

### 4.4.5  THE IDEAL SEDIMENTATION BASIN

Settling is assumed to occur in exactly the same way as in a quiescent settling basin in an ideal horizontal flow basin (Figure 4.5). Thus, for optimum performance, the tank ought to have the following characteristics (Huisman 1977, Rowe and Abdel-Magid 1995, Metcalf and Eddy 2013):

- Flow within the settling zone is uniform.
- All particles entering the sludge zone are removed, and they remain removed.
- Particles are uniformly distributed in the flow as they enter the settling zone.
- All particles entering the effluent zone leave the tank.
- Concentration of particles at inlet, for each size, is the same at all points in the vertical plane perpendicular to the direction of flow.

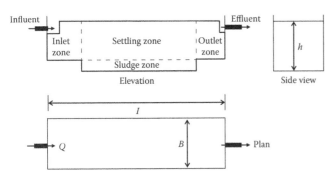

**FIGURE 4.5**  Schematic diagram of an ideal settling basin.

### 4.4.6  ELEMENTS THAT REDUCE THE EFFICIENCY IN THE PERFORMANCE OF SEDIMENTATION BASINS

The sedimentation efficiency is affected by turbulence, bottom scour, nonuniform velocity distribution, and short-circuiting.

#### 4.4.6.1  Turbulence

With turbulent flow, transverse velocity components exist. This condition scatters the discrete particles and reduces the efficiency of the tank. The horizontal flow occurs under laminar flow conditions when Reynolds numbers (Re) are less than 580–2000, depending upon the following:

- The construction of the inlet zone
- The kind and condition of the sludge removal equipment
- The presence of columns, cross beams, and rough walls
- The other obstacles that might interfere with the flow

Reynolds numbers signify the ratio between inertial forces and viscous forces. It can be shown that Reynolds number may be expressed as shown in the following equation (see Figure 4.5):

$$\text{Re} = v_H * \frac{r_H}{v} = \frac{Q}{\left[ v*(B+2*h) \right]} = \frac{v_s*B*l}{\left[ v*(B+2*h) \right]} \qquad (4.10)$$

where:

Re is the Reynolds number

$r_H$ is the hydraulic radius [$= A/w_p = B*h/(B + 2*h)$], m

$A$ is the area perpendicular to horizontal flow, m$^2$

$w_p$ is the wetted perimeter, m

$v$ is the kinematic viscosity (= $\mu/\rho$), m$^2$/s

$\mu$ is the dynamic (absolute) viscosity, N*s/m$^2$

$\rho$ is the density, kg/m$^3$

$v_H$ is the velocity of horizontal water movement [= $Q/(B*h)$], m/s

$v_s$ is the settling rate, displacement velocity, loading rate, overflow rate (= $Q/Bl$), m/s

$Q$ is the flow rate, m$^3$/s

$B$ is the tank width, m

$h$ is the tank depth, m

$l$ is the length of tank, m

In order to decrease Re numbers, one of the following conditions must be considered:

- Reduce the flow rate entering tank (less flow).
- Increase the width of tank (wider tank).
- Increase the tank depth (deeper tank).
- Decrease the length of the tank (shorter tank).

Therefore, to prevent reduction in basin efficiency by turbulence and bottom scour, the tank must be short, wide, and deep. Equation 4.11 can be used to evaluate Re numbers for various tank conditions as follows:

$$\text{Re} = \frac{v_{so} * r_H}{v} < 2000 \qquad (4.11)$$

where $v_{so}$ is the design settling velocity.

### 4.4.6.2 Bottom Scour

Scouring of settled material starts at a specific velocity due to an increase in the displacement velocity. The scouring velocity develops when the hydraulic shear, between the flowing water and the sludge deposits, equals the mechanical friction of the deposits at the bottom of the tank.

The ratio between scour and settling velocities may be determined for laminar settling conditions as presented in the following equation:

$$\frac{v_{sco}}{v_s} = 36 * \upsilon * \sqrt{\frac{10 * \rho}{\left[3 * g\left(\rho_s - \rho\right)d^3\right]}} \qquad (4.12)$$

where:

$v_{sco}$ is the scour velocity, m/s

$v_s$ is the settling velocity, m/s

$g$ is the gravitational acceleration, m/s$^2$

$\rho_s$ is the density of the particle, kgm$^3$

$\rho$ is the density of fluid, kgm$^3$

For turbulent flow, Equation 4.13 holds:

$$\frac{v_{sco}}{v_s} = 2 \qquad (4.13)$$

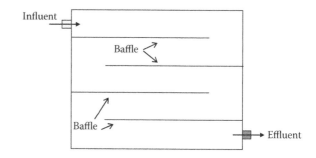

**FIGURE 4.6**   Round-the-end baffles in a sedimentation tank.

The reduction in settling basin efficiency by scour does not create problems as far as the design settling velocity is considered, as long as it is less than the scour velocity (i.e., $v_{so} < v_{sco}$). The ratio between the horizontal water velocity and the settling velocity in a rectangular horizontal flow basin can be calculated from the following equation:

$$\frac{v_H}{v_{so}} = \frac{\left(Q/B*h\right)}{\left(Q/B*l\right)} = \frac{1}{h} \qquad (4.14)$$

Equation 4.15 gives the ratio between the length of the tank and its depth:

$$\frac{l}{h} < \frac{v_{sco}}{v_{so}} \qquad (4.15)$$

Resuspension of settled matter can be hindered by using baffles. Figure 4.6 illustrates a sketch of round-the-end baffles.

### 4.4.6.3 Nonuniform Velocity Distribution and Short-Circuiting

The displacement velocity is not constant due to the frictional drag along the walls and floor of the basin, which slows down the flow of the water. As such, the velocity is slowed near the walls and floor and has a value larger than average in the center of the basin. Nonetheless, the variation in velocity over the depth of the tank has little effect on the sedimentation efficiency.

Due to the variation in the velocities of the horizontal water movement, some of the particles at the inflow reach the basin outlet in a shorter time than the theoretical detention period, whereas some other particles take a much longer time period. This phenomenon is referred to as short-circuiting (Huisman 1977). A direct reduction in the sedimentation basin efficiency takes place due to stagnant areas or eddying currents. This may be induced by an unequal flow distribution of the incoming water or by wind action that produces surface currents. Methods for reducing short-circuiting include the following:

- Proficient settling tank design
- Equal inflow and outflow over the full width and depth of the basin
- Avoidance of tank inlets with high-flow velocities
- Good mixing of tank contents to prevent density currents.

Tank stability can be approached by increasing the ratio between inertia forces and gravity forces, that is, Froude number. The Froude number, for a rectangular horizontal flow basin, can be expressed mathematically as presented in the following equation:

$$\text{Fr} = \frac{v_H^2}{(g * r_H)} = \frac{Q^2 * (B + 2 * h)}{(g * B^3 * h^3)} = \frac{\left\{ v_s^2 * l^2 * \left[ 1 + \left( \frac{2 * h}{B} \right) \right] \right\}}{g * h^3} \quad (4.16)$$

where:

$\text{Fr}$ is the Froude number, dimensionless
$v_H$ is the average displacement velocity, m/s
$g$ is the gravitational acceleration, m/s$^2$
$r_H$ is the hydraulic radius, m
$Q$ is the flow rate, m$^3$/s
$B$ is the breadth of the basin, m
$h$ is the tank depth, m
$v_s$ is the settling velocity, m/s
$l$ is the length of the tank, m

High Froude numbers can be achieved if one or all of the following conditions are considered:

- Increasing the incoming flow to the tank
- Reducing the width of the sedimentation tank (narrower tank)
- Reducing the depth of tank (shallower basin)
- Increasing the length of the basin (longer tank)

High Froude numbers are described but should not be so high as to endanger the basin efficiency by turbulence or bottom scour. Froude numbers should satisfy the parameters in Equation 4.17. They are empirical in nature and are based on experimental finding used in designing a settling tank (Huisman 1977).

$$\text{Fr} = \frac{v_{so}^2}{g * r_H} > 10^{-5} \quad (4.17)$$

Baffles guide and regulate the water flow; any structural material (e.g., wood) may be used for construction of baffles, as the water pressures on both sides of the baffles are the same.

## Example 4.3

1. Write a computer program that can be used to check the stability and effects of turbulence (Froude and Reynolds numbers) in a settling basin for horizontal water movement, given the flow rate (m$^3$/s), tank dimensions, and temperature.
2. A suspension enters a horizontal sedimentation tank at a flow rate of 6 m$^3$/min. The tank is 4 m wide, 10 m long, and 1.5 m deep. Compute, under ideal settling conditions, the values for the Froude and the Reynolds numbers for the settling basin, given that the kinematic viscosity of the water is 1.004 * 10$^{-6}$ m$^2$/s.

## Solution

1. For solution to Example 4.3 (1), see the listing of Program 4.3.
2. Solution to Example 4.3 (2):
   a. Given: $Q = 6/60 = 0.1$ m$^3$/s, $B = 4$ m, $l = 10$ m, $h = 1.5$ m, kinematic viscosity = 1.004 * 10$^{-6}$ m$^2$/s.
   b. Determine the hydraulic radius as follows: $T_H = A/w_p = B * h/[B + 2 * h] = 4 * 1.5/(4 + 2 * 1.5) = .857$ m.
   c. Find the displacement or horizontal tank velocity as follows: $V_H = Q/B*h = 0.1/4*1.5 = 0.017$ m/s.
   d. Determine Froude's number for horizontal water movement as follows: $\text{Fr} = v_H^2/g * r_H$. $\text{Fr} = (0.017)^2/(9.81 * 0.857) = 3.3 * 10^{-5}$. This value is > 10$^{-5}$; thus, it is all right.
   e. Find the Reynolds number for the horizontal water movement as follows: $\text{Re} = v_H * r_H/v = 0.017 * 0.857/1.004 * 10^{-6} = 14,229$, which is > 2000, which is not all right. To remedy this situation and avoid erratic tank behavior, baffles should be introduced.

---

**LISTING OF PROGRAM 4.3   (CHAP4.3\FORM1.VB): STABILITY AND EFFECTS OF TURBULENCE IN A SETTLING TANK**

```
'***********************************************************************************
'Program 4.3: Turbulence
'Checks the stability and effects of turbulence
'***********************************************************************************

Public Class Form1
    Dim Q, B, h, L, u As Double
    'acceleration due to gravity
    Const g = 9.81
```

```
    Private Sub Form1_Load(ByVal sender As System.Object, ByVal e As System.EventArgs)
Handles MyBase.Load
        Me.Text = "Program 4.3: Checks the stability and effects of turbulence"
        Me.MaximizeBox = False
        Me.FormBorderStyle = Windows.Forms.FormBorderStyle.FixedSingle
        Label1.Text = "Enter the flow rate Q (m3/s):"
        Label6.Text = "Enter the dimensions of the tank:"
        Label2.Text = "Width of tank B (m):"
        Label3.Text = "Depth of tank h (m):"
        Label4.Text = "Length of tank L (m):"
        Label5.Text = "Kinematic viscosity of water (m2/s):"
        Label7.Text = "Output:"
        TextBox6.Multiline = True
        TextBox6.Height = 50
        Button1.Text = "&Calculate"
    End Sub

    Sub calculateResults()
      Q = Val(TextBox1.Text)
      B = Val(TextBox2.Text)
      h = Val(TextBox3.Text)
      L = Val(TextBox4.Text)
      u = Val(TextBox5.Text)
      Dim Rh, VH, Fr, Re As Double
      'compute hydraulic radius
      Rh = B * h / (B + 2 * h)
      VH = Q / (B * h)
      Fr = VH ^ 2 / (Rh * g) 'Froude number
      Re = VH * Rh / u
      TextBox6.Text = "Froude number =" + Fr.ToString
      If Fr >= 0.00001 Then
         TextBox6.Text += " O.K."
      Else : TextBox6.Text += " NOT O.K., too low Froude number!"
      End If
      TextBox6.Text += vbCrLf + "Reynolds number =" + Re.ToString
      If Re <= 2000 Then TextBox6.Text += " O.K., No baffles needed"
      If Re > 2000 Then
         TextBox6.Text += " NOT O.K., to avoid erratic tank behavior, use baffles"
      End If
    End Sub

    Private Sub Button1_Click(ByVal sender As System.Object, ByVal e As System.EventArgs)
Handles Button1.Click
        calculateResults()
    End Sub
End Class
```

### 4.4.7 DESIGN OF THE SETTLING ZONE

The following steps may be used in the design of the settling zone in a sedimentation basin:

- The required surface area can be computed as follows:

$$A = \frac{Q}{v_{so}} \qquad (4.18)$$

where:

$A$ is the surface area, m$^2$

$Q$ is the amount of water to be treated, m$^3$/s

$v_{so}$ is the design surface loading to be applied, m/s

- Depth of tank may be computed from the following relationship:
- For circular tanks:

$$h = 0.17 * A^{1/3} \qquad (4.19)$$

where $h$ is the depth of the tank, m.

- For rectangular tanks:

$$h = \frac{l^{0.8}}{12} \qquad (4.20)$$

- The ratio of length of the tank to its width can be calculated from the following empirical equation:

$$\frac{\text{Length}}{\text{width}} = \frac{l}{B} = 6 \text{ to } 10 \qquad (4.21)$$

- The flow rate per unit length (weir loading) must not be too high. This is to prevent disturbing the settling of suspended matter near the end of the tank. Settling near the end of the tank occurs due to updraft velocities created at the weir discharge. The allowable weir loading is given by the following equation:

$$\frac{Q}{B} < 5 * h * v_{so} \qquad (4.22)$$

where:
  $Q$ is the flow rate, m³s
  $B$ is the width of the tank, m
  $h$ is the depth of the tank, m
  $v_{so}$ is the overflow rate, m/s

With $v_{so} = Q/B*l$, this requirement is fulfilled in case the ratio of tank length to its depth falls below a value of 5, as shown in the following equation:

$$l/h < 5 \qquad (4.23)$$

where:
  $l$ is the length of the tank, m
  $h$ is the depth of the tank, m

### 4.4.8  Settling of Flocculent Particles

The settling of a suspension containing particles with different velocities of subsidence results in the smaller (lighter) particles being overtaken by the larger sized particles (heavier mass density). This results in a number of collisions and eventually the formation of aggregates of particles.

The removal ratio for flocculent settling increases with a decrease in the overflow rate and an increase in the basin depth. Factors affecting flocculent settling efficiency include overflow rate, detention time, and tank depth. The effects of bottom scour and short-circuiting are similar to those for discrete particles. Nonetheless, the effect of turbulence may be neglected due to dispersion. Dispersion prevents part of the suspended matter from reaching the tank bottom, but it augments the aggregation of finely divided suspended matter into larger flocs that have higher settling velocities. The net effect of turbulence on flocculent settling is minor. Thus, for the design of the sedimentation basin, for a high Froude number and stable flow conditions, the tank ought to be long, narrow, and shallow with the displacement velocity high enough to prevent bottom scour.

Any slight increase in temperature (e.g., from solar heating) results in an increase in the settling velocity of the particles (viscosity effects on settling velocity). However, an increase in temperature can also produce convection currents, which tend to scour and resuspend settled material.

### 4.4.9  Analysis of Flocculent Settling

The settling column apparatus shown in Figure 4.2 can be used for the determination of the settling characteristics of a suspension of flocculent particles. The procedure for analysis of flocculent settling in the settling column is summarized as follows:

1. See the procedure presented in Section 4.4.4 (steps 1, 2, 3, and 4).
2. The percentage of SS that have settled past the sampling point is computed as the value of $100 - (C/C_0)$.
3. The percentage of removal is to be plotted against the time and depth.
4. Curves of equal percentages of removal (iso-concentration lines or iso-removal lines) are to be developed.
5. Overall removal is determined as follows:

$$R_T = \left[ \frac{\Delta h_1 (R_1 + R_2)}{2h_t} + \frac{\Delta h_2 (R_2 + R_3)}{2h_t} + \cdots + \frac{\Delta h_n (R_n + R_{n+1})}{2h_t} \right]$$
$$= \left( \frac{1}{2h_t} \right) * \sum_{i=1}^{i=n} h_i (R_i + R_{i+1}) \qquad (4.24)$$

where:
  $h_i$ is the sampling port depth $i$, m
  $h_t$ is the height of the proposed tank, m
  $R_T$ is the total removal, %

### 4.4.10  General Design Considerations

#### 4.4.10.1  Rectangular Tanks

Rectangular tanks are characterized by their rectangular shapes with horizontal flow. For practical purposes, the following tank dimensions are considered appropriate. A maximum tank length of 100 m, with an average length of 30 m, and a maximum depth of 5 m, with an average depth of 3 m (Water Environment Federation 2009). Such tanks generally are used for primary sedimentation (Lin 2014). The sludge hopper acts as a thickener due to accumulation of sludge in it. The bottom slope may be taken as 1% on average to facilitate sludge sliding into the hopper (Water Environment Federation 2009). In the design of rectangular tanks, the following points should be considered (see Figure 4.7):

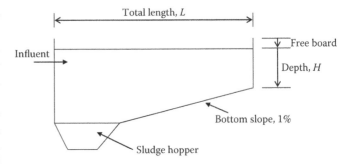

**FIGURE 4.7**  Rectangular sedimentation tank.

- The volume of the tank is computed as follows:

$$V = Q * t \qquad (4.25)$$

where:

$V$ is the volume of the tank, m³

$Q$ is the design flow, m³/day

$t$ is the detention time of tank, days (usually taken as 1.5–2 h for maximum flow)

- The minimum number of tanks is 2.
- The length:width ratio may be taken as follows (Water Environment Federation 2009):

$$4 \leq \frac{l}{b} \leq 10 \qquad (4.26)$$

where:

$l$ is the length of the tank, m

$b$ is the width of the tank, m

- The length:depth ratio may be taken as (Water Environment Federation 2009) follows:

$$25 \leq \frac{l}{H} \leq 10 \qquad (4.27)$$

where $H$ is the useful clarification depth, m (not to exceed 90 m) (Lin 2014).

- Daily sludge removal is determined as follows:

$$V_h * Q * C_0 * \text{Eff} * \frac{\left[ \dfrac{100}{(100 - \text{m.c.})} \right]}{\gamma} \qquad (4.28)$$

where:

$V_h$ is the sludge volume, m³

$Q$ is the maximum daily flow rate, m³/d

$C_0$ is the initial concentration of SS influent to tank, kg/m³

Eff is the percentage of SS removal

m.c. is the moisture content of settled sludge, %

ρ is the density of sludge, kg/m³

### 4.4.10.2 Circular Settling Tanks (Dorr Settling Tanks)

In tanks with circular shapes, the flow is horizontal. Circular tanks can be used for primary as well as secondary sedimentation purposes. The maximum diameter for circular tanks is considered to be 50 m, with an average diameter of 30 m, and the maximum water depth of 4 m, with an average water depth of 2.5 m (Water Environment Federation 2009). In the design of circular tanks, the following points merit consideration (refer to Figure 4.8):

- The ratio of diameter to clarification depth may be found as follows (Water Environment Federation 2009):
  - $6 \leq D/H \leq 20$ for circular tanks with diameters of 16 to 30 m.

- $16 \leq D/H \leq$ for circular tanks with diameters of 30 to 50 m.
- $20 \leq D/H \leq 25$ for circular tanks serving wastewaters with large amounts of organic matter (high settling velocities).
- Ratio of diameter of circular deflector to diameter of tank may be taken as (Water Environment Federation 2009):

$$b = (10\% - 20\%) * D \qquad (4.29)$$

where:

$b$ is the diameter of circular deflector (may be taken as $b \approx 0.15 * D$), m (Lin 2014)

$D$ is the diameter of the circular tank, m

Assuming a single weir in tank: surface loading * area = weir loading * length.

$$\frac{\pi}{4} * (D - b)^2 * v_s = \pi * D * q_{weir} \qquad (4.30)$$

where:

$D$ is the diameter of the tank, m

$b$ is the diameter of the deflector, m

$v_s$ is the surface loading rate, m/day

$q_{weir}$ is the weir loading (overflow rate), m³/day/m length

Taking

$$b = 0.15 * D \qquad (4.31)$$

Then,

$$D = 5.54 * \frac{q_{weir}}{v_s}$$

- Tank depth may be determined as

$$H = \frac{Q * t}{N * A} H \qquad (4.32)$$

where:

$N$ is the number of settling tanks, dimensionless

$A$ is the effective area of settlement, m²:

$$A = \frac{\pi}{4} * (D_a - b)^2 \qquad (4.33)$$

where $D_a$ is the actual diameter of tank, m.

Thus,

$$H = \frac{Q * t}{N * A} = \frac{Q * t}{N * \dfrac{\pi}{4}(D_a - b)^2} = \frac{Q * t}{N * \dfrac{\pi}{4}(D_a)^2(1 - b)^2} \qquad (4.34)$$

- Slope of settling tank bottom, β, may be taken as β = 6%–8% (or 5–10[o13]).
- Slope of sludge hopper, α, may be taken as α = 2:1[12] (≥60[o13]).

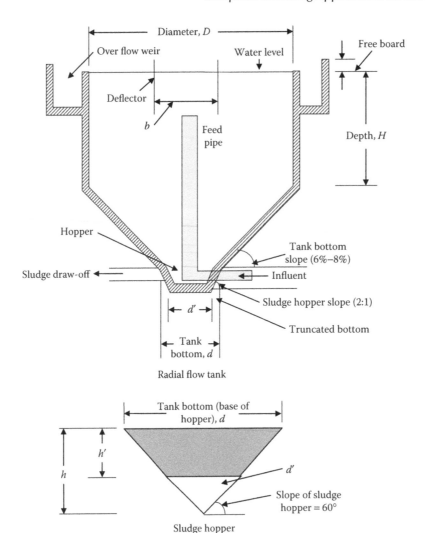

**FIGURE 4.8** Radial flow sedimentation tank.

- Detention time may be taken as 2 h at peak flow.
- Volume of sludge hopper is to hold 12–24 h of sludge production, and it may be determined as follows:

$$V_h = \frac{Q * C_0 * \text{Eff}}{a * C_e} = \frac{Q * C_0 * \text{Eff}}{a * (1 - \text{m.c.}) * 1000} \quad V_h = \frac{Q * C_0 * \text{Eff}}{a * C_c} \quad (4.35)$$

where:
  $V_h$ is the volume of sludge hopper, m³
  $Q$ is the flow rate, m³/day
  $C_0$ is the concentration of solids entering sedimentation tank, kg/m³
  Eff is the efficiency of solids removal in tank, dimensionless
  $a$ is the fraction of flow to be stored in hopper (as sludge)
  $C_c$ is the solids content of final sludge, kg/m³, which is equal to $(1 - \text{m.c.}) * 1000$, where m.c. is the moisture content of sludge (dimensionless)

Considering hopper as a truncated cone, then

$$V_h = (\pi d^2 * h / 12) - (\pi d'^2 * h' / 1) \quad (4.36)$$

where:
  $d$ is the diameter of hopper base (tank floor), m
  $h$ is the hopper imaginary depth (depth of total cone), m
  $d'$ is the diameter of truncated cone (hopper floor), m
  $h'$ is the depth of truncated cone, m

Taking a slope of sludge hopper as $\alpha = 60°$, then

$$(d/2)/h = (d'/2)/h' = \tan 30° \quad (4.37)$$

As such:

$$v_h = 0.227 * (d^3 - d'^3) \quad (4.38)$$

- The feed pipe is to be located 20–30 cm below water level.

## Example 4.4

1. Write a computer program for the hydraulic and structural design of a circular sedimentation tank, given the relevant design data of rate of flow, $Q$ (m³/s); loading rate, $v_s$ (m³/h/m²); influent solids concentration, $C_0$ (kg/m³); tank efficiency, Eff (%); sludge moisture content, m.c. (%); sludge production that can be stored by the sludge hopper; weir loading (overflow rate), $q_{weir}$ (m³/day/m weir length); retention time, $t$ (s) at peak flow; and bottom radius (truncated radius), $d'$ (m).
2. Design a circular sedimentation tank using the following information:
   a. Rate of flow, $Q$ = 15 m³/min.
   b. Loading rate (settling velocity), $v_s$ = 1.4 m³/h/m²
   c. Influent solids concentration, $C_0$ = 250 mg/L
   d. Tank efficiency, Eff = 55%
   e. Sludge moisture content, m.c. = 96%
   f. Size of sludge hopper sufficient to hold 12 h sludge production
   g. Weir loading (overflow rate), $q_{weir}$ = 200 m³/day/m
   h. Retention time, $t$ = 2 h at peak flow
   i. Bottom radius (truncated radius), $d'$ = 0.5 m

## Solution

1. For solution to Example 4.4 (1), see the listing of Program 4.4.
2. Solution to Example 4.4 (2):

a. Determine the diameter of tank using the following:
$$D = 5.54 * \frac{q_{weir}}{v_s} = 5.54 * \frac{200}{1.4*24} = 32.98 \text{ m.}$$
b. Determine the required number of sedimentation tanks: Flow from each tank
$$= \frac{\pi}{4} * D^2(1-b)^2 * v_s = \frac{\pi}{4} * 32.98^2(1-0.15)^2$$
$$* \frac{1.4*24}{4} = 20746.42 \text{ m}^3/\text{day. Flow rate to}$$
be treated = 15 m³/min = 15 * 60 * 24 = 21600 m³/day. Number of needed tanks = given flow rate/flow rate from each tank = 21600/20746.4 ≈ 2.
c. Find the actual diameter of tank: $2 * \pi * 0.85^2 * D_a^2 * 1.4 * 24/4 = 15 * 60 * 24$. Thus, $D_a$ = 23.8 m.
d. Compute the useful depth of clarification zone:
$$H = \frac{Q*t}{N*A} = \frac{Q*t}{N*\frac{\pi}{4}(D_a-b)^2} = \frac{Q*t}{N*\frac{\pi}{4}(D_a)^2(1-b)^2}$$
$$= \frac{21,600 \text{ m}^3/\text{day} * \frac{2}{24} \text{ days}}{2*\frac{\pi}{4}(23.8)^2\text{m}^2(1-0.15)^2} = 2.8 \text{ m. Check:}$$
$D/H$ = 23.8/2.8 = 8.5 (acceptable falls between 6 and 10).
e. Determine $C_c$ = (1 – m.c.)*1000 = (1 – 0.96) * 1000 = 40 kg/m³.
f. Find the volume of sludge hopper: $V_h$ = (21600 m³/d*0.25 kg/m³ * 0.55]/[(24/12 h)*40 kg/m3] = 37.125 m³. Since, $V_h$ = 0.227 ($d^3 - d'^3$) = 0.227*[$d^3$ – (0.5 * 2)³)] = 37.125, then $d$ = 5.5 m.

---

**LISTING OF PROGRAM 4.4   (CHAP4.4\FORM1.VB): DESIGN A CIRCULAR SEDIMENTATION TANK**

```
'*********************************************************************************
'Program 4.4: Tanks
'designs a circular sedimentation tank
'*********************************************************************************

Public Class Form1
  Const pi = 3.142857
  Dim Q, Vs, Co, Eff, mc, tr, Qweir, t, rdash As Double

  Private Sub Form1_Load(ByVal sender As System.Object, ByVal e As System.EventArgs)
Handles MyBase.Load
    Me.Text = "Program 4.4: Designs a circular sedimentation tank"
    Me.FormBorderStyle = Windows.Forms.FormBorderStyle.FixedSingle
    Me.MaximizeBox = False
    Button1.Text = "&Calculate"
    Label1.Text = "Enter flow rate (m3/min):"
    Label2.Text = "Enter loading rate (m3/hr/m2):"
    Label3.Text = "Enter influent solids concentration (mg/L):"
    Label4.Text = "Enter tank efficiency (%):"
    Label5.Text = "Enter sludge moisture content (%):"
    Label6.Text = "Enter period to maintain sludge in hopper (hr):"
    Label7.Text = "Enter weir loading rate (m3/d/m):"
    Label8.Text = "Enter retention time at peak flow (hr):"
    Label9.Text = "Enter bottom radius (m):"
    GroupBox1.Text = " Output: "
```

```vb
    Label10.Text = "Number of tanks needed:"
    Label11.Text = "Diameter of tank (m):"
    Label12.Text = "Depth of clarification zone (m):"
    Label13.Text = "Diameter of sludge hopper (m):"
End Sub

Sub calculateResults()
    Q = Val(TextBox1.Text)
    Vs = Val(TextBox2.Text)
    Co = Val(TextBox3.Text)
    Eff = Val(TextBox4.Text)
    mc = Val(TextBox5.Text)
    tr = Val(TextBox6.Text)
    Qweir = Val(TextBox7.Text)
    t = Val(TextBox8.Text)
    rdash = Val(TextBox9.Text)
    Dim Cc, dh, Vh, D, Q1, N, A, H, rat As Double
    Dim OK, t1 As String
    Vs = Vs * 24 'convert to m/day
    Q = Q * 60 * 24 'convert to m3/day
    Co = Co / 1000 'convert to kg/m3
    t = t / 24 'convert to days
    D = 5.54 * Qweir / Vs 'diameter of tank
    Q1 = pi / 4 * D ^ 2 * (1 - 0.15) ^ 2 * Vs 'flow from each tank
    N = CInt(Q / Q1) 'number of tanks
    Cycle:
    D = (Q / (N * pi * 0.85 ^ 2 * Vs / 4)) ^ 0.5 'actual diameter
    A = pi / 4 * (0.85 * D) ^ 2 'area of tank
    H = Q * t / (N * A) 'clarification depth
    'check D/H
    rat = D / H
    If rat > 6 And rat < 10 Then
        OK = "T"
        t1 = "D/H ratio = " + FormatNumber(rat, 2) + " is O.K."
    Else
        OK = "F"
        N = N + 1
        GoTo cycle
        End If
        'solids content in the final sludge
        Cc = (1 - mc / 100) * 1000
        'volume of sludge hopper Vh
        Eff = Eff / 100 'efficiency ratio
        Vh = Q * Co * Eff / (24 / tr * Cc)
        dh = (Vh / 0.227 + (2 * rdash) ^ 3) ^ (1 / 3)
        TextBox10.Text = FormatNumber(N, 0)
        TextBox11.Text = FormatNumber(D, 2)
        TextBox12.Text = FormatNumber(H, 2) + ": " + t1
        TextBox13.Text = FormatNumber(dh, 2)
End Sub

Private Sub Button1_Click(ByVal sender As System.Object, ByVal e As System.EventArgs)
Handles Button1.Click
    calculateResults()
End Sub
End Class
```

## 4.5 FLOCCULATION AND COAGULATION

### 4.5.1 INTRODUCTION

In water and wastewater treatment, flocculation and coagulation processes are mainly used for the removal of colloidal substance that causes color and turbidity. In relatively dilute solutions, some particles do not behave as discrete particles but rather they coalesce and combine during settling. As coalescence or flocculation occurs, the mass and settling velocity of the particle increase. The rate of flocculation depends on the opportunity for contact between particles. The factors that influence this process include surface loading, depth of settling tank, velocity gradients, and concentration and size of particles.

To enhance the process of flocculation and coagulation, a small dose of a coagulant (e.g., aluminum or ferric salts) is added to the raw water, after which flocs of coagulated colloidal matter are formed. This enables separation of particles from the fluid by sedimentation and/or filtration. The needed coagulant dose varies with the nature of the suspension. Flocculation occurs at the end of the coagulation phase. It consists of the building up and increases in the flocs with or without the use of flocculation aids or other chemical substances. The major factors that affect and influence the coagulation process include chemical factors (such as the optimum pH; inorganic ions; character of turbidity; nature, type, and concentration of the colloids involved; and type of coagulant) and physical parameters (such as temperature, mixing speed and duration of mixing, and size of colloidal particles).

### 4.5.2 ELECTROKINETICS OF COAGULATION

Electrokinetics denotes the motion of ions, molecules, or particles along a gradient of an electrical potential (Rowe and Abdel-Magid 1995, Ortega-Rivas 2011). Usually, colloidal or suspended particles in water bear a negative electrical charge. This electrical charge induces a positive charge in the vicinity of the neighboring liquid layer (electrical double layer). The electrical double layer provides a difference in potential between the particle and the bulk of the solution. This difference in potential exists across the electrical double layer and is referred to as the electrokinetic potential, zeta potential (ZP), or electrophoretic mobility.

The ZP may be defined as the potential at the surface of a particle that separates the immobile part of the double layer from the diffuse part. It is usually employed for rational control of coagulation. It also signifies a measure of the electrokinetic charge that surrounds suspended particulate matter. For most colloids, the electrokinetic potential is in the range of 30–60 mV.

Removal of colloids requires reduction of the ZP. This is due to the fact that repulsive forces of the electronegative colloid prevent the formation of a floc. The reduction of ZP is very effective in the settlement of lyophobic colloids. Equation 4.39 may be used to determine the ZP (Rowe and Abdel-Magid 1995, Ortega-Rivas 2011):

$$ZP = \frac{4\pi * B * q^+}{D_I} \qquad (4.39)$$

where:

ZP is the zeta potential
$B$ is the thickness of the boundary layer
$q^+$ is the charge on the colloid
$D_I$ is the dielectric constant

Electrophoretic mobility (EM) refers to the rate of movement of a particle to an electrode under a given electrical potential. For natural water, EM ranges between $2 * 10^{-4}$ and $4 * 10^{-4}$ m/s/V/cm. Generally, the higher the ionic strength, the lower the ZP. The usual technique for measurement of electrophoretic mobility involves a microscopic viewing of the particle motion under the influence of an electric field. Electrophoretic mobility can be calculated by using the following equation (Ortega-Rivas 2011):

$$EM = \frac{y * A}{t * i * RES_s} \qquad (4.40)$$

where:

EM is the electrophoretic mobility, $\mu$m/s/V/cm
$y$ is the distance covered in $t$ s, $\mu$m
$A$ is the cross-sectional area, cm$^2$
$i$ is the current, $A$
$RES_s$ is the specific resistance of the suspension, $\Omega$*cm

In practice, ZP is assumed to be related to EM as presented in the following equation:

$$ZP = 4 * \pi * \mu * \frac{EM}{D_I} \qquad (4.41)$$

where:

ZP is the zeta potential
$\mu$ is the absolute viscosity, N*s/m$^2$
EM is the electrophoretic mobility, $\mu$m/s/V/cm
$D_I$ is the dielectric constant

### Example 4.5

Write a short computer program that can be used to determine the ZP and electrophoretic mobility, given viscosity, dielectric constant, cross-sectional area, current density, specific resistance of the suspension, thickness of the boundary layer, and charge on the colloid. Use the relevant equation in each case.

### Solution

For solution to Example 4.5, see the listing of Program 4.5.

### LISTING OF PROGRAM 4.5   (CHAP4.5\FORM1.VB): ELECTROKINETICS OF COAGULATION

```vb
'*********************************************************************************
'Program 4.5: Coagulation
'computes the electrokinetics of coagulation
'*********************************************************************************

Public Class Form1
  Const pi = 3.14152654
  Dim u, Di, A, i, Rs, B, q As Double

  Private Sub Form1_Load(ByVal sender As System.Object, ByVal e As System.EventArgs)
Handles MyBase.Load
    Me.Text = "Program 4.5: Computes the electrokinetics of coagulation"
    Me.FormBorderStyle = Windows.Forms.FormBorderStyle.FixedSingle
    Me.MaximizeBox = False
    Label1.Text = "Viscosity coefficient (N*s/m2):"
    Label2.Text = "dielectric constant:"
    Label3.Text = "Cross sectional area (cm2):"
    Label4.Text = "Current density (A):"
    Label5.Text = "The specific resistance (Ω*cm):"
    Label6.Text = "Thickness of boundary layer:"
    Label7.Text = "Charge on the colloid:"
    groupbox1.text = " Output: "
    Label8.Text = "Zeta potential:"
    Label9.Text = "Electrophoretic mobility (μ/s/V/cm):"
    Button1.Text = "&Calculate"
  End Sub

  Sub calculateResults()
    u = Val(TextBox1.Text)
    Di = Val(TextBox2.Text)
    A = Val(TextBox3.Text)
    i = Val(TextBox4.Text)
    Rs = Val(TextBox5.Text)
    B = Val(TextBox6.Text)
    q = Val(TextBox7.Text)
    Dim zp, EM As String
    'zeta potential
    zp = 4 * pi * B * q / Di
    'EQUATION: EM = y*A/(t*i*Rs)
    'electrophoretic mobility
    EM = zp * Di / (4 * pi * u)
    textbox8.text = zp.ToString
    textbox9.text = EM.ToString
  End Sub

  Private Sub Button1_Click(ByVal sender As System.Object, ByVal e As System.EventArgs)
Handles Button1.Click
    calculateResults()
  End Sub
End Class
```

### 4.5.3 DESIGN PARAMETERS

Parameters important for design of a coagulation system are

1. Flow rate
2. Flocculation time (residence time) in the range of 15–30 min
3. Power input per unit volume (usually from 0.5 to 1.5 kWh/m³)
4. Mean velocity gradient: for quick circulation in the system, the velocity gradient lies between 20 and 75 s⁻¹. Equation 4.42 shows the relationship between the velocity gradient and the power:

$$G = \left(\frac{POW_v}{\mu}\right)^{\frac{1}{2}} \qquad (4.42)$$

where:

POW$_v$ is the power input per unit volume, W/m³
$G$ is the velocity gradient, s⁻¹

When impellers are used, velocity gradient may be calculated from the following equation:

$$G = \left(\frac{2\pi * g * \omega * TOR}{60 * V * \mu}\right)^{\frac{1}{2}} \qquad (4.43)$$

where:

$G$ is the velocity gradient, s⁻¹
$g$ is the gravitational acceleration, m/s²
$\omega$ is the impeller rotation, number of rotations/min
TOR is the net torque input, dyne/cm
$V$ is the volume, m³
$\mu$ is the dynamic viscosity, N*s/m²

5. The power (kW) requirement to be of values:

$$0.5 < POW < 2.5 \qquad (4.44)$$

6. Depth:

$$\frac{(Volume)^{\frac{1}{3}}}{2} = \frac{(Flow * detention\,time)^{\frac{1}{3}}}{2} = \frac{(Q * t)^{\frac{1}{3}}}{2} \qquad (4.45)$$

7. Surface area:

$$A = \frac{Q}{v} = \frac{Q}{(h/t)} = Q * \frac{t}{h} = \frac{Q * t}{\left(Q * \dfrac{t^{1/3}}{2}\right)} = 2 * (Q * t)^{\frac{2}{3}} \qquad (4.46)$$

### 4.5.4 DESIGN OF FLOCCULATOR

The velocity gradients are brought about by rotating paddles in a flocculator; therefore, particles in a rapidly moving

stream can overtake and collide with others in a slowly moving stream. The power needed for driving the paddle (Punmia 1979) through the fluid is given by the following equation:

$$POW = \rho * C_D * A * v^3 / 2 \qquad (4.47)$$

where:

POW is the power needed for driving the paddle through the fluid, J/s
$\rho$ is the fluid density, kg/m³
$C_D$ is the drag coefficient
$A$ is the paddle area, m²
v is the velocity of paddle relative to fluid, m/s

The velocity gradient initiated as a result of the input power is given by the following equation:

$$G = \left(\frac{POW}{\mu * V}\right)^{\frac{1}{2}} \qquad (4.48)$$

where:

$G$ is the velocity gradient, s⁻¹
$\mu$ is the viscosity, N*s/m²
$V$ is the tank volume, m³

The generally accepted design standard requires that $G$ should be between 30 and 60 s⁻¹ (Pepper and Gerba 2006). Time is also an important variable in flocculation. The term $G*t$ often is used in the design ($t$ = hydraulic retention time in a flocculator). Typical $G*t$ values range from $1 * 10^{14}$ to $1 * 10^5$ (Pepper and Gerba 2006).

#### Example 4.6

1. Write a computer program to determine the power needed for a flocculator system and the area of the paddle required for effective flocculation, given temperature, coefficient of drag for rectangular paddles, paddle tip velocity, velocity gradient, flocculator volume, and the relative velocity of the paddle.
2. A treatment plant incorporates coagulation and sedimentation processes. The existing flocculator has the following properties:

| Item | Value |
| --- | --- |
| Fluid temperature ($T$) | 20°C |
| Coefficient of drag for rectangular paddles ($C_D$) | 1.8 |
| Paddle tip velocity ($v_p$) | 0.7 m/s |
| Velocity gradient ($G$) | 70 s |
| Flocculator volume ($V$) | 4000 m³ |

Using the aforementioned information, compute the following:

a. Power needed for the system
b. Area of the paddle required for effective flocculation (Hint: Use a relative velocity of the paddle equal to 75% of that of the paddle tip velocity.)

**Solution**

1. For solution to Example 4.6 (1), see the listing of Program 4.6.
2. Solution to Example 4.6 (2):
   a. Given: $T = 20°C$, $G = 70$ s, $V = 4000$ m$^3$, $v_p = 0.7$ m/s, $C_D = 1.8$.
   b. Find the viscosity and density from Table A.1 for a temperature of 20°C as follows: viscosity $\mu = 1.002 * 10^{-3}$ N*s/m$^2$ and density $\rho = 998.2$ kg/m$^3$.
   c. Determine the theoretical power requirements as follows: $POW = \mu * G^2 * V$.
   d. Substitute the values given in the previous equation: $POW = 1.002 * 10^{-3}*(70)^2 * 4000 = 19.6$ kW.
   e. Determine the velocity of the paddle: $v = 0.75 * 0.7 = 0.525$ m/s.
   f. Find the required paddle area, $A$, as follows: $A = 2*POW/C_D * \rho * v^3 = (2 * 19.6 * 10^3)/[1.8 * 998.2 * (0.525)^3] = 151$ m$^2$.

---

### LISTING OF PROGRAM 4.6   (CHAP4.6\FORM1.VB): POWER NEEDED FOR A FLOCCULATOR SYSTEM

```
'***************************************************************************************
'Program 4.6: Coagulation 2
'computes the power needed for a flocculator system
'***************************************************************************************

Public Class Form1
    Const pi = 3.141592654
    Dim Cd, vp, G, v, Rv As Double
    Dim T As Integer
    Dim mu, Rho As Double
    Dim density(34), viscosity(34) As Double
    Dim temp(34) As Integer
    '***************************************************************************************
    'Fill the tables with default values,
    'taken from Appendix A1.
    '***************************************************************************************

Private Sub fill_tables()
    temp(0)  = 5
    temp(1)  = 6
    temp(2)  = 7
    temp(3)  = 8
    temp(4)  = 9
    temp(5)  = 10
    temp(6)  = 11
    temp(7)  = 12
    temp(8)  = 13
    temp(9)  = 14
    temp(10) = 15
    temp(11) = 16
    temp(12) = 17
    temp(13) = 18
    temp(14) = 19
    temp(15) = 20
    temp(16) = 21
    temp(17) = 22
    temp(18) = 23
    temp(19) = 24
    temp(20) = 25
    temp(21) = 26
    temp(22) = 27
    temp(23) = 28
    temp(24) = 29
    temp(25) = 30
    temp(26) = 31
```

```
temp(27) = 32
temp(28) = 33
temp(29) = 34
temp(30) = 35
temp(31) = 36
temp(32) = 37
temp(33) = 38
density(0) = 999.956
density(1) = 999.941
density(2) = 999.902
density(3) = 999.849
density(4) = 999.781
density(5) = 999.7
density(6) = 999.605
density(7) = 999.498
density(8) = 999.377
density(9) = 999.244
density(10) = 999.099
density(11) = 998.943
density(12) = 998.774
density(13) = 998.595
density(14) = 998.405
density(15) = 998.203
density(16) = 997.992
density(17) = 997.77
density(18) = 997.538
density(19) = 997.296
density(20) = 997.044
density(21) = 996.783
density(22) = 996.512
density(23) = 996.232
density(24) = 995.944
density(25) = 995.646
density(26) = 995.34
density(27) = 995.025
density(28) = 994.702
density(29) = 994.371
density(30) = 994.03
density(31) = 993.68
density(32) = 993.33
density(33) = 992.96
viscosity(0) = 1.5189
viscosity(1) = 1.4727
viscosity(2) = 1.4289
viscosity(3) = 1.3874
viscosity(4) = 1.3479
viscosity(5) = 1.3101
viscosity(6) = 1.274
viscosity(7) = 1.2396
viscosity(8) = 1.2069
viscosity(9) = 1.1757
viscosity(10) = 1.1457
viscosity(11) = 1.1168
viscosity(12) = 1.0889
viscosity(13) = 1.0618
viscosity(14) = 1.0357
viscosity(15) = 1.0105
viscosity(16) = 0.9863
viscosity(17) = 0.9629
viscosity(18) = 0.9403
viscosity(19) = 0.9186
```

```
      viscosity(20) = 0.8976
      viscosity(21) = 0.8774
      viscosity(22) = 0.8581
      viscosity(23) = 0.8395
      viscosity(24) = 0.8214
      viscosity(25) = 0.8039
      viscosity(26) = 0.7871
      viscosity(27) = 0.7708
      viscosity(28) = 0.7551
      viscosity(29) = 0.7399
      viscosity(30) = 0.7251
      viscosity(31) = 0.7109
      viscosity(32) = 0.6971
      viscosity(33) = 0.6839
   End Sub


   Private Sub Form1_Load(ByVal sender As System.Object, ByVal e As System.EventArgs)
Handles MyBase.Load
      Me.Text = "Program 4.6: Computes the power needed for a flocculator system"
      Me.FormBorderStyle = Windows.Forms.FormBorderStyle.FixedSingle
      Me.MaximizeBox = False
      Label1.Text = "Fluid temperature T (Centigrade):"
      Label2.Text = "Drag coefficient Cd:"
      Label3.Text = "Paddle tip velocity Vp (m/s):"
      Label4.Text = "Velocity gradient G (s):"
      Label5.Text = "Flocculator volume V (m3):"
      Label6.Text = "Relative velocity of the paddle (%):"
      Label7.Text = "Power needed for the system POW (W):"
      Label8.Text = "Area of the paddle required:"
      Button1.Text = "&Calculate"
      GroupBox1.Text = " Output: "
      fill_tables()
   End Sub


   '********************************************************************************************
   'Search the table of densities to find
   'appropriate density according to the given temp.
   '********************************************************************************************
   Function find_rho(ByVal t As Integer) As Double
      If t < 5 Or t > 38 Then Return 0
      For i = 0 To 32
         If temp(i) = t Then Return density(i)
         If temp(i) > t And temp(i + 1) < t Then Return density(i)
      Next
      If temp(33) = t Then Return density(33)
      'no matches found?
      Return 0
   End Function


   '********************************************************************************************
   'Search the table of viscosities to find
   'appropriate viscosity according to the given temp.
   '********************************************************************************************
   Function find_mu(ByVal t As Integer) As Double
      If t < 5 Or t > 38 Then Return 0
      For i = 0 To 32
         If temp(i) = t Then Return viscosity(i) / 1000
         If temp(i) > t And temp(i + 1) < t Then Return viscosity(i) / 1000
      Next
      If temp(33) = t Then Return viscosity(33) / 1000
```

```
      'no matches found?
      Return 0
    End Function

    Sub calculateResults()
      T = Val(TextBox1.Text)
      Cd = Val(TextBox2.Text)
      vp = Val(TextBox3.Text)
      G = Val(TextBox4.Text)
      v = Val(TextBox5.Text)
      Rv = Val(TextBox6.Text) / 100
      mu = find_mu(T)
      Rho = find_rho(T)
      If mu = 0 Or Rho = 0 Then
        MsgBox("Please enter valid temp. between 5 and 38.",
          vbOKOnly Or vbInformation)
        Exit Sub
      End If
      Dim POW, vp1, A As Double
      'power requirement
      POW = mu * G ^ 2 * v
      'velocity of the paddle
      vp1 = Rv * vp
      'required paddle area
      A = 2 * POW / (Cd * Rho * vp1 ^ 3)
      TextBox7.Text = POW.ToString
      TextBox8.Text = A.ToString
    End Sub

    Private Sub Button1_Click(ByVal sender As System.Object, ByVal e As System.EventArgs)
Handles Button1.Click
        calculateResults()
    End Sub
End Class
```

## 4.6 AERATION AND GAS TRANSFER

The process of gas transfer (aeration) signifies the intimate contact between the air and the water for a limited period for the purpose of promoting gas transfer. The advantages of this artificially induced gas process include the following (Mueller et al. 2002):

- Addition of oxygen to groundwater
- Addition of oxygen to sewage to assist in biological reactions
- Removal of carbon dioxide, $CO_2$
- Removal of hydrogen sulfide, $H_2S$
- Removal of methane, $CH_4$
- Reduction of taste and odor problems in water
- Removal of volatile oils and similar odor- and taste-producing substances
- Removal of ammonia, $NH_3$

There are many types of aerators such as gravity aerators (e.g., weir aeration and cascades, inclined plane, and vertical stacks of perforated pans or tower cascades), spray aerators, air diffusers (bubble aeration), and mechanical aerators. The factors that influence the solubility of a gas in water include the nature of the gas, the concentration of the respective gas in the gas phase (i.e., as related to partial pressure of gas in gas phase), the temperature of the water ($C_s$ decreases as $T$ increases), and the level of impurities contained in the water ($C_s$ decreases with an increase in the impurities).

The higher the gas concentration in the gaseous phase, the greater the saturation concentration in the liquid phase. The saturation concentration can be determined by using Henry's law as presented in the following equation:

$$C_s = \frac{K_D * MW * P}{R * T} = k_H * P \qquad (4.49)$$

where:

$C_s$ is the saturation concentration, g/m$^3$
$K_D$ is the distribution coefficient
MW is the molecular weight of the gas
$P$ is the partial pressure of the respective gas in the gas phase, Pa
$R$ is the universal gas constant (8314.3 J/kg*mol)
$T$ is the absolute temperature, K
$k_H$ is the Henry's constant, g/m$^3$*Pa, g/J (=$K_D$*MW/$R$*$T$)

Solubilities also can be estimated by the Bunsen absorption coefficient, as shown in Equation 4.50:

$$C_s = \frac{K_b * MW * P}{R * T_0} \qquad (4.50)$$

where:

$K_b$ is the Bunsen absorption coefficient, g/J ($= K_D T_0 T$)
$T_0$ is the standard temperature, °C

During the process of gas transfer from the gas phase into water, the concentration in the gas phase is decreased. This decrease is due to the absorption of the gas into the liquid phase (process of diffusion, i.e., the tendency of any substance to spread uniformly throughout the space available to it). Some of the theories describing the mechanism of gas transfer include film, penetration, surface renewal, and film surface renewal.

In some gas transfer operations (e.g., cascades, weir aeration), the efficiency coefficient for the system may be calculated as presented in the following equation (Mueller et al. 2002):

$$C_e - C_0 = K(C_s - C_0) \qquad (4.51)$$

where:

$C_e$ is the effluent concentration, g/m³
$C_0$ is the influent concentration, g/m³
$K$ is the efficiency coefficient of the system
$C_s$ is the saturation concentration, g/m³

For a cascade aerator, Equation 4.52 can be used to calculate the gas saturation concentration in the liquid phase:

$$\frac{(C_s - C_N)}{(C_s - C_0)} = (1 - k_N)^N \qquad (4.52)$$

where:

$C_s$ is the saturation concentration, g/m³
$C_N$ is the effluent concentration from step $N$ of the cascade aerator, g/m³
$C_0$ is the influent concentration, g/m³
$k_N$ is the efficiency coefficient of the system
$N$ is the number of steps of the cascade aerator

## Example 4.7

1. Write a computer program that can be used to determine the efficiency coefficient for a one-step cascade aerator. Also include in the program the procedure to calculate the concentration of gases sorbed into an $N$-step cascade aerator.
2. Water at a temperature of 20°C has an oxygen content of 15% saturation with zero chloride concentration. A cascade aerator is to be used. The cascade aerator is composed of three steps. Each step can raise the oxygen content of an anaerobic water to 40% saturation. Estimate the oxygen concentration of the effluent from a three-step cascade aerator.
3. Use the computer program developed in (1) to verify the solution of the problem presented in (2).

## Solution

1. For solution to Example 4.7 (1), see the listing of Program 4.7.
2. Solution to Example 4.7 (2):
   a. Given: influent concentration $C_0 = 0.15 * C_e$, $T = 20$°C, $C_0$ for each step = 0, $C_e$ for each step = $0.4 * C_s$, $N = 3$ steps.
   b. Find the saturation concentration from Table A.2 corresponding to temperature of 20°C as follows: $C_s = 9.2$ mg/L.
   c. Determine the initial concentration of water as follows: $C_0 = 0.15 * 9.2 = 1.38$ mg/L.
   d. Find the efficiency coefficient for each step as follows: $K = (C_e - C_0)/(C_s - C_0) = (0.4(C_s - 0)/(C_s - 0) = 0.4$.
   e. Find the effluent oxygen concentration from each step of the three-step aerator as follows:
      i. $C_{e1} = C_0 + K*(C_s - C_0) = 1.38 + 0.4*(9.2 - 1.38) = 4.508$ mg/L.
      ii. $C_{e2} = C_{e1} + K*(C_s - C_{e1}) = 4.508 + 0.4*(9.2 - 4.508) = 6.385$ mg/L.
      iii. $C_{e3} = C_{e2} + K*(C_s - C_{e2}) = 6.385 + 0.4*(9.2 - 6.385) = 7.51$ mg/L.
   f. Otherwise $(C_s - C_N)/(C_s - C_0) = (1 - k_=)^N$ and $(9.2 - C_{e3})/(9.2 - 1.38) = (1 - 0.4)^3$, which yields $C_{e3} = 7.51$ mg/L.

---

### LISTING OF PROGRAM 4.7　(CHAP4.7\FORM1.VB): EFFICIENCY COEFFICIENT OF A CASCADE AERATOR

```
'****************************************************************************
'Program 4.7: Cascade
'Computes the efficiency coefficient
'of a cascade aerator
'****************************************************************************


Public Class Form1
   Dim Ce(20)
   Const gammaw = 9.806 'KN/m3
```

```
Const G = 9.81 'Gravity constant
Dim N As Double
Dim T As Integer
Dim PCo, Cc, ps As Double
Dim Cs(31) As Double
'********************************************************************************
'Fill the O2 saturation table with values
'from Appendix A2.
'********************************************************************************
Sub fill_table()
  Cs(0) = 14.6
  Cs(1) = 14.2
  Cs(2) = 13.8
  Cs(3) = 13.5
  Cs(4) = 13.1
  Cs(5) = 12.8
  Cs(6) = 12.5
  Cs(7) = 12.2
  Cs(8) = 11.9
  Cs(9) = 11.6
  Cs(10) = 11.3
  Cs(11) = 11.1
  Cs(12) = 10.8
  Cs(13) = 10.6
  Cs(14) = 10.4
  Cs(15) = 10.2
  Cs(16) = 10
  Cs(17) = 9.7
  Cs(18) = 9.5
  Cs(19) = 9.4
  Cs(20) = 9.2
  Cs(21) = 9
  Cs(22) = 8.8
  Cs(23) = 8.7
  Cs(24) = 8.5
  Cs(25) = 8.4
  Cs(26) = 8.2
  Cs(27) = 8.1
  Cs(28) = 7.9
  Cs(29) = 7.8
  Cs(30) = 7.6
End Sub

  Private Sub Form1_Load(ByVal sender As System.Object, ByVal e As System.EventArgs)
Handles MyBase.Load
    Me.Text = "Program 4.7: computes the efficiency coefficient of a cascade aerator"
    Me.FormBorderStyle = Windows.Forms.FormBorderStyle.FixedSingle
    Me.MaximizeBox = False
    label1.text = "Enter number of steps in the aerator:"
    label2.text = "Enter the temperature (C):"
    label3.text = "Enter influent concentration as percent saturation:"
    label4.text = "Enter chloride concentration in water mg/L:"
    label5.text = "Enter percentage of saturation caused by each step:"
    groupbox1.text = " Output: "
    textbox6.multiline = True
      TextBox6.Height = 150
      TextBox6.ScrollBars = ScrollBars.Vertical
      Button1.Text = "&Calculate"
      fill_table()
  End Sub
```

```
Sub calculateResults()
  n = Val(textbox1.text)
  T = Val(TextBox2.Text)
  If T < 0 Or T > 30 Then
     MsgBox("Enter Temperature between 0 and 30", vbOKOnly Or vbInformation)
     Exit Sub
  End If
  PCo = Val(textbox3.text)
  Cc = Val(textbox4.text)
  ps = Val(textbox5.text)
  Dim Ce1, Cs_val, Co, K As Double
  'get saturation concentration at temperature T
  Cs_val = Cs(T)
  textbox6.text = ""
  If PCo > 1 Then PCo = PCo / 100 'convert percentage to ratio
  If PS > 1 Then PS = PS / 100 'convert percentage to ratio
  Co = 0
  Ce1 = ps * Cs_val
  K = (Ce1 - Co) / (Cs_val - Co) 'for each step
  'compute effluent oxygen concentration for each step
  Ce(0) = PCo * Cs_val 'initial concentration of water
  For i = 1 To N
     Ce(i) = Ce(i - 1) + K * (Cs_val - Ce(i - 1))
     textbox6.text += "Effluent oxygen concentration for step" + i.ToString + " = " +
Ce(i).ToString _
     + " mg/L" + vbCrLf
  Next i
End Sub

Private Sub Button1_Click(ByVal sender As System.Object, ByVal e As System.EventArgs)
Handles Button1.Click
   calculateResults()
End Sub
End Class
```

## 4.7  FILTRATION

### 4.7.1  INTRODUCTION

Filtration is a process used for separating suspended or colloidal impurities from a liquid by passing the liquid through a porous medium (Davis and Cornwell 2006). Filtration objectives include the following (Rowe and Abdel-Magid 1995):

- Partial removal of SS and colloidal solids
- Alteration of chemical properties of constituents
- Reduction in the number of disease-causing agents
- Reduction of color, tastes, and odor
- Removal of iron and manganese

Effective filtration depends upon several mechanisms, which include mechanical sieving or straining, precipitation, adsorption, chemical actions, and biological processes (Huisman 1977, IRC Rep. Intl. Appraisal Meeting 1981, Van Dijk and Oomen 1982, Huisman et al 1986, Rowe and Abdel-Magid 1995, Logsdon 2008, Viessman et al 2008, Hammer and Hammer 2011, Metcalf and Eddy 2013, Pescod, Abouzaid and Sundaresan).

The porous media preferred for filtration must be (Rowe and Abdel-Magid 1995)

- Inexpensive.
- Readily available in sufficient quantities.
- An inert material.
- Easily cleaned for reuse.
- Able to withstand existing pressures.

Examples of suitable materials used as filter media include ordinary sand, anthracite coal, broken stones, glass beads, plastic substances, concrete, diatomaceous earth, and so on.

In water treatment, sand filters are often used. Sand filters can be classified as single media, dual media, or multimedia filters. Also, they can be classified according to their allowable loading rate, such as slow or rapid sand filters. Another

## TABLE 4.1
## Comparison of Rapid and Slow Sand Filters

| Parameter | Rapid Sand Filter | Slow Sand Filter |
|---|---|---|
| Reason for filtration | Removal of SS and reduction of pathogens | Reduction in finalizing treatment |
| Location in treatment work | After coagulation or sedimentation | Without or after coagulation, or after RSF |
| Efficiency | Depends on raw water quality and design | Depends on raw water quality and design |
| Raw water turbidity needed | High | Moderate (<15 NTU) |
| Design period | 10–15 years | 10–15 years |
| Life span | Relatively long | Relatively long |
| Filtration rate ($m^3/m^2/h$) | 5–15 | 0.1–0.2 |
| Total area ($A_t$) | Flow rate/filtration velocity | Flow rate/filtration velocity |
| Area of each filter | ($A_t/N - 1$), ($A_t/N - 2$) | ($A_t/N - 1$), ($A_t/N - 2$) |
| Dimensions[20] | | $1 = [2A_t/(N + 1)]^{1/2}$ |
| | | $B = (N + 1)L/2N$ |
| Effective grain size (mm) | 0.4–3 | 0.15–0.35 |
| Uniformity coefficient | >1.2–1.5 | <3–5 (2.5) |
| Bed thickness (m) | 0.6–3 | 0.8–1.2 |
| Supernatant water level (m) | 1–1.5 | 1–1.5 |
| Minimum depth before resanding (m) | Depends on treatment | 0.5 |
| Number of filters[1] (minimum of two filters) | $12(Q)^{1/2}$ ($Q$ in $m^3/s$) | $15(Q)^{1/2}$ ($Q$ in $m^3/s$) |
| Operation period | 24 h/day (intermittent not recommended) | 24 h/day (intermittent not recommended) |
| Interval between successive cleanings | 12–72 h | 20–60 days or more |
| Filter bed resistance | 1.5–4 m | |
| Method of cleaning | Backwashing (water and/or air) | Scraping top 0.5–2 cm layer |
| Sludge removal | Manual, mechanical, and hydraulic | Manual, mechanical, and hydraulic |
| Filter material | Concrete, brick, plastics, and so on | Concrete, brick, plastics, ans so on |
| Maintenance | Continuous | Continuous |
| Hazards | Algal growth, change in water quality, and clogging | Algal growth, change in water quality, and clogging |
| Control measures | Head loss, flow rate, and turbidity | Head loss, flow rate, and turbidity |
| Important quality parameters to be analyzed | Turbidity and bacteriological quality | Turbidity and bacteriological quality |

*Source:* Rowe, D. R. and Abdel-Magid, I. M., *Handbook of Wastewater Reclamation and Reuse,* CRC Press/Lewis Publishers, Boca Raton, FL, 1995. With permission.

SS, suspended solids; RSF, rapid sand filter; NTU, neplometric turbidity unit; $Q$, flow.

---

classification is based on water movement through the filter, such as gravity or pressure filters.

Table 4.1 provides a general comparison between slow and rapid sand filters.

## Example 4.8

1. Write a computer program to find the required number of filters to be used, the filtration area, and the required area for each filter, given the filtration rate and the amount of water to be filtered when using
   a. Slow sand filters.
   b. Rapid sand filters.
2. A sedimentation tank effluent is applied to rapid sand filters at a rate of 20 $m^3$/min. The rate of filtration is to be 200 $m^3/m^2$/day. Calculate the following:

    a. Required number of filters
    b. Total filtration area
    c. Unit area of each filter
3. Use the program developed in (1) to verify the results obtained in (2).

## Solution

1. For solution to Example 4.8 (1), see the listing of Program 4.8.
2. Solution to Example 4.8 (2):
   a. Given: $Q = 20$ $m^3$/min, $v_f = 200$ m/day.
   b. Find the required number of filters needed by using the empirical equation: $N = 12 * (Q)^{1/2} = 12 * (20/60)^{1/2} = 7$ filters (take nine filters, two to serve as standby).
   c. Determine the total filtration surface area as follows: $A = Q/v_f = 20 * 60 * 24/200 = 144$ $m^2$.
   d. Compute the unit filter area for each filter as follows: $A_n = A/(N - 2) = 144/(9 - 2) = 20.6$ $m^2$.

**LISTING OF PROGRAM 4.8   (CHAP4.8\FORM1.VB): NUMBER OF FILTERS, FILTRATION AREA, AND FILTER AREAS**

```vb
'********************************************************************************
'Program 4.8: Filters
'computes number of filters, filtration area, and filter areas
'********************************************************************************

Public Class Form1
  Dim Q, Vf As Double

  Private Sub Form1_Load(ByVal sender As System.Object, ByVal e As System.EventArgs)
Handles MyBase.Load
    Me.Text = "Program 4.8: computes number of filters, filtration area, and filter
areas"
    Me.FormBorderStyle = Windows.Forms.FormBorderStyle.FixedSingle
    Me.MaximizeBox = False
    label1.text = "Enter rate of flow in m3/min:"
    label2.text = "Enter the rate of filtration in m3/m2/day:"
    label3.text = "Select the type of the filter:"
    radiobutton1.text = "Slow"
    radiobutton2.text = "Rapid"
    RadioButton1.Checked = True
    textbox3.multiline = True
    textbox3.height = 100
    GroupBox1.Text = " Output: "
    Button1.Text = "&Calculate"
  End Sub

  Sub calculateResults()
    Q = Val(textbox1.text)
    Vf = Val(textbox2.text)
    Q = Q / 60 'm3/s
    Dim N, Nadd, N1 As Integer
    Dim A, An As Double
    If radiobutton1.checked Then
      N = Math.Ceiling(12 * (Q ^ 0.5))
    Else
      N = Math.Ceiling(15 * (Q ^ 0.5))
    End If
    'stand by filters
    If N > 3 Then Nadd = 2 Else Nadd = 1
    'adding standby filters
    N1 = N + Nadd
    'Q per day
    Q = Q * 3600 * 24
    'total filtration surface area
    A = Q / Vf
    'unit filter area for each filter
    An = A / N
    TextBox3.Text = "The number of filters needed: " + N.ToString
    textbox3.text += " filters + " + Nadd.tostring + " standby = " + N1.tostring
    textbox3.text += vbCrLf + "Total filtration surface area (m2): " + A.tostring
    TextBox3.Text += vbCrLf + "Unit filter area for each filter (m2): " + An.ToString
  End Sub

  Private Sub Button1_Click(ByVal sender As System.Object, ByVal e As System.EventArgs)
Handles Button1.Click
    calculateResults()
  End Sub
End Class
```

## 4.7.2 FILTRATION THEORY

Filter clogging results in an increase in the head loss (see Figure 4.9). The Carman–Kozeny equation can be used to estimate the head loss for clean unsized sand:

$$h_L = f * \frac{l}{d} * \frac{(1-e) * v_f^2}{\phi * d * e^3 * g} \tag{4.53}$$

where:

$h_L$ is the head loss through the filter, m
$f$ is the friction factor, dimensionless, which is given as follows:

$$f = \left\{ \frac{150(1-e)}{\text{Re}} \right\} + 1.75 \tag{4.54}$$

where:

$l$ is the filter depth, m
$e$ is the porosity of bed, dimensionless
$v_f$ is the filtration velocity, m/s
$\phi$ is the shape factor, dimensionless
$d$ is the diameter of particle, m
$g$ is the gravitational acceleration, m/s²
Re is the Reynolds number, dimensionless

Another equation that can be used to predict the head loss through a sand filter was developed by Rose and is presented as follows:

$$h_L = 1.067 * C_D * \frac{l}{d} * \frac{v_f^2}{\left( g * d * \phi * e^4 \right)} \tag{4.55}$$

where:

$h_L$ is the head loss in the filter, m
$C_D$ is the Newton's drag coefficient, which is given as follows:

$$C_D = \left( \frac{24}{\text{Re}} \right) + \left( \frac{3}{\sqrt{\text{Re}}} \right) + 0.34 \tag{4.56}$$

$v_f$ is the filtration velocity, m/s
$l$ is the filter depth, m
$g$ is the gravitational acceleration, m/s²
$d$ is the particle diameter, m

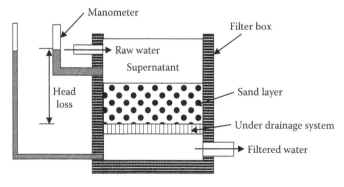

**FIGURE 4.9**   Schematic diagram of a sand filter.

$\phi$ is the particle shape factor, dimensionless
$e$ is the bed porosity, dimensionless

### Example 4.9

1. Write a computer program to compute the head loss by using the Rose or Carman–Kozeny equation. The given information includes the filtration velocity, water viscosity, filter depth, average particle size, and specific gravity for sand, the particle shape factor, and the bed porosity. The program needs to be designed so that the head loss can be determined for various temperatures.
2. A trifilter media (third layer on top) is used for filtering water at 20°C. The properties of the filter are as shown below:

| Item | Sand Layers | | Anthracite Layer |
|---|---|---|---|
| | First | Second | Third |
| Depth (m) | 0.6 | 0.6 | 0.4 |
| Average particle size (mm) | 0.5 | 1.0 | 1.4 |
| Specific gravity | 2.65 | 2.0 | 1.5 |
| Particle shape factor | 0.92 | 0.83 | 0.9 |
| Porosity of bed (%) | 50 | 45 | 45 |
| Filtration rate (m³/m²/day) | 200 | 200 | 200 |

Determine the head loss using both the Carman–Kozeny and Rose equations. Find the percentage error between the two methods.
3. Use the computer program developed in (1) to verify the results determined in (2).

### Solution

1. For solution to Example 4.9 (1), see the listing of Program 4.9.
2. Solution to Example 4.9 (2):
   a. Given: the properties of media, $v_f = 200/(24 * 60 * 60) = 2.32 * 10^{-3}$ m/s, $T = 20°C$, which gives from Table A.1 $\mu = 1.0087 * 10^{-3}$ and $\rho = 998.203$ kg/m³.
   b. Rose equation
      i. For the first sand layer:
         A. Determine Reynolds number: Re = $\rho v_f * d/v = 998.203 * 2.32 * 10^{-3} * 0.5 * 10^{-3}/1.0087 * 10^{-3} = 1.148$.
         B. Compute Newton's drag coefficient as follows: $C_D = (24/\text{Re}) + [3/(\text{Re})^{1/2}] + 0.34 = (24/1.148) + [3/(1.148)^{1/2}] + 0.34 = 24.05$.
         C. Determine the head loss for the first layer as follows: $h_L = 1.067*C_D*v_f^2 * 1/g*d*\phi*e^4$. $h_L = [1.067 * 24.05* (2.32 * 10^{-3})^2 * 0.6]/[9.81 * 0.5 * 10^{-3} * 0.92 * (0.5)^4] = 0.293$ m.

ii. For the second sand layer:
   A. Determine Reynolds number: Re = $998.203 * 2.32 * 10^{-3}*1 * 10^{-3}$ = 2.3.
   B. Find Newton's drag coefficient as follows: $C_D = (24/2.3) + [3/(2.3)^{1/2}] + 0.34 = 12.75$.
   C. Estimate the head loss for the second layer as follows: $h_L = [1.067 * 12.75*(2.32 * 10^{-3})^2 * 0.6]/[9.81 * 1*10^{-3} * 0.83 * (0.45)^4] = 0.132$ m.

iii. For the third anthracite layer:
   A. Determine Reynolds number: Re = $998.203 * 2.32 * 10^{-3}*1.4 * 10^{-3}/1.0087 * 10^{-3} = 3.21$.
   B. Find Newton's drag coefficient as follows: $C_D = (24/3.21) + [3/(3.21)^{1/2}] + 0.34 = 9.49$.
   C. Estimate the head loss for the third layer as follows: $h_L = [1.067 * 9.49*(2.32 * 10^{-3})^2 * 0.4]/[9.81 * 1.4 * 10^{-3}*0.9* (0.45)^4] = 0.043$ m.

iv. Determine the total head loss = head loss through the first layer + head loss through the second layer + head loss through anthracite layer = 0.293 + 0.132 + 0.043 = 0.468 m.

c. Carman–Kozeny equation
   i. For the first sand layer:
      A. Determine Reynolds number: Re = $\rho v_i*d/\mu = 998.203 * 2.32 * 10^{-3}*0.5 * 10^{-3}/1.0087 * 10^{-3} = 1.148$.
      B. Find the coefficient: $f = [150(1 - e)/Re] + 1.75$. $f = [150(1 - 0.5)/1.148] + 1.75 = 67.08$.

C. Determine the head loss for first layer as follows: $h_L = f*1*(1 - e)* v_f^2/\phi*d*e^3*g$. $h_L = 67.08 * 0.6* (1 - 0.5)*(2.32 * 10^{-3})^2/[0.92 * 0.5 * 10^{-3}*(0.5)^3 * 9.81] = 0.192$ m.

ii. For the second sand layer:
   A. Determine Reynolds number: Re = $998.203 * 2.32 * 10^{-3}*1 * 10^{-3}/1.0087 * 10^{-3} = 2.3$.
   B. Find the coefficient: $f = (150(1 - e)/Re) + 1.75$. $f = [150(1 - 0.45)/2.3] + 1.75 = 37.62$.
   C. Estimate the head loss for the second layer as follows: $h_L = 37.62 * 0.6* (1 - 0.45)*(2.32 * 10^{-3})^2/[0.83 * 1*10^{-3}*(0.45)^3 * 9.81] = 0.09$ m.

iii. For the third anthracite layer:
   A. Determine Reynolds number: Re = $998.203 * 2.32 * 10^{-3}*1.4 * 10^{-3}/1.0087 * 10^{-3} = 3.21$.
   B. Find the coefficient: $f = (150(1 - e)/Re) + 1.75$. $f = [150(1 - 0.45)/3.21] + 1.75 = 27.45$.
   C. Estimate the head loss for the third layer as follows: $h_L = 27.45 * 0.4* (1 - 0.45)*(2.32 * 10^{-3})^2/[0.9 * 1.4 * 10^{-3}*(0.45)^3 * 9.81] = 0.029$ m.

iv. Determine the total head loss = head loss through the first sand layer + head loss through the second sand layer + head loss through anthracite layer = 0.192 + 0.09 + 0.029 = 0.311 m.

v. Determine the percentage error: (0.468 – 0.311)*100/0.468 = 34%.

---

### LISTING OF PROGRAM 4.9   (CHAP4.9\FORM1.VB): HEAD LOSS IN FILTERS

```
'*********************************************************************************
'Program 4.9: HEAD
'Computes the Head loss in filters
'*********************************************************************************

Public Class Form1
  Dim temp(34), density(34), viscosity(34)
  Dim layer As Integer
  Dim total_hL1, total_hL2 As Double
  Dim vf, frate As Double
  Dim N, L(), d(), S(), phi(), e1(), H_Rose(), H_Carman(), T As Double
  Dim is_calculated() As Boolean
  'acceleration due to gravity
  Const g = 9.81
  '*********************************************************************************
  'Fill the tables with default values,
  'taken from Appendix A1.
  '*********************************************************************************

  Private Sub fill_tables()
    temp(0) = 5
    temp(1) = 6
```

```
temp(2)  = 7
temp(3)  = 8
temp(4)  = 9
temp(5)  = 10
temp(6)  = 11
temp(7)  = 12
temp(8)  = 13
temp(9)  = 14
temp(10) = 15
temp(11) = 16
temp(12) = 17
temp(13) = 18
temp(14) = 19
temp(15) = 20
temp(16) = 21
temp(17) = 22
temp(18) = 23
temp(19) = 24
temp(20) = 25
temp(21) = 26
temp(22) = 27
temp(23) = 28
temp(24) = 29
temp(25) = 30
temp(26) = 31
temp(27) = 32
temp(28) = 33
temp(29) = 34
temp(30) = 35
temp(31) = 36
temp(32) = 37
temp(33) = 38
density(0)  = 999.956
density(1)  = 999.941
density(2)  = 999.902
density(3)  = 999.849
density(4)  = 999.781
density(5)  = 999.7
density(6)  = 999.605
density(7)  = 999.498
density(8)  = 999.377
density(9)  = 999.244
density(10) = 999.099
density(11) = 998.943
density(12) = 998.774
density(13) = 998.595
density(14) = 998.405
density(15) = 998.203
density(16) = 997.992
density(17) = 997.77
density(18) = 997.538
density(19) = 997.296
density(20) = 997.044
density(21) = 996.783
density(22) = 996.512
density(23) = 996.232
density(24) = 995.944
density(25) = 995.646
density(26) = 995.34
density(27) = 995.025
density(28) = 994.702
```

```
        density(29) = 994.371
        density(30) = 994.03
        density(31) = 993.68
        density(32) = 993.33
        density(33) = 992.96
        viscosity(0) = 1.5188
        viscosity(1) = 1.4726
        viscosity(2) = 1.4288
        viscosity(3) = 1.3872
        viscosity(4) = 1.3476
        viscosity(5) = 1.3097
        viscosity(6) = 1.2735
        viscosity(7) = 1.239
        viscosity(8) = 1.2061
        viscosity(9) = 1.1748
        viscosity(10) = 1.1447
        viscosity(11) = 1.1156
        viscosity(12) = 1.0875
        viscosity(13) = 1.0603
        viscosity(14) = 1.034
        viscosity(15) = 1.0087
        viscosity(16) = 0.9843
        viscosity(17) = 0.9608
        viscosity(18) = 0.938
        viscosity(19) = 0.9161
        viscosity(20) = 0.8949
        viscosity(21) = 0.8746
        viscosity(22) = 0.8551
        viscosity(23) = 0.8363
        viscosity(24) = 0.8181
        viscosity(25) = 0.8004
        viscosity(26) = 0.7834
        viscosity(27) = 0.767
        viscosity(28) = 0.7511
        viscosity(29) = 0.7357
        viscosity(30) = 0.7208
        viscosity(31) = 0.7064
        viscosity(32) = 0.6925
        viscosity(33) = 0.6791
    End Sub

    Private Sub Form1_Load(ByVal sender As System.Object, ByVal e As System.EventArgs)
Handles MyBase.Load
        Me.Text = "Program 4.9: Computes the Head loss in filters"
        Me.FormBorderStyle = Windows.Forms.FormBorderStyle.FixedSingle
        Me.MaximizeBox = False
        Label1.Text = "Number of layers in the filter:"
        Label2.Text = "Temperature (Centigrade):"
        Label3.Text = "Filter depth (m):"
        Label4.Text = "Particle diameter (mm):"
        Label5.Text = "Specific gravity:"
        Label6.Text = "Particle shape factor (phi):"
        Label7.Text = "Bed porosity e (%):"
        Label8.Text = "Filtration rate (m3/m2/day):"
        Label9.Text = "Head loss hL (m):"
        Label10.Text = "Total head loss in the filter (m):"
        Label11.Text = "(1) Rose Equation:"
        Label12.Text = "(2) Carman-Kozeny Equation:"
        Label13.Text = "Head loss hL (m):"
        Label14.Text = "Total head loss in the filter (m):"
        Button1.Text = "&Calculate"
```

```
      Button3.Text = "Prev. layer"
      Button2.Text = "Next layer"
      Button3.Enabled = False
      Button2.Enabled = False
      fill_tables()
      total_hL1 = 0
      total_hL2 = 0
      N = 1 : layer = 1
      GroupBox1.Text = "LAYER NO. " + layer.ToString
      TextBox1.Text = N.ToString
      ReDim H_Rose(N)
      ReDim H_Carman(N)
      ReDim L(N)
      ReDim d(N)
      ReDim S(N)
      ReDim phi(N)
      ReDim e1(N)
      ReDim is_calculated(N)
    End Sub

    Sub calculateResults()
      'temp
      T = Val(TextBox2.Text)
      'filter depth
      L(layer) = Val(TextBox3.Text)
      'particle diameter
      d(layer) = Val(TextBox4.Text)
      'specific gravity
      S(layer) = Val(TextBox5.Text)
      'particle shape factor
      phi(layer) = Val(TextBox6.Text)
      'bed porosity
      e1(layer) = Val(TextBox7.Text)
      'filtration rate
      frate = Val(TextBox8.Text)
      vf = frate / (24 * 60 * 60)
      Dim kr, Ro, Vs, Re, Cd As Double
      'determine the density and the viscosity
      kr = -1
      For i = 0 To 32
        If T = temp(i) Then
          kr = i
          GoTo found
        End If
        If T > temp(i) And T < temp(i + 1) Then
          kr = i
          GoTo found
        End If
      Next i
      If T = temp(33) Then
        kr = 33
      Else
        MsgBox("Please enter temp. between 5 and 38", vbOKOnly Or vbInformation)
        Exit Sub
      End If
found:
      Ro = density(kr)
      Vs = viscosity(kr) / 1000
      '*****************************************************************************
      '*** 1. Calculate using Rose Equation:
      '*****************************************************************************
```

```
     Dim _e, _d As Double
     If e1(layer) > 1 Then _e = e1(layer) / 100 Else _e = e1(layer)
     'Convert to m
     _d = d(layer) / 1000
     'Reynolds number
     Re = Ro * vf * _d / Vs
     'Drag coefficient
     Cd = (24 / Re) + (3 / (Re ^ 0.5)) + 0.34
     'Compute head loss hL
     H_Rose(layer) = 1.067 * Cd * (vf ^ 2) * L(layer) / (g * _d * phi(layer) * (_e ^ 4))
     '********************************************************************************
     '*** 2. Calculate using Carman-Kozeny Equation:
     '********************************************************************************
     Dim f As Double
     f = ((150 * (1 - _e)) / Re) + 1.75
     H_Carman(layer) = (f * L(layer) * (1 - _e) * (vf ^ 2)) / (phi(layer) * _d * (_e ^ 3)
  * g)
     total_hL1 = 0
     total_hL2 = 0
     For i = 1 To N
       total_hL1 += H_Rose(i)
       total_hL2 += H_Carman(i)
     Next
     Label9.Text = "Head loss hL (m): " + H_Rose(layer).ToString
     Label10.Text = "Total head loss in the filter (m): " + total_hL1.ToString
     Label13.Text = "Head loss hL (m): " + H_Carman(layer).ToString
     Label14.Text = "Total head loss in the filter (m): " + total_hL2.ToString
     is_calculated(layer) = True
  End Sub

  Private Sub Button1_Click(ByVal sender As System.Object, ByVal e As System.EventArgs)
Handles Button1.Click
     calculateResults()
  End Sub

  Private Sub TextBox1_TextChanged(ByVal sender As System.Object, ByVal e As System.
EventArgs) Handles TextBox1.TextChanged
     If Val(TextBox1.Text) > 1 Then
       If Val(TextBox1.Text) > N Then
          Button2.Enabled = True
          N = Val(TextBox1.Text)
          ReDim H_Rose(N)
          ReDim H_Carman(N)
          ReDim L(N)
          ReDim d(N)
          ReDim S(N)
          ReDim phi(N)
          ReDim e1(N)
          ReDim is_calculated(N)
       Else
          N = Val(TextBox1.Text)
       End If
     Else
       Button2.Enabled = False
       Button3.Enabled = False
     End If
  End Sub

  Private Sub Button2_Click(ByVal sender As System.Object, ByVal e As System.EventArgs)
Handles Button2.Click
     If layer >= N Then
```

```
         Button2.Enabled = False
         Button3.Enabled = True
      Else
         layer += 1
         GroupBox1.Text = "LAYER NO. " + layer.ToString
         TextBox3.Text = L(layer).ToString
         TextBox4.Text = d(layer).ToString
         TextBox5.Text = S(layer).ToString
         TextBox6.Text = phi(layer).ToString
         TextBox7.Text = e1(layer).ToString
         If is_calculated(layer) = True Then
            Label9.Text = "Head loss hL (m): " + H_Rose(layer).ToString
            Label10.Text = "Total head loss in the filter (m): " + total_hL1.ToString
            Label13.Text = "Head loss hL (m): " + H_Carman(layer).ToString
            Label14.Text = "Total head loss in the filter (m): " + total_hL2.ToString
         Else
            Label9.Text = "Head loss hL (m): "
            Label10.Text = "Total head loss in the filter (m): "
            Label13.Text = "Head loss hL (m): "
            Label14.Text = "Total head loss in the filter (m): "
         End If
         Button3.Enabled = True
         If layer >= N Then
            Button2.Enabled = False
         End If
      End If
   End Sub

   Private Sub Button3_Click(ByVal sender As System.Object, ByVal e As System.EventArgs)
Handles Button3.Click
      If layer <= 1 Then
         Button3.Enabled = False
         Button2.Enabled = True
      Else
         layer -= 1
         GroupBox1.Text = "LAYER NO. " + layer.ToString
         TextBox3.Text = L(layer).ToString
         TextBox4.Text = d(layer).ToString
         TextBox5.Text = S(layer).ToString
         TextBox6.Text = phi(layer).ToString
         TextBox7.Text = e1(layer).ToString
         If is_calculated(layer) = True Then
            Label9.Text = "Head loss hL (m): " + H_Rose(layer).ToString
            Label10.Text = "Total head loss in the filter (m): " + total_hL1.ToString
            Label13.Text = "Head loss hL (m): " + H_Carman(layer).ToString
            Label14.Text = "Total head loss in the filter (m): " + total_hL2.ToString
         Else
            Label9.Text = "Head loss hL (m): "
            Label10.Text = "Total head loss in the filter (m): "
            Label13.Text = "Head loss hL (m): "
            Label14.Text = "Total head loss in the filter (m): "
         End If
            Button2.Enabled = True
         If layer <= 1 Then
            Button3.Enabled = False
         End If
      End If
   End Sub
End Class
```

**Example 4.10**

1. Write a short computer program that can be used to estimate the filtration rate when given the head loss, viscosity, specific gravity, particle shape factor, and other elements included in the Rose and Carman–Kozeny equations. Plan the computer program so that the suitability of the filtration rate can be evaluated.
2. A single, unisized sand bed filter of depth 1.2 m and a porosity of 0.6 with uniform sand grains of 0.8 mm is used to treat water at 22°C. Find the range of the filtration rate (in m/day) for a head loss of 0.12 m. (Use both the Rose and Carman–Kozeny equations.)
3. Use computer programs developed in (1) to verify computations in (2).

**Solution**

1. For solution to Example 4.10 (1), see the listing of Program 4.10.
2. Solution to Example 4.10 (2):
   a. Given: $l = 1.2$ m, $e = 0.6$, $d = 0.8 * 10^{-3}$ m, $h_f = 0.12$ m, $T = 22°C$. From Table A.1, $\mu = 0.9608 * 10^{-3}$, $\rho = 997.77$.
      i. Rose equation:
         A. Find Reynolds number as follows: $Re = 997.77*(v/60 * 60 * 24)* 0.8 * 10^{-3}/ 0.9608 * 10^{-3} = 9.616 * 10^{-3}v$.
         B. Find Newton's drag coefficient as follows: $C_D = (24/9.616 * 10^{-3}/v) +$ $[3/(9.616 * 10^{-3}v)^{1/2}] + 0.34 = 2495.8/v + 30.59/v^{1/2} + 0.34$.
      C. Use the head loss equation: $h_L = 0.12 = 1.067*(2495.8/v+30.59/v^{1/2}+ 0.34)*(v/60 * 60 * 24)^2 * 1.2)/(9.81 * 0.8 * 10^{-3}*1*(0.6)^4)$.
      D. Estimate the filtration velocity by trial and error, which yields $v = 234$ m/day.
      ii. Carman–Kozeny equation:
         a. Reynolds number as determined before is $Re = 9.616 * 10^{-3}v$.
         b. Find the coefficient: $f = [150(1 – e)/ Re] + 1.75. f = [150(1 – 0.6)/9.616 * 10^{-3}v] + 1.75 = 6239.6/v + 1.75$.
         c. Estimate the filtration rate from the head loss equation as follows: $h_L = 0.12 = (6239.6/v + 1.75)*1.2*(1 – 0.6)*(v/60 * 60 * 24)^2/[1 * 0.8 * 10^{-3}*(0.6)^3 * 9.81]$, which yields $v = 450$ m/day.
      iii. Determine the percentage difference between the two methods = (450 – 234)*100/450 = 48%.
3. In the development of Program 4.10, the roots of the equation were determined using the numerical method of Newton–Raphson. For a function such as $f(x) = a_0 + a_1x + a_2x^2 + \cdots + a_nx^n$, if an estimate of the root is chosen to be $x_r$, an improved estimate can be obtained by taking the derivate at $x = x_r$. Accordingly, the improved estimate for the root can be obtained as $x_{r+1} = x_r + f(x_r)/[df(x_r)/dx]$.

---

**LISTING OF PROGRAM 4.10   (CHAP4.10\FORM1.VB): FILTRATION RATE FOR A GIVEN HEAD LOSS**

```
'*******************************************************************************
'Program 4.10: FILTRATION
'Computes the filtration rate for a given head loss
'REQUIRES ZEROS OF A FUNCTION - USES NEWTON RAPHSON METHOD
'*******************************************************************************

Imports System.Math
Public Class Form1
    Dim density(34), viscosity(34)
    Dim H1, L, d, phi, _e As Double
    Dim temp(34), T As Integer
    Const g = 9.81 'acceleration due to gravity
    '*******************************************************************************
    'Fill the tables with default values,
    'taken from Appendix A1.
    '*******************************************************************************

    Private Sub fill_tables()
        temp(0) = 5
        temp(1) = 6
        temp(2) = 7
        temp(3) = 8
        temp(4) = 9
```

```
temp(5)  = 10
temp(6)  = 11
temp(7)  = 12
temp(8)  = 13
temp(9)  = 14
temp(10) = 15
temp(11) = 16
temp(12) = 17
temp(13) = 18
temp(14) = 19
temp(15) = 20
temp(16) = 21
temp(17) = 22
temp(18) = 23
temp(19) = 24
temp(20) = 25
temp(21) = 26
temp(22) = 27
temp(23) = 28
temp(24) = 29
temp(25) = 30
temp(26) = 31
temp(27) = 32
temp(28) = 33
temp(29) = 34
temp(30) = 35
temp(31) = 36
temp(32) = 37
temp(33) = 38
density(0)  = 999.956
density(1)  = 999.941
density(2)  = 999.902
density(3)  = 999.849
density(4)  = 999.781
density(5)  = 999.7
density(6)  = 999.605
density(7)  = 999.498
density(8)  = 999.377
density(9)  = 999.244
density(10) = 999.099
density(11) = 998.943
density(12) = 998.774
density(13) = 998.595
density(14) = 998.405
density(15) = 998.203
density(16) = 997.992
density(17) = 997.77
density(18) = 997.538
density(19) = 997.296
density(20) = 997.044
density(21) = 996.783
density(22) = 996.512
density(23) = 996.232
density(24) = 995.944
density(25) = 995.646
density(26) = 995.34
density(27) = 995.025
density(28) = 994.702
density(29) = 994.371
density(30) = 994.03
density(31) = 993.68
```

```
        density(32) = 993.33
        density(33) = 992.96
        viscosity(0) = 1.5188
        viscosity(1) = 1.4726
        viscosity(2) = 1.4288
        viscosity(3) = 1.3872
        viscosity(4) = 1.3476
        viscosity(5) = 1.3097
        viscosity(6) = 1.2735
        viscosity(7) = 1.239
        viscosity(8) = 1.2061
        viscosity(9) = 1.1748
        viscosity(10) = 1.1447
        viscosity(11) = 1.1156
        viscosity(12) = 1.0875
        viscosity(13) = 1.0603
        viscosity(14) = 1.034
        viscosity(15) = 1.0087
        viscosity(16) = 0.9843
        viscosity(17) = 0.9608
        viscosity(18) = 0.938
        viscosity(19) = 0.9161
        viscosity(20) = 0.8949
        viscosity(21) = 0.8746
        viscosity(22) = 0.8551
        viscosity(23) = 0.8363
        viscosity(24) = 0.8181
        viscosity(25) = 0.8004
        viscosity(26) = 0.7834
        viscosity(27) = 0.767
        viscosity(28) = 0.7511
        viscosity(29) = 0.7357
        viscosity(30) = 0.7208
        viscosity(31) = 0.7064
        viscosity(32) = 0.6925
        viscosity(33) = 0.6791
    End Sub

    Private Sub Form1_Load(ByVal sender As System.Object, ByVal e As System.EventArgs)
Handles MyBase.Load
        Me.Text = "Program 4.10: Computes the filtration rates in filters"
        Me.FormBorderStyle = Windows.Forms.FormBorderStyle.FixedSingle
        Me.MaximizeBox = False
        Label1.Text = "Head loss hL (m):"
        label2.text = "Filter depth (m):"
        label3.text = "Particle diameter (mm):"
        label4.text = "Particle shape factor (phi):"
        Label5.Text = "Bed porosity e (%):"
        Label6.Text = "Temperature (between 5-38°C):"
        GroupBox1.Text = " Output: "
        TextBox7.Multiline = True
            TextBox7.Height = 150
            TextBox7.ScrollBars = ScrollBars.Vertical
            Button1.Text = "&Calculate"
            fill_tables()
    End Sub

    Sub calculateResults()
        H1 = Val(textbox1.text)
        L = Val(textbox2.text)
```

```
    d = Val(textbox3.text)
    phi = Val(textbox4.text)
    _e = Val(TextBox5.Text)
    T = Val(textbox6.text)
    If T < 5 Or T > 38 Then
      MsgBox("Please enter temperature between 5 and 38", vbOKOnly, "Error")
      Exit Sub
    End If
    Dim kr, Ro, Vs, k, C11, C12, C13 As Double
    Dim Vc, B, C1, C2, C3, C4, M2, M1, f, fprevious, Xprevious, Re As Double
    Dim Xn, X, Epsilon, fdash, deltaX, V, difference As Double
    'determine the density and the viscosity
    For i = 0 To 33
      If T = temp(i) Then kr = i
    Next i
    Ro = density(kr)
    Vs = viscosity(kr) / 1000
    If _e > 1 Then _e = _e / 100 'convert to ratio
    d = d / 1000 'convert to meter units
    'compute velocity
      k = 60.0! * 60.0! * 24.0! 'conversion factor to /day
    'set up Carman-Kozeny equation
    C11 = 1.75
    C12 = 12960000.0 * (1 - _e) * Vs / (Ro * d)
    C13 = -H1 * g * d * phi * _e ^ 3 * k ^ 2 / ((1 - _e) * L)
    Vc = (-c12 + (c12 ^ 2 - 4 * c13 * c11) ^ 0.5) / (2 * c11)
    'set up Rose equation
    B = Ro * d / (k * vs)
    C1 = 0.34
    C2 = 3 / B ^ 0.5
    C3 = 24 / B
    C4 = H1 * g * d * phi * _e ^ 4 * k ^ 2 / (1.067 * L)
    M2 = (c4 / c1) ^ 0.25
    M1 = M2 / 10
    'zeroes of function - search for solution interval
    Dim status = 0
    For X = M1 To M2 Step M1
      f = C1 * X ^ 4 + C2 * X ^ 3 + C3 * X ^ 2 - C4
      If X > M1 Then Call check(f, fprevious, X, M1, status)
      If status > 0 Then GoTo jmp
      fprevious = f
      Xprevious = X
      Re = Ro * X ^ 2 * d / Vs
    Next X
jmp:
    'Newton Raphson approximation of roots
    Xn = X
    Epsilon = 0.000001
cycle:
    X = Xn
    f = (C1 * X ^ 4) + (C2 * X ^ 3) + (C3 * X ^ 2) - C4
    fdash = (4 * C1 * X ^ 3) + (3 * C2 * X ^ 2) + (2 * C3 * X)
    V = X ^ 2
    deltaX = f / fdash
    Xn = Xn - deltaX 'improved estimate
    If ABS(deltaX) > Epsilon Then GoTo cycle
    TextBox7.Text = "Solution using Rose equation:" + vbCrLf
    Call RESULTS(V, 6)
```

```
            TextBox7.Text += "Solution using Carman-Kozeny equation:" + vbCrLf
            Call RESULTS(Vc, 12)
            difference = ABS(Vc - V) / Vc * 100
            TextBox7.Text += "Difference between the two methods (%): " + CInt(difference).
    tostring
        End Sub

        Sub check(ByVal f, ByVal fprevious, ByVal X, ByVal m1, ByVal status)
            If X > m1 Then
                If Sign(f) <> Sign(fprevious) Then status = 1
            End If
        End Sub

        Sub RESULTS(ByVal V, ByVal R)
            'output the result of the analysis
            TextBox7.Text += "The filtration velocity is (m/day): " + V.ToString + vbCrLf
            'select a suitable type of sand filter:
            V = V / 24 'in m/hr
            Dim tt = "Not suitable for slow nor for rapid sand filter"
            If V >= 0.1 And V <= 0.2 Then tt = "SLOW SAND FILTER"
            If V <= 5 And V <= 15 Then tt = "RAPID SAND FILTER"
            TextBox7.Text += "Recommendation: " + tt + vbCrLf
        End Sub

        Private Sub Button1_Click(ByVal sender As System.Object, ByVal e As System.EventArgs)
    Handles Button1.Click
            calculateResults()
        End Sub
    End Class
```

### 4.7.3 Clogging of the Filter Bed

The quality of the effluent water from a filter can be evaluated by using the following equation:

$$C_e = C_0 * e^{-\lambda_0 * l} \tag{4.57}$$

where:
$C_e$ is the concentration of contaminant in the effluent, mg/L
$C_0$ is the concentration of contaminant at the filter inlet surface, mg/L
$\lambda_0$ is the filtration coefficient, m$^{-1}$
$l$ is the filter depth, m

The rate of deposition of particulate solids in the filter bed can be estimated from the following equation (Huisman 1977, Rowe and Abdel-Magid 1995):

$$\psi = v * \lambda_0 * C_0 * e^{-l_0 y} * t \tag{4.58}$$

where:
$\psi$ is the the rate of deposition
$v$ is the filtration rate, m/s
$\lambda_0$ is the filtration coefficient, m$^{-1}$
$C_0$ is the concentration of solid particles at the filter inlet surface, mg/L
$y$ is the depth of deposition of solids from the filter inlet surface, m
$t$ is the time taken for deposition of solids within the filter bed, s

### Example 4.11

1. Write a computer program that will estimate the concentration of a contaminant in the effluent from a filter bed as well as the rate of deposition of the contaminant at various depths throughout the filter. Given the concentration of contaminant at the filter inlet surface, temperature, filtration coefficient, filter depth, filtration rate, and time taken for deposition of solids within the filter bed. Plan the computer program so that the percentage reduction of the contaminant by the filter is also determined.

2. Water at a temperature of 20°C with a contaminant concentration of 200 mg/L is to be treated by a filter bed of 1 m depth with a coefficient of filtration of 5 m$^{-1}$. Find the concentration of the contaminant in the effluent if the filtration rate is 10 m$^3$/m$^2$/h.

3. Use the computer program developed in (1) to verify the computations in (2).

#### Solution

1. For solution to Example 4.11 (1), see the listing of Program 4.11.
2. Solution to Example 4.11 (2):
   a. Given: $C_0 = 200$ mg/L, $l = 1$ m, $\lambda = 5$/m, $v_f = 10$ m/h $= 2.78 * 10^{-3}$ m/s.
   b. Find the effluent contaminant concentration using Equation 3.57: $C_0 * e^{-\lambda_0 L} = 200 * e^{-5 * l} = 1.3$ mg/L.

**LISTING OF PROGRAM 4.11    (CHAP4.11\FORM1.VB): CONCENTRATION
OF IMPURITIES OF EFFLUENTS OF A FILTER**

```vb
'*********************************************************************************
'Program 4.11: Effluents
'computes the concentration of impurities of
'the effluents
'*********************************************************************************

Imports System.Math
Public Class Form1
  Dim Co, lamda, L, v As Double

  Private Sub Form1_Load(ByVal sender As System.Object, ByVal e As System.EventArgs)
Handles MyBase.Load
    Me.Text = "Program 4.11: Computes the concentration of impurities of effluents"
    Me.FormBorderStyle = Windows.Forms.FormBorderStyle.FixedSingle
    Me.MaximizeBox = False
    Label1.Text = "Concentration of impurities in the effluent (mg/L):"
    Label2.Text = "Filtration coefficient (/m):"
    Label3.Text = "Filter depth (m):"
    Label4.Text = "Filtration rate (m/s):"
    GroupBox1.Text = " Output: "
    TextBox5.Multiline = True
    TextBox5.Height = 150
    TextBox5.ScrollBars = ScrollBars.Vertical
    Button1.Text = "&Calculate"
  End Sub

  Sub calculateResults()
    Co = Val(TextBox1.Text)
    lamda = Val(TextBox2.Text)
    L = Val(TextBox3.Text)
    v = Val(TextBox4.Text)
    Dim y, Cx, delta, percent As Double
    Cx = Co * EXP(-lamda * L)
    TextBox5.Text = "Effluent concentration of impurities (mg/L): " + Cx.ToString + vbCrLf
    percent = 100 * (Co - Cx) / Co
    TextBox5.Text += "Percentage reduction of impurities (%): " + percent.ToString + vbCrLf
    y = L / 2
    Cx = Co * EXP(-lamda * y)
    TextBox5.Text += "Concentration of impurities at half depth (mg/L): " + Cx.ToString
+ vbCrLf
    delta = v * lamda * Co * EXP(-lamda * L)
    TextBox5.Text += "Rate of deposition of impurities:" + delta.ToString
  End Sub

  Private Sub Button1_Click(ByVal sender As System.Object, ByVal e As System.EventArgs)
Handles Button1.Click
    calculateResults()
  End Sub
End Class
```

### 4.7.4 Backwashing a Rapid Sand Filter

Rapid filters are cleaned by backwashing. Backwashing refers to reversal of the direction of flow in order to expand the bed at the end of a filter run. Expansion of the bed produces a situation in which the accumulated contaminants are scoured off the sand particle surface. Filter bed expansion may be estimated from the following empirical equation:

$$\frac{l_e}{l} = \frac{(1-e)}{(1-e_e)} = \frac{(1-e)}{\left[1-\left(v/v_s\right)^{0.22}\right]} \qquad (4.59)$$

where:

$l_e$ is the expanded bed depth, m

$l$ is the bed depth, m

$e$ is the porosity of bed, dimensionless

$e_e$ is the expanded bed porosity, dimensionless

v is the face velocity of the backwash water (rise rate), m/s

$v_s$ is the sand particles settling velocity, m/s

## Example 4.12

1. Write a short computer program to find the filter bed expansion during backwashing, given the rate of water backwashing, the sand particle shape factor, the sand bed porosity, the settling velocity of the sand particles, and the depth of the filter.

2. A uniform sand filter bed of depth 0.6 m with a porosity of 0.4 is used for filtering water at a filtration rate of 15 $m^3/m^2/h$ with a measured settling velocity of sand particles of 0.05 m/s. When backwashing is required, it is carried out at a rate of 20 $m^3/m^2h$. Determine the depth of bed expansion.

3. Use computer program developed in (1) to verify computations in (2).

**Solution**

1. For solution to Example 4.12 (1), see the listing of Program 4.12.
2. Solution to Example 4.12 (2):
   a. Given: $l = 0.6$ m, $e = 0.4$, $v = 20$ m/h, $v_s = 0.05$ m/s $= 180$ m/h.
   b. Determine the depth of bed expansion as follows: $l_e/l = (1 - e)/[1 - (v/v_s)^{0.22}]$. $l_e/0.6 = (1 - 0.4)/[1 - (20/180)^{0.22}]$. This yields a value of $l_e = 0.94$ m.

---

### LISTING OF PROGRAM 4.12   (CHAP4.12\FORM1.VB): DEPTH OF BED EXPANSION

```
'***********************************************************************************
'Program 4.12: Backwashing
'Computes the depth of bed expansion
'***********************************************************************************

Public Class Form1
    Dim Lx, e, L, v, Vs As Double

    Private Sub Form1_Load(ByVal sender As System.Object, ByVal e As System.EventArgs)
Handles MyBase.Load
        Me.Text = "Program 4.12: Computes the depth of bed expansion"
        Me.FormBorderStyle = Windows.Forms.FormBorderStyle.FixedSingle
        Me.MaximizeBox = False
        label1.text = "Enter porosity of bed:"
        label2.text = "Enter depth of bed (m):"
        label3.text = "Enter backwashing rate (m/min):"
        label4.text = "Enter particle settling velocity (m/min):"
        label5.text = "The expanded bed depth (m):"
        Button1.Text = "&Calculate"
    End Sub

    Sub calculateResults()
        e = Val(textbox1.text)
        L = Val(textbox2.text)
        v = Val(textbox3.text)
        Vs = Val(TextBox4.Text)
        'Convert from m/s to m/h
        Vs *= 60 * 60
        Lx = L * (1 - e) / (1 - (v / Vs) ^ 0.22)
        TextBox5.Text = FormatNumber(Lx, 2)
    End Sub

    Private Sub Button1_Click(ByVal sender As System.Object, ByVal e As System.EventArgs)
Handles Button1.Click
        calculateResults()
    End Sub
End Class
```

## 4.8 METHODS OF DESALINATION

### 4.8.1 Introduction

Desalination or desalting is a term applied to removal of dissolved salts from waters with the objective of producing water suitable for human consumption (Rowe and Abdel-Magid 1995, Smith and Scott 2002). Table 4.2 gives an estimate of the dissolved solids concentration of various types of water.

---

**TABLE 4.2**
**Dissolved Solids Content for Different Waters**

| Type of Water | Total Dissolved Solids (mg/L) |
|---|---|
| Brackish water | 1,500–12,000 |
| Seawater (Middle East) | 50,000 |
| Seawater (North Sea) | 35,000 |

*Source:* Rowe, D. R. and Abdel-Magid, I. M., *Handbook of Wastewater Reclamation and Reuse,* CRC Press/Lewis Publishers, Boca Raton, FL, 1995. With permission.

---

Desalination processes include the following:

1. *Distillation*: This process concerns the boiling of water and the consequent condensation of vapor produced (examples are to be found in solar stills; multistage flash, multiple-effect boiling, and vapor compression distillation).
2. *Freezing*: Freezing of a solution of saline water enables the separation of dissolved salts and the production of pure water.
3. *Membrane process*: Reverse osmosis (RO) applies pressure to force water molecules to flow through a semipermeable membrane while retaining dissolved substances in the solution. Electrodialysis uses ion-selective membranes to separate dissolved salts under the action of an electric current.

Figure 4.10 summarizes the various types of desalination techniques (Rowe and Abdel-Magid 1995).

### 4.8.2 Distillation

In distillation, saline water is boiled or evaporated to produce two separate streams: freshwater and concentrate or a

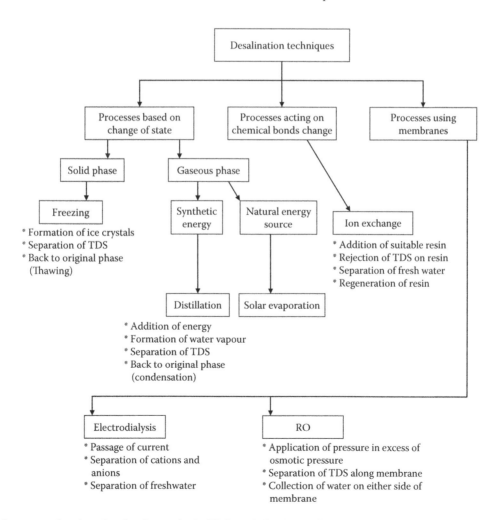

**FIGURE 4.10** Summary of various desalination methods. TDS, total dissolved solids.

brine stream. The first stream has a low concentration of dissolved salts, whereas the second stream contains the remaining dissolved salts. Vapors or steam is then condensed to supply pure water. Benefits of the distillation process include the following (Rowe and Abdel-Magid 1995, Buros 2000):

- Removal of harmful organisms from feed water (e.g., bacteria, viruses)
- Elimination of nonvolatile matter from feed water (e.g., dissolved gases such as carbon dioxide and ammonia)

In the distillation of water, at least two heat exchangers are needed for evaporating the feed water and for aiding vapor in the condensation.

In a multiple-effect distillation unit, the feed water in the first phase is boiled under high pressure, whereas the last phase operates at atmospheric pressure. Figure 4.11 outlines the pressure operation in a multiple-effect distillation process. Heating tubes of the second stage aids the condensation of water vapor emerging from the first phase. The energy released (by latent heat of condensation and by temperature decrease) is employed by the liquid water in the second stage. Cooling water is used in the last stage to finalize the condensation (Rowe and Abdel-Magid 1995).

Heat unexchanged in any phase can be determined by using the following equation:

$$HE_i = u_i * A_i * \Delta T_i \tag{4.60}$$

where:

$HE_i$ is the heat exchanged to effect number $i$, J
$u_i$ is the heat transfer coefficient for heat exchanger number $i$
$A_i$ is the area of heat exchanger number $i$, m$^2$
$\Delta T_i$ is the temperature difference between water in effect and steam entering heat exchanger, °C

$$\Delta T_1 = T_0 - T_1$$

where:

$T_0$ is the temperature of heating steam supplied to first phase, °C
$T_1$ is the boiling temperature in first phase, °C

$$\Delta T_2 = T_1 - T_2$$

where $T_2$ is the boiling temperature in second phase, and so on, °C.

For similar areas of heat exchangers and similar transfer rates of heat in each phase, the relationship between the heat generated in each phase and the temperature difference may be related as illustrated in the following equations:

$$HE_1 = HE_2 = \cdots = HE_i \tag{4.61}$$

$$A_1 = A_2 = A_3 = \cdots = A_i \tag{4.62}$$

$$u_1 * \Delta T_1 = u_2 * \Delta T_2 = \cdots = u_i * \Delta T_i \tag{4.63}$$

### Example 4.13

1. Write a computer program that determines the temperature difference in each phase of a multiphase still composed of $N$ number of stills with a drop in both temperature (from $T_1$ to $T_n$) and heat transfer coefficients (from $u_1$ to $u_n$). Assume the stills have equal areas in the heat exchanger.
2. A triple-phase still of equal areas in the heat exchanger is employed for water distillation. Dry steam enters the first still at a temperature of 120°C and leaves the triple-phase still at a temperature of 40°C with a drop in the pressure in the last phase. Heat transfer coefficients for the three stills are in the ratio of 4:3:2, respectively. Find the temperature difference in each phase in the triple distiller.

### Solution

1. For the solution to Example 4.13 (1), see the listing of Program 4.13.
2. Solution to Example 4.13 (2):
   a. Given: $u_1{:}u_2{:}u_3 = 4{:}3{:}2$, $T_1 = 120$, $T_3 = 40$.
   b. Find the temperature differences for the different phases of the still as follows: $u_1*\Delta T_1 = u_2*\Delta T_2 = u_3*\Delta T_3$.
   c. Determine the total temperature drop in the three phases: $\Delta T_1 + \Delta T_2 + \Delta T_3 = \Delta T = 120 - 40 = 80$.
   d. Let $u_1*\Delta T_1 = u_2*\Delta T_2 = u_3*\Delta T_3 = a$, then $\Delta T_1/\Delta T = (a/u_1 + a/u_2 + a/u_3)$.
   e. Multiply by $u_1/a$, then $\Delta T_1/\Delta T = (1)/(1 + u_1/u_2 + u_1/u_3)$.
   f. Find temperature differences as follows: $\Delta T_1/80 = (1)/[1 + (4/3) + (4/2)]$. This yields $\Delta T_1 = 18.5$°C.
   g. Similarly, $\Delta T_2/\Delta T = (1)/(u_2/u_1 + 1 + u_2/u_3)$, which gives $\Delta T_2 = 24.6$°C and $\Delta T_3/\Delta T = (1)/(u_3/u_1 + u_3/u_2 + 1)$, which gives $\Delta T_3 = 36.9$°C.
3. Use the computer program developed in (1) to verify the manual solution presented in (2).

**FIGURE 4.11**  Pressure regulation with a multiple-effect distiller.

## LISTING OF PROGRAM 4.13 (CHAP4.13\FORM1.VB): TEMPERATURE DIFFERENCE IN MULTIPHASE STILL

```vb
'*********************************************************************************
'Program 4.13: Desalination
'computes the temperature difference in
'multi-phase still
'*********************************************************************************

  Public Class Form1
  Dim rat(), deltaT() As Double
  Dim N As Integer = 1
  Dim T1, T3, dT As Double

  Private Sub Form1_Load(ByVal sender As System.Object, ByVal e As System.EventArgs)
Handles MyBase.Load
    Me.Text = "Program 4.13: Computes the temperature difference in multi-phase still"
    Me.FormBorderStyle = Windows.Forms.FormBorderStyle.FixedSingle
    Me.MaximizeBox = False
    label1.text = "Enter number of stills:"
    textbox1.text = N.ToString
    label2.text = "Enter " + N.ToString + " heat transfer coefficient ratios:"
    datagridview1.columns.clear()
    datagridview1.rows.clear()
    datagridview1.columns.add("col", "ratio")
    DataGridView1.Rows.Add(N)
    DataGridView1.AllowUserToAddRows = False
    label3.text = "Enter initial temperature T1:"
    Label4.Text = "Enter final temperature T" + N.ToString + ":"
    datagridview2.columns.clear()
    datagridview2.columns.add("col", "dT")
    datagridview2.rows.clear()
    DataGridView2.AllowUserToAddRows = False
    Button1.Text = "&Calculate"
    Label5.Text = "Temperature difference in the " + N.ToString + " stills are as
    follows:"
  End Sub

  Sub calculateResults()
    N = Val(textbox1.text)
    t1 = Val(textbox2.text)
    t3 = Val(textbox3.text)
    ReDim rat(N)
    ReDim deltaT(N)
    For i = 1 To N
      rat(i) = datagridview1.rows(i - 1).cells(0).value
    Next
    dT = T1 - T3
    DataGridView2.Rows.Clear()
    Dim X As Double
    For i = 1 To N
      X = 0
      For j = 1 To N
        X = X + rat(i) / rat(j)
      Next j
      deltaT(i) = dT / X
      datagridview2.rows.add(deltaT(i))
    Next i
  End Sub
```

```
    Private Sub Button1_Click(ByVal sender As System.Object, ByVal e As System.EventArgs)
Handles Button1.Click
        calculateResults()
    End Sub

    Private Sub TextBox1_TextChanged(ByVal sender As System.Object, ByVal e As System.
EventArgs) Handles TextBox1.TextChanged
        If Val(TextBox1.Text) <= 0 Then Exit Sub
        If N > Val(TextBox1.Text) Then
          'remove some cells
          For i = N To Val(TextBox1.Text) + 1 Step -1
            DataGridView1.Rows.RemoveAt(i - 1)
          Next
        ElseIf N < Val(TextBox1.Text) Then
          For i = 1 To Val(TextBox1.Text) - N
            DataGridView1.Rows.Add()
          Next
        End If
        N = Val(TextBox1.Text)
        Label4.Text = "Enter final temperature T" + N.ToString + ":"
        Label5.Text = "Temperature difference in the " + N.ToString + " stills are as
        follows:"
    End Sub
End Class
```

### 4.8.3 OSMOSIS

Osmosis concerns the tendency of a solvent to diffuse through a membrane to a solute. The flow is directed from the more diluted to the more concentrated solution. To prevent diffusion toward the more concentrated solution, osmotic pressure is applied. Figure 4.12 illustrates the concepts of normal and RO processes.

Osmotic pressure, a measure of the force that brings together the solvent molecules, depends upon the number of particles of solute in solution (Rowe and Abdel-Magid 1995). The passage of the solvent through a membrane produces a driving force, which may be estimated by the difference in vapor pressure of the solvent on the membrane sides. The transfer of solvent through the membrane from the dilute to the concentrated solution continues until hydrostatic pressure exceeds the driving force of the vapor pressure differential (Rowe and Abdel-Magid 1995, Sawyer and McCarty 2002).

Equation 4.64 may be used to estimate the osmotic pressure at equilibrium for an incompressible solvent:

$$P_{osm} = \left(\frac{R*T}{V}\right) * \ln\left(\frac{P_0}{P}\right) \quad (4.64)$$

where:

$P_{osm}$ is the osmotic pressure, Pa or atm

$R$ is the universal gas constant (= 0.08206 L*atm/mol*K = 9.314 J/K*mol = 8314 J/kg*K), J/kg*K

$T$ is the temperature, K

$P_0$ is the vapor pressure of solvent in dilute solution, Pa or atm

$P$ is the vapor pressure of solvent in concentrated solution, Pa or atm

$V$ is the volume per mole of solvent (= 0.018 L for water), m³ or L

Generally, the existence of a nonvolatile solute in a liquid decreases the vapor pressure of the solution. Raoult's law assumes that for dilute solutions, the reduction in vapor pressure of a solvent is directly proportional to the concentration of particle in solution (Rowe and Abdel-Magid 1995, Sawyer and McCarty 2002). Equation 4.65 signifies Raoult's law for dilute solution relating osmotic pressure to molar concentration of particles in the concentrated solution:

$$P_{osm} = C * R * T \quad (4.65)$$

where:

$P_{osm}$ is the osmotic pressure, atm

$C$ is the molar concentration of particles

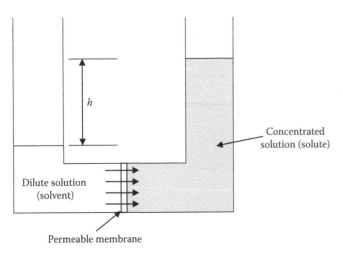

**FIGURE 4.12** Osmosis process.

*R* is the universal gas constant for all gases, L*atm/mol*K

*T* is the temperature, K

In the RO process, the dissolved solids are separated by a semipermeable membrane. In this case, pressure in excess of the natural osmotic pressure is applied to the feed water (see Figure 4.13).

The advantages of the RO process include the following (Rowe and Abdel-Magid 1995, Buros 2000, Cotruvo et al. 2010, American Water Works Association 2011, Riffat 2012):

- Reduction of content of total dissolved solids (up to 99%)
- Removal of biological and colloidal materials (up to 98%)

- Removal of harmful microorganisms (up to 100%)
- Removal of dissolved organic substances (up to 97%)
- Production of treated water suitable for human consumption

### Example 4.14

1. Write a short computer program to compute the osmotic pressure per unit volume for a salt solution, given the temperature and the vapor pressure of solvent in dilute and concentrated solutions.
2. A salt solution has a vapor pressure (concentrated solution) of 2.26 kPA and a temperature of 20°C. Assume pure water (dilute solution in this case) has a vapor pressure of 2.34 kPa at this temperature. Estimate the osmotic pressure per unit volume.
3. Use the computer program developed in (1) to verify the manual solution presented in (2).

### Solution

1. For solution to Example 4.14 (1), see the listing of Program 4.14.
2. Solution to Example 4.14 (2):
   a. Given: $T = 20°C$, $P_0$ is the vapor pressure of solvent in dilute solution = 2.34 kPa (may be found from Appendix A1 corresponding to the given temperature in case of the dilute solution being water). $P$ is the vapor pressure of the solvent in the concentrated solution = 2.26 kPa.
   b. Determine the temperature as follows: $T = 273.16 + 20 = 293.16$ K.
   c. Find the osmotic pressure as follows: $(P_{osm} = (R*T/V)*\ln(P_0/P)$. $P_{osm} = (8.314$ J/K/mol*293.16 K)*$\ln(2.34/2.26) = 84.8$ Pa/unit volume.

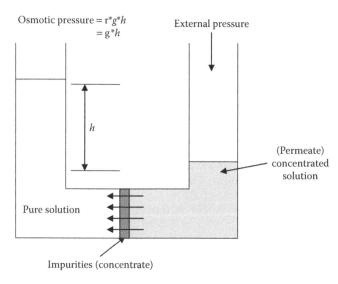

Osmotic pressure = r*g*h
                 = g*h

External pressure

*h*

(Permeate) concentrated solution

Pure solution

Impurities (concentrate)

**FIGURE 4.13** RO process.

---

### LISTING OF PROGRAM 4.14 (CHAP4.14\FORM1.VB): OSMOTIC PRESSURE PER UNIT VOLUME FOR SALT SOLUTIONS

```
'********************************************************************************
'Program 4.14: Osmosis
'Computes the osmotic pressure per unit volume for
'salt solutions
'********************************************************************************

Public Class Form1
    Dim T, Po, P, Posm As Double

    Private Sub Form1_Load(ByVal sender As System.Object, ByVal e As System.EventArgs)
Handles MyBase.Load
        Me.Text = "Program 4.14: Computes the osmotic pressure for salt solutions"
        Me.FormBorderStyle = Windows.Forms.FormBorderStyle.FixedSingle
        Me.MaximizeBox = False
```

```
      label1.text = "Enter temperature T (C):"
      label2.text = "Enter vapor pressure of solvent in dilute solution (kPa):"
      label3.text = "Enter vapor pressure of solvent in concentrated solution (kPa):"
      label4.text = ""
      button1.text = "&Calculate"
    End Sub

    Sub calculateResults()
      T = Val(textbox1.text)
      Po = Val(textbox2.text)
      P = Val(textbox3.text)
      T = T + 273.16 'Kelvin
      Dim R = 8.314 'J/K/mol
      Dim V = 1 'unit volume
      Posm = (R * T / V) * Math.Log(Po / P)
      label4.text = "Osmotic pressure (kPa/unit volume): " + Posm.ToString
    End Sub

    Private Sub Button1_Click(ByVal sender As System.Object, ByVal e As System.EventArgs)
Handles Button1.Click
      calculateResults()
    End Sub
End Class
```

## Example 4.15

1. Write a computer program to find the difference in osmotic pressure across a semipermeable membrane, given the concentrations (in molar, or mg/L) of cations and anions and the prevailing temperature.
2. Analysis of a water sample at a temperature of 23°C indicated the following concentration of ions (in mg ion/L). Cations: $Mg^{2+}$ = 0.5, $Ca^{2+}$ = 0.8, $Na^+$ = 0.3, $K^+$ = 0.2. Anions: $SO_4^{--}$ = 1.2, $Cl^-$ = 0.1, $HCO_3^-$ = 0.25, $NO_3^-$ = 0.25. The water sample is to be treated by RO. Determine the osmotic pressure across the semipermeable membrane.

## Solution

1. For solution to Example 4.15 (1), see the listing of Program 4.15.

2. Solution to Example 4.15 (2):
   a. Determine the molar concentration of ions in the water sample as follows: molar ion concentration = concentration of ion (mg/L)/MW, where MW = molecular weight. Thus (and similarly), $Mg^{2+}$ = 0.5/24.3 = 0.0206; $Ca^{2+}$ = 0.8/40 = 0.02; $Na^+$ = 0.3/23 = 0.013; $K^+$ = 0.2/39 = 0.0051; $SO_4^-$ = 1.2/96 = 0.0125; $Cl^-$ = 0.1/35.5 = 0.0028; $HCO_3^-$ = 0.25/61 = 0.0041; $NO_3^-$ = 0.25/62 = 0.004.
   b. Find the molar ion concentration of the sample of water as follows: $C$ = 0.0206 + 0.02 + 0.013 + 0.0051 + 0.0125 + 0.0028 + 0.0041 + 0.004 = 0.0821 M.
   c. Determine the osmotic pressure as follows: $P_{osm} = C*R*T$ = 0.0821 * 0.082*(273.16 + 23) = 1.99 atm.
3. Use the computer program developed in (1) to verify the manual solution presented in (2).

---

**LISTING OF PROGRAM 4.15   (CHAP4.15\FORM1.VB): DIFFERENCE
IN OSMOTIC PRESSURE ACROSS MEMBRANES**

```
'*********************************************************************************
'Program 4.15: Osmosis-2
'Computes difference in osmotic pressure
'across membranes
'*********************************************************************************


Public Class Form1
   Dim mw(8), conc(8)
   Dim T, mol, Posm As Double
```

```vb
     Private Sub Form1_Load(ByVal sender As System.Object, ByVal e As System.EventArgs)
Handles MyBase.Load
     '** symbols
     Label4.Text = "Mg2+"
     Label5.Text = "Ca2+"
     Label6.Text = "Na+ "
     Label7.Text = "K+ "
     Label11.Text = "SO4-"
     Label10.Text = "Cl-"
     Label19.Text = "HCO3-"
     Label8.Text = "NO3"
     '** molecular weights
     mw(1) = 24.3
     mw(2) = 40
     mw(3) = 23
     mw(4) = 39
     mw(5) = 96
     mw(6) = 35.5
     mw(7) = 61
     mw(8) = 62
     Me.Text = "Program 4.15: Computes difference in osmotic pressure across membranes"
     Me.FormBorderStyle = Windows.Forms.FormBorderStyle.FixedSingle
     Me.MaximizeBox = False
     Label1.Text = "Enter temperature T (C):"
     GroupBox1.Text = "Enter ion concentration in mg ion/L:"
     Label2.Text = "CATIONS:"
     Label3.Text = "ANIONS:"
     Label12.Text = ""
     Button1.Text = "&Calculate"
   End Sub

   Sub calculateResults()
     mol = 0
     T = Val(TextBox1.Text)
     conc(1) = Val(TextBox2.Text)
     conc(2) = Val(TextBox3.Text)
     conc(3) = Val(TextBox4.Text)
     conc(4) = Val(TextBox5.Text)
     conc(5) = Val(TextBox9.Text)
     conc(6) = Val(TextBox8.Text)
     conc(7) = Val(TextBox7.Text)
     conc(8) = Val(TextBox6.Text)
     For I = 1 To 8
        'molar ion concentration
        mol = mol + conc(I) / mw(I)
     Next
     Dim R = 0.082
     'osmotic pressure
     Posm = mol * R * (T + 273.16)
     Label12.Text = "Osmotic pressure: " + FormatNumber(Posm, 2) + " atm."
   End Sub

   Private Sub Button1_Click(ByVal sender As System.Object, ByVal e As System.EventArgs)
Handles Button1.Click
     calculateResults()
   End Sub
End Class
```

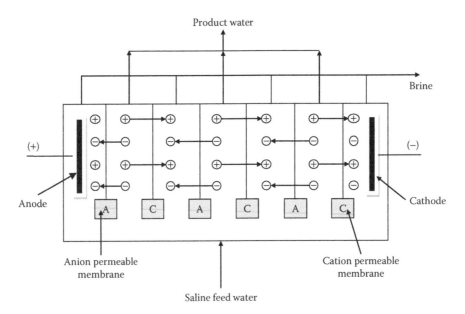

**FIGURE 4.14**   Schematic representation of an electrodialysis process.

### 4.8.4   ELECTRODIALYSIS

In electrodialysis, ions are transferred through an ion-selective membrane from one solution to another under the action of a direct electrical potential (Rowe and Abdel-Magid 1995). Usually, the electrodialysis device is an array of alternating anion- and cation-selective membranes across which an electric potential is applied (see Figure 4.14) (Rowe and Abdel-Magid 1995, American Water Works Association 2011).

The efficiency of the system to transport counterions can be determined by using the following equation (American Water Works Association 2011):

$$i * \frac{\left(\text{Eff}' - \text{Eff}\right)}{100} * \text{Far} = \frac{\text{Diff} * \left(C_0 - C\right)}{B} \qquad (4.66)$$

where:
$i$ is the current density, A/cm$^2$
Eff$'$ is the current efficiency for transport of counterions through the membrane, %
Eff is the current efficiency for transport of the same ions in solution in contact with the membrane, %
Far is the Faraday constant (=26.8 A*h)
Diff is the diffusion constant for electrolyte at the temperature of electrolysis
$C_0$ is the concentration of electrolyte in the depletion compartment
$C$ is the concentration of electrolyte at the membrane–solution interface
$B$ is the thickness of diffusion layer at the interface

The maximum value of the current density can be determined by using the following equation (American Water Works Association 2011):

$$i_{\max} = \frac{\left(\text{Far} * \text{Diff} * C_0\right)}{\left[B * \left(\text{Eff}' - \text{Eff}\right)\right]} * 100 \qquad (4.67)$$

where $i_{\max}$ is the maximum value of the current density, A/cm$^2$.

Problems associated with the electrodialysis process include the following (Rowe and Abdel-Magid 1995):

- High costs involved in treating seawater
- Expensive pretreatment
- Problems encountered with high-sulfate waters
- Limitation of process of removing only common mineral ions
- Required skilled maintenance and supervision

### Example 4.16

1. Write a computer program to determine the maximum current density, given the total dissolved solids, the current efficiency for transport of counterions through the membrane, the current efficiency for transport of the same ions in solution in contact with the membrane, the thickness of diffusion layer at the interface, the diffusion constant for electrolyte at the temperature of electrolysis, and Faraday's constant.
2. Brackish water with a dissolved solids concentration of 1800 mg/L is introduced to an electrodialysis unit. The unit operates under the following conditions:
   a. Current efficiency for transport of counterions through the membrane = 80%.
   b. Current efficiency for transport of the same ions in solution in contact with the membrane = 58%.
   c. Diffusion constant for electrolyte at the temperature of electrolysis = 1.4 * 10$^{-5}$ cm$^2$/s.
   d. Thickness of diffusion layer at the interface = 0.5 mm.

Compute the maximum current density (take Faraday's constant as 26.8 A*h).

3. Use the computer program developed in (1) to verify the manual solution presented in (2).

## Solution

1. For solution to Example 4.16 (1), see the listing of Program 4.16.

2. Solution to Example 4.16 (2):
   a. Given: $C_0 = 1800$ mg/L, Eff = 58%, Eff' = 80%, Diff = $1.4 * 10^{-5}$ cm$^2$/s, $B = 0.5$ mm = $5 * 10^{-2}$ cm, Far = 26.8 A*h.
   b. Determine the maximum current density as follows: $i_{max} = 100*Far*Diff*C_0/[B*(Eff'-Eff)]$ = $(100 * 26.8 * 3600 * 1.4 * 10^{-5}*1800 * 10^{-6})/[5 * 10^{-2}*(80 - 58)] = 0.22$ A*g/cm$^2$.

---

### LISTING FOR PROGRAM 4.16   (CHAP4.16\FORM1.VB): ELECTRODIALYSIS MAXIMUM CURRENT DENSITY

```
'******************************************************************************
'Program 4.16: Electrodialysis
'computes maximum current density
'******************************************************************************

Public Class Form1
    Dim Co, eff1, eff2, Diff, B As Double
    Dim imax As Double
    Private Sub Form1_Load(ByVal sender As System.Object, ByVal e As System.EventArgs)
Handles MyBase.Load
        Me.Text = "Program 4.16: Computes maximum current density"
        Me.MaximizeBox = False
        Me.FormBorderStyle = Windows.Forms.FormBorderStyle.FixedSingle
        Label1.Text = "Enter dissolved solids concentration (mg/L):"
        Label2.Text = "Enter efficiency for transport of ions through the membrane (%):"
        Label3.Text = "Enter efficiency for transport of ions in the solution (%):"
        Label4.Text = "Enter diffusion constant for electrolyte (cm2/s):"
        Label5.Text = "Enter thickness of diffusion layer at the interface (cm):"
        Label6.Text = ""
        Label7.Text = "Decimal places:"
        NumericUpDown1.Maximum = 10
        NumericUpDown1.Minimum = 0
        NumericUpDown1.Value = 2
        button1.text = "&Calculate"
    End Sub


    Sub calculateResults()
        Co = Val(textbox1.text)
        eff1 = Val(textbox2.text)
        eff2 = Val(textbox3.text)
        Diff = Val(textbox4.text)
        B = Val(textbox5.text)
        Dim Far, i As Double
        Far = 26.8 * 3600 'Amp*hr
        i = Far * Diff * Co * 0.000001 * 100 / (B * (eff1 - eff2))
        i = Math.Abs(i)
        imax = i
    End Sub


    Private Sub Button1_Click(ByVal sender As System.Object, ByVal e As System.EventArgs)
Handles Button1.Click
        calculateResults()
        show_results()
    End Sub
```

```
Private Sub show_results()
   Label6.Text = "Maximum current density (Amp.gm/cm2): " + FormatNumber(imax,
   NumericUpDown1.Value)
End Sub

Private Sub NumericUpDown1_ValueChanged(ByVal sender As System.Object, ByVal e As
System.EventArgs) Handles NumericUpDown1.ValueChanged
   show_results()
End Sub
End Class
```

## 4.9 DISINFECTION PROCESS

### 4.9.1 INTRODUCTION

Disinfection involves the killing of pathogenic organisms with the objective of preventing the spread of water- and wastewater-borne disease. Disinfection methods can be classified as either physical or chemical. The former methods include heat treatment, exposure to ultraviolet rays, and metal ions (e.g., silver, copper). The latter methods include the usage of oxidants such as chlorine gas, ozone, iodine, and chlorinated compounds (e.g., chlorine dioxide), and potassium permanganate.

The preferred characteristics for an effective chemical disinfectant include the following (Committee Report 1978, Dyer-Smith et al. 1983, El-Hassan and Abdel-Magid 1986, Sawyer and McCarty 2002, Degremont 2007, Nemerow et al. 2009, Shammas et al. 2010):

- Effective in killing pathogenic microorganisms
- Ease of detection and measurement of its concentration in water
- Solubility in water in doses required for disinfection
- Absence of toxicity to humans and animals
- Ease of handling, transporting, and controlling it
- Capable of producing a residual
- Absence of a taste, odor, or color when used
- Available at a reasonable cost

The disinfection process is a function of type and concentration of microorganisms, type and concentration of disinfectant, presence of oxidant-consuming compounds, temperature, dose of chemical, contact time, and pH.

### 4.9.2 CHLORINATION

Chlorination is the process of adding chlorine to water with the purpose of killing pathogens. When chlorine is added to water, part of it reacts to produce hypochlorous acid (HOCl), which is the more effective disinfectant (available chlorine); another part forms hypochlorite ion (OCl⁻); a third part reacts with ammonia to form chloramines; a fourth part oxidizes inorganic matter (e.g., hydrogen sulfide, iron, manganese); and a fifth part reacts with organic compounds to form trihalomethanes (THMs) and other chlorinated organics. The benefits of using chlorine include effectiveness, reliability, and establishment of a residual in the system (Rowe and Abdel-Magid 1995).

The rate of kill of organisms is a function of many interacting parameters, such as the concentration of disinfectant, the ability and time needed to penetrate the bacterial cell wall, the distribution of disinfectant and microorganisms, and the contact time between pathogens and disinfectant according to Chick's law.

Equation 4.68 is a mathematical expression for Chick's law, which is not valid for all disinfectants nor for all microorganisms:

$$\frac{N}{N_0} = e^{-k*t} \qquad (4.68)$$

where:

$N$ is the number of viable microorganisms of one type at time $t$

$N_0$ is the number of viable microorganisms of one type at time $t = 0$

$t$ is the time, days

$k$ is the constant, day⁻¹

Another relationship between disinfectant concentration and contact time is given in the following equation:

$$C^a * t = k \qquad (4.69)$$

where:

$C$ is the concentration of chlorine, mg/L

$t$ is the contact time or time required for a given percentage of microbial kill, min

$a$ and $k$ are the experimental constants that are valid for a particular system

The Van't Hoff–Arrhenius equation can be used to relate the effects of temperature on the disinfection process (Peavey et al. 1985, Metcalf and Eddy 2013).

$$\ln\left(\frac{t_1}{t_2}\right) = \frac{\left[E'(T_2 - T_1)\right]}{R} \qquad (4.70)$$

**TABLE 4.3**

**Activation Energies (E′) for Aqueous Chlorine**

| pH | E′(cal) |
|---|---|
| 7.0 | 8,200 |
| 8.5 | 6,400 |
| 9.8 | 12,000 |
| 10.7 | 15,000 |

*Source:* Rowe, D. R. and Abdel-Magid, I. M., *Handbook of Wastewater Reclamation and Reuse,* CRC Press/Lewis Publishers, Boca Raton, FL, 1995. With permission.

where:

$t_1$ and $t_2$ are the time required for the given kills, s

$E′$ is the activation energy (see Table 4.3), cal

$T_1$ and $T_2$ are the temperatures corresponding to $t_1$ and $t_2$ respectively, K

$R$ is the gas constant

Problems associated with chlorination include the following:

- Safety considerations during handling, transportation, and storage of chemicals

- Formation of chlorinated organic compounds (THMs) that can be injurious to human health

**Example 4.17**

1. Using Chick's law, write a computer program to find the contact time required for a disinfectant to achieve a percent kill for a microbial system that has a rate constant of $k$ (to base e) or $k′$ (to base 10).
2. Determine the contact time needed for a disinfectant to achieve a 99.99% kill for a pathogenic microorganism that has a rate constant (to base 10) of 0.1 $s^{-1}$.
3. Use the program developed in (1) to verify computations of part (2).

**Solution**

1. For solution to Example 4.17 (1), see the listing of Program 4.17.
2. Solution to Example 4.17 (2):
   a. Given: the rate of kill = 99.99%, $k′ = 0.1$ $s^{-1}$.
   b. Find the contact time from Chick's law as follows: $t = -(1/k′)*\log(N/N_0)$. $t = -(1/0.1)*\log(100 - 99.99/100) = 40$ s.

---

**LISTING OF PROGRAM 4.17   (CHAP4.17\FORM1.VB): CHLORINATION TIME OF KILL**

```
'*********************************************************************************
'Program 4.17: Chlorination
'computes the time needed by a disinfectant to
'achieve a certain kill
'*********************************************************************************

Imports System.Math
Public Class Form1
  Dim k, N1 As Double

  Private Sub Form1_Load(ByVal sender As System.Object, ByVal e As System.EventArgs)
Handles MyBase.Load
      Me.Text = "Program 4.17: Computes the time needed by a disinfectant"
      Me.FormBorderStyle = Windows.Forms.FormBorderStyle.FixedSingle
      Me.MaximizeBox = False
      Label1.Text = "Enter the rate constant k (/s):"
      Label2.Text = "Is it to base 10?"
      Label3.Text = "Enter the percentage kill to achieve:"
      radiobutton1.text = "Yes"
      RadioButton2.Text = "No"
      RadioButton1.Checked = True
      Button1.Text = "&Calculate"
      Label4.Text = ""
      Label4.AutoSize = False
      Label4.Height = 40
      Label4.Width = 290
  End Sub

  Sub calculateResults()
    k = Val(textbox1.text)
```

```
    N1 = Val(textbox2.text)
    Dim N2, lgn, t As Double
    If N1 < 1 Then N1 = N1 * 100
    N2 = (100 - N1) / 100
    'logarithm base 10 of N2
    lgn = Log10(N2)
    'logarithm base e of N2
    If RadioButton2.Checked Then lgn = Log(N2)
    t = -(1 / k) * lgn
    Label4.Text = "time needed to achieve this level of kill (seconds): " +
FormatNumber(t, 2)
  End Sub

  Private Sub Button1_Click(ByVal sender As System.Object, ByVal e As System.EventArgs)
Handles Button1.Click
    calculateResults()
  End Sub
End Class
```

## 4.10  CORROSION

### 4.10.1  INTRODUCTION

Water with corrosive characteristics can cause problems in distribution networks and residential plumbing systems. Generally, problems encountered may be classified as follows (Eliassen and Shrindle 1958, Hoyt et al 1979, American Water Works Association 1980, Narayan 1983, American Water Works Association 1984, American Water Works Association 2010):

- *Health problems*: These problems originate from dissolution of certain substances into water from transmission and distribution pipelines or plumbing system.
- *Economic problems*: These problems result from reduced service life of materials due to deterioration within transmission, distribution, and plumbing systems.
- *Aesthetic problems*: These problems are due to dissolution of certain metallic substances into the water.

### 4.10.2  CORROSION POTENTIAL INDICATORS

The main corrosion potential indicators include Langelier index (LI), Ryznar index (RI), and aggressiveness index (AI) (Eliassen and Shrindle 1958, Hoyt et al 1979, American Water Works Association 1980, American Water Works Association 1984, American Water Works Association 2010). These indicators are briefly outlined in Sections 4.10.2.1 through 4.10.2.3.

#### 4.10.2.1  Langelier Index

The LI illustrates the tendency for deposition (noncorrosiveness) or dissolution of calcium carbonate scale in pipe. It is defined as the difference between pH of the water and its saturated pH (when water is in equilibrium with calcium carbonate) (Smith and Scott 2002). Equations 4.71 through 4.73 present mathematical expressions for determining LI:

$$LI = pH_a - pH_s \tag{4.71}$$

where:

LI is the Langelier index
$pH_a$ is the actual (measured) pH of the water
$pH_s$ is the saturation pH

$$pH_s = pK_2' - pK_s' + pCa^{2+} + pAlk + S \tag{4.72}$$

where:

$pK_2' - pK_s'$ is the dissolution constant estimates based on the temperature and total dissolved solids or ionic strength
$pCa^{2+}$ is the negative logarithm of the calcium ion concentration ($Ca^{2+}$ in mol/L)
pAlk is the negative logarithm of the total alkalinity (alkalinity in equivalents of $CaCO_3$/L)
$S$ is the salinity correction term:

$$S = \frac{\left(2.5\,\xi^{1/2}\right)}{\left(1 + 5.3\,\xi^{1/2} + 5.5\,\xi\right)} \tag{4.73}$$

where $\xi$ is the ionic strength.

Positive LI values indicate a supersaturated condition that will deposit calcium carbonate on the pipe. Negative LI values indicate an undersaturated condition that will dissolve calcium carbonate and will not deposit a protective film of the salt on pipe to stop corrosion (Smith and Scott 2002).

#### 4.10.2.2  Ryznar Index

The RI provides an indication of the rate of scale formation or the tendency of the water to be aggressive. Equations 4.74 and 4.75 can be used to determine the RI:

$$RI = 2pH_s - pH_a \qquad (4.74)$$

where:

RI is the Ryznar index

$pH_s$ is the saturation pH

$pH_a$ is the actual (measured) pH

The relationship between the RI and the LI is presented in the following equation:

$$RI = pH_a - 2LI \qquad (4.75)$$

where:

RI < 6, calcium carbonate scale deposition increases proportionately.

RI > 6, corrosion increases.

RI > 10, conditions are extremely aggressive.

### 4.10.2.3 Aggressiveness Index

The AI defines water quality that can be transported through asbestos–cement pipes without adverse effects. The relationship between the AI, pH, total alkalinity, and calcium hardness is presented in the following equation:

$$AI = pH + \log_{10}\left[(Alk)*(HCa)\right] \qquad (4.76)$$

where:

AI is the aggressiveness index

pH is the negative logarithm of the hydrogen ion concentration

Alk is the total alkalinity, mg/L $CaCO_3$

HCa is the calcium hardness, mg/L $CaCO_3$

Table 4.4 gives a classification of water according to the value of the AI.

**TABLE 4.4**

**Aggressiveness Index**

| Value | Description |
|---|---|
| <10 | Highly aggressive |
| 10–11.9 | Moderately aggressive |
| >12 | Nonaggressive |

---

**LISTING OF PROGRAM 4.18 (CHAP4.18\FORM1.VB): CORROSION POTENTIAL INDICATORS**

```
'******************************************************************************
'Program 4.18: Corrosion
'determines whether a sample is corrosive or precipitative
'******************************************************************************
Imports System.Math
Public Class Form1
    Dim H, ALK, HCa, pHa, pHs As Double
    Dim compute As Char 'whether or not to compute Langelier & Ryznar indices
    Private Sub Form1_Load(ByVal sender As System.Object, ByVal e As System.EventArgs)
Handles MyBase.Load
        Me.Text = "Program 4.18: Determines whether a sample is corrosive or precipitative"
        Me.MaximizeBox = False
        Me.FormBorderStyle = Windows.Forms.FormBorderStyle.FixedSingle
        Me.Height = 256 'hide the lower part of the form
        label1.text = "Enter the hydrogen ion concentration (mol/L):"
        label2.text = "Enter the total alkalinity (mg/K CaCO3):"
        label3.text = "Enter the calcium hardness (mg/L CaCO3):"
        label4.text = ""
        Label4.AutoSize = False
        Label4.Height = 40
        Label4.Width = 297
        Button1.Text = "&Calculate"
        Button2.Text = "Compute Langelier & Ryznar indices >>"
        Button3.Text = "Calculate indices"
        compute = "N"
        Label5.Text = "Enter measured pH of water:"
        Label6.Text = "Enter saturation pH:"
        Label7.AutoSize = False
        Label7.Height = 40
```

```
      Label7.Width = 297
      Label7.Text = ""
   End Sub

   Sub calculateResults()
     H = Val(textbox1.text)
     ALK = Val(textbox2.text)
     HCa = Val(textbox3.text)
     Dim pH, AI As Double
     Dim type As String
     pH = -Log10(H)
     AI = pH + Log10(ALK * HCa)
     Select Case AI
       Case Is < 10
       type = "highly aggressive"
       Case Is > 12
       type = "Non aggressive"
     Case Else
       type = "moderately aggressive"
     End Select
     Label4.Text = "Aggressiveness Index: " + AI.ToString + vbCrLf + type
   End Sub

   Private Sub Button2_Click(ByVal sender As System.Object, ByVal e As System.EventArgs)
Handles Button2.Click
     If compute = "N" Then
       compute = "Y"
       Me.Height = 426
       Button2.Text = "Hide <<"
     Else
       compute = "N"
       Me.Height = 256
       Button2.Text = "Compute Langelier & Ryznar indices >>"
     End If
   End Sub

   Private Sub Button3_Click(ByVal sender As System.Object, ByVal e As System.EventArgs)
Handles Button3.Click
     pHa = Val(TextBox4.Text)
     pHs = Val(TextBox5.Text)
     Dim LI, RI As Double
     LI = pHa - pHs
     RI = 2 * pHs - pHa
     Label7.Text = "Langelier Index = " + LI.ToString
     Label7.Text += vbCrLf + "Ryznar Index = " + RI.ToString
   End Sub

   Private Sub Button1_Click(ByVal sender As System.Object, ByVal e As System.EventArgs)
Handles Button1.Click
     calculateResults()
   End Sub
End Class
```

## Example 4.18

1. Write a computer program to determine whether a water sample is corrosive or scale forming, using the different indexes (LI, RI, AI). Data given include the parameters involved in the computations of each of the relative indexes.

2. Analysis of a water sample revealed the following: calcium hardness = 125 mg/L as $CaCO_3$, alkalinity = 195 mg/L as $CaCO_3$, hydrogen ion concentration of $2.6 * 10^{-7}$ mol/L. Using the AI, determine whether this water is aggressive to asbestos–cement pipes.

3. Use the computer program involved in (1) to verify computations conducted in (2).

## Solution

1. For solution to Example 4.18 (1), see the listing of Program 4.18.
2. Solution to Example 4.18 (2):
   1. Given: HCa = 125, Alk = 195, $[H^+]$ = 2.6 * $10^{-7}$ mol/L.
   2. Determine the pH as follows: pH = $-\log[H^+]$ = $-\log 2.6 * 10^{-7}$ = 6.57.
   3. Find the AI as follows: AI = pH + $\log_{10}$(Alk)*(HCa) = 6.75 + log(195 * 125) = 10.96. Thus, AI = 10.96 < 11.9. This value shows that the water is moderately aggressive to asbestos–cement pipes (Table 4.4).

### 4.10.3 RATE OF CORROSION

Corrosion may be defined as an attack on the surface of a metal or other solid (Smith and Scott 2002). Figures 4.15 and 4.16 give a general classification for the corrosion processes. The rate of metal removal due to corrosion can be calculated by using the following equation (Narayan 1983):

$$R_{mr} = \frac{K * wt}{A * t * \rho} \qquad (4.77)$$

where:

$R_{mr}$ is the rate of corrosion, m/s
$K$ is the constant
wt is the weight loss, kg
$A$ is the area of specimen, $m^2$
$t$ is the time of exposure, s
$\rho$ is the density, $kg/m^3$

## Example 4.19

1. Write a short computer program that determines the rate of corrosion of a metallic surface, given the weight loss of material, wt (kg)—or electrochemical equivalent, $E_e$ (mg/C); current, $I$ (A); the area of specimen, $A$ ($m^2$); time of exposure, $t$(s); and the density of material, $\rho$ ($kg/m^3$).
2. In an electrochemical cell, the rate of corrosion of a 0.5 cm radius zinc anode is found to be 1.2 * $10^{-6}$ mm/s for an electric current of 90 mA. Determine the depth to which the anode is immersed in the electrolyte to produce the aforementioned corrosion rate. For zinc, the electrochemical equivalent = 0.3387 mg/C, the density = 7000 $kg/m^3$, and $K$ = 1.

## Solution

1. For solution to Example 4.19 (1), see the listing of Program 4.19.
2. Solution to Example 4.19 (2):
   1. Given: $r$ = 0.5 cm, $R_{mr}$ = 1.2 * $10^{-6}$ mm/s, $I$ = 90 mA, $E_e$ = 0.3387 mg/C, $\rho$ = 7000 $kg/m^3$.
   2. Use the corrosion rate equation (4.77) to find the corroded area as follows: A = $E_e$*$I$*$K$/$R_{mr}$ *$\rho$ = (0.3387 * $10^{-3}$ g/C*90 * $10^{-3}$ A)/(1.2 * $10^{-7}$*7 $g/cm^3$) = 36.29 $cm^2$. Area exposed to corrosion = $\pi d^2/4 + \pi dh$, where $d$ is the diameter of anode and $h$ is the depth of immersion.
   3. Find the depth of immersion to produce given corrosion rate as follows: h = $(A - \pi*r^2)$/$2\pi$ = (36.29 – $\pi$*$0.5^2$)/2*$\pi$*0.5 = 11.3 cm.

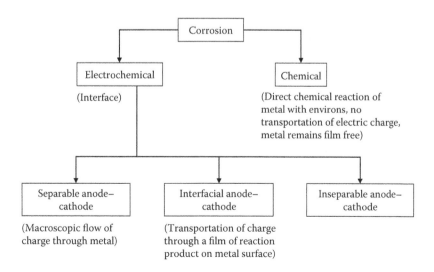

**FIGURE 4.15** Types of corrosion.

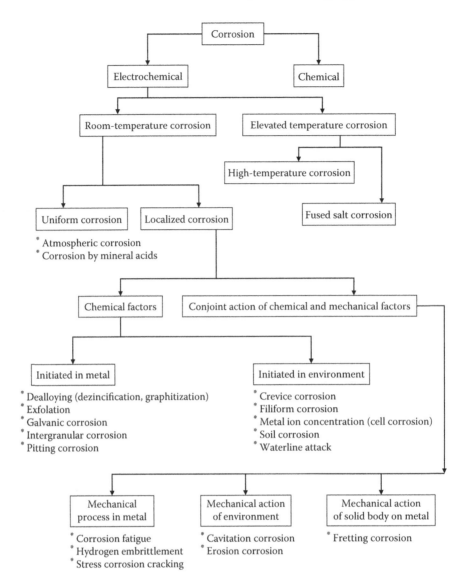

**FIGURE 4.16** Classification of corrosion.

---

**LISTING OF PROGRAM 4.19 (CHAP4.19\FORM1.VB): RATE OF CORROSION OF A METALLIC SURFACE**

```
'*******************************************************************************
'Program 4.19: corrosion
'Determines the rate of corrosion of a metallic surface
'*******************************************************************************

Public Class Form1
    Const PI = 3.14159
    Dim Rmr, wt, A, t, k As Double
    Dim I, Ee, rho, r, h As Double

    Private Sub Form1_Load(ByVal sender As System.Object, ByVal e As System.EventArgs)
Handles MyBase.Load
        RadioButton1.Text = "Determine rate of corrosion"
        RadioButton2.Text = "Determine the depth of immersion for a given rate"
        RadioButton1.Checked = True
        Me.Text = "Program 4.19: Determines the rate of corrosion of a metallic surface"
```

```vbnet
      Me.FormBorderStyle = Windows.Forms.FormBorderStyle.FixedSingle
      Me.MaximizeBox = False
      Button1.Text = "&Calculate"
      Label7.Text = ""
      TextBox6.Visible = False
      Label6.Visible = False
   End Sub

   Private Sub RadioButton1_CheckedChanged(ByVal sender As System.Object, ByVal e As
System.EventArgs) Handles RadioButton1.CheckedChanged
      If RadioButton1.Checked Then
         Label1.Text = "Constant, k:"
         Label2.Text = "Weight loss, wt (kg):"
         Label3.Text = "Area of specimen, A (m2):"
         Label4.Text = "Time of exposure, t (s):"
         Label5.Text = "Density of material, Rho (kg/m3):"
         Label6.Visible = False
         TextBox6.Visible = False
      Else
         Label1.Text = "Radius of specimen, r (cm):"
         Label2.Text = "Rate of corrosion, Rmr (mm/s):"
         Label3.Text = "The electric current, I (mA):"
         Label4.Text = "The electrochemical equivalent, Ee (mg/C):"
         Label5.Text = "Density of material, Rho (kg/m3):"
         Label6.Text = "Constant, k:"
         Label6.Visible = True
         TextBox6.Visible = True
      End If
   End Sub

   Sub calculateResults()
      If RadioButton1.Checked Then
         k = Val(TextBox1.Text)
         wt = Val(TextBox2.Text)
         A = Val(TextBox3.Text)
         t = Val(TextBox4.Text)
         rho = Val(TextBox5.Text)
         Rmr = k * wt / (A * t * rho)
         Label7.Text = "Rate of corrosion: " + FormatNumber(Rmr, 2) + " m/s"
      Else
         r = Val(TextBox1.Text)
         'Convert from mm/s to cm/s
         Rmr = Val(TextBox2.Text) / 10
         'Convert from mA to A
         I = Val(TextBox3.Text) / 1000
         'Convert from mg/C to g/C
         Ee = Val(TextBox4.Text) / 1000
         'Convert from kg/m3 to g/cm3
         rho = Val(TextBox5.Text) / 1000
         k = Val(TextBox6.Text)
         A = (Ee * I * k) / (Rmr * rho) 'corroded area
         h = (A - (PI * (r ^ 2))) / (2 * PI * r) 'depth of immersion
         Label7.Text = "Depth of immersion: " + FormatNumber(h, 2) + " cm"
      End If
   End Sub

   Private Sub Button1_Click(ByVal sender As System.Object, ByVal e As System.EventArgs)
Handles Button1.Click
         calculateResults()
   End Sub
End Class
```

## 4.11 HOMEWORK PROBLEMS IN COMPUTER MODELING APPLICATIONS FOR WATER TREATMENT

### 4.11.1 Discussion Problems

1. What are the various ways in which water treatment can be classified?
2. Discuss the objectives of water treatment.
3. What are the functions of screens at water treatment works?
4. Why should a water suspension have a minimum velocity in order to pass through a screen opening?
5. Define the following terms: sedimentation, flotation, and plain sedimentation.
6. Outline the objective of the sedimentation process.
7. Why are square and circular settling tanks preferred in water treatment?
8. Discuss the types of settling and outline their main differences.
9. Give examples for each type of settling outlined in Problem 8.
10. Name the factors that affect class I settling.
11. For what purposes are cumulative frequency distribution curves drawn?
12. What are the factors that influence the efficiency of sedimentation basins?
13. Explain the kinetics of the filtration process.
14. Define the terms: aeration, aerator, and saturation concentration.
15. What are the advantages of aeration in water and wastewater treatment?
16. Give some examples of aerators.
17. Outline the factors that affect the solubility of a gas in a liquid.
18. Differentiate between coagulation and flocculation.
19. Indicate how to enhance the process of coagulation and filtration.
20. Define the following terms: electrophoretic mobility, zeta potential, and velocity gradient. State the values for each.
21. What are the parameters that govern the design of a flocculator?
22. Define filtration and indicate its objectives in water treatment plants.
23. Differentiate between "slow sand filters" and "rapid sand filters".
24. What are the characteristics of a good filter medium?
25. Why is a rapid filter backwashed?
26. What are the main practical desalination methods? Indicate the advantages and disadvantages of each method.
27. Define the following terms: osmosis, desalination, distillation, and electrodialysis.
28. What is the purpose of disinfection?
29. What are the desirable properties of an effective disinfectant?
30. Give examples for physical and chemical disinfectants.
31. Present the reactions that can occur if water is chlorinated.
32. What are the problems associated with corrosive waters?
33. Differentiate between a corrosive and a scale-forming water suspension.
34. What are the factors that govern the rate of corrosion of metallic substances.
35. Give a general classification scheme for the type of corrosion expected in a sewerage collection system.

### 4.11.2 Specific Mathematical Problems

1. Write a short computer program to compute the diameter, $D$ (m), and the depth of a circular sedimentation tank, $h$ (m), given the influent rate of flow, $Q$ (m$^3$/s); tank's overflow rate, $v_s$ (m$^3$/m$^2$/s); and detention time of tank, $t$ (s). Verify your computations for the data given: $Q = 38$ m$^3$/h, $v_s = 5$ m$^3$/m$^2$/d, $t = 2$ h.
2. Write a short computer program to compute the diameter, $d$ (m), of spherical particles allowed to settle in an ideal sedimentation tank, given the settling velocity of particles, $v_s$ (m$^3$/m$^2$/s); the specific weight of settling particles, s.g.; and the temperature of mixture during the settling process. Assume Stoke's law is valid. Design the program so that a check can be made as to the Reynolds number. Verify your computations for the data given here: $v_s = 0.03$ m/min, s.g. $= 2.6$, and $T = 20°C$.
3. Write a short computer program to evaluate the drag force, $F$ ($N$), acting on a moving spherical particle with a small Reynolds number and its terminal settling velocity, given that $F = 6\pi*r*\mu*v$, where $\mu$ is the dynamic viscosity (N*s/m$^2$), v is the settling velocity of the sphere (m/s), and $r$ is the radius (m).
4. Develop a computer program that can determine the concentration of the SS in the effluent, the Froude and Reynolds numbers for horizontal water movement, the percentage of SS in the effluent having a settling velocity smaller than a certain given value, and the detention time of the water in the tank. The information given includes the temperature of the settling suspension, $T$ (°C); the population number, POP; the average per-capita domestic water consumption, $Q$ (L/day); SS content of influent, $C_0$ (mg/L); tank dimensions (i.e., width, $B$; length $l$; and depth, $h$ [m]); and settling column test results, which showed a straight-line relationship for the cumulative frequency distribution of the settling velocities of the particles with characteristics of the following:
   a. $x_1$ percentage of the particles having a settling velocity larger than $v_1$ (mm/s)
   b. $x_2$ percentage of the particles having a settling velocity smaller than $v_2$ (mm/s)
   Verify your program using the following data: $T = 22°C$; POP $= 71,280$; $Q = 400$ L/day; $C_0 = 125$ mg/L. The settling column test results in the following:

c. Ten percent of the particles having a settling velocity larger than 0.7 mm/s

d. Ten percent of the particles having a settling velocity smaller than 0.2 mm/s

e. $B = 10$ m, $l = 60$ m, and $i = 1.5$ m.

5. Write a computer program to estimate the average grain diameter (mm) needed for a sand filter bed consisting of unsized grains of a porosity, $e$ (%). The bed depth is $l$ (m) operating at a temperature of $T$ (°C) with a filtration velocity of $v_f$ (m³/m²/day) at a maximum bed head loss of $h_f$ (m). Assume that the head loss has been determined by the Rose equation. Design the program so that the suitability of the grain diameter is evaluated. Verify your program by using the following data: $l = 1$ m, $e = 50\%$, $v_f = 0.2$ m³/m²/h, $h_f = 0.24$ m, and $v = 1.0105 * 10^{-6}$ m²/s.

6. Write a computer program to find the porosity of a filter bed after being backwashed with a filter bed expansion of $x$ (%). Take the filter bed thickness to be $l$ (m), bed porosity of $e$ (%), and a grain size diameter of $i$ (mm). Check the validity of your program for the following data: $l = 120$ cm, $e = 40\%$, $d = 0.95$ mm, and $x = 25\%$.

7. Write a computer program to design a slow sand filter intended to treat a flow of $Q$ (m³/s), using the relationships found in Table 4.1. Use the following data to verify your program: $Q = 1000$ m³/day, $v_f = 8$ m/day.

8. Write a program that provides for the determination of the amount of a disinfectant (in kg/day) needed for disinfecting a water flow of $Q$ (m³/s) to obtain $x$ mg/L residual, given chlorine contained in the compound as $y$ percent. Use your program to find the amount of chlorine compound needed under the following conditions: $Q = 150$ m³/h, $x = 0.3$ mg/L, $y = 65\%$.

9. The following laboratory data were obtained from a settling column test. The initial solids concentration equals 550 mg/L. Estimate the overall removal efficiency and concentration of effluent of a settling basin with a depth of 2 m and a detention time of 2 h.

### Settling Data

| Time (min) | Removal (Initial Solids Concentration 550 mg/L) (%) Depth (m) | | | |
| --- | --- | --- | --- | --- |
| | 0.5 | 1.0 | 1.5 | 2.0 |
| 10 | 20 | 11 | 9 | 8 |
| 20 | 32 | 22 | 17 | 15 |
| 30 | 36 | 31 | 26 | 23 |
| 40 | 46 | 34 | 31 | 30 |
| 50 | 59 | 38 | 34 | 33 |
| 60 | 64 | 46 | 38 | 35 |
| 70 | 70 | 55 | 45 | 34 |
| 90 | 81 | 66 | 60 | 53 |
| 110 | 89 | 76 | 89 | 63 |
| 120 | 91 | 80 | 73 | 69 |

## REFERENCES

American Water Works Association. 1980. Internal corrosion, *J. Am Water Works Assoc.*, 72(5): 267–279.

American Water Works Association. 1984. Determining integral corrosion potential in water supply system. *J. Am Water Works Assoc.*, 76(8):83.

American Water Works Association. 2011. *Desalination of Seawater (M61): AWWA Manual of Water Supply.* Denver, CO: American Water Works Association.

American Water Works Association and Edzwald, J. 2010. *Water Quality & Treatment: A Handbook on Drinking Water* (Water Resources and Environmental Engineering Series), 6th ed. New York: McGraw-Hill Professional.

Buros, O. K. 2000. *The ABCs of Desalting.* Topsfield, MA: International Desalination Association.

Committee Report. 1978. Disinfection. *Am. Water Works Assoc.* 70(4): 219.

Cotruvo, J., Voutchkov, N., Fawell, J., Payment, P., Cunliffe, D., and Lattemann, S. 2010. *Desalination Technology: Health and Environmental Impacts.* Boca Raton, FL: CRC Press.

Davis, M. and Cornwell, D. 2006. *Introduction to Environmental Engineering*, 4th ed. New York: McGraw-Hill Science/Engineering/Math.

Degremont. 2007. *Water Treatment Handbook*, 7th ed. Paris, France: Lavoisier.

Dyer-Smith, P., Brown, Beveri and Co. 1983. Water disinfection status and trends. *J. Water Sewage Treat.* 2(4):13.

El-Hassan, B. M. and Abdel-Magid, I. M. 1986. *Environment and Industry: Treatment of Industrial Wastes.* Khartoum, Sudan: Institute of Environmental Studies, Khartoum University.

Eliassen, R. R. T. and Shrindle, W. B. D. 1958. Experimental performance of miracle water condition. *J. Am Water Works Assoc.* 50:1371.

Hammer, M. J. Sr. and Hammer, M. J. Jr. 2011. *Water and Wastewater Technology*, 7th ed. Upper Saddle River, NJ: Prentice Hall.

Hoyt, B. P. et al. 1979. Evaluating home plumbing corrosion problems. *J. Am Water Works Assoc.* 71(12):720.

Huisman, L. 1977. *Sedimentation and Flotation: Sedimentation and Flotation—Mechanical Filtration—Slow Sand Filtration—Rapid Sand Filtration.* Herdruk, NL: Delft University of Technology.

Huisman, L., Sundaresan, B. B., Netto, J. M. D., Lanoix, J. N., and Hofkes, E. H. 1986. *Small Community Water Supplies.* New York: IRC.

IRC Rep. Intl. Appraisal Meeting, Nagpur, India, September 15–19, 1980. 1981. *Slow Sand Filtration for Community Water Supply in Developing Countries*, BS 16. The Hague, the Netherlands: IRC.

Lin, S. D. 2014. *Water and Wastewater Calculations Manual*, 3rd ed. New York: McGraw-Hill Education.

Logsdon, G. 2008. *Water Filtration Practice: Including Slow Sand Filters and Precoat Filtration.* Denver, CO: American Water Works Association.

Mueller, J. A., Boyle, W. C., and Popel, H. J. 2002. *Aeration: Principles and Practice*, Vol. 11 (Water Quality Management Library). Boca Raton, FL: CRC Press.

Narayan, R. M. 1983. *An Introduction to Metallic Corrosion and Its Prevention.* New Delhi, India: Oxford.

Nemerow, N., L., Agardy, F. J., and Salvato, J. A. 2009. *Environmental Engineering*, 6th ed. Hoboken, NJ: Wiley.

Peavey, H. S., Rowe, D. R., and Tchobanoglous, G. 1985. *Environmental Engineering.* New York: McGraw-Hill.

Pepper, I. L. and Gerba, C. P. 2006. *Environmental and Pollution Science*, 2nd ed. Amsterdam, the Netherlands: Academic Press.

Pescod, M. B., Abouzaid, H., and Sundaresan, B. B. 1986. *Slow Sand Filtration: A Low Cost Treatment for Water Supplies in Developing countries*. Published for the World Health Organization Regional Office for Europe by the Water Research Center, UK, in collaboration with the IRC, the Netherlands.

Punmia, B. C. 1979. *Environmental Engineering Water Supply*, Vol 1. New Delhi, India: Standard Book House.

Riffat, R. 2012. *Fundamentals of Wastewater Treatment and Engineering*. Boca Raton, FL: CRC Press.

Rowe, D. R. and Abdel-Magid, I. M. 1995. *Handbook of Wastewater Reclamation and Reuse*. Boca Raton, FL: Lewis Publishers.

Sawyer, C. N. and McCarty, P. L. 2002. *Chemistry for Environmental Engineering and Science*, 5th ed. Boston, MA: McGraw-Hill.

Shammas, N. K., Fair, G. M., Geyer, J. C., and Okun D. A. 2010. *Water and Wastewater Engineering: Water Supply and Wastewater*, 3rd ed. Hoboken, NJ: Wiley.

Smith, P. G. and Scott, J. S. 2002. *Dictionary of Water and Waste Management*. London: IWA.

Steel, E. W. and McGhee, T. J. 1991. *Water Supply and Sewerage*, 6th ed. New York: McGraw-Hill.

Van Dijk, J. C. and Oomen, J. H. C. M. 1982. *Slow Sand Filtration for Community Water Supply in Developing Countries—A Design and Construction Manual*. Technical Paper No. 11. The Hague, the Netherlands: IRC.

Viessman, W., Hammer, M. J., Perez, E. M., and Chadik, P. A. 2008. *Water Supply and Pollution Control*, 8th ed. Upper Saddle River, NJ: Prentice Hall.

Wang, K. L. and Hung, Y. T. 2009. *Handbook of Advanced Industrial and Hazardous Wastes Treatment (Advances in Industrial and Hazardous Wastes Treatment)*. Boca Raton, FL: CRC Press.

Water Environment Federation. 2009. *Design of Municipal Wastewater Treatment Plants MOP 8*, 5th ed. (Wef Manual of Practice 8: Asce Manuals and Reports on Engineering Practice, No. 76). McGraw-Hill Education.

# 5 Computer Modeling Applications for Wastewater Collection System and Treatment Technology and Disposal

## 5.1 INTRODUCTION

The collection and disposal of wastewater from its sources (domestic, commercial, industrial, agricultural, etc.) are of paramount importance to safeguarding public health and hygiene. Sewage ought to be collected and conveyed as soon as possible from the point of production to a treatment facility or an approved final disposal location with a minimum of cost. This chapter presents information dealing with the terminology used in the wastewater collection field, such as the classification of sewers (sanitary, combined, and storm), and the advantages and disadvantages of each of these systems. The methods used to estimate the design or peak wastewater flows including extraneous water entering the sewers due to infiltration and inflow are included. The rational method to estimate peak flows from storm water runoff is covered, and a computer program to estimate the flow and the required storm sewer size is presented. Also discussed are the fundamentals of sewer design considering the following items: uniform flow, incompressible flow, Bernoulli's equation, application of the momentum equation, continuity equation, and determination of the hydraulic radius. The most frequently used equations, formulas, and models used in the design of sewers are documented. Dry sewerage collection system and nano-bioreactor wastewater technology are not discussed. The reader is advised to consult relevant sources.

This information lays the foundation for the development of computer programs for each of these models. The computer programs can then be used to make rapid and repeated calculations for design and evaluation of the various elements involved in sanitary sewer design. Equations to calculate the self-cleaning velocities in sewers are presented, and a computer program is included that can be used for the same purpose. A summary of the information needed and the steps involved in the design, layout, construction, and installation of sanitary sewers are documented. The basic information needed includes topographical, geological, aerial, and developmental maps. The next step in the process is the sizing and required gradients for the sewers as well as the location for manholes. Other important elements in the design of sanitary sewers are included here.

Equations that can be applied in the evaluation of sulfide buildup in sewers flowing both full and partially full are presented, and a computer program is incorporated for rapid and easy determinations for the various variables involved. The last section deals with the design, operation, and construction of waste disposal systems for rural areas. Septic tanks are the most often used systems in this situation; however, they are also the avenue of last resort. A computer program for septic tank sizing and layout that incorporates the structural design of septic tanks is displayed. A brief discussion of Imhoff tanks concludes this chapter.

Wastewater may be described as "A combination of the liquid or water-carried wastes removed from residences (domestic), institutions, commercial, and industrial establishments together with such groundwater, surface water, and storm water as may be present" (Rowe and Abdel-Magid 1995; Metcalf and Eddy 2013). Wastewater is a complex heterogeneous solution that can pollute or contaminate our environment, namely, water, air, food, and soil. Such conditions demand appropriate collection, proper treatment, and final disposal of wastewater in order to safeguard the environment and protect human health and welfare (Rowe and Abdel-Magid 1995). The first section deals with the environmental problems associated with the discharge of inadequately treated wastewater. The sources of wastewater, as well as the evaluation and selection of the wastewater design flow rates (minimum, maximum, average, and peak), are presented. Computer programs are developed to calculate these various flow rates. A discussion of the relationship between $BOD_5$ and the population equivalent (PE) is accompanied by a computer program that correlates these two factors.

The next section gives the reasons for treating wastewater and the classification and selection of treatment processes (physical, chemical, and biological). The basic principles involved in the design of grit chambers and a computer program that can be used for this purpose are provided. The most commonly used biological processes such as activated sludge (and its many modifications), trickling filters (high and low rates), and waste stabilization ponds (WSPs) are presented. In all cases, computer programs to aid in the evaluation and design of each system are formulated. The operation and design of systems that can be used to render sludge suitable for disposal or use are considered. A computer program for the design of anaerobic digesters is presented here. Also discussed is the specific resistance test used to evaluate the dewatering characteristics of a sludge, which is accompanied by a computer program for determination of both specific resistance and the compressibility coefficient for sludge. The design and scaling up of centrifugal sludge dewatering units are considered, and computer programs that can be used for this purpose are developed. The last section is devoted to the ultimate disposal of treated wastewater, including "dilution" (which, of course, is no solution to pollution) and discharge to surface water such as rivers, lakes, reservoirs, and estuaries. Computer programs that can be used to evaluate the effects of wastewater discharges in each of these situations are included.

The problems associated with inadequately treated wastewater discharges include the following (Barnes et al. 1981; Peavy et al. 1988; Rowe and Abdel-Magid 1995; Wang and Hung 2005; Abdel-Magid and Abdel-Magid 2015; Metcalf and Eddy 2013):

- Transmission of diseases (disease-causing agents such as bacteria, protozoa, viruses, and helminthes) and other public health long-term physiological effects (such as newly created organic substances)
- Accumulation of highly persistent detergents, pesticides, and other toxic substances
- Generation of odors (e.g., hydrogen sulfide, ammonia, mercaptans, and other trace gases containing sulfurs, hydrogen, and nitrogen)
- Pollution of bathing sites with oil, grease, and other undesirable debris
- Establishment of eutrophic conditions (enrichment of water by plant nutrients, etc.)
- Production of objectionable and dangerous levels of solids in wastewater along the banks of rivers and streams, which leads to degradation of water quality both for groundwater and surface water
- Destruction of fish and wildlife
- Contamination of water supplies
- Aesthetic objections

## 5.2 SEWERS AND SEWERAGE SYSTEMS

A sewer is a pipe or conduit, generally closed but normally not flowing full, for conveying sewage. The functions of a sewer include the following:

- Collection of wastewater from its sources
- Transportation of wastewater to treatment works and points of final disposal
- Protection of public health and welfare, especially in areas of concentrated population and/or development

Socioeconomic factors of the system affect the selection of one of the following sewer systems:

1. *Separate sanitary sewer*: This system is used for collection and conveyance of wastes from domestic, commercial, and industrial establishments. In this system, surface waters or runoffs are disposed of in a storm sewer system, whereas domestic sewage, commercial, and industrial wastes are carried in another set of sewers.
2. *Pseudoseparate (partially separate) system*: It is a combination of the separate and combined systems. One system receives the sewage and a part of the storm water (the runoff from the roofs of the buildings, which have a sanitary connection to the system); the other system takes care of the remaining storm water.
3. *Combined sewer system*: In this system, the same sewer serves to carry domestic, commercial, and industrial wastes, as well as surface and storm water flow.

Here the terms sewage and wastewater are often used interchangeably. In some cases, however, the term sewage indicates that human body wastes are present. The term wastewater is more inclusive and indicates not only the presence of discharges from individual homes but also commercial, institutional, and industrial discharges.

### 5.2.1 ADVANTAGES AND DISADVANTAGES OF SEWER SYSTEMS

Sections 5.2.1.1 through 5.2.1.3 outline the main advantages and disadvantages of sewer systems:

#### 5.2.1.1 Separate System

Advantages of the separate sewer system include the following:

- System is economical, because it is smaller in size.
- There is a lower risk of stream pollution, as storm overflows are excluded.
- Quantity of sewage to be treated is smaller.
- Pumping of sewage (when needed) is less costly than in the combined sewer system.

Disadvantages of the separate sewer system include the following:

- System requires flushing (expensive) because it is difficult to assure a self-cleansing velocity in a sewer (unless laid on a steep gradient or slope).
- Double house-plumbing (two sewers in a street) leads to obstruction of traffic during maintenance and repairs.
- Maintenance costs of the two systems may be greater than those for one.

#### 5.2.1.2 Combined System

Advantages of the combined sewer system include the following:

- Rain water keeps sewage fresh, making it easier and more economical to treat.
- Sewage is diluted.
- Water provides automatic flushing.
- Cleaning is easier because sewers are bigger in size.
- House plumbing is economical.

Disadvantages of the combined system include the following:

- With much higher flows, a larger and more expensive wastewater treatment facility is needed.
- Operational problems occur when the capacity of the treatment facility is exceeded (lowered efficiency).

- Grit, sand, and other debris interfere with operation of the treatment facility.
- With increased flows during storms, the sewer lines can be surcharged. Sewer lines normally are not designed to handle flows under pressure.
- Washout of solids in the secondary treatment system can wipe out the efficiency of the treatment plant and result in pollution of the receiving rivers or streams.
- Oil, grease, rubber tire debris, and toxic trace metals can be present in the runoff from roads and streets.

### 5.2.1.3 Other Considerations

Misuse of a sanitary sewer system may create problems, such as explosions and creation of fire hazards, clogging (e.g., grease, bed load, and other debris), physical damage (e.g., discharge of corrosive or abrasive wastes), overloading (e.g., improper connections), pollution of waterways, and interference with treatment (e.g., peak flows, unbiodegradable wastes). Investigations carried out during sewer design include the following (Bizier 2007): topographical surveys (e.g., maps, photographs), surface and subsurface investigations (e.g., runoff coefficient, permeability, depth of water table, infiltration), existing sewerage system (e.g., shortcomings, difficulties, percent loaded), water supply (e.g., water consumption, percentage returned to sewer), public services (e.g., layout of distribution networks, electricity, telephone, and gas), existing structures and their elevations, development plan of the area, industries, population (e.g., maxima, minima, average and distribution of rainfall, river flow, sea level, current prevailing wind, temperature, and evaporation), history of the town, political information (e.g., ordinances and laws affecting sewer connections regulations, rates, and sewerage authority), financial information, and miscellaneous (e.g., tourism, reuse of effluent for irrigation). Figure 5.1 shows the various phases in a sanitary sewer project.

### 5.2.2 Flow Rates of Sanitary Wastewater

The rate of flow of wastewater varies through the day. Relatively steady flow from industrial and commercial sources occurs mostly during the day. The design should be based on

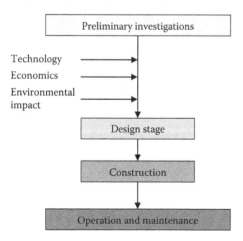

**FIGURE 5.1** Phases of projects of a sanitary sewer.

flows estimated at some future time (design period), usually taken to be between 25 and 50 years (Metcalf and Eddy 2013).

When estimating the design sewage flows for a sewer, allowance must be made for extraneous water entering the sewer as infiltration and inflow. Infiltration is water other than wastewater that enters the sewerage system through cracked pipes, defective joints, and cracks in the walls of manholes. Inflow is also water other than wastewater entering the sewerage system from sources such as roof basement, yard, and foundation drains; from springs and swampy areas and perforated manhole covers; through cross connections between storm and sanitary sewers, cooling towers, street wash water, and surface runoff.

Velocities and hydraulic forces should not exceed maximum allowable flow rates as dictated by the sewer material. A minimum velocity ought to be maintained in the sewer to prevent settling of grit and suspended solids (SS) and accumulations of grease and slime on sewer walls. For most sanitary sewers, flows occur under the action of gravitational forces. Sanitary sewers should be designed for peak flow rates and for areas with a saturated density of people. Estimations of flow rates may be based on maximum water use (assume average sewage flow = average water consumption rates) anticipated through population density or conducted in accordance with the number of structures or extent of the area served. The peak wastewater flow may be estimated from the following equation:

$$Q_p = (2-3)Q_a \tag{5.1}$$

where:

$Q_p$ is the peak flow, m³/s
$Q_a$ is the average flow, m³/s

### 5.2.3 Storm Water

The primary source of storm water flow is precipitation. Factors that may affect the production of storm water include intensity and duration of storms; distance traveled by water before reaching sewers; topography and slope of drainage area; or shape and size of drainage area. The rational method (Lloyd-Davies method) may be employed to estimate the rate of runoff. The method is valid for suburban areas up to 15 km² (Bizier 2007; Abdel-Magid and Abdel-Magid 2015). The mathematical presentation of this method is provided in the following equation:

$$Q_p = 0.278 * c * I * A \tag{5.2}$$

where:

$Q$ is the peak flow rate at point, m³/s
$A$ is the drainage area upstream of point, km²
$I$ is the mean rainfall intensity for a duration equal to the time of concentration (the time required for the entire tributary area to contribute runoff from heavy rainfall to the flow at the point, i.e., flow time from the most remote part of the area to the point), mm/h
$c$ is the coefficient of runoff, dimensionless ($0 < c < 1$, see Table 5.1)

## TABLE 5.1
### Typical Values of *c* for the Rational Method

| Area | C |
|---|---|
| City and commercial | 0.70–0.95 |
| Industrial | |
|     Light areas | 0.50–0.80 |
|     Heavy areas | 0.60–0.90 |
| Residential | |
|     Single family | 0.30–0.50 |
|     Multiunits, detached | 0.40–0.60 |
|     Multiunits, attached | 0.60–0.75 |
|     Suburban | 0.25–0.40 |
| Parks and undeveloped land | 0.10–0.30 |
| Pavement | |
|     Asphaltic | 0.70–0.95 |
|     Concrete | 0.80–0.95 |
|     Brick | 0.70–0.85 |
| Playgrounds | 0.20–0.35 |
| Lawns | 0.05–0.35 |
| Watertight roofs | 0.70–0.95 |
| Woodland areas | 0.01–0.20 |

*Sources:* Abdel-Magid, I. M., and Abdel-Magid, M. I. M., *Problem Solving in Environmental Engineering*, CreateSpace Independent Publishing Platform, 2015; Davis, M. L., and Cornwell, D. A., *Introduction to Environmental Engineering*, 4th ed, McGraw-Hill Science/Engineering/Math, 2006; Hammer, M. J. Sr., and Hammer, M. J. Jr., *Water and Wastewater Technology*, 7th ed, Prentice Hall, Upper Saddle River, NJ, 2011; Viessman, W., and Lewis, G. L., *Introduction to Hydrology*, 5th ed, Prentice Hall, Upper Saddle River, NJ, 2002; Viessman, W. Jr., et al. *Water Supply and Pollution Control*, 8th ed, Prentice Hall, Upper Saddle River, NJ, 2008; Nathanson, J. A., *Basic Environmental Technology: Water Supply, Waste Management & Pollution Control*, 5th ed, Prentice Hall, Upper Saddle River, NJ, 2007.

In storm sewers, greater velocities are required than in sanitary sewers because of the heavy sand and grit, which are washed into them. The minimum allowable velocity is 0.75 m/s, and 0.9 m/s is desirable. Because of abrasive character of the solids, excessively high velocities should be avoided, 2.4 m/s being considered the desirable upper limit (McGhee and Steel 1991).

### Example 5.1

1. Write a computer program to estimate the diameter of a sewer for a suburb of any area, $A$ (km$^2$); given the peak flow rate at point, $Q$ (m$^3$/s); the drainage area upstream of point, $A$ (km$^2$); the mean rainfall intensity, $I$ (mm/h); and the coefficient of runoff, $c$ (as presented in Table 5.1).
2. A circular storm sewer is required to serve an area of 900 ha; the average ground slope is 1 in 500. Determine the size of the sewer at point of discharge knowing that the runoff coefficient is 0.4, the time of concentration is 1000 s, and $n = 0.013$. Take $I = 750/(t+10)$.
3. Use the computer program developed in (1) to support manual computations conducted in (2).

### Solution

1. For solution to Example 5.1 (1), see the listing of Program 5.1.
2. Solution to Example 5.2 (2):
   a. Given: $A = 900$ ha $= $ km$^2$, $j = 1/500 = 0.002$, $c = 0.4$, $t = 1000$ s $= 1000/60 = 16.7$ min.
   b. Find the rainfall intensity as follows: $I = 750/(16.7 + 10) = 28.1$ mm/h.
   c. Find $Q$ as $0.278 * c * I * A = 0.278 * 0.4 * 28.1 * 9 = 28.1$ m$^3$/s $= 1687.3$ m$^3$/min.
   d. The diameter of the sewer and flow rate through it may be determined from a nomograph corresponding to the specified slope (Fair et al. 1966; Peavy et al. 1988; McGhee and Steel 1991; Jonathan et al. 2003; Bizier 2007; Mays 2010; Hicks 2014; Alderdiri et al. 2015; Alderdiri and Abdel-Magid 2015). Alternatively, the following procedure can be used: $v = (1/n)(D/4)^{2/3}*j^{1/2}$, and with $Q = (\pi/4)*D^2*v$, it gives: $D = (4^{5/3}*Q*n/\pi*j^{1/2})^{3/8} = (4^{5/3}*28.1 * 0.013/\pi * 0.002^{1/2})^{3/8} = 3.4$ (take 3.5) m, which gives a discharge of $Q = (\pi/4) * (3.4)^2 * (3.4/4)^{2/3} * (0.002)^{1/2}/0.013 = 28$ m$^3$/s.

---

### LISTING OF PROGRAM 5.1   (CHAP5.1\FORM1.VB): WASTEWATER COLLECTION

```
'******************************************************************************
'Program 5.1: Water Collection
'******************************************************************************

Public Class Form1
    Const pi = 3.1415962#
    Const g = 9.81
    Dim A, n, c, j, t As Double
    Dim I, Q, D As Double
    Dim c_table() As Double =
```

```
    {
      0.95, 0.8, 0.9, 0.5, 0.6, 0.75, 0.4, 0.3,
      0.95, 0.95, 0.85, 0.35, 0.35, 0.95, 0.2
    }

'*******************************************************************************
'Code in Form_Load sub will be executed when the form is loading,
'i.e., first thing after program loads into the memory
'*******************************************************************************

  Private Sub Form1_Load(ByVal sender As System.Object, ByVal e As System.EventArgs)
Handles MyBase.Load
    Me.MaximizeBox = False
    Me.Text = "Program 5.1: Water Collection"
    Me.FormBorderStyle = Windows.Forms.FormBorderStyle.FixedSingle
    Me.MaximizeBox = False
    Label1.Text = "Drainage area, A (Km2):"
    Label2.Text = "Enter the value of n:"
    Label3.Text = "Type of area for runoff coefficient, c:"
    Label4.Text = "Average ground slope, j:"
    Label5.Text = "Time of concentration, t (sec):"
    GroupBox1.Text = " Output: "
    Label6.Text = "The required diameter of sewer D (m):"
    Label7.Text = "This delivers a discharge of Q (m3/s):"
    Label8.Text = "Decimal Places:"
    NumericUpDown1.Maximum = 10
    NumericUpDown1.Minimum = 0
    NumericUpDown1.Value = 2
    ComboBox1.Items.Clear()
    ComboBox1.Items.Add("City and commercial")
    ComboBox1.Items.Add("Industrial: Light area")
    ComboBox1.Items.Add("Industrial: Heavy area")
    ComboBox1.Items.Add("Residential: Single family")
    ComboBox1.Items.Add("Residential: Detached multi units")
    ComboBox1.Items.Add("Residential: Attached multi units")
    ComboBox1.Items.Add("Residential: Suburban")
    ComboBox1.Items.Add("Parks and undeveloped land")
    ComboBox1.Items.Add("Pavement: Asphaltic")
    ComboBox1.Items.Add("Pavement: Concrete")
    ComboBox1.Items.Add("Pavement: Brick")
    ComboBox1.Items.Add("Playgrounds")
    ComboBox1.Items.Add("Lawns")
    ComboBox1.Items.Add("Watertight roofs")
    ComboBox1.Items.Add("Woodland areas")
    ComboBox1.SelectedIndex = 0
  End Sub

  Function get_c(ByVal index As Integer) As Double
    If index < 0 Or index > 14 Then Return 0
    Return c_table(index)
  End Function

  Sub calculateResults()
    A = Val(TextBox1.Text)
    n = Val(TextBox2.Text)
    j = Val(TextBox3.Text)
    t = Val(TextBox4.Text)
    If ComboBox1.SelectedIndex = -1 Then
      MsgBox("Please select type of area from list.",
        vbOKOnly Or vbInformation)
      Exit Sub
```

```
      End If
      c = get_c(ComboBox1.SelectedIndex)
      t = t / 60 'convert to minutes
      I = 750 / (t + 10) 'rainfall intensity
      Q = 0.278 * c * I * A
      D = (4 ^ (5 / 3) * Q * n / (pi * j ^ 0.5)) ^ (3 / 8)
      Q = pi / 4 * D ^ 2 * (D / 4) ^ (2 / 3) * j ^ 0.5 / n
      showResults()
   End Sub

   Sub showResults()
      TextBox5.Text = FormatNumber(D, NumericUpDown1.Value)
      TextBox6.Text = FormatNumber(Q, NumericUpDown1.Value)
   End Sub

'*********************************************************************************
'Code in TextChanged member of the TextBox will be executed
'when the user changes the text.
'The following code will recalculate the results and show them when ever
'the user changes the input.
'*********************************************************************************

   Private Sub TextBox1_TextChanged(ByVal sender As System.Object, ByVal e As System.
EventArgs) Handles TextBox1.TextChanged
      calculateResults()
   End Sub

   Private Sub TextBox2_TextChanged(ByVal sender As System.Object, ByVal e As System.
EventArgs) Handles TextBox2.TextChanged
      calculateResults()
   End Sub

   Private Sub TextBox3_TextChanged(ByVal sender As System.Object, ByVal e As System.
EventArgs) Handles TextBox3.TextChanged
      calculateResults()
   End Sub

   Private Sub TextBox4_TextChanged(ByVal sender As System.Object, ByVal e As System.
EventArgs) Handles TextBox4.TextChanged
      calculateResults()
   End Sub

'*********************************************************************************
'Whenever a different number of decimals is selected, re-show the results.
'*********************************************************************************

   Private Sub NumericUpDown1_ValueChanged(ByVal sender As System.Object, ByVal e As
System.EventArgs) Handles NumericUpDown1.ValueChanged
      showResults()
   End Sub
End Class
```

### 5.2.4 Hydraulics of Sewers

In the design of sanitary sewers, the following assumptions are made:

- One-dimensional flow conditions (velocity of flow is uniform across each section of flow)
- Incompressible flow (fluid motion with negligible changes in density), where pressure flow totally fills a closed conduit, except for the possible occurrence of water hammer in pressure conduits
- Steady-state flow (flow conditions at any point in fluid do not change with time) with constant rate of flow; actually flow is quasi-steady (varies significantly from hour to hour)
- Valid continuity principle ($Q = A*v$)

- *Valid Bernoulli equation*: Total energy (at a point in a flowing fluid) per unit mass = Σ(potential energy + pressure energy + kinetic energy), or

$$H = (Z) + \left(\frac{P}{\gamma}\right) + \left(\frac{v^2}{2g}\right) \qquad (5.3)$$

where:

$H$ is the total energy (total head, energy head), m
$Z$ is the elevation of point on a streamline above horizontal datum, m
$P$ is the pressure at point, Pa
$\gamma$ is the fluid specific weight, N/m³
$v$ is the average velocity for all streamlines, m/s

- Momentum principle is applied as presented in the following equation:

$$\Sigma F = \frac{\gamma \left[Q(v_2 - v_1)\right]}{g} \qquad (5.4)$$

where:

$\Sigma F$ is the sum of all forces acting on fluid contained between two cross sections (pressure, weight, constraints-friction), N
$v$ is the average velocity, m/s
$\gamma$ is the density

- Hydraulic radius is given by

$$r_H = \frac{A}{W_p} \qquad (5.5)$$

where:

$r_H$ is the hydraulic radius (= $D/4$ for circular closed conduits), m
$A$ is the area, m²
$W_p$ is the wetted perimeter of flow cross sections, m

### 5.2.5 FLOW FRICTION FORMULAE

The most frequently used flow equations are briefly outlined as follows (Fair et al. 1966; Peavy et al. 1988; McGhee and Steel 1991; Peirce et al. 1997; Jonathan et al. 2003; Davis and Cornwell 2006; Bizier 2007; Mays 2010; Hammer and Hammer 2011; Hicks 2014; Abdel-Magid and Abdel-Magid 2015; Abdel-Magid et al. 2015):

Darcy–Weisbach equation for pipe flow

$$h_f = f * \frac{v^2}{2g} * \frac{L}{D} \qquad (5.6)$$

where:

$h_f$ is the friction head, m
$f$ is the friction factor
$v$ is the velocity of flow, m/s
$L$ is the length of conduit, m
$g$ is the gravitational acceleration, m/s²
$D$ is the diameter of circular conduit, m

Chezy equation

$$v = c_f \sqrt{(r_H * j)} \qquad (5.7)$$

where:

$c_f$ is the friction coefficient, or coefficient of DeChezy, m$^{1/2}$/s
$r_H$ is the hydraulic radius, m
$j$ is the slope of invert or bed, m/m

Manning's formula
Metric units:

$$v = \frac{1}{n} * (r_H)^{2/3} * j^{1/2} \qquad (5.8)$$

English units:

$$v = \frac{1.49}{n} * r_H^{2/3} * j^{1/2} \qquad (5.9)$$

where $n$ is the Manning's coefficient.

Kutter–Manning's formula

$$v = \left[\frac{23 + (0.00155/j) + (1/n)}{1 + \frac{\left[23 + (0.00155/j)\right]n}{\sqrt{r_H}}}\right] * \sqrt{r_H * j} \qquad (5.10)$$

where:

$v$ is the mean velocity of flow, m/s
$j$ is the slope of energy grade line, dimensionless
$n$ is the Kutter coefficient of roughness, which is equal to the Manning's coefficient for types of pipe commonly used in sewer construction

Hazen–Williams formula
Metric units:

$$v = 0.85 * c * r_H^{0.63} * j^{0.54} \qquad (5.11)$$

English units:

$$v = 1.32 * c * r_H^{0.63} * j^{0.54} \qquad (5.12)$$

where $c$ is the coefficient that depends on the roughness of conduit, $100 < c < 140$.

Prandtl–Colebrook formula

$$v = -2 * \log\left[\frac{2.51v}{D\sqrt{2g * d * j}} + \frac{k}{3.71 * D}\right]\sqrt{2g * D * j} \qquad (5.13)$$

where:

$v$ is the kinematic viscosity, m²/s
$D$ is the diameter of sewer, mm
$K$ is the factor, usually taken as 0.025 mm

v is the velocity of flow, m/s

$j$ is the slope, m/km

Strickler formula (Manning's)

$$v = c * r_H^{2/3} * j^{1/2} \qquad (5.14)$$

where $c$ is the factor (= 100).

Scimemi formula (Hazen)

$$v = c * D^{0.68} * j^{0.56} \qquad (5.15)$$

where $c$ = factor (= 61.5).

The flow velocity should not exceed a maximum value to avoid excessive erosion and dissipation of energy at the point of discharge. Recommended limits are as follows:

- For clean water, velocities up to 12 m/s are recommended.
- For sanitary sewers, velocities up to 3 m/s are recommended (possibly up to 5 m/s).
- For storm sewers, velocities up to 5 m/s are allowed (possibly up to 10 m/s).

A minimum velocity should be maintained to avoid accumulation of deposits and their biodegradation (to avoid production of hydrogen sulfide, which promotes crown corrosion).

## Example 5.2

1. Write a computer program that would allow the computation of velocity (m/s) of flow in a pipe using one of the following formulae, given the relevant data for each: Darcy–Weisbach equation, Chezy equation, Manning's formula, Kutter formula, Hazen–Williams formula, Prandtl–Colebrook formula, Strickler formula, or Scimemi formula.
2. Using Manning's equation, find the velocity of flow and sewer capacity for a concrete sewer of diameter 1.52 m laid at a slope of 0.0008, when sewer is flowing full. (Take Manning's friction coefficient equal to 0.013.)

### Solution

1. For solution to Example 5.2 (1), see the listing of Program 5.2.
2. Solution to Example 5.2 (2):
   a. Given: $D = 1.52$ m, $j = 0.008$.
   b. Determine the hydraulic radius as follows: $r_H = (\pi D^2/4)/(\pi D) = D/4 = 1.52/4 = 0.38$ m.
   c. Find the average velocity of flow in the pipe as follows: $v = r_H^{2/3} * j^{1/2}/n = 0.38^{2/3} * 0.0008^{1/2}/0.013 = 1.14$ m/s.
   d. Determine the capacity of the sewer as follows: $Q = v * A = 1.14 * (\pi * 1.52^2)/4 = 2.07$ m³/s.

---

**LISTING OF PROGRAM 5.2   (CHAP5.2\FORM1.VB): VELOCITY OF FLOW IN PIPES**

```
'*****************************************************************************************
'Program 5.2: Friction Flow in Pipes
'*****************************************************************************************

Imports System.Math
Public Class Form1
    Const pi = 3.1415962#
    Const g = 9.81
    Dim D, hf, L, f, v, n, j, Cf, rH, c, fact As Double
    Dim vis, k, z As Double

'*****************************************************************************************
'Code in Form_Load sub will be executed when the form is loading,
'i.e., first thing after program loads into the memory
'*****************************************************************************************

    Private Sub Form1_Load(ByVal sender As System.Object, ByVal e As System.EventArgs)
Handles MyBase.Load
        Me.Text = "Program 5.2: Friction Flow in Pipes"
        Me.MaximizeBox = False
        Me.FormBorderStyle = Windows.Forms.FormBorderStyle.FixedSingle
        ListBox1.Items.Clear()
        ListBox1.Items.Add("Darcy-Weisbach equation")
        ListBox1.Items.Add("Chezy equation")
        ListBox1.Items.Add("Manning's equation")
        ListBox1.Items.Add("Kutter-Manning's formula")
```

```
        ListBox1.Items.Add("Hazen-Williams formula")
        ListBox1.Items.Add("Prandtl-Colebrook formula")
        ListBox1.Items.Add("Strickler formula")
        ListBox1.Items.Add("Scimemi formula (Hazen)")
        ListBox1.SelectedIndex = 0 'Select the first option
        Label1.Text = "To compute the velocity of flow using:"
        Label2.Text = "Enter the diameter of pipe (m):"
        GroupBox1.Text = "Use Metric Units:"
        RadioButton1.Checked = True
        RadioButton1.Text = "Yes"
        RadioButton2.Text = "No"
        GroupBox1.Visible = False
        GroupBox2.Text = " Output: "
        Label6.Text = "The velocity, v (m/s):"
        Label7.Text = "Sewer Capacity, A (m3/s):"
    End Sub

    Private Sub ListBox1_SelectedIndexChanged(ByVal sender As System.Object, ByVal e As
    System.EventArgs) Handles ListBox1.SelectedIndexChanged
        If ListBox1.SelectedIndex < 0 Then Exit Sub
        Select Case ListBox1.SelectedIndex
            Case 0
                'Darcy-Weisbach equation
                Label3.Text = "Enter the head loss, hf (m):"
                Label4.Text = "Enter length of pipe, L (m):"
                Label5.Text = "Enter friction factor, f:"
                GroupBox1.Visible = False
                Label3.Visible = True
                Label4.Visible = True
                Label5.Visible = True
                TextBox2.Visible = True
                TextBox3.Visible = True
                TextBox4.Visible = True
            Case 1
                'Chezy equation
                Label3.Text = "Enter Chezy coefficient Cf:"
                Label4.Text = "Enter slope of the pipe, j:"
                GroupBox1.Visible = False
                Label3.Visible = True
                Label4.Visible = True
                Label5.Visible = False
                TextBox2.Visible = True
                TextBox3.Visible = True
                TextBox4.Visible = False
            Case 2
                'Manning's equation
                Label3.Text = "Enter slope of the pipe, j:"
                Label4.Text = "Enter Manning's coefficient n:"
                GroupBox1.Visible = True
                Label3.Visible = True
                Label4.Visible = True
                Label5.Visible = False
                TextBox2.Visible = True
                TextBox3.Visible = True
                TextBox4.Visible = False
            Case 3
                'Kutter Manning's formula
                Label3.Text = "Enter slope of the pipe, j:"
                Label4.Text = "Enter Kutter's coefficient, n:"
                GroupBox1.Visible = False
                Label3.Visible = True
```

```
          Label4.Visible = True
          Label5.Visible = False
          TextBox2.Visible = True
          TextBox3.Visible = True
          TextBox4.Visible = False
      Case 4
          'Hazen-Williams formula
          Label3.Text = "Enter slope of the pipe, j:"
          Label4.Text = "Enter Hazen-William coefficient, c:"
          GroupBox1.Visible = True
          Label3.Visible = True
          Label4.Visible = True
          Label5.Visible = False
          TextBox2.Visible = True
          TextBox3.Visible = True
          TextBox4.Visible = False
      Case 5
          'Prandtl-Colebrook formula
          Label3.Text = "Enter slope of the pipe, j:"
          Label4.Text = "Enter kinematic viscosity, v (m2/s):"
          Label5.Text = "Enter factor k:"
          GroupBox1.Visible = False
          Label3.Visible = True
          Label4.Visible = True
          Label5.Visible = True
          TextBox2.Visible = True
          TextBox3.Visible = True
          TextBox4.Visible = True
      Case 6
          'Strickler formula
          Label3.Text = "Enter slope of the pipe, j:"
          GroupBox1.Visible = False
          Label3.Visible = True
          Label4.Visible = False
          Label5.Visible = False
          TextBox2.Visible = True
          TextBox3.Visible = False
          TextBox4.Visible = False
      Case 7
          'Scimemi formula (Hazen)
          Label3.Text = "Enter slope of the pipe, j:"
          GroupBox1.Visible = False
          Label3.Visible = True
          Label4.Visible = False
          Label5.Visible = False
          TextBox2.Visible = True
          TextBox3.Visible = False
          TextBox4.Visible = False
    End Select
End Sub

Sub calculateResults()
    If ListBox1.SelectedIndex < 0 Then Exit Sub
    D = Val(TextBox1.Text)
    rH = D / 4
    Select Case ListBox1.SelectedIndex
      Case 0
          'Darcy-Weisbach equation
          hf = Val(TextBox2.Text)
```

```
        L = Val(TextBox3.Text)
        f = Val(TextBox4.Text)
        v = (hf * D * 2 * g / (L * f)) ^ 0.5
    Case 1
        'Chezy equation
        j = Val(TextBox2.Text)
        Cf = Val(TextBox3.Text)
        v = Cf * ((rH * j) ^ 0.5)
    Case 2
        'Manning's equation
        n = Val(TextBox2.Text)
        j = Val(TextBox3.Text)
        'Using metric units?
        If RadioButton1.Checked Then
            fact = 1
        Else
            fact = 1.49
        End If
        v = fact * rH ^ (2 / 3) * (j ^ 0.5) / n
    Case 3
        'Kutter Manning's formula
        j = Val(TextBox2.Text)
        n = Val(TextBox3.Text)
        Dim v1 As Double = (23 + (0.00155 / j) + (1 / n))
        Dim v2 As Double = 1 + (((23 + (0.00155 / j)) * n) / Sqrt(rH))
        Dim v3 As Double = Sqrt(rH * j)
        v = (v1 / v2) * v3
    Case 4
        'Hazen-Williams formula
        j = Val(TextBox2.Text)
        c = Val(TextBox3.Text)
        'Using metric units?
        If RadioButton1.Checked Then
            fact = 0.85
        Else
            fact = 1.32
        End If
        v = fact * c * (rH ^ 0.63) * (j ^ 0.54)
    Case 5
        'Prandtl-Colebrook formula
        j = Val(TextBox2.Text)
        vis = Val(TextBox3.Text)
        k = Val(TextBox4.Text)
        z = (2 * g * D * j) ^ 0.5
        v = -2 * log10((2.51 * vis / (D * z)) + (k / (3.71 * D))) * z
    Case 6
        'Strickler formula
        j = Val(TextBox2.Text)
        c = 100
        v = c * (rH ^ (2 / 3)) * (j ^ 0.5)
    Case 7
        'Scimemi formula (Hazen)
        j = Val(TextBox2.Text)
        c = 61.5
        v = c * (D ^ 0.68) * (j ^ 0.56)
End Select
TextBox5.Text = v.ToString
Dim A As Double = pi * (D ^ 2) / 4
Dim Q As Double = v * A
TextBox6.Text = Q.ToString
End Sub
```

```
    Private Sub Button1_Click(ByVal sender As System.Object, ByVal e As System.EventArgs)
Handles Button1.Click
        calculateResults()
    End Sub
End Class
```

### 5.2.6 DESIGN COMPUTATIONS

Design computations include conduit capacity, self-cleansing velocities, maximum or minimum slopes, significant water-surface elevations, and changes in conduit size.

#### 5.2.6.1 Capacity of Flow Estimates

1. For pipes flowing full, the size and slope may be determined by using the aforementioned pipe flow equations.
2. For partially full pipes, the hydraulic element chart (see Appendix A4) gives, $Q/Q_f$, $A/A_f$, $r_H/r_{Hf}$, $n/n_f$, and $D/D_f$; otherwise, the following equation may be used (see Figure 5.2):

$$\phi = 2\cos^{-1}\left(1 - \frac{2d}{D}\right) \tag{5.16}$$

where:

$D$ is the diameter of the sewer
$d$ is the depth of the flow
$\phi$ is the angle subtended between the two radii formed at the ends of chord of partial depth

From Figure 5.2, the following equations can be developed:

1. For a pipe flowing full:
   Area: $A_f = \pi D^2/4$
   Wetted perimeter: $(w_p)_f = \pi D$
   Hydraulic radius:

$$(r_H)f = \frac{D}{4} \tag{5.17}$$

Velocity of flow:

$$v_f = \left(\frac{1}{n}\right)r_H^{2/3}j^{1/2} = \left(\frac{1}{n}\right)\left(\frac{D}{4}\right)^{2/3}j^{1/2} \tag{5.18}$$

where $j$ is the slope.

Flow:

$$Q_f = A_f v_f = \frac{1}{n}\left(\frac{D}{4}\right)^{2/3}j^{1/2}\pi\frac{D^2}{4} = \frac{1}{n}(D)^{8/3}j^{1/2}\pi\frac{1}{4^{5/3}} \tag{5.19}$$

2. For a pipe flowing partially full:
   Area:

$$A_p = \pi\frac{D^2}{4}\frac{\phi}{360} - \left(\frac{D}{2} - d\right)\sqrt{\left(Dd - d^2\right)}$$

$$= D^2\left[\frac{\pi\phi}{1440} - \left(\frac{1}{2} - \frac{d}{D}\right)\sqrt{\left[d - \left(\frac{d}{D}\right)^2\right]}\right] \tag{5.20}$$

Wetted perimeter:

$$\left(w_p\right)_p = \frac{\pi D\phi}{360} \tag{5.21}$$

Hydraulic radius:

$$\left(r_H\right)_p = \frac{A_p}{(w_p)_p} = \frac{D}{4} - \frac{360D}{\pi\phi}\left[\frac{1}{2} - \frac{d}{D}\sqrt{\frac{d}{D} - \left(\frac{d}{D}\right)^2}\right]$$

Velocity of flow:

$$v_p = \left(\frac{1}{n}\right)\left(r_H\right)_p^{2/3}j^{1/2} \tag{5.22}$$

Flow: $Q_p = A_p * v_p$

The hydraulic elements can be specified or formulated by the following equations:

$$\frac{\left(r_H\right)_p}{D} = \frac{1}{4} - \frac{360}{\pi\phi}\left[\frac{1}{2} - \frac{d}{D}\sqrt{\frac{d}{D} - \left(\frac{d}{D}\right)^2}\right] \tag{5.23}$$

$$\frac{A_p}{A_f} = \frac{\phi}{360} - 4\pi\left(\frac{1}{2} - \frac{d}{D}\right)\sqrt{\frac{d}{D} - \left(\frac{d}{D}\right)^2} \tag{5.24}$$

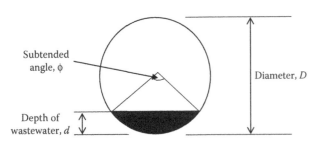

**FIGURE 5.2** Cross section of a sewer.

$$\frac{(r_H)_p}{(r_H)_f} = 1 - 1440\pi\phi \left[ \frac{1}{2} - \frac{d}{D}\sqrt{\frac{d}{D} - \left(\frac{d}{D}\right)^2} \right] \quad (5.25)$$

$$\frac{v_p}{v_f} = \left[ \frac{(r_H)_p}{(r_H)_f} \right]^{2/3} = \left\{ \frac{1 - \left(\frac{1440}{\pi\phi}\right)*\left[0.5 - \left(\frac{d}{D}\right)\right]}{*\left[(d/D) - (d/D)^2\right]^{1/2}} \right\}^{2/3} \quad (5.26)$$

$$\frac{Q_p}{Q_f} = \frac{v_p A_p}{v_f A_f} = \left\{ 1 - \frac{1440}{\pi\phi}\left[ \frac{1}{2} - \frac{d}{D}\sqrt{\frac{d}{D} - \left(\frac{d}{D}\right)^2} \right] \right\}^{2/3}$$

$$\left\{ \frac{\phi}{360} - 4\pi\left(\frac{1}{2} - \frac{d}{D}\right)\sqrt{\left[\frac{d}{D} - \left(\frac{d}{D}\right)^2\right]} \right\} \quad (5.27)$$

Table 5.2 can be developed from the above Equations 5.20 through 5.27.

---

**TABLE 5.2**
**Design Computations for Partial Pipe Flow**

| Depth Ratio (d/D) | Hydraulic Radius/ Diameter[(r_H)_p/D] | Area Ratio (A_p/A_f) | Hydraulic Radius Ratio (r_H)_p/(r_H)_f | For the Same n | |
|---|---|---|---|---|---|
| | | | | Velocity Ratio (v_p/v_f) | Discharge Ratio (Q_p/Q_f) |
| 0.10 | 0.06352 | 0.052023 | 0.254081 | 0.401157 | 0.020869 |
| 0.11 | 0.069522 | 0.059825 | 0.278086 | 0.426042 | 0.025488 |
| 0.12 | 0.075458 | 0.067945 | 0.301833 | 0.449964 | 0.030573 |
| 0.13 | 0.08133 | 0.076363 | 0.32532 | 0.473014 | 0.036121 |
| 0.14 | 0.087136 | 0.08506 | 0.348546 | 0.495268 | 0.042128 |
| 0.15 | 0.092878 | 0.094022 | 0.37151 | 0.51679 | 0.04859 |
| 0.16 | 0.098553 | 0.103234 | 0.394212 | 0.537633 | 0.055502 |
| 0.17 | 0.104162 | 0.112682 | 0.416649 | 0.557845 | 0.062859 |
| 0.18 | 0.109705 | 0.122353 | 0.438821 | 0.577464 | 0.070654 |
| 0.19 | 0.115182 | 0.132237 | 0.460727 | 0.596526 | 0.078882 |
| 0.20 | 0.120591 | 0.142321 | 0.482365 | 0.61506 | 0.087536 |
| 0.21 | 0.125934 | 0.152597 | 0.503735 | 0.633094 | 0.096608 |
| 0.22 | 0.131209 | 0.163054 | 0.524835 | 0.650652 | 0.106091 |
| 0.23 | 0.136416 | 0.173683 | 0.545664 | 0.667755 | 0.115977 |
| 0.24 | 0.141555 | 0.184475 | 0.566221 | 0.684422 | 0.126259 |
| 0.25 | 0.146626 | 0.195422 | 0.586503 | 0.70067 | 0.136927 |
| 0.26 | 0.151628 | 0.206517 | 0.606511 | 0.716516 | 0.147973 |
| 0.27 | 0.156561 | 0.21775 | 0.626242 | 0.731973 | 0.159387 |
| 0.28 | 0.161424 | 0.229116 | 0.645696 | 0.747054 | 0.171162 |
| 0.29 | 0.166218 | 0.240606 | 0.66487 | 0.761771 | 0.183287 |
| 0.30 | 0.170941 | 0.252214 | 0.683764 | 0.776135 | 0.195752 |
| 0.31 | 0.175594 | 0.263933 | 0.702375 | 0.790156 | 0.208549 |
| 0.32 | 0.180176 | 0.275757 | 0.720703 | 0.803842 | 0.221665 |
| 0.33 | 0.184686 | 0.287679 | 0.738745 | 0.817203 | 0.235092 |
| 0.34 | 0.189125 | 0.299693 | 0.7565 | 0.830244 | 0.248819 |
| 0.35 | 0.193492 | 0.311793 | 0.773967 | 0.842975 | 0.262834 |
| 0.36 | 0.197786 | 0.323973 | 0.791143 | 0.855401 | 0.277127 |
| 0.37 | 0.202007 | 0.336228 | 0.808026 | 0.867528 | 0.291687 |
| 0.38 | 0.206154 | 0.348551 | 0.824616 | 0.879362 | 0.306502 |
| 0.39 | 0.210227 | 0.360937 | 0.84091 | 0.890908 | 0.321561 |
| 0.40 | 0.214226 | 0.37338 | 0.856905 | 0.90217 | 0.336852 |
| 0.41 | 0.21815 | 0.385875 | 0.872601 | 0.913154 | 0.352363 |
| 0.42 | 0.221999 | 0.398417 | 0.887995 | 0.923862 | 0.368082 |
| 0.43 | 0.225771 | 0.411 | 0.903085 | 0.934299 | 0.383997 |
| 0.44 | 0.229467 | 0.423619 | 0.917868 | 0.944467 | 0.400094 |
| 0.45 | 0.233086 | 0.436269 | 0.932343 | 0.954371 | 0.416362 |
| 0.46 | 0.236627 | 0.448944 | 0.946506 | 0.964012 | 0.432787 |
| 0.47 | 0.240089 | 0.46164 | 0.960356 | 0.973393 | 0.449357 |
| 0.48 | 0.243473 | 0.474351 | 0.973891 | 0.982517 | 0.466058 |

*(Continued)*

**TABLE 5.2 (*Continued*)**
**Design Computations for Partial Pipe Flow**

| Depth Ratio (d/D) | Hydraulic Radius/ Diameter[(r_H)_p/D] | Area Ratio (A_p/A_f) | Hydraulic Radius Ratio (r_H)_p/(r_H)_f | For the Same n | |
|---|---|---|---|---|---|
| | | | | Velocity Ratio (v_p/v_f) | Discharge Ratio (Q_p/Q_f) |
| 0.49 | 0.246776 | 0.487072 | 0.98106 | 0.991385 | 0.482876 |
| 0.50 | 0.25 | 0.499799 | 1 | 1 | 0.499799 |
| 0.51 | 0.253142 | 0.512525 | 1.01257 | 1.008362 | 0.516811 |
| 0.52 | 0.256203 | 0.525247 | 1.024812 | 1.016474 | 0.533899 |
| 0.53 | 0.259181 | 0.537958 | 1.036725 | 1.024336 | 0.551049 |
| 0.54 | 0.262076 | 0.550654 | 1.048304 | 1.031949 | 0.568246 |
| 0.55 | 0.264886 | 0.563329 | 1.059546 | 1.039313 | 0.585475 |
| 0.56 | 0.267612 | 0.575979 | 1.070448 | 1.04643 | 0.602722 |
| 0.57 | 0.270251 | 0.588598 | 1.081005 | 1.0533 | 0.61997 |
| 0.58 | 0.272804 | 0.601181 | 1.091216 | 1.059922 | 0.637205 |
| 0.59 | 0.275269 | 0.613723 | 1.101074 | 1.066296 | 0.65441 |
| 0.60 | 0.277644 | 0.626218 | 1.110577 | 1.072422 | 0.67157 |
| 0.61 | 0.27993 | 0.638661 | 1.119719 | 1.0783 | 0.688668 |
| 0.62 | 0.282124 | 0.651047 | 1.128497 | 1.083927 | 0.705688 |
| 0.63 | 0.284226 | 0.66337 | 1.136905 | 1.089305 | 0.722612 |
| 0.64 | 0.286234 | 0.675624 | 1.144938 | 1.09443 | 0.739423 |
| 0.65 | 0.288148 | 0.687804 | 1.15259 | 1.099301 | 0.756104 |
| 0.66 | 0.289964 | 0.699904 | 1.159857 | 1.103917 | 0.772636 |
| 0.67 | 0.291683 | 0.711918 | 1.166732 | 1.108275 | 0.789001 |
| 0.68 | 0.293302 | 0.72384 | 1.173209 | 1.112372 | 0.80518 |
| 0.69 | 0.29482 | 0.735664 | 1.17928 | 1.116207 | 0.821153 |
| 0.70 | 0.296235 | 0.747383 | 1.184939 | 1.119774 | 0.836901 |
| 0.71 | 0.297544 | 0.758991 | 1.190177 | 1.123072 | 0.852402 |
| 0.72 | 0.298747 | 0.770482 | 1.194986 | 1.126096 | 0.867636 |
| 0.73 | 0.299839 | 0.781847 | 1.199358 | 1.12884 | 0.882581 |
| 0.74 | 0.300821 | 0.793081 | 1.203282 | 1.131301 | 0.897213 |
| 0.75 | 0.301687 | 0.804175 | 1.206748 | 1.133473 | 0.911511 |
| 0.76 | 0.302436 | 0.815123 | 1.209745 | 1.135349 | 0.925448 |
| 0.77 | 0.303065 | 0.825915 | 1.21226 | 1.136922 | 0.939001 |
| 0.78 | 0.30357 | 0.836544 | 1.21428 | 1.138184 | 0.952141 |
| 0.79 | 0.303947 | 0.847001 | 1.21579 | 1.139128 | 0.964842 |
| 0.80 | 0.304193 | 0.857276 | 1.216773 | 1.139742 | 0.977074 |
| 0.81 | 0.304303 | 0.867361 | 1.217211 | 1.140015 | 0.988805 |
| 0.82 | 0.304271 | 0.877245 | 1.217084 | 1.139936 | 1.000003 |
| 0.83 | 0.304092 | 0.886916 | 1.216369 | 1.139489 | 1.010631 |
| 0.84 | 0.30376 | 0.896364 | 1.215041 | 1.138659 | 1.020653 |
| 0.85 | 0.303267 | 0.905575 | 1.213068 | 1.137427 | 1.030026 |
| 0.86 | 0.302605 | 0.914537 | 1.210419 | 1.13577 | 1.038704 |
| 0.87 | 0.301763 | 0.923235 | 1.207054 | 1.133664 | 1.046638 |
| 0.88 | 0.300731 | 0.931653 | 1.202925 | 1.131077 | 1.053771 |
| 0.89 | 0.299495 | 0.939772 | 1.197978 | 1.127975 | 1.060039 |
| 0.90 | 0.298037 | 0.947575 | 1.192147 | 1.124311 | 1.065369 |
| 0.91 | 0.296337 | 0.955037 | 1.185347 | 1.120032 | 1.069672 |
| 0.92 | 0.294369 | 0.962135 | 1.177476 | 1.115068 | 1.072846 |
| 0.93 | 0.292099 | 0.968838 | 1.168397 | 1.109329 | 1.07476 |
| 0.94 | 0.289482 | 0.975111 | 1.157926 | 1.102691 | 1.075247 |
| 0.95 | 0.286452 | 0.980912 | 1.145806 | 1.094983 | 1.074082 |
| 0.96 | 0.282912 | 0.986186 | 1.131647 | 1.085944 | 1.070942 |
| 0.97 | 0.278702 | 0.99086 | 1.114807 | 1.075143 | 1.065316 |
| 0.98 | 0.273515 | 0.994827 | 1.094058 | 1.061762 | 1.056269 |
| 0.99 | 0.266576 | 0.997906 | 1.066304 | 1.043728 | 1.041542 |
| 1.00 | 0.25 | 0.999598 | 1 | 1 | 0.999598 |

## Example 5.3

1. Write a computer program to determine the velocity, the gradient (slope), and the rate of flow for a sewer flowing partially full at depth, $d$ (m), given the sewer diameter, $D$ (m), and the velocity of the sewer when flowing full, $v_f$ (m/s).
2. A sewer of diameter 150 mm is to flow at a depth of 40% on a grade that enables self-cleaning, similar to that at full depth at a velocity of 0.85 m/s. Find the needed grades, associated velocities, and flow rates at full depth and at 40% depth. Take Manning's coefficient of friction to be equal to 0.013.

## Solution

1. For solution to Example 5.3 (1), see the listing of Program 5.3.
2. Solution to Example 5.3 (2):
   a. Given: $D = 0.15$ m, $d = 0.4 * D$, $v_f = 0.85$ m/s, $n = 0.013$.
   b. Find the slope as follows: $j = (v_f * n/r_H^{2/3})^2 = [0.85 * 0.13/(0.15/4)^{2/3}]^2 = 9.7 * 10^{-3}$.
   c. Compute $Q_f$ as $\pi * D^2 * v_f/4 = (\pi * 0.15^2/4) * 0.85 = 0.015$ m³/s.
   d. Find the central angle as follows:

$$\phi = 2\cos^{-1}\left(1 - \frac{2d}{D}\right) = 2\cos^{-1}(1 - 2*0.4)$$

$$= 2\cos^{-1}(0.2) = 1.369438 \text{ radians}$$

$$\phi = 1.369438 * \frac{360}{2\pi} = 156.93°$$

e. Find the ratio of areas at partial and full flow as follows:

$$\frac{A_p}{A_f} = \frac{\phi}{360} - \frac{\sin(\phi)}{2\pi} = \frac{156.93}{360} - \frac{\sin(156.93)}{2\pi} = 0.37$$

f. Determine $A_p$ as $(\pi * 0.15^2/4)(0.37 = 6.5 * 10^{-3}$ m² or

$$A_p = \left(\frac{D^2}{4}\right)\left(\pi * \phi/360 - (\sin\phi)/2\right)$$

$$= (0.15^2/4)(\pi * 156.93/360 - ((\sin 156.93)/2))$$

$$= 6.6 * 10 - 3$$

g. Determine the hydraulic radius as follows:

$$r_h = \left(\frac{D}{4}\right) * \left(1 - \frac{360 * \sin(\phi)}{2\pi\phi}\right)$$

$$= \left(\frac{0.15}{4}\right) * \left(1 - \frac{360 * \sin(156.93)}{2\pi * 156.93}\right) = 0.032$$

Thus, velocity

$$v_p = \frac{1}{n}r_H^{2/3}\sqrt{j} = \frac{1}{0.013}0.032^{2/3}\sqrt{9.7 * 10^{-3}} = 0.76 \text{ m/s}$$

---

### LISTING OF PROGRAM 5.3   (CHAP5.3\FORM1.VB): CAPACITY OF FLOW ESTIMATES

```
'*********************************************************************************
'Program 5.3: Capacity of flow estimates
'*********************************************************************************

Imports System.Math
Public Class Form1
    Const g = 9.81
    Dim D, dD, vf, n, rH, j, Q, z As Double
    Dim phi, Ap, vp As Double
    Dim ApAf As Double


    '*********************************************************************************
    'Code in Form_Load sub will be executed when the form is loading,
    'i.e., first thing after program loads into the memory
    '*********************************************************************************

    Private Sub Form1_Load(ByVal sender As System.Object, ByVal e As System.EventArgs)
Handles MyBase.Load
        Me.MaximizeBox = False
        Me.Text = "Program 5.3: Capacity of flow estimates"
        Me.FormBorderStyle = Windows.Forms.FormBorderStyle.FixedSingle
        Button1.Text = "&Calculate"
        Label1.Text = "Enter the diameter of sewer, D (m):"
```

```vbnet
        Label2.Text = "Enter velocity of full flow, vf (m/s):"
        Label3.Text = "Enter depth of flow in the pipe, d/D (%):"
        Label4.Text = "Enter Manning's coefficient, n:"
        GroupBox1.Text = "Results:"
        Label5.Text = "Ratio of areas at partial and full flow:"
        Label6.Text = "The central angle, phi (degrees):"
        Label7.Text = "The full flow, Qf (m3/s):"
        Label8.Text = "The slope:"
        Label9.Text = "The velocity of flow (m/s):"
        Label11.Text = "Area of partial flow Ap (m2):"
        Label10.Text = "The hydraulic radius (m):"
        Label12.Text = "Decimal points:"
        NumericUpDown1.Value = 2
        NumericUpDown1.Maximum = 10
        NumericUpDown1.Minimum = 0
    End Sub

    Sub calculateResults()
        D = Val(TextBox1.Text)
        vf = Val(TextBox2.Text)
        dD = Val(TextBox3.Text)
        n = Val(TextBox4.Text)
        If dD > 1 Then dD /= 100
        rH = D / 4 'hydraulic radius
        j = (vf * n / rH ^ (2 / 3)) ^ 2 'the slope
        Q = pi * D ^ 2 / 4 * vf 'the flow
        phi = 2 * Acos(1 - (2 * dD))
        z = (phi * 360) / (2 * PI)
        ApAf = (z / 360) - (Sin(phi) / (2 * PI))
        Ap = PI * (D ^ 2) / 4
        Ap *= ApAf
        rH = (D / 4) * (1 - (360 * Sin(phi) / (2 * PI * z)))
        vp = (rH ^ (2 / 3)) * (j ^ 0.5) / n
    End Sub

    Sub showResults()
        'Show results in the respective textboxes
        Dim decimals As Integer = NumericUpDown1.Value
        TextBox5.Text = FormatNumber(ApAf, decimals)
        TextBox6.Text = FormatNumber(z, decimals)
        TextBox7.Text = FormatNumber(Q, decimals)
        TextBox8.Text = FormatNumber(j, decimals)
        TextBox9.Text = FormatNumber(vp, decimals)
        TextBox11.Text = FormatNumber(Ap, decimals)
        TextBox10.Text = FormatNumber(rH, decimals)
    End Sub

    Private Sub Button1_Click(ByVal sender As System.Object, ByVal e As System.EventArgs) _
Handles Button1.Click
        calculateResults()
        showResults()
    End Sub

    Private Sub NumericUpDown1_ValueChanged(ByVal sender As System.Object, ByVal e As _
System.EventArgs) Handles NumericUpDown1.ValueChanged
        showResults()
    End Sub
End Class
```

## 5.2.6.2 Self-Cleansing Velocities

Self-cleansing velocity is referred to as that velocity sufficient to prevent deposition of suspended matter. The velocity needed to carry sediment in a pipe flowing full is given by the following equation:

$$v_{sc} = \frac{r_H^{1/6}}{n} \sqrt{b(\text{s.g.} - 1)d} = \sqrt{\frac{8b,(\text{s.g.} - 1)d}{f}} \qquad (5.28)$$

where:

$v_{sc}$ is the self-cleansing velocity, m/s

$r_H$ is the hydraulic radius, m

$b'$ is the dimensionless constant (sediment-scouring characteristics, see Table 5.3)

s.g. is the specific gravity, dimensionless

$d$ is the diameter of sediment grain, m

$n$ is the roughness factor (Manning and Kutter factor), $m^{1/6}$

$f$ is the friction (Darcy–Weisbach) factor, dimensionless (usually 0.02–0.03) (McGhee and Steel 1991)

A minimum velocity is often accepted as the design criterion. As such, the recommended velocities of Table 5.4 may be adopted. Generally, it is desirable to have a minimum velocity of 0.91 m/s or more whenever possible. As such, 0.6 m/s < minimum design velocity < 3.5 m/s at peak flow.

### 5.2.7 Summary of Sewer System Design

1. The first information needed to design and lay out a sanitary sewer system is a topographical map (contour map), not only for the immediate area under consideration but also for the surrounding area (Health Education Services 1990; McGhee And Steel 1991; Bizier 2007). With this information regarding the various drainage basins, it is then possible to evaluate the number and location for the lift stations. A geological map indicating the soil type and rock formations is also necessary. Aerial photographs are very helpful if available.

2. Details as to the zoning in the design and layout area are required, indicating which areas are industrial, commercial, and residential. Essential information such as lot sizes, population density, and street layouts must be provided.

3. With the information indicated in (1) and (2), the size and gradient of the sewers can be determined. The sewer must be on an adequate slope to prevent deposition of solids (a slope of 2% for house connections, with a minimum slope of 1% being used). Sanitary sewers with diameters of 375 mm (15″) are designed to flow half full; large sewers are designed to flow three quarters full.

4. The sanitary sewers must be installed deep enough to receive flows from the tributary areas. Sufficient depth must be provided to prevent freezing in cold climates. In most cases, a minimum depth from the ground surface to the crown of the pipe should not be less than 1.5 m. The sanitary sewer also should be at least 1 m below the first-floor elevation of any adjacent dwellings.

5. Manhole locations are at
   a. Junctions of sanitary sewers.
   b. Changes in gradient or gradient or alignment (except in curved alignments).
   c. Locations that provide easy access to the sewer for preventive maintenance and emergency service.

   A minimum diameter for a manhole is 1.22 m. A straight sewer is to be used between manholes. Large accessible sewers may be curved, and manholes may be placed at 100–200 m intervals. The spacing for manholes is in the range of 90–150 and 150–300 m for large diameter sewers.

6. Sanitary sewers should not be installed in the same trench as water mains in order to protect public health (Health Education Services 1990).

7. No sewers line should pass under a building.

8. Sewer lines should be installed down the center line of the lanes rather than the centerline of the streets. In this way, one lane is always available to keep traffic moving. In wide streets, sewers can be laid outside the curb between the curb and the sidewalk.

9. Coordination between utilities responsible for power poles, gas meters, telephone junction boxes television wires, and so on is necessary to plan for construction and layout of the sewer lines. This planning can help avoid conflicts and problems that may arise among the various organizations involved.

10. Planting of trees and shrubs and construction of fences or retaining walls that interfere with access to the sewer lines should be controlled or prohibited.

## TABLE 5.3
## Values of $b'$

| Value of $b'$ | Significance |
| --- | --- |
| 0.04 | To start motion of clean granular particles |
| 0.8 | For adequate self-cleansing of cohesive material |

*Source:* McGhee, T. J., and Steel, E. W., *Water Supply and Sewerage*, 5th ed. McGraw-Hill College, 1991.

## TABLE 5.4
## Recommended Minimum Velocities

| Sewer | $(v_{sc})_{min}$ (m/s) |
| --- | --- |
| Sanitary | <0.61 |
| Storm | <0.75 |

11. Integrity of a sewer line (leakage) is tested for by a low-pressure air test. Water testing generally has been replaced by low-pressure air testing (Health Education Services 1990).
12. Combined sewers should now be prohibited in order to prevent present and future wastewater treatment problems (see Section 5.2.1 for disadvantages).

### 5.2.8 Corrosion in Sanitary Sewers

Corrosion in a sanitary sewer is caused by the production of hydrogen sulfide, $H_2S$. Some of the problems encountered with $H_2S$ are as follows (Bizier 2007):

- Generating odors
- Creating health risk to maintenance workers
- Developing corrosion on unprotected sewer pipes made of concrete materials or metals
- Interfering with treatment processes (it affects activated sludge and increases disinfectant demand, and public complaint may arise due to generation of odor issuing from influent structure)
- $H_2S$ toxicity (numerous workers are killed each year by entering manholes without the proper precautions)

Generation of sulfides in sanitary sewers depends on wastewater temperature, dilution rate, hydraulic flow conditions, and agitation of deposited waste solids. When insufficient dissolved oxygen is present in the wastewater, anaerobic bacteria produce sulfides. Generation of sulfide is encountered on the interior walls of sanitary sewers. Sulfide in municipal wastewater may exist, in part, as insoluble sulfides of various metals; however, the concentration of metal sulfide is generally low (Bizier 2007). The major part of sulfides is normally retained in solution as a mixture (referred to as dissolved sulfide) of hydrogen sulfide and ionic $HS^-$.

#### 5.2.8.1 Sulfide Buildup Estimates

An indicator of the potential of sulfide buildup in relatively small gravity sewers (not over 600 mm or 24″ in diameter) can be estimated by using the following equation (Bizier 2007):

$$f = \frac{BOD_e * w_p}{\sqrt{j} * Q^{0.33} * B} \qquad (5.29)$$

where:
$f$ is the defined function (see Table 5.5)
$BOD_e$ is the effective biochemical oxygen demand, mg/L
$w_p$ is the wetted perimeter, ft
$j$ is the hydraulic slope, dimensionless

### TABLE 5.5
### Sulfide Condition for Different Values of $f$

| $f$ Value | Sulfide Condition |
|---|---|
| $f < 5,000$ | Sulfide rarely generated |
| $\geq 5,000$ to $\leq 10,000$ | Marginal condition of sulfide generation |
| $> 10,000$ | Sulfide generation common |

*Source:* From Joint Task Force of the American Society of Civil Engineers and the Water Pollution Control Federation, *Gravity Sanitary Sewer Design and Construction*, ACSE Manuals and Reports on Engineering, Practice Number 60, ASCE WPCF, New York, 1982. Reprinted by permission of the publisher, ASCE.

$Q$ is the discharge volume, ft³/s
$B$ is the surface width, ft

### Example 5.4

1. Write a computer program to estimate the likelihood of sulfide buildup in a sewer, given the pipe diameter of sanitary sewer, $D$ (mm); the flow rate, $Q$ (m³/s); the slope, $j$ (m/m); the wetter perimeter, $w_p$ (ft); the surface width of flow, $B$ (ft); and the effective biochemical oxygen demand, $BOD_e$ (mg/L). Let the program indicate the level of sulfide generation based on Table 5.5.
2. A sanitary sewer of pipe diameter 600 mm (24″) carries a flow = 2.2 m³/min at a slope = 0.0015. Taking wetted perimeter = 0.82 m and surface width of flow = 1 m, give an indicator of the likelihood of sulfide buildup in the sewer, knowing that the effective biochemical oxygen demand = 295 mg/L.

#### Solution

1. For solution to Example 5.4 (1), see the listing of Program 5.4.
2. Solution to Example 5.4 (2):
   a. Given: $d = 600$ m, $Q = 2.2$ m3/min = 2.2 * 35.3147/60 = 1.295 cfs, $j = 0.0015$, $w_p = 0.82$ m = 0.82 * 3.2808 = 2.69 ft, width = $B = 1$ m = 1 * 3.2808 = 3.2808 ft, BOD = 295mg/L.
   b. Determine the indicator of sulfide buildup as: $f = BOD_e * w_p/(j^{1/2} * Q^{0.33} * B) = 295 * 2.69/[(0.0015)^{1/2} * 1.2950.33 * 3.2808] = 5734.6$.
   c. By comparing this value with values outlined in Table 5.5, it can be concluded that the value 5735 indicates a marginal condition for sulfide generation.

**LISTING OF PROGRAM 5.4   (CHAP5.4\FORM1.VB): SULFIDE BUILD UP INDICATOR**

```vb
'*****************************************************************************
'Program 5.4: Sulfide Build Up Indicator
'*****************************************************************************

Imports System.Math
Public Class Form1
   Const g = 9.81
   Dim D, Q, S, Wp, B, BOD, Z As Double
   '*****************************************************************************
   'Code in Form_Load sub will be executed when the form is loading,
   'i.e., first thing after program loads into the memory
   '*****************************************************************************

   Private Sub Form1_Load(ByVal sender As System.Object, ByVal e As System.EventArgs)
Handles MyBase.Load
      Me.Text = "Program 5.4: Sulfide Build up Indicator"
      Me.FormBorderStyle = Windows.Forms.FormBorderStyle.FixedSingle
      Me.MaximizeBox = False
      Label1.Text = "Enter the diameter of the sewer, D (m):"
      Label2.Text = "Enter the flow rate in the sewer, Q (m3/s):"
      Label3.Text = "Enter the slope of the pipe, S:"
      Label4.Text = "Enter the wetter perimeter, Wp (m):"
      Label5.Text = "Enter the surface width of flow, b (m):"
      Label6.Text = "Enter the effective biochemical demand, BOD (mg/L):"
      GroupBox1.Text = " Output: "
      Label7.Text = ""
      Button1.Text = "&Calculate"
   End Sub

   Sub calculateResults()
      D = Val(textbox1.text)
      Q = Val(textbox2.text)
      S = Val(textbox3.text)
      Wp = Val(textbox4.text)
      B = Val(textbox5.text)
      BOD = Val(textbox6.text)
      D = D * 0.3048 'convert to ft
      Q = Q / (0.3048 ^ 3) / 60 'convert to cfs
      Wp = Wp / 0.3048 'convert to ft
      B = B / 0.3048 'convert to ft
      Z = BOD * Wp / (S ^ 0.5 * Q ^ 0.33 * B)
      Dim R As String
      If Z < 5000 Then
        R = "Sulfide rarely generated"
      ElseIf Z <= 10000 Then
        R = "Marginal condition of sulfide generation"
      Else
        R = "Sulfide generation common"
      End If
      Label7.Text = "Sulfide buildup indicator Z = " + FormatNumber(Z, 2)
      Label7.Text += vbCrLf
      Label7.Text += R
   End Sub

   Private Sub Button1_Click(ByVal sender As System.Object, ByVal e As System.EventArgs)
Handles Button1.Click
      calculateResults()
   End Sub
End Class
```

### 5.2.8.2 Filled Pipe Conditions

Equation 5.30 may be used to estimate the level of sulfide buildup in completely filled sewer pipes (force mains) (Bizier 2007):

$$\frac{dS_u}{dt} = \frac{\left[ 3.28 * a * BOD_e \left( \frac{(1+0.12 * D_i)}{0.12 * D_i} \right) \right]}{0.25 * D_i} \quad (5.30)$$

where:

$dS_u/dt$ is the increase in sulfide concentration, mg/L*h
$a$ is the coefficient (= 0.3 mm/h) (Bizier 2007)
$D_i$ is the internal diameter of sewer pipe, mm
$BOD_e$ is the effective biochemical oxygen demand:     (5.31)
$BOD(\theta)^{T-20}$

where:

BOD is the climate BOD, mg/L
$T$ is the temperature, °C
$\theta$ is the temperature coefficient (may be taken as equal to 1.07) (Bizier 2007)

### 5.2.8.3 Partially Filled Pipe Conditions

Equation 5.32 may be used to estimate the sulfide buildup in gravity flow in a partially filled sanitary sewer:

$$\frac{dS_u}{dt} = \left( \frac{a' * BOD_e}{r_H} \right) - \left[ b \left( j * v \right)^{0.375} * S_u / D_m \right] \quad (5.32)$$

where:

$dS_u/dt$ is the rate of change of total sulfide concentration, mg/L*h
$a'$ is the effective sulfide flux coefficient ($a' = 0.4 * 10^{-3}$ m/h for DO < 0.5 mg/L; $a' = 0$ when DO is high) m/h (Bizier 2007)
$r_H$ is the hydraulic radius of stream, m
$b$ is the empirical coefficient, sulfide loss coefficient (= 0.64 [conservative] – 0.96 [less conservative value]) (Bizier 2007)
$S_u$ is the total sulfide concentration, mg/L
$j$ is the energy gradient of stream
$v$ is the velocity of flow, m/s
$D_m$ is the mean hydraulic depth, ft

Equation 5.32 should not be used for systems when DO > 0.5 mg/L or for sanitary sewers with an effective slope > 0.6% (Bizier 2007). Sulfide concentrations approach a limiting value ($S_{ulim}$), when Equation 5.32 approaches zero (Bizier 2007). This can be represented by Equation 5.33, taking the value of 0.96 for the sulfide loss coefficient, $b$ (Bizier 2007).

$$S_{ulim} = \left[ \frac{\left( 0.33 * 10^{-3} BOD_e * D_m \right)}{\left( j * v \right)^{0.375} r_H} \right]$$

$$= \left[ \frac{\left( 0.52 * 10^{-3} BOD_e * w_p \right)}{\left( j * v \right)^{0.375} B} \right] \quad (5.33)$$

Sulfide generation in a long reach of sewer line with uniform slope and flow can be determined from the following equation (Bizier 2007):

$$S_{u2} = S_{ulim}$$
$$- \left\{ \left( S_{ulim} - S_{u1} \right) / \log^{-1} \left[ \left( j * v \right)^{0.375} * \Delta t / 1.15 D_m \right] \right\} \quad (5.34)$$

where:

$S_{u1}$ is the sulfide concentration at upstream location, mg/L
$S_{u2}$ is the sulfide concentration at downstream location, mg/L
$\Delta t$ is the time of flow from upstream location to downstream location, h

## 5.3 WASTEWATER DISPOSAL FOR RURAL INHABITANTS

Small communities in rural areas often need inexpensive, robust, compact sewage treatment works that require little maintenance and can be installed and operated with unskilled labor. Examples of these units include septic tanks, Imhoff tanks, ventilated-improved pit (VIP) latrines, and so on. The most widely used system is the septic tank. Following is information that can be used in the design, construction, and operation of septic tanks.

### 5.3.1 SEPTIC TANKS

#### 5.3.1.1 Introduction

A septic tank is an underground (covered) box (Fair et al. 1966; Mara 1980; Peirce et al. 1997; Jonathan Ricketts et al. 2003; Nemerow et al. 2009; Metcalf and Eddy 2013; Hicks 2014; Abdel-Magid and Abdel-Magid 2015). The tank receives wastewater and treats it in a horizontal flow pattern. It removes solids and promotes partial deposition. Functions taking place in a septic tank include settling of solids, flotation of grease, anaerobic decomposition of solids, and storage of sludge. The retention of large solids and grease is essential to avoid plugging final disposal localities, such as absorption fields. The liquor within the tank has a considerable detention time to maximize solid deposition. For large installations serving multiple families or institutions, a shorter detention time may be adhered to.

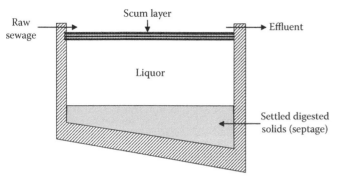

**FIGURE 5.3**   Regions in a septic tank.

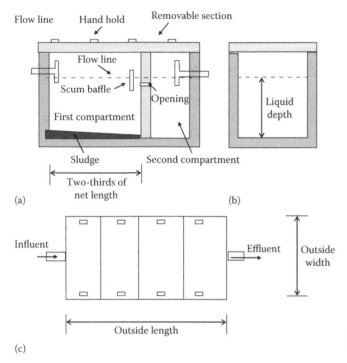

**FIGURE 5.4**   A schematic diagram of a septic tank: (a) elevation; (b) side view; (c) plan.

Figures 5.3 and 5.4 illustrate the zones that are found in a septic tank:

- The scum layer zone, which forms a crust on surface of tank
- The liquor layer zone, where there is solids deposition and diffusion of solubilized material from the third layer
- The settling and digestion-of-solids zone

The materials that usually are used in the construction of a septic tank include concrete, perforated reinforced concrete, concrete blocks, steel, brick, stone masonry, and fiberglass. The septic tank may be of a rectangular or circular (prefabricated concrete pipes) shape.

### 5.3.1.2   Designing Septic Tanks

In designing a septic tank, the following points must be considered:

- Selection of the location of the tank (U.S. Agency for International Development (a) 1982)
- To be located downhill
- To be located at least 15 m from the nearest water supply system (including neighbors)
- To be located at least 3 m from the nearest building
- Rain or surface water not to flow over or stand on the tank
- Vehicles not to drive over the tank
- Determination of tank capacity and dimensions
- Determination of necessary labor, materials, and tools.
- Maintenance of the septic tank
- Sludge removed every 2–5 years (under normal loading conditions)
- Sludge discharged to a nearby wastewater treatment plant for treatment or pumped into the soil (avoiding groundwater pollution)
- Use of strong bactericidal cleansers by householders or use of unnecessary amounts of detergents discouraged
- With good maintenance, a tank can last for ≥20 years
- Tank inspected at least once a year to determine if it needs cleaning; tank needs to be cleaned if the depth of sludge exceeds one-third of liquid depth or when the bottom of the scum layer is within 75 mm of the bottom of the outlet pipe "T" fitting (U.S. Agency for International Development (a) 1982)
- Removing crust (black-colored digesting sludge) to be avoided when desludging the tank

General information that can be helpful in designing a septic tank is as follows:

1. Tank should receive all sewage from the building to be served (this includes excreta and wash water and excludes rainwater, surface water, or subsurface drainage). Determine the amount of sewage entering the tank during each 24 h period = $Q$.
2. Determine the desired retention time $t$.
    a.  A minimum value of $t$ of 1 day is taken (a smaller tank implies less initial cost).
    b.  A maximum value of $t$ of 3 days is taken (a large tank signifies high initial cost and smaller cleaning intervals, and it treats more sewage, and thus increases the life of subsurface absorption system).
3. Determine the capacity of tank $V = Q*t$. Otherwise, tank capacity may be determined from Equations 5.35 through 5.37.
    a.  Volume of septic tank:

$$V = 180 * P + 2000 \qquad (5.35)$$

where:

    $V$ is the minimum volume of septic tank, L

    $P$ is the number of people served

b. Where garbage grinders are used, Equation 5.36 can be used to determine the needed capacity:

$$V = 250 * P + 2000 \tag{5.36}$$

c. When septic tank is used for serving schools and intermittent uses, its volume can be determined from the following equation:

$$V = 90 * P + 2000 \tag{5.37}$$

4. Determine the tank dimensions as follows:
   a. Tank length:width ratio = 3:1 to 2:1.

$$L/B = 3 : 1 \text{ to } 2 : 1 \tag{5.38}$$

where:

    $L$ is the internal length of tank, m

    $B$ is the internal width of tank, m (width of tank not to be less than 1 m)

b. Find the inside and outside tank dimensions. Recommended liquid depth in tank = 1.2 m (usually varies from 1.1 to 1.8 m). For a tank serving up to 50 people, use an outlet liquid depth of 1.2 m; when it serves up to 150 people, use a liquid depth of 1.5 m; and when it serves a larger number of people, use a liquid depth of 1.8.
c. Inlet and outlet pipes should be fitted with open T sewer pipe fittings.
d. The bottom of an inlet pipe 300 mm should be below the top of the tank.
e. The bottom of an outlet pipe 75 mm should be below the bottom of the inlet (375 mm below the top of the tank).
f. Downward T connection of inlet should be 20% of outlet liquid depth, whereas that of the outlet should be 40% of liquid depth.
5. The tank floor can be sloped toward the inlet at a slope of 10%.
6. When two compartments are considered, one compartment (near inlet) can be twice the capacity of the second compartment. This yields a first compartment length as determined from the following equation:

$$L_c = \frac{2L}{3} \tag{5.39}$$

where:

    $L_c$ is the internal length of first compartment, m

    $L$ is the internal length of tank, m

7. Walls must be watertight (25-mm-thick inside coating of cement plaster is applied, usually two coats of 12 mm each are applied) (U.S. Agency for International Development (a) 1982).
8. The floor of tank should be of reinforced concrete (usually 100–150 mm thick) and should rest on a bed of gravel or sand 75 mm thick (U.S. Agency for International Development (a) 1982)
9. Top should be watertight (usually reinforced concrete is used). Generally, all or most of the sections can be removed to clean tank, and one or two sections over outlet can be removed to inspect the tank.
10. Inlet and outlet pipes should be fitted with open T sewer pipe fittings.
11. A proper plan view (top, side, and/or end views) of the tank is needed.

### 5.3.1.3 Septic Tank Effluents

The effluent from a septic tank is offensive and potentially dangerous (BOD$_5$, 120–270 mg/L; SS$_{mean}$, 44–69 mg/L). Usually, the effluent is discharged to subsurface tile field (leaching field, drain field), absorption and evapotranspiration mounds (beds of imported porous soil covered with a top soil that is planted with vegetation to appear as a part of the landscape), seepage pits, or intermittent sand filters. The ability of the ground to absorb the effluent ought to be determined. Percolation tests, used to measure the suitability of ground for the fields, are conducted as follows (Peirce et al. 1997):

1. A hold about ≥100 mm (4″) in diameter is dug as deep as the proposed tile field trench.
2. The hole sides are scratched, and all loose soil is removed.
3. A 500-mm (2″) fine gravel or coarse sand is placed in the bottom of pit.
4. The hole is filled with water to a depth of 300 mm and allowed to stand overnight (at least 4 h).
5. Next day, pit is filled with water to about 150 mm (6″) above gravel, and the drop in water level in 30 min is measured.
6. The percolation rate is calculated as mm/min ($t$).
7. Tables are used to determine the required trench area, or it can be computed according to the following equation:

$$Q = 204 * t_{25} \tag{5.40}$$

where:

    $Q$ is the flow that can be applied per unit area, L/m$^2$*day

    $t_{25}$ is the time required for the water surface to fall 25 mm = 25/subsidence rate, min

## Example 5.5

1. Write a computer program to design a septic tank for any number of people. Design the program so that the necessary structural design of the tank is included.
2. Design a septic tank to serve a family dwelling of 20 people.
3. Use the above data with the program developed in (1) in order to produce the full design of the tank.

## Solution

1. For solution to Example 5.5 (1), see the listing of Program 5.5.
2. Solution to Example 5.5 (2):
   a. Given: $P = 20$.
   b. Determine the tank volume: $V = 180 * P + 2000 = 180 * 20 + 2000 = 5600$ L.
   c. Take an outlet liquid depth $d = 1.2$ m.
   d. Take a length:width ratio of 2.5, that is, $L/B = 2.5$, and determine the plan area as follows:

   $$\text{Area} = \frac{\text{Volume}}{\text{Depth}}$$

   $V/d = 5600 * 10^{-3}/1.2 = 4.67$ m$^2$ $= L * B = 2.5 * B,^2$ which yields a width $B = 1.36$ m; therefore, $L = 2.5 * B = 3.4$ m.
   e. Determine the liquid depth at inlet ($D$) from relationship of slope: slope $= (D - d)/L$; thus,

$D = $ length * floor slope + outlet depth $= 0.1 * 3.4 - 1.2 = 1.54$ m.

   f. Determine the total depth of tank as follows: total depth = liquid depth at inlet + distance of bottom of inlet pipe from liquid surface + distance between inlet and tank cover = $1.54 + 0.075 + 0.3 = 1.92$ m.
   g. Find the distance of the bottom of the T sewer inlet pipe from the top of the tank as $0.3 + 0.075 + 20\%$ of liquid depth $= 0.3 + 0.075 + 0.2 * 1.2 = 0.62$ m.
   h. Find the distance of the bottom of the T sewer outlet pipe from the top of the tank as $0.3 + 0.75 + 40\%$ of liquid depth $= 0.3 + 0.075 + 0.4 * 1.2 = 0.86$ m.
   i. When considering a two-compartment tank, the length of the first compartment can be determined as follows: $L_c = 2 * L/3 = 2 * 3.4/3 = 2.27$ m.
3. As an example, the following data can be taken to run the program: number of people = 20, no garbage grinder wastes, not a public place, grade of concrete = 30, yield stress for steel = 250 N/mm$^2$. Severe exposure is assumed, as is a modulus of subgrade soil 0.
The results of the program are as follows:
   a. Vertical wall thickness = 200 mm; steel in all directions will be 1040 mm$^2$ everywhere.
   b. Base slab thickness = 200 mm; steel in both directions will be 1300 mm$^2$.
   c. Cover slab thickness = 150 mm; steel in both directions will be 780 mm$^2$.

---

### LISTING OF PROGRAM 5.5    (A) LISTING OF FORM1 CODE (CHAP5.5\FORM1.VB): DESIGN OF SEPTIC TANKS

```vb
'*********************************************************************************
'Program 5.5: Design of Septic Tanks
'*********************************************************************************


Imports System.Math
Public Class Form1
  Const pi = 3.1415962#
  Const g = 9.81
  Dim P, fcu, fy, V As Double
  Dim d0, rat, Ly, Lx, S, di, dt, dti, dt0, Lc As Double
  Dim tw, tb, tc, H, D, Alpha, cv As Double
  Dim Ast1, Ast2, Ast3, Ast4, Ast5, Ast6, Ast7, Ast8 As Double
  Dim B, qx, qy As Double 'used in GetCoverPlateMoments()
  'The following variables are used in Designsection()
  Dim k, Z, TN, Ast As Double
  Dim stress, Ra, d1, Vc As Double
  Dim Mmax, dr, Ec, Es, Asm, r, x, fs, fsa, fact As Double
  'The following variables are used in GetBasePlateMoments()
  Dim E, mu, udl, w, wxx, wyy As Double
  Dim y, m1, n1, c, cx, cy, cd As Double
  'The following variables are used in GetWallsMoments()
  Dim NDIVX, NDIVY, NX, NY, DX, DY, NDOF As Double
```

```
Dim RIGIDITY, pload, MM As Double
Dim L3, L4, F, A As Double
Dim K1, K2, K3, K4, K5, K6, K7, K8, K9, K10, K11, K12, K13, K14, K15, K16 As Double
Dim r1, r2, r3, mmt As Double
Dim f1 As Form2 'used in GetCover()
Dim f2 As Form3 'used in calculateResults() to display the results

Private Sub Form1_Load(ByVal sender As System.Object, ByVal e As System.EventArgs)
Handles MyBase.Load
    Me.Text = "Program 5.5: Design of Septic Tanks"
    Me.MaximizeBox = False
    Label1.Text = "Enter the number of people to be served:"
    Label2.Text = "Will garbage grinder wastes be used?"
    Label3.Text = "Is the tank in a school/public utility?"
    Label4.Text = "Enter the grade of concrete N/mm2:"
    Label5.Text = "Enter the yield stress of steel N/mm2:"
    Label6.Text = "Select a sides ratio (L/B) between 2 and 3:"
    RadioButton1.Text = "Yes"
    RadioButton2.Text = "No"
    RadioButton3.Text = "Yes"
    RadioButton4.Text = "No"
    RadioButton2.Checked = True
    RadioButton4.Checked = True
    Button2.Text = "Calculate dimensions"
    GroupBox1.Text = " OVERALL DIMENSIONS: "
    Label7.Text = "Length of tank (m):"
    Label8.Text = "Width of tank (m):"
    Label9.Text = "Depth at inlet (m):"
    Label10.Text = "Depth at outlet (m):"
    Label11.Text = "Depth of T-sewer inlet pipe from top (m):"
    Label12.Text = "Depth of T-sewer outlet pipe from top (m):"
    Label13.Text = "Length of the second compartment (m):"
    Label14.Text = "If the dimensions are acceptable, click on Design button"
    Button1.Text = "&Design"
    Button1.Enabled = False
End Sub

Sub calculateResults()
    P = Val(TextBox1.Text)
    fcu = Val(TextBox2.Text)
    fy = Val(TextBox3.Text)
    V = 180 * P + 2000 'volume of tank in liters
    If RadioButton1.Checked Then V = 250 * P + 2000
    If RadioButton3.Checked Then V = 90 * P + 2000
    V = V / 1000 'convert to m3
    'Compute liquid depth at outlet (m)
    If P <= 50 Then d0 = 1.2
    If P > 50 And P <= 150 Then d0 = 1.5
    If P > 150 Then d0 = 1.8
    'Compute depth at inlet (m) pipe
    rat = Val(TextBox4.Text)
    If rat < 2 Or rat > 3 Then
        MsgBox("Please select a sides ratio (L/B) between 2 and 3.", vbOKOnly, "Error")
        Exit Sub
    End If
    Ly = (V / (d0 * rat)) ^ 0.5 'width of tank
    Lx = Ly * rat 'length of tank
    'Compute liquid depth at inlet di
    S = 0.1 'slope of base
    di = d0 + S * Lx 'liquid depth at inlet pipe
    dt = di + 0.075 + 0.3 'total depth of tank at inlet
```

```
        dti = 0.375 + 0.2 * d0 'depth from top of tank to bottom of t-sewer at inlet
        dt0 = 0.375 + 0.4 * d0 'depth from top of tank to bottom of t-sewer at inlet
        Lc = 2 * Lx / 3 'length of the second compartment
        'Output results of overall dimensioning
        TextBox5.Text = (CInt(Lx * 100) / 100).ToString
        TextBox6.Text = (CInt(Ly * 100) / 100).ToString
        TextBox7.Text = (CInt(di * 100) / 100).ToString
        TextBox8.Text = (CInt(d0 * 100) / 100).ToString
        TextBox9.Text = (CInt(dti * 100) / 100).ToString
        TextBox10.Text = (CInt(dt0 * 100) / 100).ToString
        TextBox11.Text = (CInt(Lc * 100) / 100).ToString
        Button1.Enabled = True
    End Sub


    Sub Designsection(ByVal t, ByVal cv, ByVal Mh, ByVal Mv, ByVal n, ByVal V, ByVal fcu,
ByVal fy, ByVal Astx, ByVal Astv)

        '**************************************************************************
        'Structural design of a section
        '**************************************************************************

        Dim br = 1000 '1m width
        D = t - cv - 10 'effective depth, assuming 20mm bars
        Dim ft = 1.3 'tensile strength of concrete for crack control
        Dim rcrit = ft / fy 'critical steel percentage for crack control
        Dim fst = 0.87 * fy 'design steel stress
        Dim Astmin = rcrit * br * t 'minimum area of steel for grade 250 steel
RECYCLE:
        'X-direction steel
        Dim MaxMu = 0.156 * fcu * br * D ^ 2 / 10 ^ 6 'in KNm
        Do While Mh > MaxMu
            D = D + 25
            t = t + 25
            MaxMu = 0.156 * fcu * br * D ^ 2 / 10 ^ 6
        Loop
        k = Mh * 10 ^ 6 / (fcu * br * D ^ 2)
        Z = D * (0.5 + (0.25 - k / 0.9) ^ 0.5)
        If Z > 0.94 * D Then Z = 0.94 * D
        TN = ABS(n)
        Ast = TN * 10 ^ 3 / fst
        Ast2 = ABS((Mh * 10 ^ 6 - TN * 10 ^ 3 * (D - t / 2)) / (fst * Z))
        Astx = Ast + Ast2
        If Astx < Astmin Then Astx = Astmin
        'Y-direction steel
        Do While Mv > MaxMu
            D = D + 25
            t = t + 25
            MaxMu = 0.156 * fcu * br * D ^ 2 / 10 ^ 6
        Loop
        k = Mv * 10 ^ 6 / (fcu * br * D ^ 2)
        Z = D * (0.5 + (0.25 - k / 0.9) ^ 0.5)
        If Z > 0.94 * D Then Z = 0.94 * D
        'for base slab, there is tension in this direction as well
        If n < 0 Then TN = ABS(n) Else TN = 0.0!
        Ast = TN * 10 ^ 3 / fst
        Ast2 = ABS((Mv * 10 ^ 6 - TN * 10 ^ 3 * (D - t / 2)) / (fst * Z))
        Astv = Ast + Ast2
        If Astv < Astmin Then Astv = Astmin
        'check for shear stress
        Ast = MAX2(Astx, Astv)
        stress = V * 10 ^ 3 / (br * D)
```

```
    Ra = 100 * Ast / (br * D)
    d1 = D
    If d1 > 400 Then d1 = 400
    Vc = 0.79 * Ra ^ (1 / 3) * (400 / d1) ^ 0.25 / 1.25
    If fcu > 25 Then Vc = Vc * (fcu / 25) ^ (1 / 3)
    If stress > Vc Then
      D = D * stress / Vc * 1.1
      t = D + cv + 10
      GoTo RECYCLE
    End If
    'check for cracking at serviceability limit state
    n = n / 1.5 'service ring tension
    Mmax = Mh
    Asm = Ast
    dr = 1
    If Mmax < Mv Then
      Mmax = Mv
      Asm = Astv
      dr = 2
    End If
    Mmax = Mmax / 1.5 'service maximum moment
    Ec = 5500 * fcu ^ 0.5 'Young's modulus of concrete N/mm2
    Ec = 0.5 * Ec 'Service Young's modulus of concrete
    Es = 2 * 10 ^ 5 'Young's modulus for steel
    Alpha = Es / Ec 'modular ratio
    r = Alpha * Asm / (br * D) 'steel ratio
    x = D * r * ((1 + 2 / r) ^ 0.5 - 1)
    Z = D - x / 3 'lever arm
    fs = Mmax / (Asm * Z) 'service stress in steel
    fsa = 0.58 * fy 'allowable service stress
    If fs > fsa Then
      'cracking is excessive, so increase area of steel
      Fact = fs / fsa
      If dr = 1 Then Ast = Ast * fact Else Astv = Astv * fact
    End If
End Sub

Sub GetCover(ByRef fcu, ByRef cv)

    '*********************************************************************************
    'Determine the cover for a septic tank
    '*********************************************************************************

    Try
      f1.ShowDialog()
    Catch ex As Exception
      f1 = New Form2
      f1.ShowDialog()
    End Try
    Select Case f1.choice
      Case 0
        If fcu <= 30 Then : cv = 25
        ElseIf fcu >= 35 Then : cv = 20
        End If
      Case 1
        If fcu <= 35 Then : cv = 35
        ElseIf fcu = 40 Then : cv = 30
        ElseIf fcu = 45 Then : cv = 25
        ElseIf fcu = 50 Then : cv = 20
        End If
```

```
         Case 2
           If fcu <= 40 Then : cv = 40
           ElseIf fcu = 45 Then : cv = 30
           ElseIf fcu = 50 Then : cv = 25
           End If
         Case 3
           If fcu <= 40 Then : cv = 50
           ElseIf fcu = 45 Then : cv = 40
           ElseIf fcu = 50 Then : cv = 30
           End If
         Case 4
           If fcu <= 45 Then : cv = 60
           ElseIf fcu = 50 Then : cv = 50
           End If
      End Select
   End Sub


   Sub GetBasePlateMoments(ByRef t, ByRef Lx, ByRef Ly, ByRef H, ByRef fcu, ByRef Mx,
ByRef Myy)

      '*******************************************************************************
      'Routine: PLATES ON ELASTIC FOUNDATIONS.
      'Originally written by Dr. A. W. Hago, Muscat, April 1995
      'Updated by Dr. M. I. Abdel-Magid, Muscat, 2013
      '*******************************************************************************

      Dim kmodulus = Val(InputBox("Enter the modulus of subgrade soil (KN/m2) (0 if
unknown):", _
         "STRUCTURAL DESIGN OF TANK BASE SLAB", "0"))
      If kmodulus = 0 Then kmodulus = 39000 'assumed clay soil
      E = 5500000.0! * (fcu) ^ 0.5 'Young's modulus of concrete, KN/m2
      mu = 0.16 'Poisson's ratio for concrete
      udl = 10 * H 'Uniform Pressure on the base slab
      D = E * t ^ 3 / (12 * (1 - mu ^ 2))
      x = 0.5 * Lx : y = 0.5 * Ly 'coordinates of center point
      w = 0.0!
      wxx = 0.0!
      wyy = 0.0!
      m1 = 15
      n1 = 15
      For m = 1 To m1 Step 2
        For n = 1 To n1 Step 2
           c = 16 * udl / (m * n * pi ^ 2)
           cx = Sin(m * pi * x / Lx)
           cy = Sin(n * pi * y / Ly)
           cd = pi ^ 4 * D * ((m / Lx) ^ 2 + (n / Ly) ^ 2) ^ 2 + kmodulus
           w = w + c * cx * cy / cd
           wxx = wxx - c * (m * pi / Lx) ^ 2 * cx * cy / cd
           wyy = wyy - c * (n * pi / Ly) ^ 2 * cx * cy / cd
        Next n
      Next m
      Mx = -D * (wxx + mu * wyy)
      Myy = -D * (wyy + mu * wxx)
   End Sub


   Sub GetCoverPlateMoments(ByVal Lx, ByVal Ly, ByVal udl, ByVal Mx, ByVal Myy)

      '*******************************************************************************
      'Routine to compute moments in a simple supported slab
      'using the deflection method (Dr. Marcus)
      '*******************************************************************************
```

```
      Alpha = Ly / Lx 'sides ratio
      B = Alpha ^ 4
      qx = udl * B / (1 + B) 'load carried by x strips
      qy = udl / (1 + B) 'load carried by y strips
      Mx = qx * Lx ^ 2 / 8
      Myy = qy * Ly ^ 2 / 8
   End Sub

   Sub GetWallsMoments(ByRef Lx, ByRef Ly, ByRef H, ByRef fcu, ByRef MAXMX, ByRef MAXMY,
ByRef MAXSX, ByRef MAXSY, ByRef MAXTN)

      '*********************************************************************************
      'routine to compute the moments on the walls of a tank, using the
      'finite difference method. The side walls are assumed as
      'FIXED ON THREE SIDES, SIMPLY SUPPORTED ON THE FOURTH TOP SIDE,
      'under linearly varying Hydrostatic Pressure of a liquid.
      '*********************************************************************************

      NDIVX = 4 'four divisions of one half for symmetry
      NDIVY = 4 'for divisions on full height
      NX = NDIVX
      NY = NDIVY - 1
      DX = 0.5 * Lx / NDIVX 'consider half wall for symmetry
      DY = Ly / NDIVY
      NDOF = NX * NY 'total number of degrees of freedom
      udl = 10.0! * Ly 'maximum liquid pressure in KN/m2
      mu = 0.16 'Poisson's ratio for concrete
      E = 5500000.0! * fcu ^ 0.5 'Young's modulus for concrete, KN/m2
      RIGIDITY = E * H ^ 3 / (12 * (1 - mu ^ 2)) 'flexural rigidity
      pload = udl / RIGIDITY
      MM = NDOF + 1
      Dim S(NDOF, MM), w(NDOF), Mx(NDOF), Myy(NDOF), MXV(NY), MYH(NX), AXIAL(NDIVY) As
Double
      Dim Q(NDOF) As Double
      L3 = DY ^ 3
      L4 = DX ^ 4
      Z = (DX / DY) ^ 2
      A = (6 + 6 * Z ^ 2 + 8 * Z) / L4
      B = -4 * (1 + Z) / L4
      c = -4 * Z * (1 + Z) / L4
      D = 2 * Z / L4
      E = Z ^ 2 / L4
      F = 1 / L4
      For I = 1 To NDOF
        For J = 1 To NDOF + 1
          S(I, J) = 0.0!
        Next J
      Next I
      'CONNECTIVITY
      k = 0
      For I = 1 To NX
        For J = 1 To NY
        K13 = 0 : K14 = 0 : K15 = 0 : K16 = 0
        k = k + 1
        K1 = k - NY
        K2 = k + NY
        K3 = k - 1
        K4 = k + 1
        K5 = k + NY + 1
        K6 = k - NY + 1
```

```
         K7 = k - NY - 1
         K8 = k + NY - 1
         K9 = k - 2 * NY
         K10 = k + 2 * NY
           K11 = k - 2
           K12 = k + 2
           'SPECIAL CASES
           If I = 1 Then
             K1 = 0
             K6 = 0
             K7 = 0
             K9 = k
           End If
           If I = 2 Then K9 = 0
           If I = NX Then
             K2 = k - NY
             K5 = K2 + 1
             K8 = K2 - 1
             K10 = k - 2 * NY
           End If
           If I = NX - 1 Then K10 = k
           If J = I Then
             K7 = 0
             K3 = 0
             K8 = 0
             K11 = k
           End If
           If J = 2 Then K11 = 0
           If J = NY - 1 Then
             K4 = 0
             K5 = 0
             K6 = 0
             K12 = k
           End If
           S(k, k) = S(k, k) + A
           If K1 > 0 Then S(k, K1) = S(k, K1) + B
           If K2 > 0 Then S(k, K2) = S(k, K2) + B
           If K3 > 0 Then S(k, K3) = S(k, K3) + C
           If K4 > 0 Then S(k, K4) = S(k, K4) + C
           If K5 > 0 Then S(k, K5) = S(k, K5) + D
           If K6 > 0 Then S(k, K6) = S(k, K6) + D
           If K7 > 0 Then S(k, K7) = S(k, K7) + D
           If K8 > 0 Then S(k, K8) = S(k, K8) + D
           If K9 > 0 Then S(k, K9) = S(k, K9) + F
           If K10 > 0 Then S(k, K10) = S(k, K10) + F
           If K11 > 0 Then S(k, K11) = S(k, K11) + E
           If K12 > 0 Then
             If J = NY - 1 Then S(k, K12) = S(k, K12) - E
             If J < NY - 1 Then S(k, K12) = S(k, K12) + E
           End If
           'LOAD VECTOR
           Q(k) = pload * (1 - J / NDIVY) 'HYDROSTATIC PRESSURE
         Next J
       Next I
       Call SOLVE(S, Q, w, NDOF)
       'COMPUTE MOMENT Mx ON INTERIOR POINTS
       For I = 1 To NDOF
         For J = 1 To NDOF
           S(I, J) = 0.0!
         Next J
       Next I
```

```
            r1 = 2 * RIGIDITY * (1 / DX ^ 2 + mu / DY ^ 2)
            r2 = -RIGIDITY / DX ^ 2
            r3 = -mu * RIGIDITY / DY ^ 2
            k = 0
            For I = 1 To NX
            For J = 1 To NY
              k = k + 1
              K1 = k - NY
              K2 = k + NY
              K3 = k - 1
              K4 = k + 1
              'SPECIAL CASES
              If I = 1 Then K1 = 0
              If I = NX Then K2 = k - NY
              If J = 1 Then K3 = 0
              If J = NY - 1 Then K4 = 0
              S(k, k) = S(k, k) + r1
              If K1 > 0 Then S(k, K1) = S(k, K1) + r2
              If K2 > 0 Then S(k, K2) = S(k, K2) + r2
              If K3 > 0 Then S(k, K3) = S(k, K3) + r3
              If K4 > 0 Then S(k, K4) = S(k, K4) + r3
            Next J
          Next I
          For I = 1 To NDOF
            mmt = 0.0!
            For J = 1 To NDOF
              mmt = mmt + S(I, J) * w(J)
            Next J
            Mx(I) = mmt
          Next I
          'COMPUTE SUPPORT MOMENT MXV OF THE VERTICAL SIDE Ly
          For I = 1 To NY
            MXV(I) = -2 * RIGIDITY * w(I) / DX ^ 2
          Next I
          'COMPUTE MOMENT My ON INTERIOR POINTS
          For I = 1 To NDOF
            For J = 1 To NDOF
              S(I, J) = 0.0!
            Next J
          Next I
          r1 = 2 * RIGIDITY * (1 / DY ^ 2 + mu / DX ^ 2)
          r2 = -mu * RIGIDITY / DX ^ 2
          r3 = -RIGIDITY / DY ^ 2
          k = 0
          For I = 1 To NX
            For J = 1 To NY
              k = k + 1
              K1 = k - NY
              K2 = k + NY
              K3 = k - 1
              K4 = k + 1
              'SPECIAL CASES
              If I = 1 Then K1 = 0
              If I = NX Then K2 = k - NY
              If J = 1 Then K3 = 0
              If J = NY - 1 Then K4 = 0
              S(k, k) = S(k, k) + r1
              If K1 > 0 Then S(k, K1) = S(k, K1) + r2
              If K2 > 0 Then S(k, K2) = S(k, K2) + r2
```

```
            If K3 > 0 Then S(k, K3) = S(k, K3) + r3
            If K4 > 0 Then S(k, K4) = S(k, K4) + r3
          Next J
        Next I
        For I = 1 To NDOF
          mmt = 0.0!
          For J = 1 To NDOF
            mmt = mmt + S(I, J) * w(J)
          Next J
          Myy(I) = mmt
        Next I
        'COMPUTE SUPPORT MOMENT MYH ON HORIZONTAL SIDE Lx
        For I = 1 To NX
          k = (I - 1) * NY + 1
          MYH(I) = -2 * RIGIDITY * w(k) / DY ^ 2
        Next I
        'COMPUTE AXIAL TENSION
        For I = 1 To NDIVY
          AXIAL(I) = 10 * Ly * (1 - I / NDIVY) * Lx / 2.0!
        Next I
        'GET MAXIMUM VALUES
        MAXMX = MAX(Mx, NDOF) 'MAXIMUM +VE x MOMENT
        MAXMY = MAX(Myy, NDOF) 'MAXIMUM +VE y MOMENT
        MAXSX = MAX(MXV, NY) 'MAIMUM -VE x MOMENT
        MAXSY = MAX(MYH, NX) 'MAXIMUM -VE y MOMENT
        MAXTN = MAX(AXIAL, NDIVY) 'MAXIMUM TENSION
End Sub

Function MAX(ByVal A() As Double, ByVal n As Integer)

    '*****************************************************************************************
    'Function to determine the maximum value in an array
    '*****************************************************************************************

    Dim x = 0
    For I = 1 To n
      A(I) = Abs(A(I))
      If x < A(I) Then x = A(I)
    Next I
    MAX = x
End Function

Function MAX2(ByVal x, ByVal y)

    '*****************************************************************************************
    'This function returns with the maximum of two numbers
    '*****************************************************************************************

    Dim Z = Abs(x)
    If x > y Then Z = x Else Z = y
    MAX2 = Z
End Function

  Sub SOLVE(ByVal GSTIF(,) As Double, ByVal GCON() As Double, ByVal HEAD() As Double,
ByVal n As Integer)

    '*****************************************************************************************
    'SOLVING EQUATIONS BY GAUSSIAN ELIMINATION
    '*****************************************************************************************

    Dim O(n), S(n), x(n)
    Dim PIVOT, DUMMY, BIG, FACTOR, SUM As Double
```

```
'********** ORDERING THE EQUATIONS
For I = 1 To n
  O(I) = I
  S(I) = Abs(GSTIF(I, 1))
  For J = 2 To n
    If Abs(GSTIF(I, J)) > S(I) Then
      S(I) = Abs(GSTIF(I, J))
    Else
    End If
  Next J
Next I
'ELIMINATION *********
For K = 1 To n - 1
  PIVOT = K
  BIG = Abs(GSTIF(O(K), K) / S(O(K)))
  For II = K + 1 To n
    DUMMY = Abs(GSTIF(O(II), K) / S(O(II)))
    If DUMMY > BIG Then
      BIG = DUMMY
      PIVOT = II
    Else
    End If
  Next II
  DUMMY = O(PIVOT)
  O(PIVOT) = O(K)
  O(K) = DUMMY
  For I = K + 1 To n
  FACTOR = GSTIF(O(I), K) / GSTIF(O(K), K)
  For J = K + 1 To n
    GSTIF(O(I), J) = GSTIF(O(I), J) - FACTOR * GSTIF(O(K), J)
  Next J
    GCON(O(I)) = GCON(O(I)) - FACTOR * GCON(O(K))
  Next I
Next K
'********* BACK SUBSTITUTION **********
x(n) = GCON(O(n)) / GSTIF(O(n), n)
HEAD(n) = x(n)
For I = n - 1 To 1 Step -1
  SUM = 0
  For J = I + 1 To n
    SUM = SUM + GSTIF(O(I), J) * x(J)
  Next J
  x(I) = (GCON(O(I)) - SUM) / GSTIF(O(I), I)
  HEAD(I) = x(I)
Next I
End Sub

Function fN(ByVal n As Double) As String
  'formats a number into 6 decimal places, e.g. ##.######
  'returns the result as a string
  Return FormatNumber(n, 6).ToString
End Function

Sub designTank()
  Dim Mx, Myy, Mxb, Myb, Mxc, Myc, V1, V2 As Double
  Dim M11, M12, M13, M14, TN1 As Double
  Dim M21, M22, M23, M24, TN2 As Double
  'Structural design of the tank walls
  'Assure thickness as follows:
  tw = 200 'Thickness of walls in mm
  tb = 200 'thickness of base slab in mm
```

```
tc = 150 'thickness of cover slab in mm
H = dt 'total depth of tank in mm
Call GetCover(fcu, cv) 'determine cover to steel
'Design the walls
Call GetWallsMoments(Lx, H, tw, fcu, M11, M12, M13, M14, TN1)
Call GetWallsMoments(Lx, H, tw, fcu, M21, M22, M23, M24, TN2)
'Midspan section
Mx = MAX2(M11, M21)
Myy = MAX2(M12, M22)
TN = MAX2(TN1, TN2)
Call Designsection(tw, cv, Mx, Myy, TN, TN, fcu, fy, Ast1, Ast2)
'Section at the joint of the walls
Mx = MAX2(M13, M23)
Myy = MAX2(M14, M24)
TN = MAX2(TN1, TN2)
Call Designsection(tw, cv, Mx, Myy, TN, TN, fcu, fy, Ast3, Ast4)
'Design the bottom base slab
Call GetBasePlateMoments(tb, Lx, Ly, H, fcu, Mxb, Myb)
TN = -10 * dt 'tension in the base slab made -ve for identification only
Call Designsection(tb, cv, Mxb, Myb, TN, 0, fcu, fy, Ast5, Ast6)
'Design the cover slab
udl = 1.6 * 1.5 + 1.4 * 24 * tc / 1000 'Ultimate design load in KN/m2
Call GetCoverPlateMoments(Lx, Ly, udl, Mxc, Myc)
V1 = udl * Lx / 2 'shear on short support
V2 = udl * Ly / 2 'shear on long support
V = MAX2(V1, V2)
Call Designsection(tc, cv, Mxc, Myc, 0, V, fcu, fy, Ast7, Ast8)
'Output results of the structural design
Dim res As String 'the output string
res = "Structural Design of Septic Tanks:"
res += Chr(13) + Chr(10)
res += Chr(13) + Chr(10)
res += "Walls Forces: "
res += fN(M11) + " " + fN(M12) + " " + fN(M13) + " " + fN(M14) + " " + fN(TN1)
res += Chr(13) + Chr(10)
res += "Walls Forces: "
res += fN(M21) + " " + fN(M22) + " " + fN(M23) + " " + fN(M24) + " " + fN(TN2)
res += Chr(13) + Chr(10)
res += "Bottom slab: "
res += fN(Mxb) + " " + fN(Myb)
res += Chr(13) + Chr(10)
res += "Cover slab: "
res += fN(Mxc) + " " + fN(Myc)
res += Chr(13) + Chr(10)
res += Chr(13) + Chr(10)
res += "Vertical walls:"
res += Chr(13) + Chr(10)
res += "Thickness of walls: " + CInt(tw).ToString + " mm"
res += Chr(13) + Chr(10)
res += "Vertical steel: " + CInt(Ast1).ToString + " mm2/m"
res += Chr(13) + Chr(10)
res += "Horizontal steel:" + CInt(Ast2).ToString + " mm2/m"
res += Chr(13) + Chr(10)
res += "Horizontal steel at support: " + CInt(Ast3).ToString + " mm2/m"
res += Chr(13) + Chr(10)
res += "Vertical steel at support: " + CInt(Ast3).ToString + " mm2/m"
res += Chr(13) + Chr(10)
res += Chr(13) + Chr(10)
res += "Base slab:"
res += Chr(13) + Chr(10)
res += "Thickness of base: " + CInt(tb).ToString + " mm"
```

```vb
      res += Chr(13) + Chr(10)
      res += "Long direction steel: " + CInt(Ast5).ToString + " mm2/m"
      res += Chr(13) + Chr(10)
      res += "Short direction steel: " + CInt(Ast6).ToString + " mm2/m"
      res += Chr(13) + Chr(10)
      res += Chr(13) + Chr(10)
      res += "Cover slab:"
      res += Chr(13) + Chr(10)
      res += "Thickness of cover slab: " + CInt(tc).ToString + " mm"
      res += Chr(13) + Chr(10)
      res += "Long direction steel: " + CInt(Ast7).ToString + " mm2/m"
      res += Chr(13) + Chr(10)
      res += "Short direction steel: " + CInt(Ast8).ToString + " mm2/m"
      Try
        f2.showText = res
        f2.ShowDialog()
      Catch ex As Exception
        f2 = New Form3
        f2.showText = res
        f2.ShowDialog()
      End Try
    End Sub

  Private Sub Button1_Click(ByVal sender As System.Object, ByVal e As System.EventArgs)
Handles Button1.Click
      designTank()
    End Sub

  Private Sub Button2_Click(ByVal sender As System.Object, ByVal e As System.EventArgs)
Handles Button2.Click
      calculateResults()
    End Sub
End Class
```

## (B) LISTING OF FORM2 CODE (CHAP5.5\FORM2.VB):

```vb
Public Class Form2
  Public choice As Integer

  Private Sub Form2_Load(ByVal sender As System.Object, ByVal e As System.EventArgs)
Handles MyBase.Load
    Me.Text = "THE EXPOSURE CONDITION"
    Me.FormBorderStyle = Windows.Forms.FormBorderStyle.FixedDialog
    Label1.Text = "Select exposure condition:"
    ListBox1.Items.Clear()
    ListBox1.Items.Add("Mild")
    ListBox1.Items.Add("Moderate")
    ListBox1.Items.Add("Severe")
    ListBox1.Items.Add("Very severe")
    ListBox1.Items.Add("Extreme")
    ListBox1.SelectedIndex = 0
    Button1.Text = "&OK"
    Button2.Text = "&Cancel"
    Me.CancelButton = Button2
    Me.AcceptButton = Button1
  End Sub
```

```
   Private Sub ListBox1_SelectedIndexChanged(ByVal sender As System.Object, ByVal e As
System.EventArgs) Handles ListBox1.SelectedIndexChanged
      choice = ListBox1.SelectedIndex
   End Sub

   Private Sub Button1_Click(ByVal sender As System.Object, ByVal e As System.EventArgs)
Handles Button1.Click
      choice = ListBox1.SelectedIndex
      Me.Hide()
   End Sub

   Private Sub Button2_Click(ByVal sender As System.Object, ByVal e As System.EventArgs)
Handles Button2.Click
      choice = -1
      Me.Hide()
   End Sub
End Class
```

### (C) LISTING OF FORM3 CODE (CHAP5.5\FORM3.VB):

```
Public Class Form3
   Private txt As String

Public Property showText
   Get
   showText = txt
   End Get
   Set(ByVal value)
   txt = value
   TextBox1.Text = value
   Button1.Select()
   End Set
End Property

   Private Sub Form3_Load(ByVal sender As System.Object, ByVal e As System.EventArgs)
Handles MyBase.Load
      TextBox1.Multiline = True
      TextBox1.Height = 259
      TextBox1.ScrollBars = ScrollBars.Vertical
      Me.FormBorderStyle = Windows.Forms.FormBorderStyle.FixedDialog
      Button1.Text = "&OK"
      Me.CancelButton = Button1
      Me.AcceptButton = Button1
   End Sub

   Private Sub Button1_Click(ByVal sender As System.Object, ByVal e As System.EventArgs)
Handles Button1.Click
      Me.Hide()
   End Sub
End Class
```

### 5.3.2 IMHOFF TANKS

An Imhoff tank is a form of a septic tank. It is a two-storey tank that performs the same functions of sedimentation and anaerobic digestion of sludge. The two processes are achieved in separate compartments. The incoming wastewater flows through the upper compartment, allowing solids to settle to the bottom of the chamber, which is in the shape of a hopper. At the bottom of the hopper, the solids pass through a baffled outlet into the lower chamber in which anaerobic digestion takes place. Because there is no intimate contact between sewage and the digesting sludge, a better effluent is obtained. Gases from the sludge compartment are directed away from the falling solids so as not to hinder their descent (Smith and Scott 2002). Although Imhoff tanks function better than septic tanks (especially in warm climates), they are more elaborate and more costly to operate.

The advantages of the Imhoff tank are as follows:

- It yields good results without skilled attention.
- It has minimum problems with sludge disposal.

The disadvantages of the Imhoff tank are as follows:

- Greater depth means greater costs.
- It is unsuitable when high acidic conditions prevail.
- It has no adequate control over its operation.
- Insufficient area of vents causes the gas to lift the fluffy scum, a condition known as "foaming."

For the design of an Imhoff tank, the following points must be considered:

- Detention time should be 1–2 h.
- The length-to-width ratio should be between 2:1 and 4:1.
- Depth should be kept shallow, 10–12 m.
- Vent area for gas release and accumulation of scum should be 20%–30% of the total Imhoff tank surface area.

## 5.4 SOURCES AND EVALUATION OF WASTEWATER FLOW RATES

Sources of generation of wastewater are as follows:

- *Domestic wastewater*—discharges from residences, commercial, institutional, and related units
- *Industrial wastewater*—discharges from factories and industrial establishments
- *Storm water*—water that result from rainfall runoff
- *Infiltration/inflow*—extraneous water that enters the sewer from the ground via different routes and storm water that is discharged from sources such as roof leaders, foundation drains, and storm sewers (Rowe and Abdel-Magid 1995)

The domestic wastewater rate of flow is normally estimated by finding the dry weather flow, DWF, which is defined as the total average discharge of sanitary sewage and is the normal flow in a sewer during the dry weather (Rowe and Abdel-Magid 1995). DWF also may be described as the average daily flow in the sewer after several days during which rainfall has not exceeded 2.5 mm in the previous 24 h (Lin and Lee 2007). The DWF for a certain locality can be determined from the following equation:

$$DWF = (P * Q) + I_r + T_w - E_v \qquad (5.41)$$

where:
DWF is the dry weather flow, L/day
$P$ is the number of people served by the sewer, dimensionless

$Q$ is the average water consumption, L/c * day
$I_r$ is the average infiltration into the sewer, L/day (also estimated as L/day/km length of sewer for different ages of the sewer). Usually, $I_r$ varies between 0% and 30% of DWF (Rowe and Abdel-Magid 1995; Lin and Lee 2007)
$T_w$ is the average trade waste discharge, L/day
$E_v$ is the rate of evaporation

The DWF can also be estimated from the data of water consumption as presented in the following equation:

$$DWF = \chi * Q * P \qquad (5.42)$$

where:
DWF is the dry weather flow, m³/s
$\chi$ is the percentage of water entering the sewerage system (usually ranging from 80% to 90% of water consumption)
$Q$ is the water consumption, m³/cs
$P$ is the population served, dimensionless

Maximum wastewater flow may be determined by using the following equation (Mullick 1987):

$$Q_{max} = a * DWF \qquad (5.43)$$

where:
$Q_{max}$ is the maximum daily wastewater flow, m³/day
$a$ is the factor (= 2–4)
DWF is the dry weather flow, m³/s

Equation 5.44 gives an estimate of the minimum wastewater flow (Rowe and Abdel-Magid 1995; American Water Works Association and American Society of Civil Engineers 2012):

$$Q_{min} = b' * Q_a \qquad (5.44)$$

where:
$Q_{min}$ is the minimum wastewater flow, m³/day
$b'$ is the constant (with $b' = 30\%–50\%$ for small communities, and $b' = 66\%–80\%$ for large communities that are greater than 100,000 persons)
$Q_a$ is the average wastewater flow, m³/day

Equation 5.45 gives the relationship between the maximum wastewater flows and the average flows called peaking factor (Mullick 1987):

$$pf = \frac{Q_{max}}{Q_a} \qquad (5.45)$$

where $pf$ is the peaking factor, given by the following equation:

$$pf = \frac{5}{(P/1000)^{0.167}} \qquad (5.46)$$

where $P$ is the number of people served in thousands, dimensionless

Equation 5.46 is valid for a population size of 200 up to 1000.

## Example 5.6

1. Write a computer program to determine the DWF for a sanitary sewer that is required to serve an area with a population number of $P$, for an average daily sewage flow of $Q$ (m³/day). Take the daily infiltration in the area to be $I$ (m³/km * length of sewer) and total length of sewer as l (km).
2. An area with a population of 12,000 is to be served by a sanitary sewer. The average wastewater flow is 15 m³/h per capita. The daily infiltration in the area is judged to be 55 m³/km length of sewer. For a total sewer length of 7 km, compute the DWF.

3. Use the computer program developed in (1) to verify the manual solution obtained in (2).

### Solution

1. For solution to Example 5.6 (1), see the listing for Program 5.6.
2. Solution to Example 5.6 (2):
   a. Given: $P = 12,000$; $Q = 15$ m³/h * c, $I_r = 55$ m³/length of sewer, length of sewer = 7 km.
   b. Use equation of DWF as DWF = $(P * Q) + I_r + T_w - E_v$.
   c. Find the average dry weather infiltration into the sewer as follows: $I_r = 55$ (m3) * 7 (km of sewer length) = 385 m³/day.
   d. Assume both average trade waste discharge, $T_w$ and evaporation rate, $E_v$ to be negligible, and find the DWF as follows: DWF = 12000 * 15/24 + 385 = 7885 m³/day.

---

### LISTING OF PROGRAM 5.6   (CHAP5.6\FORM1.VB): DRY WEATHER FLOW, DWF

```
'********************************************************************************
'Program 5.6: Determination of the Dry Weather Flow
'********************************************************************************

Public Class Form1
  Const Pi = 3.1415962#
  Const G = 9.81
  Dim POP, Q, I, L As Double

  Private Sub Form1_Load(ByVal sender As System.Object, ByVal e As System.EventArgs)
Handles MyBase.Load
     Me.Text = "Program 5.6: Determination of the Dry Weather Flow"
     Me.MaximizeBox = False
     Me.FormBorderStyle = Windows.Forms.FormBorderStyle.FixedSingle
     Label1.Text = "Enter the number of people to be served:"
     Label2.Text = "Enter the average daily sewage flow Q (m3/h/c):"
     Label3.Text = "Enter the daily infiltration rate I (m3/km):"
     Label4.Text = "Enter the length of the sewer L (km):"
     BUTTON1.TEXT = "&Calculate"
     label5.text = ""
  End Sub

  Sub calculateResults()
     POP = Val(textbox1.text)
     Q = Val(textbox2.text)
     I = Val(textbox3.text)
     L = Val(textbox4.text)
     Dim Ir, Tw, Ev, DWF As Double
     Ir = I * L
     Tw = 0
     Ev = 0
     DWF = POP * Q / 24 + Ir + Tw - Ev
     Label5.Text = "The dry weather flow DWF = " + FormatNumber(DWF, 2) + " m3/d"
  End Sub

  Private Sub Button1_Click(ByVal sender As System.Object, ByVal e As System.EventArgs)
Handles Button1.Click
     calculateResults()
  End Sub
End Class
```

**Example 5.7**

1. Write a computer program to determine the average, maximum, minimum, and peak wastewater flows, given population number and annual average water consumption (L/capita/d). Assume DWF amounting to any percentage, *a*, of water consumption and minimum flow equal to any percentage, *b*, of average flow.
2. Determine the different per capita wastewater flows (i.e., maximum daily, average daily, and minimum daily flow) for a population of 900 using the following data: annual average water consumption = 275 L/c/day; DWF = 90% of water consumption; and minimum flow = 50% of average flow.
3. Use the computer program developed in (1) to verify the manual solution acquired in (2).

**Solution**

1. For solution to Example 5.7 (1), see the listing of Program 5.7.
2. Solution to Example 5.7 (2):
   a. Given: $P = 900$, $Q = 275$ L/c/day, DWF = 0.9 * consumption, $Q_{min} = 0.5 * Q_a$.
   b. Determine the average annual daily DWF, or wastewater flow as 0.9 * water consumption = 0.9 * 275 = 247.5 L/c.
   c. Compute the peaking factor as follows: $pf = 5/P^{0.167} = 5/0.9^{0.167} = 5.09$.
   d. Find the maximum daily flow or the maximum 24 h flow = $pf * Q_a = 5.09 * 247.5 = 1259.5$ L/c/day. Assuming a working time activity of 16 h, the design average daily flow will be as 1259.5 * 24/16 = 1889.7 L/c/day.
   e. Determine the minimum daily flow as follows: $Q_{min} = 0.5 * Q_a = 0.5 * 247.5 = 123.75$ L/c/day.

---

**LISTING OF PROGRAM 5.7   (CHAP5.7\FORM1.VB): DIFFERENT PER CAPITA WASTEWATER FLOWS**

```
'*********************************************************************************************
'Program 5.7: Determination of Different Per Capita Wastewater Flows
'*********************************************************************************************

Imports System.Math
Public Class Form1
  Const G = 9.81
  Dim POP, DWF, Q, Qamp As Double
  Dim Pf, Qmax, Qd, Qmin As Double

  Private Sub Form1_Load(ByVal sender As System.Object, ByVal e As System.EventArgs)
Handles MyBase.Load
    Me.Text = "Program 5.7: Per Capita Wastewater Flows"
    Me.MaximizeBox = False
    Me.FormBorderStyle = Windows.Forms.FormBorderStyle.FixedSingle
    Label1.Text = "Enter the number of people to be served:"
    Label2.Text = "Enter the annual average water consumption L/c.d:"
    Label3.Text = "Enter the dry weather flow DWF as percentage of Q:"
    Label4.Text = "Enter the minimum flow as a percentage of Q:"
    Button1.Text = "&Calculate"
    Label5.Text = ""
    Label5.AutoSize = False
    Label5.Width = 352
    Label5.Height = 52
    Label6.Text = "Decimals:"
    NumericUpDown1.Value = 2
    NumericUpDown1.Maximum = 10
    NumericUpDown1.Minimum = 0
  End Sub

  Sub calculateResults()
    POP = Val(TextBox1.Text)
    Q = Val(TextBox2.Text)
    DWF = Val(TextBox3.Text)
    Qamp = Val(TextBox4.Text)
    If DWF > 1 Then DWF /= 100
    If Qamp > 1 Then Qamp /= 100
    DWF = DWF * Q
    'peak factor
```

```
        Pf = 5 / (POP / 1000) ^ 0.167
        'maximum daily flow
        Qmax = Pf * DWF
        'design average daily flow
        Qd = Qmax * 24 / 16
        'minimum daily flow
        Qmin = Qamp * DWF
    End Sub

    Sub showResults()
        Dim decimals As Integer = NumericUpDown1.Value
        Label5.Text = "Average annual daily DWF (L/c): " + FormatNumber(DWF, decimals)
        Label5.Text += vbCrLf + "Maximum daily flow (L/c.d): " + FormatNumber(Qmax, decimals)
        Label5.Text += vbCrLf + "Minimum daily flow (L/c.d): " + FormatNumber(Qmin, decimals)
    End Sub

    Private Sub Button1_Click(ByVal sender As System.Object, ByVal e As System.EventArgs)
Handles Button1.Click
        calculateResults()
        showResults()
    End Sub

    Private Sub NumericUpDown1_ValueChanged(ByVal sender As System.Object, ByVal e As
System.EventArgs) Handles NumericUpDown1.ValueChanged
        showResults()
    End Sub
End Class
```

## 5.5 CONCEPT OF PE

PE of sewage or wastewater refers to a certain quality parameter—for example, BOD or SS—as a per capita contribution, compared to some quality parameter of the per capita contribution of a standard sewage. Domestic sewage is adopted as the standard. Thus, the PE of any sewage is the number of persons to the standard sewage. The relationship between PE and $BOD_5$ is presented in the following equation (Smith and Scott 2002; Lin and Lee 2007):

$$PE = \frac{BOD_5\, Q}{BOD_s} \qquad (5.47)$$

where:
PE is the population equivalent
$BOD_5$ is the 5-day BOD of wastewater, mg/L
$Q$ is the wastewater flow rate, m³/s

$BOD_s$ is the BOD of standard sewage (= 60 g of $BOD_5$ produced by an average person per day in the United Kingdom and 80 g of $BOD_5$ produced by an average person in the United States per day).

The main benefits of the PE concept include the following:

• Estimation of the strength of industrial sewage for the purposes of waste treatment

• Measurement of changes of wastewater treatment for various industries
• Assessment of charges for factories and industrial firms

### Example 5.8

1. Write a short computer program to find the PE of an industrial establishment that produces wastewater of $Q$ (m³/day). The wastewater has a 5-day BOD of $L$ (mg/L).
2. The daily wastewater production from a certain industry amounts to 6 * 10⁶ L with a 5-day BOD of 370 mg/L. Determine the PE for this industry.
3. Use the computer program developed in (1) to verify the manual solution in (2).

**Solution**

1. For solution to Example 5.8 (1), see the listing of Program 5.8.
2. Solution to Example 5.8 (2):
   a. Given: wastewater flow, $Q$ = 6 * 10⁶ L/day, BOD = 370 mg/L.
   b. Assume the average person exerts a 5-day BOD load of 0.06 kg/day.
   c. Determine the PE as follows: PE = $BOD_5$ of waste * flow rate/BOD of standard sewage = (370 * 10⁻³ g/L * 6 * 10⁶ L/d)/(0.06 * 10³ g/d) = 37,000.

### LISTING OF PROGRAM 5.8 (CHAP5.8\FORM1.VB): POPULATION EQUIVALENT OF INDUSTRIAL ESTABLISHMENTS

```vbnet
'*****************************************************************************
'Program 5.8: Population Equivalent of Industrial establishments
'*****************************************************************************

Imports System.Math
  Public Class Form1
     Const G = 9.81
     Dim Q, BOD5 As Double
     Dim Ex, PE As Double

  Private Sub Form1_Load(ByVal sender As System.Object, ByVal e As System.EventArgs)
Handles MyBase.Load
     Me.Text = "Program 5.8: Population Equivalent of Industrial establishments"
     Me.MaximizeBox = False
     Me.FormBorderStyle = Windows.Forms.FormBorderStyle.FixedSingle
     Label1.Text = "Daily wastewater production, Q (L/d):"
     Label2.Text = "The 5-day BOD (mg/L):"
     Button1.Text = "&Calculate"
     Label3.Text = ""
     Label4.Text = "Decimals:"
     NumericUpDown1.Value = 2
     NumericUpDown1.Maximum = 10
     NumericUpDown1.Minimum = 0
  End Sub


  Sub calculateResults()
     Q = Val(TextBox1.Text)
     bod5 = Val(TextBox2.Text)
     'assumed per capita 5-d BOD load in kg/d
     Ex = 0.06
     'convert to mg/L
     BOD5 = BOD5 / 1000
     'population equivalent
     PE = BOD5 * Q / (Ex * 10 ^ 3)
  End Sub


  Sub showResults()
     Label3.Text = "Population equivalent PE: " + FormatNumber(PE, NumericUpDown1.Value)
  End Sub

  Private Sub Button1_Click(ByVal sender As System.Object, ByVal e As System.EventArgs)
Handles Button1.Click
     calculateResults()
     showResults()
  End Sub

  Private Sub NumericUpDown1_ValueChanged(ByVal sender As System.Object, ByVal e As
System.EventArgs) Handles NumericUpDown1.ValueChanged
     showResults()
  End Sub
End Class
```

## 5.6 REASONS FOR TREATING WASTEWATER

The main reasons for treatment of wastewater include the following:

- Stabilization of organic pollutants
- Reduction of the number of disease-causing agents found in sewage
- Prevention of pollutants from entering water sources
- Reduction of odors and other nuisances resulting from sewage
- Water reclamation and reuse
- By-product recovery and use

## 5.7 WASTEWATER TREATMENT UNIT OPERATIONS AND PROCESSES

Wastewater treatment units can be classified simply into the following:

1. *In situ treatment works* are small treatment plants (package plants) that handle the disposal of wastewater emerging from individual households or small rural communities.
2. *Large treatment works* are centrally located wastewater plants that treat the wastewater discharges from large metropolitan areas.

Table 5.6 gives a quick review of the most essential unit operations and processes used in water and wastewater treatment plants.

## 5.8 PRELIMINARY TREATMENT: GRIT REMOVAL

Grit is composed of sand, gravel, cinders, ashes, metal fragments, glass, inert inorganic solids, pebbles, eggshells, bone chips, seeds, coffee and tea grounds, earth from vegetable washing, large organic particles such as food

---

**TABLE 5.6**
**Water and Wastewater Treatment Units**

**Water Treatment**

| | |
|---|---|
| • Screening | Bars, mesh, or strainer to remove large solids |
| • Flocculation | Mechanical mixing to favor agglomeration of solid particles |
| • Coagulation | Addition of chemicals to coagulate SS |
| • Sedimentation | Settling out of flocculated substances |
| • Sand filtration | Filtering out remaining suspended matter |
| • Disinfection | Addition of disinfectant to kill harmful disease-causing agents |
| • Sludge treatment | Thickening by brevity and then disposal |
| • Other | Water softening, pH adjustment, and fluoridation |

**Preliminary Treatment**

| | |
|---|---|
| • Screening | As above |
| • Grit removal | Removing grit and inorganic matter (e.g., sand) but not organic matter |
| • Storm overflow | Diverting sewage in excess of treatment plant capacity to storm water-holding tanks |

**Primary Treatment**

| | |
|---|---|
| • Primary sedimentation | Settling of SS (only 40%– 60% removed), no chemicals added |

**Secondary Treatment**

| | |
|---|---|
| • Anaerobic oxidation of organic matter | Biodegrading organic matter through the action of microorganisms in a biological treatment unit such as activated sludge or trickling filter plant |
| • Secondary sedimentation | Settling out of sludge containing microorganisms to produce a treated effluent |

**Tertiary Treatment (Effluent polishing, advanced treatment)**

| | |
|---|---|
| • Finalizing treatment | Polishing of effluent by operations such as sand filters and microstrainers |

**Sludge Treatment**

| | |
|---|---|
| • Anaerobic digestion | Decomposing thickened sludge in the absence of oxygen |
| • Gravity thickening | Thickening of primary and secondary sludge |
| • Mechanical dewatering | Removing water from sewage sludges by methods such as centrifuges and pressure or vacuum filters |
| • Drying beds | Drying sewage sludge in open atmosphere |

**Sludge Disposal**

- Composted to be used as a soil conditioner (digested only)
- Dumped at sea (undigested)
- Incinerated (normally undigested but thickened)
- Landfilled (preferably digested and dewatered or dried)

*Source:* Rowe, D. R. and Abdel-Magid, I. M., *Handbook of Wastewater Reclamation and Reuse*, CRC Press/Lewis Publishers, Boca Raton, FL, 1995. With permission.

wastes, or other heavy solid material with specific gravities greater than those of the organic putrescent solids found in wastewater.

Problems in treating wastewater containing grit are as follows:

- Grit is an abrasive material.
- It will produce wear and tear on pumps and other treatment works.
- It absorbs oil and grease on its surface (it becomes sticky and solidifies) in pipes, sumps, sedimentation tanks, or other areas of low hydraulic shear.
- It can occupy a valuable volume of the sludge digester.

Grit chambers serve the following functions:

- Protecting the moving mechanical equipment from abrasion and development of any abnormal wear and tear
- Reducing the formation of heavy deposits inside pipelines, channels, and conduits
- Reducing the frequency of digester cleaning
- Facilitating the operation of treatment units

The quantity of grit in wastewater differs according to the status of the sewerage system, the quantity of storm water present, and the ratio of industrial to other sources of wastewater. Separation of grit particles in grit removal chambers relies on the difference in the specific gravity between the organic and inorganic solids. All particles are assumed to settle with a terminal velocity in agreement with Newton's law of motion as presented in the following Equation (Steel and McGhee 1991):

$$v = \sqrt{\frac{4}{3}\left(\rho_S - \rho\right)\frac{gd}{\rho C_D}} = \sqrt{\frac{4}{3}\left(\text{s.g.} - \rho\right)\frac{gd}{C_D}} \qquad (5.48)$$

where:
  v is the terminal settling velocity, m/s
  $g$ is the gravitation acceleration, m/s$^2$
  $\rho_s$ is the mass density of particle, kg/m$^3$
  $\rho$ is the mass density of fluid, kg/m$^3$
  $d$ is the diameter of particle, m
  s.g. is the specific gravity of settling particle, dimensionless
  $C_D$ is the drag coefficient (dimensionless)

$$C_D = \frac{24}{\text{Re}} + \frac{3}{\sqrt{\text{Re}}} + 0.34 \qquad (5.49)$$

where Re is the Reynolds number, dimensionless.

Particles within the grit chamber are scoured at a horizontal scouring velocity given by the following equation (Steel and McGhee 1991):

$$v_{sco} = \sqrt{\frac{8a}{f}g\left(\text{s.g.}-1\right)d} \qquad (5.50)$$

where:
  $v_{sco}$ is the horizontal velocity of flow (scour velocity), m/s
  s.g. is the specific gravity of the particle, dimensionless
  $f$ is the Darcy–Weisbach friction factor, dimensionless (usually, $f = 0.02$–0.003) (Steel and McGhee 1991)
  $a$ is a constant, dimensionless (usually $a = 0.04$–0.06) (Steel and McGhee 1991)

Generally, a grit removal system possesses the following properties:

- A long constant-velocity channel
- A length-to-depth ratio of 10 (i.e., $l/h = 10$). In practice (owing to turbulence at inlet and outlet), values of $l/h$ up to 25 are employed
- A width-to-depth ratio of 2, that is, $B/h = 2$
- The width of the channel at any point above the invert has a settling velocity of 30 cm/s (to avoid removal of organic particles) and is given by the following equation (Abdel-Magid 1986; Lin and Lee 2007):

$$B = 4.92 * Q_{max}/d_{max} \qquad (5.51)$$

where:
  $B$ is the width of the channel, m
  $Q_{max}$ is the maximum flow rate, m$^3$/s
  $d_{max}$ is the maximum depth of flow, m

Velocity control within the grit chamber may be made by a flow control device such as a standing wave flume, vertical throat, and proportional flow weir. For standing wave flumes, the flow rate is presented by the following equation (Abdel-Magid 1986; Lin and Lee 2007):

$$\begin{aligned} Q &= 1.71 * y * d^{1.5} \\ &= 1.14 * B * d^{1.5} \end{aligned} \qquad (5.52)$$

where:
  $Q$ is the flow rate, m$^3$/s
  $d$ is the depth of flow, m
  $y$ is the throat width, m

The throat width may be computed as given in the following equation:

$$y = 2 * B/3 \qquad (5.53)$$

where $B$ is the width of channel at any point above invert, m.

Table 5.7 presents typical design parameters of grit chambers.

## TABLE 5.7
### Grit Chamber Typical Design Parameters

| Parameter | Value |
|---|---|
| Detention time (s) | 60 |
| Horizontal design velocity (cm/s) | 30 |
| Equivalent diameter of grit removed (mm) | 0.2 |
| Specific gravity of particles captured | 2.65 |
| Length of chamber (m) | $>18 d_{max}$ |

*Sources:* Metcalf and Eddy, Inc., Tchobanoglous, G., Stensel, H. D., Tsuchihashi R., and Burton, F., *Wastewater Engineering: Treatment and Resource Recovery*, 5th ed, McGraw-Hill Science/Engineering/Math, 2013; Peavy, Rowe, and Tchobanoglous, 1985; Abdel-Magid, I. M., *Selected Problems in Wastewater Engineering*, National Research Council, Khartoum University Press, Khartoum, Sudan, 1986; Peirce, J. J. et al. *Environmental Pollution and Control*, 4th ed, Butterworth-Heinemann, Oxford, 1997.

## Example 5.9

1. Write a computer program for designing a grit chamber that consists of $N$ mechanically cleaned channels. Each channel of the chamber is required to carry a maximum flow of $Q_{max}$ (m³/s) with a maximum depth of $d_{max}$ (m).
2. Design a grit chamber that consists of four mechanically cleaned channels. Each channel carries a maximum flow of 725 L/s with a maximum depth of 0.6 m (use l = 18 * $d_{max}$).
3. Use the computer program developed in (1) to verify the manual solution in (2).

## Solution

1. For the solution to Example 5.9 (1), see the listing of Program 5.9.
2. Solution to Example 5.9 (2):
   a. Given: $Q_{max} = 725$ L/s, $d_{max} = 0.6$ m, $N = 4$.
   b. Determine the width of the channel as follows: $B = 4.92 * Q_{max}/d_{max} = 4.92 * 0.725/0.6 = 5.95$ m.
   c. Width of each channel = total width/number of channels = $B/N = 5.95/4 = 1.5$ m.
   d. Find the throat width of the standing wave flume in channels that maintain the velocity as follows: $y = (2/3) * B = (2/3) * 1.5 = 0.99$ m.
   e. Determine the length of the channel as follows: l = 18 * $d_{max}$ = 19 * 0.99 = 17.8 m.

---

**LISTING OF PROGRAM 5.9   (CHAP5.9\FORM1.VB): DESIGNING A GRIT CHAMBER**

```
'****************************************************************************
'Program 5.9: Designing a Grit Chamber
'****************************************************************************

Imports System.Math
Public Class Form1
    Const G = 9.81
    Dim N, Q, dmax As Double
    Dim B, W, Y, L As Double

    Private Sub Form1_Load(ByVal sender As System.Object, ByVal e As System.EventArgs)
Handles MyBase.Load
        Me.Text = "Program 5.9: Grit Chamber Design"
        Me.MaximizeBox = False
        Me.FormBorderStyle = Windows.Forms.FormBorderStyle.FixedSingle
        Label1.Text = "Number of mechanically cleaned channels, N:"
        Label2.Text = "Flow per channel, Q (m3/s):"
        Label3.Text = "Required maximum depth of flow, dmax (m):"
        Button1.Text = "&Calculate"
        Label4.Text = ""
```

```
      Label4.AutoSize = False
      Label4.Width = 352
      Label4.Height = 52
      Label5.Text = "Decimals:"
      NumericUpDown1.Value = 2
      NumericUpDown1.Maximum = 10
      NumericUpDown1.Minimum = 0
   End Sub

   Sub calculateResults()
     N = Val(TextBox1.Text)
     Q = Val(TextBox2.Text) / 1000
     dmax = Val(TextBox3.Text)
     'width of channel
     B = 4.92 * Q / dmax
     'width of each channel
     W = B / N
     'throat width
     Y = 2 / 3 * W
     'length of channel
     L = 18 * Y
   End Sub

   Sub showResults()
     Dim decimals As Integer = NumericUpDown1.Value
     Label4.Text = "Width of each channel (m): " + FormatNumber(W, decimals)
     Label4.Text += vbCrLf + "Throat width (m): " + FormatNumber(Y, decimals)
     Label4.Text += vbCrLf + "Length of channel (m): " + FormatNumber(L, decimals)
   End Sub

   Private Sub Button1_Click(ByVal sender As System.Object, ByVal e As System.EventArgs)
Handles Button1.Click
     calculateResults()
     showResults()
   End Sub

   Private Sub NumericUpDown1_ValueChanged(ByVal sender As System.Object, ByVal e As
System.EventArgs) Handles NumericUpDown1.ValueChanged
     showResults()
   End Sub
End Class
```

## 5.9 SECONDARY TREATMENT (AEROBIC AND BIOLOGICAL)

### 5.9.1 INTRODUCTION

Secondary wastewater treatment serves the following functions:

- Stabilization of organic matter
- Coagulation and removal of nonsettleable colloidal solids
- Reduction of organic matter usually found in sewage sludge
- Reduction of growth nutrients in sewage, such as nitrogen and phosphorus

Secondary treatment facilities can be broadly classified as:

- *Attached growth processes*: In these processes, microorganisms are fixed or attached to a solid surface or media. Organic matter is brought in contact with the mircoorganisms by various mechanisms (Berger 1987).
- *Suspended growth processes*: In these processes, microorganisms (bacteria, fungi, and protozoa) and small organisms (rotifers and nematode worms) are free to move within the reactor and utilize the organic material for energy, growth, and reproduction.

### 5.9.2 Suspended Growth Systems (Aerobic Suspended Growth Process): Activated Sludge Process

Activated sludge is a form of a suspended growth system. This process is an aerobic, continuous (or semicontinuous, fill-and-draw) method used for biological wastewater treatment. It includes carbonaceous oxidation and nitrification (see Figure 5.5). The activated sludge process utilizes all or part of the following:

- Dissolved and colloidal biodegradable organic substances
- Unsettled suspended matter
- Mineral nutrients (e.g., phosphorus and nitrogen compounds)
- Volatile organic materials
- Other materials which can be sorbed on, or entrapped by, the activated sludge floc

Oxygen is provided to the activated sludge to

- Supply needed oxygen for microbial oxidation and synthesis reactions (Vernick and Walker 1981).
- Properly mix the contents of the aeration reactor.
- Maintain microorganisms in suspension.

The activated sludge floc adsorbs suspended and colloidal solids on its surface and, to some extent, soluble organic substances found in wastewater. Microbial activity transforms part of the available organic matter into a reserve food, thus rapidly reducing the BOD. Growth and multiplication of the aerobic microorganisms in the tank form an active biomass termed activated sludge. The activated sludge and wastewater in the aeration unit is called a mixed liquor (American Water Works Association and American Society of Civil Engineers 2012). The rate of removal of organic constituents depends upon the remaining BOD and the concentration of the activated sludge. The mixed culture of microbial species usually found in an activated sludge system includes viruses, bacteria, protozoa, and other organisms, singly or jointly grouped. Generally, the microbes are contained in a fabric of organic debris, dead cells, or other waste products. Factors that influence operation of the process include sensitivity of organisms to nutrients, the wastewater composition, and other conditions (e.g., inorganic salt content, turbulence, pH, temperature, presence of other competing microorganisms, etc.).

### 5.9.3 Activated Sludge Process Kinetics

Kinetics of the activated sludge may be represented by the Monod equation for evaluation of the rate of limiting substrate biological growth. The equation relates the growth rate of organisms to the growth-limiting substrate concentration, as shown in the following equation:

$$\mu = (\mu_s)_{\max} \frac{S^*}{\left(K_S + S^*\right)} \tag{5.54}$$

$\mu_s$ is the growth rate of microorganisms, per day $= \dfrac{DR}{R_u}$

$$\tag{5.55}$$

where:
  DR is the dilution rate
  $R_u$ is the recycling ratio, dimensionless
  $(\mu_s)_{\max}$ is the maximum growth rate of microorganisms, per day
  $K_s$ is the half-velocity constant (substrate concentration expressed in mg/L at half of the maximum growth rate)
  $s^*$ is the growth-limiting substrate concentration in solution, mg/L

When the substrate concentration is low, compared to the half-velocity constant, Equation 5.54 may be simplified into a first-order equation, as presented in the following equation:

$$\mu_s = \frac{(\mu_s)_{\max} * s^*}{K_s} = K * s^* \tag{5.56}$$

where:
  $\mu_s$ is the growth rate of microorganisms
  $(\mu_s)_{\max}$ is the maximum growth rate of microorganisms
  $K_s$ is the half-velocity constant
  $K$ is the constant

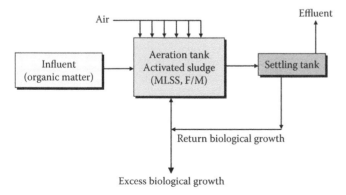

**FIGURE 5.5**  Activated sludge process. MLSS, mixed liquor SS; F/M, food-to-microorganism ratio.

$$K = \frac{(\mu_s)_{max}}{K_s} \qquad (5.57)$$

concentration of the growth-limiting substrate in the system for a dilution rate of 0.75/h.

3. Use the computer program developed in (1) to verify solution obtained in (2).

**Example 5.10**

1. Write a computer program to determine the growth rate of microorganisms (per day) using Monod equation, given the incorporated variables.

2. A fully mixed continuous activated culture with partial feedback of cells is adopted for wastewater treatment. The recycling ratio of the process is 2. Values of the half-velocity constant and maximum growth rate of microorganisms are 1.5 mg/L and 1/h, respectively. Evaluate the

**Solution**

1. For the solution to Example 5.10 (1), see the listing of Program 5.10.

2. Solution to Example 5.10 (2):
   a. Given: $R_u = 2$, $DR = 0.75$, $K_s = 1.5$, $\mu_{max} = 1$.
   b. Find $\mu_s$ as $DR/R_u = 0.75/2 = 0.375$.
   c. Find the concentration of growth-limiting substrate from the Monod equation as follows: $0.375 = 1 * s*/(1.5 + s*)$: therefore, $s* = 0.9$ mg/L.

3. Use the computer program written in (1) to verify computations attained in (2).

---

### LISTING OF PROGRAM 5.10  (CHAP5.10\FORM1.VB): DETERMINATION OF GROWTH RATE OF MICROORGANISMS

```
'***************************************************************************
'Program 5.10: Determination of Growth Rate of Microorganisms
'***************************************************************************

Imports System.Math
Public Class Form1
  Const G = 9.81
  Dim DR, Ru, Kx, Umax As Double
  Dim Ux, S As Double

  Private Sub Form1_Load(ByVal sender As System.Object, ByVal e As System.EventArgs)
Handles MyBase.Load
      Me.Text = "Program 5.10: Growth Rate of Microorganisms"
      Me.MaximizeBox = False
      Me.FormBorderStyle = Windows.Forms.FormBorderStyle.FixedSingle
      Label1.Text = "Enter the dilution rate, DR (/h):"
      Label2.Text = "Enter the recycling ratio, Ru:"
      Label3.Text = "Enter half-velocity constant, Kx (mg/L):"
      Label4.Text = "Enter the maximum growth rate, Umax (/h):"
      Button1.Text = "&Calculate"
      Label5.Text = ""
      Label6.Text = "Decimals:"
      NumericUpDown1.Value = 2
      NumericUpDown1.Maximum = 10
      NumericUpDown1.Minimum = 0
  End Sub

  Sub calculateResults()
      DR = Val(TextBox1.Text)
      Ru = Val(TextBox2.Text)
      kx = Val(TextBox3.Text)
      Umax = Val(TextBox4.Text)
      Ux = DR / Ru
      S = Ux * Kx / (Umax - Ux)
  End Sub

  Sub showResults()
      Label5.Text = "The growth limiting substrate conc. (mg/L): " + FormatNumber(S,
      NumericUpDown1.Value)
  End Sub
```

```
    Private Sub Button1_Click(ByVal sender As System.Object, ByVal e As System.EventArgs)
Handles Button1.Click
        calculateResults()
        showResults()
    End Sub

    Private Sub NumericUpDown1_ValueChanged(ByVal sender As System.Object, ByVal e As
System.EventArgs) Handles NumericUpDown1.ValueChanged
        showResults()
    End Sub
End Class
```

### 5.9.4 Factors Affecting the Activated Sludge Process

The main factors affecting the activated sludge process include volume and sludge loading, sludge age, sludge volume index (SVI), sludge density index (SDI), wastewater aeration time, wastewater flow and quality, mixed liquor SS, dissolved oxygen in reactor, conditions of mixing and turbulence, wastewater temperature, and wastewater solids content (Fair et al. 1966; Mara 1980; Vernick and Walker 1981; James 1984; Peavy et al. 1985; Abdel-Magid 1986; Negulescu 1986; Berger 1987; Steel and McGhee 1991; Rowe and Abdel-Magid 1995; Peirce et al. 1997; Jonathan Ricketts et al. 2003; Jerry and Nathanson 2007; Nemerow et al. 2009; Hammer and Hammer 2011; Abdel-Magid 2012; Howe et al. 2012; Metcalf and Eddy 2013).

#### 5.9.4.1 Volume and Sludge Loadings

The volume and sludge loadings are given by the ratio of food to microorganisms, or *F/M* (also referred to as sludge-loading rate, SLR, or substrate loading). The following equation defines the *F/M* ratio:

$$\frac{F}{M} = \text{SLR}$$

$$= \text{mass of BOD}_5 \text{ input to aeration basin/}$$

$$(\text{MLVSS} * \text{tank volume}) \qquad (5.58)$$

$$= L_i * Q/\text{MLVSS} * V$$

$$= L/\text{MLVSS} * t$$

where:
  *F/M* is the food-to-microorganism ratio, per day
  $L_i$ is the influent BOD concentration, mg/L
  *Q* is the wastewater flow, m³/s
  MLVSS is the mixed liquor volatile suspended solids concentration, mg/L
  *V* is the tank volume, m³
  *t* is the tank retention time, *d*

Operation of an activated sludge unit at a high *F/M* ratio results in an incomplete metabolism of organic matter, inadequate BOD removal, and a lowering of the settling rate. On the contrary, operation of the process at a low *F/M* ratio increases the efficiency of organic matter removal, improves settleability of the activated sludge, and augments BOD removal. The volumetric organic loading, VOL rate, is defined by the following equation:

$$\text{VOL} = Q * L_i/V = \text{SLR} * \text{MLVSS} \qquad (5.59)$$

where:
  VOL is the volumetric organic loading rate
  SLR is the sludge-loading rate, *d*

Table 5.8 outlines suitable values for sludge-loading rate for particular treatment systems.

#### 5.9.4.2 Sludge Age, Mean Cell Residence Time, Solids Retention Time, or Cell Age

Sludge age is defined as the total sludge in the biological treatment process divided by the daily waste sludge. This is presented as follows:

$$\text{SA} = \frac{\left[\text{Mass sludge solids in aeration tank } (\text{kg})\right]}{\left[\text{Mass sludge solids wasted } (\text{kg/day})\right]}$$

**TABLE 5.8**
**SLR for Certain Treatment Units**

| Unit | Reasonable Value of SLR (day⁻¹) |
|---|---|
| Conventional plants | 0.3–0.35 |
| Extended aeration | 0.05–0.2 |
| Step aeration | 0.2–0.5 |

*Sources:* Metcalf and Eddy, Inc., Tchobanoglous, G., Stensel, H. D., Tsuchihashi R., and Burton, F., *Wastewater Engineering: Treatment and Resource Recovery*, 5th ed, McGraw-Hill Science/Engineering/Math, 2013; Abdel-Magid, I. M., *Selected Problems in Wastewater Engineering*, National Research Council, Khartoum University Press, Khartoum, Sudan, 1986; Peirce, J. J. et al. *Environmental Pollution and Control*, 4th ed, Butterworth-Heinemann, Oxford 1997.

SLR, sludge-loading rate.

$$SA = \frac{V * MLSS}{Q_w * SS} \qquad (5.60)$$

where:

SA is the sludge age, days

$V$ is the volume of aeration tank, m$^3$

MLVSS is the concentration of mixed liquor volatile suspended solids, mg/L

$Q_w$ is the waste sludge flow, m$^3$/day

SS is the suspended solids in waste sludge, mg/L

The sludge age also is related to the reciprocal of the net microorganism's specific growth rate, as depicted in the following equation:

$$\frac{1}{SA} = \left[\frac{a(F/M)}{Eff}\right] - k \qquad (5.61)$$

where:

SA is the sludge age, days

$a$ is the cell yield coefficient, mg cell per mg substrate

$F/M$ is the food-to-microorganisms ratio, day$^{-1}$

Eff is the efficiency of BOD removal

$k$ is the microorganisms endogenous decay coefficient, day$^{-1}$

For a completely mixed process operating under steady-state conditions, the sludge age usually is calculated by the relationship presented in the following equation:

$$SA = \frac{V * SS}{Q_w * SS_R + (Q - Q_w) * SS_e} \qquad (5.62)$$

where:

SA is the sludge age, days

$V$ is the volume of aeration tank, m$^3$

$Q_w$ is the waste sludge flow, m$^3$/day

$SS_R$ is the suspended solids in recycle line, mg/L

$Q$ is the influent volumetric flow rate, m$^3$/s

SS is the suspended solids in aeration tank, mg/L

$SS_e$ is the suspended solids in final effluent, mg/L

Table 5.9 gives the value of the mean cell residence time for the various modifications of the activated sludge process.

## Example 5.11

1. Write a computer program to determine the detention time (h), the volumetric organic loading rate (kg BOD/m$^3$/day), and the sludge-loading ratio (kg BOD per kg MLVSS per day), given the flow rate (m$^3$/day), BOD of incoming waste (mg/L), the number of aeration tanks (activated sludge units), and their dimensions (m). Design the program to determine the sludge age for the activated sludge system, given the daily sludge wasted (kg/day).
2. Settling sewage flowing at a daily rate of 5150 m$^3$ is introduced to an activated sludge plant. The sewage has a 5-day BOD of 245 mg/L.

**TABLE 5.9**

**Mean Cell Residence Time for Various Modifications of the Activated Sludge Process**

| Process | (SA) (days) |
| --- | --- |
| Contact stabilization | 5–15 |
| Conventional activated sludge | 5–15 |
| Extended aeration | 20–30 |
| High-rate aeration | 5–10 |
| Modified aeration | 0.2–0.5 |
| Step aeration | 5–15 |

*Sources:* Metcalf and Eddy, Inc., Tchobanoglous, G., Stensel, H. D., Tsuchihashi R., and Burton, F., *Wastewater Engineering: Treatment and Resource Recovery*, 5th ed, McGraw-Hill Science/Engineering/Math, 2013; Abdel-Magid, I. M., *Selected Problems in Wastewater Engineering*, National Research Council, Khartoum University Press, Khartoum, Sudan,1986.

SA, sludge age.

The aeration unit is composed of two aeration tanks each of dimensions: 3 m depth, 8 width, 40 m length, with an MLVSS concentration of 2100 mg/L.

  a. Determine the aeration unit detention time.
  b. Find the volumetric organic loading rate.
  c. Compute the sludge-loading ratio of the aeration unit.
3. Determine the sludge age for the activated sludge system mentioned in (2), given that the daily sludge wasted was 495 kg.
4. Use the program developed in (1) to verify the manual computations of (2) and (3).

## Solution

1. For solution to Example 5.11 (1), see the listing of Program 5.11.
2. Solution to Example 5.11 (2):
    a. Given: $Q$ = 5150 m$^3$/d, $L_1$ = 245 mg/L, $V$ for each tank = 3 * 8*40, MLVSS = 2100, number of tanks = 2.
    b. Determine the volume of the aeration unit = volume of each tank * number of tanks = 3 * 8 * 40 * 2 = 1920 m$^3$.
    c. Find the detention time: $t$ = $V/Q$ = 1920/5150 = 0.37 d = 8.9 h.
    d. Compute the volumetric organic loading rate as follows: VOL = $Q$ * $L/V$ = (5150 m$^3$/day * 245 * 10$^{-3}$ kg/m$^3$)/1920 = 0.66 kg BOD/m$^3$/day.
    e. Determine the sludge-loading ratio, SLR, as follows: $F/M$ = SLR = $L_i$ * $Q$/MLVSS * $V$ = (245 * 10$^{-3}$ kg/m$^3$ * 5150 m$^3$/day)/(2100 * 10$^{-3}$ kg/m$^3$ * 1920 m$^3$) = 0.31 kg BOD/kg MLVSS/day.
3. Solution to Example 6.6 (3):
    a. Given: $Q$ = 5150 m$^3$/day, $L_i$ = 245 mg/L, $V$ for the plant = 1920 m$^3$, MLVSS = 2100 mg/L, $Q_w$ * SS = 495 kg/day.
    b. Compute the SA as follows: SA = $V$ * MLVSS/$Q_w$ * SS = 1920 * 2100 * 10$^{-3}$/495 = 8.1 days.

**LISTING OF PROGRAM 5.11   (CHAP5.11\FORM1.VB): ACTIVATED SLUDGE PROCESS**

```vb
'******************************************************************************
'Program 5.11: Activated Sludge Processes
'******************************************************************************

Imports System.Math
Public Class Form1
  Const G = 9.81
  Dim Q, N, Li, MLVSS As Double
  Dim L(), D(), W() As Double

  Private Sub Form1_Load(ByVal sender As System.Object, ByVal e As System.EventArgs)
Handles MyBase.Load
    Me.Text = "Program 5.11: Activated Sludge Processes"
    Me.MaximizeBox = False
    Me.FormBorderStyle = Windows.Forms.FormBorderStyle.FixedSingle
    Label1.Text = "Enter the flow rate, Q (m3/d):"
    label2.text = "Enter the number of aeration tanks:"
    DataGridView1.Columns.Clear()
    DataGridView1.Columns.Add("LCol", "Length (m)")
    DataGridView1.Columns.Add("WCol", "Width (m)")
    DataGridView1.Columns.Add("DCol", "Depth (m)")
    DataGridView1.AllowUserToAddRows = False
    label3.text = "Enter the 5-day BOD (mg/L):"
    label4.text = "Enter the MLVSS concentration (mg/L):"
    Button1.Text = "&calculate"
  End Sub

  Sub calculateResults()
    Q = Val(textbox1.text)
    N = Val(textbox2.text)
    li = Val(TextBox3.Text)
    mlvss = Val(TextBox4.Text)
    For i = 1 To N
      L(i) = Val(DataGridView1.Rows(i - 1).Cells("LCol").Value)
      D(i) = Val(DataGridView1.Rows(i - 1).Cells("DCol").Value)
      W(i) = Val(DataGridView1.Rows(i - 1).Cells("WCol").Value)
    Next
    Dim V, t, VOL, SLR As Double
    'determine total volume of units
    V = 0.0!
    For I = 1 To N
    V = V + L(I) * W(I) * D(I)
    Next I
    'CONVERT TO KG/L
    Li = Li / 1000
    MLVSS = MLVSS / 1000
    'DETENTION TIME IN HOURS
    t = V / Q * 24
    'VOLUMETRIC ORGANIC LOADING RATE
    VOL = Q * Li / V
    'SLUDGE LOADING RATIO
    SLR = Li * Q / (MLVSS * V)
    Dim s As String
    s = "The detention time = " + FormatNumber(t, 2) + " h" + vbCrLf
    s += "The volumetric organic loading rate = " + FormatNumber(VOL, 2) + " kg
BOD/m3/d" + vbCrLf
    s += "The sludge loading ratio = " + FormatNumber(SLR, 2) + " Kg BOD/Kg MLVSS/d"
    MsgBox(s, vbOKOnly, "Program 6.6: Activated Sludge Processes")
    Dim r = MsgBox("Do you want to compute the sludge age?", vbYesNo, "Prompt")
```

```
      If r = vbYes Then
         Dim DSL = InputBox("Enter the daily sludge wasted (kg):", "Enter DSL", "0")
         Dim SA = V * MLVSS / DSL
         MsgBox("The sludge age for this daily sludge = " + FormatNumber(SA, 2) + " days",
vbOKOnly)
      End If
   End Sub


   Private Sub TextBox2_TextChanged(ByVal sender As System.Object, ByVal e As System.
EventArgs) Handles TextBox2.TextChanged
      If N <> Val(TextBox2.Text) Then
         If Val(TextBox2.Text) <= 0 Then Exit Sub
         N = Val(TextBox2.Text)
         DataGridView1.Rows.Clear()
         DataGridView1.Rows.Add(CInt(N))
         ReDim L(N), D(N), W(N)
      End If
   End Sub


   Private Sub Button1_Click(ByVal sender As System.Object, ByVal e As System.EventArgs)
Handles Button1.Click
      calculateResults()
   End Sub
End Class
```

### 5.9.4.3 Effects of SVI (Mohlman Sludge Volume Index)

The SVI estimates the settleability of the activated sludge, and it monitors the performance and operation of the aeration unit. It is also defined as the volume in milliliters occupied by 1 g of activated sludge mixed liquor solids, dry weight, after settling for 30 min in a 1 L graduated cylinder. Equation 5.63 expresses mathematically the concept of the SVI:

$$SVI = \frac{V_S * 1000}{MLSS} \qquad (5.63)$$

where:
  SVI is the sludge volume index, mL/g
  $V_s$ is the settled volume of sludge in a 1000 mL graduated cylinder in 30 min (mL/L or %)
  MLVSS is the mixed liquor volatile suspended solids, mg/L

Table 5.10 gives SVI values relating to the activated sludge.

### 5.9.4.4 SDI or Donaldson Index

SDI is the reciprocal of the SVI multiplied by 100 as given by the following equation:

$$SDI = \frac{100}{SVI} \qquad (5.64)$$

where:
  SDI is the sludge density index, g/mL
  SVI is the sludge volume index, mL/g

**TABLE 5.10**

**Activated Sludge Settleability according to SVI**

| Value (mL/g) | Criteria |
|---|---|
| <40 | Excellent settling properties |
| 40–75 | Good settling properties |
| 76–120 | Fair settling properties |
| 121–200 | Poor settling properties |
| >200 | Bulking sludge |

*Sources:* Metcalf and Eddy, Inc., Tchobanoglous, G., Stensel, H. D., Tsuchihashi R., and Burton, F., Wastewater Engineering: Treatment and Resource Recovery, 5th ed, McGraw-Hill Science/Engineering/Math, 2013; Abdel-Magid, I. M., Selected Problems in Wastewater Engineering, National Research Council, Khartoum University Press, Khartoum, Sudan, 1986; Ganczarczyk, J. J., Activated Sludge Process: Theory and Practices, Pollution Engineering and Technology Series No. 23, Marcel Dekker, New York, 1983.

SVI, sludge volume index.

SDI varies from around 2 for a good sludge to about 0.3 for a poor sludge, that is, 0.3 < SDI < 2.

### Example 5.12

1. Write a computer program to compute the SVI (mL/g) and the SDI (g/mL) of wastewater in a step aeration plant, given the concentration of the mixed liquor suspended solids as MLSS (mg/L) and the volume of sludge settled after 30 min in a 1 L graduated cylinder, $V_s$ (mL).

2. In a step aeration plant, the concentration of the MLSS was 2400 mg/L. A sample of the wastewater

was withdrawn for a SVI test. The volume of sludge settled after 0.5 min in a 1 L graduated cylinder was found to be 285 mL. Compute the SVI of the wastewater and the SDI.

3. Use the computer program of section (1) to verify the results obtained by manually solving (2).

**Solution**

1. For solution to Example 5.12 (1), see the listing of Program 5.12.
2. Solution to Example 5.12 (2):
   a. Given: $V_s = 285$ mL, MLVSS = 2400 mg/L.
   b. Compute the sludge volume index of the sample as follows: SVI = $V_s$ * 1000/MLVSS = 285 * 1000/2400 = 119 mL/g.
   c. From Table 5.10, it can be concluded that this SVI value signifies fair settling properties.
   d. Determine the SDI as follows: SDI = 100/SVI = 100/119 = 0.84.

Table 5.11 gives the general design criteria of the conventional activated sludge process.

---

### LISTING OF PROGRAM 5.12 (CHAP5.12\FORM1.VB): SLUDGE VOLUME INDEX

```vb
'*******************************************************************************
'Program 5.12: Sludge Volume Index
'*******************************************************************************

Imports System.Math
Public Class Form1
  Const G = 9.81
  Dim MLVSS, Vx As Double
  Dim SVI, SDI As Double
  Dim Comment = ""

  Private Sub Form1_Load(ByVal sender As System.Object, ByVal e As System.EventArgs)
Handles MyBase.Load
    Me.Text = "Program 5.12: Sludge Volume Index"
    Me.MaximizeBox = False
    Me.FormBorderStyle = Windows.Forms.FormBorderStyle.FixedSingle
    Label1.Text = "The MLVSS (mg/L):"
    Label2.Text = "Volume of sludge settling in 1/2 h, Vs (mL):"
    Button1.Text = "&calculate"
    Label3.Text = ""
    Label4.Text = "Decimals:"
    NumericUpDown1.Value = 2
    NumericUpDown1.Minimum = 0
    NumericUpDown1.Maximum = 10
  End Sub

  Sub calculateResults()
    MLVSS = Val(textbox1.text)
    Vx = Val(textbox2.text)
    SVI = Vx * 1000 / MLVSS
    SDI = 100 / SVI
    If SVI <= 40 Then Comment = "Excellent settling properties"
    If SVI > 40 And SVI <= 75 Then Comment = "Good settling properties"
    If SVI > 75 And SVI <= 120 Then Comment = "Fair settling properties"
    If SVI > 120 And SVI <= 200 Then Comment = "Poor settling properties"
    If SVI > 200 Then Comment = "Bulking sludge"
  End Sub

  Sub showResults()
    Label3.Text = "The sludge volume index = " + FormatNumber(SVI, NumericUpDown1.Value)
    + " " + Comment
    Label3.Text += vbCrLf + "The sludge density index = " + FormatNumber(SDI,
    NumericUpDown1.Value)
  End Sub
```

```
    Private Sub Button1_Click(ByVal sender As System.Object, ByVal e As System.EventArgs)
Handles Button1.Click
        calculateResults()
        showResults()
    End Sub

    Private Sub NumericUpDown1_ValueChanged(ByVal sender As System.Object, ByVal e As
System.EventArgs) Handles NumericUpDown1.ValueChanged
        showResults()
    End Sub
End Class
```

**TABLE 5.11**
**Design Criteria for Conventional Activated Sludge**

| Item | Value |
|---|---|
| MLSS (mg/L) | 1500–3000 |
| Volumetric organic loading (g BOD per m³ per day) | 500–700 |
| Aeration detention time (h) (based on average daily flow) | 4–8 |
| F/M ratio (g BOD per g MLSS per d) | 0.1–0.6 |
| Sludge retention time (d) | 5–15 |
| Sludge age (days) | 3–4 |
| Recycling ratio | 0.25–0.5 |
| Sludge yield index (kg solids per kg BOD removed) | 0.7–0.9 |
| BOS5 removal (%) | 85–95 |
| Optimum pH for aerobic bacterial growth | 6.5–7.5 |
| Depth of aeration tanks (m) | 3 |

*Sources:* Metcalf and Eddy, Inc., Tchobanoglous, G., Stensel, H. D., Tsuchihashi R., and Burton, F., *Wastewater Engineering: Treatment and Resource Recovery*, 5th ed, McGraw-Hill Science/Engineering/Math, 2013; Barnes, D. et al. *Water and Wastewater Engineering Systems*, Pitman Publishing, Marshfield, MA, 1981; Abdel-Magid, I. M., Selected Problems in Wastewater Engineering, National Research Council, Khartoum University Press, Khartoum, Sudan, 1986; Ganczarczyk, J. J., Activated Sludge Process: Theory and Practices, Pollution Engineering and Technology Series No. 23, Marcel Dekker, New York, 1983; Vernick, A. S., and Walker, F. C., *Handbook of Wastewater Treatment Processes*, Pollution Engineering and Technology Series No. 19, Marcel Dekker, New York, 1981.

MLSS, mixed liquor suspended solids; BOD, biological oxygen demand; *F/M*, food-to-microorganism ratio.

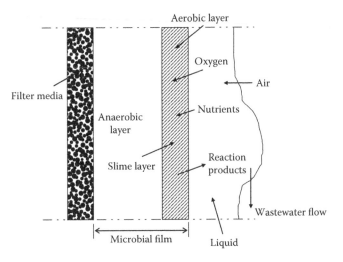

**FIGURE 5.6**   Schematic diagram of a trickling filter process.

## 5.10   ATTACHED GROWTH TREATMENT PROCESSES: TRICKLING FILTER

### 5.10.1   INTRODUCTION

An attached growth culture is used to treat wastewater that comes in contact with microorganisms attached to surfaces of the media of the reactors. The trickling filtration process is a form of an attached growth system. In this process, settled wastewater is brought in contact with microorganisms attached to the filter bed media (see Figure 5.6). Wastewater is sprayed over a fixed media composed of a bed often packed with rocks or plastic structures. Thus, biological slimes develop and coat the filter surface area. These slimes provide a film consisting of heterotrophic organisms, facultative bacteria, protozoa, fungi, algae, rotifers, sludge worms, insect larvae, and snails. As the wastewater trickles over the media, microbes in the slime layer extract organic and inorganic matter from the liquid film. Suspended and colloidal particles are retained on the surfaces to be further converted to soluble products. Aerobic reactions are sustained by oxygen from the gas phase in the pores of the media through the liquid film to the slime layer. The digested waste products diffuse outward with the water flow or air currents through the voids in the filter medium. With time, the biological film increases in thickness. This increase in film thickness is accompanied by a lowering of the content of oxygen and food. This condition eventually leads to anaerobic and endogenous metabolism at the slime-media interface, with increased detachment of organisms from the media. The shearing action of the flowing wastewater across the file then detaches and washes away the slime from the media. This process is referred to as filter sloughing off. Sloughing off develops because the microorganisms grow

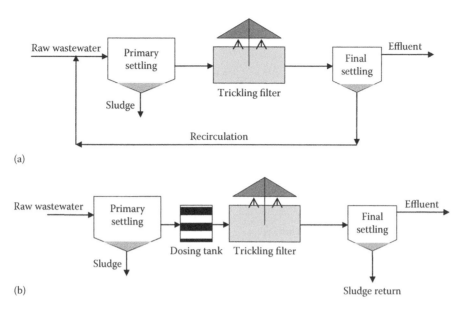

**FIGURE 5.7**   (a) High- and (b) low-rate trickling filters.

and multiply, and the slime layer gets thicker until it is washed from the media by the incoming wastewater.

Trickling filters are classified based on the hydraulic loading rate (shear velocities) and organic loading rate (metabolism in slime layer). As such, trickling filters are grouped into low-rate and high-rate filters (see Figure 5.7). Low-rate filters are rather simple, having low hydraulic loading rates, and do not incorporate the recycling of wastewater effluent. A constant hydraulic loading is maintained in low-rate filters by a dosing siphon or by a suction level-controlled pump. However, high-rate filters have a high hydraulic loading maintained by recirculating part of the effluent from the filter or part of the final effluent.

### 5.10.2   Recirculation to the Trickling Filter

Recirculation of a portion of the treated effluent to the trickling filter is performed to

- Increase biological solids in the system.
- Provide continuous seeding with recirculated sloughed-off solids.
- Maintain uniform hydraulic load through filter.
- Attain continuous rotation of the distribution arm.
- Establish uniform organic load.
- Dilute influent for production of better quality effluent.
- Decrease thickness of biological slime layer.
- Improve removal efficiency of pollutants.

The recirculation ratio ($R_u$) lies between a value of 50% and 1000% of the raw wastewater flow. Usually, the recirculation ratio is between 50% and 300% (Steel and McGhee 1991).

### 5.10.3   Efficiency of a Trickling Filter (BOD Removal Efficiency)

Computations of efficiency of a trickling filter are generally based on mathematical models and empirical formulae, which are described as follows:

1. *Mathematical models*: Mathematical models usually assume the existence of uniform biological layer and even distribution of load within the trickling filter bed media. From an operational viewpoint, mathematical models are impractical for design purposes.
2. *Empirical formulae*: Empirical formulae are based on operational data collected from existing treatment plants. The collected data were analyzed to produce a design procedure.

Some of the most important formulae used in the design of trickling filters include the National Research Council (NRC), Velz, Rankin, and Rumpf formulae.

### 5.10.3.1   NRC Formula

The NRC formula is based on wastewater treatment plants in U.S. military installations. Assumptions incorporated in the formula are as follows:

- Oxygen is the rate-limiting factor with a soluble BOD in excess of 40 mg/L.
- Wastewater is not extremely dilute.
- The trickling filter is followed by a final settling tank or a secondary clarifier.
- Trickling filters treat settled domestic wastewater at a temperature of 20°C (temperatures different from this value have to be corrected for).

- The NRC formula includes the effect of secondary clarifiers.
- The maximum practical rate of recirculation predicted through the formula is 800%.

Equation 5.65 presents the NRC formula for a first-stage trickling filter:

$$\text{Eff}_1 = \left(\frac{L_e - L_i}{L_i}\right) = \frac{100}{1 + 100\sqrt{(W_1/V_1 F_1)}} \quad (5.65)$$

where:

Eff$_1$ is the efficiency of the first-stage trickling filter, %
$L_i$ is the influent BOD, mg/L
$L_e$ is the effluent BOD from the first stage, mg/L
$W_1$ is the BOD load of the first-stage filter, kg/day

$$W_1 = Q * L_i = \text{BOD}_{\text{load}} \quad (5.66)$$

where:

$Q$ is the flow of wastewater, m³/day
$V_1$ is the volume of first-stage filter media, m³
$F_1$ is the first-stage recirculation factor

$$F_1 = \frac{(1 + R_u)}{(1 + 0.1 * R_u)^2} \quad (5.67)$$

$$R_u = \frac{\text{Recirculated flow }(Q_R)}{\text{wastewater flow }(Q)}$$

$$= \frac{Q_R}{Q} \quad (5.68)$$

Equation 5.69 gives the NRC equation for a second-stage trickling filter:

$$\text{Eff}_2 = \left(\frac{L_i - L_e}{L_i}\right)_2 = \frac{100}{1 + \dfrac{0.44\sqrt{[W_2/(V_2 F_2)]}}{1 - \text{Eff}_1}} \quad (5.69)$$

where:

Eff$_2$ is the efficiency of the second-stage trickling filter, %
$W_2$ is the BOD load of the second-stage filter = $Q*L_i$ = BOD$_{\text{load}}$, kg/day
$V_2$ is the volume of the second-stage filter, m³
$F_2$ is the second-stage recirculation factor
$L_{i2}$ is the influent BOD to the second-stage filter, mg/L
$L_{e2}$ is the effluent BOD of the second-stage filter, mg/L

### 5.10.3.2 Velz Formula

The Velz formula is valid for BOD removals of ≤ 90%. The formula can be expressed as indicated in the following equation:

$$L_e = \frac{(L_i + R_u * L_e)e^{-kh}}{1 + R_u} \quad (5.70)$$

where:

$L_i$ is the influent BOD, mg/L
$L_e$ is the effluent BOD, mg/L
$k$ is the experimental coefficient
= 0.49 for high-rate filters
= 0.57 for low-rate filters
$h$ is the filter depth, m
$R_u$ is the recirculated flow/wastewater flow

### 5.10.3.3 Rankin Formula

The Rankin formula (Equation 5.71) may be used for design of a single-stage filter:

$$L_e = \frac{L_i}{3 + 2R_u} \quad (5.71)$$

where:

$L_e$ is the BOD of the settled filter effluent, mg/L
$L_i$ is the BOD of the settled sewage, mg/L
$R_u$ is the recirculation ratio

The Rankin formula applies to all plants treating sewage that have presedimentation, high-rate filtration, and final settling tanks. This applies when the BOD does not exceed the value of 0.7 kg/m³/day and where recirculation, if applied, maintains a dosing rate between 93 and 244 * 10³/ha/day (O'Connor and Dobbins 1956).

### 5.10.3.4 Rumpf Formula

The Rumpf equation can be used to evaluate a trickling filter's efficiency, as presented in the following equation:

$$E = 93 - \frac{0.017W}{V} \quad (5.72)$$

where:

E is the trickling filter efficiency, %
W is the organic loading of the trickling filter, g BOD per day
V is the volume of the filter media, m³

### 5.10.3.5 Eckenfelder Equation

Eckenfelder equation is depicted as follows:

$$\frac{L_i}{L_e} = \frac{1}{\left(1 + \dfrac{18.6h^{0.67}}{\sqrt{Q/A_s}}\right)} \quad (5.73)$$

where:

$L_i$ is the influent BOD including return flow, mg/L
$L_e$ is the nonsettled BOD of flow to filter, mg/L
$h$ is the filter depth, m
$Q$ is the influent flow, mL/day
$A_s$ is the area, ha

### 5.10.3.6 Galler and Gotaas Equation

An equation has been developed to design filter based on multiple regression analysis for data adopted from existing stations taking in to consideration return flow, hydraulic load, filter depth, and wastewater temperature as represented in the following equation:

$$L_e = \frac{0.182\,K\,(Q_i * L_i + Q_r * L_e)^{1.19}}{(1+r)^{0.78} * (h+0.305)^{0.67}\,(d/2)^{0.25}} \quad (5.74)$$

where:

$L_e$ is the nonsettling BOD for effluent from filter, mg/L
$K$ is the constant

$$K = \frac{9.731}{Q_i^{0.28}\,T^{0.15}} \quad (5.75)$$

$Q_i$ is the inflow, mL/day
$L_i$ is the influent BOD to filter, mg/L
$Q_r$ is the recirculation or return flow, mL/day
$r$ is the ratio of recirculation (maximum value is 4:1)
$h$ is the filter depth, m
$(d/2)$ is the radius of filter, m
$T$ is the wastewater temperature, °C

### 5.10.4 Overall Treatment Plant Efficiency

Many factors affect the operation and performance of a trickling filter, including organic loading, hydraulic flow rate, wastewater characteristics, rate of diffusability of food and air to the biological slime layer, temperature, and so on. The overall treatment plant efficiency of a system comprising a two-stage filter can be determined by using the following equation (Viessman et al. 2008; Hammer and Hammer 2011):

$$\text{Eff}_{over} = 100 - 100 * \left[ (1 - \text{Eff}_{sed})(1 - \text{Eff}_1)(1 - \text{Eff}_2) \right] \quad (5.76)$$

where:

$\text{Eff}_{over}$ is the overall treatment plant efficiency, %
$\text{Eff}_{sed}$ is the percentage of BOD removed in primary settling (usually is equal to 35%)
$\text{Eff}_1$ is the BOD efficiency of the first-stage filter and intermediate clarifier corrected for the operating temperature, %
$\text{Eff}_2$ is the BOD efficiency of the second-stage filter and final clarifier corrected for the operating temperature, %

Temperature correction can be carried out as shown by the following equation:

$$\text{Eff}_T = \text{Eff}_{20} * (T_c)^{T-20} \quad (5.77)$$

where:

$\text{Eff}_T$ is the efficiency at temperature, $T\%$
$\text{Eff}_{20}$ is the efficiency at a temperature of 20°C, %
$T$ is the temperature, °C
$T_c$ is the temperature correction factor, usually taken as equal to 1.035

### Example 5.13

1. Write a computer program to compute the trickling filter efficiency using different formulae (NRC, Velz, Rumpf, and Rankin), given the population number, the rate of population growth, the clarifier efficiency, the rate of wastewater flow to plant (m³/s), 5-day BOD (mg/L), the recirculation ratio, and the filter diameter (m). Design the program so that the trickling filters determine the volume, area, diameter, or depth, and efficiency can be estimated at any temperature.
2. A treatment work includes primary sedimentation and trickling filtration to treat wastewater of 8,500 inhabitants. The sedimentation unit has an efficiency of 35%. The wastewater flows at a rate of 325 L/c/day with a 5-day BOD of 310 mg/L. The trickling filter has the following characteristics: removal efficiency = 82%, recirculation ratio = 3:1, and filter depth = 2.5 m. Determine the trickling filter diameter using the NRC formula.
3. Use the computer program of (1) to confirm computations in (2).

### Solution

1. For solution to Example 5.13 (1), see the listing of Program 5.13.
2. Solution to Example 5.13 (2):
   a. Given: POP = 8500, efficiency of sedimentation = 35%, q = 325 L/c/day, $L_i$ = 310 mg/L, efficiency of filter = 82%, $R_u$ = 3/l, h = 2.5 m.
   b. Find the recirculation factor as follows: $F = (1 + R_u)/(1 + 0.1*R_u)^2 = (1 + 3)/(1 + 0.1 * 3)^2 = 2.37$.
   c. Find the wastewater flow: $Q = q * \text{POP} = 325 * 10^{-3} * 8500 = 2762.5$ m³/day.
   d. Find the BOD concentration emerging from the primary sedimentation and entering the trickling filter as follows: $L_i = \text{BOD}_i * (1 - \text{Eff}_{clarifier}) = 310 * (1 - 0.35) = 201.5$ mg/L.
   e. Determine the influent BOD load to the filter as follows: $W = Q * L_i = 2762.5 * 201.5 * 10^{-3} = 556.6$ kg/day.
   f. Find the volume of the filter using the NRC efficiency equation as follows: $\text{Eff} = 100/[1 + 0.44(W/V * F)^{0.5}] = 82 = 100/[1 + 0.44(556.6/V * 2.37)^{0.5}]$. This yields $V = 944$ m³.
   g. Compute the surface area as follows: $A = V/h = 944/2.5 = 377.6$ m².
   h. Find the diameter of the filter as follows: $D = (4 * A/\pi)^{0.5} = (4 * 377.6/\pi)^{0.5} = 22$ m.

### LISTING OF PROGRAM 5.13 (CHAP5.13\FORM1.VB): TRICKLING FILTER EFFICIENCY

```vb
'*******************************************************************************
'Program 5.13: Calculation of Trickling Filter Efficiency
'*******************************************************************************

Imports System.Math
Public Class Form1
    Const G = 9.81
    Dim Li, Le, Q, Qr, Ru, V, Eff, W, F1 As Double
    Dim K, h, A, POP, rate, BOD5, Qc, Effr As Double

    Private Sub Form1_Load(ByVal sender As System.Object, ByVal e As System.EventArgs)
Handles MyBase.Load
        Me.Text = "Program 5.13: Trickling Filter Efficiency"
        Me.MaximizeBox = False
        Me.FormBorderStyle = Windows.Forms.FormBorderStyle.FixedSingle
        Label1.Text = "To compute Trickling Filter Efficiency using:"
        ListBox1.Items.Clear()
        ListBox1.Items.Add("National Research Council formula")
        ListBox1.Items.Add("Velz formula")
        ListBox1.Items.Add("Rankin formula")
        ListBox1.Items.Add("Rumpf formula")
        ListBox1.Items.Add("Design a filter using NRC formula")
        ListBox1.SelectedIndex = 0
        RadioButton1.Text = "Yes"
        RadioButton2.Text = "No"
        RadioButton2.Checked = True
        Label11.Text = ""
        Button1.Text = "&Calculate"
    End Sub

    Private Sub ListBox1_SelectedIndexChanged(ByVal sender As System.Object, ByVal e As
System.EventArgs) Handles ListBox1.SelectedIndexChanged
        Select Case ListBox1.SelectedIndex
            Case 0
                'National Research Council formula
                Label2.Text = "Enter the influent BOD, Li (mg/L):"
                Label3.Text = "Enter the effluent BOD, Le (mg/L) (0 if unknown):"
                Label4.Text = "Enter the wastewater flow, Q (m3/day):"
                Label5.Text = "Enter the recirculation ratio, Ru:"
                Label6.Text = "Enter the volume of filter, V (m3):"
                RadioButton1.Visible = False
                RadioButton2.Visible = False
                TextBox3.Visible = True
                TextBox4.Visible = True
                TextBox5.Visible = True
                TextBox6.Visible = False
                TextBox7.Visible = False
                TextBox8.Visible = False
                Label4.Visible = True
                Label5.Visible = True
                Label6.Visible = True
                Label7.Visible = False
                Label8.Visible = False
                Label9.Visible = False
                Label10.Visible = False
```

```
       Case 1
         'Velz formula
         Label2.Text = "Enter the influent BOD, Li (mg/L):"
         Label3.Text = "Enter the recirculated flow, Ru (m3/day):"
         Label4.Text = "Enter the filter depth, h (m):"
         Label7.Text = "Is it a high-rate filter?"
         Label7.Visible = True
         RadioButton1.Visible = True
         RadioButton2.Visible = True
         Label5.Visible = False
         Label6.Visible = False
         Label8.Visible = False
         Label9.Visible = False
         Label10.Visible = False
         TextBox4.Visible = False
         TextBox5.Visible = False
         TextBox7.Visible = False
         TextBox6.Visible = False
         TextBox8.Visible = False
       Case 2
         'Rankin formula
         Label2.Text = "Enter the influent BOD, Li (mg/L):"
         Label3.Text = "Enter the recirculated flow, Ru (m3/day):"
         RadioButton1.Visible = False
         RadioButton2.Visible = False
         Label4.Visible = False
         Label5.Visible = False
         Label6.Visible = False
         Label7.Visible = False
         Label8.Visible = False
         Label9.Visible = False
         Label10.Visible = False
         TextBox3.Visible = False
         TextBox4.Visible = False
         TextBox5.Visible = False
         TextBox6.Visible = False
         TextBox7.Visible = False
         TextBox8.Visible = False
       Case 3
         'Rumpf formula
         Label2.Text = "Enter the organic load of filter, W (g BOD/day):"
         Label3.Text = "Enter the volume of the filter, V (m3):"
         RadioButton1.Visible = False
         RadioButton2.Visible = False
         Label4.Visible = False
         Label5.Visible = False
         Label6.Visible = False
         Label7.Visible = False
         Label8.Visible = False
         Label9.Visible = False
         Label10.Visible = False
         TextBox3.Visible = False
         TextBox4.Visible = False
         TextBox5.Visible = False
         TextBox6.Visible = False
         TextBox7.Visible = False
         TextBox8.Visible = False
```

```
      Case 4
        Label2.Text = "Enter the population number:"
        Label3.Text = "Enter the rate of population growth, 0 if unknown:"
        Label4.Text = "Enter the 5-day BOD, mg/L:"
        Label5.Text = "Enter the recirculation ratio Ru:"
        Label6.Text = "Enter the flow rate of wastewater, Q (m3/c/day):"
        Label8.Text = "Enter the efficiency of sedimentation (%):"
        Label9.Text = "Enter the removal efficiency (%):"
        Label10.Text = "Enter the filter depth, h (m):"
        RadioButton1.Visible = False
        RadioButton2.Visible = False
        TextBox3.Visible = True
        TextBox4.Visible = True
        TextBox5.Visible = True
        TextBox6.Visible = True
        TextBox7.Visible = True
        TextBox8.Visible = True
        Label4.Visible = True
        Label5.Visible = True
        Label6.Visible = True
        Label7.Visible = False
        Label8.Visible = True
        Label9.Visible = True
        Label10.Visible = True
    End Select
End Sub

Sub calculateResults()
    Select Case ListBox1.SelectedIndex
      'National Research Council formula
      Case 0
        Li = Val(TextBox1.Text)
        Le = Val(TextBox2.Text)
        If Le = 0 Then
          Q = Val(TextBox3.Text)
          Ru = Val(TextBox4.Text)
          V = Val(TextBox5.Text)
          W = Q * Li
          F1 = (1 + Ru) / ((1 + 0.1 * Ru) ^ 2)
          Eff = 100 / (1 + 0.44 * (W / (V * F1)) ^ 0.5)
        Else
          Eff = 100 * (Li - Le) / Li
        End If
      Case 1
        'Velz formula
        Li = Val(TextBox1.Text)
        Ru = Val(TextBox2.Text)
        'is it a high rate filter?
        If RadioButton1.Checked Then
          K = 0.49
        Else : K = 0.57
        End If
        h = Val(TextBox3.Text)
        A = (1 + Ru) * Exp(K * h)
        Le = Li / (A - Ru)
        Eff = 100 * (Li - Le) / Li
      Case 2
        'Rankin formula
        Li = Val(TextBox1.Text)
```

```
                  Ru = Val(TextBox2.Text)
                  Le = Li / (3 + (2 * Ru))
                  Eff = 100 * (Li - Le) / Li
              Case 3
                'Rumpf formula
                W = Val(TextBox1.Text)
                V = Val(TextBox2.Text)
                Eff = 93 - (0.017 * W / V)
              Case 4
                POP = Val(TextBox1.Text)
                rate = Val(TextBox2.Text)
                BOD5 = Val(TextBox3.Text)
                Ru = Val(TextBox4.Text)
                Qc = Val(TextBox5.Text) / 1000
                Eff = Val(TextBox6.Text)
                Effr = Val(TextBox7.Text)
                h = Val(TextBox8.Text)
                If Eff < 1 Then Eff = Eff * 100
                If Effr < 1 Then Effr = Effr * 100
                'recirculation factor
                Dim F = (1 + Ru) / (1 + 0.1 * Ru) ^ 2
                'wastewater flow
                Q = Qc * POP
                'BOD concentration entering trickling filter
                Li = BOD5 * (1 - (Eff / 100))
                'influent BOD load to the filter
                W = Q * Li / 1000
                'volume of filter by NCR formula
                V = W / (F * ((100 / Effr - 1) / 0.44) ^ 2)
                'surface area
                A = V / h
                'diameter of filter
                Dim D = (4 * A / PI) ^ 0.5
                Label11.Text = "The diameter of the filter (m) = " + FormatNumber(D, 2)
          End Select
          If ListBox1.SelectedIndex <> 4 Then
            Label11.Text = "Efficiency = " + FormatNumber(Eff, 2) + " %"
          End If
      End Sub


    Private Sub Button1_Click(ByVal sender As System.Object, ByVal e As System.EventArgs)
Handles Button1.Click
        calculateResults()
      End Sub
End Class
```

## 5.10.5 Trickling Filter Clarifier

Final or secondary clarifiers are required subsequent to trickling filtration in order to collect the relatively large particles of the sloughed off biological slime solids or humus. In the design of secondary clarifiers, the following parameters need to be considered (Steel and McGhee 1991):

- Neither thickening nor hindered settling is assumed to develop in the secondary clarifier.

- Settling design criteria are based upon the particle size and density.
- Surface overflow rate ($v_s$) lies between 25 and 33 m/day at average flow rates, not to exceed the value of 50 m/day at peak flow rates.
- Weir loading rates are in the range of 120–370 $m^3$/day/m of the weir length at peak flow rate.

Table 5.12 shows the typical design information for trickling filter.

**TABLE 5.12**
**Typical Design Information for Trickling Filter**

| Parameter | Low-Rate Filter | High-Rate Filter | Roughing Filter |
|---|---|---|---|
| Organic loading (kg/m$^3$/day) | 0.07–0.32 | 0.32–1.0 | 0.8–6.0 |
| Hydraulic loading (m$^3$/m$^2$/day) | 1–4 | 10–40 | 30–200 |
| Depth (m) | 1.5–3(2) | 1–2(2) | 4.5–12 |
| Recirculation ratio | 0 | 0.5–3 | Varies |
| Filter media | Crushed stone, gravel, slag, synthetic material, and so on | Crushed stone slag, gravel, rock, and so on | Synthetic materials, red wood, and so on |
| Power requirement (kW/1000 m$^3$) | 2–4 | 6–10 | 10–20 |
| Sloughing | Intermittent | Continuous | Continuous |
| Dosing intervals | Not greater than 5 min (generally intermittent) | Not greater than 15 s (Continuous) | Continuous |
| Filter flies | Many | Few, larvae are washed away | |

## 5.11 COMBINED SUSPENDED AND ATTACHED GROWTH SYSTEMS WSP, LAGOON, OR OXIDATION POND

### 5.11.1 INTRODUCTION

WSPs are shallow earthen basins with a controlled shape or a natural depression that receives wastewater and retains it so that biological processes can proceed at acceptable levels. They can be classified according to the type of biological activity taking place, such as anaerobic, facultative, or aerobic.

Anaerobic WSPs are deep ponds devoid of oxygen except for a relatively thin surface layer. This type of pond receives wastewater with a high organic load or wastewater with a high level of solids (i.e., not presettled in a settling unit operation such as a septic tank). The objectives of an anaerobic WSP include settling of solids, partial treatment of wastewater, and discharge of the effluent to a facultative pond. The main biological action here is achieved by anaerobic organisms. Usually, anaerobic ponds are used in series with facultative (or aerobic–anaerobic) ponds to provide complete treatment.

In facultative WSPs, aerobic conditions prevail in the upper surface mainly by using oxygen produced by algal organisms and to a lesser extent oxygen from the atmosphere. Anaerobic conditions develop at the bottom of the pond due to stagnant conditions. The depth of the aerobic and anaerobic zones is a function of mixing conditions, wind action, sunlight penetration, and the hours of daylight. Algae and bacteria are responsible for biodegradation of organic matter in the aerobic layers of the pond. Bacteria use oxygen for oxidation of waste organics to synthesize new cells and produce stable end products (such as carbon dioxide, nitrates, and phosphates). These inorganic compounds are used by algae, in the presence of sunlight, to yield new cells with end products such as oxygen. The oxygen produced is then used by the bacteria. This mutual benefit between bacteria and algae is termed

symbiotic relationship. The biological solids in the aerobic zone, together with heavy solids, settle to the anaerobic zone in the pond. These solids provide the needed food for the benthic anaerobic organisms. Solids in the anaerobic zone are then converted to organic acids and gases yet to be released in a soluble form to organisms in the aerobic zone of the pond. A facultative pond receives wastewater from a sewer system or from an anaerobic pond. Effluent is then retained for some days, and the treated wastewater is discharged to a polishing pond. Figure 5.8 illustrates a schematic diagram of a facultative WSP.

Tertiary (maturation) WSPs receive treated wastewater from facultative ponds and retain it for a period of time to improve its quality under aerobic conditions.

The effluent from a maturation pond can be used for landscape or crop irrigation. Maturation ponds can be used to grow fish and aquatic birds. Aerobic ponds are used primarily for the treatment of soluble organic wastes and effluents from wastewater treatment plants.

WSP can be operated in series or in parallel or can function on an individual basis. In the series arrangement of ponds, the wastewater or effluent is introduced from one pond to the next. Each pond in series discharges an effluent of a better quality than the one preceding it. Thus, this arrangement helps in treating unsettled raw sewage and improves the quality of the final effluent. In a parallel arrangement, ponds are placed adjacent to each other. The ponds simultaneously receive effluent from the same source and concurrently discharge the treated sewage to the same outlet or discharge point. The quality of the discharge effluent from parallel ponds is comparable. The advantage of such an arrangement is that during maintenance and repair programs, one pond can remain in operation.

An individual WSP system of ponds can be used, but a series of ponds are preferred. The series system of ponds can be selected for different reasons, such as obtaining higher treatment level, growing fish, or cleaning ponds without shutting off the collecting sewer system.

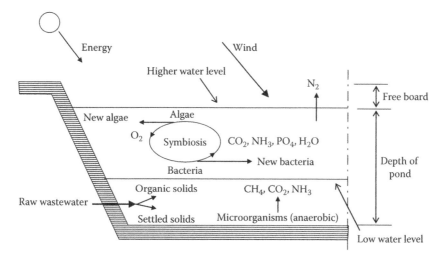

**FIGURE 5.8**  Schematic diagram of a facultative WSP.

## 5.11.2  WSP Design

The design of a WSP is based on such various interrelated parameters as follows:

1. *Site selection*: Gravity flow is practical in order to minimize pumping. Pumping of wastewater will increase the cost, use energy, and also require added operational and maintenance efforts.
2. *Soil*: The soil needs to be fairly impermeable, not sandy or gravely (to avoid seepage and groundwater contamination), and must be easy to excavate. Adequate soil provides material for constructing the embankments.
3. *Drainage*: Good drainage is required to provide for the discharge of the treated wastewater from the pond.
4. *Size*: The size provided for a WSP must be large enough to accommodate the anticipated flows.
5. *Layout*: Ponds must be located at an appropriate distance from the neighboring houses and dwellings. The distance between the pond site and the nearest dwelling must be in excess of 200 m. The greater the distance between the ponds and the community, the better the situation.
6. *Wind direction*: Ponds must be located in a down-wind direction from a community to avoid airborne pollution problems.

In the design of WSP, the following parameters can be used:

### 5.11.2.1  General

- The expected daily flow of wastewater into the WSP is reported as $Q$, L/day.
- The average annual water temperature in the WSP is reported as $T$, °C.
- The organic loading rate of the wastewater effluent is given as $OL$, g/L.

- The minimum required area of WSP can then be calculated from the relationship given in the following equation:

$$A = OL * Q / OL_{max} \tag{5.78}$$

where:
   $A$ is the minimum required area of WSP, m²
   $OL$ is the organic loading rate of the effluent, g/L
   $Q$ is the daily flow of sewage into the WAP, L/day

$$OL_{max} = \text{Maximum allowable organic load, g/m2/d} \tag{5.79}$$
$$= 2 * T - 12$$

$T$ is the average annual water temperature, °C

- For a rectangular WSP, the dimensional relationship of the following equation can be used:

$$l = [2 - 3] * B \tag{5.80}$$

where:
   $l$ is the length of the pond, m
   $B$ is the width of the pond, m

- The depth of a WSP usually lies between 1 and 3 m, depending on the type and volume of wastewater, climatic conditions, and WSP classification.

### 5.11.2.2  Completely Mixed Conditions

- Assuming complete mixing conditions without solid recycling (i.e., a facultative pond), the relationship of the mass balance between the BOD entering and that leaving a WSP is given by the following equation:

$$BOD_{in} = BOD_{out} + BOD_{consumed} \qquad (5.81)$$

$$Q * L_i = Q * L_e + V * k_p L_e$$

where:

Q is the wastewater flow rate to the pond, m³/s
$L_i$ is the influent BOD, mg/L
$L_e$ is the effluent BOD, mg/L
V is the volume, m³
$k_p$ is the removal rate constant for WSP, day⁻¹

- The mathematical relationship for a single WSP may be derived upon rearrangement of Equation 6.41, as presented in the following equation:

$$\frac{L_e}{L_i} = \frac{1}{\left[1+\left(k_p * \dfrac{V}{Q}\right)\right]} = \frac{1}{\left(1+k_p * t\right)} \qquad (5.82)$$

where:

$L_e$ is the effluent BOD, mg/L
$L_i$ is the influent BOD, mg/L
$K_p$ is the removal rate constant for WSP, day⁻¹
V is the volume, m³
Q is the wastewater flow rate to the pond, m³/s
t is the detention time, days

- For more than one WSP arranged in series, the effluent of one pond becomes influent to the next. Thus, a substrate balance written across a series of N ponds yields the following equation:

$$\frac{L_e}{L_i} = \frac{1}{\left(1+\dfrac{k_p t}{N}\right)^N} \qquad (5.83)$$

where:

$L_e$ is the effluent BOD, mg/L
$L_i$ is the influent BOD, mg/L
t is the retention time, days
$k_p$ is the removal rate constant for WSP, day⁻¹
N is the number of ponds

The variation of the removal rate constant with the temperature can be calculated by the following equation:

$$\left(k_p\right)_T = \left(k_p\right)_{20} * \left(T_c\right)^{(T-20)} \qquad (5.84)$$

where:

$(k_p)_T$ is the removal rate constant for WSP at temperature T, day⁻¹
$(k_p)_{20}$ is the removal rate constant for WSP at a temperature of 20°C, day⁻¹. (It relates to

degradability of waste organics, temperature, and completeness of aeration mixing. It ranges between 0.3 and 1 per day to base e and at 20°C.) (Hammer and Hammer 2011)
$T_c$ is the temperature constant (a function of biodegradability, ranging from 1.03 to 1.12, usually taken equal to 1.035)
n is the number of WSP

- The value of the detention time can be found from the relationship shown in the following equation:

$$t = \frac{V}{Q} \qquad (5.85)$$

where:

V is the volume of the pond, m³
Q is the rate of wastewater flow, m³/s

- The depth of the WSP may be assumed, and the surface area may be determined from the following equation:

$$A = \frac{V}{h} \qquad (5.86)$$

where:

A is the area of WSP, m²
h is the depth of WSP, m

## Example 5.14

1. Write a computer program for the design of a WSP given the 5-day BOD of wastewater (mg/L), the flow rate (m³/s), the lowest temperature (°C), the removal rate constant for the pond (d⁻¹), at any temperature (°C), the temperature correction factor, the effluent 5-day BOD from the pond (mg/L), and the depth of pond (m).
2. Wastewater with a 5-day BOD of 200 mg/L and flowing at a rate of 1.6 m³/min is to be treated in a WSP. Determine the surface area of the pond, given the following data:
   a. Lowest temperature = 23°C
   b. Removal rate constant for the pond = 0.25/day at 20°C, temperature correction factor = 1.05
   c. Effluent 5-day BOD from the pond = 25 mg/L
   d. Depth of pond = 1.4 m
3. Use the program established in (1) to verify computations obtained in (2).

## Solution

1. For solution to Example 5.14 (1), see the listing of Program 5.14.
2. Solution to Example 5.14 (2):
   a. Given: Q = 1.6 m³/min, $L_i$ = 200 mg/L, $L_e$ = 25 mg/L, $(k_p)_{20}$ = 0.25/day, h = 1.4 m, $T_{min}$ = 23°C, $T_c$ = 1.05.

b. Find the removal rate constant for the pond at the lowest temperature of 25°C as follows: $(k_p)_{25} = (k_p)_{20}(1.05)^{(T-20)} = 0.25(1.05)^{(23-20)} = 0.29$/day.

c. Find the detention time of the pond as follows: $L_e/L_i = 1/(1 + k_p * t)$. Use equation to find detention time: $25/200 = 1/(1 + 0.29 * t)$. This yields detention time of $t = 24.1875$ day.

d. Compute the volume of the WSP as follows: $V = t * Q = 24.1875 * 1.6 * 60 * 24 = 55727.9$ m³.

e. Determine the surface area of the pond as follows: $A = V/h = 55727.9/1.4 = 39805$ m².

### 5.11.2.3 Maturation WSP

A maturation WSP is a polishing pond that enables the production of an effluent with improved bacteriological quality. The following points provide a summary of the most important relationships in connection with maturation ponds:

- The rate of die-off of fecal organisms in a single maturation pond can be estimated by using the following equation:

$$\frac{N_e}{N_i} = \frac{1}{1 + k't} \tag{5.87}$$

---

**LISTING OF PROGRAM 5.14  (CHAP5.14\FORM1.VB): DESIGN OF A WASTE STABILIZATION POND**

```vb
'*********************************************************************************
'Program 5.14: Design of a Waste Stabilization Pond
'*********************************************************************************
Imports System.Math
Public Class Form1
  Const G = 9.81
  Dim Q, Li, Tmin, Kp, Tc, Lo, h As Double
  Dim A As Double

  Private Sub Form1_Load(ByVal sender As System.Object, ByVal e As System.EventArgs)
Handles MyBase.Load
      Me.Text = "Program 5.14: Design of a Waste Stabilization Pond"
      Me.MaximizeBox = False
      Me.FormBorderStyle = Windows.Forms.FormBorderStyle.FixedSingle
      label1.text = "Enter the flow rate, Q (m3/min):"
      label2.text = "Enter the 5-day BOD (mg/L):"
      label3.text = "Enter the lowest temperature, T (degree C):"
      label4.text = "Enter the removal rate constant for the pond (/day):"
      label5.text = "Enter the temperature correction factor:"
      label6.text = "Enter the effluent 5-day BOD, (mg/L):"
      label7.text = "Enter the depth of the pond, d (m):"
      button1.text = "&Calculate"
      Label8.Text = ""
      Label9.Text = "Decimal places:"
      NumericUpDown1.Value = 2
      NumericUpDown1.Maximum = 10
      NumericUpDown1.Minimum = 0
  End Sub

  Sub calculateresults()
      Q = Val(textbox1.text)
      Li = Val(textbox2.text)
      Tmin = Val(textbox3.text)
      Kp = Val(textbox4.text)
      Tc = Val(textbox5.text)
      Lo = Val(textbox6.text)
      h = Val(textbox7.text)
      Dim Kpt, dt, V As Double
      'removal rate constant at T
      Kpt = Kp * Tc ^ (Tmin - 20)
      'detention time
      dt = (Li / Lo - 1) / Kpt
      'volume of stabilization pond
```

```
        V = dt * Q * 60 * 24
        'surface area of the pond
        A = V / h
    End Sub

    Sub showResults()
        Label8.Text = "The required surface area of the pond (m2) = " + FormatNumber(A,
NumericUpDown1.Value)
    End Sub

    Private Sub Button1_Click(ByVal sender As System.Object, ByVal e As System.EventArgs)
Handles Button1.Click
        calculateresults()
        showResults()
    End Sub

    Private Sub NumericUpDown1_ValueChanged(ByVal sender As System.Object, ByVal e As
System.EventArgs) Handles NumericUpDown1.ValueChanged
        showResults()
    End Sub
End Class
```

where:

$N_e$ is the effluent bacterial number, the number of bacteria per 100 mL

$N_i$ is the influent bacterial number, the number of bacteria per 100 mL

$k'$ is the bacterial die-off rate, day$^{-1}$

$t$ is the retention time, days

- The rate of die-off of fecal organisms in a multicelled maturation WSP can be estimated by the following equation:

$$\frac{N_e}{N_i} = \frac{1}{\left(1+k'*t\right)^N} \quad (5.88)$$

where $N$ is the number of WSPs in series.

WSPs can be round, square, or rectangular in shape.

- The detention time of the maturation pond may be determined from the relationship between volume and flow as presented in the following equation:

$$t = \frac{V}{Q} \quad (5.89)$$

or

$$V = t*Q = A*h$$

For which the area, $A$, can be calculated.

## Example 5.15

1. Write a computer program to find the retention time and volume of a pond required to reduce the bacterial number by any percentage (a) with a bacterial die-off rate of $k'$ (per day), given the rate of wastewater flow to the pond as $Q$ (m$^3$/day).
2. A polishing pond reduces the bacterial number by 99% with a bacterial die-off rate of 0.5/day. Wastewater is introduced to the pond at the rate of 20 m$^3$/day. Find the necessary retention time and the volume of the pond.
3. Use the program established in (1) to verify computations obtained in (2).

### Solution

1. For solution to Example 5.15 (1), see the listing of Program 5.15.
2. Solution to Example 5.15 (2):
   a. Given: $k'$ is the 0.5/day, $N_e/N_i = (100-99)/100 = 0.01$, $Q = 20$ m$^3$/day.
   b. Find the detention time of the pond as follows: $N_e/N_i = 1/(1 + k' * t) = 0.01 = 1/(1 + 0.5 * t)$. This yields $t = 198$ days.
   c. Determine the volume of the pond as follows: $V = t * Q = 198 * 20 = 3960$ m$^3$.

Table 5.13 gives the typical design information for WSPs.

**LISTING OF PROGRAM 5.15   (CHAP5.15\FORM1.VB): RETENTION TIME AND VOLUME OF A POND**

```vb
'********************************************************************************************
'Program 5.15: Retention Time and Volume of a Pond
'********************************************************************************************
Imports System.Math
Public Class Form1
  Const G = 9.81
  Dim Q, k, kil, rat, t, V As Double

  Private Sub Form1_Load(ByVal sender As System.Object, ByVal e As System.EventArgs)
Handles MyBase.Load
    Me.Text = "Program 5.15: Retention Time and Volume of a Pond"
    Me.MaximizeBox = False
    Me.FormBorderStyle = Windows.Forms.FormBorderStyle.FixedSingle
    Label4.AutoSize = False
    Label4.Height = 52
    Label4.Width = 372
    Label4.Text = ""
    Label1.Text = "Enter flow rate, Q (m3/day):"
    Label2.Text = "Enter bacterial die away rate, k (/day):"
    Label3.Text = "Enter bacterial number reduction (%):"
    Button1.Text = "&Calculate"
    Label5.Text = "Decimals:"
    NumericUpDown1.Value = 2
    NumericUpDown1.Maximum = 10
    NumericUpDown1.Minimum = 0
  End Sub

  Sub calculateResults()
    Q = Val(TextBox1.Text)
    k = Val(TextBox2.Text)
    kil = Val(TextBox3.Text)
    If kil < 1 Then kil *= 100
    rat = (100 - kil) / 100
    'detention time
    t = (1 / rat - 1) / k
    'volume of pond
    V = t * Q
  End Sub

  Sub showResults()
    Label4.Text = "The detention time of the pond (days): " + FormatNumber(t,
    NumericUpDown1.Value) + vbCrLf
    Label4.Text += "The volume of the pond (m3): " + FormatNumber(V, NumericUpDown1.Value)
  End Sub

  Private Sub Button1_Click(ByVal sender As System.Object, ByVal e As System.EventArgs)
Handles Button1.Click
    calculateResults()
    showResults()
  End Sub

  Private Sub NumericUpDown1_ValueChanged(ByVal sender As System.Object, ByVal e As
System.EventArgs) Handles NumericUpDown1.ValueChanged
    showResults()
  End Sub
End Class
```

**TABLE 5.13**

**Design Information for WSPs**

| | Anaerobic | Facultative | Aerobic |
|---|---|---|---|
| Influent | Sewage with high organic load and high degree of solids | Sewage from a sewerage system or anaerobic pond | Sewage from facultative ponds |
| Treatment | Partial | | |
| Disposal of effluent | To facultative pond | Maturation pond | Agricultural irrigation, fish farming, and aquatic birds |
| Depth (m) | 2–4 | 1–1.5 | 1 |
| Detention time (days) | 8–20 | 20–180 | 5–10 |
| Main biological action | Organisms that do not require DO for feeding and reproduction | Anaerobic and aerobic | Aerobic organisms |
| Operation | Parallel or series connection | At least three ponds in series, parallel useful for large ponds | One or more, series or parallel |
| Color | Grayish black | Green or brownish green | Green |
| Frequency of sludge removal (years) | 2–12 | 8–20 | Probably never |
| Optimum temperature (°C) | 30 | 20 | 20 |
| Oxygen requirement | – | – | 0.7–1.4 times removed BOD |
| pH | 6.8–7.2 | 6.5–9.0 | 6.5–8.0 |
| Chemicals needed | Nutrients when there is a deficiency; no other chemicals | Nutrients when there is a deficiency; no other chemicals | |
| Expected problems | Odors, large land requirements, groundwater pollution | Odors when loading is high, groundwater pollution, reduction in biological activity in cold climates | Reduction in problems associated with biological activity under cold weather conditions |

*Sources:* Metcalf and Eddy, Inc., Tchobanoglous, G., Stensel, H. D., Tsuchihashi R., and Burton, F., Wastewater Engineering: Treatment and Resource Recovery, 5th ed, McGraw-Hill Science/Engineering/Math, 2013; Abdel-Magid, I. M., Selected Problems in Wastewater Engineering, National Research Council, Khartoum University Press, Khartoum, Sudan, 1986; Vernick, A. S., and Walker, F. C., *Handbook of Wastewater Treatment Processes*, Pollution Engineering and Technology Series No. 19, Marcel Dekker, New York, 1981.

## 5.12 SLUDGE TREATMENT AND DISPOSAL

### 5.12.1 SLUDGE DIGESTION

Sludge digestion is the controlled decomposition of organic matter present in sludges. It facilitates volume reduction, conversion of solids to inert compounds, and removal of pathogenic microorganisms. The digestion process may be aerobic or anaerobic.

The aerobic process used to treat waste-activated humus or primary sludges, or a mixture thereof, in an open tank can be regarded as a modification of the activated sludge process (Vernick and Walker 1981; Smith and Scott 2002). The dominating reaction in aerobic digestion is endogenous respiration. The main advantages of aerobic digestion include reduction in odor and biodegradable material, improvement of the dewatering process, and reduction in the level of pathogenic microorganisms.

Anaerobic digestion is a fermentation process of sewage sludge in which facultative and anaerobic bacteria break down organic matter primarily into the gases such as carbon dioxide and methane. The process is carried down by acidic bacteria and methanogenic bacteria, respectively (Vernick and Walker 1981; Smith and Scott 2002). The anaerobic sludge digestion process is affected by many interrelated factors such as pH, temperature, nutrients present, toxic substances, volatile acids, ammonia, type and characteristics of decomposed materials, shock loads, and mixing conditions (Rowe and Abdel-Magid 1995).

Acid formers convert solids to soluble organic acids and alcohols. Formation of acids results in a decrease in pH of the system and may end the action of acid formers. These species are then replaced by anaerobic bacteria (methane formers), which function within a narrow pH range of from 6.5 to 7.5. Methane formers convert the formed acids and alcohols to carbon dioxide and methane gas with traces of other gases such as hydrogen sulfide (Rowe and Abdel-Magid 1995).

The volumetric gas production, or specific yield, from an anaerobic sludge digester can be determined from the relationship presented in the following equation (Gunnerson and Stuckey 1986):

$$V_g = \frac{(Y_t * \text{VS})}{t}\left(1 - \frac{k_n}{t * \mu_{\text{smax}} - 1 + k_n}\right) \quad (5.90)$$

where:

$V_g$ is the volumetric gas production rate, or the specific yield, m³ gas per m³ digester per day

$Y_t$ is the ultimate gas yield, m³ gas per kg VS added

VS is the concentration of influent volatile solids, kg/m³

$K_n$ is the kinetic coefficient, dimensionless

$t$ is the hydraulic detention time, day

$\mu_{\text{smax}}$ is the maximum specific growth rate of microorganisms, day⁻¹

Table 5.14 gives the general design information for a conventional anaerobic digester.

## TABLE 5.14
### Design Information for a Conventional Anaerobic Digester

| Parameter | Value |
|---|---|
| Volatile solids loading (kg/m$^3$/day) | 0.3–2 |
| Volatile solids destruction (%) | 40–50 |
| Gas production (m$^3$ gas per kg VS) | 0.2–1.5 |
| Influent sludge solids (kg/m$^3$/day) | 2–5 |
| Total solids decomposition (%) | 30–40 |
| pH | 6.5–7.4 |
| Alkalinity concentration (mg/L) | 2000–3500 |
| Solids retention time (days) | 30–90 |
| Digester capacity, m$^3$/capita | 0.1–0.17 |
| Gas composition (%) | |
| Methane | 65–70 |
| Carbon dioxide | 32–35 |
| Hydrogen sulfide | Trace |
| Temperature (°C) | 30–55 |

*Sources:* Rowe, D. R. and Abdel-Magid, I. M., *Handbook of Wastewater Reclamation and Reuse*, CRC Press/Lewis Publishers, Boca Raton, FL, 1995; Lin, S., and Lee, C., *Water and Wastewater Calculations Manual*, 2nd ed, McGraw-Hill Professional, 2007; Abdel-Magid, I. M., Selected Problems in Wastewater Engineering, National Research Council, Khartoum University Press, Khartoum, Sudan, 1986; Gunnerson, C. G., and Stuckey, D. C., *Integrated Resources Recovery: Anaerobic Digestion Principles and Practice for Biogas Systems*, Technical Paper Number 49, World Bank, Washington, DC, 1986.

## Example 5.16

1. Develop a computer program to find the volumetric gas production rate in an anaerobic digester, given the specific yield (m$^3$ gas per m$^3$ digester per day), the ultimate gas yield (m$^3$ gas per kg VS added), the concentration of influent volatile solids (kg/m$^3$), the kinetic coefficient, the hydraulic detention time (d), and the maximum specific growth rate of microorganisms (day$^{-1}$).
2. Compute the volumetric gas production rate from a beef waste, given that the influent volatile solids concentration to the digester amounts to 180 kg/m$^3$ for a hydraulic retention time of 15 days. The maximum specific growth rate of microorganisms is 0.15/day at a temperature of 20°C. Take the ultimate methane yield equal to 0.3 m$^3$ methane per kg VS added, the kinetic coefficient equal to 1.5, and the digester volume equal to 6 m$^3$.
3. Use the computer program of (1) to verify computations obtained in (2).

### Solution

1. For the solution to Example 5.16 (1), see the listing of Program 5.16.
2. Solution to Example 5.16 (2):
   a. Given: $Y_t = 0.3$, $VS = 180$, $k_n = 1.5$, $t = 15$, $\mu_{smax} = 0.15$.
   b. Determine the volumetric gas production rate from the beef manure as follows: $V_g = (Y_t * VS * \{1-[k_n/(t * \mu_{smax}-1 + k_n)]\})/t = (0.3 * 180 * \{1-[1.5/(15 * 0.15 + 1.5)]\})/15 = 1.64$ m$^3$ methane gas per m$^3$ digester volume per day.
   c. Determine the daily amount of gas production as 1.64 * 6 = 9.84 m$^3$.

---

### LISTING OF PROGRAM 5.16   (CHAP5.16\FORM1.VB): VOLUMETRIC GAS PRODUCTION RATE IN DIGESTERS

```vb
'*****************************************************************************************
'Program 5.16: Volumetric Gas Production Rate in Digesters
'*****************************************************************************************
Public Class Form1
    Dim Yt, VS, Uxmax, t, Kn, Vd, Vx, Px As Double

    Private Sub Form1_Load(ByVal sender As System.Object, ByVal e As System.EventArgs)
Handles MyBase.Load
        Me.Text = "Program 5.16: Volumetric Gas Production Rate in Digesters"
        Me.MaximizeBox = False
        Me.FormBorderStyle = Windows.Forms.FormBorderStyle.FixedSingle
        Label7.AutoSize = False
        Label7.Height = 52
        Label7.Width = 372
        Label7.Text = ""
        Label1.Text = "Enter the specific yield, Yt (m3 gas/kg VS):"
        Label2.Text = "Enter the influent volatile solids concentration, VS (kg/m3):"
        Label3.Text = "Enter the max specific growth rate of microorganisms (/day):"
        Label4.Text = "Enter the retention time, t (days):"
        Label5.Text = "Enter the kinetic coefficient, Kn:"
        Label6.Text = "Enter the volume of the digester (m3):"
```

```
      Button1.Text = "&Calculate"
      Label8.Text = "Decimals:"
      NumericUpDown1.Value = 2
      NumericUpDown1.Maximum = 10
      NumericUpDown1.Minimum = 0
  End Sub

  Sub calculateResults()
      Yt = Val(TextBox1.Text)
      VS = Val(TextBox2.Text)
      Uxmax = Val(TextBox3.Text)
      t = Val(TextBox4.Text)
      Kn = Val(TextBox5.Text)
      Vd = Val(TextBox6.Text)
      Vx = Yt * VS * (1 - (Kn / (t * Uxmax - 1 + Kn))) / t
      Px = Vx * Vd
  End Sub

  Sub showResults()
      Label7.Text = "The volumetric gas production rate (m3/m3/day): " + FormatNumber(Vx,
NumericUpDown1.Value) + vbCrLf
      Label7.Text += "The daily amount of gas production (m3): " + FormatNumber(Px,
NumericUpDown1.Value)
  End Sub

  Private Sub Button1_Click(ByVal sender As System.Object, ByVal e As System.EventArgs)
Handles Button1.Click
      calculateResults()
      showResults()
  End Sub

  Private Sub NumericUpDown1_ValueChanged(ByVal sender As System.Object, ByVal e As
System.EventArgs) Handles NumericUpDown1.ValueChanged
      showResults()
  End Sub
End Class
```

### 5.12.2 Sludge Dewatering

#### 5.12.2.1 Introduction

Sludge dewatering is a physical unit operation utilized in rendering the moisture content of sludge with the following objectives:

- Reduction of sludge volume
- Promotion of the sludge-handling process
- Retardation of biological decomposition
- Increase the sludge calorific value by removal of excess moisture
- Removal of sludge odors
- Reduction of leachate production at sanitary landfills
- Safe use of sludge for landscape fertilizer or as a farmland fertilizer

The methods used for sludge dewatering include land disposal (drying beds, injection, ridge and furrow, spray irrigation, lagooning), vacuum filters, pressure filters, centrifuges, composting, elutration, and compaction. These methods are of a physical rather than chemical nature.

In the drying bed operation, sludge is spread in an open area to assist in the moisture reduction through evaporation or drainage according to prevailing climatic conditions, soil topography, hydrology of the area, meteorological factors, type and properties of sludge, nature and size of solid particles, bed dimensions especially the depth, shape of drying surface as related to air flow, land value, and management (Vernick and Walker 1981; Rowe and Abdel-Magid 1995; Viessman et al. 2008; U.S. Environmental Protection Agency 2012). Collected filtrate is usually returned to the treatment plant, and the sludge is removed from the drying bed after it has drained and dried sufficiently to be stable (Smith and Scott 2002).

Vacuum filtration is a continuous operation carried out at pressures ~70% of the atmospheric pressure (Rowe and Abdel-Magid 1995). Chemical conditioning is used before vacuum filtration to agglomerate small particles. This method is used in larger facilities where space is limited, or when incineration is necessary for maximum volume reduction (Vernick and Walker 1981). The factors that affect the vacuum filter yield include the type and properties of sludge, the type, and the concentration of the conditioner and conditioning procedure, the type and condition of the filter cloth, and the operational parameters of the filter (Vernick and Walker 1981; Rowe and Abdel-Magid 1995).

Pressure filtration is a batch process that dewaters a conditioned sludge. Application of pressure drains the liquor and retains the sludge cake. Factors influencing operation include the pressure applied, the type and life of filter, the sludge properties, and the conditioning and collection and final disposal of the cake.

Centrifugation (solid bowl, disc, basket type) uses centrifugal and Stoke's frictional forces to increase the settling rate of sludge solids. The method is used in large facilities where space is limited or when incineration is required (Vernick and Walker 1981). Factors of importance include solids content, particle size and shape, electrostatic factors, viscosity effect, differential density between solid particles and liquid, disposal of centrate, availability of power, properties of sludge, rotational speed, hydraulic loading rate, depth of liquid pool in bowl, and use of polyelectrolyte to improve performance (Vernick and Walker 1981; Rowe and Abdel-Magid 1995; Metcalf and Eddy 2013).

In sludge composting, the sludge is mixed with a bulking material (e.g., wood chips, leaves, garbage, rags, paper, cardboard, etc.) to form an aerated, porous pile. Decomposition is achieved by aerobic and thermophilic bacteria. The method may be used to convert digested and undigested sludge cake (Vernick and Walker 1981). Factors affecting the process include weather conditions, properties of sludge, and the aeration rate.

In elutriation, the sludge is washed with treated effluent to yield a porous sludge. During mixing of elutriant and sludge, substances such as carbonates and phosphates are removed from the sludge, together with products of decomposition and nonsettleable fine solids (Rowe and Abdel-Magid 1995; U.S. Environmental Protection Agency 2012).

In general, the main factors that affect dewatering practice include the nature and content of the fine particles, sludge solids content, particle size and charge, shearing strength, protein content, pH, moisture content, anaerobic digestion, and the type and concentration of the filter aids and conditioners used.

### 5.12.2.2 Filtration of Sludge

Filtration may be defined as the separation of a solid from a suspension of a liquid. It retains the solid on a screen or membrane and allows the filtrate to drain off. The methods used for determining how well sludge dewaters include the cracking time test, the capillary suction time (CST) test, the filter leaf test, and the specific resistance test.

The cracking time is the time necessary for a formed cake to crack. The CST is the time required by water from a sludge to travel a given distance along a filter paper (Smith and Scott 2002). The shorter the CST, the higher the dewaterability of the sludge. Long CST values indicate problems in the filtration of the sludge analyzed (Rowe and Abdel-Magid 1995). The filter leaf test stimulates the principle of operation of a vacuum filter, whereby the filter yield is estimated based on the drying cycle period of the sludge (Rowe and Abdel-Magid 1995; Peirce, Vesilind and Weiner 1997). Coackley (Coackley

1955; Coackley 1965) advocated the use of the specific resistance concept whereby the ease of dewatering is given by the Carman and Coackley (Carman 1938; Newitt et al. 1949; Coackley 1953; Coackley and Allos 1962; Coackley 1965; Coackley 1967; Coackley 1975; Sanin et al. 2010) equation for constant pressure, as presented in the following equation:

$$\frac{t}{V} = \left( \frac{\mu * r_S * C}{2P * A^2} \right) V + \frac{\mu * R_m}{P * A} \tag{5.91}$$

where:
  $t$ is the time of filtration, s
  $V$ is the volume of filtrate, m$^3$
  $\mu$ is the viscosity of filtrate, N*s/m$^2$
  $r_s$ is the specific resistance of sludge cake, m/kg
  $C$ is the solids content, kg/m$^3$
  $P$ is the pressure applied, N/m$^2$
  $A$ is the area of filtration, m$^2$
  $R_m$ is the resistance of filter medium, m$^{-1}$

Equation 6.48 may be put in the form of the following equation:

$$\frac{t}{V} = b * V + a \tag{5.92}$$

where:

$$b = \frac{\mu r_s C}{2PA^2} \tag{5.93}$$

$$a = \frac{\mu R_m}{PA} \tag{5.94}$$

Plotting $t/V$ versus $V$ yields a straight line of gradient $b$, and $r_s$ is found as indicated in the following equation:

$$r_S = \frac{2 b PA^2}{\mu C} \tag{5.95}$$

where:
  $r_s$ is the specific resistance of sludge cake, m/kg
  $b$ is the slope of the straight line of $t/V$ versus $V$, s/m$^6$
  $P$ is the pressure applied, N/m$^2$
  $A$ is the area of filtration, m$^2$
  $\mu$ is the viscosity of filtrate, N*s/m$^2$
  $C$ is the solids content, kg/m$^3$

The specific resistance to filtration may be defined as follows: "The resistance to filtrate flow caused by a cake of unit weight of dry solids per unit filter area" (Abdel-Magid 1982). Table 5.15 gives a general outline of the values of the specific resistance and gives an estimation of the degree of the sewage sludge dewaterability.

The value of $r_s$ for most wastewater sludges changes with pressure as indicated in the following equation:

$$r_s = r_s' * P^a \tag{5.96}$$

## TABLE 5.15

### Sludge Characteristics as Related to Specific Resistance

| Specific Resistance Value | Sludge Characteristics (m/kg) |
|---|---|
| $10^{11}$–$10^{12}$ | Easily filtered sludge |
| $10^{14}$–$10^{15}$ | Poorly filtered sludge |

*Sources:* Rowe, D.R. and Abdel-Magid, I.M., *Handbook of Wastewater Reclamation and Reuse*, CRC Press/Lewis Publishers, Boca Raton, FL, 1995; Abdel-Magid, I. M., Selected Problems in Wastewater Engineering, National Research Council, Khartoum University Press, Khartoum, Sudan, 1986; Fair, G. M. et al. *J. Am. Water Works Assoc.*, 40,1051, 1966.

where:

$r_s$ is the specific resistance to filtration at applied pressure *P*, m/kg

$r_s'$ is a constant

*a* is the coefficient of compressibility (varies between 0 and 1)

Equation 5.96 can be arranged in a straight line relationship form as presented in the following equation:

$$\log(r_s) = a * \log(P) + \log(r_s') \qquad (5.97)$$

The compressibility coefficient is taken to be the slope, *a*, of the straight line obtained by plotting $\log(r_s)$ versus $\log(P)$. The greater the value of *a*, the more compressible is the sludge. Figure 5.9 shows a sketch of the apparatus used in determining the specific resistance of sludge to filtration.

## Example 5.17

1. Write a computer program to compute the specific resistance of a sample of a sludge, given the variation of a collected volume of filtrate *V* (mL) with *t* (s), the vacuum applied *P* (N/m²), the filtrate viscosity μ (N*s/m²), the solids concentration *C* (kg/m³), and the area of filtration *A* (m²). Design the computer program to show whether the sludge under consideration is amenable to dewatering by vacuum filtration.

2. The following data were obtained in an experiment in order to determine its specific resistance to filtration:

| Volume of Filtrate Collected (mL) | Time Taken to Collect Volume (min) |
|---|---|
| 6.8 | 1 |
| 9.5 | 2 |
| 11.6 | 3 |
| 13.4 | 4 |
| 14.9 | 5 |

a. Find the specific resistance of the sludge, given the vacuum pressure used = 60 kPa, the dynamic viscosity of filtrate = 1.139 * $10^{-3}$ N*s/m², the volume of sample used in filtration = 50 mL, the solids concentration = 0.084 g/mL, and the diameter of filter paper used in the experiment = 7 cm.

b. Is this sludge amenable to dewatering by vacuum filtration? State your reasons.

3. Use the computer program of (1) to verify computations obtained in (2).

**Solution**

1. For the solution to Example 5.17 (1), see the listing of Program 5.17.

2. Solution to Example 5.17 (2):

a. Given: the variation of volume of filtrate collected with time of filtration, $P = 60 * 10^3$ N/m², μ = 1.139 * $10^{-3}$ N * s/m², C = 0.084 g/mL, D = 7 cm.

b. Find the ratio of time to filtrate volume, *t/V*, as follows:

| | Time (s) | | | | |
|---|---|---|---|---|---|
| | 60 | 120 | 180 | 240 | 300 |
| Time/volume of filtrate collected (s/m²) | 8.82 | 12.63 | 15.52 | 17.91 | 20.13 |
| Volume of filtrate collected (mL) | 6.8 | 9.5 | 11.6 | 13.4 | 14.9 |

c. Draw the straight line of the plot of *t/V* versus *V*, and determine the slope of the straight line as follows: $b = 1.387 * 10^{12}$ s/m⁶.

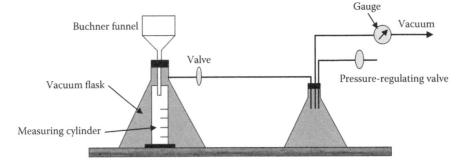

**FIGURE 5.9**   Specific resistance apparatus.

d. Find the area of filtration as follows: $A = \pi(0.07)^2/4 = 38.48 * 10^{-4}$ m$^2$.

e. Find the sludge solids concentration as follows: $C = 0.084 * 10^{-3}/10^{-6} = 84$ kg/m$^3$.

f. Compute the sludge specific resistance as follows: $r_s = 2 * b * P * A^2/\mu * C = (2 * 1.387$

$* 10^{12} * 60 * 10^3 * (38.48 * 10^{-4})^2)/(1.139 * 10^{-3} * 84) = 2.58 * 10^{13}$ m/kg.

g. Because the specific resistance of this sludge exceeds the value of $1 * 10^{12}$ m/kg (with $r_s = 2.58 * 10^{13}$ m/kg), it will not filter well in a vacuum filter.

---

### LISTING OF PROGRAM 5.17   (CHAP5.17\FORM1.VB): SPECIFIC RESISTANCE OF SLUDGE

```
'*************************************************************************************
'Program 5.17: Specific Resistance of Sludge
'*************************************************************************************
Imports System.Math
Public Class Form1
  Dim Rx, P, mu, c, d, N As Double
  Dim t(), v(), tov() As Double
  Dim assess As String

  Private Sub Form1_Load(ByVal sender As System.Object, ByVal e As System.EventArgs)
Handles MyBase.Load
    Me.Text = "Program 5.17: Specific Resistance of Sludge"
    Me.MaximizeBox = False
    Me.FormBorderStyle = Windows.Forms.FormBorderStyle.FixedSingle
    Label1.Text = "Enter the vacuum applied, P (N/m2):"
    Label2.Text = "Enter the filtrate viscosity, mu (Ns/m2):"
    Label3.Text = "Enter the solids concentration, C (g/mL):"
    Label4.Text = "Enter the diameter of the filter paper used, d (m):"
    Label5.Text = "Enter the number of measurements of volume and time:"
    Label6.Text = "Enter the data collected for volume versus time:"
    Label7.Text = ""
    Label7.AutoSize = False
    Label7.Width = 215
    Label7.Height = 45
    Button1.Text = "&Calculate"
    DataGridView1.Columns.Clear()
    DataGridView1.Columns.Add("VCol", "Volume of filtrate (mL)")
    DataGridView1.Columns.Add("TCol", "Time (minutes)")
    DataGridView1.AllowUserToAddRows = False
    Label8.Text = "Decimals:"
    NumericUpDown1.Value = 2
    NumericUpDown1.Maximum = 10
    NumericUpDown1.Minimum = 0
  End Sub

  Sub calculateResults()
    P = Val(TextBox1.Text)
    mu = Val(TextBox2.Text)
    c = Val(TextBox3.Text)
    d = Val(TextBox4.Text)
    N = Val(TextBox5.Text)
    If N <= 0 Then Exit Sub
    If DataGridView1.ColumnCount < 3 Then
      DataGridView1.Columns.Add("TVCol", "t/V")
      DataGridView1.Columns("TVCol").ReadOnly = True
    End If
    ReDim t(N), v(N), tov(N)
    Dim b, A, X1, Y1, sum1, sum2 As Double
    Y1 = 0.0!
```

```vb
    X1 = 0.0!
    For i = 1 To N
        v(i) = Val(DataGridView1.Rows(i - 1).Cells("VCol").Value)
        t(i) = Val(DataGridView1.Rows(i - 1).Cells("TCol").Value)
        t(i) = t(i) * 60 'convert to seconds
        tov(i) = t(i) / v(i)
        DataGridView1.Rows(i - 1).Cells("TVCol").Value = FormatNumber(tov(i), 2)
        Y1 = Y1 + tov(i)
        X1 = X1 + v(i)
    Next i
    'compute averages for regression analysis
    Y1 = Y1 / N 'average t/v
    X1 = X1 / N 'average v
    Sum1 = 0.0!
    Sum2 = 0.0!
    For i = 1 To N
        Sum1 = sum1 + (tov(i) - Y1) * (v(i) - X1)
        Sum2 = sum2 + (v(i) - X1) ^ 2
    Next i
    b = sum1 / sum2 'the slope of the line
    b *= (10 ^ 12)
    A = Pi / 4 * d ^ 2 'area of filtration
    c = c * 10 ^ 3 'convert to kg/m3
    Rx = 2 * b * P * (A ^ 2) / (mu * c) 'specific resistance
    assess = "case not covered by Table 5.15"
    If Rx < (10 ^ 12) Then assess = "Easily filtered sludge"
    If Rx > (10 ^ 12) Then assess = "Poorly filtered sludge"
End Sub


Sub showResults()
    Label7.Text = "The specific resistance of the sludge (m/kg): " + FormatNumber(Rx,
NumericUpDown1.Value)
    Label7.Text += vbCrLf + assess
End Sub


Private Sub TextBox5_TextChanged(ByVal sender As System.Object, ByVal e As System.
EventArgs) Handles TextBox5.TextChanged
    If N <> Val(TextBox5.Text) Then
        If Val(TextBox5.Text) <= 0 Then Exit Sub
        N = Val(TextBox5.Text)
        DataGridView1.Rows.Clear()
        DataGridView1.Rows.Add(CInt(N))
    End If
End Sub


Private Sub Button1_Click(ByVal sender As System.Object, ByVal e As System.EventArgs)
Handles Button1.Click
    calculateResults()
    showResults()
End Sub


Private Sub NumericUpDown1_ValueChanged(ByVal sender As System.Object, ByVal e As
System.EventArgs) Handles NumericUpDown1.ValueChanged
    showResults()
End Sub
End Class
```

## Example 5.18

1. Write a computer program to determine the compressibility of a sludge sample for a set of specific resistance values and corresponding pressures.
2. For a sample of sewage sludge, the following specific resistance values were obtained for the corresponding applied vacuum pressures:

| Pressure Applied (kPa) | Specific Resistance, $r*10^{-13}$ (m/kg) |
|---|---|
| 293.04 | 57.85 |
| 586.075 | 104.45 |
| 1172.15 | 186.54 |
| 1758.225 | 256.84 |
| 2344.3 | 331.89 |

Determine the compressibility coefficient of the sludge.
3. Use program developed in (1) to verify solution obtained in (2).

## Solution

1. For the solution to example 5.18 (1), see the listing of Program 5.18.
2. Solution to Example 5.18 (2):
   a. Given: the specific resistance values corresponding to the applied pressures.
   b. Determine $\log(P)$ and $\log(r)$ as presented in the following table:

| Pressure Applied (kPa) | Specific Resistance, $r*10^{-13}$ m/kg | $\log(P)$ | $\log(r)$ |
|---|---|---|---|
| 293.04 | 57.85 | 5.466927 | 14.7623 |
| 586.075 | 104.45 | 5.767953 | 15.01891 |
| 1172.15 | 186.54 | 6.068983 | 15.27077 |
| 1758.225 | 256.84 | 6.245074 | 15.40966 |
| 2344.3 | 331.89 | 6.370013 | 15.52099 |

   c. Plot $\log(r)$ as a function of $\log(P)$. Find the slope of the drawn line to correspond to the compressibility coefficient of the sample as $s = 0.84$.
3. Use program developed in (1) to verify solution obtained in (2).

---

### LISTING OF PROGRAM 5.18   (CHAP5.18\FORM1.VB): COMPRESSIBILITY OF SLUDGES

```
'********************************************************************************************
'Program 5.18: Compressibility of Sludges
'********************************************************************************************
Imports System.Math
Public Class Form1
  Dim N, X1, Y1, P, R, b As Double
  Dim logP(), logR() As Double

  Private Sub Form1_Load(ByVal sender As System.Object, ByVal e As System.EventArgs)
Handles MyBase.Load
    Me.Text = "Program 5.18: Compressibility of Sludges"
    Me.MaximizeBox = False
    Me.FormBorderStyle = Windows.Forms.FormBorderStyle.FixedSingle
    Label1.Text = "Enter the number of measurements of data of volume and time:"
    DataGridView1.Columns.Clear()
    DataGridView1.Columns.Add("PCol", "Pressure (kPa)")
    DataGridView1.Columns.Add("RCol", "S.R. (/10^13) m/kg")
    DataGridView1.AllowUserToAddRows = False
    Label2.Text = ""
    button1.text = "&Calculate"
    Label3.Text = "Decimals:"
    NumericUpDown1.Value = 2
    NumericUpDown1.Maximum = 10
    NumericUpDown1.Minimum = 0
  End Sub

  Sub calculateResults()
    N = Val(TextBox1.Text)
    If N <= 0 Then Exit Sub
    ReDim logP(N), logR(N)
    Y1 = 0.0!
```

```vb
      X1 = 0.0!
      If DataGridView1.ColumnCount < 4 Then
         DataGridView1.Columns.Add("LPCol", "Log(P)")
         DataGridView1.Columns.Add("LRCol", "Log(r)")
         DataGridView1.Columns("LPCol").ReadOnly = True
         DataGridView1.Columns("LRCol").ReadOnly = True
      End If
      For i = 1 To N
         P = Val(DataGridView1.Rows(i - 1).Cells("PCol").Value)
         R = Val(DataGridView1.Rows(i - 1).Cells("RCol").Value)
         R *= Math.Pow(10, 13)
         P *= 1000
         logR(i) = Log10(R)
         logP(i) = Log10(P)
         DataGridView1.Rows(i - 1).Cells("LPCol").Value = FormatNumber(logP(i), 2)
         DataGridView1.Rows(i - 1).Cells("LRCol").Value = FormatNumber(logR(i), 2)
         Y1 = Y1 + logR(i)
         X1 = X1 + logP(i)
      Next i
      Dim sum1, sum2 As Double
      'compute averages for regression analysis
      Y1 = Y1 / N 'average t/v
      X1 = X1 / N 'average v
      Sum1 = 0.0!
      Sum2 = 0.0!
      For i = 1 To N
         sum1 = sum1 + (logR(i) - Y1) * (logP(i) - X1)
         sum2 = sum2 + (logP(i) - X1) ^ 2
      Next i
      b = sum1 / sum2 'the slope of the line
   End Sub


   Sub showResults()
      Label2.Text = "The compressibility coefficient of the sample = " + FormatNumber(b,
NumericUpDown1.Value)
   End Sub


   Private Sub Button1_Click(ByVal sender As System.Object, ByVal e As System.EventArgs)
Handles Button1.Click
      calculateResults()
      showResults()
   End Sub


   Private Sub TextBox1_TextChanged(ByVal sender As System.Object, ByVal e As System.
EventArgs) Handles TextBox1.TextChanged
      If N <> Val(TextBox1.Text) Then
         If Val(TextBox1.Text) <= 0 Then Exit Sub
         DataGridView1.Rows.Clear()
         DataGridView1.Rows.Add(CInt(Val(TextBox1.Text)))
         N = Val(TextBox1.Text)
      End If
   End Sub


   Private Sub NumericUpDown1_ValueChanged(ByVal sender As System.Object, ByVal e As
System.EventArgs) Handles NumericUpDown1.ValueChanged
      showResults()
   End Sub
End Class
```

### 5.12.2.3  Centrifugation

In centrifugation, settling and rejection of solids are the parameters that are considered in the modeling and scale-up between two geometrically similar centrifuges. Settling of particles is estimated by the sigma equation, whereas rejection of solids is measured by the beta equation (Rowe and Abdel-Magid 1995; Peirce et al. 1997).

The sigma equation is illustrated in Equation 5.98 for two centrifuges having the same settling characteristics within the bowl (Rowe and Abdel-Magid 1995; Peirce et al. 1997):

$$\frac{Q_1}{\Sigma_1} = \frac{Q_2}{\Sigma_2} \tag{5.98}$$

where:

$Q_1$ is the liquid flow rate into the first centrifuge, m³/s
$\Sigma_1$ is a parameter related to the properties of the first centrifuge
$Q_2$ is the liquid flow rate into the second centrifuge, m³/s
$\Sigma_2$ is a parameter related to the properties of the second centrifuge

The sigma value for a solid bowl centrifuge can be calculated by using the following equation:

$$\Sigma = \frac{v_r^2 * V}{g * \ln(r_2/r_1)} \tag{5.99}$$

where:

$\Sigma$ is a parameter related to the properties of the centrifuge (Rowe and Abdel-Magid 1995; Peirce et al. 1997)
$v_r$ is the rotational velocity of the bowl, rad/s
$V$ is the liquid volume in the pool, m³
$g$ is the local gravitational acceleration, m/s²
$r_1$ is the radius from the centerline to the surface of the sludge, m
$r_2$ is the radius from the centerline to inside the bowl wall, m

Solids movement in a centrifugation system may be estimated by the beta equation (Rowe and Abdel-Magid 1995; Peirce et al. 1997):

$$\frac{W_1}{\beta_1} = \frac{W_2}{\beta_2} \tag{5.100}$$

where:

$W_1$ is the solids loading rate for the first centrifuge, kg/h
$\beta_1$ is the beta function for the first centrifuge
$W_2$ is the solids loading rate for the second centrifuge, kg/h
$\beta_2$ is the beta function for the second centrifuge

The beta function may be computed by using the following equation:

$$\beta = v_w * d_p * n * \pi * Z * D \tag{5.101}$$

where:

$\beta$ is the beta function for a centrifuge
$v_w$ is the difference in the rotational velocity between the bowl and the conveyor, rad/s
$d_p$ is the distance between blades, or the scroll pitch, m
$n$ is the number of leads
$Z$ is the depth of sludge in the bowl, m
$D$ is the bowl diameter, m

### Example 5.19

1. Develop a computer program to find the flow rate at which a solid bowl centrifuge will perform similarly as an old one that is to be replaced in order to increase sludge production. Data given include the solids concentration (%); the rate of flow of sludge to the old centrifuge that needs dewatering (m³/d); and the characteristics of both centrifuges, in terms of bowl length (cm), bowl diameter (cm), bowl speed (rpm), bowl depth (cm), scroll pitch (cm), number of leads, and conveyor velocity (rpm). Assume that the new centrifuge is scaled up geometrically similar to larger.
2. The method of centrifugation is used to dewater a sludge of 3.5% solids content. The sludge enters the solid bowl centrifuge at an hourly rate of 0.5 m³. Due to an increase in quantity of sludge requiring dewatering, the centrifuge is scaled up to another geometrically similar and larger one. The properties of the two centrifuges are tabulated as follows. Determine the flow rate at which the second centrifuge will operate similarly to the first one.

| Characteristic | First Centrifuge | Second Centrifuge |
|---|---|---|
| Bowl length (cm) | 24 | 60 |
| Bowl diameter (cm) | 20 | 40 |
| Bowl speed (rpm) | 4500 | 4000 |
| Bowl depth (cm) | 3 | 5 |
| Scroll pitch (cm) | 5 | 10 |
| Number of leads | 1 | 1 |
| Conveyor velocity (rpm) | 4450 | 3950 |

3. Use the computer program developed in (1) to verify computations obtained in (2).

### Solution

1. For the solution to Example 5.19 (1), see the listing of Program 5.19.
2. Solution to example 5.19 (2):
   a. Given: $C = 3.5\%$, $Q = 0.5$ m³/h = 12 m³/day, properties of the two centrifuges.
   b. Construct the following table using the given data:

| Parameter | First Centrifuge | Second Centrifuge |
|---|---|---|
| $l$ (cm) | 24 | 60 |
| $D$ (cm) | 20 | 40 |
| $r_2$ (cm) | 20/2 = 10 | 40/2 = 20 |
| $v_r$ (rad/s) | $(2\pi/60) * 4500 = 471$ | $(2\pi/60) * 4000 = 419$ |
| $Z$ (cm) | 3 | 5.0 |
| $r_1 = r_2 - Z$ (cm) | 10 − 3 = 7 | 20 − 5 = 15 |
| $d_p$ (cm) | 5 | 10 |
| $n$ (dimensionless) | 1 | 1 |
| $V_w$ (rad/s) | 4500 − 4450 = 50 | 4000 − 3950 = 50 |

c. Estimate the settling of each centrifuge by using the sigma equation as follows: $(Q_1/\Sigma_1) = (Q_2/\Sigma_2)$.

i. Find the volume as follows: $V = 2\pi([r_1 + r_2]/2) * (r_2 − r_1) * 1$.

ii. Compute the sigma as follows: $\Sigma = [(v_r)^2 * V]/[g * \ln (r_2/r_1)]$.

| Parameter | First Centrifuge | Second Centrifuge |
|---|---|---|
| $V$ (cm³) | 3,847 | 33,000 |
| $v_r$ (rad/s) | 471 | 419 |
| $\ln(r_2/r_1)$ | ln(10/7) = 0.35667 | ln(20/15) = 0.28768 |
| $g$ (cm²/s) | 981 | 981 |
| $\Sigma$ | 2,443,409 | 20,533,306 |
| $Q$ (m³/day) | 12 | ? |

iii. Compute the liquid flow rate into the second centrifuge as follows: $Q_2 = (Q_1/\Sigma_1) * \Sigma_2 = (12/2443409) * 20533306 = 101$ m³/day. With respect to settling of solids, the second centrifuge will achieve equal dewaterability when the flow rate is 101 m³/day.

d. Estimate the movement of solids out of centrifuge by finding the beta equation as follows: $W_1/\beta_1 = W_2/\beta_2$.

i. Determine the solids loading rate for the two centrifuges as follows: $W$ = flow rate (m³/d) * solids concentration * density. (Assume density of water = 1000 kg/m³.)

ii. Determine the beta value for each centrifuge as follows: $\beta = v_w * d_p * n * \pi * Z * D$.

| Parameter | First Centrifuge | Second Centrifuge |
|---|---|---|
| $W$ (kg/d) | 420 | ? |
| $V_w$ (rad/s) | 50 | 50 |
| $d_p$ (cm) | 5 | 10 |
| $n$ (dimensionless) | 1 | 1 |
| $Z$ (cm) | 3.0 | 5.0 |
| $D$ (cm) | 20 | 40 |
| $B$ | 47,124 | 314,159 |

e. Estimate the solids movement of the second centrifuge as follows: $W_2 = (W_1/\beta_1) * \beta_2 = (420/47,124) * 314,159 = 2800$ kg/day. This is equivalent to a value of 2800/(0.035 * 1000) = 80 m³/day (assuming the same solids concentration of 3.5% and the same density).

f. In summary, and to have similar dewaterability to the first centrifuge, the following conditions must apply to the second centrifuge:

i. Sigma equation advocates the settling properties for the second centrifuge when the liquid flow rate attains a value of 101 m³/day.

ii. Beta equation reveals the movement of solids out of the second centrifuge when the solids loading rate attains a value of 80 m³/d.

iii. Because the scale-up procedure indicates that the lower value governs the centrifuge capacity, in this case, the solids loading rate governs. As such, the new centrifuge is not to be operated at a solids loading rate exceeding 80 m³/day.

---

**LISTING OF PROGRAM 5.19   (CHAP5.19\FORM1.VB): SLUDGE DEWATERING BY CENTRIFUGATION**

```
'********************************************************************************
'Program 5.19: Centrifugation
'********************************************************************************
Imports System.Math
Public Class Form1
    Dim C, Q As Double
    Const G = 9.81
    Dim L1, L2, D1, D2, vs1, vs2, z1, z2, dp1, dp2, n1, n2, Vw1, Vw2 As Double

    Private Sub Form1_Load(ByVal sender As System.Object, ByVal e As System.EventArgs)
Handles MyBase.Load
        Me.Text = "Program 5.19: Centrifugation"
        Me.MaximizeBox = False
        Me.FormBorderStyle = Windows.Forms.FormBorderStyle.FixedSingle
        Label1.Text = "Enter the solids concentration (%):"
        Label2.Text = "Enter the flow rate of sludge to be dewatered Q (m3/h):"
        label3.text = "Enter the characteristics of the two centrifuges:"
```

```
        Label4.AutoSize = False
        Label4.Height = 68
        Label4.Width = 390
        Label4.Text = ""
        Button1.Text = "&Calculate"
        DataGridView1.Columns.Clear()
        DataGridView1.Columns.Add("CCol", "Characteristic")
        DataGridView1.Columns.Add("FCol", "First centrifuge")
        DataGridView1.Columns.Add("SCol", "Second centrifuge")
        DataGridView1.Rows.Add(7)
        DataGridView1.AllowUserToAddRows = False
        DataGridView1.Columns("CCol").ReadOnly = True
        DataGridView1.Rows(0).Cells("CCol").Value = "Bowl Length, cm"
        DataGridView1.Rows(1).Cells("CCol").Value = "Diameter, cm"
        DataGridView1.Rows(2).Cells("CCol").Value = "Speed, rpm"
        DataGridView1.Rows(3).Cells("CCol").Value = "Depth, cm"
        DataGridView1.Rows(4).Cells("CCol").Value = "Scroll pitch, cm"
        DataGridView1.Rows(5).Cells("CCol").Value = "Number of leads"
        DataGridView1.Rows(6).Cells("CCol").Value = "Conveyor velocity, rpm"
End Sub

Sub calculateResults()
    C = Val(TextBox1.Text)
    Q = Val(TextBox2.Text)
    L1 = Val(DataGridView1.Rows(0).Cells("Fcol").Value)
    L2 = Val(DataGridView1.Rows(0).Cells("Scol").Value)
    D1 = Val(DataGridView1.Rows(1).Cells("Fcol").Value)
    D2 = Val(DataGridView1.Rows(1).Cells("Scol").Value)
    vs1 = Val(DataGridView1.Rows(2).Cells("Fcol").Value)
    vs2 = Val(DataGridView1.Rows(2).Cells("Scol").Value)
    z1 = Val(DataGridView1.Rows(3).Cells("Fcol").Value)
    z2 = Val(DataGridView1.Rows(3).Cells("Scol").Value)
    dp1 = Val(DataGridView1.Rows(4).Cells("Fcol").Value)
    dp2 = Val(DataGridView1.Rows(4).Cells("Scol").Value)
    n1 = Val(DataGridView1.Rows(5).Cells("Fcol").Value)
    n2 = Val(DataGridView1.Rows(5).Cells("Scol").Value)
    Vw1 = Val(DataGridView1.Rows(6).Cells("Fcol").Value)
    Vw2 = Val(DataGridView1.Rows(6).Cells("Scol").Value)
    Dim R11, R12, R21, R22 As Double
    Dim V1, V2, Vr1, Vr2, sig1, sig2 As Double
    Dim Q1, Q2, Ro, b1, b2, C1, W1, W2 As Double
    R11 = D1 / 2 : r12 = r11 - z1
    R21 = D2 / 2 : r22 = r21 - z2
    V1 = volume(r12, r11, L1)
    V2 = volume(r22, r21, L2)
    Vr1 = 2 * Pi / 60 * vs1
    Vr2 = 2 * Pi / 60 * vs2
    Sig1 = sigma(vr1, v1, r12, r11)
    Sig2 = sigma(vr2, v2, r22, r21)
    Q1 = Q * 24 'flow in first centrifuge, m3/d
    Q2 = Q1 * sig2 / sig1 'flow in second centrifuge
    Ro = 1000 'density of water (assumed), kg/m3
    Vw1 = vs1 - Vw1
    Vw2 = vs2 - Vw2
    B1 = beta(Vw1, dp1, n1, z1, D1) 'beta for first centrifuge
    B2 = beta(Vw2, dp2, n2, z2, D2) 'beta for second centrifuge
    C1 = C / 100 'concentration of solids in first centrifuge
    W1 = Q1 * C1 * Ro 'solids loading rate in first centrifuge
    W2 = W1 / b1 * b2 'solids loading rate in second centrifuge
    Dim s As String
    s = "Results:" + vbCrLf
```

```
       += "The flow rate in the second centrifuge = " + FormatNumber(Q2, 2) + " m3/d" + vbCrLf
       s += "The solids movement in second centrifuge = " + FormatNumber(W2, 2) + " kg/d,"
+ vbCrLf
       s += "equivalent to " + FormatNumber(W2 / ((C / 100) * 1000), 2) + " m3/d."
       Label4.Text = s
    End Sub

    Function BETA(ByVal Vw, ByVal dp, ByVal n, ByVal z, ByVal D)
       BETA = Vw * dp * n * Pi * z * D
    End Function

    Function SIGMA(ByVal Vr, ByVal V, ByVal R1, ByVal R2)
       SIGMA = Vr ^ 2 * V / (G * LOG(R2 / R1))
    End Function

    Function VOLUME(ByVal R1, ByVal R2, ByVal L)
       VOLUME = 2 * Pi * ((R1 + R2) / 2) * (R2 - R1) * L
    End Function

    Private Sub Button1_Click(ByVal sender As System.Object, ByVal e As System.EventArgs)
Handles Button1.Click
       alculateResults()
    End Sub
End Class
```

## 5.13 WASTEWATER DISPOSAL

### 5.13.1 DILUTION

Discharges of small quantities of relatively dilute sludges and treated effluents find their way into water courses. Such wastewater discharges can affect the water quality in rivers, streams, lakes, estuaries, and oceans. To predict the effects imposed by these discharges on water courses, mathematical models have been formulated. Factors affecting natural purification processes include the rate of flow and characteristics of body of water, reoxygenation processes, water use downstream from the disposal point, and the quantity of the wastewater discharge (Rowe and Abdel-Magid 1995).

In Figure 5.10, application of the principle of mass balance between two points, upstream and downstream of the discharge point, gives an estimate of the degree of dilution in the river as presented in the following equation (dilution law):

$$C_w * Q_w + C_r * Q_r = C_m * Q_m \qquad (5.102)$$

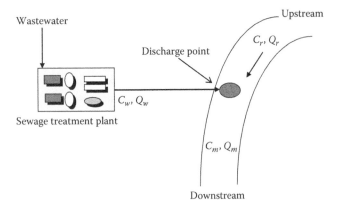

**FIGURE 5.10**  Wastewater disposal into a stream (dilution).

where:

$C_w$ is the concentration of pollutant in effluent from wastewater treatment works, mg/L

$Q_w$ is the wastewater flow rate to adjacent river, m³/s

$C_r$ is the concentration of same pollutant in river upstream of the discharge point, mg/L

$Q_r$ is the river flow rate, m³/s

$C_m$ is the concentration of the pollutant in mixture of river water and wastewater discharge, or concentration of pollutant in the river downstream of point of discharge, mg/L

$Q_m$ is the rate of flow of mixture of river water and wastewater discharge or the rate of flow of river downstream of point of discharge, m³/s

**Example 5.20**

1. Write a short computer program to determine the concentration of a pollutant in the mixture of effluent and river, $C_m$ (mg/L), given the concentration, $C_w$ (mg/L); the rate of flow, $Q_w$ (m³/s), of pollutant introduced to the river from the treatment facility; the river flow, $Q_r$ (m³/s); and the amount of pollutant initially found in it, $C_r$ (mg/L).

2. A sewage treatment work discharges its wastewater at the rate of 8 m³/s with a pollutant concentration of 15 mg/L to a receiving stream. The concentration of a particular pollutant upstream of the sewage treatment plant is 2 mg/L, and the river flows at a rate of 720 m³/min. Find the pollutant concentration downstream from the sewage treatment plant.

3. Use the program indicated in (1) to verify computations of (2).

**Solution**

1. For the solution to Example 5.20 (1), see the listing of Program 5.20.
2. Solution to example 5.20 (2):
   a. Given: $Q_w = 8$ m³/s, $C_w = 15$ mg/L, $C_r = 2$ mg/L, $Q_r = 720$ m³/min $= 12$ m³/s.
   b. Compute the pollutant concentration of the mixture downstream of the treatment

plant as follows: $C_m*Q_m = C_w * Q_w + C_{st} * Q_{st}$.

c. Find the discharge of the mixture of river water flow and the treatment plant effluent as follows: $Q_m = Q_w + Q_{st} = 8 + 12 = 20$ m³/s.
d. Find the concentration of the pollutant of mixture as follows: $C_m = (15 * 8 + 2 * 12)/20 = 7.2$ mg/L.

---

**LISTING OF PROGRAM 5.20   (CHAP5.20\FORM1.VB): CONCENTRATION OF POLLUTANTS IN RIVERS**

```vb
'*******************************************************************************
'Program 5.20: Concentration of Pollutants in Rivers
'*******************************************************************************

Imports System.Math
Public Class Form1
  Dim Qw, Qr, Cw, Cr As Double
  Dim Qm, Cm As Double

  Private Sub Form1_Load(ByVal sender As System.Object, ByVal e As System.EventArgs)
Handles MyBase.Load
    Me.Text = "Program 5.20: Concentration of Pollutants in Rivers"
    Me.MaximizeBox = False
    Me.FormBorderStyle = Windows.Forms.FormBorderStyle.FixedSingle
    Label1.Text = "Enter the wastewater flow rate, Qw (m3/s):"
    Label2.Text = "Enter the river flow, Qr (m3/s):"
    Label3.Text = "Enter the effluent pollutant concentration, Cw (mg/L):"
    Label4.Text = "Enter the concentration of pollutants in river, Cr (mg/L):"
    Label5.Text = ""
    Label6.Text = "Decimal places:"
    Button1.Text = "&Calculate"
    NumericUpDown1.Value = 2
    NumericUpDown1.Maximum = 10
    NumericUpDown1.Minimum = 0
  End Sub

  Sub calculateResults()
    Qw = Val(TextBox1.Text)
    Qr = Val(TextBox2.Text)
    Cw = Val(TextBox3.Text)
    Cr = Val(TextBox4.Text)
    Qm = Qw + Qr
    Cm = (Cw * Qw + Cr * Qr) / Qm
  End Sub

  Sub showResults()
    Label5.Text = "The concentration of pollutants in the river = " + FormatNumber(Cm,
NumericUpDown1.Value) + " mg/L"
  End Sub

  Private Sub Button1_Click(ByVal sender As System.Object, ByVal e As System.EventArgs)
Handles Button1.Click
    calculateResults()
    showResults()
  End Sub

  Private Sub NumericUpDown1_ValueChanged(ByVal sender As System.Object, ByVal e As
System.EventArgs) Handles NumericUpDown1.ValueChanged
    showResults()
  End Sub
End Class
```

## 5.13.2 Disposal into Natural Waters

### 5.13.2.1 Introduction

Natural bodies of waters may be used for discharge of treated wastewater when discharge regulations and standards are met. The reduction of pollutant concentrations exceeding permit discharge regulations may be achieved by adoption of appropriate treatment unit operations and processes. Table 5.16 gives general classification patterns for rivers in terms of biochemical oxygen demand, suspended solids, and dissolved oxygen.

The adverse effects of improperly treated wastewater discharges to rivers or water courses include the following (Rowe and Abdel-Magid 1995):

- Prevention of usage of waster course for beneficial purposes
- Deterioration in quality of a water course
- Endangering aquatic life
- Generation of odors, tastes, and nuisances
- Introduction of public epidemiological health hazards in the form of classical diseases or chronic chemical poisoning
- Delaying the mechanisms of self-purification in a water course

### 5.13.2.2 Oxygen Renewal and Depletion in Rivers

The main sources of oxygen renewal or reoxygenation in a river are reaeration from the atmosphere and photosynthesis of aquatic plants and algae (Rowe and Abdel-Magid 1995; Metcalf and Eddy 2013). Atmospheric reaeration may be evaluated by using the following equation:

$$r_r = k''(C_s - C) \qquad (5.103)$$

where:

$r_r$ is the rate of reaeration
$C_s$ is the dissolved oxygen saturation concentration, mg/L
$C$ is the dissolved oxygen concentration, mg/L
$K''$ is the reaeration constant, day$^{-1}$ (to base e)

$k''$ can be estimated from Equation 5.104 (O'Connor and Dobbins 1956; Rowe and Abdel-Magid 1995; Metcalf and Eddy 2013) or chosen from tables such as Table 5.17:

$$k'' = \frac{294 * \text{Diff}_T * \sqrt{v}}{\sqrt{H^3}} \qquad (5.104)$$

where:

v is the mean river velocity, m/s
$h$ is the average depth of flow, m
$\text{Diff}_T$ is the molecular diffusion coefficient for oxygen, m$^2$/d

$$(\text{Diff})_T = (\text{Diff}_c) * (T_c)(T - 20) \qquad (5.105)$$

where:

$\text{Diff}_T$ is the molecular diffusion coefficient for oxygen at a temperature of $T°C$, m$^2$/day
$\text{Diff}_c$ is the molecular diffusion coefficient for oxygen at temperature of 20°C, m$^2$/day; this may be taken to be equal to 1.76 * 10$^{-4}$
$T_c$ is the temperature correction coefficient that may be taken to be equal to 1.037
$T$ is the temperature, °C

For temperature other than 20°C, $k''$ can be determined by using the following equation:

$$(k'')_T = (k'')_{20} * (1.024)^{(T-20)} \qquad (5.106)$$

## TABLE 5.16
### River Classification Patterns

| Classification Scheme | BOD$_5{}^{20}$ (mg/L) | SS (mg/L) | DO, as a Percentage of Saturation Value |
|---|---|---|---|
| Very clean | ≤1 | ≤4 | |
| Clean | 2 | 10 | ≥90 |
| Fairly clean | 3 | 15 | 75–90 |
| Doubtful | 5 | 21 | 50–75 |
| Poor | 7.5 | 30 | <50 |
| Bad | 10 | 35 | |
| Very bad | ≥20 | ≥40 | |

*Sources:* Abdel-Magid 2012; Rowe, D.R. and Abdel-Magid, I.M., *Handbook of Wastewater Reclamation and Reuse*, CRC Press/Lewis Publishers, Boca Raton, FL, 1995; Peirce, J. J. et al. *Environmental Pollution and Control*, 4th ed, Butterworth-Heinemann, Oxford, 1997; Abdel-Magid, I. M., Selected Problems in Wastewater Engineering, National Research Council, Khartoum University Press, Khartoum, Sudan, 1986. With permission.

## TABLE 5.17
### Reaeration Constants

| Water Body | Range of $k''$ at 20°C (to base e) |
|---|---|
| Small ponds and back waters | 0.1–0.23 |
| Sluggish streams and large lakes | 0.23–0.35 |
| Large streams of low velocity | 0.35–0.46 |
| Large streams of normal velocity | 0.46–0.69 |
| Swift streams | 0.69–1.15 |
| Rapids and waterfalls | >1.15 |

*Sources:* Metcalf and Eddy, Inc., Tchobanoglous, G., Stensel, H. D., Tsuchihashi R., and Burton, F., Wastewater Engineering: Treatment and Resource Recovery, 5th ed, McGraw-Hill Science/Engineering/Math, New York, 2013; Rowe, D.R. and Abdel-Magid, I.M., *Handbook of Wastewater Reclamation and Reuse*, CRC Press/Lewis Publishers, Boca Raton, FL, 1995. With permission.

where:

$(k'')_T$ is the reaeration constant at a temperature of $T°C$
$(k'')_{20}$ is the reaeration constant at a temperature of $20°C$

Oxygen depletion (deoxygenation) in a river is due primarily to microbial metabolic activities and benthic deposits (Rowe and Abdel-Magid 1995; Metcalf and Eddy 2013). Deoxygenation by microbial decomposition of organic matter may be estimated as shown in the following equation:

$$r_D = -k' * L_o * e^{-k'*t} \qquad (5.107)$$

where:

$r_D$ is the rate of deoxygenation
$k'$ is the first-order reaction rate constant, day$^{-1}$
$L_o$ is the ultimate BOD at point of discharge, mg/L
$t$ is the time, days

The effect of organic solids and sediments accumulated in the benthic layer may be estimated by the Fair et al. empirical equation (Fair et al. 1948; Rowe and Abdel-Magid 1995):

$$L_m = 3.14\left(10^{-2} L_o\right)T_c * VS * \frac{5+160VS}{1+160VS} \sqrt{t} \qquad (5.108)$$

where:

$L_m$ is the maximum daily benthal oxygen demand, g/m$^2$
$L_0$ is the BOD$_5{}^{20}$ of benthal deposit, g/kg volatile matter
$T_c$ is the temperature correction factor
VS is the daily rate of volatile solids deposition, kg/m$^2$
$t$ is the time during which settling takes place, day

### 5.13.2.3  Dissolved Oxygen Sag Curves in Rivers

Equation 5.109 gives the simple river oxygenation model of Streeter and Phelps (Rowe and Abdel-Magid 1995; Metcalf and Eddy 2013):

$$DO_t = \left[\frac{k'}{\left(k''-k'\right)} * L_o \left(e^{-k'*t} - e^{-k''*t}\right)\right] + DO_o * e^{-k'*t} \qquad (5.109)$$

where:

$DO_t$ is the oxygen deficit at time $t$, mg/L
$k'$ is the first-order reaction rate constant, day$^{-1}$
$L_o$ is the ultimate BOD at point of discharge, mg/L
$k''$ is the reaeration constant, day$^{-1}$
$t$ is the time of travel in river from point of discharge, days
$DO_o$ is the initial oxygen deficit at point of waste discharge, at time $t = 0$

The limitations of the model are as follows (Peavy et al. 1985; Steel and McGhee 1991; Rowe and Abdel-Magid 1995; Masters and Ela 2007; Viessman et al. 2008; Abdel-Magid 2012; Metcalf and Eddy 2013):

- It ignores the effects of oxygen production by algae.
- It omits oxygen depletion by benthic deposits.

- It assumes one pollution source or a point source.
- It neglects factors influencing organic load apart from BOD.
- It assumes steady-state conditions along each river reach.

In spite of these limitations, the model provides reasonable estimation and is used extensively.

The point of lowest dissolved oxygen concentration is an important point in the sag curve. This point represents the maximum impact on the dissolved oxygen deficit due to organic waste disposal. The critical oxygen deficit may be estimated as indicated in the following equation:

$$DO_c = \frac{k'}{k''} L_o\, e^{-k't_c} \qquad (5.110)$$

where:

$DO_c$ is the critical oxygen deficit, mg/L
$k'$ is the first-order reaction rate constant, day$^{-1}$
$L_o$ is the ultimate carbonaceous BOD at point of discharge, mg/L
$t_c$ is the critical time required to reach the critical distance

The critical time can be found from d(DO)/d$t$ = 0, as found in the following equation:

$$t_c = \frac{1}{k''-k'} \ln\left\{\frac{k''}{k'}\left[1 - \frac{DO_o\left(k''-k'\right)}{k'L_o}\right]\right\} \qquad (5.111)$$

where:

$t_c$ is the critical time, days
$k''$ is the reaeration constant, day$^{-1}$
$k'$ is the first-order reaction rate constant, day$^{-1}$
$DO_o$ is the initial oxygen deficit at the point of waste discharge, at time $t = 0$
$L_o$ is the ultimate BOD at point of discharge, mg/L

The critical distance may be found from the following equation:

$$X_c = t_c * v \qquad (5.112)$$

where:

$x_c$ is the critical distance, m
$t_c$ is the critical time, days
v is the velocity of flow in the river, m/day

### Example 5.21

1. Develop a computer program to find the temperature, DO, and 5-day BOD of a mixture of river water and wastewater; the initial dissolved oxygen deficit of the river just below wastewater treatment plant outfall; the distance downstream to critical DO; and the minimum DO in the river below the wastewater treatment work, given the daily production of wastewater (m$^3$/s),

the treatment plant effluent BOD (mg/L), the temperature of the wastewater (°C), dissolved oxygen in effluent of plant as the percentage of saturation concentration (%), river flow ($m^3/s$), the velocity of flow in river (m/s), the temperature of river water before wastewater addition to river (°C), dissolved oxygen content in river as the percentage of saturation concentration (%), river BOD (mg/L), the first-order reaction constant $k'$ ($day^{-1}$), the reaeration constant $k''$ ($day^{-1}$), and the temperature correction for $k'$ and $k''$.

2. A river flows at a rate of 12,600 $m^3/h$ with a 5-day BOD of 1 mg/L, and it is saturated with oxygen. Wastewater effluent is discharged to the river at the rate of 12 $m^3/min$ with a 5-day BOD of 200 mg/L. The wastewater is 10% saturated with dissolved oxygen. The velocity of the river is 3 km/h. Assume the temperatures to be constant at 20°C, and $k' = 0.1$, $k'' = 0.4$/day. Compute for the mixture: the dissolved oxygen concentration, the 5-day BOD, the initial deficit in the river, the minimum dissolved oxygen content, and its location downstream from the wastewater treatment plant.

### Solution

1. For the solution to Example 5.21 (1), see the listing of Program 5.21.
2. Solution to Example 5.21 (2):

a. Given: $Q_r = 12600/60 * 60 = 3.5$ $m^3/s$, $BOD_r = 1$ mg/L, $DO_r = C_s$, $Q_w = 12/60 = 0.2$ $m^3/s$, $BOD_w = 200$ mg/L, $DO_w = 0.1 * C_s$.

b. Find the saturation value of oxygen, $C_s$, from Appendix A2 as equal to 9.2 for a temperature $T$ of 20°C and $DO_w = 0.1 * 9.2 = 0.92$ mg/L.

c. Determine the BOD of the mixture of river water and wastewater effluent by using the dilution law as follows: $BOD_m = (C_r * Q_r + C_w * Q_w)/Q_m = (1 * 3.5 + 200 * 0.2)/(3.5 + 0.2) = 11.76$ mg/L.

d. Determine the dissolved oxygen content of the mixture: $DO_m = (9.2 * 3.5 + 0.2 * 9.2)/(3.5 + 0.2) = 8.7524$ mg/L.

e. Determine the ultimate BOD of the mixture: $L_o = BOD_m/(1 - e^{-k'*5})$; $L_{om} = 11.76/(2 - e^{-0.1 * 5}) = 29.88$ mg/L.

f. Compute the initial dissolved oxygen concentration as follows: $DO_o = C_s - DO_m = 9.2 - 8.7524 = 0.4476$ mg/L.

g. Find critical time: $t_c = [1/(k'' - k')]*\ln[(k''/k')(1 - \{(DO_m L_{om})*[(k'' - k'')/k'']\})] = 4.468$ days.

h. Determine the critical oxygen as follows: $DO_c = (k' * L_o * e^{-k' * tc})/k''$; $DO_c = (0.1/0.4) * 29.88 * e^{-0.1 * 4.468} = 4.39$ mg/L.

i. Determine the position downstream where $DO_c$ occurs: $X_c = v * tc = 3* (24 \text{ h}) * 4.468$ $d = 321.7$ km.

---

### LISTING OF PROGRAM 5.21 (CHAP5.21\FORM1.VB): DO AND BOD OF RIVER–WASTEWATER MIXTURES

```
'*****************************************************************************
'Program 5.21: DO & BOD of River-Wastewater Mixtures
'*****************************************************************************

Imports System.Math
Public Class Form1
  Dim Cs(31) As Double
  Dim v, Qw, PDOw, BODw, Qr, Tr, PDOr, BODr, kd, kdd, ckd, ckdd As Double
  Dim Tw As Integer
  Sub loadCsTable()
    Cs(0)  = 14.6
    Cs(1)  = 14.2
    Cs(2)  = 13.8
    Cs(3)  = 13.5
    Cs(4)  = 13.1
    Cs(5)  = 12.8
    Cs(6)  = 12.5
    Cs(7)  = 12.2
    Cs(8)  = 11.9
    Cs(9)  = 11.6
    Cs(10) = 11.3
    Cs(11) = 11.1
    Cs(12) = 10.8
    Cs(13) = 10.6
    Cs(14) = 10.4
    Cs(15) = 10.2
    Cs(16) = 10.0
    Cs(17) = 9.7
```

```
        Cs(18) = 9.5
        Cs(19) = 9.4
        Cs(20) = 9.2
        Cs(21) = 9.0
        Cs(22) = 8.8
        Cs(23) = 8.7
        Cs(24) = 8.5
        Cs(25) = 8.4
        Cs(26) = 8.2
        Cs(27) = 8.1
        Cs(28) = 7.9
        Cs(29) = 7.8
        Cs(30) = 7.6
    End Sub

    Private Sub Form1_Load(ByVal sender As System.Object, ByVal e As System.EventArgs)
Handles MyBase.Load
        'data from Table A2
        loadCsTable()
        Me.Text = "Program 5.21: DO & BOD of River-Wastewater Mixtures"
        Me.MaximizeBox = False
        Me.FormBorderStyle = Windows.Forms.FormBorderStyle.FixedSingle
        Label1.Text = "Enter the daily production of wastewater, Qw (m3/s):"
        Label2.Text = "Enter the temperature of wastewater, Tw (C):"
        Label3.Text = "Enter the percentage of saturation dissolved oxygen in waste, %:"
        Label4.Text = "Enter the 20 C 5-day BOD of the waste, BOD205w (mg/L):"
        Label5.Text = "Enter the river flow, Qr (m3/s):"
        Label6.Text = "Enter the temperature of the river before the waste, Tr:"
        Label7.Text = "Enter the percentage of saturation dissolved oxygen in river, %:"
        Label8.Text = "Enter the 20 C 5-day BOD of the river, BOD205r(mg/L):"
        Label9.Text = "Enter the velocity of the river, v(km/h):"
        Label10.Text = "Enter the first order reaction constant (at 20C), k' /day:"
        Label11.Text = "Enter the reaeration constant (at 20C), k'' /day:"
        Label12.Text = "Enter the temperature correction of k':"
        Label13.Text = "Enter the temperature correction for k'':"
        Button1.Text = "&Calculate"
        TextBox14.Multiline = True
        TextBox14.Height = 135
        TextBox14.ScrollBars = ScrollBars.Vertical
    End Sub

    Sub calculateResults()
        Qw = Val(TextBox1.Text)
        Tw = Val(TextBox2.Text)
        PDOw = Val(TextBox3.Text)
        BODw = Val(TextBox4.Text)
        Qr = Val(TextBox5.Text)
        Tr = Val(TextBox6.Text)
        PDOr = Val(TextBox7.Text)
        BODr = Val(TextBox8.Text)
        V = Val(TextBox9.Text)
        kd = Val(TextBox10.Text)
        kdd = Val(TextBox11.Text)
        ckd = Val(TextBox12.Text)
        ckdd = Val(TextBox13.Text)
        If PDOw < 1 Then PDOw = PDOw * 100
        If PDOr < 1 Then PDOr = PDOr * 100
        Dim Csw, DOw, Csr, DOr, BOD5m, LOm, Csm, DOi, Kd20, Tc, Xc, Dc, DOc As Double
        Dim BOD5, BOD205c, Qm, Tm, DOm As Double
        'DO concentration in wastewater at Tw
        Csw = GetCs(Tw)
```

```
        'DO of wastewater at Tw
        DOw = PDOw / 100 * Csw
        'DO concentration in river at Tr
        Csr = GetCs(Tr)
        'DO of river at Tr
        DOr = PDOr / 100 * Csr
        'Mixture flow
        Qm = Qw + Qr
        'mixture temp.
        Tm = (Qw * Tw + Qr * Tr) / Qm
        'mixture DO
        DOm = (Qw * DOw + Qr * DOr) / Qm
        'mixture BOD5
        BOD5m = (Qw * BODw + Qr * BODr) / Qm
        'mixture Lo
        LOm = BOD5m / (1 - Exp(-kd * 5))
        'DO concentration in mixture at Tm
        Csm = GetCs(Tm)
        'initial oxygen deficit in mixture
        DOi = Csm - DOm
        'store value of kd at 20C
        Kd20 = kd
        'corrected kd at temperature of mixture
        kd = kd * ckd ^ (Tm - 20)
        'corrected kdd at temperature of mixture
        kdd = kdd * ckdd ^ (Tm - 20)
        'determine critical time tc
        Tc = 1 / (kdd - kd) * Log(kdd / kd * (1 - DOi / LOm * (kdd - kd) / kd))
        'convert to km/d
        v = v * 24
        'position downstream where critical conditions exist
        Xc = v * Tc
        Dc = kd / kdd * LOm * Exp(-kd * Tc)
        'DO at xc
        DOc = Csm - Dc
        'BOD5 at xc at temp. Tm
        BOD5 = LOm * Exp(-kd * Tc)
        'BOD5 at xc at temp. 20C
        BOD205c = BOD5 * (1 - Exp(-Kd20 * 5))
        Dim s As String
        s = "RESULTS:" + vbCrLf
        s += "temperature of the mixture = " + FormatNumber(Tm, 2) + " C" + vbCrLf
        s += "Dissolved oxygen of the mixture = " + FormatNumber(DOm, 2) + " mg/L" + vbCrLf
        s += "5-day BOD of the mixture = " + FormatNumber(BOD5m, 2) + " mg/L" + vbCrLf
        s += "Initial DO deficit in river = " + FormatNumber(DOi, 2) + " mg/L" + vbCrLf
        s += "Critical time tc for minimum DO = " + FormatNumber(Tc, 2) + " day" + vbCrLf
        s += "Position where minimum DO occurs, xc = " + FormatNumber(Xc, 2) + " km" +
vbCrLf
        s += "Minimum DO in the river (critical) = " + FormatNumber(DOc, 2) + " mg/L" +
vbCrLf
        s += "5-day BOD at xc at 20C = " + FormatNumber(BOD205c, 2) + " mg/L"
        TextBox14.Text = s
    End Sub

    Function GetCs(ByVal T) As Double
        If T < 0 Or T > 30 Then Return 0.0
        Return Cs(T)
    End Function
```

```
   Private Sub Button1_Click(ByVal sender As System.Object, ByVal e As System.EventArgs)
   Handles Button1.Click
        calculateResults()
     End Sub
End Class
```

### 5.13.3  Disposal into Lakes

Treated wastewater can find its way into lakes and reservoirs, especially in inland locations where nearby streams are not present. Lakes and reservoirs are often subject to a significant mixing due to wind-induced currents (Metcalf and Eddy 2013). The theoretical model for analysis assumes complete mixing conditions (for small lakes and reservoirs), constant flow rates, and biodegradation of the pollutant follows the first-order reaction ($r_c = K' * C$). Figure 5.11 is a schematic of a lake system receiving a contaminant. The material mass balance for this lake is given by Equation 5.113, which estimates the concentration of pollutant in the lake[1]:

$$C = \left[ W * \frac{\left(1 - e^{\beta * t}\right)}{\beta * V} \right] + C_0 * e^{-\beta * t} \qquad (5.113)$$

where:

$C$ is the pollutant concentration in the lake and in effluent, $kg/m^3$

$$W = Q_r * C_r + Q_w * C_w \qquad (5.114)$$

where:

$Q_r$ is the river flow rate into lake, $m^3/s$
$C_r$ is the concentration of pollutant in river, $kg/m^3$
$Q_w$ is the wastewater flow rate into lake, $m^3/s$
$C_w$ is the concentration of pollutant in wastewater, $kg/m^3$

$$\beta = \frac{1}{t} + k' \qquad (5.115)$$

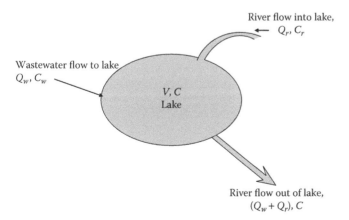

**FIGURE 5.11**  Disposal into a lake. $V$, lake volume; $C$, concentration of pollutant.

where:

$t$ is the detention time $= V/Q$
$k'$ is the first-order decay constant (base e), $day^{-1}$
$V$ is the lake volume
$C_0$ is the concentration in the lake at time $t = 0$

The equilibrium concentration can be found by letting $t$ equal $\infty$, as presented in the following equation:

$$C_e = \frac{W}{\beta V} \qquad (5.116)$$

where $C_e$ is the equilibrium waste concentration in the lake.

**Example 5.22**

1. Write a computer program to find the equilibrium concentration of a radioactive pollutant discharged into a small lake with average dimensions (m) $L$, $B$, $d$ for length, breadth, depth, respectively, given the concentration of pollutant entering the lake $C_p$ (mg/L), half-life of pollutant $t_{1/2}$, at a flow rate of $q_p$ (m³/s). Assume that wind currents enable complete mixing of the lake contents.
2. Use the program developed in (1) to determine the equilibrium concentration of a radioactive substance of a 12 h half-life, discharged at a rate of 0.05 m³/s with a concentration of 4 mg/L to a small lake of average dimensions 60 * 25 * 2 m for length, width, and depth, respectively.

**Solution**

1. For solution to Example 5.22 (1), see the listing of Program 5.22.
2. Solution to Example 5.22 (2):
   a. Given: $t_{1/2} = 12/24 = 0.5$ d, $Q_r = 0$ (no river flow out of lake), $q_w = 0.05$ m³/s, $C_w = 4$, $L = 60$ m, $B = 25$ m, $d = 2$ m.
   b. Determine $W$ as follows: $W = Q_r * C_r + q_w * C_w = 0 + 0.05 * 4 = 0.2$ g/s = 17280 g/day.
   c. Determine the detention time: $t = V/Q = L * B * d/(q_r + q_w) = 60 * 25 * 2/0.05 * 60 * 60 * 24 = 0.694$ day.
   d. Find the first-order decay rate as follows: $L_t = L_o * e^{-k't}$, and when $t = 0.5$ day, $L_{1/2} = 1/2 L_o$ or $k' = -\ln(1/2) = 0.693$/day.
   e. Determine the factor $\beta$ as follows: $\beta = 1/t_o + k' = (1/0.694) + 0.693 = 2.133$/day.
   f. Find the equilibrium concentration as follows: $C_e = W/\beta V = 17,280/(2.133 * 60 * 25 * 2) = 2.7$ mg/L.

**LISTING OF PROGRAM 5.22   (CHAP5.22\FORM1.VB): CONCENTRATION OF POLLUTANTS IN LAKES**

```vb
'********************************************************************************
'Program 5.22: Equilibrium concentration of pollutants in lakes
'********************************************************************************

Imports System.Math
Public Class Form1
  Dim Ceq As Double

  Private Sub Form1_Load(ByVal sender As System.Object, ByVal e As System.EventArgs)
Handles MyBase.Load
      Me.Text = "Program 5.22: Equilibrium concentration of pollutants in lakes"
      Me.MaximizeBox = False
      Me.FormBorderStyle = Windows.Forms.FormBorderStyle.FixedSingle
      Label1.Text = "Enter the length of the lake, L (m):"
      Label2.Text = "Enter the breadth of the lake, B (m):"
      Label3.Text = "Enter the depth of the lake, d (m):"
      Label4.Text = "Enter the concentration of pollutants entering, Cw (mg/L):"
      Label5.Text = "Enter the half-life of pollutants, tw (h):"
      Label6.Text = "Enter the flow rate of pollutants, Qw (m3/s):"
      Label7.Text = ""
      Label8.Text = "Decimals:"
      Button1.Text = "&Calculate"
      NumericUpDown1.Value = 2
      NumericUpDown1.Maximum = 10
      NumericUpDown1.Minimum = 0
  End Sub

  Sub calculateResults()
  Dim L, B, d, tw, Qw, Cw, Qr, Q, Cr, W, V, t, k, beta As Double
    L = Val(TextBox1.Text)
    B = Val(TextBox2.Text)
    d = Val(TextBox3.Text)
    Cw = Val(TextBox4.Text)
    tw = Val(TextBox5.Text)
    Qw = Val(TextBox6.Text)
    'convert to days
    tw = tw / 24
    'convert to m3/d
    Qw = Qw * 3600 * 24
    'no flow out of the lake
    Qr = 0
    Cr = 0
    W = Qr * Cr + Qw * Cw
    'volume
    V = L * B * d
    Q = Qr + Qw
    'detention time
    t = V / Q
    k = -Log(tw)
    beta = (1 / t) + k
    Ceq = W / (beta * V)
  End Sub

  Sub showResults()
      Label7.Text = "The equilibrium concentration (mg/L): " + FormatNumber(Ceq,
NumericUpDown1.Value)
  End Sub
```

```
   Private Sub Button1_Click(ByVal sender As System.Object, ByVal e As System.EventArgs)
Handles Button1.Click
      calculateResults()
      showResults()
   End Sub


   Private Sub NumericUpDown1_ValueChanged(ByVal sender As System.Object, ByVal e As
System.EventArgs) Handles NumericUpDown1.ValueChanged
      showResults()
   End Sub
End Class
```

## 5.13.4  Disposal in Estuaries

An estuary signifies the zone where a river meets the sea. Tidal action is assumed to result in a significant lateral mixing in reaches of the river adjacent to the estuary and increases the amount of mixing and dispersion of waste disposed along the length of channel. For continuous discharge of waste to an estuary, the concentration may be determined as presented in the following equation (Metcalf and Eddy 2013):

$$C = C_0 * e^{j*X} \qquad (5.117)$$

where:

$C$ is the concentration of waste discharged at time $t$
$C_0$ is the initial concentration in estuary

$$C_0 = \frac{W}{\left[ Q(1+4k'*E/v^2)^{0.5} \right]} \qquad (5.118)$$

$$J = \left( \frac{v}{2E} \right) * \left[ (1 \pm (1+4k'*E/v^2)^{0.5} \right] \qquad (5.119)$$

where:

$E$ is the coefficient of eddy diffusion or turbulent mixing, $m^2/s$
v is the velocity of flow in river ($= Q/A$), m/s
$A$ is the cross-sectional area, $m^2$
$k'$ is the first-order decay constant, $day^{-1}$
$W$ is the pollutant load as determined by Equation 5.114

The positive root for $J$ refers to the upstream $(-x)$ direction, and the negative root refers to the downstream $(+x)$ direction.

### Example 5.23

Write a computer program that enables determination of the concentration of a pollutant at any time $t$ after its disposal to a neighboring estuary, given the relevant parameters.

**Solution**

See the listing of Program 5.23.

---

**LISTING OF PROGRAM 5.23   (CHAP5.23\FORM1.VB): WASTE CONCENTRATION IN ESTUARIES**

```
'****************************************************************************************
'Program 5.23: Waste concentration in estuaries
'****************************************************************************************

Imports System.Math
Public Class Form1
  Dim A, Q, k, E, W, x, C As Double

  Private Sub Form1_Load(ByVal sender As System.Object, ByVal e As System.EventArgs)
Handles MyBase.Load
     Me.Text = "Program 5.23: Waste concentration in estuaries"
     Me.MaximizeBox = False
     Me.FormBorderStyle = Windows.Forms.FormBorderStyle.FixedSingle
     Label1.Text = "Enter the cross-sectional area of the river, A (m2):"
     Label2.Text = "Enter the flow rate in river, Q (m3/s):"
     Label3.Text = "Enter the decay constant, k:"
     Label4.Text = "Enter the coefficient of turbulent mixing, E (m2/s):"
     Label5.Text = "Enter the pollutant load, W:"
```

```
        Label6.Text = "Enter the distance where concentration is required, x (m):"
        Label7.Text = ""
        Label18.Text = "Decimals:"
        Button1.Text = "&Calculate"
        NumericUpDown1.Value = 2
        NumericUpDown1.Maximum = 10
        NumericUpDown1.Minimum = 0
    End Sub

    Sub calculateResults()
        A = Val(TextBox1.Text)
        Q = Val(TextBox2.Text)
        k = Val(TextBox3.Text)
        E = Val(TextBox4.Text)
        W = Val(TextBox5.Text)
        x = Val(TextBox6.Text)
        Dim v, J, Co As Double
        v = Q / A
        If x < 0 Then
            J = (v / (2 * E)) * (1 + (1 + 4 * k * E / (v ^ 2)) ^ 0.5)
        Else
            J = (v / (2 * E)) * (1 - (1 + 4 * k * E / (v ^ 2)) ^ 0.5)
        End If
        Co = W / (Q * (1 + 4 * k * E / (v ^ 2)) ^ 0.5)
        C = Co * Exp(J * x)
    End Sub

    Sub showResults()
        Label7.Text = "The concentration of wastes: " + FormatNumber(C, NumericUpDown1.
Value)
    End Sub

    Private Sub Button1_Click(ByVal sender As System.Object, ByVal e As System.EventArgs)
Handles Button1.Click
        calculateResults()
        showResults()
    End Sub

    Private Sub NumericUpDown1_ValueChanged(ByVal sender As System.Object, ByVal e As
System.EventArgs) Handles NumericUpDown1.ValueChanged
        showResults()
    End Sub
End Class
```

## 5.14 HOMEWORK PROBLEMS IN WASTEWATER COLLECTION SYSTEM, WASTEWATER TREATMENT TECHNOLOGY, AND DISPOSAL

### 5.14.1 DISCUSSION PROBLEMS

1. What are the various sources of wastewater?
2. Outline the problems associated with sewer systems.
3. Define the following terms: sewer, sewerage system, and storm sewer.
4. What are the functions of sewers?
5. Indicate the classes of sewer systems. Point out the advantages and disadvantages of each system. Which system would you recommend in your community? State your reasons.
6. What is the difference between the terms sewage and wastewater?
7. Briefly discuss the investigations that need to be carried out during sewer design.
8. Why do industrial and commercial rates of flow of wastewater experience diurnal variations?
9. Indicate the factors influencing design of sewers.
10. What are the benefits of maintaining a minimum flow rate in a sewer?
11. Outline the factors affecting the volume of storm water that must be controlled.
12. Point out the limitations of the rational method.

13. Why is there a need to have greater velocities in storm sewers compared to sanitary sewers?

14. Briefly explain the assumptions used in the design of sanitary sewers.

15. Define the following terms: one-dimensional flow, uniform flow, incompressible flow, steady flow, quasi-steady flow, continuity equation, Bernoulli's equation, momentum principle, hydraulic radius, laminar flow, and self-cleaning velocity.

16. What are the serious effects of generation of sulfides in sewer operations?

17. Indicate the factors that govern the generation of sulfide in a sanitary sewer.

18. Show how to avoid and control sulfide buildup in a sanitary sewer.

19. Write briefly about the methods of wastewater disposal in rural areas.

20. Define the following terms: septic tank, septage, Imhoff tank, Imhoff cone, and VIP latrine.

21. Outline the differences between a septic tank and an Imhoff tank.

22. Discuss the following statement: "The effluent of a septic tank is offensive and potentially dangerous."

23. What are the factors that affect the design of percolation fields?

24. State which system gives a better effluent, a septic tank or an Imhoff tank? Why? State your reasons.

25. Discuss the factors that affect the structural design of a reinforced concrete septic tank.

26. What are the pros and cons of using an Imhoff tank?

27. Define the following terms: wastewater, dry weather flow, population equivalent, and grit.

28. What are some of the problems associated with improper disposal of wastewater?

29. Outline the main sources of wastewater.

30. What are the reasons for the treatment of wastewaters?

31. Differentiate between wastewater treatment unit operations and processes. Give examples of each.

32. What are the problems associated with grit?

33. What are the reasons for installing secondary treatment units in a sewage treatment facility?

34. Differentiate between fixed-growth and suspended growth systems. Give examples for each.

35. Define the following terms: activated sludge, food-to-microorganisms ratio, volumetric loading rate, hydraulic loading rate, cell age, sludge volume index, and SDI.

36. What are the factors that affect the activated sludge process?

37. Briefly describe the trickling filter process.

38. What are the merits of recirculating a portion of a treated effluent to a trickling filter unit?

39. Indicate how to estimate the efficiency of the performance of a trickling filter unit.

40. Indicate the assumptions made in the NRC formula for estimating efficiency of a trickling filter.

41. Define the following terms: waste stabilization pond and symbiotic relationship.

42. Differentiate between anaerobic, aerobic, facultative, and maturation waste stabilization ponds. Indicate the factors affecting design and operation of each.

43. Differentiate between aerobic and anaerobic digestion. Indicate the factors affecting each.

44. Outline the major methods that can be used to dewater sewage sludges. Indicate the factors that govern the design and operation of each system. What demerits are to be expected in each system?

45. Outline the major methods that can be used to dewater sewage sludges. Indicate the factors that govern the design and operation of each system. What demerits are to be expected in each system?

46. What are the major factors that influence filterability of sewage sludges?

47. Define the following terms: sludge filtration, specific resistance to filtration, compressibility of sludge, and sigma equation.

48. What methods can be used to determine the filterability of a sludge sample? What are the limitations of each method?

49. What are the factors that affect the natural purification processes in a water course?

50. What are the limitations of the dilution law?

51. Show how to differentiate between a clean and a polluted river.

52. What are the effects of untreated wastewater discharges on surface water?

53. What are the main sources for reoxygenation of a river?

54. What are the limitations of the Streeter–Phelps model?

## 5.14.2 Specific Mathematical Problems

1. Using the equations mentioned in this chapter, write a computer program to find the limiting total sulfide, amount, and dissolved sulfide concentration that can develop in a sanitary sewer. Data given: the diameter of pipe (ft), the length of sanitary sewer (ft), the flow rate (cfs), the relative depth of flow, slope, Manning's coefficient, wastewater characteristics, BOD (mg/L), climatic temperature (°C), pH, insoluble sulfide (mg/L), total sulfide (mg/L), and hydraulic characteristics—surface width of flow (ft), wetted perimeter (ft), exposed perimeter (ft), velocity of flow (fps), and mean hydraulic depth (ft).

2. Write a computer program to design a two-component septic tank. The volume of the tank can be computed from the following equation: $V = 180 * POP + 2000$, where $V$ is volume of tank in liters and POP is the number of people to be served by the unit.

3. Write a computer program to design a suitable percolation field for a septic tank serving a certain number of people.

4. Determine the size of a septic tank and the percolation field for a mobile home park, which has 187 residents. Percolation tests indicate an average percolation rate of 6 mm/min.

5. Expand the computer program of Example 5.5 to enable plotting the plan, side view, and elevation of the septic tank.

6. a. Write a short computer program to determine the solids content for an activated sludge plant, given tank aeration volume, $V$ (L/c), MLVSS (mg/L), and sludge wasting rate, $Q_w$ (kg/c/day).

   b. Using this program, computer the solids content of the plant, given $V = 48$ L/c, MLVSS = 2400 mg/L, and $Q_w = 30$ g/c/day.

7. a. Write a computer program that can be used to estimate the NRC efficiency of a two-stage trickling filter unit, given the wastewater flow rate to the plant being $Q$ (m³/s); filter depth, $h$ (m); influent $BOD_5$, $L_i$ (mg/L); recirculation to the first filter, $Ru_1$; recirculation to the second filter, $Ru_2$; volume of the first filter, $V_1$ (m³); and volume of the second filter, $V_2$ (m³).

   b. Use this program to determine the efficiency of a two-stage trickling filter, given the following data: $Q = 3$ m³/min, $h = 2$ m, $L_i = 215$ mg/L, $Ru_1 = 150\%$ of $Q$, $Ru_2 = 125\%$ of $Q$, $V_1 = 250$ m³, and $V_2 = 250$ m³. Use the NRC formula to compute the effluent $BOD_5$ of the two-stage trickling filter unit.

8. a. Write a computer program to design a trickling filter to treat a wastewater flow of $Q$ (m³/s) with an average BOD of $L_i$ (mg/L). The treatment works are assumed to have one single-stage, high-rate trickling filter with a BOD loading, $BOD_1$ (kg/m³*s), and depth, $h$ (m). The 5-day $BOD^{20}$ removal ratio in the primary sedimentation tank and trickling filter is $E_s$ and $E_f$, respectively.

   b. Use this program to design a trickling filter, given the following data: $Q = 1600$ m³/day $L_i = 230$ mg/L, $h = 2$ m, $BOD_1 = 330$ kg/m³/day, $E_s = 35\%$, and $E_f = 62\%$.

9. a. Use the NRC equation to write a short computer program to calculate the influent concentration of BOD of a wastewater, $Y$, in mg/L, introduced to a trickling filter, given the filter diameter, $D$ (m); the depth, $h$ (m); the wastewater effluent, $Q$ (m³/s); recirculated flow of $Q_r$ (m³/s); and the final effluent $BOD_5$, $L_e$ (mg/L).

   b. Use this program to determine $Y$, given $D = 10$ m, $h = 2.5$ m, $Q = 0.02$ m³/s, $Q_r = 1.6$ m3/min, and $L_e = 25$ mg/L.

10. a. Write a computer program to compare the dewaterability of two sludges, A and B, which have the same coefficient of compressibility "s." The value of the specific resistance of sludge A at a pressure of $P_1$ (Pa) is given as $r_A$ m/kg. The specific resistance of sludge B is reported as $r_B$ (s²/g) at a pressure of $P_2$ (Pa).

    b. Use this program to compare the dewaterability of the two sludges A and B, given s = 0.75, $r_A = 2.3 * 10^{14}$ m/kg, $P_1 = 60$ kPa, $r_B = 3.7 * 10^8$ s²/g, $P_2 = 0.2$ MPa.

11. Expand the computer program of Example 5.21 so that the dissolved oxygen sag curve can be plotted for a river receiving the treated waste discharges indicated in this example.

12. a. Develop a computer program to determine the effluent 5-day BOD of a treated wastewater $Y$ (mg/L). The river flows at a rate of $q_r$ (m³/h) with a 5-day BOD of $BOD_r$ (mg/L) and is $x\%$ saturated with oxygen. The wastewater effluent is discharged to the river at a rate of $Q_w$ (m³/s) with a 5-day BOD of $Y$ (mg/L), and it is $z\%$ saturated with dissolved oxygen. The critical oxygen deficit amounts to $DO_c$ (mg/L). Temperatures are $T_r$ and $T_w$ degrees Celsius for the river water and wastewater effluent, respectively.

    b. Use the program developed in 12a to determine the value of $Y$ and the dissolved oxygen deficit after 3 days for the following data: $Q_r = 1800$ m³/h, $BOD_r = 1$ mg/L, $x = 100\%$, $Q_w = 2$ m³/min, $z = 10\%$, $DO_c = 2.5$ mg/L, $T_r = T_w = 20°C$, $k' = 0.1$/d, and $k'' = 0.4$/day.

## REFERENCES

Abdel-Magid, I. M. 1982. The role of filter aids in sludge dewatering. PhD thesis, University of Strathclyde, Glasgow.

Abdel-Magid, I. M. 1986. *Selected Problems in Wastewater Engineering.* Khartoum, Sudan: National Research Council, Khartoum University Press.

Abdel-Magid, I. M., and Abdel-Magid, M. I. M. 2015. *Problem Solving in Environmental Engineering.* North Charleston, SC: CreateSpace Independent Publishing Platform, Second version, improved, updated and revised.

Abdel-Magid, I. M., Alderdiri, A. M., and Abdel-Magid, M. I. M. 2015. *Wastewater* (Arabic Edition), 2nd ed. North Charleston, SC: CreateSpace Independent Publishing Platform.

American Water Works Association and American Society of Civil Engineers. 2012. *Water Treatment Plant Design,* 5th ed. New York: McGraw-Hill Professional.

Barnes, D., Bliss, P. J., Gould, B. W., and Vallentine, H. R. 1981. *Water and Wastewater Engineering Systems.* Marshfield, MA: Pitman Publishing.

Berger, B. B. (Ed.). 1987. *Control of Organic Substances in Water and Wastewater.* Park Ridge, NJ: Noyes Publishing.

Bizier, P. 2007. *Gravity Sanitary Sewer Design and Construction* (ASCE Manuals and Reports on Engineering Practice No. 60), 2nd ed. American Society of Civil Engineers.

Carman, P. C. 1938. Fundamental principles of industrial filtration. *Trans. Inst. Chem. Eng.* 16:168–188.

Coackley, P. 1953. The dewatering treatment. PhD thesis, London University.

Coackley, P. 1955. Research on sewage sludge carried out in the Civil Engineering Department of the University College London. *J. Proc. Instit. Sewage Purification* 1:59–72.

Coackley, P. 1965. The theory and practice of sludge dewatering. *J. Instit. Public Health Eng.* 64(1):34.

Coackley, P. 1967. Sludge dewatering treatment. *Proc. Biochem.* 2(3):17.

Coackley, P. 1975. *Development in Our Knowledge of Sludge Dewatering Behavior.* 8th Public Health Engineering Conference held in the Dept. of Civil Engineering, Loughborough University of Technology, 5.

Coackley, P., and Allos, R. 1962. The drying characteristics of some sewage sludges. *J. Proc. Inst. Sewage Purification* 6:557.

Davis, M. L., and Cornwell, D. A. 2006. *Introduction to Environmental Engineering,* 4th ed. New York: McGraw-Hill Science/Engineering/Math.

Fair, G. M., Geyer, J. C., and Okun, D. A. 1966. *Water and Wastewater Engineering,* Vols I and II. New York: John Wiley & Sons.

Fair, G. M., Morris, F. C., Chang, S. L., Weil, I., and Burden, R. A. 1948. The behavior of chlorine as a water disinfectant. *J. Am. Water Works Assoc.* 40:1051.

Ganczarczyk, J. J. 1983. *Activated Sludge Process: Theory and Practices.* Pollution Engineering and Technology Series No. 23. New York: Marcel Dekker.

Gunnerson, C. G., and Stuckey, D. C. 1986. *Integrated Resources Recovery: Anaerobic Digestion Principles and Practice for Biogas Systems.* Technical Paper Number 49. Washington, DC: World Bank.

Hammer, M. J. Sr., and Hammer, M. J. Jr. 2011. *Water and Wastewater Technology,* 7th ed. Upper Saddle River, NJ: Prentice Hall.

Health Education Services. 1990. *Recommended Standards for Wastewater Facilities: Policies for the Design, Review, and Approval of Plans and Specifications for Wastewater Collection and Treatment Facilities.* Report of the Wastewater Committee of the Great Lakes-Upper Mississippi River Board of State Public Health and Environmental Managers, Albany.

Hicks, T. 2014. *Standard Handbook of Engineering Calculations,* 5th ed. New York: McGraw-Hill Education.

Howe, K. J., Hand, D. W., Crittenden, J. C., and Trussell, R. R. 2012. *Principles of Water Treatment,* 1st ed. Hoboken, NJ: John Wiley & Sons.

James, A. 1984. *An Introduction to Water Quality Modeling.* New York: John Wiley & Sons.

Jerry, A., and Nathanson, P.E. 2007. *Basic Environmental Technology: Water Supply, Waste Management & Pollution Control,* 5th ed. Upper Saddle River, NJ: Prentice Hall.

Jonathan Ricketts, J., Loftin, M., and Frederick, M. 2003. *Standard Handbook for Civil Engineers,* 5th ed. New York: McGraw-Hill Professional.

Lin, S., and Lee, C. 2007. *Water and Wastewater Calculations Manual,* 2nd ed. New York: McGraw-Hill Professional.

Mara, D. 1980. *Sewage Treatment in Hot Climates.* New York: John Wiley & Sons.

Masters, G. M., and Ela, W. P. 2007. *Introduction to Environmental Engineering and Science,* 3rd ed. Upper Saddle River, NJ: Prentice Hall.

Mays, L. W. 2010. *Water Resources Engineering,* 2nd ed. Hoboken, NJ: John Wiley & Sons.

McGhee, T. J., and Steel, E. W. 1991. *Water Supply and Sewerage,* 5th ed. New York: McGraw-Hill College.

Metcalf and Eddy, Inc., Tchobanoglous, G., Stensel, H. D., Tsuchihashi R., and Burton, F. 2013. *Wastewater Engineering: Treatment and Resource Recovery,* 5th ed. New York: McGraw-Hill Science/Engineering/Math.

Mullick, M. A. 1987. *Wastewater Treatment Processes in the Middle East.* Sussex: The Book Guild.

Nathanson, J. A. 2007. *Basic Environmental Technology: Water Supply, Waste Management & Pollution Control,* 5th ed. Upper Saddle River, NJ: Prentice Hall.

Negulescu, M. 1986. *Municipal Wastewater Treatment.* Developments in Water Science Series No. 23. Amsterdam, the Netherlands: Elsevier.

Nemerow, N. L., Agardy, F. J., and Salvato, J. A. 2009. *Environmental Engineering,* 6th ed. Hoboken, NJ: John Wiley & Sons.

Newitt, D. M., Oliver, T. R., and Pearse, J. F. 1949. The mechanism of the drying of solids. *Trans. Inst. Chem. Eng.* 27:1.

O'Connor, D., and Dobbins, W. 1956. The mechanism of re-aeration in natural streams. *J. Sanitary Eng.* SA6.

Peavy, H. S., Rowe, D. R., and Tchobanoglous, G. 1988. *Environmental Engineering.* New York: McGraw-Hill.

Peirce, J. J., Vesilind, P. A., and Weiner, R. 1997. *Environmental Pollution and Control,* 4th ed. Oxford: Butterworth-Heinemann.

Rowe, D. R., and Abdel-Magid, I. M. 1995. *Handbook of Wastewater Reclamation and Reuse.* Boca Raton, FL: CRC Press/Lewis Publishers.

Sanin, F. D., Clarkson, W. W. and Vesilind P. A. 2010. *Sludge Engineering: The Treatment and Disposal of Wastewater Sludges.* Lancaster, PA: DEStech Publications.

Smith, P. G., and Scott, J. S. 2002. Dictionary of water and waste management. 2nd ed. London, UK: International Water Association (IWA) Publishing.

U.S. Agency for International Development. 1982. *Water for the World,* Technical notes produced under contract to the U.S. Agency for International Development by National Demonstration Water Project, Institute for Rural Water and National Environmental Health Association. (a) *Designing Septic Tanks,* Technical Notes No. SAN2.D.3. (b) *Operating and Maintaining Septic Tanks,* Technical Note No. SAN.2.0.3.

U.S. Environmental Protection Agency. 2012. *Design Manual: Dewatering Municipal Wastewater Sludges.* BiblioGov.

Viessman, W., and Lewis, G. L. 2002. *Introduction to Hydrology,* 5th ed. Upper Saddle River, NJ: Prentice Hall.

Viessman, W. Jr., Hammer, M. J., Perez, E. M., and Chadik, P. A. 2008. *Water Supply and Pollution Control,* 8th ed. Upper Saddle River, NJ: Prentice Hall.

Vernick, A. S., and Walker, F. C. 1981. *Handbook of Wastewater Treatment Processes.* Pollution Engineering and Technology Series No. 19. New York: Marcel Dekker.

Wang, L. K., and Hung, Y. 2005. *Waste Treatment in the Process Industries.* Boca Raton, FL: CRC Press.

# 6 Computer Modeling Applications for Municipal Solid Waste Classification, Quantities, Properties, Collection, Processing, Material Separation, and Cost Estimates

## 6.1 INTRODUCTION

Municipal solid waste (MSW) could be defined as a heterogeneous collection of residential throwaways from a certain community, as well as the more homogeneous accumulations of commercial, agricultural, industrial, and mineral wastes. Otherwise it may be defined as materials that do not represent major outputs of market product and serve no productive or consumptive purpose to its producer, which justifies its disposal. Generally, a solid waste represents those things that are unwanted, useless, of no added value, and not needed by someone (Abdel-Magid and Abdel-Magid 2015a; Worrell and Vesilind 2012).

The sources of solid waste, garbage, and sweeping vary, and they include agricultural, building and construction, commercial and offices, demolition, domestic from homes, educational institutional hospitals, housing, industrial, mining, open markets, restaurants, shops, and so on. Table 6.1 gives examples of the main sources and types of production units of solid waste and garbage. Variation in solid waste generation depends on many intervening factors such as location, season, climate and meteorological parameters, socioeconomic conditions and behavior, population and demographical aspects, and market for waste materials.

Equation 6.1 illustrates a rough estimate for the amount of MSW or refuse generated from a certain community:

$$MSW = Refuse$$
$$+ \text{construction and demolition waste} \quad (6.1)$$
$$+ \text{leaves} + \text{bulky items}$$

or

$$MSW = Refuse + C \& D \text{ waste}$$
$$+ \text{leaves} + \text{bulky items} \quad (6.2)$$

where C&D is the construction and demolition.

Refuse can be defined in terms of as-generated and as-collected solid waste. The refuse generated includes all of the wastes produced by a household. Often some part of the refuse, especially organic matter and yard waste, is composted on premises. The fraction of refuse that is generated but not collected is called diverted refuse. The as-generated refuse is always larger than the as-collected refuse, and the difference is the diverted refuse:

$$As\text{-generated refuse}$$
$$= As\text{-collected refuse} + \text{diverted refuse} \quad (6.3)$$

### Example 6.1

1. Given annually generated fractions of a MSW from a certain community, write a short computer program that calculates the percentage of diversion of the solid waste produced.
2. A certain community generates annually the quantities of MSW presented in the following table:

| Fraction of Waste Produced | Tons per Year |
|---|---|
| Commercial waste | 60 |
| C&D debris | 120 |
| Leaves and miscellaneous | 50 |
| Mixed house waste | 200 |
| Recyclables | 40 |

Generated recyclables are collected separately and processed at a materials recovery facility (MRF). Both mixed household and commercial wastes are dumped at the municipality landfill, as do the leaves and miscellaneous solid wastes. The C&D wastes are used to fill a large ravine located in the area. Calculate the percentage of diversion.
3. Use the program developed in (1) to verify the manual computations conducted in (2).

### Solution

1. For the solution of Example 6.1 (1), see the listing of Program 6.1.
2. Given: MSW annual quantities in tons.

**TABLE 6.1**

**Sources and Types of Production Units of Solid Waste**

| Type of Solid Waste | Production Units | Source of Waste |
|---|---|---|
| Spoiled food wastes, agricultural waste, rubbish, hazardous materials | Field, raw crops, and different farm types from planting farms, harvesting of fields, livestock farms producing dairy, butter and cheese, meat, slaughterhouses, fruit orchards vineyards squares, laboratories, experimental fields, feedlots, and so on. | Agricultural |
| Food waste, rubbish, ashes, demolition and construction waste, special waste, hazardous waste | Many of the items as generated by household waste that may arise from design health facilities, hotels, maintenance workshops and warehouses, markets, offices and office buildings, stores, shops, printing institutions, rest houses, restaurants, hotels and motels, auto repair shops, medical facilities, and so on | Commercial |
| Pathological and infectious, remains of experimental animals, corpses, remnants of drugs, poisons and chemicals, and containers, and so on | Radioactive materials, chemical, biomaterial, hospitals and medical waste, household hazardous waste, and so on | Hazardous waste |
| Food waste, rubbish, ashes, demolition and construction wastes, special waste, hazardous waste, bricks, concrete, dust, stones, mortar, outputs of refrigeration and air conditioning, plumbing, electricity, water, phone and networks connections, and so on | Refuse from construction, demolition and ruins of buildings, light and heavy manufacturing fabrications, oil refineries, chemical plants, power plants, sources of metals production and processing, mining, logging, and so on | Industrial |
| Paper, cardboard, plastics, wood, food wastes, glass, metals, special wastes, hazardous wastes. | Schools, hospitals, prisons, government centers and departments | Institutional |
| Damaged computers, peripherals and software, CD plastic, DVDs, and so on | Industrial domains, hardware units of information revolution, technology updates | Other waste products |
| Food waste, rubbish, garbage, ashes, special waste | Single and multifamily dwellings, houses, low medium, and high-rise apartments, villas, and so on having mixed household and residential waste, garbage, food waste, and recyclables (such as newspapers, aluminum cans, steel cans, milk cartons, plastic soft drink bottles, corrugated cardboard, and other material collected by the community) | Residential or domestic MSW from residential homes, community |
| Special waste, rubbish | Sweeping (thrown on the side of road from users), damaged cars placed on both sides of road, cleanliness of streets, alleys, parks, playgrounds, beaches, bathing and recreational areas, squares, highways, gardens litter, and waste from community trash cans produced by individuals, municipal containers, debris and rubble, dead animals (small animals—such as pets like cats, dogs, rabbits, and turtles—and big animals—such as horses, sheep, donkeys, and cattle | Square and open areas Street refuse |
| Treatment plant wastes, principally composed of residuals, sludge, and so on | Water and wastewater treatment processes, solid waste treatment, purification plants, industrial wastewater treatment processes, and air pollution and control plants | Treatment plant sites |
| Bulky refuse items, large household applications such as refrigerators, stoves, air conditioners, washing machines, and so on | Bicycles, furniture, old and used cars, damaged vehicles, refrigerators and gas and electric stoves, rugs, and so on | White waste |
| Tree branches, leaves, seeds and so on | Originating with individual households and animal waste | Yard (or green) waste |

*Sources:* From Blackman, W. C. 2001. *Basic Hazardous Waste Management*, 3rd ed. CRC Press; de Bertoldi, M., (Ed.). 1996. *Science of Composting*. Springer; CEHA. 1995. Solid waste management in some countries of the Eastern Mediterranean region. CEHA Document No., Special studies, ss-4. Amman, Jordan: WHO, Eastern Mediterranean Regional Office, Regional Centre for Environmental Health Activities; Ojovan, M. I., (Ed.). 2011. *Handbook of Advanced Radioactive Waste Conditioning Technologies*. Cambridge: Woodhead Publishing; Walsh, P. and O'Leary, P. 1986. *Implementing Municipal Solid Waste to Energy Systems*. University of Wisconsin—Extension for Great Lakes Regional Biogas Energy Program; Worrell, W. A. and Vesilind, P. A. 2012. *Solid Waste Engineering*. CL-Engineering Pub.

## LISTING OF PROGRAM 6.1 (CHAP6.1\FORM1.VB): DIVERSION OF SOLID WASTE

```vb
'*****************************************************************************
'Example 6.1: Calculates percentage of diversion of solid waste.
'*****************************************************************************
Public Class Form1
  Private Sub Form1_Load(ByVal sender As System.Object, ByVal e As System.EventArgs)
Handles MyBase.Load
    Me.Text = "Example 6.1: Calculates percentage of diversion of solid waste."
    Me.FormBorderStyle = BorderStyle.FixedSingle
    Me.MaximizeBox = False
    RadioButton1.Text = "Calculate MSW (municipal solid waste):"
    RadioButton2.Text = "Calculate as fraction of refuse (mixed household and commercial
waste):"
    RadioButton1.Checked = True
    Label1.Text = "Recyclables (ton per year):"
    Label2.Text = "Construction and demolition debris (ton per year):"
    Label3.Text = "Leaves and miscellaneous (ton per year):"
    Label4.Text = "Mixed house waste (ton per year):"
    Label5.Text = "Commercial waste (ton per year):"
    Label6.Text = "Recyclables (ton per year):"
    Label8.Text = "Mixed house waste (ton per year):"
    Label9.Text = "Commercial waste (ton per year):"
    Label7.Text = "The formula is:" + vbCrLf
    Label7.Text += "MSW, municipal solid waste = (refuse) + construction and demolition
waste "
    Label7.Text += vbCrLf + " + leaves + bulky items"
    Label10.Text = ""
    Button1.Text = "&Calculate"
  End Sub


  Private Sub RadioButton1_CheckedChanged(ByVal sender As System.Object, ByVal e As
System.EventArgs) Handles RadioButton1.CheckedChanged
    changeDisplay()
  End Sub


  Private Sub RadioButton2_CheckedChanged(ByVal sender As System.Object, ByVal e As
System.EventArgs) Handles RadioButton2.CheckedChanged
    changeDisplay()
  End Sub

  '*****************************************************************************
  'Enables/disables the textboxes and their
  'labels according to which radiobutton is selected.
  '*****************************************************************************
  Sub changeDisplay()
    If RadioButton1.Checked Then
    Label1.Enabled = True
    Label2.Enabled = True
    Label3.Enabled = True
    Label4.Enabled = True
    Label5.Enabled = True
    Label6.Enabled = False
    Label8.Enabled = False
    Label9.Enabled = False
    TextBox1.Enabled = True
    TextBox2.Enabled = True
    TextBox3.Enabled = True
    TextBox4.Enabled = True
    TextBox5.Enabled = True
    TextBox6.Enabled = False
```

```vb
        TextBox7.Enabled = False
        TextBox8.Enabled = False
      Else
       Label1.Enabled = False
       Label2.Enabled = False
       Label3.Enabled = False
       Label4.Enabled = False
       Label5.Enabled = False
       Label6.Enabled = True
       Label8.Enabled = True
       Label9.Enabled = True
       TextBox1.Enabled = False
       TextBox2.Enabled = False
       TextBox3.Enabled = False
       TextBox4.Enabled = False
       TextBox5.Enabled = False
       TextBox6.Enabled = True
       TextBox7.Enabled = True
       TextBox8.Enabled = True
      End If
    End Sub

  Private Sub Button1_Click(ByVal sender As System.Object, ByVal e As System.EventArgs)
Handles Button1.Click
        calculateResults()
    End Sub

  Sub calculateResults()
      Dim tot, rec, CD, leaves, mixed, comm, div As Double
      If RadioButton1.Checked Then
      'calculate MSW
      rec = Val(TextBox1.Text)
      CD = Val(TextBox2.Text)
      leaves = Val(TextBox3.Text)
      mixed = Val(TextBox4.Text)
      comm = Val(TextBox5.Text)
      tot = rec + CD + leaves + mixed + comm
      div = (rec + CD + leaves) / tot * 100
      Label10.Text = "Diversion = (" + rec.ToString + "+" + CD.ToString + "+" + _
        leaves.ToString + ")/" + tot.ToString + "*100 = " + Format(div, "##.#").ToString +
"%"
    Else
      'calculate fraction of refuse
      rec = Val(TextBox6.Text)
      mixed = Val(TextBox7.Text)
      comm = Val(TextBox8.Text)
      tot = mixed + comm
      div = rec / tot * 100
      Label10.Text = "Diversion = " + rec.ToString + "/(" + mixed.ToString + "+" + _
        comm.ToString + ")" + "*100 = " + Format(div, "##.#").ToString + "%"
    End If
  End Sub
End Class
```

a. If the calculation is on basis of MSW, the total waste generated is (200 + 40 + 60 + 120 + 50) = 470 tons per year. If everything not going to the landfill is counted as having been diverted, the diversion is calculated as

$$\text{Diversion} = \frac{(40+120+50)}{470} \times 100 = 45\%$$

b. If the diversion is calculated as that fraction of the refuse (mixed household and commercial waste) that has been kept out of the landfill by the recycling program, the diversion is

$$\text{Diversion} = \frac{(40)}{200+60} \times 100 = 15\%$$

(a reasonable diversion)

Properties and characteristics of MSW have a significant impact and govern the design of solid waste collection systems, treatment and disposal plants, operation and maintenance schedules, and management and performance of units and facilities. Notable MSW properties constitute physical, chemical, and biological aspects. Physical and material properties of solid waste affect the design of storage equipment and waste collection, separation, transfer, transpiration, and treatment. Physical properties may include grain size and distribution, material components, use of solid waste material fractions, degree of purity, contents of solid waste, moisture content, grain size calorific value, weight, density, and mechanical properties. Chemical properties may include chemical composition and content, chemical properties and characteristics, toxicity, and so on. Biological properties may refer to biodegradability, degree of decomposition, nutrients content, and so on.

Properties of MSW of significance and interest include the following:

1. Physical properties:
   a. Composition by identifiable items (steel cans, office paper, etc.)
   b. Weight
   c. Density
   d. Moisture content
   e. Particle size and grain size distribution
   f. Heat and calorific values
   g. Angle of stability
   h. Mechanical properties to evaluate alternative processes and options for energy recovery by focusing on pressure stress, stress–strain curve for some materials, and modulus of elasticity
2. Chemical properties:
   a. Chemical composition: carbon, hydrogen, and concentration of metals
   b. Proximate analysis
   c. Fusing point of ash
   d. Ultimate analysis (major elements)
   e. Compositional analysis
   f. Calorimetry
   g. Energy content
   h. Volatile solids lost upon ignition
   i. pH value
   j. Toxic elements
   k. Nutrients (carbon, nitrogen, and phosphorus)
3. Biological properties (biodegradability)

Benefits of MSW properties may be summarized as follows:

- Knowledge of *hazardous and harmful substances* that may be present for appropriate sorting and ultimate disposal
- Estimation of combustible materials for incineration and *energy recovery*
- Evaluation of *recyclables* for augmentation of recycling plans
- Quantifying tons of *special (hazardous)* waste *materials* generated for controlled disposal or sanitary landfilling
- Assessment of useful organic materials for *production of gas* from a landfill for some beneficial use

## 6.2 PHYSICAL PROPERTIES OF SOLID WASTE

### 6.2.1 Moisture Content

The moisture content becomes important when the refuse is processed into fuel or when it is fired directly. Moisture content influences many MSW properties of importance. The extent of this effect depends on the material. When the moisture level exceeds 50%, the high organic fraction can undergo spontaneous combustion if the material is allowed to stand undisturbed (Abdel-Magid and Abdel-Magid 2015b; Worrell and Vesilind 2012).

Moisture content, on a wet basis, is found as presented in the following equation:

$$M = \frac{W_w - W_d}{W_w} \times 100 \qquad (6.4)$$

where:
$M$ is the moisture content (on a wet basis), %
$W_w$ is the initial (wet) weight of sample
$W_d$ is the final (dry) weight of the sample

Moisture content may be evaluated on a dry weight basis as shown in the following equation:

$$M_d = \frac{W_w - W_d}{W_d} \times 100 \qquad (6.5)$$

where:
$M_d$ is the moisture content (on a dry basis), %

Table 6.2 gives the moisture content of uncompacted refuse components

## TABLE 6.2
## Moisture Content of Uncompacted Refuse Components

| Component | Moisture Content | |
|---|---|---|
| | Range | Typical |
| **Residential** | | |
| Aluminum cans | 2–4 | 3 |
| Cardboard | 4–8 | 5 |
| Ferrous metals | 2–6 | 3 |
| Fines (dirt, etc.) | 6–12 | 8 |
| Food waste | 50–80 | 70 |
| Garden trimmings | 30–80 | 60 |
| Glass | 1–4 | 2 |
| Grass | 40–80 | 60 |
| Leather | 8–12 | 10 |
| Leaves | 20–40 | 30 |
| Nonferrous metal | 2–4 | 2 |
| Paper | 4–10 | 6 |
| Plastics | 1–4 | 2 |
| Rubber | 1–4 | 2 |
| Steel cans | 2–4 | 3 |
| Textiles | 6–15 | 10 |
| Wood | 15–40 | 20 |
| Yard waste | 30–80 | 60 |
| **Commercial** | | |
| Construction (mixed) | 2–15 | 8 |
| Dirt, ashes, bricks, and so on | 6–12 | 8 |
| Food waste | 50–80 | 70 |
| Mixed organics | 10–60 | 25 |
| Mixed | 10–25 | 15 |
| Municipal waste | 15–40 | |
| Wooden shipping crates and plant scales | 10–30 | 30 |

*Sources:* From de Bertoldi, M., (Ed.). 1996. *Science of Composting.* Springer; CEHA. 1995. Solid waste management in some countries of the Eastern Mediterranean region. CEHA Document No., Special studies, ss-4. Amman, Jordan: WHO, Eastern Mediterranean Regional Office, Regional Centre for Environmental Health Activities; Ojovan, M. I., (Ed.). 2011. *Handbook of Advanced Radioactive Waste Conditioning Technologies.* Cambridge: Woodhead Publishing; Walsh, P. and O'Leary, P. 1986. *Implementing Municipal Solid Waste to Energy Systems.* University of Wisconsin—Extension for Great Lakes Regional Biogas Energy Program; Worrell, W. A. and Vesilind, P. A. 2012. *Solid Waste Engineering.* CL-Engineering Pub.

## Example 6.2

1. Write a short computer program that enables calculating moisture concentration of a certain solid waste using the typical values.
2. A residential waste has the components presented in the table. Estimate its moisture concentration using the typical values.

| | |
|---|---|
| Food | 30% |
| Garden trimmings | 10% |
| Leather | 10% |
| Paper | 40% |
| Steel cans | 10% |

3. Use the program developed in (1) to verify the manual computations conducted in (2).

**Solution**

Assume a wet sample weighing 100 lb. Set up the tabulation:

| Component | % | Moisture from Table | Dry Weight (based on 100 lb.) |
|---|---|---|---|
| Food | 30 | 70 | 9 |
| Garden trimmings | 10 | 60 | 4 |
| Leather | 10 | 10 | 9 |
| Paper | 40 | 6 | 37.5 |
| Steel cans | 10 | 3 | 9.7 |
| Total | 100 | | 69.2 |

The moisture content (on a wet basis) would then be $(100 - 69.2)/100 = 31\%$.

### 6.2.2 PARTICLE SIZE

Particle size of solid waste is of paramount importance to the recovery of materials especially with mechanical means (trammel* screens and magnetic separators). This is besides its significant role in treatment, reuse, and recycling operations. Particle size of solid waste has its effect on physicomechanical characteristics of particles such as compressibility, flowability, packing, and porosity (Abdel-Magid 2012). Particle size distribution may be accurately expressed in a graphical format. Nonetheless, several mathematical expressions are used for its exemplification. In water engineering, the particle size of a sand filter is usually expressed using the uniformity coefficient concept, defined as presented in the following equation:

$$UC = \frac{D_{60}}{D_{10}} \qquad (6.6)$$

where:

$UC$ is the uniformity coefficient

$D_{60}$ is the particle (sieve) size where 60% of the particles are smaller than that size

$D_{10}$ is the particle (sieve) size where 10% of the particles are smaller than that size

---

* Trammel: A hollow cylindrical screen. Usually, it is a horizontal perforated rotating cylinder used to break open trash bags, large particles of glass, and small abrasive items (e.g., stones, dirt, etc.).

## LISTING OF PROGRAM 6.2   (CHAP6.2\FORM1.VB): MOISTURE CONCENTRATION OF SOLID WASTE

```vb
'*******************************************************************************
'Program 6.2: Moisture concentration of solid waste
'using the typical values.
'*******************************************************************************
Public Class Form1
  Dim comp(27) As String
  Dim moist(27) As Integer
  Dim moistC As Double
  Private Sub Form1_Load(ByVal sender As System.Object, ByVal e As System.EventArgs)
Handles MyBase.Load
      Me.Text = "Program 6.2: Moisture concentration of solid waste"
      Me.FormBorderStyle = Windows.Forms.FormBorderStyle.FixedSingle
      Me.MaximizeBox = False
      Label1.Text = "Select component:"
      Label2.Text = "Percentage:"
      Label3.Text = "Moisture (from Table):"
      Label4.Text = "Dry weight (based on 100lb.):"
      Label5.Text = ""
      Label6.Text = "Decimal places:"
      Button1.Text = "&Calculate"
      NumericUpDown1.Value = 2
      NumericUpDown1.Maximum = 10
      NumericUpDown1.Minimum = 0
      'DATA FROM TABLE 6.2
      comp(0) = "Residential"
      comp(1) = " Aluminum cans"
      comp(2) = " Cardboard"
      comp(3) = " Fines (dirt, etc.)"
      comp(4) = " Food waste"
      comp(5) = " Glass"
      comp(6) = " Grass"
      comp(7) = " Leather"
      comp(8) = " Nonferrous metal"
      comp(9) = " Leaves"
      comp(10) = " Paper"
      comp(11) = " Plastics"
      comp(12) = " Ferrous metals"
      comp(13) = " Rubber"
      comp(14) = " Steel cans"
      comp(15) = " Textiles"
      comp(16) = " Wood"
      comp(17) = " Yard waste"
      comp(18) = " Garden trimmings"
      comp(19) = " Commercial"
      comp(20) = " Food waste"
      comp(21) = " Mixed organics"
      comp(22) = " Mixed"
      comp(23) = " Wooden shipping crates and plant scales"
      comp(24) = " Construction (mixed)"
      comp(25) = " Dirt, ashes, bricks, etc."
      comp(26) = " Municipal waste"
      comp(27) = "Select component"
      'MOISTURE CONTENT FROM TABLE 6.2
      moist(0) = 0
      moist(1) = 3
      moist(2) = 5
      moist(3) = 8
```

```
    moist(4) = 70
    moist(5) = 2
    moist(6) = 60
    moist(7) = 10
    moist(8) = 2
    moist(9) = 30
    moist(10) = 6
    moist(11) = 2
    moist(12) = 3
    moist(13) = 2
    moist(14) = 3
    moist(15) = 10
    moist(16) = 20
    moist(17) = 60
    moist(18) = 60
    moist(19) = 0
    moist(20) = 70
    moist(21) = 25
    moist(22) = 15
    moist(23) = 30
    moist(24) = 8
    moist(25) = 8
    moist(26) = 0
    'ADD THE ITEMS INTO THE COMBOBOXES
    ComboBox1.Items.Clear()
    ComboBox2.Items.Clear()
    ComboBox3.Items.Clear()
    ComboBox4.Items.Clear()
    ComboBox5.Items.Clear()
    ComboBox6.Items.Clear()
    ComboBox1.Items.AddRange(comp)
    ComboBox2.Items.AddRange(comp)
    ComboBox3.Items.AddRange(comp)
    ComboBox4.Items.AddRange(comp)
    ComboBox5.Items.AddRange(comp)
    ComboBox6.Items.AddRange(comp)
    'DISABLE THE TEXTBOXES
    TextBox2.Enabled = False
    TextBox3.Enabled = False
    TextBox4.Enabled = False
    TextBox5.Enabled = False
    TextBox7.Enabled = False
    TextBox8.Enabled = False
    TextBox10.Enabled = False
    TextBox11.Enabled = False
    TextBox13.Enabled = False
    TextBox14.Enabled = False
    TextBox16.Enabled = False
    TextBox17.Enabled = False
End Sub

Sub calculateResults()
    Dim i, M, Ww, Wd As Double
    Dim totalWW, totalWd As Double
    totalWd = 0
    totalWW = 0
    'Calculate the dry weights of the components
    i = ComboBox1.SelectedIndex
    If i <> 0 And i <> 19 And i <> 27 Then
        Ww = Val(TextBox1.Text)
        M = Val(TextBox2.Text)
```

```
         Wd = Ww - (M * Ww / 100)
         TextBox3.Text = Wd.ToString
         totalWd += Wd : totalWW += Ww
      End If
      i = ComboBox2.SelectedIndex
      If i <> 0 And i <> 19 And i <> 27 Then
         Ww = Val(TextBox6.Text)
         M = Val(TextBox5.Text)
         Wd = Ww - (M * Ww / 100)
         TextBox4.Text = Wd.ToString
         totalWd += Wd : totalWW += Ww
      End If
      i = ComboBox3.SelectedIndex
      If i <> 0 And i <> 19 And i <> 27 Then
         Ww = Val(TextBox9.Text)
         M = Val(TextBox8.Text)
         Wd = Ww - (M * Ww / 100)
         TextBox7.Text = Wd.ToString
         totalWd += Wd : totalWW += Ww
      End If
      i = ComboBox4.SelectedIndex
      If i <> 0 And i <> 19 And i <> 27 Then
         Ww = Val(TextBox12.Text)
         M = Val(TextBox11.Text)
         Wd = Ww - (M * Ww / 100)
         TextBox10.Text = Wd.ToString
         totalWd += Wd : totalWW += Ww
      End If
      i = ComboBox5.SelectedIndex
      If i <> 0 And i <> 19 And i <> 27 Then
         Ww = Val(TextBox15.Text)
         M = Val(TextBox14.Text)
         Wd = Ww - (M * Ww / 100)
         TextBox13.Text = Wd.ToString
         totalWd += Wd : totalWW += Ww
      End If
      i = ComboBox6.SelectedIndex
      If i <> 0 And i <> 19 And i <> 27 Then
         Ww = Val(TextBox18.Text)
         M = Val(TextBox17.Text)
         Wd = Ww - (M * Ww / 100)
         TextBox16.Text = Wd.ToString
         totalWd += Wd : totalWW += Ww
      End If
      moistC = ((totalWW - totalWd) / totalWW) * 100
   End Sub


   Sub showResults()
      Label5.Text = "The moisture content (wet basis) = " + FormatNumber(moistC,
NumericUpDown1.Value) + "%"
   End Sub


   Private Sub ComboBox1_SelectedIndexChanged(ByVal sender As System.Object, ByVal e As
System.EventArgs) Handles ComboBox1.SelectedIndexChanged
      'enter the moisture content (From Table 6.2) into the 'moisture' field
      If ComboBox1.SelectedIndex = 0 Or ComboBox1.SelectedIndex = 19 Or ComboBox1.
SelectedIndex = 27 Then
         'These are items not to be selected by the user, so clear the textbox and exit
sub
         TextBox2.Text = ""
      Else
```

```
            TextBox2.Text = moist(ComboBox1.SelectedIndex).ToString
        End If
    End Sub

    Private Sub ComboBox2_SelectedIndexChanged(ByVal sender As System.Object, ByVal e As
System.EventArgs) Handles ComboBox2.SelectedIndexChanged
        'enter the moisture content (From Table 6.2) into the 'moisture' field
        If ComboBox2.SelectedIndex = 0 Or ComboBox2.SelectedIndex = 19 Or ComboBox2.
SelectedIndex = 27 Then
            'These are items not to be selected by the user, so clear the textbox and exit
sub
            TextBox5.Text = ""
        Else
            TextBox5.Text = moist(ComboBox2.SelectedIndex).ToString
        End If
    End Sub

    Private Sub ComboBox3_SelectedIndexChanged(ByVal sender As System.Object, ByVal e As
System.EventArgs) Handles ComboBox3.SelectedIndexChanged
        'enter the moisture content (From Table 6.2) into the 'moisture' field
        If ComboBox3.SelectedIndex = 0 Or ComboBox3.SelectedIndex = 19 Or ComboBox3.
SelectedIndex = 27 Then
            'These are items not to be selected by the user, so clear the textbox and exit
sub
            TextBox8.Text = ""
        Else
            TextBox8.Text = moist(ComboBox3.SelectedIndex).ToString
        End If
    End Sub

    Private Sub ComboBox4_SelectedIndexChanged(ByVal sender As System.Object, ByVal e As
System.EventArgs) Handles ComboBox4.SelectedIndexChanged
        'enter the moisture content (From Table 6.2) into the 'moisture' field
        If ComboBox4.SelectedIndex = 0 Or ComboBox4.SelectedIndex = 19 Or ComboBox4.
SelectedIndex = 27 Then
            'These are items not to be selected by the user, so clear the textbox and exit
sub
            TextBox11.Text = ""
        Else
            TextBox11.Text = moist(ComboBox4.SelectedIndex).ToString
        End If
    End Sub

    Private Sub ComboBox5_SelectedIndexChanged(ByVal sender As System.Object, ByVal e As
System.EventArgs) Handles ComboBox5.SelectedIndexChanged
        'enter the moisture content (From Table 6.2) into the 'moisture' field
        If ComboBox5.SelectedIndex = 0 Or ComboBox5.SelectedIndex = 19 Or ComboBox5.
SelectedIndex = 27 Then
            'These are items not to be selected by the user, so clear the textbox and exit
sub
            TextBox14.Text = ""
        Else
            TextBox14.Text = moist(ComboBox5.SelectedIndex).ToString
        End If
    End Sub

    Private Sub ComboBox6_SelectedIndexChanged(ByVal sender As System.Object, ByVal e As
System.EventArgs) Handles ComboBox6.SelectedIndexChanged
        'enter the moisture content (From Table 6.2) into the 'moisture' field
        If ComboBox6.SelectedIndex = 0 Or ComboBox6.SelectedIndex = 19 Or ComboBox6.
SelectedIndex = 27 Then
```

```
            'These are items not to be selected by the user, so clear the textbox and exit sub
            TextBox17.Text = ""
        Else
            TextBox17.Text = moist(ComboBox6.SelectedIndex).ToString
        End If
    End Sub


    Private Sub Button1_Click(ByVal sender As System.Object, ByVal e As System.EventArgs)
Handles Button1.Click
        calculateResults()
        showResults()
    End Sub


    Private Sub NumericUpDown1_ValueChanged(ByVal sender As System.Object, ByVal e As
System.EventArgs) Handles NumericUpDown1.ValueChanged
        showResults()
    End Sub
End Class
```

Correlation between the quantity of a particulate material and its particle size may be determined via Gates–Gaudin–Schuhmann function as shown in Equation 6.7. A distribution following this equation gives a straight line in a logarithmic plot (Masato Nakamura, Castaldi and Themelis 2005):

$$D = 100 \left( \frac{d}{d_o} \right)^k \qquad (6.7)$$

where:

$D$ is the cumulative passing weight, %
$d$ is the particle size
$d_o$ is the maximum particle size
$k$ is the dimensionless exponent

Rosin–Rammler–Sperling–Bennett (RRSB) function likewise describes cumulative weight fraction passing a certain sieve size as presented in Equation 6.8. RRSB function is particularly appropriate for materials that do not have a defined upper size limit (Masato Nakamura, Castaldi and Themelis 2005):

$$D = 1 - e^{-(d/d')^n} \qquad (6.8)$$

where:

$D$ is the cumulative passing weight fraction
$d$ is the particle size
$d'$ and $n$ are the characteristics of the material studied and must be determined by analyzing the data obtained. This can conveniently be done by data regression after appropriate manipulation or using especially constructed charts, from which the two parameters could be read off after plotting size data obtained by sieving or other means (Masato Nakamura, Castaldi and Themelis 2005).

### 6.2.3 PERMEABILITY OF COMPACTED WASTE

Permeability is an important parameter in the treatment of solid waste as in the design of leachate collection and recirculation system. Hydraulic conductivity of compacted wastes governs the movement of fluids, liquids, and gases in a solid waste facility such as a sanitary landfill. It permits the passage of fluid through its interconnected pore spaces (von Blottnitz, Pehlken and Pretz 2002). Coefficient of permeability may be determined as depicted in the following equation:

$$K = C_d^2 \frac{\gamma}{\mu} = k \frac{\gamma}{\mu} \qquad (6.9)$$

where:

$K$ is the coefficient of permeability
$C_d$ is the constant or shape factor, dimensionless
$\gamma$ is the specific weight of water
$\mu$ is the dynamic viscosity of water
$k$ is the intrinsic permeability (or specific permeability) = $C_d^2$ (Typical values for the intrinsic permeability for compacted solid waste in a landfill are in the range between $\sim 10^{-11}$ and $10^{-12}$ m$^2$ in the vertical direction and $\sim 10^{-10}$ m$^2$ in the horizontal direction).

### 6.2.4 APPARENT DENSITY

Apparent density may be used in estimating the amount of solid waste in some cases. Likewise, the property is used to assess requirements of a sanitary landfill cover mate-rial. The apparent density of solid waste and garbage varies greatly with exerted pressure, degree of compaction, level of economic development, concentration of produced waste products, geographic location, and season of the year and solid waste storage time.

Overall bulk density for a mixture of materials in a con-tainer may be estimated by knowing the bulk density of each substance, separately. For a mixture of two materials (A) and (B), the bulk density of the mixture can be estimated as shown in the following equation:

$$\rho_C = \rho_{A+B} = \frac{\rho_A V_A + \rho_B V_B}{V_A + V_B} \qquad (6.10)$$

where:

$\rho_C = \rho_{A+B}$ is the bulk density of the mixture of material (A) and material (B)
$\rho_A$ is the bulk density of material (A)
$\rho_B$ is the bulk density of material (B)
$V_A$ is the volume of material (A)
$V_B$ is the volume of material (B)

Bulk density of the mixture of materials can also be estimated by the mass of materials from the following equation:

$$\rho_{A+B} = \frac{M_A + M_B}{\left(M_A/\rho_A\right) + \left(M_B/\rho_B\right)} \qquad (6.11)$$

where $M$ is the mass of the material (pounds or tons in the American Standard System or kilograms or tons in the SI system).[*]

## 6.2.5   ANGLE OF REPOSE

The angle of repose is (see Figure 6.1) the angle to the horizontal to which the material will stack without sliding or slumping. Sand, for example, depending on its moisture content, has an angle of repose of ~35°. Due to variable den-sity, moisture, and particle size, the angle of repose of shred-ded refuse can vary from 45° to >90° (Walsh and O'Leary 1986).

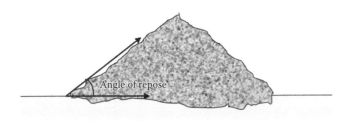

**FIGURE 6.1**   Angle of repose.

## 6.2.6   SIZE OF REDUCTION IN VOLUME (REDUCTION VOLUME)

In design and operation when packaging or compacting solid waste in a landfill, it is of value computing size of reduction in volume as outlined in the following equation:

$$F = \frac{V_c}{V_0} \qquad (6.12)$$

where:

$F$ is the volume of reduction (remaining ratios of original size as a result of compaction)
$V_0$ is the original size (initial)
$V_c$ is the volume after compaction

Relationship of reduction volume to apparent density can be found from the following equation:

$$F = \frac{V_c}{V_o} = \frac{M/\rho_c}{M/\rho_o} = \frac{\rho_c}{\rho_o} \qquad (6.13)$$

where:

$\rho_o$ is the original apparent density
$\rho_c$ is the apparent density after compaction

### Example 6.3

1. Write a short computer program that enables estimating the percent age of volume reduction achieved during compaction of the waste.
2. Assume that a certain solid waste has the components and bulk densities depicted in the following table:

#### Table of Solid Waste Components and Bulk Densities

| Component | Percentage (by Weight) | Apparent Density before Compaction (g/cm³) |
|---|---|---|
| Aluminum | 10 | 0.04 |
| Glass | 5 | 0.30 |
| Various papers | 15 | 0.06 |
| Food waste | 70 | 0.4 |

  a. For a compaction of landfill of 700 kg/m³, find the percentage of volume reduction achieved upon compacting this solid waste.
  b. Determine the total apparent density before compaction by removing various paper components. Estimate the percentage of volume reduction achieved.
  c. Comment on your results.
3. Use the program developed in (1) to verify the manual computations conducted in (2).

*Hint*: Use the following equations for estimating bulk density and percentage of volume reduction of a mixture of solid waste, respectively.

$$F = \frac{\rho_0}{\rho_c} = \frac{232.3\,\text{lb/yd}^3}{1300\,\text{lb/yd}^3} = 0.179$$

**Solution**

1. The overall bulk density prior to compaction is

$$\rho_{(A+B+C+D)} = \frac{M_A + M_B + M_C + M_D}{\left(\dfrac{M_A}{\rho_A}\right) + \left(\dfrac{M_B}{\rho_B}\right) + \left(\dfrac{M_C}{\rho_C}\right) + \left(\dfrac{M_D}{\rho_D}\right)}$$

$$= \frac{10 + 5 + 15 + 70}{\left(\dfrac{10}{0.04}\right) + \left(\dfrac{5}{0.3}\right) + \left(\dfrac{15}{0.06}\right) + \left(\dfrac{70}{0.4}\right)}$$

$$= 0.145\,\text{g/cm}^3 = 145\,\text{kg/m}^3$$

a. The volume reduction achieved during compaction is

$$F = \frac{\rho_0}{\rho_c} = \frac{0.214\,\text{g/cm}^3}{0.7\,\text{g/cm}^3} = 0.31$$

b. So the required landfill volume is ~31% of the volume required without compaction. If the mixed paper is removed, the uncompacted density would be as follows:

$$\rho_{(A+B+C+D)} = \frac{M_A + M_B + M_C + M_D}{\left(\dfrac{M_A}{\rho_A}\right) + \left(\dfrac{M_B}{\rho_B}\right) + \left(\dfrac{M_C}{\rho_C}\right) + \left(\dfrac{M_D}{\rho_D}\right)}$$

$$= \frac{10 + 5 + 0 + 70}{\left(\dfrac{10}{0.04}\right) + \left(\dfrac{5}{0.3}\right) + 0 + \left(\dfrac{70}{0.4}\right)}$$

$$= 0.192\,\text{g/cm}^3 = 19.2\,\text{kg/m}^3$$

$$F = \frac{\rho_0}{\rho_c} = \frac{0.192\,\text{g/cm}^3}{0.7\,\text{g/cm}^3} = 0.275$$

---

### LISTING OF PROGRAM 6.3   (CHAP6.3\FORM1.VB): VOLUME REDUCTION OF COMPACTED SOLID WASTE

```vb
'*********************************************************************************
'Program 6.3: Estimates percent volume reduction achieved
'during compaction of the waste
'*********************************************************************************
Public Class Form1
  Dim comp() As String
  Dim perc(), dens() As Double
  Dim rho, rhoc, F As Double

  Private Sub Form1_Load(ByVal sender As System.Object, ByVal e As System.EventArgs)
Handles MyBase.Load
    Me.Text = "Program 6.3: Volume reduction"
    Me.FormBorderStyle = Windows.Forms.FormBorderStyle.FixedSingle
    Me.MaximizeBox = False
    Label1.Text = "Program 6.3: Estimates percent volume reduction achieved "
    Label1.Text += vbCrLf + "during compaction of the waste"
    Label2.Text = "Enter each component's name, percentage, and bulk density:"
    Label3.Text = "Enter the compaction in the landfill (kg/cm3):"
    Label4.Text = ""
    Label5.Text = "Decimals:"
    NumericUpDown1.Value = 2
    NumericUpDown1.Maximum = 10
    NumericUpDown1.Minimum = 0
    Button1.Text = "&Calculate"
    DataGridView1.Columns.Clear()
    DataGridView1.Columns.Add("CCol", "Component")
    DataGridView1.Columns.Add("PCol", "% (by weight)")
    DataGridView1.Columns.Add("DCol", "Bulk density (gm/cm3)")
    DataGridView1.AutoResizeColumns()
  End Sub
```

```vb
Sub calculateResults()
  If DataGridView1.RowCount <= 1 Then
    MsgBox("Enter at least one component!", vbOKOnly, "Error")
    Exit Sub
  End If
  Dim count As Integer = DataGridView1.RowCount - 1
  Dim nom, denom As Double
  ReDim comp(count), perc(count), dens(count)
  For i = 0 To count - 1
    comp(i) = DataGridView1.Rows(i).Cells("CCol").Value
    perc(i) = Val(DataGridView1.Rows(i).Cells("PCol").Value)
    dens(i) = Val(DataGridView1.Rows(i).Cells("DCol").Value)
  Next
  nom = 0
  denom = 0
  For i = 0 To count - 1
    nom += perc(i)
    denom += perc(i) / dens(i)
  Next
  rho = nom / denom
  rhoc = Val(TextBox1.Text) / 1000
  F = rho / rhoc
End Sub

Sub showResults()
  Dim decimals = NumericUpDown1.Value
  If decimals < 0 Or decimals > 10 Then decimals = 2
  Label4.Text = "The overall bulk density prior to compaction:"
  Label4.Text += FormatNumber(rho, decimals) + (" gm/cm3")
  Label4.Text += vbCrLf + "The volume reduction achieved during compaction: "
  Label4.Text += FormatNumber(F, decimals)
End Sub

Private Sub Button1_Click(ByVal sender As System.Object, ByVal e As System.EventArgs) Handles Button1.Click
  calculateResults()
  showResults()
End Sub

Private Sub NumericUpDown1_ValueChanged (ByVal sender As System.Object, ByVal e As System.EventArgs) Handles NumericUpDown1.ValueChanged
  showResults()
End Sub
End Class
```

### 6.2.7 MATERIAL ABRASIVENESS

MSW and refuse consist of different types of abrasive particles and grains such as sand, glass, metals, and rocks. Removal of this abrasive material is often necessary before some operations (such as pneumatic conveying) can become practical (Walsh and O'Leary 1986).

## 6.3 CHEMICAL PROPERTIES OF SOLID WASTE

Chemical properties of solid waste are of value in economics of material or energy recovery. Chemical components of solid waste have a significant variability and change due to the heterogeneity of solid waste, geographical location, and temporal changes. Typically, solid wastes represent a combination of semimoist combustible and noncombustible materials. When using solid waste as a fuel, its chemical properties of significance include the fusing point of ash, proximate analysis, ultimate analysis (major elements), and energy content.

### 6.3.1 FUSION POINT OF ASH (ASH FUSIBILITY)

The fusion point of ash may be defined as the temperature at which the ash resulting from the burning of waste will form a solid (clinker) by fusion and agglomeration. Typical fusion temperatures for the formation of clinker from solid waste range from 1100°C to 1200°C.

## 6.3.2 Proximate Analysis

Proximate analyses are to determine percentage (fraction) of volatile organics and fixed carbon in solid waste and garbage (fuel).

## 6.3.3 Ultimate Analysis

Ultimate analysis uses the chemical makeup of the fuel to approximate its heat value, and it depends on elemental composition.

## 6.3.4 Volatile Solids

Volatile solids can be estimated upon ignition at a temperature of 550°C for 4 h and then cooling in a dryer. Loss in weight represents volatile organics, which includes disintegrating organic material and nondecomposable material as reflected in the following equations:

$$\text{Loss in weight} = \text{Volatile organics} \qquad (6.14)$$

$$\text{VO} = \text{D} + \text{ND} \qquad (6.15)$$

where:
VO is the volatile organics
D is the disintegrating organic material
ND is the nondecomposable material

## 6.3.5 Heat Value of Refuse

Heat value of refuse is of paramount importance in resource recovery. Heat value is expressed as British thermal unit per pound, Btu*/lb, of refuse, or kJ/kg in the SI system of units. Heat value of refuse and other heterogeneous materials may be measured with a calorimeter. A calorimeter is a device in which a sample is combusted, and the temperature rise is recorded. Knowing the mass of the sample and the heat generated by the combustion, the Btu/lb is calculated.

The most popular method using ultimate analysis is the DuLong equation, which originally was developed for estimating the heat value of coal.

Energy values of solid waste and garbage can be estimated by using DuLong formula as shown in the following equation:

$$\frac{\text{kJ}}{\text{kg}} = 337\text{C} + 1428\left(\text{H} - \frac{\text{O}}{8}\right) + 9\text{S} \qquad (6.16)$$

where:
C is carbon, %
H is hydrogen, %
O is oxygen, %
S is sulfur, %

The DuLong formula is cumbersome to use in practice, and it does not give acceptable estimates of heat value for materials other than coal. Total energy content may be determined using the modified DuLong formula as presented in the following equation:

$$\frac{\text{Btu}}{\text{lb}} = 145\text{C} + 610\left(\text{H}_2 - \frac{\text{O}_2}{8}\right) + 40\text{S} + 10\text{N} \qquad (6.17)$$

where:
Btu/lb† is the total energy
C is carbon, % by weight
$\text{H}_2$ is hydrogen, % by weight
$\text{O}_2$ is oxygen, % by weight
S is sulfur, % by weight
N is nitrogen, % by weight

Table 6.3 is an illustration of ideal data for final analysis of components of a combustible MSW.

Another equation for estimating the heat value of refuse using ultimate analysis is illustrated in the following equation:

$$\frac{\text{Btu}}{\text{lb}} = 144\text{C} + 672\text{H} + 6.2\text{O} + 41.4\text{S} - 10.8\text{N} \qquad (6.18)$$

## TABLE 6.3
## Ideal Data for Final Analysis of Components of a Combustible MSW

| Component | Percentages by Mass (on Dry Bases) | | | | | |
|---|---|---|---|---|---|---|
| | Carbon | Hydrogen | Oxygen | Nitrogen | Sulfur | Ash |
| Food waste | 0.048 | 6.4 | 37.6 | 2.6 | 0.4 | 0.05 |
| Paper | 43.5 | 0.06 | 0.044 | 0.3 | 0.2 | 0.06 |
| Cardboard | 0.044 | 5.9 | 44.6 | 0.3 | 0.2 | 0.05 |
| Plastics | 0.060 | 7.2 | 22.8 | – | – | 10.0 |
| Textiles | 0.055 | 6.6 | 31.2 | 4.6 | 0.15 | 2.5 |
| Rubber | 0.078 | 10.0 | – | 0.02 | – | 0.010 |
| Leather | 0.060 | 8.0 | 11.6 | 0.010 | 0.4 | 10.0 |
| Garden trimmings | 47.8 | 6.0 | 0.038 | 3.4 | 0.3 | 4.5 |
| Timber | 49.5 | 0.06 | 42.7 | 0.2 | 0.1 | 1.5 |
| Mixture of organic materials | 48.5 | 6.5 | 37.5 | 2.2 | 0.3 | 0.05 |
| Dirt, ash, bricks, and so on | 26.3 | 0.03 | 0.02 | 0.5 | 0.2 | 68.0 |

*Sources:* From Blackman, W. C. 2001. *Basic Hazardous Waste Management*, 3rd ed. CRC Press; de Bertoldi, M., (Ed.). 1996. *Science of Composting*. Springer; CEHA. 1995. Solid waste management in some countries of the Eastern Mediterranean region. CEHA Document No., Special studies, ss-4. Amman, Jordan: WHO, Eastern Mediterranean Regional Office, Regional Centre for Environmental Health Activities; Ojovan, M. I., (Ed.). 2011. *Handbook of Advanced Radioactive Waste Conditioning Technologies*. Cambridge: Woodhead Publishing; Walsh, P. and O'Leary, P. 1986. *Implementing Municipal Solid Waste to Energy Systems*. University of Wisconsin—Extension for Great Lakes Regional Biogas Energy Program; Worrell, W. A. and Vesilind, P. A. 2012. *Solid Waste Engineering*. CL-Engineering Pub.

---

* 1 Btu = heat necessary to raise the temperature of 1 lb of water 1°F.
† (Btu/1b) × 2.326 = kJ/kg.

where C, H, O, S, and N are the weight percentages (on a dry basis) of carbon, hydrogen, oxygen, sulfur, and nitrogen, respectively, in the combustible fraction of the fuel. The sum of all of these percentages has to add to 100%.

Formulas based on compositional analyses are an improvement over formulas based on ultimate analyses. One such formula is indicated in the following equation:

$$\frac{Btu}{lb} = 49R + 22.5(G + P) - 3.3W \qquad (6.19)$$

where:

  $R$ is the plastics, % by weight of total MSW, on a dry basis
  $G$ is the food waste, % by weight of total MSW, on a dry basis
  $P$ is the paper, % by weight of total MSW, on a dry basis
  $W$ is the water, % by weight, on a dry basis

Using regression analysis and comparing the results to actual measurements of heat value, an improved form of a compositional model is suggested by the following equation:

$$\frac{Btu}{lb} = 1238 + 15.6R + 4.4P + 2.7G - 20.7W \qquad (6.20)$$

where:

  $R$ is the plastics, % by weight, on a dry basis
  $P$ is the paper, % by weight, on a dry basis
  $G$ is the food waste, % by weight, on a dry basis
  $W$ is the water, % by weight, on a dry basis.

## Example 6.4

1. Write a short computer program that enables calculating the approximate chemical formula of the organic component of a sample composition of a solid waste using chemical composition obtained to estimate the energy content of this solid waste.
2. Find the approximate chemical formula of the organic component of the sample composition of a solid waste as set out in the following table. Use chemical composition obtained to estimate the energy content of this solid waste.

| Component | Percentage by Mass |
|---|---|
| Garden trimmings | 10 |
| Food waste | 25 |
| Timber | 5 |
| Paper | 33 |
| Cardboard | 15 |
| Rubber | 7 |
| Tin cans | 5 |
| Total sum | 100 |

Determine the total organic composition of the solid waste assuming a mass of 100 kg of the sample as shown in the following table:

| Component | Mass (kg) |
|---|---|
| Carbon | 34 |
| Hydrogen | 5 |
| Oxygen | 26 |
| Nitrogen | 0.6 |
| Sulfur | 0.1 |
| Ash | 4 |

3. Use the program developed in (1) to verify the manual computations conducted in (2).

**Solution**

1. For the computer program that enables calculating the approximate chemical formula of Example 6.4 (1), see the listing of Program 6.4: chemical composition and energy content of solid waste.
2. Determine the dry mass to solid waste sample assuming a mass of 100 kg of solid waste as shown in the following table (column 4):

| Composition (1) | Percentage by Mass (2) | Moisture Content (%) (from Table) (3) | Dry Mass (kg) (4) |
|---|---|---|---|
| Garden trimmings | 10 | 60 | 4 |
| Food waste | 25 | 75 | 7.5 |
| Wood | 5 | 20 | 4 |
| Paper | 33 | 6 | 31.02 |
| Cardboard | 14.25 | 5 | 15 |
| Rubber | 7 | 2 | 6.86 |
| Tin cans | 5 | 3 | 4.85 |
| Total | 100 | | |

Determine the dry mass of mixture without tin cans: 67.63%.
Determine the moisture content of mixture: 27.37.

3. The following table illustrates a summary of data of total organic composition of the solid waste assuming a mass of 100 kg of the sample.
Change the moisture content ($H_2O$) in the previous step to hydrogen and oxygen.
Hydrogen = 3.04; oxygen = 24.33

| Component | Mass (kg) |
|---|---|
| Moisture content | 28.62% |
| Carbon | 34 |
| Hydrogen | 8.04 |
| Oxygen | 50.33 |
| Nitrogen | 0.6 |
| Sulfur | 0.1 |
| Ash | 4 |
| Total | 98.32 |

4. Find the percentage by mass as in the following table:

| Composition | Mass (kg) | % (by Mass) |
|---|---|---|
| Carbon | 34 | 35.03 |
| Hydrogen | 8.04 | 8.28 |
| Oxygen | 50.33 | 51.85 |
| Nitrogen | 0.6 | 0.62 |
| Sulfur | 0.1 | 0.10 |
| Ash | 4 | 4.12 |
| Total | 98.32 | 100 |

5. Find the energy value for waste from DuLong formula and the following table:

$$\frac{kJ}{kg} = 337C + 1428\left(H - \frac{O}{8}\right) + 9S$$
$$= 14{,}379\, kJ/kg$$

| Element | Mass (kg) | Atomic Weight | Number of Moles | Chemical Formula with Sulfur | Chemical Formula without Sulfur |
|---|---|---|---|---|---|
| Carbon | 33.73 | 12 | 2.8 | 906.7 | 66.1 |
| Hydrogen | 7.46 | 1 | 8.04 | 2357.2 | 187.6 |
| Oxygen | 50.05 | 16 | 3.15 | 1006.6 | 73.4 |
| Nitrogen | 0.61 | 14 | 0.043 | 13.7 | 1 |
| Sulfur | 0.13 | 32 | 0.003 | 1 | 0.072917 |

Then, the chemical formula with sulfur is $C_{906.7}H_{2573.2}O_{1006.6}N_{13.7}S$.

The chemical formula without sulfur is $C_{66.1}H_{187.6}O_{73.4}N$.

---

### LISTING OF PROGRAM 6.4 (CHAP6.4\FORM1.VB): CHEMICAL COMPOSITION AND ENERGY CONTENT OF SOLID WASTE

```
'******************************'***********************************************************
'Program 6.4: Calculates chemical composition and estimates
'energy content of solid waste.
'*********************************************************************************************
Public Class Form1
    Dim comp(27) As String
    Dim moist(27) As Integer
    Dim PERC(10, 5) As Double
    Private Sub Form1_Load(ByVal sender As System.Object, ByVal e As System.EventArgs)
Handles MyBase.Load
        Me.Text = "Program 6.4: Chemical composition of solid waste"
        Me.FormBorderStyle = Windows.Forms.FormBorderStyle.FixedSingle
        Me.MaximizeBox = False
        Label1.Text = "Select component:"
        Label2.Text = "Percentage:"
        Label3.Text = "Moisture (from Table):"
        Label4.Text = "Dry weight (based on 100lb.):"
        Label5.Text = "Component:"
        Label6.Text = "Mass, kg:"
        Label7.Text = "Carbon"
        Label8.Text = "Hydrogen"
        Label9.Text = "Oxygen"
        Label10.Text = "Nitrogen"
        Label11.Text = "Sulfur"
        Label12.Text = "Ash"
        Label13.Text = ""
        Button1.Text = "&Calculate"
        'DATA FROM TABLE 6.2
        comp(0) = "Residential"
        comp(1) = " Aluminum cans"
        comp(2) = " Cardboard"
        comp(3) = " Fines (dirt, etc.)"
        comp(4) = " Food waste"
        comp(5) = " Glass"
        comp(6) = " Grass"
```

```
comp(7) = " Leather"
comp(8) = " Non-ferrous Metal"
comp(9) = " Leaves"
comp(10) = " Paper"
comp(11) = " Plastics"
comp(12) = " Ferrous metals"
comp(13) = " Rubber"
comp(14) = " Steel cans"
comp(15) = " Textiles"
comp(16) = " Wood"
comp(17) = " Yard waste"
comp(18) = " Garden trimmings"
comp(19) = "Commercial"
comp(20) = " Food waste"
comp(21) = " Mixed organics"
comp(22) = " Mixed"
comp(23) = " Wooden shipping crates and plant scales"
comp(24) = " Construction (mixed)"
comp(25) = " Dirt, ashes, bricks ... etc."
comp(26) = " Municipal waste"
comp(27) = "Select component"
'MOISTURE CONTENT FROM TABLE 6.2
moist(0) = 0
moist(1) = 3
moist(2) = 5
moist(3) = 8
moist(4) = 70
moist(5) = 2
moist(6) = 60
moist(7) = 10
moist(8) = 2
moist(9) = 30
moist(10) = 6
moist(11) = 2
moist(12) = 3
moist(13) = 2
moist(14) = 3
moist(15) = 10
moist(16) = 20
moist(17) = 60
moist(18) = 60
moist(19) = 0
moist(20) = 70
moist(21) = 25
moist(22) = 15
moist(23) = 30
moist(24) = 8
moist(25) = 8
moist(26) = 0
'ADD THE ITEMS INTO THE COMBOBOXES
ComboBox1.Items.Clear()
ComboBox2.Items.Clear()
ComboBox3.Items.Clear()
ComboBox4.Items.Clear()
ComboBox5.Items.Clear()
ComboBox6.Items.Clear()
ComboBox7.Items.Clear()
ComboBox1.Items.AddRange(comp)
ComboBox2.Items.AddRange(comp)
ComboBox3.Items.AddRange(comp)
ComboBox4.Items.AddRange(comp)
```

```
        ComboBox5.Items.AddRange(comp)
        ComboBox6.Items.AddRange(comp)
        ComboBox7.Items.AddRange(comp)
        'DISABLE THE TEXTBOXES
        TextBox2.Enabled = False
        TextBox3.Enabled = False
        TextBox4.Enabled = False
        TextBox5.Enabled = False
        TextBox7.Enabled = False
        TextBox8.Enabled = False
        TextBox10.Enabled = False
        TextBox11.Enabled = False
        TextBox13.Enabled = False
        TextBox14.Enabled = False
        TextBox16.Enabled = False
        TextBox17.Enabled = False
        TextBox19.Enabled = False
        TextBox20.Enabled = False
        'PREPARE THE RADIOBUTTONS AND THE LAST COMBOBOX
        RadioButton1.Text = "Select values from Table 6.3:"
        RadioButton2.Text = "Enter new values:"
        RadioButton2.Checked = True
        ComboBox8.Items.Clear()
        ComboBox8.Items.Add("--Select waste--")
        ComboBox8.Items.Add("Food waste")
        ComboBox8.Items.Add("Paper")
        ComboBox8.Items.Add("Cardboard")
        ComboBox8.Items.Add("Plastics")
        ComboBox8.Items.Add("Textile")
        ComboBox8.Items.Add("Rubber")
        ComboBox8.Items.Add("Leather")
        ComboBox8.Items.Add("Garden trimmings")
        ComboBox8.Items.Add("Timber")
        ComboBox8.Items.Add("Mixture of organic materials")
        ComboBox8.Items.Add("Dirt, ash, bricks etc.")
        'DATA FROM TABLE 6.3
        'PERCENTAGES OF Carbon, Hydrogen, Oxygen, Nitrogen, Sulfur, Ash
        Dim percStr As String
        percStr = "48,6.4,37.6,2.6,0.4,5.0,"
        percStr += "43.5,6,44,0.3,0.2,6,"
        percStr += "44,5.9,44.6,0.3,0.2,5,"
        percStr += "60,7.2,22.8,0,0,10,"
        percStr += "55,6.6,31.2,4.6,0.15,2.5,"
        percStr += "78,10,0,2,0,10,"
        percStr += "60,8,11.6,10,0.4,10,"
        percStr += "47.8,6,38,3.4,0.3,4.5,"
        percStr += "49.5,6,42.7,0.2,0.1,1.5,"
        percStr += "48.5,6.5,37.5,2.2,0.3,5,"
        percStr += "26.3,3,2,0.5,0.2,68,"
        Dim last As Integer = 0
        For i = 0 To 10
            For j = 0 To 5
            PERC(i, j) = Val(percStr.Substring(last, percStr.IndexOf(",", last) - last))
            last = percStr.IndexOf(",", last) + 1
        Next
    Next
End Sub


Sub calculateResults()
    Dim i, M, Ww, Wd As Double
    Dim totalWW, totalWd, moistC As Double
```

```
Dim cansWd, cansWw As Double
totalWd = 0
totalWW = 0
'Calculate the dry weights of the components
i = ComboBox1.SelectedIndex
If i <> 0 And i <> 19 And i <> 27 Then
    Ww = Val(TextBox1.Text)
    M = Val(TextBox2.Text)
    Wd = Ww - (M * Ww / 100)
    If i = 1 Then 'if selection is 'Tin Cans', save it to be subtracted later
        cansWd = Wd
        cansWw = Ww
    End If
    TextBox3.Text = Wd.ToString
    totalWd += Wd : totalWW += Ww
End If
i = ComboBox2.SelectedIndex
If i <> 0 And i <> 19 And i <> 27 Then
    Ww = Val(TextBox6.Text)
    M = Val(TextBox5.Text)
    Wd = Ww - (M * Ww / 100)
    If i = 1 Then 'if selection is 'Tin Cans', save it to be subtracted later
        cansWd = Wd
        cansWw = Ww
    End If
    TextBox4.Text = Wd.ToString
    totalWd += Wd : totalWW += Ww
End If
i = ComboBox3.SelectedIndex
If i <> 0 And i <> 19 And i <> 27 Then
    Ww = Val(TextBox9.Text)
    M = Val(TextBox8.Text)
    Wd = Ww - (M * Ww / 100)
    If i = 1 Then 'if selection is 'Tin Cans', save it to be subtracted later
        cansWd = Wd
        cansWw = Ww
    End If
    TextBox7.Text = Wd.ToString
    totalWd += Wd : totalWW += Ww
End If
i = ComboBox4.SelectedIndex
If i <> 0 And i <> 19 And i <> 27 Then
    Ww = Val(TextBox12.Text)
    M = Val(TextBox11.Text)
    Wd = Ww - (M * Ww / 100)
    If i = 1 Then 'if selection is 'Tin Cans', save it to be subtracted later
        cansWd = Wd
        cansWw = Ww
    End If
    TextBox10.Text = Wd.ToString
    totalWd += Wd : totalWW += Ww
End If
i = ComboBox5.SelectedIndex
If i <> 0 And i <> 19 And i <> 27 Then
    Ww = Val(TextBox15.Text)
    M = Val(TextBox14.Text)
    Wd = Ww - (M * Ww / 100)
    If i = 1 Then 'if selection is 'Tin Cans', save it to be subtracted later
        cansWd = Wd
        cansWw = Ww
    End If
End If
```

```
            TextBox13.Text = Wd.ToString
            totalWd += Wd : totalWW += Ww
        End If
        i = ComboBox6.SelectedIndex
        If i <> 0 And i <> 19 And i <> 27 Then
            Ww = Val(TextBox18.Text)
            M = Val(TextBox17.Text)
            Wd = Ww - (M * Ww / 100)
            If i = 1 Then 'if selection is 'Tin Cans', save it to be subtracted later
                cansWd = Wd
                cansWw = Ww
            End If
            TextBox16.Text = Wd.ToString
            totalWd += Wd : totalWW += Ww
        End If
        i = ComboBox7.SelectedIndex
        If i <> 0 And i <> 19 And i <> 27 Then
            Ww = Val(TextBox21.Text)
            M = Val(TextBox20.Text)
            Wd = Ww - (M * Ww / 100)
            If i = 1 Then 'if selection is 'Tin Cans', save it to be subtracted later
                cansWd = Wd
                cansWw = Ww
            End If
            TextBox19.Text = Wd.ToString
            totalWd += Wd : totalWW += Ww
        End If
        totalWd -= cansWd
        totalWW -= cansWw
        moistC = totalWW - totalWd
        'FORM TABLE (B) AS IN THE EXAMPLE
        Dim total, Carbon, Hydrogen, Oxygen, Nitrogen, Sulfur, Ash As Double
        Carbon = Val(TextBox22.Text)
        Hydrogen = Val(TextBox23.Text)
        Oxygen = Val(TextBox24.Text)
        Nitrogen = Val(TextBox25.Text)
        Sulfur = Val(TextBox26.Text)
        Ash = Val(TextBox27.Text)
        'Change moisture content (H2O) in previous step to hydrogen and oxygen
        Hydrogen += (2 / 18) * moistC
        Oxygen += (16 / 18) * moistC
        total = Carbon + Hydrogen + Oxygen + Nitrogen + Sulfur + Ash
        'Calculate the percentages for the gases
        Carbon /= total
        Hydrogen /= total
        Oxygen /= total
        Nitrogen /= total
        Sulfur /= total
        Ash /= total
        'Find energy value for waste from Dulong formula
        Dim Dulong As Double
        Dulong = (337 * Carbon) + (1428 * (Hydrogen - (Oxygen / 8))) + (9 * Sulfur)
        'Calculate moles for each gas
        Dim molC, molH, molO, molN, molS As Double
        molC = Carbon / 12 'Moles = Mass/Atomic weight
        molH = Hydrogen / 1 'Moles = Mass/Atomic weight
        molO = Oxygen / 16 'Moles = Mass/Atomic weight
        molN = Nitrogen / 14 'Moles = Mass/Atomic weight
        molS = Sulfur / 32 'Moles = Mass/Atomic weight
        'Approximate chemical formula with sulfur
        'sulfur will be 1 mole = 0.004 * 250.. so multiply all moles by 250
```

```
        Dim molC2, molH2, molO2, molN2, molS2 As Double
        molC2 = molC * 250 * 100
        molH2 = molH * 250 * 100
        molO2 = molO * 250 * 100
        molN2 = molN * 250 * 100
        molS2 = molS * 250 * 100
        Label13.Text = "Chemical formula with sulfur is: C" + Format(molC2, "n") + _
            " H" + Format(molH2, "n") + " O" + Format(molO2, "n") + " N" + Format(molN2,
"n") + " S"
        'Approximate chemical formula without sulfur
        'nitrogen will be 1 mole = 0.044 * 22.73.. so multiply all moles by 22.73
        molC2 = molC * 22.73 * 100
        molH2 = molH * 22.73 * 100
        molO2 = molO * 22.73 * 100
        molN2 = molN * 22.73 * 100
        molS2 = molS * 22.73 * 100
        Label13.Text += vbCrLf + "Chemical formula without sulfur is: C" + Format(molC2,
"n") + _
            " H" + Format(molH2, "n") + " O" + Format(molO2, "n") + " N" + Format(molN2,
"n")
    End Sub


    Private Sub ComboBox1_SelectedIndexChanged(ByVal sender As System.Object, ByVal e As
System.EventArgs) Handles ComboBox1.SelectedIndexChanged
        'enter the moisture content (From Table 6.2) into the 'moisture' field
        If ComboBox1.SelectedIndex = 0 Or ComboBox1.SelectedIndex = 19 Or ComboBox1.
SelectedIndex = 27 Then
            'These are items not to be selected by the user, so clear the textbox and
exit sub
            TextBox2.Text = ""
        Else
            TextBox2.Text = moist(ComboBox1.SelectedIndex).ToString
        End If
    End Sub


    Private Sub ComboBox2_SelectedIndexChanged(ByVal sender As System.Object, ByVal e As
System.EventArgs) Handles ComboBox2.SelectedIndexChanged
        'enter the moisture content (From Table 6.2) into the 'moisture' field
        If ComboBox2.SelectedIndex = 0 Or ComboBox2.SelectedIndex = 19 Or ComboBox2.
SelectedIndex = 27 Then
            'These are items not to be selected by the user, so clear the textbox and
exit sub
            TextBox5.Text = ""
        Else
            TextBox5.Text = moist(ComboBox2.SelectedIndex).ToString
        End If
    End Sub


    Private Sub ComboBox3_SelectedIndexChanged(ByVal sender As System.Object, ByVal e As
System.EventArgs) Handles ComboBox3.SelectedIndexChanged
        'enter the moisture content (From Table 6.2) into the 'moisture' field
        If ComboBox3.SelectedIndex = 0 Or ComboBox3.SelectedIndex = 19 Or ComboBox3.
SelectedIndex = 27 Then
            'These are items not to be selected by the user, so clear the textbox and
exit sub
            TextBox8.Text = ""
        Else
            TextBox8.Text = moist(ComboBox3.SelectedIndex).ToString
        End If
    End Sub
```

```vb
    Private Sub ComboBox4_SelectedIndexChanged(ByVal sender As System.Object, ByVal e As
System.EventArgs) Handles ComboBox4.SelectedIndexChanged
        'enter the moisture content (From Table 6.2) into the 'moisture' field
        If ComboBox4.SelectedIndex = 0 Or ComboBox4.SelectedIndex = 19 Or ComboBox4.
SelectedIndex = 27 Then
            'These are items not to be selected by the user, so clear the textbox and
exit sub
            TextBox11.Text = ""
        Else
            TextBox11.Text = moist(ComboBox4.SelectedIndex).ToString
        End If
    End Sub

    Private Sub ComboBox5_SelectedIndexChanged(ByVal sender As System.Object, ByVal e As
System.EventArgs) Handles ComboBox5.SelectedIndexChanged
        'enter the moisture content (From Table 6.2) into the 'moisture' field
        If ComboBox5.SelectedIndex = 0 Or ComboBox5.SelectedIndex = 19 Or ComboBox5.
SelectedIndex = 27 Then
            'These are items not to be selected by the user, so clear the textbox and
exit sub
            TextBox14.Text = ""
        Else
            TextBox14.Text = moist(ComboBox5.SelectedIndex).ToString
        End If
    End Sub

    Private Sub ComboBox6_SelectedIndexChanged(ByVal sender As System.Object, ByVal e As
System.EventArgs) Handles ComboBox6.SelectedIndexChanged
        'enter the moisture content (From Table 6.2) into the 'moisture' field
        If ComboBox6.SelectedIndex = 0 Or ComboBox6.SelectedIndex = 19 Or ComboBox6.
SelectedIndex = 27 Then
            'These are items not to be selected by the user, so clear the textbox and
exit sub
            TextBox17.Text = ""
        Else
            TextBox17.Text = moist(ComboBox6.SelectedIndex).ToString
        End If
    End Sub

    Private Sub ComboBox7_SelectedIndexChanged(ByVal sender As System.Object, ByVal e As
System.EventArgs) Handles ComboBox7.SelectedIndexChanged
        'enter the moisture content (From Table 6.2) into the 'moisture' field
        If ComboBox7.SelectedIndex = 0 Or ComboBox7.SelectedIndex = 19 Or ComboBox7.
SelectedIndex = 27 Then
            'These are items not to be selected by the user, so clear the textbox and
exit sub
            TextBox20.Text = ""
        Else
            TextBox20.Text = moist(ComboBox7.SelectedIndex).ToString
        End If
    End Sub

    Private Sub Button1_Click(ByVal sender As System.Object, ByVal e As System.EventArgs)
Handles Button1.Click
        calculateResults()
    End Sub

    Private Sub RadioButton1_CheckedChanged(ByVal sender As System.Object, ByVal e As
System.EventArgs) Handles RadioButton1.CheckedChanged
        ComboBox8.Enabled = True
    End Sub
```

```
    Private Sub RadioButton2_CheckedChanged(ByVal sender As System.Object, ByVal e As
System.EventArgs) Handles RadioButton2.CheckedChanged
        ComboBox8.Enabled = False
        TextBox22.Text = ""
        TextBox23.Text = ""
        TextBox24.Text = ""
        TextBox25.Text = ""
        TextBox26.Text = ""
        TextBox27.Text = ""
    End Sub

    Private Sub ComboBox8_SelectedIndexChanged(ByVal sender As System.Object, ByVal e As
System.EventArgs) Handles ComboBox8.SelectedIndexChanged
        Dim i = ComboBox8.SelectedIndex
        If i <= 0 Then Exit Sub
        TextBox22.Text = PERC(i - 1, 0).ToString
        TextBox23.Text = PERC(i - 1, 1).ToString
        TextBox24.Text = PERC(i - 1, 2).ToString
        TextBox25.Text = PERC(i - 1, 3).ToString
        TextBox26.Text = PERC(i - 1, 4).ToString
        TextBox27.Text = PERC(i - 1, 5).ToString
    End Sub
End Class
```

## Example 6.5

1. Write a short computer program that enables calculating the percentage of heat value to be generated from a municipal refuse, given its fractions.
2. Based on the given typical heat values, estimate the heat value of a processed refuse-derived fuel (RDF) with the composition set along the following table:

**Table of Fractions by Weight of MSW Components**

| Component | Fraction by Weight, Dry Basis |
|---|---|
| Paper | 0.5 |
| Plastics | 0.3 |
| Glass | 0.1 |
| Wood | 0.1 |

**Typical Heat Values of MSW Components**

| Component | Heat Value (Btu/lb), Dry Weight |
|---|---|
| Cardboard | 7,000 |
| Dirt, ashes, and other fines | 3,000 |
| Ferrous metals | 300 |
| Food waste | 2,000 |
| Garden trimmings | 2,800 |
| Glass | 60 |
| Leather | 7,500 |
| Nonferrous metals | 300 |
| Paper | 7,200 |
| Plastics | 14,000 |
| Rubber | 10,000 |
| Textiles | 7,500 |
| Wood | 8,000 |

## Solution

1. For the solution of Example 6.5 (1), see the listing of Program 6.5. Given: Refuse composition.
2. Form the following table:

| Component | Fraction by Weight, Dry Basis | Heat Value |
|---|---|---|
| Paper | 0.5 | 7,200 |
| Plastics | 0.3 | 14,000 |
| Glass | 0.1 | 60 |
| Wood | 0.1 | 8,000 |

Estimate the heat value based on the typical values in table: 0.5 * 7200 + 0.3 * 14000 + 0.1 * 60 + 0.1 * 8000 = 8606 Btu/lb.

3. Use the program developed in (1) to verify the manual computations conducted in (2).

## LISTING OF PROGRAM 6.5   (CHAP6.5\FORM1.VB): HEAT VALUE

```vb
'**************************************************************************
'Program 6.5: calculates percentage of heat value
'**************************************************************************
Public Class Form1
    Dim MSW_Component() As String =
        {"Cardboard", "Dirt, ashes and other fines",
        "Ferrous metals", "Food waste",
        "Garden trimmings", "Glass",
        "Leather", "Nonferrous metals",
        "Paper", "Plastics", "Rubber",
        "Textiles", "Wood", "--Select comp.--"}
    Dim MSW_Value() As Double =
        {7000, 3000, 300, 2000,
        2800, 60, 7500, 300,
        7200, 14000, 10000,
        7500, 8000}

    Private Sub Form1_Load(ByVal sender As System.Object, ByVal e As System.EventArgs)
Handles MyBase.Load
        Me.Text = "Program 6.5: Calculates percentage of heat value"
        Me.FormBorderStyle = Windows.Forms.FormBorderStyle.FixedSingle
        Me.MaximizeBox = False
        Label1.Text = "Select component:"
        Label2.Text = "Fraction by weight (dry basis):"
        Label3.Text = "Heat value:"
        Label4.Text = ""
        Button1.Text = "&Calculate"
        'prepare the selection boxes
        ComboBox1.Items.Clear()
        ComboBox1.Items.AddRange(MSW_Component)
        ComboBox1.SelectedIndex = 13
        ComboBox2.Items.Clear()
        ComboBox2.Items.AddRange(MSW_Component)
        ComboBox2.SelectedIndex = 13
        ComboBox3.Items.Clear()
        ComboBox3.Items.AddRange(MSW_Component)
        ComboBox3.SelectedIndex = 13
        ComboBox4.Items.Clear()
        ComboBox4.Items.AddRange(MSW_Component)
        ComboBox4.SelectedIndex = 13
        ComboBox5.Items.Clear()
        ComboBox5.Items.AddRange(MSW_Component)
        ComboBox5.SelectedIndex = 13
        ComboBox6.Items.Clear()
        ComboBox6.Items.AddRange(MSW_Component)
        ComboBox6.SelectedIndex = 13
        TextBox2.Enabled = False
        TextBox3.Enabled = False
        TextBox5.Enabled = False
        TextBox7.Enabled = False
        TextBox9.Enabled = False
        TextBox11.Enabled = False
    End Sub

    Sub calculateResults()
        Dim heatV As Double = 0
```

```
            'check each combobox to see if the user selected a component, if yes, add it to
the total.
        If ComboBox1.SelectedIndex <> 6 Then heatV += addHeatValue(1)
        If ComboBox2.SelectedIndex <> 6 Then heatV += addHeatValue(2)
        If ComboBox3.SelectedIndex <> 6 Then heatV += addHeatValue(3)
        If ComboBox4.SelectedIndex <> 6 Then heatV += addHeatValue(4)
        If ComboBox5.SelectedIndex <> 6 Then heatV += addHeatValue(5)
        If ComboBox6.SelectedIndex <> 6 Then heatV += addHeatValue(6)
        Label4.Text = "Heat value = " + heatV.ToString + " Btu/Lb."
    End Sub

    Function addHeatValue(ByVal n As Integer) As Double
        Dim fraction, heat As Double
        Select Case n
            'which combobox is it?
            Case 1
                fraction = Val(TextBox1.Text)
                heat = Val(TextBox2.Text)
                Return fraction * heat
            Case 2
                fraction = Val(TextBox4.Text)
                heat = Val(TextBox3.Text)
                Return fraction * heat
            Case 3
                fraction = Val(TextBox6.Text)
                heat = Val(TextBox5.Text)
                Return fraction * heat
            Case 4
                fraction = Val(TextBox8.Text)
                heat = Val(TextBox7.Text)
                Return fraction * heat
            Case 5
                fraction = Val(TextBox10.Text)
                heat = Val(TextBox9.Text)
                Return fraction * heat
            Case 6
                fraction = Val(TextBox12.Text)
                heat = Val(TextBox11.Text)
                Return fraction * heat
            Case Else
                Return 0
        End Select
    End Function

    Private Sub ComboBox1_SelectedIndexChanged(ByVal sender As System.Object, ByVal e As
System.EventArgs) Handles ComboBox1.SelectedIndexChanged
        'change the 'heat value' according to the selected component from the combobox.
        If ComboBox1.SelectedIndex = -1 Or ComboBox1.SelectedIndex = 13 Then
            TextBox2.Text = ""
            Exit Sub
        End If
        TextBox2.Text = MSW_Value(ComboBox1.SelectedIndex).ToString
    End Sub
```

```vb
    Private Sub ComboBox2_SelectedIndexChanged(ByVal sender As System.Object, ByVal e As
System.EventArgs) Handles ComboBox2.SelectedIndexChanged
        'change the 'heat value' according to the selected component from the combobox.
        If ComboBox2.SelectedIndex = -1 Or ComboBox2.SelectedIndex = 13 Then
            TextBox3.Text = ""
            Exit Sub
        End If
        TextBox3.Text = MSW_Value(ComboBox2.SelectedIndex).ToString
    End Sub

    Private Sub ComboBox3_SelectedIndexChanged(ByVal sender As System.Object, ByVal e As
System.EventArgs) Handles ComboBox3.SelectedIndexChanged
        'change the 'heat value' according to the selected component from the combobox.
        If ComboBox3.SelectedIndex = -1 Or ComboBox3.SelectedIndex = 13 Then
            TextBox5.Text = ""
            Exit Sub
        End If
        TextBox5.Text = MSW_Value(ComboBox3.SelectedIndex).ToString
    End Sub

    Private Sub ComboBox4_SelectedIndexChanged(ByVal sender As System.Object, ByVal e As
System.EventArgs) Handles ComboBox4.SelectedIndexChanged
        'change the 'heat value' according to the selected component from the combobox.
        If ComboBox4.SelectedIndex = -1 Or ComboBox4.SelectedIndex = 13 Then
            TextBox7.Text = ""
            Exit Sub
        End If
        TextBox7.Text = MSW_Value(ComboBox4.SelectedIndex).ToString
    End Sub

    Private Sub ComboBox5_SelectedIndexChanged(ByVal sender As System.Object, ByVal e As
System.EventArgs) Handles ComboBox5.SelectedIndexChanged
        'change the 'heat value' according to the selected component from the combobox.
        If ComboBox5.SelectedIndex = -1 Or ComboBox5.SelectedIndex = 13 Then
            TextBox9.Text = ""
            Exit Sub
        End If
        TextBox9.Text = MSW_Value(ComboBox5.SelectedIndex).ToString
    End Sub

    Private Sub ComboBox6_SelectedIndexChanged(ByVal sender As System.Object, ByVal e As
System.EventArgs) Handles ComboBox6.SelectedIndexChanged
    'change the 'heat value' according to the selected component from the combobox.
    If ComboBox6.SelectedIndex = -1 Or ComboBox6.SelectedIndex = 13 Then
        TextBox11.Text = ""
        Exit Sub
    End If
        TextBox11.Text = MSW_Value(ComboBox6.SelectedIndex).ToString
    End Sub

    Private Sub Button1_Click(ByVal sender As System.Object, ByVal e As System.EventArgs)
Handles Button1.Click
        calculateResults()
    End Sub
End Class
```

**TABLE 6.4**

**General Overview of Solid Waste Properties and Characteristics of Concern**

| Property | Influencing Parameters | Typical Values | Measuring Device |
|---|---|---|---|
| Specific weight | Location, season, storage time, equipment used, processing (compaction, shredding, etc.) | 600–900 lb/yd$^3$ as delivered (50–300 kg/m$^3$) | Pycnometry, density meter |
| Permeability | Type of waste, compaction, soil characteristics | Compacted solid wastes in a landfill are in the range between ~10$^{-11}$ and 10$^{-12}$ m$^2$ | Oedometer, consolidation equipment |
| Moisture content | Unit weight, decomposition stage, temperature, waste composition, pore fluid composition | 40%–65% | Moisture gage, moisture analyzer, electrical resistance sensor |
| Angle of repose | Density, surface area and shapes of the particles, the coefficient of friction of the material, gravity | 45°–90° | Shear test method, tilting box, fixed funnel and revolving cylinder method |
| Fusion point of ash (ash fusibility) | Acidic and alkaline composition and content, type of waste and fuel, type of reactor and burning equipment, rate of burning, temperature and thickness of fire bed or ball, distribution of ash, viscosity | 1100°C–1200°C for the formation of clinker | Analytical instrument, furnace |
| Heat value | Waste composition, organic matter, moisture content | | Calorimeter, analytical equipment |

Table 6.4 gives a general overview of solid waste properties and characteristics of concern.

## 6.4  SOLID WASTE COLLECTION

MSW collection systems are customarily person/truck systems. MSW collectors traverse generation sites and source production in trucks to transport the collected refuse to a site at which the truck is emptied. This process may be an intermediate stopover where the refuse is transferred from a small truck into trailers, larger vans, barges, or railway cars for long-distance transport or a selected final site such as the landfill, compost site, or MRF. During this cycle, some of the useful MSW may be isolated, sorted out, or segregated for reuse or recycling or conversion into other useful products.

The process of refuse collection is a multiphase process, and it can be divided into separate phases: house-to-can (transferring MSW from home to dustbin inside or outside the house), can-to-truck (for movement of MSW from dustbin to MSW and garbage car by MSW workers or owner of housing), truck from house-to-house (collection phase of MSW from different sources by best and efficient ways and its transfer to collection areas and to areas of intermediate or final disposal), truck routing (stage of path of truck through the city's road network), and truck-to-disposal (stage of final disposal or recovery of materials).

### Example 6.6

1. Write a short computer program that enables calculating the number of cans and compacted blocks produced by a family given the number of people, the rate of per capita daily generation of solid waste, the bulk density of refuse in a typical garbage can, and the collection frequency.

2. A family of four people generates MSW at a rate of 2.5 lb/c/day, and the bulk density of refuse in a typical garbage can is ~250 lb/yd$^3$. If collection is once a week, how many 30-gal garbage cans will they need, or the alternative, how many compacted 20-lb blocks would the family produce if they had a home compactor? How many cans wovuld they need in that case?

3. Use the program developed in (1) to verify the manual computations conducted in (2).

**Solution**

1. For the solution of Example 6.6 (1), see the listing of Program 6.6.

2. Solution to Example 6.6 (2):

   a. Given: $P = 4$, generated waste = 2.5 lb/c/day, $\rho = 250$ lb/yd$^3$.

   b. Weight of SW generated = 2.5 lb/c/day × 4 persons × 7 days/week = 70 lb.

   c. Volume of solid waste = weight/density = 70 lb/250 lb/yd$^3$ = 0.28 yd$^3$

      i. Volume (convert to gallons) = 0.28 yd$^3$ × 202 gal/yd$^3$ = 56.6 gal.

      ii. They will require *two* 30-gal cans.

   d. If the refuse is compacted into 20-lb blocks, they would need to produce such compacted blocks to take care of the week's refuse.

      i. If each block of compacted refuse is 1400 lb/yd$^3$, the necessary volume is 70 lb./1400 lb./yd$^3$) × 202 gal/yd$^3$ = 10.1 gal.

      ii. They would need *only one* 30-gal can.

**LISTING OF PROGRAM 6.6   (CHAP6.6\FORM1.VB): NUMBER OF CANS
AND COMPACTED BLOCKS PRODUCED BY A FAMILY**

```vb
'*********************************************************************************
'Program 6.6: Calculates number of cans and
' compacted blocks produced by a family
'*********************************************************************************
Public Class Form1
    Private Sub Form1_Load(ByVal sender As System.Object, ByVal e As System.EventArgs)
Handles MyBase.Load
        Me.Text = "Program 6.6: Cans and compacted blocks"
        Me.FormBorderStyle = BorderStyle.FixedSingle
        Me.MaximizeBox = False
        Label1.Text = "Number of people in the family:"
        Label2.Text = "Rate of daily generation of solid waste (lb/c/day):"
        Label3.Text = "Bulk density of refuse in a typical garbage can (lb/yd3):"
        Label4.Text = "Collection frequency (per week):"
        Label5.Text = ""
        button1.text = "&Calculate"
        Me.Height = 188
    End Sub

    Sub calculateResults()
        Dim s As String = ""
        Dim P, rate, dens, freq As Double
        Dim weight, vol, vol2, vol3 As Double
        P = Val(TextBox1.Text)
        rate = Val(TextBox2.Text)
        dens = Val(TextBox3.Text)
        freq = Val(TextBox4.Text)
        weight = rate * P * (7 / freq) 'convert frequency per week to days
        s += vbCrLf + "Weight of SW generated = " + rate.ToString + " lb/c/day x " +
P.ToString + _
            " persons x " + (7 / freq).ToString + " days/week = " + weight.ToString + "
Lb."
        vol = weight / dens
        s += vbCrLf + "Volume of solid waste = weight/density = " + weight.ToString + "
lb./" + dens.ToString _
            + " lb./yd3 = " + vol.ToString + " yd3"
        vol2 = vol * 202 'convert to gallons
        s += vbCrLf + "Volume (convert to gallons) = " + vol.ToString + " yd3 x 202
gal/yd3 = " _
            + vol2.ToString + " gal"
        s += vbCrLf + vbCrLf + "They will require " + numberInLetters(Math.Ceiling(vol2 /
30)) + " 30-gal cans. "
        s += vbCrLf + "If the refuse is compacted into 20-lb blocks, they would need to "
        s += vbCrLf + "produce such compacted blocks to take care of the week's refuse"
vol3 = (weight / 1400) * 202
        s += vbCrLf + vbCrLf + "If each block of compacted refuse is 1400 lb/yd3, the
necessary volume is "
        s += vbCrLf + "(" + weight.ToString + " lb./1400 lb./yd3) x 202 gal/yd3 = " + _
Format(vol3, "##.##").ToString + " gal"
        s += vbCrLf + "They would need only " + numberInLetters(Math.Ceiling(vol3 / 30)) +
" 30-gal can."
        Label5.Text = s
        Me.Height = 372
    End Sub
```

```
'*******************************************************************************
'takes a number as integer, and returns it in letters.
'e.g. 1 -> 'One', 2 -> 'Two' and so one.
'*******************************************************************************
Function numberInLetters(ByVal n As Integer) As String
    Select Case n
        Case 0 : Return "Zero"
        Case 1 : Return "One"
        Case 2 : Return "Two"
        Case 3 : Return "Three"
        Case 4 : Return "Four"
        Case 5 : Return "Five"
        Case 6 : Return "Six"
        Case 7 : Return "Seven"
        Case 8 : Return "Eight"
        Case 9 : Return "Nine"
        Case 10 : Return "Ten"
        Case 11 : Return "Eleven"
        Case 12 : Return "Twelve"
        Case 13 : Return "Thirteen"
        Case 14 : Return "Fourteen"
        Case 15 : Return "Fifteen"
        Case Else : Return ""
    End Select
End Function

Private Sub Button1_Click(ByVal sender As System.Object, ByVal e As System.EventArgs)
Handles Button1.Click
    calculateResults()
End Sub
End Class
```

The number of collection vehicles needed for a community may be determined from the following equation:

$$N = \frac{S * F}{X * W} \qquad (6.21)$$

where:

$N$ is the number of collection vehicles needed

$S$ is the total number of customers serviced

$F$ is the collection frequency, number of collections per week

$X$ is the number of customers a single truck can service per day

$W$ is the number of workdays per week

## Example 6.7

1. Write a short computer program that enables calculating the number of collection vehicles that a community would need for a total of services (customers) that are to be collected once per week during working days in a city.

2. Use the equation for estimating the number of MSW collection vehicles needed for a certain community:

$$N = \frac{S * F}{X * W}$$

The number of collection vehicles a community would need for solid waste collection in a certain municipality is six trucks. Find the total number of services (customers) that are to be collected once per week during working days in a Al-Doha in KSA. (Realistically, most trucks can service only ~200–300 customers before the truck is full, and a trip to the landfill is necessary).

3. Use the program developed in (1) to verify the manual computations conducted in (2).

## Solution

1. Given:

   $N$ is the number of collection vehicles needed = 6

   $S$ is the total number of customers serviced = ?

   $F$ is the collection frequency, number of collections per week = 1.

   $X$ is the number of customers a single truck can service per day (A single truck can service 300 customers in a single day and still have time to take the full loads to the landfill) = 300.

2. $W$ is the number of workdays per week (The town wants to collect on Saturdays, Sundays, Mondays, and Tuesdays leaving Wednesdays for special projects and truck maintenance) = 4 days.

3. Thus, $S = N * X * W/F = (300 * 4 * 6)/(1) = 7200$.

4. The total number of customers to be serviced = 7200.

## LISTING OF PROGRAM 6.7 (CHAP6.7\FORM1.VB): NUMBER OF COLLECTION VEHICLES NEEDED BY A COMMUNITY

```vbnet
'****************************************************************************
'Program 6.7: Calculates number of collection vehicles needed by a community
'****************************************************************************
Public Class Form1
    Private Sub Form1_Load(ByVal sender As System.Object, ByVal e As System.EventArgs)
Handles MyBase.Load
        Me.Text = "Program 6.7: Collection vehicles"
        Me.FormBorderStyle = BorderStyle.FixedSingle
        Me.MaximizeBox = False
        Label6.Text = "Formula:" + vbCrLf + "N = (S* F)/(X * W)"
        Label7.Text = "Solving for:"
        Button1.Text = "&Calculate"
        ComboBox1.Items.Clear()
        ComboBox1.Items.Add("N (Number of collection vehicles needed)")
        ComboBox1.Items.Add("S (Total number of customers serviced)")
        ComboBox1.Items.Add("F (Collection frequency per week)")
        ComboBox1.Items.Add("X (Number of customers a single truck can service)")
        ComboBox1.Items.Add("W (Number of workdays per week)")
        ComboBox1.SelectedIndex = 0
    End Sub

    Sub calculateResults()
        Dim str As String = ""
        Dim S, F, X, W, N As Double
        If ComboBox1.SelectedIndex = 0 Then
            'Solving for N
            S = Val(TextBox1.Text)
            F = Val(TextBox2.Text)
            X = Val(TextBox3.Text)
            W = Val(TextBox4.Text)
            N = (S * F) / (X * W)
            str = "N = " + N.ToString
            str += vbCrLf + "The community will need" + numberInLetters(Math.Ceiling(N))
+ " trucks."
        ElseIf ComboBox1.SelectedIndex = 1 Then
            'Solving for S
            N = Val(TextBox1.Text)
            F = Val(TextBox2.Text)
            X = Val(TextBox3.Text)
            W = Val(TextBox4.Text)
            S = (N * X * W) / (F)
            str = "The total number of customers to be serviced = " + CInt(S).ToString + "
customer"
        ElseIf ComboBox1.SelectedIndex = 2 Then
            'Solving for F
            N = Val(TextBox1.Text)
            S = Val(TextBox2.Text)
            X = Val(TextBox3.Text)
            W = Val(TextBox4.Text)
            F = (N * X * W) / (S)
            str = "The collection frequency = " + CInt(F).ToString + " per week"
        ElseIf ComboBox1.SelectedIndex = 3 Then
            'Solving for X
            N = Val(TextBox1.Text)
            S = Val(TextBox2.Text)
            F = Val(TextBox3.Text)
            W = Val(TextBox4.Text)
            X = (S * F) / (N * W)
```

```
            str = "The number of customers a single truck can service = " + CInt(X).
ToString + " customer(s)"
        ElseIf ComboBox1.SelectedIndex = 4 Then
        'Solving for W
        N = Val(TextBox1.Text)
        S = Val(TextBox2.Text)
        F = Val(TextBox3.Text)
        X = Val(TextBox4.Text)
        W = (S * F) / (N * X)
        str = "The number of workdays per week = " + CInt(W).ToString + " day(s)"
        End If
        Label5.Text = str
    End Sub

    '**********************************************************************************
    'takes a number as integer, and returns it in letters.
    'e.g. 1 -> 'One', 2 -> 'Two' and so one.
    '**********************************************************************************
    Function numberInLetters(ByVal n As Integer) As String
        Select Case n
            Case 0 : Return "Zero"
            Case 1 : Return "One"
            Case 2 : Return "Two"
            Case 3 : Return "Three"
            Case 4 : Return "Four"
            Case 5 : Return "Five"
            Case 6 : Return "Six"
            Case 7 : Return "Seven"
            Case 8 : Return "Eight"
            Case 9 : Return "Nine"
            Case 10 : Return "Ten"
            Case 11 : Return "Eleven"
            Case 12 : Return "Twelve"
            Case 13 : Return "Thirteen"
            Case 14 : Return "Fourteen"
            Case 15 : Return "Fifteen"
            Case 16 : Return "Sixteen"
            Case 17 : Return "Seventeen"
            Case 18 : Return "Eighteen"
            Case 19 : Return "Nineteen"
            Case 20 : Return "Twenty"
            Case Else : Return ""
        End Select
    End Function

    Private Sub Button1_Click(ByVal sender As System.Object, ByVal e As System.EventArgs)
Handles Button1.Click
        calculateResults()
    End Sub

    Private Sub ComboBox1_SelectedIndexChanged(ByVal sender As System.Object, ByVal e As
System.EventArgs) Handles ComboBox1.SelectedIndexChanged
        Label5.Text = ""
        If ComboBox1.SelectedIndex = 0 Then
            Label1.Text = "Total number of customers serviced, S:"
            Label2.Text = "Collection frequency, F (per week):"
            Label3.Text = "Number of customers a single truck can service per day, X:"
            Label4.Text = "Number of workdays, W (per week):"
        ElseIf ComboBox1.SelectedIndex = 1 Then
            Label1.Text = "Number of collection vehicles needed, N:"
            Label2.Text = "Collection frequency, F (per week):"
```

```
            Label3.Text = "Number of customers a single truck can service per day, X:"
            Label4.Text = "Number of workdays, W (per week):"
        ElseIf ComboBox1.SelectedIndex = 2 Then
            Label1.Text = "Number of collection vehicles needed, N:"
            Label2.Text = "Total number of customers serviced, S:"
            Label3.Text = "Number of customers a single truck can service per day, X:"
            Label4.Text = "Number of workdays, W (per week):"
        ElseIf ComboBox1.SelectedIndex = 3 Then
            Label1.Text = "Number of collection vehicles needed, N:"
            Label2.Text = "Total number of customers serviced, S:"
            Label3.Text = "Collection frequency, F (per week):"
            Label4.Text = "Number of workdays, W (per week):"
        ElseIf ComboBox1.SelectedIndex = 4 Then
            Label1.Text = "Number of collection vehicles needed, N:"
            Label2.Text = "Total number of customers serviced, S:"
            Label3.Text = "Collection frequency, F (per week):"
            Label4.Text = "Number of customers a single truck can service per day, X:"
        End If
    End Sub
End Class
```

## 6.5 SOLID WASTE COLLECTION, PROCESSING, AND MATERIAL SEPARATION

MSW refuse is a heterogeneous material with unpredictable and time-variable characteristics and amounts. These conditions interfere negatively with design MSW processing and MRFs when attempting to utilize the material in producing a desirable product. Some types of MSW refuse are easily processed, yet certain other types are difficult and/or dangerous to handle. MSW processing operations incorporate many interlinked factors at operational, organizational, and safety levels. Therefore, the material ought to be designed for extraordinary contingencies. Such a requirement often results in overdesign and underutilization of resources when processing all of the feed material.

Primary treatment operations are targeted to prepare MSW and garbage in a sustainable management system format, increasing efficiency of operation, extracting useful materials and resources, and restoring products and energy. This could be achieved through size reduction by mechanical means (e.g., compaction or fragmentation) or chemical means (e.g., incineration and burning) or by automatic and mechanical means (e.g., separation of components or extraction of moisture content and drying).

The basic types of conveyors used primarily to move MSW and refuse include rubber-belted conveyors, live bottom feeders, and pneumatic conveyors. Other conveyors used to feed or meter refuse to a load-sensitive device (such as a combustor) incorporate vibratory feeders, screw feeders, and drag chains.

Power requirements of belt conveyors can be estimated by a number of empirical equations. Equation 6.22 represents the power necessary to move a load horizontally and vertically together with power loss due to friction.

$$\text{Horsepower} = \frac{LSF}{1000} + \frac{LTC}{990} + \frac{TH}{990} + P \quad (6.22)$$

where:
$L$ is the length of conveyor belt, ft
$S$ is the speed of belt, ft/min
$F$ is the speed factor, dimensionless
$T$ is the capacity, tons/h
$C$ is the idle resistance factor, dimensionless
$H$ is the lift, ft
$P$ is the pulley friction, horsepower

Screw conveyors are used to meter shredded refuse into a furnace. The volume of material moved by a screw conveyor in a flooded condition can be estimated by the following equation:

$$Q = CNRV \quad (6.23)$$

where:
$Q$ is the delivery of refuse, m³/min
$C$ is the efficiency factor
$N$ is the number of conveyor leads or the number of blades that are wrapped around the conveyor hub
$R$ is the rotational speed of screw, rpm
$V$ is the volume of refuse between each pitch, m³

Crushing devices or rolls break fragile materials such as glass while unfolding rout iron such as iron cans, which would facilitate separation by sieving. Crushing rolls strongly hold the raw material entering between two rollers operating in opposite directions. The maximum volume that can be squeezed by crushing cylinders can be estimated by the following equation:

$$C = kvDLs\rho \quad (6.24)$$

where:

C is the capacity, tons/h

k is the constant, dimensionless = 60, when taking units and dimensions described in Equation 6.24

v is the speed of cylinders, rpm

D is the cylinder diameter, m

L is the length of cylinder, m

s is the distance of separation between cylinders (disks), m

ρ is the density of the material, g/cm$^3$

## Example 6.8

1. Write a short computer program that enables calculating the capacity of a crushing cylinder given the speed of cylinder, cylinder diameter and length, the density of material, and the distance between disks.

2. Find the capacity of a crushing cylinder noting that the speed of cylinder is 4600 cycles per hour, cylinder diameter is 200 mm, its length is 45 cm, the density of material is 2400 kg/m$^3$, and the distance between disks is 5 mm.

3. Use the program developed in (1) to verify the manual computations conducted in (2).

### Solution

1. For the solution of Example 6.8 (1), see the listing of Program 6.8.
2. Solution to Example 6.8 (2):
   a. Given: v = 4600 cycles/h = 4600/60 cycles/min, D = 200 mm = 20 cm, L = 0.45 m, ρ = 2400/1000 = 2.4 g/cm$^3$, s = 5 mm.
   b. Then, capacity C = kvDLs ρ or C = kv. D.L.s. ρ = 60 × (4600/60) rpm × 0.2 m × 0.45 m × 0.005 m × 2.4 (g/cm$^3$) = 5 tons/h.

---

### LISTING OF PROGRAM 6.8   (CHAP6.8\FORM1.VB): CAPACITY OF A CRUSHING CYLINDER

```
'********************************************************************************
'Program 6.8: Calculates capacity of a crushing cylinder
'********************************************************************************
Public Class Form1
    Dim C, k, v, D, L, s, rho As Double
    Private Sub Form1_Load(ByVal sender As System.Object, ByVal e As System.EventArgs)
Handles MyBase.Load
        Me.Text = "Program 6.8: Capacity of a crushing cylinder"
        Me.MaximizeBox = False
        Me.FormBorderStyle = Windows.Forms.FormBorderStyle.FixedSingle
        Label1.Text = "Formula:" + vbCrLf
        Label1.Text += "C = kvDLsrho"
        Label2.Text = "Speed of cylinders (v), cycles/hr:"
        Label3.Text = "Cylinder diameter (D), mm:"
        Label4.Text = "Length of cylinder (L), cm:"
        Label5.Text = "Distance of separation between discs (s), mm:"
        Label6.Text = "Density of the material (rho), kg/m3:"
        Label7.Text = ""
        Label8.Text = "Decimals:"
        Button1.Text = "&Calculate"
        NumericUpDown1.Value = 2
        NumericUpDown1.Maximum = 10
        NumericUpDown1.Minimum = 0
    End Sub

    Sub calculateResults()
        v = Val(TextBox1.Text)
        D = Val(TextBox2.Text)
        L = Val(TextBox3.Text)
        s = Val(TextBox4.Text)
        rho = Val(TextBox5.Text)
        k = 60
        v = v / 60       'convert to cycles/min
        D = D / 1000     'convert to m.
        L = L / 100      'convert to m.
```

```
        s = s / 1000    'convert to m.
        rho = rho / 1000 'convert to g/cm3
        C = k * v * D * L * s * rho
    End Sub

    Sub showResults()
        Label7.Text = "Capacity, C = " + FormatNumber(C, NumericUpDown1.Value) + "
tones/hr"
    End Sub

    Private Sub Button1_Click(ByVal sender As System.Object, ByVal e As System.EventArgs)
Handles Button1.Click
        calculateResults()
        showResults()
    End Sub

    Private Sub NumericUpDown1_ValueChanged(ByVal sender As System.Object, ByVal e As
System.EventArgs) Handles NumericUpDown1.ValueChanged
        showResults()
    End Sub
End Class
```

In separating various pure materials from a mixture, the separation can be either binary (two output streams) or polynary (more than two output streams). A binary separator is designed to extract one type of material from a waste stream (e.g., magnet drawing off ferrous materials as the desired output or product or extract). A polynary separator separates components from each other through multiple paths (e.g., screen with a series of different sized holes, producing several products) (Rietema 1957, Rodic-wiersma 2005).

Suppose in a binary separator, the input stream is composed of a mixture of $x$ and $y$ to be separated. The mass per time (e.g., tons/h) of $x$ and $y$ fed to the separator is $x_0$ and $y_0$, respectively. The mass per time of $x$ and $y$ exiting in the first output stream is $x_1$ and $y_1$, and the second output stream is $x_2$ and $y_2$. The device separates $x$ into the first output stream and $y$ into the second. The effectiveness of the separation then can be expressed in terms of recovery (Figure 6.2).

The recovery of component $x$ in the first output stream is $R_{x1}$, defined as shown in the following equation:

$$R_{x_1} = \frac{x_1}{x_0} \times 100 \qquad (6.25)$$

where:

$R_{x1}$ is the recovery of component $x$ in the first output stream (1)

$x_1$ is the first component emerging of the first output stream (1), mass/time

$x_0$ is the $x$ component entering to the binary separator, mass/time

Similarly, the recovery of $y$ in the second output stream may be found from the following equation:

$$R_{y_2} = \frac{y_2}{y_0} \times 100 \qquad (6.26)$$

Purity of output can be determined from the following equation:

$$P_{x_1} = \frac{x_1}{(x_1 + x_2)} \times 100 \qquad (6.27)$$

where $P_{x1}$ is the purity of the first output stream in terms of $x$, which is expressed as a percentage.

Similarly, the purity of the second output stream in terms of $y$ is as presented in the following equation:

$$P_{y_2} = \frac{y_2}{(x_2 + y_2)} \times 100 \qquad (6.28)$$

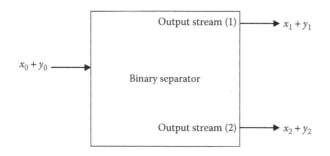

**FIGURE 6.2** Binary separator.

Overall recovery is useful only for process design, such as sizing conveyor belts. Nevertheless, it is not a measure of separation effectiveness. It should not be used in describing the operation of materials separation. Overall recovery may be defined as in the following equation:

$$OR_{x,y} = \left( \frac{x_1 + y_1}{x_0 + y_0} \right) \times 100 \qquad (6.29)$$

where $OR_{x,y}$ is the overall recovery for components $x$ and $y$.

Rietema separation effectiveness measure for a binary separation with inputs of $x_0$ and $y_0$ yields the following equation:

$$E_{x,y} = 100 \left| \frac{x_1}{x_0} - \frac{y_1}{y_0} \right| = 100 \left| \frac{x_2}{x_0} - \frac{y_2}{y_0} \right| \qquad (6.30)$$

Worrell–Stessel equation for finding the separation performance of a binary separator is reflected in the following equation:

$$E_{x,y} = \left| \frac{x_1}{x_0} \frac{y_2}{y_0} \right|^{1/2} \times 100 \qquad (6.31)$$

## Example 6.9

1. Write a short computer program that enables calculating the recoveries and effectiveness of separation of a binary separator, given the feed rate, the hourly output (1) and output (2), the $x$ constituent, and the amount of $x$ ending up in output 2 using different methods.
2. A binary separator has a feed rate of 1 ton/h. It is operated so that during any 1 h, 750 kg reports as output 1 and 250 kg as output 2. Of the 750 kg, the $x$ constituent is 700 kg, whereas 70 kg of $x$ ends up in output 2. Calculate the recoveries and effectiveness of the separation using the methods of recovery, purity, Rietema effectiveness, and Worrell–Stessel equation. Comment on your results.
3. Use the program developed in (1) to verify the manual computations conducted in (2).

## Solution

1. For the solution of Example 6.9 (1), see the listing of Program 6.9.
2. Solution to Example 6.9 (2):
   Given: Data: $x_0 + y_0 = 1000$ kg (1 ton), $x_1 = 750$ kg, $x_2 = 70$ kg (see figure).

From the data, find $x_0 = x_1 + x_2 = 700 + 70 = 770$ kg.
Then, find $y_0$ = total value $- x_0 = 1000 - 770 = 230$ kg.
From track (1), the value of $y_1 = 750 - x_1 = 750 - 700 = 50$ kg.
From track (2), the value of $y_2 = 250 - x_2 = 250 - 70 = 180$ kg.
The recovery of $x$ in the first output is

$$R_{x_1} = \frac{x_1}{x_0} \times 100 = \frac{700}{770} = 100 = 91\%$$

The purity of this output stream is

$$P_{x_1} = \frac{x_1}{(x_1 + y_1)} \times 100 = \frac{700}{(700 + 180)} \times 100 = 80\%$$

Using Rietema's definition of effectiveness,

$$E_{x,y} = 100 \left| \frac{x_1}{x_0} - \frac{y_1}{y_0} \right|$$

$$= 100 \left| \frac{x_2}{x_0} - \frac{y_2}{y_0} \right|$$

$$= 100 \left| \frac{700}{770} - \frac{50}{230} \right|$$

$$= 100 \left| \frac{70}{770} - \frac{180}{230} \right|$$

$$= 69\%$$

And according to Worrell–Stessel effectiveness equation:

$$E_{x,y} = \left| \frac{x_1}{x_0} \frac{y_2}{y_0} \right|^{1/2} \times 100 - \left| \frac{700}{770} \frac{180}{230} \right|^{1/2} \times 100 = 84\%$$

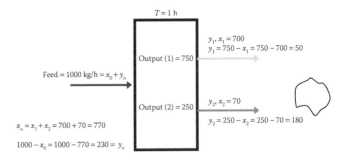

## LISTING OF PROGRAM 6.9 (CHAP6.9\FORM1.VB): RECOVERIES AND THE EFFECTIVENESS OF SEPARATION OF A BINARY SEPARATOR

```vb
'*******************************************************************************
'Program 6.9: Calculates the recoveries and the effectiveness
' of separation of a binary separator.
'*******************************************************************************
Imports System.Math
Public Class Form1
    Private Sub Form1_Load(ByVal sender As System.Object, ByVal e As System.EventArgs)
Handles MyBase.Load
        Me.Text = "Program 6.9:"
        Me.FormBorderStyle = Windows.Forms.FormBorderStyle.FixedSingle
        Me.MaximizeBox = False
        Label1.Text = "Program 6.9: Calculates the recoveries and the effectiveness "
        Label1.Text += vbCrLf + "of separation of a binary separator."
        Label2.Text = "Feed rate (kg/h):"
        Label3.Text = "Hourly output (1) (kg):"
        Label4.Text = "Hourly output (2) (kg):"
        Label5.Text = "The x constituent (x1) (kg):"
        Label6.Text = "The amount of x ending up in output2 (x2) (kg):"
        Label7.Text = ""
        button1.text = "&Calculate"
    End Sub

    Sub calculateResults()
        Dim feed, x0, x1, x2, y0, y1, y2, Rx, Px, Exy, Exy2 As Double
        Dim out1, out2 As Double
        feed = Val(TextBox1.Text)
        out1 = Val(TextBox2.Text)
        out2 = Val(TextBox3.Text)
        x1 = Val(TextBox4.Text)
        x2 = Val(TextBox5.Text)
        x0 = x1 + x2
        y0 = feed - x0
        y1 = out1 - x1
        y2 = out2 - x2
        'Find the recovery of x in the first output:
        Rx = (x1 / x0) * 100
        'Find purity of this output stream:
        Px = x1 / (x1 + y1) * 100
        'Using Rietema's definition of effectiveness:
        Exy = 100 * abs((x1 / x0) - (y1 / y0))
        'Worrell-Stessel effectiveness equation:
        Exy2 = sqrt(abs((x1 / x0) * (y2 / y0))) * 100
        Dim str As String
        str = "x0 = x1 + x2 = " + fN(x0) + "kg"
        str += vbCrLf + "y0 = total - x0 = " + fN(y0) + "kg"
        str += vbCrLf + "y1 = output1 - x1 = " + fN(y1) + "kg"
        str += vbCrLf + "y2 = output2 - x2 = " + fN(y2) + "kg"
        str += vbCrLf + "Recovery of x in the first output Rx1 = " + fN(Rx) + "%"
        str += vbCrLf + "Purity of output stream Px1 = " + fN(Px) + "%"
        str += vbCrLf + "Effectiveness (using Rietema's definition) = " + fN(Exy) + "%"
        str += vbCrLf + "Effectiveness (using Worrell-Stessel eqn.) = " + fN(Exy2) + "%"
        Label7.Text = str
    End Sub
```

```
    Function fN(ByVal n As Double) As String
        'formats a number 'n' as ##.# and returns result as string
        Return Format(n, "##.#")
    End Function
    Private Sub Button1_Click(ByVal sender As System.Object, ByVal e As System.EventArgs)
Handles Button1.Click
        calculateResults()
    End Sub
End Class
```

## 6.6  SANITARY LANDFILL

A sanitary landfill is an engineered method for land disposal of solid or hazardous wastes in a manner that protects the environment. Within the landfill, biological, chemical, and physical processes occur in biodegrading wastes, resulting in the production of leachate and gaseous substances (Schubeler et al. 1996, Senate et al. 2003). Leachate production and quantity can be estimated using empirical data or a water balance technique.

Water balance system in the landfill facilitates estimating the amount of percolating water production by establishing a mass balance among precipitation, evapotranspiration, surface runoff, and soil moisture storage as presented in the following equation:

$$C = P(1 - r) - S - E \qquad (6.32)$$

where:
C is the total amount percolating within the top layer of soil, mm/year
P is the precipitation, mm/year
r is the coefficient of runoff (can be estimated for different types of soil)
S is the storage in the soil or solid waste, mm/year
E is the evapotranspiration, mm/year

Gas production over time may be estimated from the EPA landfill gas emissions model (LandGEM) model[*] based on the following equation:

$$Q_T = \sum_{i=1}^{n} 2k\,L_o M_i e^{-kt_i} \qquad (6.33)$$

where:
$Q_T$ is the total gas emission rate from a landfill, volume/time
$n$ is the total time periods of waste placement
$k$ is the landfill gas emission constant, time$^{-1}$
$L_o$ is the methane generation potential, volume/mass of waste
$t_i$ is the age of the $i$th section of waste, time
$M_i$ is the mass of wet waste, placed at time $i$

## Example 6.10

1. Write a short computer program that enables calculating the peak gas production for a number of years, given the landfill opening period, the annual waste received, the gas emission constant, and the methane generation potential.
2. A landfill cell is open for 4 years, receiving 90,000 tons of waste per year (recall that 1 ton = 1000 kg). Find the peak gas production for the first year if the landfill gas emission constant is 0.03 per year, and the methane generation potential is 140 m³/ton.
3. Use the program developed in (1) to verify the manual computations conducted in (2).

### Solution

1. For the solution of Example 6.10 (1), see the listing of Program 6.10.
2. Solution to Example 6.10 (2):
   a.  Given: $M_i$ = 90000 tons/year, $k$ = 0.03, $L_o$ = 140, $t_i$ = 1

   $$Q_T = \sum_{i=1}^{n} 2kL_o M_i e^{-kt_i}$$

   b.  For the first year, $Q_T$ = 2 (0.03) (140) (90000) $(e^{-0.0307(1)})$ = 733657 m³

---

[*] This model can be downloaded at http://www.epa.gov/ttn/catc/products.html#software.

**LISTING OF PROGRAM 6.10   (CHAP6.10\FORM1.VB): PEAK GAS PRODUCTION FROM A LANDFILL**

```vb
'*********************************************************************************************
'Program 6.10: Calculates the peak gas production for a number of years
'*********************************************************************************************
Public Class Form1
    Dim Mi(), k(), Lo() As Double
    Dim Qt, t As Double
    Dim n As Integer
    Private Sub Form1_Load(ByVal sender As System.Object, ByVal e As System.EventArgs)
Handles MyBase.Load
        Me.Text = "Program 6.10: Peak gas production"
        Me.FormBorderStyle = Windows.Forms.FormBorderStyle.FixedSingle
        Me.MaximizeBox = False
        Label1.Text = "For each year, enter the annual waste received (Mi), gas emission
constant (k),"
        Label1.Text += vbCrLf + "and the methane generation potential (Lo):"
        Label2.Text = ""
        Label3.Text = "Decimals:"
        DataGridView1.Columns.Clear()
        DataGridView1.Columns.Add("MCol", "Mi (tones/yr)")
        DataGridView1.Columns.Add("KCol", "K (time-1)")
        DataGridView1.Columns.Add("LCol", "Lo (m3/ton)")
        Button1.Text = "&Calculate"
        NumericUpDown1.Value = 2
        NumericUpDown1.Maximum = 10
        NumericUpDown1.Minimum = 0
    End Sub


    Private Sub Button1_Click(ByVal sender As System.Object, ByVal e As System.EventArgs)
Handles Button1.Click
        calculateResults()
        showResults()
    End Sub

    Sub calculateResults()
        n = DataGridView1.Rows.Count - 1
        If n <= 0 Then
            MsgBox("Enter at least one period's data!", vbOKOnly, "Error")
            Exit Sub
        End If
        '*********************************************************************************
        'Formula is:
        '      n
        ' QT = SUM 2k * Lo * Mi * e^(-kti)
        '      i=1
        '*********************************************************************************

        Qt = 0
        t = 0
        ReDim Mi(n), k(n), Lo(n)
        For i = 0 To n - 1
            Mi(i) = Val(DataGridView1.Rows(i).Cells("MCol").Value)
            k(i) = Val(DataGridView1.Rows(i).Cells("KCol").Value)
            Lo(i) = Val(DataGridView1.Rows(i).Cells("LCol").Value)
            t += 1
            Qt += 2 * k(i) * Lo(i) * Mi(i) * (Math.E ^ (-k(i) * t))
        Next
    End Sub


    Sub showResults()
```

```
        Dim decimals As Integer = NumericUpDown1.Value
        If decimals < 0 Or decimals > 10 Then decimals = 2
        If n = 1 Then
            Label2.Text = "For one year, QT = " + FormatNumber(Qt, decimals) + " m3"
        Else
            Label2.Text = "For " + n.ToString + " years, QT = " + FormatNumber(Qt,
decimals) + " m3"
        End If
    End Sub

    Private Sub NumericUpDown1_ValueChanged(ByVal sender As System.Object, ByVal e As
System.EventArgs) Handles NumericUpDown1.ValueChanged
        showResults()
    End Sub
End Class
```

## Example 6.11

1. Use the EPA LandGEM model used for estimating methane production from a sanitary landfill.

   A landfill cell is open for 3 years, receiving 200,000 tons of waste per year (recall that 1 ton 1000 kg). Calculate the peak gas production if the landfill gas emission constant is 0.05 year$^{-1}$, and the methane generation potential is 170 m$^3$/ton.

   *Hint:* For the second year, this waste produces less gas, but the next new layer produces more, and the two layers are added to yield the total gas production for the second year.

### Solution

$Q_T$ is the total gas emission rate from a landfill, volume/time
$n$ is the total time periods of waste placement
$k$ is the landfill gas emission constant, time$^{-1}$
$L_o$ is the methane generation potential, volume/mass of waste
$t_i$ is the age of the $i$th section of waste, time
$M_i$ is the mass of wet waste, placed at time $i$
Given: $M_i = 200,000$ tons/year, $k = 0.05$ year$^{-1}$, $L_o = 170$ m$^3$/ton, $t_i = 1$ year

$$Q_T = \sum_{i=1}^{n} 2kL_0 M_i e^{-kt_i}$$

For the first year,
$QT_1 = 2 (0.05) (170) (200,000)(e-0.05 * 1) = 3,234,180$ m$^3$.
Determine the gas produced for other years as shown in the following table:

| T | Mass, M | Mass, M | Constant, k | Emission, L | Total Gas, Q |
|---|---------|---------|-------------|-------------|--------------|
| 1 | 200,000 | 200,000 | 0.05 | 170 | 3,234,180 |
| 2 | 200,000 | 400,000 | 0.05 | 170 | 6,152,894 |
| 3 | 200,000 | 600,000 | 0.05 | 170 | 8,779,221 |
| 4 | 200,000 | 497,100 | 0.05 | 170 | 6,918,848 |
| 5 | 200,000 | 497,100 | 0.05 | 170 | 6,581,412 |
| 6 | 200,000 | 497,100 | 0.05 | 170 | 6,260,433 |
| 7 | 200,000 | 497,100 | 0.05 | 170 | 5,955,108 |
| 8 | 200,000 | 497,100 | 0.05 | 170 | 5,664,674 |
| 9 | 200,000 | 497,100 | 0.05 | 170 | 5,388,404 |
| 10 | 200,000 | 497,100 | 0.05 | 170 | 5,125,609 |

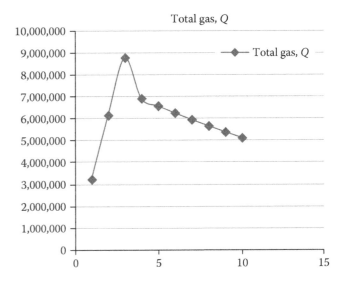

Total gas, Q

Procedure for using LandGEM
- Given: $M_i = 200,000$ tons/year, $k = 0.05$ year$^{-1}$, $L_o = 170$ m$^3$/ton, $t_i = 1$ year.
- Make a working copy of the file landfill gas emission model, LandGEM v3.02 (.xls MS Excel file).
- Start MS Excel opening file landGEM-v302.xls to have the introduction sheet.
- Select "Enable content" to have a working copy. Select continue.

- Select Sheet 2 for "user inputs," and insert your values.
  - Provide landfill characteristics: landfill open year = 0, landfill closure year = 4.
  - Determine model parameters: methane generation rate = 0.03 (select user-specified and type your value), potential methane generation capacity = 150 (select user-specified and type your value).
  - Enter waste acceptance rates = 90,000, 90,000, 90,000, and 90,000 for 4 years.
- Select Sheet 3: Pollutants if you have any.
- Select Sheet 4: Input review to review your data.
- Select Sheet 5: Methane to review findings.
- Select Sheet 6: Results for results of computations.
- Select Sheet 7: Graphs to view results graphically.
- Select Sheet 8: Inventory.
- Select Sheet 9: Report to view report of the model.
- Save your file.

Depth of leachate in the lining can be estimated using Darcy's law and continuity equation, depending on the rates of filtration, the permeability of drainage materials, the distance from discharge tube, and the slope of drainage system. Equation 6.34 illustrates the depth of leachate estimates:

$$Y_{max} = \frac{P}{2}\left(\frac{q}{k}\right)\left[\frac{k\tan^2\alpha}{q} + 1 - \frac{k\tan\alpha}{q}\left(\tan^2\alpha + \frac{q}{k}\right)^{\frac{1}{2}}\right] \quad (6.34)$$

where:
  $Y_{max}$ is the maximum saturated depth over the liner, ft
  $P$ is the distance between collection pipes, ft
  $q$ is the vertical inflow (infiltration), defined in this equation as from a 25-year, 24-h storm, ft/day
  $\alpha$ is the inclination of liner from horizontal, degrees
  $K$ is the hydraulic conductivity of the drainage layer, ft/day

Needed land area can be found from the estimates of the required volume from the following equation:

$$V = \frac{W}{\rho}\left(1 - \frac{x}{100}\right) + v_r \quad (6.35)$$

where:
  $V$ is the volume of sanitary landfill area
  $W$ is the weight of solid waste to be buried
  $\rho$ is the average density of solid waste and garbage
  $x$ is the percentage of compressed solid waste volume, %
  $v_r$ is the volume of a layer of coverage required (thickness of 15–30 cm for medium layers, temporary edge and front and overhead slope, and at least 60 cm in the final layer), and that this volume

ranges between 17% of volume of solid waste for deep burial to 33% for surface burial and on average 25%.

Average sanitary landfill volume can be estimated as shown in the following equation:

$$V = 1.25\frac{W}{\rho}\left(1 - \frac{x}{100}\right) \quad (6.36)$$

## Example 6.12

1. Write a computer program to estimate the design storm (vertical flow) in cm/s using a coarse discharge material assuming that rainwater from 25 years and a 24-h storm entering the leachate drainage system, given the maximum design depth of leachate above lining in a sanitary landfill, the distance between leachate collection tubes, hydraulic conductivity, and the slope of discharge.
2. The maximum design depth of leachate above lining in a sanitary landfill is 16 cm, and the distance between leachate collection tubes is 19m. The hydraulic conductivity is 0.02 cm/s for a slope of discharge 2%. Estimate the design storm (vertical flow) in cm/s using a coarse discharge material and assuming that rainwater from 25 years and a 24-h storm entering the leachate drainage system.
3. Use program developed in (1) to verify your computations in (2).

### Solution

Given: $Y_{max} = 16$ cm, $P = 1900$ cm, $k = 0.02$ cm/s, $q = ?$ cm/s, $\tan \alpha = 2/100 = 0.02$.

$$Y_{max} = \frac{P}{2}\left(\frac{q}{k}\right)\left[\frac{k\tan^2\alpha}{q} + 1 - \frac{k\tan\alpha}{q}\left(\tan^2\alpha + \frac{q}{k}\right)^{\frac{1}{2}}\right]$$

$$= \frac{1900}{2}\left(\frac{q}{0.02}\right)$$

$$\left[\frac{0.02*(0.02^2)}{q} + 1 - \frac{0.02*(0.02)}{q}\left(0.02^2 + \frac{q}{0.02}\right)^{\frac{1}{2}}\right] = 16$$

By trial and error, $q = 0.00038$ cm/s (http://www.wolframalpha.com/).

## LISTING OF PROGRAM 6.12   (CHAP6.12\FORM1.VB): VERTICAL FLOW

```vb
'*********************************************************************************
'Example 6.12: Estimates the design storm (vertical flow)
'*********************************************************************************
Public Class Form1
    Dim Ymax, P, k, tanA, Q As Double
    Private Sub Form1_Load(ByVal sender As System.Object, ByVal e As System.EventArgs)
Handles MyBase.Load
        Me.Text = "Example 6.12"
        Me.FormBorderStyle = Windows.Forms.FormBorderStyle.FixedSingle
        Me.MaximizeBox = False
        Label1.Text = "The maximum design depth, Ymax (cm) ="
        Label2.Text = "The distance between leachate tubes, P (m) ="
        Label3.Text = "The hydraulic conductivity, k (cm/s) ="
        Label4.Text = "The slope of discharge, tan a ="
        Label5.Text = "The design storm (vertical flow), Q (cm/s) ="
        Label6.Text = "Decimal places:"
        Button1.Text = "&Calculate Q"
        NumericUpDown1.Value = 2
        NumericUpDown1.Maximum = 10
        NumericUpDown1.Minimum = 0
    End Sub


    Private Sub Button1_Click(ByVal sender As System.Object, ByVal e As System.EventArgs)
Handles Button1.Click
        Ymax = Val(TextBox1.Text)
        P = Val(TextBox2.Text) * 100
        k = Val(TextBox3.Text)
        tanA = Val(TextBox4.Text)
        Dim Z As Double
        Dim b As Double = P / (2 * k)
        Dim c As Double = k * (tanA ^ 2)
        Dim d As Double = k * tanA
        Dim _e As Double = (tanA ^ 2)
        Z = ((Ymax ^ 2) - (((b ^ 2) * (c ^ 2)) + ((b ^ 2) * (d ^ 2) * _e))) / (b ^ 2)
        Dim X As Double = (d ^ 2) / k
        Dim Y As Double = (0.5 * X) ^ 2
        'Calculate the roots of the equation
        Dim Q1 As Double = -Y + Math.Sqrt(Z + Y)
        Dim Q2 As Double = -Y - Math.Sqrt(Z + Y)
        If Q1 < 0 Then
            Q = Q2
        Else
            Q = Q1
        End If
        showResult()
    End Sub


    Sub showResult()
        TextBox5.Text = FormatNumber(Q, NumericUpDown1.Value)
    End Sub


    Private Sub NumericUpDown1_ValueChanged(ByVal sender As System.Object, ByVal e As
System.EventArgs) Handles NumericUpDown1.ValueChanged
        showResult()
    End Sub
End Class
```

## 6.7 BIOCHEMICAL PROCESSES, COMBUSTION, AND ENERGY RECOVERY

In MSW, the components reliable for bioconversion processes are garbage (food waste), paper products, and yard wastes for their cellulose content. All methods of biochemical conversion (anaerobic digestion, composting) use the organic fraction of refuse. Decay of organic matter under anaerobic conditions produces end products that include gases such as methane ($CH_4$), carbon dioxide ($CO_2$), small amounts of hydrogen sulfide ($H_2S$), ammonia ($NH_3$), and a few others.

Ideally, production of methane and carbon dioxide can be calculated using Equation 6.37 if chemical composition of a material is known:

$$C_aH_bO_cN_d + \left(\frac{4a-b-2c+3d}{4}\right)H_2O \rightarrow$$

$$\left(\frac{4a+b-2c-3d}{8}\right)CH_4 + \left(\frac{4a-b+2c+3d}{8}\right) \quad (6.37)$$

$$CO_2 + dNH_3$$

### Example 6.13

1. Write a short computer program that enables estimating the production of $CO_2$ and $CH_4$ during the anaerobic decomposition of a certain material.
2. Estimate the production of $CO_2$ and $CH_4$ during the anaerobic decomposition of acetic acid, $CH_3COOH$.

   Estimate the percentage of landfill gas that is methane for a rapidly decomposing waste with the following composition: $C_{68}H_{111}O_{50}N$. Assume that the maximum theoretical landfill gas may be estimated using the following equation (University of Dammam, 2015):

$$C_aH_bO_cN_d + \left(\frac{4a-b-2c+3d}{4}\right)H_2O \rightarrow$$

$$\left(\frac{4a+b-2c-3d}{8}\right)CH_4 + \left(\frac{4a-b+2c+3d}{8}\right)CO_2 + dNH_3$$

3. Use the program developed in (1) to verify the manual computations conducted in (2).

### Solution

1. For the solution of Example 6.13 (1), see the listing of program 6.13.
2. Solution to Example 6.13 (2):
   a. Given: Reactants and products: $a = 68$, $b = 111$, $c = 50$, $d = 1$.
   b. The formula for acetic acid is $CH_3COOH$.

$$C_aH_bO_cN_d + \left(\frac{4a-b-2c+3d}{4}\right)H_2O \rightarrow$$

$$\left(\frac{4a+b-2c-3d}{8}\right)CH_4 + \left(\frac{4a-b+2c+3d}{8}\right)CO_2 + dNH_3$$

   c. The formula for the reaction is $C_{68}H_{111}O_{50}N + 16H_2O = 35CH_4 + 33CO_2 + NH_3$.
   d. Determine the water coefficient as $(4a - b - 2c + 3d)/4 = (4 * 68 - 111 - 2 * 50 + 3 * 1)/4 = 16$.
   e. Determine the methane coefficient as $(4a + b - 2c - 3d)/4 = (4 * 68 + 111 - 2 * 50 - 3 * 1)/8 = 35$.
   f. Determine carbon dioxide coefficient as $(4a - b + 2c + 3d)/4 = (4 * 68 - 111 + 2 * 50 + 3 * 1)/8 = 33$.
   g. Determine ammonia dioxide coefficient as $d = 1$.
   h. Total of methane, carbon dioxide, and ammonia $= 35 + 33 + 1 = 69$ moles.
   i. Percentage of methane $= 35$ moles/69 moles $= 51\%$.

---

### LISTING OF PROGRAM 6.13   (CHAP6.13\FORM1.VB): PRODUCTION OF $CO_2$ AND $CH_4$

```
'*********************************************************************************
'Program 6.13: Estimates the production of CO2 and CH4
'*********************************************************************************
Imports System.Math
Public Class Form1
    Dim a, b, c, d As Double
    Dim H2O, CH4, CO2, NH3
    Private Sub Form1_Load(ByVal sender As System.Object, ByVal e As System.EventArgs)
Handles MyBase.Load
        Me.Text = "Program 6.13: The production of CO2 and CH4"
        Me.FormBorderStyle = Windows.Forms.FormBorderStyle.FixedSingle
        Me.MaximizeBox = False
        Label1.Text = "Enter the chemical formula of the material, for example, acetic
acid is CH3COOH"
        Label2.Text = ""
        Button1.Text = "&Calculate"
    End Sub
```

```
    Sub calculateResults()
        readChemicalFormula()
        H2O = ((4 * a) - b - (2 * c) + (3 * d)) / 4
        CH4 = ((4 * a) + b - (2 * c) - (3 * d)) / 8
        CO2 = ((4 * a) - b + (2 * c) + (3 * d)) / 8
        NH3 = d
        H2O = abs(H2O)
        CH4 = abs(CH4)
        CO2 = abs(CO2)
        NH3 = abs(NH3)
        Label2.Text = "a=" + a.ToString + ", b=" + b.ToString + ", c=" + c.ToString + ",
d=" + d.ToString
        Label2.Text += vbCrLf + TextBox1.Text + " + " + H2O.ToString + "H2O = " + CH4.
ToString + "CH4 + " _
              + CO2.ToString + "CO2 + " + NH3.ToString + "NH3"
        Dim molCH4, molCO2, totMol
        molCH4 = CH4 * 16 'molecular wt of CH4 is 16
        molCO2 = CO2 * 44 'molecular wt of CO2 is 44
        totMol = molCH4 + molCO2
        molCH4 = molCH4 / totMol
        molCO2 = molCO2 / totMol
        Label2.Text += vbCrLf + vbCrLf + "1 kg of " + TextBox1.Text.ToUpper + _
              " produces " + FormatNumber(molCH4, 2) + " kg CH4 and " +
FormatNumber(molCO2, 2) + " kg CO2."
    End Sub

    '*******************************************************************************************
    'Reads the chemical formula entered in TextBox1, goes through it letter-by-letter.
    'if the letter is 'C', it adds one to variable 'a', if it is 'H', it adds to 'b',
    'and so on. If there is a number after the letter, e.g, 'H3' in 'CH3COOH', it adds
    'the value accordingly, in the above example it adds 3 to variable 'c', and so on...
    '*******************************************************************************************
    Sub readChemicalFormula()
        Dim i As Integer = 0
        Dim formula As String = TextBox1.Text.ToUpper + " "
        Dim tmp As Double
        a = 0 : b = 0 : c = 0 : d = 0
        While i < formula.Length - 1
            Select Case formula.Chars(i)
                Case "C"
                  If Val(formula.Chars(i + 1)) > 0 Then
                     tmp = Val(formula.Chars(i + 1))
                     i += 2
                     While i < formula.Length - 1
                       If formula.Chars(i) >= CChar("0") And formula.Chars(i) <=
CChar("9") Then
                          tmp = (tmp * 10) + Val(formula.Chars(i))
                       Else : Exit While
                       End If
                       i += 1
                     End While
                     a += tmp
                     Continue While
                  Else : a += 1
                  End If
                Case "H"
                  If Val(formula.Chars(i + 1)) > 0 Then
                     tmp = Val(formula.Chars(i + 1))
                     i += 2
                     While i < formula.Length - 1
```

```
                            If formula.Chars(i) >= CChar("0") And formula.Chars(i) <=
CChar("9") Then
                                tmp = (tmp * 10) + Val(formula.Chars(i))
                            Else : Exit While
                            End If
                            i += 1
                        End While
                        b += tmp
                        Continue While
                    Else : b += 1
                    End If
                Case "O"
                    If Val(formula.Chars(i + 1)) > 0 Then
                        tmp = Val(formula.Chars(i + 1))
                        i += 2
                        While i < formula.Length - 1
                          If formula.Chars(i) >= CChar("0") And formula.Chars(i) <=
CChar("9") Then
                                tmp = (tmp * 10) + Val(formula.Chars(i))
                          Else : Exit While
                          End If
                          i += 1
                        End While
                        c += tmp
                        Continue While
                    Else : c += 1
                    End If
                Case "N"
                    If Val(formula.Chars(i + 1)) > 0 Then
                        tmp = Val(formula.Chars(i + 1))
                        i += 2
                        While i < formula.Length - 1
                          If formula.Chars(i) >= CChar("0") And formula.Chars(i) <=
                          CChar("9") Then
                                tmp = (tmp * 10) + Val(formula.Chars(i))
                          Else : Exit While
                          End If
                          i += 1
                        End While
                        d += tmp
                        Continue While
                    Else : d += 1
                    End If
            End Select
            i += 1
        End While
    End Sub

    Private Sub Button1_Click(ByVal sender As System.Object, ByVal e As System.EventArgs)
Handles Button1.Click
        calculateResults()
    End Sub
End Class
```

In the beginning of composting process, mesophilic[*] microorganisms are frequent with most of the occurring biochemical reactions being attributed to them. The increase in these organisms, after about a week, increases the temperature of compost, which limits the growth of these organisms to be replaced by thermophilic[†] organisms. When the temperature drops, it usually means that the compost needs to be aerated, or watered, or that composting is complete. It is desirable to work at a temperature between 60°C and 75°C for complete digestion.

---

[*] Mesophiles live in medium temperature ranging from 25°C to 45°C.
[†] Thermophiles are microorganisms that live at high temperatures, >45°C.

A critical variable in composting is the moisture content. If the mixture is too dry, the microorganisms cannot survive, and composting stops. If there is too much water, the oxygen from the air is not able to penetrate to where the microorganisms are, and the mixture becomes anaerobic. The right amount of moisture, whether wastewater sludge or other sources of water, that needs to be added to the solids to achieve just the right moisture content can be calculated from a simple mass balance as presented in the following equation:

$$M_P = \frac{M_a x_a + 100 x_s}{x_s + x_a} \qquad (6.38)$$

where:

$M_P$ is the moisture in mixed pile (heap) ready to begin the process of composting, %

$M_a$ is the moisture in solids as in shredded and screened refuse, %

$x_a$ is the mass of solids, wet tons

$x_s$ is the mass of sludge or other source of water, tons. (This assumes that the solids content of the sludge is very low, a good assumption if waste activated sludge is used, which is commonly <1% solids)

## Example 6.14

1. Write a short computer program that enables calculating the amount of water or wastewater sludge to be added to the solids of an MSW to obtain the desired concentration of moisture content of the heap to start the process of composting, given composted materials mass, the amount of moisture content in it, and moisture content.

2. A mixture of solids that contains paper, newspapers, and potentially composted materials, with a mass of 8 tons, has a moisture content of 4%. It is required to make a mixture for the process of composting with the moisture content of 45% moisture. Find the amount of water or wastewater sludge to be added to the solids of this MSW to obtain the desired concentration of moisture content of the heap to start the process of composting.

3. Use the program developed in (1) to verify the manual computations conducted in (2).

## Solution

1. For the solution of Example 6.14 (1), see the listing of Program 6.14.
2. Solution to Example 6.14 (2):
   a. Given: Data: $x_a = 8$ tons, $M_a = 4\%$, $M_P = 45\%$.
   b. Use the following equation to find the amount of water needed $x_s$ from wastewater:

$$M_P = \frac{M_a x_a + 100 x_s}{x_s + x_a}$$

$$45 = \frac{(4 \times 8) + 100 x_s}{x_s + 8}$$

Then, $x_s = 6$ tons from water or from wastewater sludge.

---

### LISTING OF PROGRAM 6.14   (CHAP6.14\FORM1.VB): AMOUNT OF WATER TO BE ADDED TO MSW TO OBTAIN THE DESIRED MOISTURE CONTENT

```
'**********************************************************************************
'Program 6.14: Calculates amount of water or wastewater sludge to be added
'to the solids of a MSW to obtain desired concentration of moisture content.
'**********************************************************************************
Public Class Form1
    Dim MP, Ma, Xa, Xs As Double
    Private Sub Form1_Load(ByVal sender As System.Object, ByVal e As System.EventArgs)
Handles MyBase.Load
        Me.Text = "Program 6.14:"
        Me.FormBorderStyle = Windows.Forms.FormBorderStyle.FixedSingle
        Me.MaximizeBox = False
        Label1.Text = "The formula:" + vbCrLf
        Label1.Text += "MP = ((Ma * Xa) + (100 * Xs)) / (Xs + Xa)"
        Label2.Text = "Composted materials mass, Xa (ton):"
        Label3.Text = "The amount of moisture content in it, Ma (%):"
        Label4.Text = "Moisture content, MP (%):"
        Label5.Text = ""
        Button1.Text = "&Calculate"
    End Sub

    Sub calculateResults()
        Xa = Val(TextBox1.Text)
        Ma = Val(TextBox2.Text)
        MP = Val(TextBox3.Text)
        'Formula: "MP = ((Ma * Xa) + (100 * Xs)) / (Xs + Xa)"
        Xs = ((MP * Xa) - (Ma * Xa)) / (100 - MP)
```

```
        Label5.Text = "xs = " + Format(Xs, "###.#") + " tons from water or from
wastewater sludge."
    End Sub


    Private Sub Button1_Click(ByVal sender As System.Object, ByVal e As System.EventArgs)
Handles Button1.Click
        calculateResults()
    End Sub
End Class
```

Estimation of carbon and nitrogen levels and the *C/N* ratio is based on mass balances. If two materials such as shredded refuse and sewage sludge are mixed, the carbon of the mixture is calculated as shown in the following equation:

$$C_p = \frac{c_r x_r + c_s x_s}{x_r + x_s} \qquad (6.39)$$

where:

$C_p$ is the carbon concentration in the mixture prior to composting, % of total wet mass of mixture

$C_r$ is the carbon concentration in the refuse, % of total wet refuse mass

$C_s$ is the carbon concentration in the sludge, % of total wet sludge mass

$X_s$ is the total mass of sludge, wet tons per day

$X_r$ is the total mass of refuse, wet tons per day

MSW and refuse can be burned as is, or it can be processed to improve its heat value and to make it easier to handle in a combustor. Processed refuse also can be combined with other fuels (such as coal) and cofired in a heat recovery combustor or can be used to provide electrical power for the community. The amount of oxygen necessary to oxidize some hydrocarbon is known as stoichiometric oxygen. Usually, refuse is not burned using air as the source of oxygen. Because air contains 23.15% oxygen by weight, then stoichiometric air required can be determined from the following equation:

$$\text{Stoichiometric air} = \text{Stoichiometric oxygen}/0.2315 \qquad (6.40)$$

## Example 6.15

1. Write a short computer program that enables calculating stoichiometric oxygen and stoichiometric air required for the combustion of a gas.
2. Calculate the stoichiometric oxygen and stoichiometric air required for the combustion of methane gas.
3. Use the program developed in (1) to verify the manual computations conducted in (2).

**Solution**

1. For the solution of Example 6.15 (1), see the listing of Program 6.15.
2. Solution to Example 6.15 (2):
   a. The equation for combustion of methane is $CH_4 + 2O_2 = CO_2 + 2H_2O$.
   b. That is, it takes 16 g of methane (12 + 4) to react with $2 \times 2 \times 16 = 64$ moles of oxygen. Thus, stoichiometric oxygen required for combustion of methane is $64/16 = 4$ g $O_2$/g $CH_4$.
   c. Stoichiometric air requirement = 4/0.2315 = 17.3 g air/g methane.

In a power plant, water is heated to steam in a boiler. Steam is used to turn a turbine, which drives a generator. This process can be simplified to a simple energy balance where energy in has to equal energy out (energy wasted in the conversion

---

**LISTING OF PROGRAM 6.15   (CHAP6.15\FORM1.VB): STOICHIOMETRIC OXYGEN AND AIR REQUIRED FOR COMBUSTION OF A GAS**

```
'***********************************************************************************
'Program 6.15: Calculates stoichiometric oxygen and air
' required for the combustion of a gas
'***********************************************************************************
Imports System.Math
Public Class Form1
    Dim a, b, c, d As Double
    Dim H2O, GAS, CO2, O2 As Double
    Dim outputStr(4) As String
    Private Sub Form1_Load(ByVal sender As System.Object, ByVal e As System.EventArgs)
Handles MyBase.Load
        Me.Text = "Program 6.15:"
        Me.FormBorderStyle = Windows.Forms.FormBorderStyle.FixedSingle
```

```
            Me.MaximizeBox = False
            Label1.Text = "Enter the equation for combustion of the gas:"
            Label2.Text = ""
            CheckBox1.Checked = False
            CheckBox1.Text = "Use Example 6.15, Combustion of methane 'CH4 + 2O2 = CO2 +
2H2O' "
            Button1.Text = "&Calculate"
    End Sub

    Sub calculateResults()
            readChemicalFormula()
            Dim result, result2 As Double
            result = O2 / GAS
            result2 = result / 0.2315
            Label2.Text = "It takes " + GAS.ToString + " grams of (" + outputStr(1) + ") to
react with"
            Label2.Text += O2.ToString + " moles of oxygen"
            Label2.Text += vbCrLf + "Stoichiometric oxygen required for combustion is " +
Format(result, "###.#")
            Label2.Text += vbCrLf + "Stoichiometric air requirement = " + result.ToString +
"/0.2315 = " _
                    + Format(result2, "###.#") + " g air/g gas."
    End Sub

    '******************************************************************************
    'Reads the chemical formula entered in TextBox1, goes through it letter-by-letter.
    'it divides the formula into four components, two on the left side of the equal "="
sign
    'and two on the right side. it then calculates the weight of each one.
    'Sample formula: CH4 + 2O2 = CO2 + 2H2O
    '******************************************************************************
    Sub readChemicalFormula()
            Dim i As Integer = 0
            Dim i2 As Integer = 0
            Dim formula As String = TextBox1.Text.ToUpper + " "
            Dim f(4) As String
            Dim r(4) As Double
            Dim x, multiplier As Double
            GAS = 0 : O2 = 0 : CO2 = 0 : H2O = 0
            'break down the formula into four parts
            i = formula.IndexOf("+")
            f(1) = formula.Substring(0, i)
            i2 = formula.IndexOf("=")
            f(2) = formula.Substring(i + 1, i2 - i - 1)
            i = formula.IndexOf("+", i2)
            f(3) = formula.Substring(i2 + 1, i - i2 - 1)
            f(4) = formula.Substring(i + 1, formula.Length - i - 1)
            For j = 1 To 4
                'run each part of the formula in turn, calculating its weight
                i = 0
                multiplier = 0
                r(j) = 0
                f(j) = " " + f(j) + " "
                While i < f(j).Length - 1
                    If multiplier = 0 And IsNumeric(f(j).Chars(i)) Then
                        'the first number in the formula is the multiplier, e.g. 2H2O
                        multiplier = Val(f(j).Chars(i))
                    Else
                        Select Case f(j).Chars(i)
                            Case "C"
                                'if the next char is a number, e.g. C2, then multiply this
```

```
                                  'by the weight of carbon, then multiply by the multiplier that
                                  'was read early on if it is > 0.
                                   If Val(f(j).Chars(i + 1)) > 0 Then
                                    x = Val(f(j).Chars(i + 1)) * 12 'weight of Carbon
                                    If multiplier > 0 Then
                                      r(j) = r(j) + x * multiplier 'multiplier, e.g. 2*C
                                    Else
                                      r(j) = r(j) + x
                                      End If
                                      i += 1
                                    Else
                                      If multiplier > 0 Then
                                        r(j) = r(j) + 12 * multiplier 'multiplier, e.g. 2*C
                                      Else : r(j) += 12
                                      End If
                                    End If
                                Case "H"
                                    'if the next char is a number, e.g. H2O, then multiply this
                                    'by the weight of hydrogen, then multiply by the multiplier that
                                    'was read early on if it is > 0.
                                    If Val(f(j).Chars(i + 1)) > 0 Then
                                      x = Val(f(j).Chars(i + 1)) 'weight of Hydrogen
                                      If multiplier > 0 Then
                                        r(j) = r(j) + x * multiplier 'multiplier, e.g. 2*H
                                      Else
                                        r(j) = r(j) + x
                                        End If
                                        i += 1
                                      Else
                                        If multiplier > 0 Then
                                        r(j) = r(j) + multiplier 'multiplier, e.g. 2*H
                                      Else : r(j) += 1
                                        End If
                                    End If
                                Case "O"
                                    'if the next char is a number, e.g. O2, then multiply this
                                    'by the weight of oxygen, then multiply by the multiplier that
                                    'was read early on if it is > 0.
                                    If Val(f(j).Chars(i + 1)) > 0 Then
                                      x = Val(f(j).Chars(i + 1)) * 16 'weight of Oxygen
                                      If multiplier > 0 Then
                                        r(j) = r(j) + x * multiplier 'multiplier, e.g. 2*O2
                                      Else
                                        r(j) = r(j) + x
                                        End If
                                        i += 1
                                      Else
                                      If multiplier > 0 Then
                                        r(j) = r(j) + 16 * multiplier 'multiplier, e.g. 2*O2
                                      Else : r(j) += 16
                                      End If
                                    End If
                                End Select
                              End If
                      i += 1
            End While
            Next
            'make sure the right- and left side of the equation are balanced.
            If (r(1) + r(2)) <> (r(3) + r(4)) Then
                MsgBox("Equation is not balanced! Review and try again.", vbOKOnly, "Equation
Error")
```

```
                    Exit Sub
            End If
            'the four parts are not arranged.. arrange then so that the gas result is the
    first,
            'the oxygen is second, the CO2 is third, and the water is fourth.
            For i = 1 To 4
                If f(i).Contains("CO2") Then
                    CO2 = r(i)
                    outputStr(3) = f(i).Trim
                ElseIf f(i).Contains("O2") Then
                    O2 = r(i)
                    outputStr(2) = f(i).Trim
                ElseIf f(i).Contains("H2O") Then
                    H2O = r(i)
                    outputStr(4) = f(i).Trim
                Else
                    GAS = r(i)
                    outputStr(1) = f(i).Trim
                End If
            Next
        End Sub

        Private Sub Button1_Click(ByVal sender As System.Object, ByVal e As System.EventArgs)
    Handles Button1.Click
            calculateResults()
        End Sub

        Private Sub CheckBox1_CheckedChanged(ByVal sender As System.Object, ByVal e As
    System.EventArgs) Handles CheckBox1.CheckedChanged
            If CheckBox1.Checked Then
                TextBox1.Text = "CH4 + 2O2 = CO2 + 2H2O"
            Else
                TextBox1.Text = ""
            End If
        End Sub
    End Class
```

+ useful energy) plus energy accumulated in box (energy changed in form) as presented in the following equation:

$$\text{(Rate of energy accumulated)} = \text{(Rate of energy in)}$$
$$- \text{(Rate of energy out)} + \text{(Rate of energy produced)} \quad (6.41)$$
$$- \text{(Rate of energy consumed)}$$

Energy systems in a steady state are defined as no change occurring over time. As such, there cannot be a continuous accumulation of energy, or if some of the energy out is useful and the rest is wasted and then Equation 6.41 becomes:

$$\text{(Rate of energy in)} = \text{(Rate of energy out)} \quad (6.42)$$

$$\text{(Rate of energy in)} = \text{(Rate of energy used)}$$
$$+ \text{(Rate of energy wasted)} \quad (6.43)$$

Efficiency ($E\%$) of process can be calculated as indicted in the following equation:

$$E = \frac{\text{Energy used}}{\text{Energy in}} \times 100 \quad (6.44)$$

A thermal balance on a large combustion unit is difficult because much of the heat cannot be accurately measured. Assuming recovery of heat as steam in a combustor, input heat to a black box (see Figure 6.3) is from heat value in the fuel and heat in the water entering the water wall pipes. The output is the sensible heat in the stack gases, the latent heat of water, the heat in the ashes, the heat in the steam, and the heat lost due to radiation. If the process is in a steady state, the equation of thermal balance can be as indicated in the following equation:

$$\text{(Rate of energy accumulated)} = \text{(Rate of heat in the fuel)}$$
$$+ \text{(Rate of heat in the water)} - \text{(Rate of heat out in}$$
$$\text{the stack gases)} - \text{(Rate of heat out in the stem)}$$
$$- \text{(Rate of heat out as latent heat of vaporization)} \quad (6.45)$$
$$- \text{(Rate of heat out in the ash)} - \text{(Rate of heat loss}$$
$$\text{due to radiation)}$$

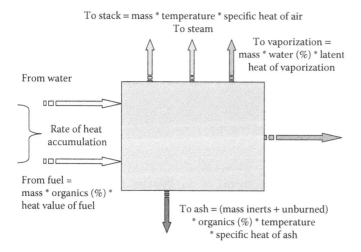

To stack = mass * temperature * specific heat of air
To steam

To vaporization =
mass * water (%) * latent
heat of vaporization

From water

Rate of heat
accumulation

From fuel =
mass * organics (%) *
heat value of fuel

To ash = (mass inerts + unburned)
* organics (%) * temperature
* specific heat of ash

**FIGURE 6.3**   Energy flow in a combustor.

## Example 6.16

1. Write a short computer program that enables calculating the temperature of the stack gases given heat value of the fuel, air flow, heat input lost due to radiation, percentage of fuel remaining uncombusted in ash, ash exit temperature from combustion chamber, specific heat of ash, specific heat of air, and latent heat of vaporization.

2. A refractory combustion unit lined with no water wall and no heat recovery is burning RDF* consisting of 85% organics, 10% water, and 5% inorganics or inerts at a rate of 950 kg/h. Determine the temperature, $T$, at which ash exits the combustion chamber (in both °C and °F), assuming the following:
   a. Heat value of the fuel = 19,000 kJ/kg on a moisture-free basis.
   b. Air flow = 9,500 kg/h and that the under- and overfire air contributes negligible heat.
   c. 5% of the heat input is lost due to radiation.
   d. 15% of the fuel remains uncombusted in the ash.
   e. Specific heat of ash = 0.837 kJ/kg/°C.
   f. Specific heat of air = 1.0 kJ/kg/°C.
   g. Latent heat of vaporization = 2575 kJ/kg.
   h. Temperature of the stack gases = 1500°C.
3. Use the program developed in (1) to verify the manual computations conducted in (2).

## Solution

1. For the solution of Example 6.16 (1), see the listing of Program 6.16.
2. Solution to Example 6.16 (2):
   Given: RDF 85% organics, 10% water, 5% inorganics or inerts.
   RDF rate = 950 kg/h.
   Heat value of the fuel = 19,000 kJ/kg.
   Air flow = 9,500 kg/h.
   Heat input is lost due to radiation = 5%.
   Fuel remaining uncombusted in the ash = 15%.
   Specific heat of ash = 0.837 kJ/kg/C.
   Temperature of the stack gases = 1500°C.
   Ash temperature = $T$°C???
   Specific heat of air is 1.0 kJ/kg/°C.
   Latent heat of vaporization = 2575 kJ/kg.

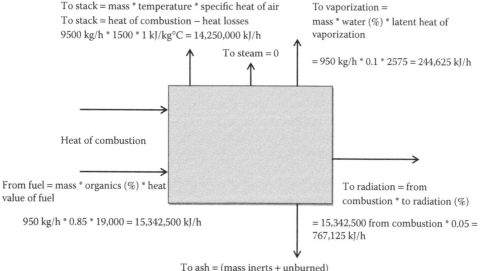

To stack = mass * temperature * specific heat of air
To stack = heat of combustion − heat losses
9500 kg/h * 1500 * 1 kJ/kg°C = 14,250,000 kJ/h

To steam = 0

To vaporization =
mass * water (%) * latent heat of vaporization

= 950 kg/h * 0.1 * 2575 = 244,625 kJ/h

Heat of combustion

From fuel = mass * organics (%) * heat value of fuel

950 kg/h * 0.85 * 19,000 = 15,342,500 kJ/h

To radiation = from combustion * to radiation (%)

= 15,342,500 from combustion * 0.05 = 767,125 kJ/h

To ash = (mass inerts + unburned)
* organics (%) * temperature * specific heat of ash

= (950 * 0.05 inerts + 950 * 0.85 organics * 0.15 uncombusted) * $T$ * 0.837 = 141.1391T kJ/h

Energy flow in a combustor

---

* Refuse Derived Fuel (RDF) refers to shredded waste as combined with fuel.

**LISTING OF PROGRAM 6.16   (CHAP6.16\FORM1.VB): TEMPERATURE OF STACK GASES**

```
'*************************************************************************************
'Program 6.16: Calculates temperature of the stack gases
'*************************************************************************************
Public Class Form1
    Dim g As Graphics
    Dim heatValue, airFlow, percRad, percAsh, exitTemp, AshSpHeat, AirSpHeat, latentHeat
As Double
    Dim RDFRate, percOrg, percWater, percInerts As Double
    Dim toStack, toVapor, toRad, toAsh, fromFuel, gasTemp As Double
    Private Sub Form1_Load(ByVal sender As System.Object, ByVal e As System.EventArgs)
Handles MyBase.Load
        Me.Text = "Program 6.16: Temperature of the stack gases"
        Me.FormBorderStyle = Windows.Forms.FormBorderStyle.FixedSingle
        Me.MaximizeBox = False
        Label1.Text = "Heat value of the fuel (kJ/kg):"
        Label2.Text = "Air flow (kg/h):"
        Label3.Text = "Heat input lost due to radiation (%):"
        Label4.Text = "Fuel remaining uncombusted in the ash (%):"
        Label5.Text = "Stack gases temp. (C):"
        Label6.Text = "Specific heat of ash (kJ/kg/C):"
        Label7.Text = "Specific heat of air (kJ/kg/C):"
        Label8.Text = "Latent heat of vaporization (kJ/kg):"
        Label9.Text = "Refuse-derived fuel (RDF) rate (kg/h):"
        Label10.Text = "Percentage of organics (%):"
        Label11.Text = "Percentage of water (%):"
        Label12.Text = "Percentage of inorganics/inerts (%):"
        Label13.Text = "Ash exit temp. from combustion chamber (C):"
        Button1.Text = "&Calculate"
    End Sub

    Sub calculateResults()
        heatValue = Val(TextBox1.Text)
        airFlow = Val(TextBox2.Text)
        percRad = Val(TextBox3.Text)
        percAsh = Val(TextBox4.Text)
        gasTemp = Val(TextBox5.Text)
        AshSpHeat = Val(TextBox6.Text)
        AirSpHeat = Val(TextBox7.Text)
        latentHeat = Val(TextBox8.Text)
        RDFRate = Val(TextBox9.Text)
        percOrg = Val(TextBox10.Text)
        percWater = Val(TextBox11.Text)
        percInerts = Val(TextBox12.Text)
        If percOrg > 1 Then percOrg /= 100 'if percentage is input as xx%, convert to 0.xx%
        If percWater > 1 Then percWater /= 100 'if percentage is input as xx%, convert
to 0.xx%
        If percInerts > 1 Then percInerts /= 100 'if percentage is input as xx%, convert
to 0.xx%
        If percRad > 1 Then percRad /= 100 'if percentage is input as xx%, convert to
0.xx%
        If percAsh > 1 Then percAsh /= 100 'if percentage is input as xx%, convert to
0.xx%
        fromFuel = RDFRate * percOrg * heatValue
        toVapor = RDFRate * percWater * latentHeat
        toRad = fromFuel * percRad
        toAsh = ((RDFRate * percInerts) + (RDFRate * percOrg * percAsh)) * AshSpHeat
        toStack = airFlow * gasTemp * AirSpHeat
        Dim ashTemp As Double = fromFuel - toRad - toVapor - toStack
        ashTemp /= toAsh
```

```vb
        TextBox13.Text = FormatNumber(ashTemp, 2)
        drawStack()
    End Sub

    Sub drawStack()
        Dim bmp As Bitmap = New Bitmap(PictureBox1.Width, PictureBox1.Height)
        g = Graphics.FromImage(bmp)
        g.Clear(Color.White)
        'draw stack box in the middle of the area
        g.DrawRectangle(Pens.Black, CInt((PictureBox1.Width / 2) - 50), _
            CInt((PictureBox1.Height / 2) - 35), 100, 70)
        'draw the 7 arrows, 2 into and 5 out of the stack
        Dim x = CInt((PictureBox1.Width / 2) - 50)
        Dim y = CInt((PictureBox1.Height / 2) - 35)
        drawArrow(g, x - 30, y + 20, x, y + 20)
        drawArrow(g, x - 30, y + 50, x, y + 50)
        drawArrow(g, x + 20, y, x + 20, y - 20)
        drawArrow(g, x + 40, y, x + 40, y - 10)
        drawArrow(g, x + 60, y, x + 60, y - 30)
        drawArrow(g, x + 60, y + 70, x + 60, y + 100)
        drawArrow(g, x + 100, y + 40, x + 130, y + 40)
        'output the results
        Dim s As String
        Dim f As Font = New Font(FontFamily.GenericSansSerif, 8)
        Dim f2 As Font = New Font(f, FontStyle.Bold Or FontStyle.Underline)
        s = "To stack = heat of combustion " + vbCrLf + "- heat losses" + vbCrLf + _
            "= " + Format(toStack, "n") + " kJ/hr"
        g.DrawString(s, f, Brushes.Black, New Point(10, y - 50))
        s = "To vaporization = mass*%water" + vbCrLf + "*latent heat of vaporization" +
        vbCrLf + _
            "= " + Format(toVapor, "n") + " kJ/hr"
        g.DrawString(s, f, Brushes.Black, New Point(x + 65, y - 50))
        s = "From fuel = mass" + vbCrLf + "*%organics*heat " + vbCrLf + "value of fuel" + _
            vbCrLf + "= " + Format(fromFuel, "n") + " kJ/hr"
        g.DrawString(s, f, Brushes.Black, New Point(10, y + 30))
        s = "To radiation = from " + vbCrLf + "combustion*%to radiation" + vbCrLf + _
            "= " + Format(toRad, "n") + " kJ/hr"
        g.DrawString(s, f, Brushes.Black, New Point(x + 110, y + 50))
        s = "To ash = (mass inerts+unburned)" + vbCrLf + "*%organics*temperature*speci
fic" + _
            vbCrLf + " heat of ash" + vbCrLf + "= " + Format(toAsh, "n") + " kJ/hr"
        g.DrawString(s, f, Brushes.Black, New Point(x + 40, y + 110))
        s = "Energy flow in a combustor"
        g.DrawString(s, f2, Brushes.Black, New Point(x - 15, y + 160))
        PictureBox1.Image = Image.FromHbitmap(bmp.GetHbitmap)
        g.Dispose()
        bmp.Dispose()
    End Sub

    Sub drawArrow(ByRef g As Graphics, ByVal x As Integer, ByVal y As Integer, ByVal x2
As Integer, ByVal y2 As Integer)
        'Draw the arrow stem
        g.DrawLine(Pens.Black, x, y, x2, y2)
        'Draw the arrow head
        If x = x2 Then 'arrow is vertical
            If y2 > y Then 'arrow is facing down
                g.DrawLine(Pens.Black, x2, y2, x2 - 4, y2 - 4)
                g.DrawLine(Pens.Black, x2, y2, x2 + 4, y2 - 4)
            Else 'arrow is facing up
                g.DrawLine(Pens.Black, x2, y2, x2 - 4, y2 + 4)
                g.DrawLine(Pens.Black, x2, y2, x2 + 4, y2 + 4)
```

```
            End If
        Else 'arrow is horizontal
            If x > x2 Then 'arrow is facing left
                g.DrawLine(Pens.Black, x2, y2, x2 + 4, y2 - 4)
                g.DrawLine(Pens.Black, x2, y2, x2 + 4, y2 + 4)
            Else 'arrow is facing right
                g.DrawLine(Pens.Black, x2, y2, x2 - 4, y2 - 4)
                g.DrawLine(Pens.Black, x2, y2, x2 - 4, y2 + 4)
            End If
        End If
    End Sub
    Private Sub Button1_Click(ByVal sender As System.Object, ByVal e As System.EventArgs)
Handles Button1.Click
        calculateResults()
    End Sub
End Class
```

Heat to ash = mass*temperature*specific heat of ash = 141.1391T kJ/h.

Temperature of ash = (15,342,500−767,125−244,625 − 14,250,000)/141.1391 = 80,750/141.1391 = 8070/141.1391 = 572°C = (572 * 9/5) +32 = 1062°F.

## 6.8   COST ESTIMATES FOR SOLID WASTE FACILITIES

Solid waste may be defined as garbage, refuse, and other solid material derived from any agricultural, commercial, consumer, or industrial operation or activity if it is both used material or residual material, and reasonably expected to be introduced into a qualified* solid waste disposal process within a reasonable time after such purchase or acquisition.

Solid waste financing or funding concerns revenues and costs, which vary with specifics of the solid waste system, ownership and contractual arrangements, and complexity of financial system. Revenue and profit for solid waste operations may be received from sale of services and goods, garbage bill paid by home or business, tipping fees at disposal site, sale of recyclables, and sale of products such as landfill gas or electricity from waste-to-energy plant. The initial cost of the facility (or its capital cost) is an important one-time investment that may be paid from the budget of the municipality

or agency, or proceeds of bank loans, or general obligation bonds, or revenue bonds, and so on.

### 6.8.1   Capital Cost and Capital Recovery Factor

The capital costs of competing facilities can be projected by determining the cost that the municipality or agency would incur if it were to pay interest on a loan of that amount and value. Computing the annual cost of a capital investment resembles computing the annual cost of a loan or mortgage on a building or land. The municipality or agency borrows the money from a moneylender or a financier or bank and then has to pay it back in a number of equal installments. If the municipality borrows ($X$) dollars and aims to pay back the loan in $n$ number of installments at an interest rate of $i$, each installment can be found as presented in the following equation:

$$Y = \left[\frac{i(1+i)^n}{(1+i)^n - 1}\right]X \qquad (6.46)$$

where:

$Y$ is the installment cost

$i$ is the annual interest rate (enter interest $i$ in decimal form, i.e., as a fraction)

$n$ is the number of installments

$X$ is the amount borrowed

A capital recovery factor (CRF) is defined as shown in the following equation:

$$CRF = \left[\frac{i(1+i)^n}{(1+i)^n - 1}\right] \qquad (6.47)$$

Equations 7.1 and 7.2 can be combined as revealed in the following equation:

$$Y = CRF * X \qquad (6.48)$$

---

* A qualified solid waste disposal process may employ any biological, engineering, industrial, or technological method. Eligible types of solid waste disposal processes include a final disposal process, an energy conversion process, and a recycling process. A final disposal process is the placement of solid waste in a landfill, the incineration of solid waste without capturing any useful energy, or the containment of solid waste with a reasonable expectation that the containment will continue indefinitely and that the solid waste has no current or future beneficial use. Energy conversion process encompasses a thermal, chemical, or other process that is applied to solid waste to create and capture synthesis gas, heat, hot water, steam, or other useful energy. The energy conversion process ends before any transfer or distribution of synthesis gas, heat, hot water, steam, or other useful energy. Recycling process regards a process reconstituting, transforming, or otherwise processing solid waste into a useful product (http://www.squiresanders.com/tax_exempt_financing_of_solid_waste_disposal_facilities/).

## Example 6.17

1. Write a short computer program that enables calculating the capital cost, CRF, and annual installments on this capital expense to purchase a refuse collection truck, given truck expected life, SAR, annual back payments, and interest rate.
2. A municipality decides to purchase a refuse collection truck that has an expected life of 10 years for SAR* 650,000. The cost of the truck is to be borrowed from the local bank and to be paid back in 10 annual payments. Determine the annual installments on this capital expense if the interest rate is 6.125%.
3. Use the program developed in (1) to verify the manual computations conducted in (2).

### Solution

1. For the solution of Example 6.17 (1), see the listing of Program 6.17.

2. Solution to Example 6.17 (2):
   a. Given: $t = 10$ years, cost of truck, $X =$ SAR 650,000, payments = 10, $i = 6.125\%$.
   b. From Table 6.6, CRF for $n$ years = 10 is 0.13667 or

$$CRF = \left[\frac{i(1+i)^n}{(1+i)^n - 1}\right] = \left[\frac{0.06125(1+0.06125)^{10}}{(1+0.06125)^{10} - 1}\right]$$

$$= 0.136674$$

   c. The annual cost to the municipality would then be $Y = CRF * X = 0.13667 * 650,000$ or $Y =$ SAR 88,836. That is, the municipality would have to pay SAR 88,836 each year for 10 years to pay back the bank loan on this truck.
   d. It is to be noted that this truck does not cost $10 * 88,836 =$ SAR 888,360, because the Saudi riyals for each year are different and cannot be augmented and added.

---

### LISTING OF PROGRAM 6.17 (CHAP6.17\FORM1.VB): CAPITAL COST, CAPITAL RECOVERY FACTOR, AND ANNUAL INSTALLMENTS

```
'*******************************************************************************
'Program 6.17: Calculates capital cost, capital recovery factor
'and annual installments
'*******************************************************************************
Public Class Form1
   Dim exLife, price, backPay, intRate As Double
   Private Sub Form1_Load(ByVal sender As System.Object, ByVal e As System.EventArgs)
Handles MyBase.Load
      Me.Text = "Program 6.17: Capital cost et al"
      Me.FormBorderStyle = Windows.Forms.FormBorderStyle.FixedSingle
      Me.MaximizeBox = False
      Label1.Text = " Calculates capital cost, capital recovery factor and annual
installments."
      Label2.Text = "Truck expected life (yr):"
      Label3.Text = "Price (SAR):"
      Label4.Text = "Annual back payments:"
      Label5.Text = "Interest rate (%):"
      Button1.Text = "&Calculate"
      Label6.Text = ""
   End Sub

   Sub calculateResults()
      exLife = Val(TextBox1.Text)
      price = Val(TextBox2.Text)
      backPay = Val(TextBox3.Text)
      intRate = Val(TextBox4.Text)
      intRate /= 100
      Dim CRF, aCost As Double
      CRF = ((intRate * ((1 + intRate) ^ exLife)) / (((1 + intRate) ^ exLife) - 1))
```

---

\* United States dollar, US$ $\approx$ 3.8 Saudi Arabia Riyal, SAR.

```
    aCost = CRF * price
    Label6.Text = "CRF for " + backPay.ToString + " years is " + Format(CRF, "n")
    Label6.Text += vbCrLf + "The annual cost would be, Y = " + Format(CRF, "n") + _
        " * " + backPay.ToString + " = SAR " + Format(aCost, "n")
    End Sub

    Private Sub Button1_Click(ByVal sender As System.Object, ByVal e As System.EventArgs)
Handles Button1.Click
        calculateResults()
    End Sub
End Class
```

### 6.8.2 Present Worth Value and Present Worth Factor

The actual cost of a capital investment also may be estimated by evaluating the present worth value or the value on a solution to example, given the date of a payment made at other times. This concerns finding the amount to be invested at the moment (present), $Y$ dollars, at a certain interest rate $i$ to have available $X$ dollars every year for $n$ number of years. The relationship can be figured as presented in the following equation:

$$Y = \left[ \frac{(1+i)^n - 1}{i(1+i)^n} \right] X \qquad (6.49)$$

where:

$Y$ is the amount that has to be invested
$i$ is the annual interest rate
$n$ is the number of years
$X$ is the amount available every year

A present worth factor, PWF can be introduced as shown in Equation 6.50:

$$PWF = \left[ \frac{(1+i)^n - 1}{i(1+i)^n} \right] \qquad (6.50)$$

Then, by combining Equations 6.49 and 6.50, the following equation is obtained:

$$Y = PWF * X \qquad (6.51)$$

### Example 6.18

1. Write a short computer program that enables calculating the amount to be invested at the moment (present), $Y$ dollars, at a certain interest rate $i$ to have available $X$ dollars every year for $n$ number of years.
2. A town wants to invest money in a bank account drawing an interest rate of 6.125% so that it can withdraw SAR 88,836 every year for the next 10 years. Compute the amount that must be invested.
3. Use the program developed in (1) to verify the manual computations conducted in (2).

### Solution

1. For the solution of Example 6.18 (1), see the listing of Program 6.18.
2. Solution to Example 6.18 (2):
   a. Given: $i = 6.125\%$, withdraw, $X = $ SAR 88,836, $n = 10$ year.
   b. From Table 6.6, the PWF for $n = 10$ is 7.316.
   c. Thus, the money required, $Y = PWF*X = 7.316 \times$ SAR 88,836 or $Y = $ SAR 649,924.

### 6.8.3 Sinking Fund and Sinking Fund Factor

Sinking fund may be defined as a fund established by a municipality or a government agency or a business for the purpose of reducing debt by repaying or purchasing outstanding loans and securities held against the entity. This indicates that a sinking fund would be a sum of money set up to collect a certain amount of money to pay for purchasing a certain commodity or paying the bill of a major work. This means that the municipality or agency is saving money by investing it so that at some later date, it would have some specified sum available. An example of such a fund in solid waste engineering is for the case of landfill entity investing money during the active life of the landfill so that, when the landfill is full, sufficient funds would be available to place the required final landfill cover.

Equation 6.52 illustrates how to determine the funds $Y$ necessary to be invested in an account that draws $i$ percent interest so that at the end of $n$ years, the fund has $X$ value in it:

$$Y = \left[ \frac{i}{(1+i)^n - 1} \right] X \qquad (6.52)$$

A sinking fund factor, SFF, may be introduced as shown in the following equation:

$$SFF = \left[ \frac{i}{(1+i)^n - 1} \right] \qquad (6.53)$$

---

### LISTING OF PROGRAM 6.18 (CHAP6.18\FORM1.VB): AMOUNT OF INVESTMENT NEEDED FOR A NUMBER OF YEARS

```vb
'*******************************************************************************
'Program 6.18: Calculates amount of investment needed
'for a number of years.
'*******************************************************************************
Public Class Form1
  Dim i, X, n, Y As Double
  Private Sub Form1_Load(ByVal sender As System.Object, ByVal e As System.EventArgs)
Handles MyBase.Load
    Me.Text = "Program 6.18:"
    Me.FormBorderStyle = Windows.Forms.FormBorderStyle.FixedSingle
    Me.MaximizeBox = False
    Label1.Text = " Calculates amount of investment needed for a number of years."
    Label2.Text = "Interest rate (%):"
    Label3.Text = "Yearly withdrawal needed (SAR):"
    Label4.Text = "Total number of years:"
    Button1.Text = "&Calculate"
    Label5.Text = ""
  End Sub

  Sub calculateResults()
    i = Val(TextBox1.Text)
    X = Val(TextBox2.Text)
    n = Val(TextBox3.Text)
    i /= 100
    Dim PWF As Double
    PWF = (((((1 + i) ^ n) - 1) / (i * ((1 + i) ^ n)))
    Y = PWF * X
    Label5.Text = "Present worth factor (PWF) for " + n.ToString + " years is " +
Format(PWF, "n")
    Label5.Text += vbCrLf + "The money required, Y = " + Format(PWF, "n") + _
      " * " + Format(X, "n") + " = SAR " + Format(Y, "n")
  End Sub

  Private Sub Button1_Click(ByVal sender As System.Object, ByVal e As System.EventArgs)
Handles Button1.Click
    calculateResults()
  End Sub
End Class
```

## Example 6.19

1. Write a short computer program that enables calculating the funds $Y$ necessary to be invested in an account that draws $i$ percent interest so that at the end of $n$ years, the fund has $X$ value in it.

2. A local solid waste enterprise aspires to have SAR 2 million available at the end of a 10-year period by investing annually into an account that gives an interest of 6.125%. Find the amount the enterprise has to invest annually.

3. Use the program developed in (1) to verify the manual computations conducted in (2).

## Solution

1. For the solution of Example 6.19 (1), see the listing of Program 6.19.
2. Solution to Example 6.19 (2).
   a. Given: $X$ = SAR 2,400,000, $t$ = 10 years, $i$ = 6.125%.
   b. From Table 6.6, the SFF at 10 years is 0.07452.
   c. The required annual investment is therefore $Y$ = SFF*$X$ = 0.07452 * SAR 2,400,000 or $Y$ = SAR 178,848.
   d. Note that the value of money of 10 * 178,848 = SAR 1,788,480 is significantly less than SAR 2,400,000. This is because the investments during the early years are drawing interest and adding to the sum available.

## LISTING OF PROGRAM 6.19   (CHAP6.19\FORM1.VB): NECESSARY FUNDS TO REACH A CERTAIN VALUE

```vb
'*******************************************************************************
'Program 6.19: Calculates necessary funds to reach a certain value
'*******************************************************************************
Public Class Form1
  Dim i, X, n, Y As Double
  Dim SFF As Double
  Private Sub Form1_Load(ByVal sender As System.Object, ByVal e As System.EventArgs)
Handles MyBase.Load
      Me.Text = "Program 6.19: Funds"
      Me.FormBorderStyle = Windows.Forms.FormBorderStyle.FixedSingle
      Me.MaximizeBox = False
      Label1.Text = "Calculates necessary funds to reach a certain value"
      Label2.Text = "Interest rate (%):"
      Label3.Text = "Total number of years:"
      Label4.Text = "Target value at the end of investment period (SAR):"
      Label5.Text = ""
      Label6.Text = "Decimals:"
      Button1.Text = "&Calculate"
      NumericUpDown1.Value = 2
      NumericUpDown1.Maximum = 10
      NumericUpDown1.Minimum = 0
  End Sub

  Sub calculateResults()
    i = Val(TextBox1.Text)
    n = Val(TextBox2.Text)
    X = Val(TextBox3.Text)
    i /= 100
    SFF = i / (((1 + i) ^ n) - 1)
    Y = SFF * X
  End Sub

  Sub showResults()
    Dim decimals As Integer = NumericUpDown1.Value
    If decimals < 0 Or decimals > 10 Then decimals = 2
    Label5.Text = "Sinking fund factor (SFF) at " + n.ToString + " years is " +
FormatNumber(SFF, decimals)
    Label5.Text += vbCrLf + "The required annual investment, Y = " + FormatNumber(SFF,
decimals)
    Label5.Text += " * " + Format(X, "n") + " = SAR " + FormatNumber(Y, decimals)
  End Sub

  Private Sub Button1_Click(ByVal sender As System.Object, ByVal e As System.EventArgs)
Handles Button1.Click
    calculateResults()
    showResults()
  End Sub

  Private Sub NumericUpDown1_ValueChanged(ByVal sender As System.Object, ByVal e As
System.EventArgs) Handles NumericUpDown1.ValueChanged
    showResults()
  End Sub
End Class
```

Table 6.5 gives a general summary of selected compounding factors.

The capital recovery factor, present worth factor, and sinking fund factor need not be computed, because they can be found in interest tables or are programmed into hand-held calculators or computer software. Table 6.6 shows these capital recovery factors for an interest rate of 6.125%.

Key:

CRF is the capital recovery factor $= \left[ \dfrac{i(1+i)^n}{(1+i)^n - 1} \right]$.

PWF is the present worth factor $= \left[ \dfrac{(1+i)^n - 1}{i(1+i)^n} \right]$.

SFF is the sinking fund factor $= \left[ \dfrac{i}{(1+i)^n - 1} \right]$.

## 6.8.4 Total Cost

Total cost to the community is the sum of the annual payback of the capital costs (fixed costs) of the investments and the labor and raw materials costs and operating and maintenance costs (variable costs).

### Example 6.20

1. Write a short computer program that enables calculating the total cost that a refuse collection truck costs the municipality every year, given the truck expected life, the truck cost, annual installments interest rate, and annual operation cost of the truck.

## TABLE 6.5
## Summary of Selected Compounding Factors

| Factor | Abbreviation | Equation | Use | Examples |
|---|---|---|---|---|
| Compound amount | CA | $\left[ (1+i)^n \right]$ | | |
| Capital recovery factor | CRF | $\left[ \dfrac{i(1+i)^n}{(1+i)^n - 1} \right]$ | Pays loan back in a number of equal installments. Converts a present value into a stream of equal annual payments over a specified time. | |
| Present worth factor | PWF | $\left[ \dfrac{(1+i)^n - 1}{i(1+i)^n} \right]$ | How much to be invested right now? | |
| Sinking fund factor | SFF | $\left[ \dfrac{i}{(1+i)^n - 1} \right]$ | Saves money by investing it so that at some later date some specified sum would be available. | When landfill owner must invest money during the active life of landfill so that, when landfill is full, there are sufficient funds to place final cover. |

## TABLE 6.6
## Capital Recovery Factors for an Interest Rate of 6.125%

| Year | CRF | PWF | SFF |
|---|---|---|---|
| 1 | 1.06,125 | 0.942,285,041 | 1 |
| 2 | 0.546,392,511 | 1.83,018,614 | 0.485,142,511 |
| 3 | 0.374,975,336 | 2.666,842,064 | 0.313,725,336 |
| 4 | 0.289,417,974 | 3.455,210,425 | 0.228,167,974 |
| 5 | 0.238,204,237 | 4.198,078,139 | 0.176,954,237 |
| 6 | 0.204,161,994 | 4.898,071,274 | 0.142,911,994 |
| 7 | 0.179,931,702 | 5.557,664,333 | 0.118,681,702 |
| 8 | 0.161,833,535 | 6.179,189,007 | 0.100,583,535 |
| 9 | 0.147,823,104 | 6.764,842,409 | 0.086,573,104 |
| 10 | 0.136,673,733 | 7.31,669,485 | 0.075,423,733 |
| 11 | 0.127,604,778 | 7.836,697,149 | 0.066,354,778 |
| 12 | 0.120,095,776 | 8.326,687,537 | 0.058,845,776 |
| 13 | 0.113,786,379 | 8.788,398,151 | 0.052,536,379 |
| 14 | 0.10,841,917 | 9.223,461,155 | 0.04,716,917 |
| 15 | 0.103,805,354 | 9.633,414,516 | 0.042,555,354 |
| 16 | 0.099,803,313 | 10.01,970,744 | 0.038,553,313 |
| 17 | 0.096,304,735 | 10.38,370,548 | 0.035,054,735 |
| 18 | 0.093,225,356 | 10.72,669,538 | 0.031,975,356 |
| 19 | 0.090,498,641 | 11.04,988,964 | 0.029,248,641 |
| 20 | 0.088,071,346 | 11.35,443,076 | 0.026,821,346 |

**LISTING OF PROGRAM 6.20   (CHAP6.20\FORM1.VB): REFUSE COLLECTION TRUCK'S COSTS**

```vb
'******************************************************************************
'Program 6.20: Calculates refuse collection truck's costs
'******************************************************************************
Public Class Form1
  Dim i, X, n, Y, cost As Double
  Dim CRF, total As Double
  Private Sub Form1_Load(ByVal sender As System.Object, ByVal e As System.EventArgs)
Handles MyBase.Load
    Me.Text = "Program 6.20: Truck costs"
    Me.FormBorderStyle = Windows.Forms.FormBorderStyle.FixedSingle
    Me.MaximizeBox = False
    Label1.Text = "Calculates refuse collection truck's costs"
    Label2.Text = "Expected life of collection truck:"
    Label3.Text = "Truck cost (SAR):"
    Label4.Text = "Annual installments interest rate (%):"
    Label5.Text = "Truck operating cost:"
    Label6.Text = ""
    Label7.Text = "Decimals:"
    Button1.Text = "&Calculate"
    NumericUpDown1.Value = 2
    NumericUpDown1.Maximum = 10
    NumericUpDown1.Minimum = 0
  End Sub

  Sub calculateResults()
    n = Val(TextBox1.Text)
    X = Val(TextBox2.Text)
    i = Val(TextBox3.Text)
    cost = Val(TextBox4.Text)
    i /= 100
    CRF = (i * ((1 + i) ^ n)) / (((1 + i) ^ n) - 1)
    Y = CRF * X
    total = Y + cost
  End Sub

  Sub showResults()
    Dim decimals As Integer = NumericUpDown1.Value
    If decimals < 0 Or decimals > 10 Then decimals = 2
    Label6.Text = "Capital recovery factor (CRF) for " + n.ToString + " years is " +
FormatNumber(CRF, decimals)
    Label6.Text += vbCrLf + "The required annual investment, Y = " + FormatNumber(CRF,
decimals)
    Label6.Text += " * " + Format(X, "n") + " = SAR " + FormatNumber(Y, decimals)
    Label6.Text += vbCrLf + "The annual cost = " + FormatNumber(cost, decimals)
    Label6.Text += " + " + FormatNumber(Y, decimals) + " = SAR " + FormatNumber(total,
decimals)
  End Sub

  Private Sub Button1_Click(ByVal sender As System.Object, ByVal e As System.EventArgs)
Handles Button1.Click
    calculateResults()
    showResults()
  End Sub

  Private Sub NumericUpDown1_ValueChanged(ByVal sender As System.Object, ByVal e As
System.EventArgs) Handles NumericUpDown1.ValueChanged
    showResults()
  End Sub
End Class
```

2. A municipality wants to purchase a refuse collection truck that has an expected life of 10 years and costs SAR 600,000. The municipality preferred to pay back the loan in 10 annual installments at an interest rate of 6.125%. The annual operation cost of the truck (gas, oil, service, and regular maintenance) amounts to SAR 80,000. How much will this truck cost the municipality every year?

3. Use the program developed in (1) to verify the manual computations conducted in (2).

### Solution

1. For the solution of Example 6.20 (1), see the listing of Program 6.20.

2. Solution to Example 6.20 (2).

   a. Given: $n$ = 10 years, $X$ = SAR 600,000, $i$ = 6.125%. truck operating cost = SAR 80,000/year.

   b. From Table 6.6, the CRF for $n$ = 10 is 0.13667.

   c. So the annual cost of the capital investment is $y$ = CRF*$X$ = 0.13667 * SAR 600,000, or $Y$ = SAR 82,002.

   d. Total annual cost to the community = operating cost + annual investment = SAR 82,002 + SAR 80,000 or total cost = SAR 160,002.

## 6.9 HOMEWORK PROBLEMS IN COMPUTER MODELING APPLICATIONS FOR MUNICIPAL SOLID WASTE CLASSIFICATION, QUANTITIES, PROPERTIES, COLLECTION, PROCESSING, MATERIAL SEPARATION, AND COST ESTIMATES

### 6.9.1 Discussion Problems

#### 6.9.1.1 Solid Waste Research

1. Select a short *research project* on solid waste for a certain locality (municipal, industrial, commercial, agricultural, hazardous, and so on). Write briefly about the following:

   a. Selection of research topic and justification

   b. Research objectives, hypothesis, and assumptions

   c. Research methodology

   d. Materials and methods for selected research area with emphasis on solid waste sources and characteristics, collection, segregation, sorting, treatment, final disposal, and reuse and recycling

   e. Results and discussions

   f. Conclusions and recommendations

   g. References

#### 6.9.1.2 Type of Waste

1. Compare between industrial and agricultural solid wastes as per indicated parameters within the following table:

### Table of Comparison between Industrial and Agricultural Solid Wastes

| Parameter | Type of Solid Waste | |
|---|---|---|
| | Industrial | Agricultural |
| Source and production unit | | |
| Type of solid waste | | |
| Quality | | |
| Reuse | | |
| Environmental impact | | |

#### 6.9.1.3 Solid Waste Types

1. Write briefly about *three* of the following:

   a. Current classification of solid waste in KSA

   b. The challenge for society is to minimize how much waste is generated and to convert waste into a resource (essence of the zero waste concept)

   c. Potential problems of solid waste, garbage, and sweeping

   d. Factors that affect quality and quantity of solid waste produced from a particular locality

   e. Most important properties of solid waste and its significance

2. Attempt writing briefly about *any three* questions of the following:

   a. "The challenge for society is to minimize how much waste is generated and to convert waste into a resource". Discuss this statement.

   b. MSW may be defined as a "heterogeneous mass of throwaways from the urban community, as well as the more homogeneous accumulations of agricultural, industrial, and mineral wastes". Based on this definition, how can you classify urban MSW? State your reasons.

   c. "Waste and garbage disposal is a big responsibility for the government. If the authority did not have good management for its disposal, it exposes itself to political and social problems." Explain why?

   d. "It is difficult to determine the relevance of diseases with waste and garbage. Nonetheless, ~50% of various diseases are transferred by flies, mosquitoes, and rodents proliferating in the waste." To take caution, what procedures would you advocate to be followed by concerned authorities?

   e. There are many sources of solid waste, garbage, and sweeping, which include agriculture, mining, building and construction, industry, housing, homes, offices, open markets, restaurants, hospitals, shops, educational institutions, hazardous, and so on. Outline the major types of hazardous solid waste. Which type would expect to be found in KSA? Why?

f. Write briefly about the most important properties of solid waste and its associated benefits.

### 6.9.1.4 Solid Waste Amount

How can you estimate the amount of solid waste in an area?

### 6.9.1.5 Solid Waste Properties

1. What is the benefit of properties of solid waste in management systems and related engineering topics?
2. What are the related effects to the physical properties of solid waste and garbage?
3. What is the benefit of knowing the *angle of stability* in a landfill?
4. Indicate the importance of *moisture content* measurements for a sample of MSW.
5. Evaluate the impact of *moisture* in a stream of solid waste of a municipality on its system of processing and final disposal.
6. Describe a method of conducting an experiment to estimate the *moisture content* of a household solid waste.
7. Why do newspapers contain higher moisture content compared to plastic materials in a domestic dustbin?
8. What is the benefit of knowing the *size of grains* of a solid waste?
9. How *bulk density* of trade solid waste is determined?
10. How can you estimate the chemical composition of garbage?
11. What is the benefit of estimating the *calorific value* of a solid waste?
12. What is the purpose of measuring *heat values* of refuse?
13. Compare between the methods used to identify the chemical components of solid waste and garbage to estimate the amount of energy or heat value in an unknown fuel as per attached table.

### Table for Methods Used to Identify Chemical Components of Solid Waste and Garbage

| Method | Description | Examples | Limitations |
| --- | --- | --- | --- |
| Ultimate analysis | | | |
| Compositional analysis | | | |
| Proximate analysis | | | |
| Calorimetry | | | |

14. Rearrange group (I) with the corresponding relative ones of group (II) in the area allocated for the answer.

| Group (I) | Rearranged Group (II) | Group (II) |
| --- | --- | --- |
| Recyclables | | Food waste |
| Refrigerators | | Hospitals |
| Bulky refuse | | Refineries |
| Yard waste | | Chemical and biological processes |
| Biomaterials | | Rubble and remnants of buildings |
| Sweeping | | Diverted refuse |
| Commercial waste | | Rest and stability |
| Municipal waste | | ASTM Standard |
| Industrial refuse | | Geotechnical engineering |
| Agriculture waste | | Generated by households |
| Hazardous waste | | Newspapers |
| Construction and demolition | | Quartering and coning |
| Refuse | | Production of gas |
| Dry weight moisture | | Thrown by users |
| Waste sampling | | Livestock farms |
| Representative samples | | Green waste |
| Not collected waste | | Furniture |
| Organic materials | | Warehouse waste |
| Angle of repose | | White waste |

### 6.9.1.6 Solid Waste Collection

1. Write briefly about the following agenda concerning solid waste as related to a selected *research project* in your residence area:
   a. Process of *collection*
   b. Sorting of solid waste components from each other
   c. Responsible personnel for collection and sorting of solid waste.
   d. Preferred method to transport in your city. Give reasons for your answer
   e. Objectives of solid waste collection
   f. Stages of solid waste collection
   g. Difference between collection of solid waste in both rural and urban areas
   h. Appropriate routes for car collecting solid waste between neighborhoods in the city
   i. Disadvantages and advantages of transfer stations.
   j. Transfer station location, divisions available, and methods to unload solid waste trucks
   k. Difference between reuse and recycling
   l. Methods of collecting recyclable materials
   m. Methods of storage of solid waste in house, apartment, and office. Harmful effects for keeping solid waste for a long time
   n. Kinds of baskets preferred for storage until transferred
   o. Antibreeding of trash flies used risks resulting from breeding of blow flies

p. Appropriate hours of collection of solid waste and related reasons

q. Obstacles to collect solid waste in your area. Give most appropriate solutions to improve the situation

2. Write briefly about the objectives of MSW *collection*.

3. Write briefly about the objectives of MSW *collection* in Al-Danaha municipality

4. Write briefly about refuse *collection* phases

5. Write briefly about the stages of MSW process *collection* in an urban area

6. Write briefly about the role of a transfer station in MSW *collection*. Indicate when it is preferred to rely on transfer stations

7. Recommend the appropriate ways for *collecting* recyclables, yard wastes, and C&D of solid wastes, as depicted in the following table:

### Table for Methods of SW Collection

| Waste | Method of Collection |
| --- | --- |
| Recyclables | |
| Yard waste | |
| C&D wastes | |

8. Compare between the different phases of MSW *collection* within a community as per the following table:

### Table of Different Phases of MSW Collection within a Community

| Stage or phase | Description | Location | Responsible Body | Limitations |
| --- | --- | --- | --- | --- |
| Phase 1: House-to-can | | | | |
| Phase 2: Can-to-truck | | | | |
| Phase 3: Truck from house-to-house | | | | |
| Phase 4: Truck routing | | | | |
| Phase 5: Truck-to-disposal | | | | |

### 6.9.1.7  Solid Waste Recycling and Reuse

1. Rank *recycling* among solid waste hierarchal system. Give an example of a material that is often recycled in your residence area.

2. Write briefly about MSW, recycling, and *reuse*.

### 6.9.1.8  Solid Waste Treatment

1. Appraise three advantages and three disadvantages of biological treatment in comparison with physical–chemical treatment.

2. Write briefly about primary treatment systems to prepare MSW.

3. The three components of MSW of greatest interest in the bioconversion processes are garbage (food waste), paper products, and yard wastes. What are the main factors that affect the variation of garbage fraction of refuse?

4. Theoretically, the combustion of refuse produced by a community is sufficient to provide ~20% of the electrical power needs for that community. Discuss this statement.

### 6.9.1.9  Solid Waste Finance and Costs

1. Write briefly about the major factors affecting the *cost* of MSW collection.

2. What innovative methods to *finance* and fund programs for collection and disposal of MSW would you propose to be adopted in KSA? Outline your reasons for your proposal.

3. Assess the effect of the factors, outlined in the following table, on the cost of MSW collection.

### Table of the Cost of MSW Collection

| Factor | Assessment |
| --- | --- |
| Time | |
| Number of collection points | |
| Distance to complete path and access final point of disposal | |
| Volume of MSW placed in each collection point and relationship to MSW collection vehicle | |
| Weight of MSW | |

4. *True/false*: Indicate whether the following statements are true (T) or false (F). In each case, justify your answer.

a. One of the goals of studying solid and hazardous waste management is transforming domestic and industrial waste into usable materials that people may reuse. ()

b. Absence of good sanitation in work areas, dump sites, and waste dumps is one of the main reasons for occurrence of risks and disease. ()

c. Ill management of waste deteriorates community public health as well as its environmental impact and pollution of water, air, and soil. ()

d. Sensory imbalance between workers may affect negatively their health. ()

e. Community opposition to landfill siting, which has exaggerated the problem of diminishing refuse disposal capacity, is known as NIMBY. ()

f. A transfer station is a site where solid waste is concentrated before it is taken to a processing facility or a sanitary landfill. ()

g. Adawha district in Dhahran serves ~80% of its population with a curbside recyclables program. ()

h. Recycling is the best approach to solid waste management. ()

i.   Reusing plastic water bottles is an example of waste recovery. ()

j.   Controlled tipping is so called because only a limited quantity of waste can be tipped at any one time. ()

k.   Packaging can sometimes reduce overall wastes. ()

l.   Minimizing waste generated by the society and converting it into a resource is the essence of the zero waste concept. ()

m.   MSW defines a heterogeneous mixture of refuse, C&D waste, leaves, and bulky items. ()

n.   It is difficult to quantify hazardous waste due to lack of real statistics, and perhaps hiding it from producers. This calls for sudden visits and regular monitoring. ()

o.   Factors that affect the quality and quantity of waste produced include standards, laws and legislation in force, living conditions, urbanization in the region, and social and economic factors. ()

p.   Low-income areas generate less waste but with higher proportion of food. ()

q.   Properties of solid waste affect the design of collection systems, treatment and disposal, operation, management, and performance of units. ()

5. *MCQs*

a.   Determine the hazardous solid waste among the following:
   i.   Agricultural waste
   ii.   Chemical waste
   iii.   Hospital waste
   iv.   Agricultural and hospital waste
   v.   Chemical and hospital waste

b.   Choose ………….. levels that can be paid by the beneficiary.
   i.   Cost
   ii.   Dust
   iii.   Leachate
   iv.   Odor
   v.   Organic matter

c.   Unpleasant and undesirable …………… resulting from bacterial decomposition of organic materials of components of the waste ought to be avoided.
   i.   Diseases
   ii.   Explosive gases
   iii.   Odors
   iv.   Toxic materials
   v.   Unregulated waste

d.   Secure landfills require all but one of the following:
   i.   A leachate collection system
   ii.   Financial guarantees for postclosure activities
   iii.   Groundwater monitoring
   iv.   Single liners

e.   Landfills have historically been the sources of all but one of the following:
   i.   Breeding grounds for insects and rodents
   ii.   Odor

iii.   Pollution to groundwater
iv.   Producers of propane gas

f.   An acronym* that indicates people's desire to locate landfills away from their communities is
   i.   MRF†
   ii.   NIMBY‡
   iii.   NIMTOO§
   iv.   NOPE**
   v.   PAYT††

g.   Biomedical waste may be disposed of by? ‡‡
   i.   Autoclaving
   ii.   Incineration
   iii.   Land filling
   iv.   Both (ii) and (iii)

h.   Which of the following is a biodegradable organic chemical/substance?§§
   i.   Garbage
   ii.   Oils
   iii.   Pesticides
   iv.   Plastics

i.   Which of the following is a practice used to reduce and manage MSW?
   i.   Waste combustion
   ii.   Source reduction
   iii.   Recycling of materials
   iv.   All of the above

j.   Which of the following is not a material in MSW?
   i.   Agricultural wastes
   ii.   Food wastes
   iii.   Glass and plastics
   iv.   Wood wastes

k.   Which of the following is not a source reduction activity?
   i.   Products package reuse
   ii.   Reducing use by modifying practices
   iii.   Saving energy by using recycled materials
   iv.   Package or product design that reduces material or toxicity

l.   Facilities that perform the function of preparing recyclables for marketing are referred to as
   _____.
   i.   WTEs (waste-to-energy)
   ii.   RDFs (refuse-derived fuel)
   iii.   TSDs (treatment, storage, and disposal)
   iv.   MRFs (materials recovery facility)

---

*  Acronym is an abbreviation formed from the initial letters of other words and pronounced as a word.
†  MRF (or Recycling or Factory)
‡  Not in My Back Yard
§  Not in My Term of Office.
**  Not on Planet Earth
††  Pay As You Throw
‡‡  Adapted from http://www.shareyouressays.com/114076/33-objective-type-questions-mcqs-with-answers-on-environmental-pollution.
§§  Adapted from http://www.shareyouressays.com/114076/33-objective-type-questions-mcqs-with-answers-on-environmental-pollution.

m. Which hierarchy is correct for the principle of waste generation
   i. Reuse-reduce-recycle-recover-resource management
   ii. Reduce-reuse-recycle-recover-resource management
   iii. Reduce-recover-recycle-reuse-resource management
   iv. Reduce-recycle-recover-reuse-resource management

6. *Complete* missing titles by using the following words and phrases (flooding, providers, odors, moisture. syringes)
   a. Garbage piles up on roads, streets, and parks producing foul ....................
   b. Bodies that deal with solid waste and garbage include: citizens on a daily basis, .................. of waste collection services, scavengers and those working on reuse.
   c. Accumulation of waste in drainage networks and waterways increases the risk of ......................... and contamination of water resources.
   d. Types of hazardous solid waste include linens, clothing, bandages and disposable, .................., needles, surgical equipment and medical devices disposed off, food and contaminated waste, and flammable materials.
   e. Common contaminants of waste items include ......................, food, and dirt.

## 6.9.2 Specific Mathematical Problems

### 6.9.2.1 Amount of Solid Waste

1. a. Write a computer program to determine the *diversion*, given the fractions of solid waste and its fate for a particular society.
   b. Verify your program for a community that produces the following in an annual basis:

| Fraction | Tons per year |
|---|---|
| Mixed house waste | 260 |
| Recyclables | 37 |
| Commercial waste | 54 |
| C&D debris | 135 |
| Leaves and miscellaneous | 65 |
| Treatment plant sludges | 8 |

The recyclables are collected separately and processed at an MRF. The mixed household waste and the commercial waste go to the landfill, as do the leaves and miscellaneous solid wastes. The sludges are dried and applied on land (not into the landfill), and the C&D wastes are used to fill a large ravine. Calculate the diversion.

### 6.9.2.2 Solid Waste Properties

1. a. Write a computer program to find the overall *moisture* content of a sample of a collected residential MSW, given the typical physical composition of the waste.
   b. Verify your program by estimating the overall moisture content of a sample of as-collected residential MSW with the typical composition given in the following table (University of Dammam, 2012).

Table of typical physical composition of residential MSW

| Component | Percentage by Weight | |
|---|---|---|
| | Range | Typical |
| **Organic** | | |
| Food wastes | 6–18 | 9.0 |
| Paper | 25–40 | 34.0 |
| Cardboard | 3–10 | 6.0 |
| Plastics | 4–10 | 7.0 |
| Textiles | 0–4 | 2.0 |
| Rubber | 0–2 | 0.5 |
| Leather | 0–2 | 0.5 |
| Yard wastes | 5–20 | 18.5 |
| Wood | 1–4 | 2.0 |
| Miscellaneous organics | – | – |
| **Inorganic** | | |
| Glass | 4–12 | 8.0 |
| Tin cans | 2–8 | 6.0 |
| Aluminum | 0–1 | 0.5 |
| Other metal | 1–4 | 3.0 |
| Dirt, ash, and so on | 0–6 | 3.0 |
| Total | | 100 |

2. a. Write a computer program to find the approximate *chemical formula* and *energy* content of organic component of a certain solid waste, given its composition and fractions.
   b. Verify your program by finding an approximate chemical formula of the organic component of the sample composition of a solid waste as set out in the following table. Use chemical composition obtained to estimate energy content of this solid waste.

| Component | Percentage by Mass |
|---|---|
| Garden trimmings | 13 |
| Food waste | 20 |
| Timber | 7 |
| Paper | 33 |
| Cardboard | 12 |
| Rubber | 8 |
| Tin cans | 7 |
| Total sum | 100 |

Assume the total organic composition of the solid waste assuming a mass of 100 kg of the sample as shown in the following table (b):

| Component | Mass (kg) |
|-----------|-----------|
| Carbon | 35 |
| Hydrogen | 5 |
| Oxygen | 28 |
| Nitrogen | 0.5 |
| Sulfur | 0.1 |
| Ash | 5 |

3. a. Write a computer program to find the *heat value* of a certain solid waste, given its fractions on a dry basis.

 b. Verify your program by estimating the heat value of a processed refuse-derived fuel having the following composition:

| Component | Fraction by Weight, Dry Basis |
|-----------|-------------------------------|
| Paper | 0.25 |
| Food waste | 0.25 |
| Plastics | 0.3 |
| Glass | 0.2 |

4. a. Write a computer program to find the comparable, *moisture*-free Btu and the moisture- and ash-free heat value for a sample of refuse, given its water content, Btu of the entire mixture, weight of sample, and ash remaining after combustion.

 b. Verify your program for a sample of refuse is analyzed and found to contain 15% water (measured as weight loss on evaporation). The Btu of the entire mixture is measured in a calorimeter and is found to be 6000 Btu/lb. A 1.0 g sample is placed in the calorimeter, and 0.3 g ash remains in the sample cup after combustion. What is the comparable, moisture-free Btu, and the moisture- and ash-free heat value?

## 6.9.2.3 Solid Waste Collection

1. a. Write a short computer program to determine the number of customers served by a truck, given each household production of refuse per week, truck capacity, and compaction density.

 b. Verify your program by assuming that each household produces 50 lb of refuse per week (as in Problem 2). How many customers can a 20-yd$^3$ truck that compacts the refuse to 580 lb/yd$^3$ collect before it has to make a trip to the landfill?

2. a. Write a computer program that allows computation of customers that can be served by a truck if it did not have to go to the landfill, given the number of crew, the time required per stop, and the number of customers to be served.

 b. Suppose a crew of two people, requires 3 min/stop, at which they can service 4 customers. If each customer generates 55 lb of refuse per week, how many customers can they service if they did not have to go to the landfill?

## 6.9.2.4 Solid Waste Treatment and Disposal

1. a. Write a computer program that enables the determination of the percolation of water through a sanitary landfill, given the annual amount of rainfall, the yearly transpiration, and the runoff coefficient.

 Verify your program by finding the percolation of water through a sanitary landfill assuming the amount of rainfall is 1300 mm per year, and transpiration is 480 mm/year. Assuming a runoff coefficient of 0.15.

2. a. Write a computer program that enables the estimation the percolation of water through a landfill, given its depth, thickness of soil cover, population, R, E, soil field capacity, and refuse field capacity as packed.

 b. Verify your program by estimating the percolation of water through a landfill 12 m deep (h), with a 1.1 m cover of sandy loam soil (d). Use the following data:
  i. Precipitation (P) = 1000 mm/yr
  ii. Runoff coefficient (R) = 0.14
  iii. Evaporation (E) = 580 mm/yr
  iv. Soil field capacity, $F_s$ = 210 mm/m
  v. Refuse field capacity, $F_r$ = 320 mm/m, as packed.

3. a. Write a computer program that allows the determination of maximum design depth above lining, given the distance between leachate collection tubes, rain water from 25 years, and a 24-h storm entering the leachate drainage system, design storm (vertical flow), hydraulic conductivity, and the slope of discharge.

 b. Verify your program by finding the maximum design depth above lining noting that the distance between leachate collection tubes is 10 m. Using a coarse discharge material and assuming that rainwater from 25 years and a storm entering the leachate drainage system, design storm (vertical flow) = 0.017 cm/min, and hydraulic conductivity 0.017 cm/s, and the slope of discharge 1.8%.

4. a. Write a computer program that allows the determination of the spacing between pipes in a leachate collection system, given design storm (25 years, 24-h), hydraulic conductivity, drainage, and the maximum design depth on liner.

 b. Verify your program by determining the spacing between pipes in a leachate collection system using granular drainage material and the following properties. Assume that in the most conservative design, all storm water from a 25-year, 24 h storm enters the leachate collection system.
  i. Design the storm (25 years, 24 h) = 8.2 in = 0.015 cm/min.
  ii. Hydraulic conductivity = 0.02 cm/s.
  iii. Drainage slope = 1.5%.
  iv. Maximum design depth on liner = 15 cm.

5. a. Write a computer program that allows writing down a balanced equation for the anaerobic decomposition of an organic compound and estimating the amount and volume produced (at STP)[*] of $CO_2$ and $CH_4$ during the anaerobic decomposition of this compound.
   b. Verify your program by writing down a balanced equation for the anaerobic decomposition of a MSW, given its chemical formula. Estimate the production of $CO_2$ and $CH_4$ during the anaerobic decomposition of MSW using the chemical composition approximation of organic fraction of refuse as described by $C_{99}H_{149}O_{59}N$.

6. a. Write a computer program that allows the determination of the amount of water or sludge that must be added to the solids to achieve a certain moisture concentration in the compost pile, given the weight of mixture and the required moisture content.
   b. Verify whether your program for 15 tons of a mixture of paper and other compostable materials has a moisture content of 8%. The intent is to make a mixture for composting of 50% moisture. How many tons of water or sludge must be added to the solids to achieve this moisture concentration in the compost pile?

7. a. Write a computer program that allows the determination of the efficiency of a certain power plant, given the amount of coal used per day, and the energy value of the coal and plant production of electricity each day.
   b. Verify whether your program for a coal-fired power plant uses 850 Mg[†] of coal per day. The energy value of the coal is 28,000 kJ/kg. The plant produces $3.4 \times 10^6$ kWh[‡] of electricity each day. What is the efficiency of the power plant?

8. a. Write a computer program that allows the determination of the temperature at which ash exits the combustion chamber (in both °C and °F) of a refractory combustion unit lined with no water wall and no heat recovery burning an RDF, given RDF organics, water, and inorganics or inerts, and their hourly rate, heat value of the fuel, air flow, heat loss due to radiation, amount of fuel remaining uncombusted in the ash, specific heat of ash, specific heat of air, latent heat of vaporization, and temperature of the stack gases.
   b. Verify your program for a refractory combustion unit lined with no water wall and no heat recovery is burning RDF consisting of 85% organics, 10% water, and 5% inorganics or inerts at a rate of 950 kg/h.

Determine the temperature, $T$, at which ash exits the combustion chamber (in both °C and °F), assuming the following:
   i. Heat value of the fuel = 18,500 kJ/kg on a moisture-free basis.
   ii. Air flow = 8,700 kg/h and that the under and overfire air contributes negligible heat.
   iii. 5% of the heat input is lost due to radiation.
   iv. 15% of the fuel remains uncombusted in the ash.
   v. Specific heat of ash = 0.837 kJ/kg/°C.
   vi. Specific heat of air is 1.0 kJ/kg/°C.
   vii. Latent heat of vaporization = 2575 kJ/kg.
   viii. Temperature of the stack gases = 1500°C.

### 6.9.2.5  Solid Waste Finance and Cost Analysis

1. a. Write a computer program that allows the determination of the annual installments on a capital expense to purchase a refuse collection truck, given its expected life, SAR, and annual payments of cost of truck to be paid back to the local bank.
   b. Verify your program for a municipality deciding to purchase a refuse collection truck that has an expected life of 15 years for SAR 650,000. The cost of the truck is to be borrowed from the local bank and to be paid back in 15 annual payments. Determine the annual installments on this capital expense if the interest rate is 6.125%.

2. a. Write a computer program that allows the determination of the rate of interest if a certain municipality chooses to purchase a refuse collection truck, given its expected life, value of SAR, and annual payments.
   b. Verify whether your program for a municipality chooses to purchase a refuse collection truck that has an expected life of 15 years for the value of SAR 700,000. The cost of the truck is to be borrowed from the local bank and to be paid back in 10 annual payments, given that the annual installments on this capital expense are SAR 68,000. Find the rate of interest.

### REFERENCES

Abdel-Magid, I.M. 2012. *Problem Solving in Solid Waste Engineering.* Dammam, KSA: University of Dammam.

Abdel-Magid, I. M., and Abdel-Magid, M. I. M. 2015a. *Problem Solving in Solid Waste Engineering*, 2nd ed. North Charleston, SC: Create Space Independent Publishing Platform.

Abdel-Magid, I. M., and Abdel-Magid, M.I.M. 2015b. *Solid Waste Engineering and Management.* North Charleston, SC: CreateSpace Independent Publishing Platform (In Arabic).

Abdel-Magid, I. M., Hago, A., and Rowe, D. R. 1995. *Modeling Methods for Environmental Engineers.* Boca Raton, FL: CRC Press/Lewis Publishers.

Blackman, W. C. 2001. *Basic Hazardous Waste Management*, 3rd ed. Boca Raton, FL: CRC Press.

---

[*] STP is used for expression of the properties and processes of ideal gases. The standard temperature is the freezing point of water and the standard pressure is one standard atmosphere. Standard temperature: 0°C = 273.15 K; standard pressure = 1 atmosphere = 760 mmHg = 101.3 kPa; standard volume of 1 mole of an ideal gas at STP: 22.4 L.

[†] Megagrams, or 1000 kg, commonly called a metric ton.

[‡] 1 kWh = $3.6 \times 10^6$ J.

CEHA. 1995. Solid waste management in some countries of the Eastern Mediterranean region. CEHA Document No., Special studies, ss-4. Amman, Jordan: WHO, Eastern Mediterranean Regional Office, Regional Centre for Environmental Health Activities.

de Bertoldi, M. (Ed.). 1996. *The Science of Composting*. London, UK: Blackie Academic & Professional.

Nakamura, M., Castaldi, M. J., and Themelis, N. J. 2005. Measurement of particle size and shape of New York city municipal solid waste and combustion residues using image analysis. http://www.seas.columbia.edu/earth/wtert/sofos/ Nakamura_JSMWE_2005.pdf (accessed on March 21, 2016).

Ojovan, M. I. (Ed.). 2011. *Handbook of Advanced Radioactive Waste Conditioning Technologies*. Cambridge: Woodhead Publishing Ltd.

Rietema, K. 1957. On the efficiency in separating mixtures of two components. *Chemical Engineering Science*, 7:89–96.

Rodic-wiersma, L. 2005. Introduction to solid waste management and engineering. *Refresher Course on Solid Waste Management and Engineering, Organized by UNESCO-IHE Institute for Water Education*, Delft, the Netherlands, October 16–22, Mombasa, Kenya.

Schubeler, P., Wehrle, K., and Christen, J. 1996. Conceptual framework for municipal solid waste management in low-income countries. Urban Management and Infrastructure, UNDP, UNCHS (Habitat), World Bank, SDC Collaborative Program on Municipal Solid Waste Management in Low-Income Countries, August 1996, Working Paper No. 9, SKAT (Swiss Centre for Development Cooperation in Technology and Management), Gallen, Switzerland.

Senate, E., Galtier, L., Bekaert, C., Lambolez-Michel, L., and Budka, A. 2003. Odor management at MSW landfill sites: Odor sources, odorous compounds and control measures. *Proceedings Sardinia, Ninth International Waste Management and Landfill Symposium*, S. Marghorita di Pula, Cagliari, Italy, October 6–10.

von Blottnitz, H., Pehlken, A., and Pretz, T. 2002. The description of solid wastes by particle mass instead of particle size distributions. *Resour, Conserv and Recy J*, 34:193–207.

Walsh, P. and O'Leary, P. 1986. *Implementing Municipal Solid Waste to Energy Systems*. Madison, WI: University of Wisconsin—Extension for Great Lakes Regional Biogas Energy Program.

Worrell, W. A., and Vesilind, P. A. 2012. *Solid Waste Engineering*. Stamford, CT: CL-Engineering Pub.

# 7 Computer Modeling Applications for Air Pollution Control Technology

The introduction to this chapter presents a brief overview of the air pollution field. This chapter deals with fundamental concepts that are needed to make calculations dealing with air pollution control. Computer programs based on mathematical equations and models relating to these fundamental concepts are also included here. The most commonly used air pollution control devices for gaseous and particulate pollutants are presented such as absorption, adsorption, and combustion for gaseous contaminants and settling chambers; cyclones; electrostatic precipitators (ESPs); venture scrubbers; and baghouse filters for controlling particulate emissions. Each control device or technique is accompanied by a computer program that can aid in the design or evaluation of air pollution control equipment.

This section concludes with mathematical models that can be used for determining effective stack heights (plume rise) as well as dispersion models that can help estimate the concentrations of air pollutants dispersed in the atmosphere. Computer programs for both stack height calculations and dispersion models are included.

## 7.1 INTRODUCTION

The first step in directing compliance and enforcement of air pollution control laws and regulations is to have a credible and acceptable definition as to what air pollution is. In the United States, the Code of Federal Regulations (40 CFR 52.741, July 1, 1994) indicates that air pollution means the presence of one or more contaminants in sufficient quantities and of such characteristics and duration as to be injurious to human, plant, or animal life; to health; or to property, or to unreasonably interfere with the enjoyment of life or property (Fed. Regist. 1994). All environmental regulations in the United States can be found in CFR Part 40.

Ambient (outdoor) unpolluted dry air consists by volume of ~78% nitrogen ($N_2$), 21% oxygen ($O_2$), 0.93% argon (Ar), 0.03% carbon dioxide ($CO_2$), and traces of other gases such as neon (Ne), helium (He), methane ($CH_4$), and krypton (Kr). The unpolluted dry air described previously exists only in theory; all air contains natural contaminants such as pollen, fungi, spores, smoke, and dust particles from forest fires and volcanic eruptions. In contrast to natural air, pollutants are contaminants of an anthropogenic (manmade) origin.

Anthropogenic sources include transportation (mobile sources), fuel combustion from electric utilities, fuel combustion from other sources, industrial processes, waste disposal and recycling, and miscellaneous sources such as open burning, agricultural burning, and recreation. The two basic physical forms of air pollutants are particulate matter (PM) and gases. PM includes small solid or liquid particles such as dust, smoke, mists, and fly ash. Gases include substances such as carbon monoxide, sulfur dioxide, and volatile organic compounds (VOCs).

In the United States, the Environmental Protection Agency (EPA) has further classified air pollutants as criteria pollutants and noncriteria pollutants. Criteria pollutants are pollutants that have been identified as being both common and detrimental to human health and welfare. The EPA currently designates six pollutants as criteria pollutants (Air Pollution Training Institute 1992):

- Carbon monoxide (CO)
- Ozone ($O_3$)
- Sulfur dioxide ($SO_2$)
- PM <10 μm in diameter
- Nitrogen dioxide ($NO_2$)
- Lead (Pb)

The U.S. Clean Air Act Amendments established a new classification of noncriteria pollutants called hazardous air pollutants (HAPs). This act listed 189 compounds as HAPs and directed the EPA to investigate possible regulation of sources emitting these pollutants (EPA 2014).

For each criteria pollutant, the EPA was required to set both a primary standard and a secondary standard. The purpose of the primary standard is to protect public health, whereas secondary standards are set at a level to protect public welfare from any adverse effects. The primary and secondary standards for each of the criteria pollutants are shown in Table 7.1. Collectively, these standards are the National Ambient Air Quality Standards (NAAQS) (Air Pollution Training Institute 1992, U.S. Environmental Protection Agency 2014).

For comparison, Table 7.2 presents air quality standards for other countries in the world, indicating the standard in micrograms per cubic meter (μg/m³) and the averaging time. Table 7.3 presents the total air pollutant emissions (million tons per year) for the six major air pollutants in the United States for 1993 (Council on Environmental Quality 1993). Of these pollutants, transportation produced ~77% of the total carbon monoxide, 36% of the VOCs, 44% of the oxides of nitrogen, 22% of the $PM_{10}$ particulates, 3% of the oxides of sulfur, and 32% of the lead. Transportation was responsible for 56% by weight of all the major contaminants emitted to the atmosphere in the United States in 1993 (Council on Environmental Quality 1993).

As mentioned earlier, air pollutants can be divided into two classes, gaseous and particulate, and they generally

**TABLE 7.1**

**National Ambient Air Quality Standards**

| Pollutant | Primary (Health-Related) Standard Level | | Secondary (Welfare-Related) Standard Level | |
|---|---|---|---|---|
| | Averaging Time | Concentration | Averaging Time | Concentration |
| Particle pollution: PM 10 μm | 24 h (not to be exceeded more than once per year on average over 3 years) | 150 μg/m³ | Same as primary | Same as primary |
| PM 2.5 μm | 24 h (98th percentile, averaged > 3 years) | 35 μg/m³ | Same as primary | Same as primary |
| $SO_2$ | 1 h (99th percentile of 1 h daily maximum concentrations, averaged > 3 years) | 75 ppb | 3 h (not to be exceeded more than once per year) | 0.5 ppm |
| CO | 8 h | 10,000 μg/m³ (9 ppm) | No secondary standard | No secondary standard |
| | 1 h | 40,000 μg/m³ (35 ppm) | No secondary standard | No secondary standard |
| $NO_2$ | Annual mean, 1 year | 53 ppb | Same as primary | Same as primary |
| | 1 h (98th percentile of 1 h daily maximum concentrations, averaged > 3 years) | 100 ppb | No secondary standard | No secondary standard |
| $O_3$ | 8 h (Annual fourth highest daily maximum 8 h concentration, averaged > 3 years) | 0.070 ppm | Same as primary | Same as primary |
| Pb | Rolling 3-month period | 0.15 μg/m³ | Same as primary | Same as primary |

*Source:* NAAQS, Air Pollution Training Institute 1992; Environmental Protection Agency, EPA. 2016. National Ambient Air Quality Standards (NAAQS). https://www3.epa.gov/ttn/naaqs/criteria.html.

**TABLE 7.2**

**Air Quality Standards for Selected Pollutants in Several Countries (μg/m³)**

| Country | Suspended Particulate | Sulfur Oxide | Carbon Monoxide | Nitrogen Oxide |
|---|---|---|---|---|
| Canada | 120/24 h<br>60–70/1 year | 450/1 h<br>150/24 h<br>30–60/1 year | 15,000/1 year<br>6000/8 h | 400/1 h<br>200/24 h<br>60–100/1 year |
| Japan | 200/1 h<br>100/24 h | 300/1 h<br>120/24 h | 11,100/8 h | 100/24 h |
| Russia | 150/24 h | 157/24 h | 5723/8 h<br>1145/24 h | 113/24 h |
| Saudi Arabia | 340/24 h<br>80/1 year | 800/1 h<br>400/24 h<br>85/1 year | 40,000/1 h<br>10,000/8 h | 660/1 h<br>100/1 year |
| West Germany | 480/1/2 h | 500/1/2 h | 40,000/1 h<br>10,000/8 h | 1,000/1/2 h |

require different prevention and control methods. The broad prevention methods to control air pollution emissions for both gaseous and particulates include the following (Air Pollution Training Institute 1992):

- Process change
- Changes in fuel
- Good operating practices
- Plant shutdowns

The basic principles involved in the control of gas emissions include the following:

- Combustion
- Adsorption
- Absorption
- Condensation

The most commonly used devices to control particulate emissions include the following:

- Settling chambers
- Cyclonics
- Electrostatic precipitation
- Venture scrubber
- Baghouse filters

Pollution control efforts utilizing these various principles and techniques have reduced the air pollution emissions of sulfur dioxide by 30% from 1970 to 1993; carbon monoxide by

**TABLE 7.3**

**Sources of Air Pollutants in the United States, 1993[a]**

| Source | CO | PM$_{10}$[b] | SO$_3$ | VOCs[c] | NO$_x$ | Pb | Total |
|---|---|---|---|---|---|---|---|
| Transportation | 75.261 | 0.592 | 0.718 | 8.301 | 10.423 | 1.589 | 96.884 |
| Fuel combustion (electric utilities) | 0.322 | 0.270 | 15.836 | 0.036 | 7.782 | 0.062 | 24.308 |
| Fuel combustion (industrial) | 0.667 | 0.219 | 2.830 | 0.271 | 3.176 | 0.018 | 7.181 |
| Fuel combustion (other) | 4.444 | 0.723 | 0.600 | 0.341 | 0.732 | 0.417 | 7.257 |
| Industrial processes | 5.219 | 0.610 | 1.868 | 11.201 | 0.911 | 2.281 | 22.090 |
| Waste disposal and recycling | 1.732 | 0.248 | 0.037 | 2.271 | 0.084 | 0.518 | 4.890 |
| Miscellaneous | 9.506 | 0.0 | 0.011 | 0.893 | 0.296 | n.a. | 10.706 |
| Total | 97.151 | 2.662 | 21.900 | 23.314 | 23.404 | 4.885 | 173.316 |

*Source:* Council on Environmental Quality, *24th Annual Report of the Council on Environmental Quality,* Washington, D.C, 1993.

[a] In million short tons (2000). (See Council on Environmental Quality 1993.)

[b] PM$_{10}$ includes only those particles with aerodynamic diameter <10 μm (see Council on Environmental Quality 1993).

[c] VOCs, volatile organic compounds (see Council on Environmental Quality 1993).

n.a., not available.

18%, and VOCs by 20%. The PM$_{10}$ particulate emissions from 1985 to 1993 were reduced 10%; however, from 1970 to 1993, nitrogen emissions have increased ~11%. One of the greatest environmental success stories has been the reduction of lead in the ambient air. From 1984 to 1993, lead concentrations at 204 sampling sites in the United States have shown an 89% reduction (Council on Environmental Quality 1993). Although there have been gains made in controlling air pollution, it must be remembered that some of these gains have been offset by an increase in all activities, especially motor vehicle registration and miles traveled.

In the past, identification and characterization of exposure of people to pollutants in the outdoor air have been emphasized. In recent years, attention has been directed at the effects of exposure to indoor air. Most people spend 80%–90% of their time indoors, and it has been shown that exposure to some pollutants can be 2–5 times higher indoors than outdoors (Air Pollution Training Institute 1992). Some of these indoor pollutants include carbon monoxide (CO), nitrogen oxides (NO$_x$), sulfur dioxides (SO$_x$), PM, asbestos, formaldehyde (HCHO), ozone, and radon gas (Ra-222). Some of the sources for these indoor air pollutants include emissions from combustion appliances, tobacco smoke, aerosol propellants, plastic furniture, rugs and curtains, refrigerants, paints, cleaners, and building materials. The broad measures that can be used to reduce exposure to indoor air pollution are careful selection and operation of household appliances, proper building design (selection of materials and construction), and, of course, proper ventilation.

Although more attention is being paid to indoor air pollution on a regional, a continental, and a global basis, some of the major air pollution problems include the following (Peavey et al. 1985):

- Global warming (greenhouse effect due to carbon dioxide [CO$_2$] emissions)

- Acid rain (sulfur dioxide, SO$_2$, and nitrogen oxides, NO$_x$, transformed into acids in the atmosphere)
- Ozone layer depletion (increased ultraviolet radiation exposure caused by depletion of the ozone layer due to reaction with chlorine released by fluorocarbons emitted to the atmosphere)

## 7.2 FUNDAMENTAL CONCEPTS IN AIR POLLUTION

### 7.2.1 Units of Measurement

In the past, many confusing and conflicting units have been used in the air pollution field. At present, the trend is to try to standardize the units by utilizing the metric system. For instance, on an international basis, it is recommended that weights be reported in grams (g), milligrams (mg), or micrograms (μg), and volumes be reported in cubic meters (m$^3$). The U.S. EPA has recommended using the units for particulates and gaseous pollutants as presented in Table 7.4 (Air Pollution Training Institute 1983).

A common practice at present is to present the concentration of the gas contaminants first in μg/m$^3$, followed by parts per million (ppm), parts per hundred million, or parts per billion by volume. For gases, ppm can be converted in to μg/m$^3$ or vice versa by using the following equation:

$$\frac{\mu g}{m^3} = \frac{ppm \times \text{molecular weight of gas} \times 10^3}{L/mol} \quad (7.1)$$

For instance,

- At 1 atm and 0°C (273 K), L/min = 22.41 L.
- At 1 atm and 25°C (298 K), or standard temperature and pressure (STP), L/mol = 24.46 L.

## TABLE 7.4
## EPA Recommended Units of Measurement

| Parameter | Unit Recommended |
|---|---|
| Particle fallout | Milligrams/square centimeter/month (or year): mg/cm²/month (or yr) |
| Outdoor airborne PM | Micrograms/cubic meter: μg/m³ |
| Gaseous materials | μg/m³ or parts per million (ppm) |
| Standard conditions for reporting gas volumes | 760 mmHg (1 atm) 25°C (STP) |
| Particle counting | Number of particles/m³ of gas |
| Temperature | °C |
| Pressure | mmHg (atm) |
| Sampling rate | m³/min or L/min |
| Visibility | Kilometers (km) |

*Source:* Air Pollution Training Institute, *Atmosphere Sampling,* (S1:435), Research Triangle Park, NC, U.S. Environmental Protection Agency, 1983.

STP, standard temperature and pressure.

## Example 7.1

1. Prepare a computer program relating the various elements in Equation 7.1 to convert ppm to μg/m³ or vice versa for gaseous air contaminants.
2. Using the computer program developed in (1), determine the concentration of $SO_2$ in μg/m³ when it has been reported to be 0.14 ppm by volume at STP.

## Solution

1. For solution to Example 7.1 (1), see the listing of Program 7.1.
2. Solution to Example 7.1 (2):

$$\frac{\mu g}{m^3} = \frac{ppm \times molecular\,weight \times 10^3}{24.46}$$

$$= \frac{0.14\,ppm \times 64.1\,g/mol \times 10^3}{24.46\,L/mol}$$

$$= 367\,\mu g/m^3$$

---

### LISTING OF PROGRAM 7.1   (CHAP7.1\FORM1.VB): CONCENTRATION OF GASEOUS AIR CONTAMINANTS

```
'********************************************************************************
'Example 7.1: Relating elements of equation 7-1:
' Converts PPM to micrograms/m{3}
'********************************************************************************
Public Class Form1
    Dim ppm, MolWgt, lMole, Ugmol, ugm As Double

    Private Sub Form1_Load(ByVal sender As System.Object, ByVal e As System.EventArgs)
Handles MyBase.Load
        Me.Text = "Exercise 7.1: converts PPM micrograms/m{3} and vice versa"
        Me.MaximizeBox = False
        Me.FormBorderStyle = Windows.Forms.FormBorderStyle.FixedSingle
        label1.text = "Convert measure to:"
        ListBox1.Items.Clear()
        ListBox1.Items.Add("ug/m{3} = (ppm * Molecular Weight * 10{3})/liter/mole")
        ListBox1.Items.Add("ppm = (ug/m{3} * liter/mole)/(Molecular Weight * 10{3})")
        ListBox1.SelectedIndex = 0
        Label2.AutoSize = False
        Label2.Height = 120
        Label2.Width = 433
        Label2.Text = "At 1 atm. And 0 degrees C (273)K, liters/mole=22.41 liters"
        Label2.Text += vbCrLf + "At 1 atm. And 25 degrees C (298K), liters/mole=24.46"
        Label2.Text += vbCrLf + "1 atm. And 25 degrees C are known as Standard Temperature
and Pressure"
        Label2.Text += vbCrLf + "u = micro"
        Label2.Text += vbCrLf + "a number in curly braces {} is a superscript number"
        Label2.Text += vbCrLf + " m{3} = cubic meters"
        Label6.Text = ""
        Button1.Text = "&Calculate"
    End Sub
```

```
Private Sub ListBox1_SelectedIndexChanged(ByVal sender As System.Object, ByVal e
As System.EventArgs) Handles ListBox1.SelectedIndexChanged
    Select Case ListBox1.SelectedIndex
      Case 0
        Label3.Text = "Enter the measurement in ppm "
        Label4.Text = "Enter the molecular weight of the gas "
        Label5.Text = "Enter the liters/mole "
      Case 1
        Label3.Text = "Enter the measure in ug/m{3} "
        Label4.Text = "Enter the liters/mole "
        Label5.Text = "Enter the molecular weight of the gas "
    End Select
End Sub


Sub calculateResults()
    Select Case ListBox1.SelectedIndex
      Case 0
        ppm = Val(TextBox1.Text)
        MolWgt = Val(TextBox2.Text)
        lMole = Val(TextBox3.Text)
        Ugmol = (ppm * MolWgt) / lMole * 1000
        Label6.Text = "micrograms/cubic meter = " + FormatNumber(Ugmol, 2)
      Case 1
        ugm = Val(TextBox1.Text)
        lMole = Val(TextBox2.Text)
        MolWgt = Val(TextBox3.Text)
        ppm = (ugm * lMole / MolWgt) * 0.001
        Label6.Text = "parts per million = " + FormatNumber(ppm, 2)
    End Select
End Sub


Private Sub Button1_Click(ByVal sender As System.Object, ByVal e As System.EventArgs)
Handles Button1.Click
    calculateResults()
  End Sub
End Class
```

## Example 7.2

Using the computer program developed in Example 7.1 (1), convert 9.0 ppm CO to $\mu g/m^3$ at 0°C (273 K), 1 atm, and at STP.

### Solution

At 0°C (273 K) and 1 atm:

$$\frac{\mu g}{m^3} = \frac{ppm \times \text{molecular weight of gas} \times 10^3}{22.41 \text{ L/mol}}$$

$$= \frac{9 \times 28 \times 10^3}{22.41}$$

$$= 11,245 \ \mu g/m^3$$

At STP, 25°C (298 K), and 1 atm:

$$\frac{\mu g}{m^3} = \frac{ppm \times \text{molecular weight of gas} \times 10^3}{24.46 \text{ L/mol}}$$

$$= \frac{9 \times 28 \times 10^3}{24.46}$$

$$= 10,302 \ \mu g/m^3$$

## 7.2.2 Mole and Mole Fraction

The mole is a practical, simple unit that has helped to make chemistry an exact and quantitative science. A mole of a substance is the molecular weight of the substance, expressed in mass units, where the molecular weight is the sum of the atomic weights of the atoms that compose the substance. For example, a molecule of hydrogen (two atoms, $H_2$) has a molecular weight of 2, and a molecule of oxygen (two atoms, $O_2$) has a molecular weight of 32. Water ($H_2O$) has a molecular weight of 18. The atomic weight expresses the ratio of the weight of one atom to that of another. Because the atomic weight is a relative weight, the numerical value must be determined by reference to some standard. In 1961, the carbon-12 atom was adopted as the atomic weight standard with a value of exactly 12. Tables are readily available giving atomic weight (see Appendix A3).

Avogadro in 1811 was the first to suggest that equal volumes of all gases at the same temperature and pressure have the same number of molecules. This was known as Avogadro's hypothesis and led to the discovery that one mole of gas, any gas, contained $6.02 \times 10^{23}$ particles or entities; this value is now called Avogadro's number. By definition, one mole of a substance contains $6.02 \times 10^{23}$ particles or constituents. It can be expressed as atoms per mole, molecules per mole, ions per mole, electrons per mole, or particles per mole:

$6.02 \times 10^{23}$ O atoms = 16.0 g O
$6.02 \times 10^{23}$ H atoms = 1.01 g H
$6.02 \times 10^{23}$ $H_2O$ molecules = 18.0 g $H_2O$
$6.02 \times 10^{23}$ $OH^-$ ions = 17.0 g $OH^-$

Also, for example, 1 mole of carbon-12 contains $6.02 \times 10^{23}$ carbon atoms. In mathematical terms, the number of moles of a gas can be expressed as

$$n = \frac{m}{MW} \qquad (7.2)$$

where:

$n$ is the number of moles of a gas
$m$ is the mass of the gas, g
MW is the molecular weight of the gas, g-mol

In a gaseous mixture, the mole fraction for each gas is the ratio of the moles of the given gas divided by the total number of moles of all gases present in the mixture, or

$$X = \frac{n}{\Sigma n_i} \qquad (7.3)$$

where:

$X$ is the mole fraction for each gas present in the gaseous mixture
$n$ is the moles of each gas present in the gaseous mixture
$\Sigma n_i = n_1 + n_2 + \cdots + n_i$ is the total number of moles present in the gaseous mixture

## Example 7.3

A gaseous mixture contains 4 g $O_2$, 10 g $N_2$, 1 g CO, and 5 g $CO_2$. Determine the moles present for each gas, as well as the mole fraction for each gas. The total number of

moles present and the mole fraction for each gas can be determined as follows:

| Component | Weight (g) | Molecular Weight | Moles | Mole Fraction |
|---|---|---|---|---|
| $O_2$ | 4.0 | 32 | 0.125 | 0.20 |
| $N_2$ | 10.0 | 28 | 0.357 | 0.56 |
| CO | 1.0 | 28 | 0.036 | 0.06 |
| $CO_2$ | 5.0 | 44 | 0.114 | 0.18 |
| Total | | | 0.632 | 1.00 |

The mole fraction has no dimensions and no units, and the sum of the mole fractions for a gas mixture equals unity.

## Example 7.4

1. Write a computer program that can be used to calculate the moles and mole fraction for each gas present in a gaseous mixture.
2. Using the computer program developed in (1), determine the moles and mole fraction for each of the gases present in the following mixture: 6.5 g methane ($CH_4$), 3 g carbon dioxide ($CO_2$), 0.1 g hydrogen sulfide ($H_2S$), 0.4 g nitrogen ($N_2$), and 0.1 g hydrogen ($H_2$). The total number of moles present and the mole fraction for each gas present can be determined as follows:

| Component | Weight (g) | Molecular Weight | Moles | Mole Fraction |
|---|---|---|---|---|
| $CH_4$ | 6.5 | 16 | 0.406 | 0.750 |
| $CO_2$ | 3.0 | 44 | 0.068 | 0.126 |
| $H_2S$ | 0.1 | 34.1 | 0.003 | 0.005 |
| $N_2$ | 0.4 | 28 | 0.014 | 0.026 |
| $H_2$ | 0.1 | 2 | 0.050 | 0.093 |
| Total | | | 0.541 | 1.000 |

---

### LISTING OF PROGRAM 7.4   (CHAP7.4\FORM1.VB): MOLES AND MOLE FRACTIONS IN A GASEOUS MIXTURE

```
'****************************************************************************
'Example 7.4: calculates moles and mole fractions
'for each gas present in a gaseous mixture
'****************************************************************************
Public Class Form1
    Dim Wgt(), MolWgt(), moles(), MolFrac() As Double
    Dim gas() As String
    Dim totmoles, totMolFrac As Double
```

```vbnet
  Private Sub Form1_Load(ByVal sender As System.Object, ByVal e As System.EventArgs)
Handles MyBase.Load
    Me.Text = "Exercise 7.4: calculates moles and mole fractions"
    Me.MaximizeBox = False

    Label1.Text = "Formulas:"
    ListBox1.Items.Clear()
    ListBox1.Items.Add("Moles = weight/molecular weight")
    ListBox1.Items.Add("Mole fraction = moles/total moles")
    Label2.Text = "Type in the names, weights, and molecular weights of the gases:"
    DataGridView1.Columns.Clear()
    DataGridView1.Columns.Add("nameCol", "Gas Name")
    DataGridView1.Columns.Add("wgtCol", "Gas Wgt")
    DataGridView1.Columns.Add("MWCol", "Gas MW")
    TextBox1.Multiline = True
    TextBox1.Height = 125
    TextBox1.ScrollBars = ScrollBars.Vertical
    TextBox1.Font = New Font(FontFamily.GenericMonospace, 9)
    Button1.Text = "&Calculate"
  End Sub

  Sub calculateResults()
    If DataGridView1.Rows.Count <= 1 Then
      MsgBox("Type in at least one gas!", vbOKOnly + vbExclamation, "Prompt")
      Exit Sub
    End If

    Dim count = DataGridView1.Rows.Count - 1

    ReDim gas(count), Wgt(count), MolWgt(count), moles(count), MolFrac(count)
    totmoles = 0
    totMolFrac = 0

    For i = 0 To count - 1
      gas(i) = DataGridView1.Rows(i).Cells("nameCol").Value
      Wgt(i) = Val(DataGridView1.Rows(i).Cells("wgtCol").Value)
      MolWgt(i) = Val(DataGridView1.Rows(i).Cells("MWCol").Value)
      moles(i) = Wgt(i) / MolWgt(i)
      Totmoles = totmoles + moles(i)
    Next

    For i = 0 To count - 1
      MolFrac(i) = moles(i) / totmoles
      totMolFrac = totMolFrac + MolFrac(i)
    Next

    Dim s As String
    s = "Gas Wgt MW Moles Mole Fraction"
    For i = 0 To count - 1
      s += vbCrLf + gas(i) + Space(8 - gas(i).Length)
      s += Format(Wgt(i), "00.000 ").ToString
      s += Format(MolWgt(i), "00.000 ").ToString
      s += Format(moles(i), "00.000 ").ToString
      s += Format(MolFrac(i), "00.000").ToString
    Next
    s += vbCrLf + Space(24) + "------"
    s += Space(2) + "------"
    s += vbCrLf + Space(24) + Format(totmoles, "00.000").ToString
    s += Space(2) + Format(totMolFrac, "00.000").ToString
    TextBox1.Text = s
  End Sub
```

```
    Private Sub Button1_Click(ByVal sender As System.Object, ByVal e As System.EventArgs)
Handles Button1.Click
        calculateResults()
    End Sub
End Class
```

### 7.2.3 Basic Gas Laws

#### 7.2.3.1 Boyle's Law

Boyle's law states that the volume of a given quantity of an ideal gas varies inversely to its absolute pressure, with the temperature being held constant. A practical and easy way to remember Boyle's law is provided by the following equation:

$$P_1V_1 = P_2V_2 \qquad (7.4)$$

where:

$P_1$ is the initial gas pressure, atm
$V_1$ is the volume of gas at $P_1$, L
$P_2$ is the final gas pressure, atm
$V_2$ is the volume of gas at $P_2$, L

**Example 7.5**

1. Prepare a computer program relating the four elements in Boyle's law (Equation 7.4).

2. Using the computer program developed in (1), determine the volume of a gas held in a 100-L gas cylinder if the initial pressure was 35 atm and the final pressure was 10 atm, with the temperature being held constant.

**Solution**

1. For solution to Example 7.5 (1), see the listing of Program 7.5.
2. Solution to Example 7.5 (2):

$$P_1V_1 = P_2V_2$$

$$35*100 = 10*V_2$$

$$V_2 = \frac{35 \times 100}{10} = 350 \text{ L}$$

---

**LISTING OF PROGRAM 7.5   (CHAP7.5\FORM1.VB): BOYLE'S LAW**

```
'****************************************************************************************
'Example 7.5: Relates the four elements in Boyle's law
'****************************************************************************************
'Equation is "P1*V1 = P2*V2"
Public Class Form1
  Dim result As Double

  Private Sub Form1_Load(ByVal sender As System.Object, ByVal e As System.EventArgs)
Handles MyBase.Load
      Me.Text = "Exercise 7.5: Boyle's law"
      Me.MaximizeBox = False
      Me.FormBorderStyle = Windows.Forms.FormBorderStyle.FixedSingle
      Label1.Text = "P1*V1 = P2*V2"
      Label2.AutoSize = False
      Label2.Height = 120
      Label2.Width = 433
      Label2.Text = "P1 = initial gas pressure, atm."
      Label2.Text += vbCrLf + "V1 = volume of gas at P1, liters"
      Label2.Text += vbCrLf + "P2 = final gas pressure, atm."
      Label2.Text += vbCrLf + "V2 = volume of gas at P2, liters"
      Label3.Text = "Enter the initial gas pressure, (P1) (in atm.)"
      Label4.Text = "Enter the volume of gas at P1, (V1) (in liters)"
      Label5.Text = "Enter the final gas pressure, (P2) (in atm.)"
      Label6.Text = ""
      Label7.Text = "Solving for:"
      Label8.Text = "Decimal points:"
      Button1.Text = "&Calculate"
```

```
      ComboBox1.Items.Clear()
      ComboBox1.Items.Add("P1")
      ComboBox1.Items.Add("V1")
      ComboBox1.Items.Add("P2")
      ComboBox1.Items.Add("V2")
      ComboBox1.SelectedIndex = 3
      NumericUpDown1.Value = 2
      NumericUpDown1.Maximum = 10
      NumericUpDown1.Minimum = 0
  End Sub

  Sub calculateResults()
    Dim V1, V2, P1, P2 As Double
    Select Case ComboBox1.SelectedIndex
      Case 0
        'Solving for P1
        V1 = Val(TextBox1.Text)
        P2 = Val(TextBox2.Text)
        V2 = Val(TextBox3.Text)
        P1 = P2 * V2 / V1
        result = P1
      Case 1
        'Solving for V1
        P1 = Val(TextBox1.Text)
        P2 = Val(TextBox2.Text)
        V2 = Val(TextBox3.Text)
        V1 = P2 * V2 / P1
        result = V1
      Case 2
        'Solving for P2
        P1 = Val(TextBox1.Text)
        V1 = Val(TextBox2.Text)
        V2 = Val(TextBox3.Text)
        P2 = P1 * V1 / V2
        result = P2
      Case 3
        'Solving for V2
        P1 = Val(TextBox1.Text)
        V1 = Val(TextBox2.Text)
        P2 = Val(TextBox3.Text)
        V2 = P1 * V1 / P2
        result = V2
    End Select
  End Sub

  Sub showResult()
    Select Case ComboBox1.SelectedIndex
      Case 0
        Label6.Text = "Initial gas pressure, P1 = " + FormatNumber(result,
NumericUpDown1.Value) + " atm."
      Case 1
        Label6.Text = "Volume of gas at P1, (V1) = " + FormatNumber(result,
NumericUpDown1.Value) + " litre(s)"
      Case 2
        Label6.Text = "Final gas pressure, P2 = " + FormatNumber(result, NumericUpDown1.
Value) + " atm."
      Case 3
        Label6.Text = "Volume of gas at P2, (V2) = " + FormatNumber(result,
NumericUpDown1.Value) + " litre(s)"
    End Select
  End Sub
```

```
    Private Sub Button1_Click(ByVal sender As System.Object, ByVal e As System.EventArgs)
Handles Button1.Click
        calculateResults()
        showResult()
    End Sub

    Private Sub ComboBox1_SelectedIndexChanged(ByVal sender As System.Object, ByVal e As
System.EventArgs) Handles ComboBox1.SelectedIndexChanged
        Select Case ComboBox1.SelectedIndex
            Case 0
                'Solving for P1
                Label3.Text = "Enter the volume of gas at P1, (V1) (in liters)"
                Label4.Text = "Enter the final gas pressure, (P2) (in atm.)"
                Label5.Text = "Enter the volume of gas at P2, (V2) (in liters)"
            Case 1
                'Solving for V1
                Label3.Text = "Enter the initial gas pressure, (P1) (in atm.)"
                Label4.Text = "Enter the final gas pressure, (P2) (in atm.)"
                Label5.Text = "Enter the volume of gas at P2, (V2) (in liters)"
            Case 2
                'Solving for P2
                Label3.Text = "Enter the initial gas pressure, (P1) (in atm.)"
                Label4.Text = "Enter the volume of gas at P1, (V1) (in liters)"
                Label5.Text = "Enter the volume of gas at P2, (V2) (in liters)"
            Case 3
                'Solving for V2
                Label3.Text = "Enter the initial gas pressure, (P1) (in atm.)"
                Label4.Text = "Enter the volume of gas at P1, (V1) (in liters)"
                Label5.Text = "Enter the final gas pressure, (P2) (in atm.)"
        End Select
    End Sub

    Private Sub NumericUpDown1_ValueChanged(ByVal sender As System.Object, ByVal e As
System.EventArgs) Handles NumericUpDown1.ValueChanged
        showResult()
    End Sub
End Class
```

### 7.2.3.2 Charles's Law

Charles' law, also known as Gay-Lussac's law, states that the volume of a given mass of an ideal gas varies directly as the absolute temperature with the pressure being held constant.

$$\frac{V}{T} = \text{constant} \tag{7.5}$$

Combining the laws of Boyle and Charles into one expression gives the following equation 7.6:

$$\frac{P_1 V_1}{T_1} = \frac{P_2 V_2}{T_2} \tag{7.6}$$

where:
$P_1$ is the initial gas pressure, atm
$V_1$ is the volume of gas at $P_1$, L
$T_1$ is the initial gas temperature, K
$P_2$ is the final gas pressure, atm

$V_2$ is the volume of gas at $P_2$, L
$T_2$ is the final gas temperature, K

### Example 7.6

1. Prepare a computer program relating the various elements in Equation 7.6, which combines Boyle's and Charles' laws.
2. One very important use of the combined Boyle's and Charles' law (Equation 7.6) is to compare an actual gas volume to its volume under a set of standard conditions. For most applications in regard to air pollution, the standard conditions are 25°C (298 K) and 1 atm (760 mmHg). The term "standard conditions for temperature and pressure" is abbreviated STP. This principal can also be extended so as to compare the actual volumetric flow rates to the volumetric flow rates under standard conditions.

## LISTING OF PROGRAM 7.6   (CHAP7.6\FORM1.VB): BOYLE'S AND CHARLES' LAWS

```vb
'********************************************************************************
'Example 7.6: Combined Charles' and Boyle's laws
'********************************************************************************
'Equation is "P1*V1/T1 = P2*V2/T2"
Public Class Form1
  Dim result As Double

  Private Sub Form1_Load(ByVal sender As System.Object, ByVal e As System.EventArgs)
Handles MyBase.Load
    Me.Text = "Exercise 7.6: Combined Charles' and Boyle's laws"
    Me.MaximizeBox = False
    Label1.Text = "(P1*V1)/T1 = (P2*V2)/T2"
    Label2.AutoSize = False
    Label2.Height = 120
    Label2.Width = 433
    Label2.Text = "P1 = initial gas pressure, atm."
    Label2.Text += vbCrLf + "V1 = volume of gas at P1, liters"
    Label2.Text += vbCrLf + "T1 = initial gas temperature, C"
    Label2.Text += vbCrLf + "P2 = final gas pressure, atm."
    Label2.Text += vbCrLf + "V2 = volume of gas at P2, liters"
    Label2.Text += vbCrLf + "T2 = final gas temperature, C"
    Label3.Text = "Enter the final gas pressure, (P2) (in atm.)"
    Label4.Text = "Enter the volume of gas at P2, (V2) (in m{3}/min)"
    Label5.Text = "Enter the initial gas temperature, (T1) (in C)"
    Label6.Text = "Enter the initial gas pressure, (P1) (in atm.)"
    Label7.Text = "Enter the final gas temperature, (T2) (in C)"
    Label8.Text = ""
    Label9.Text = "Solving for:"
    Label10.Text = "Decimal points:"
    Button1.Text = "&Calculate"
    ComboBox1.Items.Clear()
    ComboBox1.Items.Add("P1")
    ComboBox1.Items.Add("V1")
    ComboBox1.Items.Add("T1")
    ComboBox1.Items.Add("P2")
    ComboBox1.Items.Add("V2")
    ComboBox1.Items.Add("T2")
    ComboBox1.SelectedIndex = 4
    NumericUpDown1.Value = 2
    NumericUpDown1.Maximum = 10
    NumericUpDown1.Minimum = 0
  End Sub

  Sub calculateResults()
    Dim V1, V2, P1, P2, T1, T2 As Double
    Select ComboBox1.SelectedIndex
      Case 0
        'Solving for P1
        V1 = Val(TextBox1.Text)
        T1 = Val(TextBox2.Text) + 273
        P2 = Val(TextBox3.Text)
        V2 = Val(TextBox4.Text)
        T2 = Val(TextBox5.Text) + 273
        P1 = (P2 * V2 * T1) / (V1 * T2)
        result = P1
```

```vbnet
        Case 1
          'Solving for V1
          P1 = Val(TextBox1.Text)
          T1 = Val(TextBox2.Text) + 273
          P2 = Val(TextBox3.Text)
          V2 = Val(TextBox4.Text)
          T2 = Val(TextBox5.Text) + 273
          V1 = (P2 * V2 * T1) / (P1 * T2)
          result = V1
        Case 2
          'Solving for T1
          P1 = Val(TextBox1.Text)
          V1 = Val(TextBox2.Text)
          P2 = Val(TextBox3.Text)
          V2 = Val(TextBox4.Text)
          T2 = Val(TextBox5.Text) + 273
          T1 = (P1 * V1 * T2) / (P2 * V2)
          result = T1
        Case 3
          'Solving for P2
          P1 = Val(TextBox1.Text)
          V1 = Val(TextBox2.Text)
          T1 = Val(TextBox3.Text) + 273
          V2 = Val(TextBox4.Text)
          T2 = Val(TextBox5.Text) + 273
          P2 = (P1 * V1 * T2) / (V2 * T1)
          result = P2
        Case 4
          'Solving for V2
          P1 = Val(TextBox1.Text)
          V1 = Val(TextBox2.Text)
          T1 = Val(TextBox3.Text) + 273
          P2 = Val(TextBox4.Text)
          T2 = Val(TextBox5.Text) + 273
          V2 = (P1 * V1 * T2) / (P2 * T1)
          result = V2
        Case 5
          'Solving for T2
          P1 = Val(TextBox1.Text)
          V1 = Val(TextBox2.Text)
          T1 = Val(TextBox3.Text) + 273
          P2 = Val(TextBox4.Text)
          V2 = Val(TextBox5.Text)
          T2 = (P2 * V2 * T1) / (P1 * V1)
          result = T2
      End Select
   End Sub

   Sub showResult()
      Select Case ComboBox1.SelectedIndex
        Case 0
          Label8.Text = "Initial gas pressure, P1 = " + FormatNumber(result,
NumericUpDown1.Value) + " atm."
        Case 1
          Label8.Text = "Volume of gas at P1, (V1) = " + FormatNumber(result,
NumericUpDown1.Value) + " litre(s)"
        Case 2
          Label8.Text = "Initial gas temperature, T1 = " + FormatNumber(result,
NumericUpDown1.Value) + " K"
```

```
      Case 3
         Label8.Text = "Final gas pressure, P2 = " + FormatNumber(result, NumericUpDown1.
Value) + " atm."
      Case 4
         Label8.Text = "Volume of gas at P2, (V2) = " + FormatNumber(result,
NumericUpDown1.Value) + " litre(s)"
      Case 5
         Label8.Text = "Final gas temperature, T2 = " + FormatNumber(result,
NumericUpDown1.Value) + " K"
    End Select
  End Sub

  Private Sub Button1_Click(ByVal sender As System.Object, ByVal e As System.EventArgs)
Handles Button1.Click
    calculateResults()
    showResult()
  End Sub

  Private Sub ComboBox1_SelectedIndexChanged(ByVal sender As System.Object, ByVal e As
System.EventArgs) Handles ComboBox1.SelectedIndexChanged
    Select Case ComboBox1.SelectedIndex
      Case 0
        'Solving for P1
        Label3.Text = "Enter the volume of gas at P1, (V1) (in liters)"
        Label4.Text = "Enter the initial gas temperature, T1 (in C)"
        Label5.Text = "Enter the final gas pressure, (P2) (in atm.)"
        Label6.Text = "Enter the volume of gas at P2, (V2) (in liters)"
        Label7.Text = "Enter the final gas temperature, T2 (in C)"
      Case 1
        'Solving for V1
        Label3.Text = "Enter the initial gas pressure, (P1) (in atm.)"
        Label4.Text = "Enter the initial gas temperature, T1 (in C)"
        Label5.Text = "Enter the final gas pressure, (P2) (in atm.)"
        Label6.Text = "Enter the volume of gas at P2, (V2) (in liters)"
        Label7.Text = "Enter the final gas temperature, T2 (in C)"
      Case 2
        'Solving for T1
        Label3.Text = "Enter the initial gas pressure, (P1) (in atm.)"
        Label4.Text = "Enter the volume of gas at P1, (V1) (in liters)"
        Label5.Text = "Enter the final gas pressure, (P2) (in atm.)"
        Label6.Text = "Enter the volume of gas at P2, (V2) (in liters)"
        Label7.Text = "Enter the final gas temperature, T2 (in C)"
      Case 3
        'Solving for P2
        Label3.Text = "Enter the initial gas pressure, (P1) (in atm.)"
        Label4.Text = "Enter the volume of gas at P1, (V1) (in liters)"
        Label5.Text = "Enter the initial gas temperature, T1 (in C)"
        Label6.Text = "Enter the volume of gas at P2, (V2) (in liters)"
        Label7.Text = "Enter the final gas temperature, T2 (in C)"
      Case 4
        'Solving for V2
        Label3.Text = "Enter the initial gas pressure, (P1) (in atm.)"
        Label4.Text = "Enter the volume of gas at P1, (V1) (in liters)"
        Label5.Text = "Enter the initial gas temperature, T1 (in C)"
        Label6.Text = "Enter the final gas pressure, (P2) (in atm.)"
        Label7.Text = "Enter the final gas temperature, T2 (in C)"
      Case 5
        'Solving for T2
        Label3.Text = "Enter the initial gas pressure, (P1) (in atm.)"
        Label4.Text = "Enter the volume of gas at P1, (V1) (in liters)"
        Label5.Text = "Enter the initial gas temperature, T1 (in C)"
```

```
        Label6.Text = "Enter the final gas pressure, (P2) (in atm.)"
        Label7.Text = "Enter the volume of gas at P2, (V2) (in liters)"
    End Select
End Sub

Private Sub NumericUpDown1_ValueChanged(ByVal sender As System.Object, ByVal e As
System.EventArgs) Handles NumericUpDown1.ValueChanged
    showResult()
End Sub
End Class
```

## Example 7.7

Using the computer program developed in Example 7.6 section (1), determine the standard volumetric flow rate at STP for the following conditions:

Actual volumetric flow rate = 6 m³/min ($V_2$)
Actual operating temperature = 90°C ($T_2$)
Actual operating pressure = 1 atm ($P_2$)
STP = 25°C ($T_1$) and 1 atm ($P_1$)

### Solution

Substituting in Equation 7.6:

$$\frac{P_1 V_1}{T_1} = \frac{P_2 V_2}{T_2}$$

$$V_1 = \frac{P_2 V_2 T_1}{P_1 T_2} = \frac{(1 \text{ atm})(6 \text{m}^3/\text{min})(273 + 25\text{K})}{(1 \text{ atm})(273 + 90\text{K})}$$

$$= \frac{(1)(6)(298)}{(1)(363)} = 4.92 \text{m}^3/\text{min}$$

### 7.2.3.3  Ideal Gas Law

Ideal gases that obey Boyle's and Charles' laws, as well as Avogadro's hypothesis (see Section 7.2.2), comply with the ideal gas equation:

$$PV = nRT \qquad (7.7)$$

where:
  $P$ is the absolute pressure, atm
  $V$ is the volume of a gas, L
  $T$ is the absolute temperature, K
  $n$ is the number of moles of a gas
  $R$ is the ideal or universal constant, (L) atm (K)$^{-1}$ (g-mol)$^{-1}$

Typical values of $R$ are as follows:

$R$ = 0.08206 (L) atm (K)$^{-1}$ (g-mol)$^{-1}$
$R$ = 62.4 (1) (mmHg) (K)$^{-1}$ (g-mol)$^{-1}$
$R$ = 1.986 cal/g-mol-K

$R$ is chosen in order to provide appropriate units consistent with the equation used.

As indicated earlier, the number of moles of a gas, $n$, can be determined if the mass and molecular weight of a gas are known:

$$n = \frac{m}{\text{MW}} \qquad (7.2)$$

where:
  $n$ is the number of moles of gas
  $m$ is the mass of gas, g
  MW is the molecular weight of gas, g-mol

Also, the density of a gas can be determined by rearranging Equation 7.7 as follows:

$$\rho = \frac{m}{V} \qquad (7.8)$$

where:
  $\rho$ is the gas density, g/L
  $m$ is the mass of gas, g
  $V$ is the volume of gas, L

Substituting Equation 7.2 into Equation 7.7 gives

$$PV = \frac{mRT}{\text{MW}} \qquad (7.9)$$

When the above equation is rearranged,

$$\frac{m}{V} = \frac{P*\text{MW}}{RT} \qquad (7.10)$$

Substituting Equation 7.8 into the above equation gives

$$\rho_a = \frac{P*\text{MW}}{RT} \qquad (7.11)$$

The units for this equation are the same as presented earlier.

No real gas obeys the ideal gas law exactly, although the "lighter" gases (hydrogen, oxygen, air, etc.) under ambient conditions approach ideal gas law behavior. The "heavier" gases, such as sulfur dioxide and hydrocarbons, particularly at high pressures and low temperatures, deviate considerably from the ideal gas law. Despite these deviations, the ideal gas law is routinely used in air pollution calculations (Air Pollution Training Institute 1981, 1984).

## Example 7.8

1. Prepare computer programs for each of the equations dealing with the ideal gas law (Equations 7.7 through 7.10).
2. Using the computer program developed in (1), determine the volume occupied by 1 mole of an ideal gas at 0°C (273 K) and 1 atm pressure.
3. Using the computer program developed for Equation 7.9, calculate the pressure exerted by 38.0 g of carbon monoxide in a volume of 25 L at 500 K; assume that the ideal gas law applies.
4. Using the computer program developed for Equation 7.9, determine the molecular weight of a gas if 0.6 g of the gas occupies 400 mL at 25°C and 760 mmHg (1 atm) (standard conditions for temperature and pressure).
5. Using the computer program developed for Equation 7.10, determine the density of air if it has a molecular weight of 29, and the air temperature is 93°C at 1 atm of pressure.

## Solution

1. For solution to Example 7.8 (1), see the listing of Program 7.8.
2. Solution to Example 7.8 (2):

$$PV = nRT \qquad (7.7)$$

where:

$P = 1$, atm
$n = 1$, g-mol
$R = 0.08206$ (L), atm (K)$^{-1}$ (g-mol)$^{-1}$
$T = 0°C$ (273 K)
$V$ is the molar volume, L

$$(1\ atm)V = \frac{1\text{g-mol}(0.082061\text{*atm})(273\text{K})}{\text{g-mol-K}}$$

where $V = 22.4$ L.

Thus, 1 mole of an ideal gas at 273 K and 1 atm occupies 22.4 L.

3. Solution to Example 7.8 (3):

$$PV = \frac{mRT}{MW} \qquad (7.9)$$

$$P(25L) = \frac{38}{28}\left(\frac{(0.08206(1)\text{atm})}{\text{g-mol-K}}\right)(500\text{K})$$

$P = 2.22$ atm $= 1693$ mmHg (1 atm $= 760$ mmHg at 273 K)

$$PV = \frac{mRT}{MW} \qquad (7.9)$$

4. Solution to Example 7.8 (4):

$$1\text{atm}(0.41) = \frac{0.6}{MW} \frac{[0.08206\ 1(\text{atm})](273 + 25\text{K})}{\text{g-mol}}$$

which yields
MW = 36 g/g-mol

5. Solution to Example 7.8 (5):

$$\rho_a = \frac{P*MW}{RT} = \frac{1\text{atm}\ (29\ \text{g})}{0.08206\ (\text{L})\text{atm(K)}^{-1}(\text{g-mol})(273 + 93)\text{K}} \qquad (7.12)$$

$= 0.966$ g/L or 0.000966 g/cm$^3$

---

### LISTING OF PROGRAM 7.8    (CHAP7.8\FORM1.VB): IDEAL GAS LAW

```
'*********************************************************************************
'Example 7.8: Ideal gas law
'*********************************************************************************
'Equations 7.7, 7.9, and 7.10
Public Class Form1
  Dim eq(2) As String

  Private Sub Form1_Load(ByVal sender As System.Object, ByVal e As System.EventArgs)
Handles MyBase.Load
    Me.Text = "Exercise 7.8: Ideal gas law"
    Me.MaximizeBox = False
    Me.FormBorderStyle = Windows.Forms.FormBorderStyle.FixedSingle
    eq(0) = "P * V = n * R * T"
    eq(1) = "P * V = (m/MW) * R * T"
    eq(2) = "r(rho) = (P*MW)/(R * T)"
    Label1.Text = "Select option:"
    ListBox1.Items.Clear()
    ListBox1.Items.Add(eq(0))
    ListBox1.Items.Add(eq(1))
    ListBox1.Items.Add(eq(2))
    ListBox1.SelectedIndex = 0
    Label2.AutoSize = False
```

```
      Label2.Height = 120
      Label2.Width = 433
      Label2.Text = "P = 1 atmosphere"
      Label2.Text += vbCrLf + "V = molar volume, L"
      Label2.Text += vbCrLf + "n = 1 g-mol"
      Label2.Text += vbCrLf + "R = 0.08206 (L) atm, (K){-1}, (g-mol){-1} (a constant)"
      Label2.Text += vbCrLf + "T = 0 degrees C (K = 273 + degrees C)"
      Label2.Text += vbCrLf + "m = weight of gas, g"
      Label2.Text += vbCrLf + "MW = Molecular weight of gas"
      Label2.Text += vbCrLf + "r(rho) = density of a gas (in g/L or g/cm*10{3})"
      Label6.Text = ""
      Button1.Text = "&Calculate"
      ComboBox1.Visible = False
   End Sub

   Private Sub ListBox1_SelectedIndexChanged(ByVal sender As System.Object, ByVal e As
System.EventArgs) Handles ListBox1.SelectedIndexChanged
      Select Case ListBox1.SelectedIndex
        Case 0
          Label7.Text = "Equation 7.7: " + eq(0)
          Label7.Text += vbCrLf + "(1 atm.) V = n * R * T Solving for V"
          Label7.Text += vbCrLf + "1 * V = 1 (g-mol) * 0.08206 (1 atm/g-mol K)*273 (K)"
          Label7.Text += vbCrLf + "V = 1 * 0.08206 * 273"
          Label7.Text += vbCrLf + "V = 22.4 L"
          Label7.Text += vbCrLf + "1 mole of an ideal gas at 273 K and 1 atm occupies 22.4 L"
          ComboBox1.Visible = False
        Case 1
          Label7.Text = "Equation 7.9: " + eq(1)
          Label7.Text += vbCrLf + "P * V = (m/MW) * R * T Solving for (Select from the
right):"
          ComboBox1.Items.Clear()
          ComboBox1.Items.Add("Pressure, P")
          ComboBox1.Items.Add("Volume, V")
          ComboBox1.Items.Add("Mass, m")
          ComboBox1.Items.Add("Mol. Wt, MW")
          ComboBox1.SelectedIndex = 0
          ComboBox1.Visible = True
        Case 2
          Label7.Text = "Equation 7.10: " + eq(2)
          Label7.Text += vbCrLf + "r(rho) = P * (MW)/R * T Solving for r"
          Label3.Text = "Enter the number of atmospheres (P) (in atm)"
          Label4.Text = "Enter the molecular weight of the gas (MW)"
          Label5.Text = "Enter the gas temperature, T (in K)"
          ComboBox1.Visible = False
      End Select
      Label8.Text = ""
      If ListBox1.SelectedIndex = 0 Then
        Label3.Visible = False
        Label4.Visible = False
        Label5.Visible = False
        Label6.Visible = False
        TextBox1.Visible = False
        TextBox2.Visible = False
        TextBox3.Visible = False
        TextBox4.Visible = False
        Button1.Visible = False
      Else
        Label3.Visible = True
        Label4.Visible = True
        Label5.Visible = True
```

```
         If ListBox1.SelectedIndex = 1 Then
           Label6.Visible = True
           TextBox4.Visible = True
         Else
           Label6.Visible = False
           TextBox4.Visible = False
         End If
         TextBox1.Visible = True
         TextBox2.Visible = True
         TextBox3.Visible = True
         Button1.Visible = True
     End If
End Sub

Sub calculateResults()
   Dim P, V, m, MW, T, Rho As Double
   Const R As Double = 0.08206
   Select Case ListBox1.SelectedIndex
     Case 0
        Exit Sub
     Case 1
        'Equation 7.9: PV = mRT/MW
        Select Case ComboBox1.SelectedIndex
          Case 0
             'Solving for P
             V = Val(TextBox1.Text)
             m = Val(TextBox2.Text)
             MW = Val(TextBox3.Text)
             T = Val(TextBox4.Text)
             P = (m * R * T) / (MW * V)
             Label8.Text = "Gas pressure = " + FormatNumber(P, 2) + " atm"
          Case 1
             'Solving for V
             P = Val(TextBox1.Text)
             m = Val(TextBox2.Text)
             MW = Val(TextBox3.Text)
             T = Val(TextBox4.Text)
             V = (m * R * T) / (MW * P)
             Label8.Text = "Volume of gas = " + FormatNumber(V, 2) + " L"
          Case 2
             'Solving for m
             P = Val(TextBox1.Text)
             V = Val(TextBox2.Text)
             MW = Val(TextBox3.Text)
             T = Val(TextBox4.Text)
             m = (P * V * MW) / (R * T)
             Label8.Text = "Weight of gas = " + FormatNumber(m, 2) + " g"
          Case 3
             'Solving for MW
             P = Val(TextBox1.Text)
             V = Val(TextBox2.Text)
             m = Val(TextBox3.Text)
             T = Val(TextBox4.Text)
             MW = (m * R * T) / (P * V)
             Label8.Text = "Molecular weight of gas = " + FormatNumber(MW, 2) + " g/g-mol"
        End Select
     Case 2
        P = Val(TextBox1.Text)
        MW = Val(TextBox2.Text)
        T = Val(TextBox3.Text)
```

```
            Rho = (P * MW) / (R * T)
            Label8.Text = "Rho = " + FormatNumber(Rho, 2) + " g/L"
        End Select
    End Sub

    Private Sub Button1_Click(ByVal sender As System.Object, ByVal e As System.EventArgs)
Handles Button1.Click
        calculateResults()
    End Sub

    Private Sub ComboBox1_SelectedIndexChanged(ByVal sender As System.Object, ByVal e As
System.EventArgs) Handles ComboBox1.SelectedIndexChanged
        Select Case ComboBox1.SelectedIndex
          Case 0
            Label3.Text = "Enter the volume of the gas (V) (in L)"
            Label4.Text = "Enter the weight of the gas (m) (in g)"
            Label5.Text = "Enter the molecular weight of the gas (MW)"
            Label6.Text = "Enter the gas temperature, (T) (in K)"
          Case 1
            Label3.Text = "Enter the pressure (P) (in atm)"
            Label4.Text = "Enter the weight of the gas (m) (in g)"
            Label5.Text = "Enter gas's molecular weight (MW)"
            Label6.Text = "Enter the gas temperature, (T) (in K)"
          Case 2
            Label3.Text = "Enter the pressure (P) (in atm)"
            Label4.Text = "Enter the volume of the gas (V) (in L)"
            Label5.Text = "Enter the molecular weight of the gas (MW)"
            Label6.Text = "Enter the gas temperature, (T) (in K)"
          Case 3
            Label3.Text = "Enter the pressure (P) (in atm)"
            Label4.Text = "Enter the volume of the gas (V) (in L)"
            Label5.Text = "Enter the weight of the gas (m) (in g)"
            Label6.Text = "Enter the gas temperature, (T) (in K)"
        End Select
        Label8.Text = ""
    End Sub
End Class
```

### 7.2.4 Van der Waal's Equation

As indicated, the ideal gas law is not strictly applicable to real gases. Van der Waal's equation contains two discreet constants, which often provide a better representation of the actual volumetric behavior of a gas:

$$\left(P + \frac{n^2 a}{V^2}\right)(V - nb) = nRT \qquad (7.13)$$

where:

$P$ is the absolute pressure, atm

$V$ is the volume of a gas, L

$n$ is the number of moles of gas

$R$ is the ideal or universal gas constant, 0.08206 (L) atm $(K)^{-1}$ $(g\text{-mol})^{-1}$

$T$ is the absolute temperature, K

$a$ and $b$ are the constants determined experimentally for each gas; these constants are available in most handbooks of chemistry (see Table 7.5)

$a = L^2\text{-atm/mol}^2$

$b = L/mol$

**TABLE 7.5**
**Van der Waal's Constants for Some Gases**

| Gas | $a$ [L²-atm/(mol)²] | $b$ (L/mol) |
|-----|---------------------|-------------|
| $O_2$ | 1.382 | 0.03186 |
| $CO_2$ | 3.658 | 0.04286 |
| $H_2$ | 0.2453 | 0.02651 |
| $CH_4$ | 2.300 | 0.04301 |
| $N_2$ | 1.370 | 0.0387 |
| NO | 1.46 | 0.0289 |

*Source:* From Lide, D.R. and Frederikse, H.P.R., *CRC Handbook of Chemistry and Physics,* 76th ed., CRC Press, Boca Raton, FL, 1995/1996, pp. 6–48. With permission.

## Example 7.9

1. Write a computer program that relates the various elements in Van der Waal's equation (Equation 7.12).
2. Using the computer program developed in section (1), determine the pressure developed by 660 g of $CO_2$ put into an 80-L container under the conditions stated in the solutions section. Also compare the results if the ideal gas law is used.

## Solution

1. For solution to Example 7.9 (1), see the listing of Program 7.9.
2. Solution to Example 7.9 (2):

$$V = 80.01$$

$$T = 130°C$$

$$a = 3.658 \frac{L^2 - atm}{mol^2}$$

$$b = \frac{0.04286\ L}{mol}$$

Using Equation 7.2, $n = 660/44 = 15$ mol.

$$P = \frac{nRT}{V - nb} - \frac{n^2 a}{V^2}$$

$$= \frac{15(0.08206)(403)}{80 - 15(0.04286)} - \frac{[15^2(3.658)]}{(80)^2} \qquad (7.14)$$

$$= \frac{496.0527}{79.3571} - 0.1286$$

$$= 6.122\ atm$$

Using the ideal gas law,

$$PV = nRT$$

$$P = \frac{nRT}{V} = \frac{15(0.08206)(403)}{80} = 6.20\ atm \qquad (7.7)$$

Van der Waal's equation gives 6.12 atm, and the ideal gas law gives 6.20 atm. The ideal gas law gives a pressure of 0.08 atm higher than Van der Waal's equation.

---

### LISTING OF PROGRAM 7.9 (CHAP7.9\FORM1.VB): VAN DER WAAL'S EQUATION

```
'********************************************************************************
'Example 7.9: Van der Waal's equation (7.11)
'********************************************************************************
Public Class Form1
    Dim P1, P2, V, T, m, n, MW, a, b As Double
    Const R As Double = 0.08206
    Dim eq(2) As String


'********************************************************************************
'See Table 7.5 for a and b values
'********************************************************************************

    Dim a_table() As Double = {1.382, 3.658, 0.2453, 2.3, 1.37, 1.46}
    Dim b_table() As Double = {0.03186, 0.04286, 0.02651, 0.04301, 0.0387, 0.0289}

    Private Sub Form1_Load(ByVal sender As System.Object, ByVal e As System.EventArgs)
Handles MyBase.Load
        Me.Text = "Exercise 7.9: Van der Waal's equation"
        Me.MaximizeBox = False
        Me.FormBorderStyle = Windows.Forms.FormBorderStyle.FixedSingle
        Label1.Text = "Formula: P = ((n * R * T)/V - (n * b)) - (n{2} * a)/V{2}"
        Label2.AutoSize = False
        Label2.Height = 120
        Label2.Width = 433
        Label2.Text = "P = absolute pressure, atm"
        Label2.Text += vbCrLf + "V = volume of gas, L"
        Label2.Text += vbCrLf + "n = number of moles of gas"
        Label2.Text += vbCrLf + "R = 0.08206 (L) atm, (K){-1}, (g-mol){-1} (a constant)"
        Label2.Text += vbCrLf + "T = absolute temperature, K"
        Label2.Text += vbCrLf + "a,b = constants determined for each gas"
        Label2.Text += vbCrLf + "a = (L{2} - atm)/mol{2}, "
        Label2.Text += "b = L/mol"
```

```vbnet
      Label2.Text += vbCrLf + "Equation 7.11: "
      Label2.Text += "By Van der Waal's Equation"
      Label2.Text += vbCrLf + "P = (n*R*T)/(V-(n*b) - (b{2}*a)/V{2}"
      Label3.Text = "Enter the volume of gas, (V) (in liters)"
      Label4.Text = "Enter the temperature of the gas, (T), (C)"
      Label5.Text = "Enter the weight of the gas, g"
      Label6.Text = "Select the gas:"
      Label7.Text = ""
      Label10.Text = "Decimals:"
      Button1.Text = "&Calculate"
      NumericUpDown1.Value = 2
      NumericUpDown1.Maximum = 10
      NumericUpDown1.Minimum = 0
      ComboBox1.Items.Clear()
      ComboBox1.Items.Add("O2")
      ComboBox1.Items.Add("CO2")
      ComboBox1.Items.Add("H2")
      ComboBox1.Items.Add("CH4")
      ComboBox1.Items.Add("N2")
      ComboBox1.Items.Add("NO")
      ComboBox1.SelectedIndex = 0
   End Sub

   Sub calculateResults()
     V = Val(TextBox1.Text)
     T = Val(TextBox2.Text) + 273
     m = Val(TextBox3.Text)
     Dim index As Integer = ComboBox1.SelectedIndex
     If index = -1 Then
       MsgBox("Please select a gas from the list.",
         vbOKOnly Or vbInformation)
       Exit Sub
     End If
     n = m / MW
     a = a_table(index)
     b = b_table(index)
     Dim step1, step2 As Double
     'Van der Waal equ.
     step1 = (n * R * T) / (V - (n * b))
     step2 = ((n ^ 2) * a) / (V ^ 2)
     P1 = step1 - step2
     'Ideal gas equ.
     P2 = (n * R * T) / V
   End Sub

   Sub showResults()
      Label7.Text = "By Ideal Gas Law"
      Label7.Text += vbCrLf + "P = n*R*T/V"
      Label7.Text += vbCrLf + "P = " + FormatNumber(P1, NumericUpDown1.Value)
      Label7.Text += vbCrLf + "Van der Waal's equation gives " + FormatNumber(P1,
NumericUpDown1.Value) + " atm"
      Label7.Text += vbCrLf + "The ideal gas law gives " + FormatNumber(P2,
NumericUpDown1.Value) + " atm"
   End Sub

   Private Sub Button1_Click(ByVal sender As System.Object, ByVal e As System.EventArgs)
Handles Button1.Click
      calculateResults()
      showResults()
   End Sub
```

```
   Private Sub ComboBox1_SelectedIndexChanged(ByVal sender As System.Object, ByVal e As
System.EventArgs) Handles ComboBox1.SelectedIndexChanged
       Select Case ComboBox1.SelectedIndex
          Case 0 'O2
             MW = 32
          Case 1 'CO2
             MW = 44
          Case 2 'H2
             MW = 2
          Case 3 'CH4
             MW = 16
          Case 4 'N2
             MW = 28
          Case 5 'NO
             MW = 30
       End Select
   End Sub

   Private Sub NumericUpDown1_ValueChanged(ByVal sender As System.Object, ByVal e As
System.EventArgs) Handles NumericUpDown1.ValueChanged
       showResults()
   End Sub
End Class
```

## 7.2.5 Dalton's Law of Partial Pressures

When a mixture of gases (having no chemical interaction) is present in a container, the total pressure exerted by the gas mixture is equal to the sum of the pressures that each gas would exert if it alone occupied the container.

$$P_{total} = p_1 + p_2 + p_3 + \cdots \qquad (7.15)$$

The pressure exerted by each gas in the mixture is called its partial pressure:

$$\text{Partial pressure} = \frac{P_i}{P_{total}} \qquad (7.16)$$

There are two ways to express the fraction which one gaseous component contributes: either by mole fraction (Equation 7.3) or by pressure fraction (Equation 7.16).

### Example 7.10

A gas mixture contains the following components. Using Equations 7.2 and 7.3, determine the number of moles of each gas present as well as the mole fraction for each gas (Air Pollution Training Institute 1984).

| Gas | Weight (g) | Molecular Weight | Moles | Mole Fraction |
|-----|-----------|------------------|-------|---------------|
| $O_2$ | 20 | 32 | 0.625 | 0.885 |
| $SO_2$ | 2 | 64 | 0.0313 | 0.044 |
| $SO_3$ | 4.0 | 80 | 0.050 | 0.071 |
| Total | | | 0.7063 | 1.000 |

The total number of moles in this gaseous mixture is 0.7063, using the moles present for each gas and the ideal gas law, Equation 7.7. Determine the partial pressure for each gas as well as the partial volumes when the gas temperature is 25°C, and the gas is held in a 2-L container.

**Solution**

- For $O_2$,

$$P = \frac{(0.625)(0.08206)(298)}{2} = 7.64 \text{ atm}$$

- For $SO_2$,

$$P = \frac{(0.0313)(0.08206)(298)}{2} = 0.38 \text{ atm}$$

- For $SO_3$,

$$P = \frac{(0.050)(0.08206)(298)}{2} = 0.61 \text{ atm}$$

Total pressure = 8.63 atm.

- Partial pressure for $O_2 = \dfrac{7.64}{8.63} = 0.885$ atm
- Partial pressure for $SO_2 = \dfrac{0.38}{8.63} = 0.044$ atm
- Partial pressure for $SO_3 = \dfrac{0.61}{8.63} = 0.071$ atm

The partial volumes for each gas can also be calculated by using the ideal gas law, Equation 7.7.

- For $O_2$,

$$V = \frac{(0.625)(0.08206)(298)}{8.63} = 1770 \text{ L}$$

- For $SO_2$,

$$V = \frac{(0.0313)(0.08206)(298)}{8.63} = 0.089 \text{ L}$$

- For $SO_3$,

$$V = \frac{(0.0375)(0.08206)(298)}{8.63} = 0.141 \text{ L}$$

Total volume = 2.000 L.

- Partial volume for $O_2 = \dfrac{1.770}{2} = 0.885$ L

- Partial volume for $SO_2 = \dfrac{0.089}{2} = 0.044$ L

- Partial volume for $SO_3 = \dfrac{0.141}{2} = 0.071$ L

The partial pressure for each gas indicates the concentrations of that particular gas, and for most purposes, the gas concentration governs the behavior of a gaseous substance regardless of the presence of other gases in the mixture (Lide and Frederiske 1995/1996).

## 7.2.6 Henry's Law

Henry's law, simply stated, indicates that the concentration of a gas dissolved in a solvent under equilibrium conditions is proportional to the partial pressure of the gas above the solution at constant temperature. A mathematical equation describing this relationship is provided as follows:

$$C_i = K_i p_i \tag{7.17}$$

where:

$C_i$ is the concentration of gas $i$ dissolved in solution, mol/L
$K_i$ is the Henry law constant, mol/L
$p_i$ is the partial pressure of the gas, atm

Table 7.6 presents Henry's law constants for some gases in water at 25°C (Manahan 2009).

### Example 7.11

1. Write a computer program relating the elements contained in Henry's law that will make it possible to determine the concentration of a gas in water, provided Henry's gas constant is given.

**TABLE 7.6**

**Henry's Law Constants for Some Gases in Water at 25°C**

| Gas | K, mol x L$^{-1}$ x atm$^{-1}$ |
|-----|-------------------------------|
| $O_2$ | $1.28 \times 10^{-3}$ |
| $CO_2$ | $3.38 \times 10^{-2}$ |
| $H_2$ | $7.90 \times 10^{-4}$ |
| $CH_4$ | $1.34 \times 10^{-3}$ |
| $N_2$ | $6.48 \times 10^{-4}$ |
| NO | $2.0 \times 10^{-4}$ |

*Source:* Manahan, S.E., *Environmental Chemistry*, 9th ed, Boca Raton, FL, CRC Press, 2009.

2. Using the computer program prepared in (1), determine the solubility of oxygen in water at 25°C, knowing that air contains 20.95% oxygen by volume.

**Solution**

1. For a solution to Example 7.11 (1), see the listing of Program 7.11.
2. In calculating the solubility of a gas in water, a correction must be made for the partial pressure of water by subtracting it from the total pressure of the gas. At 25°C, the partial pressure of water is 0.0313 atm; values at other temperatures are readily obtained from standard handbooks. The concentration of oxygen in water saturated with air at 1 atm (760 mmHg) and 25°C (298 K) may be calculated as an example of a simple gas solubility calculation. Considering that dry air is 20.95% by volume oxygen and factoring in the partial pressure of water gives the following:

$$P_{O2} = (1.00 \text{ atm} - 0.0313 \text{ atm}) \times 0.2095 = 0.2029 \text{ atm}$$

$$C_i = K_i p_i = 1.28 \times 10^{-3} \text{mol/L/atm} \times 0.2029 \text{ atm}$$

$$= 2.60 \times 10^{-4} \text{mol/L}$$

Because the molecular weight of oxygen is 32, the concentration of dissolved oxygen in water in equilibrium with air under the conditions given above is 8.32 mg/L or 8.32 ppm.[8]

---

**LISTING OF PROGRAM 7.11    (CHAP7.11\FORM1.VB): HENRY'S LAW**

```
'********************************************************************************
'Example 7.11: relates the components in Henry's law
'to determine the concentration of a gas in water
'provided Henry's gas constant is given.
'********************************************************************************


Public Class Form1
    Dim gas As String
    Dim k, P, MW, comp, comp2 As Double
```

```
'********************************************************************************
'See Table 7.6
'********************************************************************************
Dim k_table() As Double =
    {1.28 / 1000, 3.38 / 100, 7.9 / 10000, 1.34 / 1000, 6.48 / 10000, 2.0 / 10000}

Private Sub Form1_Load(ByVal sender As System.Object, ByVal e As System.EventArgs)
Handles MyBase.Load
    Me.Text = "Exercise 7.11: Henry's law"
    Me.MaximizeBox = False
    Me.FormBorderStyle = Windows.Forms.FormBorderStyle.FixedSingle
    label1.text = "The formula is - C[i] = K[i] * P[i]"
    Label2.Text = "Select the gas"
    Label3.Text = "Enter the partial pressure of gas (in atm)"
    Label5.Text = ""
    Label6.Text = "Decimal places:"
    Button1.Text = "&Calculate"
    NumericUpDown1.Value = 4
    NumericUpDown1.Maximum = 10
    NumericUpDown1.Minimum = 0
    ComboBox1.Items.Clear()
    ComboBox1.Items.Add("O2")
    ComboBox1.Items.Add("CO2")
    ComboBox1.Items.Add("H2")
    ComboBox1.Items.Add("CH4")
    ComboBox1.Items.Add("N2")
    ComboBox1.Items.Add("NO")
    ComboBox1.SelectedIndex = 0
End Sub

Sub calculateResults()
    P = Val(TextBox1.Text)
    If ComboBox1.SelectedIndex = -1 Then
      MsgBox("Please select a gas from the list.",
        vbOKOnly Or vbInformation)
      Exit Sub
    End If
    k = k_table(ComboBox1.SelectedIndex)
    comp = k * P * MW
    comp2 = comp * 1000
End Sub

Sub showResults()
    Label5.Text = "C[i] = K[i] * p[i]"
    Label5.Text += vbCrLf + "C[i] = " + FormatNumber(k, NumericUpDown1.Value) + " * "
    Label5.Text += FormatNumber(P, NumericUpDown1.Value) + " * " + FormatNumber(MW,
NumericUpDown1.Value)
    Label5.Text += vbCrLf + "C[i] = " + FormatNumber(comp, NumericUpDown1.Value) + " g/
liter or "
    Label5.Text += FormatNumber(comp2, NumericUpDown1.Value) + " mg/L"
End Sub

Private Sub Button1_Click(ByVal sender As System.Object, ByVal e As System.EventArgs)
Handles Button1.Click
    calculateResults()
    showResults()
End Sub
```

```
    Private Sub ComboBox1_SelectedIndexChanged(ByVal sender As System.Object, ByVal e As
    System.EventArgs) Handles ComboBox1.SelectedIndexChanged
        Select Case ComboBox1.SelectedIndex
            Case 0 'O2
                MW = 32
            Case 1 'CO2
                MW = 44
            Case 2 'H2
                MW = 2
            Case 3 'CH4
                MW = 16
            Case 4 'N2
                MW = 28
            Case 5 'NO
                MW = 30
        End Select
    End Sub

    Private Sub NumericUpDown1_ValueChanged(ByVal sender As System.Object, ByVal e As
    System.EventArgs) Handles NumericUpDown1.ValueChanged
        showResults()
    End Sub
End Class
```

## Example 7.12

Use the computer program developed in (1) of Example 7.11 to determine the weight of $CO_2$ that would be dissolved in 1 L of water at 25°C under a $CO_2$ pressure of 1 atm.

### Solution

$$C_i = K_i p_i \qquad (7.17)$$

From Table 7.6,

$K_i = 3.38 \times 10^{-2}$ mol/L/atm
$C_i = 3.38 \times 10^{-2}$ mol/L/atm × 1 atm
$C_i = 3.38 \times 10^{-2}$ mol/L
1 mol $CO_2 = 44$ g
$C_i = 3.38 \times 10^{-2}$ mol/L × 44 g/mol/L = 1.48 g/L or 1487 mg/L

## 7.2.7  CLAUSIUS–CLAPEYRON EQUATION

The solubilities of gases decrease increasing temperature. This factor is taken into account with the following Clausius–Clapeyron equation:

$$\ln\left(\frac{C_2}{C_1}\right) = \frac{\Delta H}{R}\left(\frac{1}{T_1} - \frac{1}{T_2}\right) \qquad (7.18)$$

where:
$C_1$ is the gas concentration at absolute temperature $T_1$
$C_2$ is the gas concentration at absolute temperature $T_2$
$\Delta H$ is the heat of solution
$R$ is the ideal or universal gas constant (1.987 cal/deg·mol)

## Example 7.13

1. Prepare a computer program relating the various elements in Equation 7.18.
2. Using the computer program developed in (1), determine the concentration of oxygen in water at 40°C when the solubility of oxygen is 14.60 mg/L at 0°C and 7.54 mg/L at 30°C.

### Solution

1. For solution to Example 7.13 (1), see the listing of Program 7.13.
2. Using the Clausius–Clapeyron equation, determine $\Delta H$ for 0°C and 30°C oxygen concentration:

$$\ln\left(\frac{7.54}{14.60}\right) = \frac{\Delta H}{1.986}\left(\frac{1}{273} - \frac{1}{303}\right) - 0.6607$$

$$= \frac{\Delta H(0.00366 - 0.00330)}{1.986}\Delta H$$

$$= \frac{-0.6607 \times 1.986}{0.000363} = -3618$$

The estimated solubility of oxygen in water at 40°C would then be

$$\ln\left(\frac{C}{14.60}\right) = \frac{-3618}{1.986}\left(\frac{1}{273} - \frac{1}{313}\right)\ln\left(\frac{C_2}{14.60}\right)$$

$$= -1820(0.00366 - 0.003195)$$

which yields $C_2 = 6.22$ mg/L estimated oxygen concentration at 40°C.

## LISTING OF PROGRAM 7.13    (CHAP7.13\FORM1.VB): CLAUSIS–CLAPEYRON EQUATION

```vbnet
'********************************************************************************
'Example 7.13: Clausis-Clapeyron equation.
'********************************************************************************
Public Class Form1
    Dim gasc1, gasc2, gast1, gast2 As Double

    Private Sub Form1_Load(ByVal sender As System.Object, ByVal e As System.EventArgs)
Handles MyBase.Load
        Me.Text = "Exercise 7.13: Clausis-Clapeyron equation"
        Me.MaximizeBox = False
        Me.FormBorderStyle = Windows.Forms.FormBorderStyle.FixedSingle
        Label1.Text = "The formula is:"
        Label1.Text += vbCrLf + "ln(C2/C1) = ({D}H/R) * ((1/T1) - (1/T2))"
        Label2.Text = "Enter the gas concentration at absolute temp T1 (C1) (in mg/L) "
        Label3.Text = "Enter the temperature of the gas C1 (T1) (in C) (in mg/L) "
        Label4.Text = "Enter the gas concentration at absolute temp T2 (C2) (in mg/L) "
        Label5.Text = "Enter the temperature of the gas C2 (T2) (in C) (in mg/L) "
        Label6.Text = ""
        Button1.Text = "&Calculate"
    End Sub

    Sub calculateResults()
        gasc1 = Val(TextBox1.Text)
        gast1 = Val(TextBox2.Text)
        gasc2 = Val(TextBox3.Text)
        gast2 = Val(TextBox4.Text)

        Dim gasccomp, gasccompf, gastemp1, gastemp2, gastemp3, gastemp4, gastemp5, delhen As
Double
        Gasccomp = gasc2 / gasc1
        gasccompf = Math.Log(gasccomp)
        Gastemp1 = 273 + gast1 'T1
        Gastemp2 = 273 + gast2 'T2
        Gastemp3 = 1 / gastemp1
        Gastemp4 = 1 / gastemp2
        Gastemp5 = gastemp3 - gastemp4
        delhen = (gasccompf * 1.986) / gastemp5
        Label6.Text = "The Delta H (heat of solution) at " + gastemp2.ToString + " degrees
is " + FormatNumber(delhen, 2)

        Dim gastemp6, gastemp7, delhen2, step1, step2, step3, answ As Double
        Dim temp3 = InputBox("Type in the new water temperature (in C)", "Enter
Temperature")
        Gastemp6 = 273 + temp3
        Gastemp7 = gastemp3 - (1 / gastemp6)
        Delhen2 = delhen / 1.986
        Step1 = delhen2 * gastemp7
        step2 = Math.Log(gasc1)
        Step3 = step1 + step2
        answ = Math.Exp(step3)
        Label6.Text += vbCrLf + vbCrLf + "Estimated oxygen concentration at " _
            + temp3.ToString + " degrees C" + vbCrLf + "C2 = " + FormatNumber(answ, 2) + "
gm/L"
    End Sub
```

```
    Private Sub Button1_Click(ByVal sender As System.Object, ByVal e As System.EventArgs)
Handles Button1.Click
        calculateResults()
    End Sub
End Class
```

### 7.2.8  Reynolds Numbers

The Reynolds number (Re) relates inertial forces to viscous forces per unit volume for the medium involved (gas or liquid). Typical inertial force per unit volume of fluid is given as follows:

$$\frac{\rho v^2}{g_c L} \tag{7.19}$$

Typical viscous force per unit volume of fluid is given as follows:

$$\frac{\mu v}{g_c L^2} \tag{7.20}$$

The first expression divided by the second provides the dimensionless ratio known as Reynolds number:

$$\text{Re} = \frac{L v \rho}{\mu} = \frac{\text{Inertial force}}{\text{Viscous force}} \tag{7.21}$$

where:
  $\rho$ is the density of the fluid (mass/volume), $kg/m^3$
  $v$ is the velocity of the fluid, m/s
  $g$ is the dimensional constant
  $L$ is the linear dimension, m
  $\mu$ is the viscosity of the fluid, kg/m/s
  Re is the Reynolds number, dimensionless

The Reynolds number provides information on flow behavior. Laminar flow is normally encountered at a Reynolds number < 2100 in a tube, but it can persist up to Reynolds numbers of several thousands under special conditions. Under ordinary conditions of flow, the flow is turbulent at a Reynolds number ~> 4000. Between 2100 and 4000, a transition region is found where the type of flow may be either laminar or turbulent. By definition, the Reynolds number is dimensionless (Air Pollution Training Institute 1984).

### Example 7.14

1. Write a computer program that includes the elements contained in Equation 7.19 in order to calculate the Reynolds number.
2. Given the data in the Solutions section for the flow of an air stream through a circular duct, determine the Reynolds number for this air stream.

### Solution

1. For solution to Example 7.14 (1), see the listing of Program 7.14.
2. Solution to Example 7.14 (2):
   $\rho_a = 0.965$ $kg/m^3$
   $v = 10.0$ m/s
   $L = 0.50$ m (diameter of duct)
   $\mu_a = 2.15 \times 10^{-5}$ kg/m/s

$$\text{Re} = \frac{L v \rho}{\mu} = \frac{(0.50)(10.0 \text{ m/s})(0.965 \text{ kg/m}^3)}{2.15 \times 10^{-5} \text{ kg/m}} = 2.24 \times 10^5 \tag{7.22}$$

This would be turbulent flow. For most air pollution conditions, the flow is usually in the turbulent range with high Reynolds numbers.

---

**LISTING OF PROGRAM 7.14   (CHAP7.14\FORM1.VB): REYNOLDS NUMBER**

```
'****************************************************************************************
'Example 7.14: Calculates the Reynolds number
'****************************************************************************************

Public Class Form1
    Dim dimen, vel, dens, visc As Double

    Private Sub Form1_Load(ByVal sender As System.Object, ByVal e As System.EventArgs)
Handles MyBase.Load
        Me.Text = "Exercise 7.14: Reynolds number"
        Me.MaximizeBox = False
        Me.FormBorderStyle = Windows.Forms.FormBorderStyle.FixedSingle
```

```
    Label1.Text = "The formula is:"
    Label1.Text += vbCrLf + "Re = (L * v * r(rho))/u (viscosity)"
    Label2.Text = "Type in the diameter of the duct, (L) (in m)"
    Label3.Text = "Type in the velocity of the fluid, (v) (in m/s)"
    Label4.Text = "Type in the density of the gas, (r (rho)) (in kg/m{3})"
    Label5.Text = "Type in the viscosity of the fluid, (u) (in kg/m.s"
    Label5.Text += vbCrLf + " use scientific notation (x.xxxE+/-x)"
    Label6.Text = ""
    Button1.Text = "&Calculate"
  End Sub

  Sub calculateResults()
    dimen = Val(TextBox1.Text)
    vel = Val(TextBox2.Text)
    dens = Val(TextBox3.Text)
    visc = Val(TextBox4.Text)
    Dim step1, step2 As Double
    step1 = dimen * vel * dens
    step2 = step1 / visc
    Label6.Text = "The Reynolds number is " + Format(step2, "##.####").ToString
    If step2 < 2100 Then
      Label6.Text += vbCrLf + "Laminar flow"
    ElseIf step2 < 4000 Then
      Label6.Text += vbCrLf + "Transition region"
    Else
      Label6.Text += vbCrLf + "Turbulent flow"
    End If
  End Sub

  Private Sub Button1_Click(ByVal sender As System.Object, ByVal e As System.EventArgs)
Handles Button1.Click
    calculateResults()
  End Sub
End Class
```

## 7.2.9 Stoke's Law

The terminal settling velocity of spherical particles larger than the free mean path in a fluid is described by Stoke's law, provided the particles are not so large that inertial effects arise from the fluid displaced by the falling particles. Terminal velocity is a constant value of velocity reached when all forces (gravity, drag, buoyancy, etc.) are acting on a body balance. The sum of all the forces is then equal to zero (no acceleration). Particles smaller than about 50 μm reach a constant settling velocity within a fraction of a second. The terminal settling velocity is given by the following equation:

$$v_t = \frac{g d_p^2 \left(\rho_p - \rho_a\right)}{18\mu} \qquad (7.23)$$

where:

$v_t$ is the terminal settling velocity, cm/s
$g$ is the gravitational constant, cm/s$^2$
$\rho_p$ is the density of the particle, g/cm$^3$
$\rho_a$ is the density of air, g/cm$^3$
$d_p$ is the diameter of particle, μm
$\mu$ is the viscosity of air, g/cm/s

For the determination of the terminal settling velocity of particles, three size regimes are considered here, based on a dimensionless constant $k$. Terminal velocity is a constant value of velocity reached when all forces (gravity, drag, buoyancy, etc.) are acting on a body balance. The sum of all the forces is then equal to zero (no acceleration). A dimensionless constant $k$ determines the appropriate range of the fluid-particle dynamic laws which apply (Air Pollution Training Institute 1984).

$$k = d_p \left(\frac{g \rho_p \rho}{\mu^2}\right)^{1/3} \qquad (7.24)$$

where:

$k$ is the dimensionless constant, which determines the range of the fluid-particle dynamic laws
$d_p$ is the particle diameter, cm
$g$ is the gravity force, cm/s$^2$
$\rho_p$ is the particle density, g/cm$^3$
$\rho$ is the fluid (gas) density, g/cm$^3$
$\mu$ is the fluid (gas) viscosity, g/cm/s

The numerical value of $k$ determines the appropriate law, and the values of $k$ that apply for the three regimes are as follows:

- $k < 3.3$, Stoke's law range
- $3.3 < k < 43.6$, intermediate law range
- $43.6 < k < 2360$, Newton's law range

The equations for each regime are as follows:

- For Stoke's law range:

$$v_t = \frac{gd_p^2(\rho_p - \rho_a)}{18\mu} \tag{7.23}$$

- For intermediate law range:

$$v = \frac{0.153g^{0.71}d_p^{1.14}\rho_p^{0.71}}{\mu^{0.43}\rho^{0.29}} \tag{7.25}$$

- For Newton's law range:

$$v = 1.74\sqrt{\left(\frac{gd_p\rho_p}{\rho}\right)} \tag{7.26}$$

When particles approach sizes comparable to the mean free path of the fluid molecules, the medium can no longer be regarded as continuous because particles can fall between the molecules at a faster rate than predicted by aerodynamic theory. To allow for this "slip," Cunningham correction factor is introduced to Stoke' law (Air Pollution Training Institute 1984):

$$v = \left(\frac{gd_p^2\rho_p}{18\mu}\right)(C_f) \tag{7.27}$$

$C_f$ = Cunningham correction factor

$$= 1 + \left(\frac{2A\lambda}{d_p}\right) \tag{7.28}$$

where:
$A = 1.257 + 0.40e^{-1.10dp/2\lambda}$
$\lambda$ is the mean free path of the fluid molecules ($6.53 \times 10^{-6}$ cm for ambient air)

The Cunningham correction factor is usually applied to particles equal to or smaller than 1 μm (Air Pollution Training Institute 1984).

### Example 7.15

1. Write a computer program that provides for the calculation of the dimensionless constant $k$ and thus provides information as to which appropriate terminal velocity equation to use.
2. Write a computer program that relates the various elements in Stoke's law (Equation 7.23) and will determine the terminal settling velocity for spherical particles.

3. Using the computer programs developed in (1) and (2), estimate the terminal settling velocity in cm/s for a 30 μm particle with a density of 2.31 g/cm³. The room air is 24°C with a room depth of 2.7 m. Estimate the time required for this particle to settle from the ceiling to the floor.

### Solution

1. For a solution to Example 7.15 (1), see the listing of Program 7.15.
2. For a solution to Example 7.15 (2), first calculate the density of air at 24°C using Equation 7.11:

$$\rho_a = \frac{P*MW}{RT}$$

where:
$P = 1$, atm
MW = 29 g/g-mol for air
$R = 0.08206$ (L) at (g-mol)$^{-1}$ (K)$^{-1}$
$T = $ K, $(273 + °C)$

$$\rho_a = \frac{1\,atm \times 29\,g/g\text{-}mol}{0.08206\,(L)atm(g\text{-}mol)^{-1}(K)^{-1} \times (273+24)\,K}$$

$$= \frac{1.19\,g}{L} = 0.00119\,g/cm^3$$

Using Equation 7.24, calculate the value of $k$ in order to determine which equation to use to estimate the terminal settling velocity for the 30 μm particle:

$$k = d_p\left(\frac{g\rho_p\rho}{\mu^2}\right)^{1/3}$$

$\mu_a$ at 24°C $= 0.00018\,g/cm*s$

$$k = 30 \times 10^{-4}\,cm$$

$$\left[\frac{981\,cm/s^2 \times 2.31\,g/cm^3 \times 0.00119\,g/cm^3}{(0.00018\,g/cm*s)^2}\right]^{1/3}$$

If $k < 3.3$, the appropriate equation to use is Stoke's law (Equation 7.23):

$$v_t = \frac{gd_p^2(\rho_p - \rho_a)}{18\mu} \tag{7.23}$$

$$v = \frac{981\,cm/s^2\left(30 \times 10^{-4}\,cm\right)^2\left(2.31\,g/cm^3 - 0.00119\,g/cm^3\right)}{18(0.00018\,g/cm*s)}$$

$$= 6.29\,cm/s$$

The time required for this 30 μm particle to settle from the ceiling to the floor would be

$$\frac{270\,cm}{6.29\,cm/s} = 43\,s$$

## LISTING OF PROGRAM 7.15    (CHAP7.15\FORM1.VB): TERMINAL VELOCITY AND STOKE'S LAW

```vb
'*********************************************************************************
'Example 7.15: Provides for calculation of the dimensionless
'constant k and provides information on which terminal velocity
'equation to use.
'Also relates the elements of Stoke's law.
'*********************************************************************************
Public Class Form1
    Dim pres, molwgt, temp As Double

    Private Sub Form1_Load(ByVal sender As System.Object, ByVal e As System.EventArgs)
Handles MyBase.Load
        Me.Text = "Exercise 7.15"
        Me.MaximizeBox = False
        Me.FormBorderStyle = Windows.Forms.FormBorderStyle.FixedSingle
        Label1.Text = "Example 7.15: Provides for calculation of the dimensionless"
        Label1.Text += vbCrLf + " constant k and provides information on which terminal
        velocity"
        Label1.Text += " equation to use."
        Label1.Text += vbCrLf + "Also relates the elements of Stoke's law."
        Label1.Text += vbCrLf + "The initial formulas are:"
        Label1.Text += vbCrLf + "r(rho) = (P * (MW))/R * T Formula 7.10"
        Label1.Text += vbCrLf + "k = d[p] * (g * r[p] * (r/u{2})){1/3} Formula 7.20"
        Label1.Text += vbCrLf + "v[t] = (g * d[p]{2} * (r[p] - r[a]))/18 * u Formula 7.19"

        Label2.Text = "To calculate the density of air:"
        Label2.Text += vbCrLf + "Type in the pressure, (P) (in atm)"
        Label3.Text = "Type in the molecular weight of the air, (MW) (in g(g-mole){-1})"
        Label4.Text = "Type in the temperature of the air, (T) (in C)"
        Label5.Text = "Type in the particle diameter (um)"
        Label6.Text = "Type in the particle density (g/cm{3})"
        Label7.Text = "Type in the room depth (m)"
        Label8.Text = "Type in the gas viscosity (from a table) (in g/cm.s)"
        Button1.Text = "&Calculate"
    End Sub

    Sub calculateResults()
        pres = Val(TextBox1.Text)
        molwgt = Val(TextBox2.Text)
        temp = Val(TextBox3.Text)

        Dim comptemp, nom, denom, dem, dens, denscm As Double
        Dim depth, compdiama, compdiam, compdens, deptha, V, time As Double
        Dim diam, partden, visc As Double
        Dim decimals As Integer = 2
        Dim decimals2 As Integer = 4
        Dim s As String

        comptemp = 273 + temp
        nom = pres * molwgt
        dem = 0.08206 * comptemp
        dens = nom / dem
        denscm = dens * 0.001
        s = "The density of the air at " + FormatNumber(temp, decimals) + " degrees C is "
        s += FormatNumber(dens, decimals) + " g/L"
        s += " and " + FormatNumber(denscm, decimals2) + " g/cm*10{3}"
        diam = Val(TextBox4.Text)
        partden = Val(TextBox5.Text)
        depth = Val(TextBox6.Text)
        visc = Val(TextBox7.Text)
```

```
        Dim nom2, comp1, comp2, comp3, diamcomp, comp4 As Double
        Nom2 = 981 * partden * denscm
        comp1 = nom2 / visc
        comp2 = comp1 / visc
        comp3 = comp2 ^ (0.3333)
        diamcomp = diam * 0.0001
        comp4 = diamcomp * comp3
        s += vbCrLf + "k = " + FormatNumber(comp4, decimals2)
        If comp4 < 3.3 Then
            s += " (using Stoke's law range)"
        ElseIf comp4 < 43.6 Then
            s += " (using the intermediate law range)"
    Else
            s += " (using Newton's law range)"
        End If

        Compdiama = diam * diam
        Compdiam = compdiama * 0.00000001#
        Compdens = partden - denscm
        nom = 981 * compdiam * compdens
        denom = 18 * visc
        V = nom / denom
        deptha = depth * 100
        time = deptha / V
        s += vbCrLf + "Settling time = " + FormatNumber(time, decimals) + " seconds"
        s += vbCrLf + "The velocity = " + FormatNumber(V, decimals) + " cm/s"
        MsgBox(s, vbOKOnly, "Results")
    End Sub


    Private Sub Button1_Click(ByVal sender As System.Object, ByVal e As System.EventArgs)
    Handles Button1.Click
        calculateResults()
    End Sub
End Class
```

### 7.2.10  Particle Size Distribution—Log Normal

The most important single parameter in evaluating the effects of and determining the best method for control of airborne particulates is the size of the particulate. Airborne particles are never all the same size; therefore, it is necessary to try to characterize and quantify the distribution of the various sizes involved. The most often used mathematical expression to do this evaluation is the log-normal distribution. There is, however, no law that says all particles follow the log-normal classification. In a number of industrial and ambient air studies, the size distribution of the particulate followed the "normal" or "Gaussian" distribution (Rowe et al. 1985).

The log-normal distribution of suspended particulate in the atmosphere is often used, as it is a relatively simple and convenient way to analyze the data. This is done by plotting on log-probability paper the particle size diameter on the ordinate (log scale) and the cumulative percentage less than or equal to a stated size on the abscissa (probability scale). If the data are plotted as a straight line on log-probability paper, this tends to indicate that the size distribution of the particulate is log normal. The log-normal probability distribution for particulates can be described by the following mathematical expression (U.S. Department of Health and Human Services 1973):

$$F(d) = \frac{\Sigma N}{\log \sigma_g \sqrt{2\pi}} \exp\left(-\frac{\left(\log d - \log d_g\right)^2}{2\log^2 \sigma_g}\right) \tag{7.29}$$

where:
   $F(d)$ is the frequency of particle occurrence of diameter $d$
   $\Sigma N$ is the total number of particles
   $\sigma_g$ is the standard geometric deviation for the data
   $d_g$ is the geometric mean diameter of the particles

If the size distribution data do plot a straight line on log-probability paper, then two parameters that characterize these data are the geometric mean diameter $d_g = d_{50}$ and the geometric standard deviation $\sigma_g$:

$$\sigma_g = \frac{d_{84.13}}{d_{50}} = \frac{d_{50}}{d_{15.87}} \tag{7.30}$$

where:

$d_{84.13}$ is the diameter such that particles constituting 84.13% of the total mass of particles are smaller than this size

$d_{50}$ is the geometric mean diameter

$d_{15.87}$ is the diameter such that particles constituting 15.87% of the total mass of particles are smaller than this size

$\sigma_g$ is the geometric standard deviation

## Example 7.16

Determine if the following particle size distribution data are log normal. If the data indicate log-normal distribution, what then are the geometric mean diameter and the geometric standard deviation?

| Data | | Answer | |
|---|---|---|---|
| Particle Size d (μm) | Weight of Particles (μg) | Cumulative Total | Cumulative Percentage |
| 0.4 | 10 | 10 | 5.7 |
| 0.55 | 30 | 40 | 22.8 |
| 0.70 | 25 | 65 | 37.1 |
| 0.90 | 60 | 125 | 71.4 |
| 1.20 | 30 | 155 | 88.6 |
| 1.5 | 15 | 170 | 97.1 |
| 2.00 | 5 | 175 | 100 |

These data give a straight line on log-probability paper and therefore can be considered to have a log-normal distribution.

From the graph for Example 7.16:

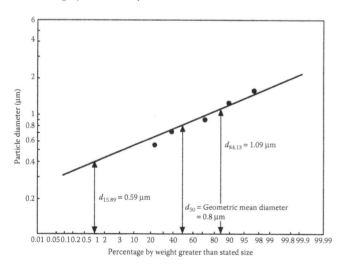

$d_{50} = 0.8$ μm or the geometric mean diameter
$d_{84.13} = 1.09$ μm
$d_{15.87} = 0.59$ μm

$$\sigma_g = \frac{0.8}{0.59} = \frac{1.09}{0.8} = 1.36, \text{ the geometric standard deviation}$$

## Example 7.17

1. Develop a computer program that can be used to determine if the data for a particle size distribution sample are log normal. If the data give a log-normal distribution, then include in the program a means, by which the geometric mean diameter, $d_{50}$, $d_{15.87}$, and $d_{84.13}$ sizes can be determined, as well as the geometric standard deviation, $\sigma_g$.

2. Using the computer program developed in (1), determine if the following data for a particle size distribution are log normal. Also indicate the $d_{15.87}$, $d_{84.13}$, and $d_{50}$ sizes as well as $\sigma_g$.

| Particle Size d (μm) | Weight of Particles (μg) |
|---|---|
| 0.4 | 30 |
| 0.7 | 123 |
| 1.1 | 130 |
| 2.1 | 540 |
| 3.3 | 230 |
| 4.0 | 99 |
| 8.5 | 128 |

### Solution

1. For a solution to Example 7.17 (1), see the listing of Program 7.17.
2. A solution to Example 7.17 (2):

| Particle Size d (μm) | Weight of Particles (μg) | Cumulative Total | Cumulative Percentage |
|---|---|---|---|
| 0.4 | 30 | 30 | 2.3 |
| 0.7 | 123 | 153 | 12.0 |
| 1.1 | 130 | 283 | 22.1 |
| 2.1 | 540 | 823 | 64.3 |
| 3.3 | 230 | 1053 | 82.3 |
| 4.0 | 99 | 1152 | 90.0 |
| 8.5 | 128 | 1280 | 100.0 |

Plot the data on a log-probability paper as shown on graph for Example 7.17. Determine from the graph the following:

$$d_{50} = 1.7 \text{ μm}$$

$$d_{15.87} = 0.8 \text{ μm}$$

$$d_{84.13} = 3.5 \text{ μm}$$

$$\sigma_g = \frac{3.5}{1.7} = 2.1$$

$$\sigma_g = \frac{1.7}{0.8} = 2.1$$

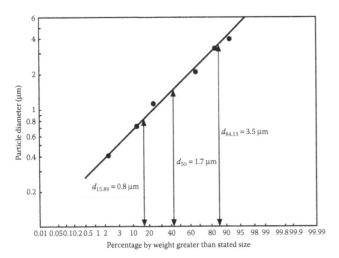

## 7.3  AIR POLLUTION CONTROL TECHNOLOGY

### 7.3.1  Settling Chambers

The oldest and probably least effective air pollution control device is the gravity settling chamber. Although simple to construct and having a low operating cost, settling chambers are still generally used only to collect large particles above 40–50 μm in size. However, if the particles have a high density, then particles as small as 10–15 μm can be collected. Settling chambers generally are simply an enlargement of the conduit carrying the contaminated airstream (see Figure 7.1).

Settling chambers are often used as a precleaning stage for treating a contaminated gas stream in order to protect the other downstream collection equipment from hot, coarse, and abrasive materials.

---

### LISTING OF PROGRAM 7.17    (CHAP7.17\FORM1.VB): PARTICLE SIZE DISTRIBUTION

```vb
'****************************************************************************************
'Program 7.17: Determine if distribution is log-normal
'and compute the geometric SD.
'****************************************************************************************
   Public Class Form1
   Dim Partsz(), Wgt(), cumtot(), cumper(), per() As Double
   Dim sum1X, sum1Y, sum2X, sum2Y As Double
   Dim m As Double
   Dim a, b As Double
   Dim g As Graphics
   '****************************************************************************************
   'get_max(): Gets the largest number of an
        'array, given a reference to
        'the array and member count.
   '****************************************************************************************
   Private Function get_max(ByRef array() As Double, ByVal count As Integer) As Double
      Dim i As Integer
      Dim max As Double = array(0)
      For i = 1 To count - 1
        If max < array(i) Then
          max = array(i)
        End If
      Next
      Return max
   End Function


   '****************************************************************************************
   'Draws a straight line from the scattered
   'point data. The algorithm used is simple:
   '(1) Divide data into two sets
   '(2) Find a mid-point in each set
   '(3) Find line equation from these two points
   '(4) Find the first and last points in this line
   '(5) Draw the line!
   '****************************************************************************************
   Private Sub draw_straight_line()
      Dim count As Integer = DataGridView1.Rows.Count - 1
      Dim w As Integer = PictureBox1.Width - 4
      Dim h As Integer = PictureBox1.Height - 4
      Dim max_Y As Double = get_max(Partsz, count)
      Dim max_X As Double = get_max(cumper, count)
```

$d_{84.13} = 3.5\ \mu m$

$d_{50} = 1.7\ \mu m$

$d_{15.89} = 0.8\ \mu m$

```
      Dim scaleY As Double = h / max_Y
      Dim scaleX As Double = w / max_X
      Dim zeroX As Integer = 2
      Dim zeroY As Integer = h + 2
      Dim mid_count As Integer = count / 2
      Dim i As Integer
      sum1X = 0 : sum2X = 0
      sum1Y = 0 : sum2Y = 0
      For i = 0 To mid_count - 1
        sum1X += cumper(i)
        sum1Y += Partsz(i)
      Next
      sum1X /= mid_count
      sum1Y /= mid_count
      For i = mid_count To count - 1
        sum2X += cumper(i)
        sum2Y += Partsz(i)
      Next
      sum2X /= (count - mid_count)
      sum2Y /= (count - mid_count)
      m = (sum2Y - sum1Y) / (sum2X - sum1X)
      'find straight line equation:
      'y = a + bx
      'y2 = mx2 - mx1 + y1
      a = sum1Y - (m * sum1X)
      b = m
      '***************************************************************************
      'find first and last points 'from the
      'equation to draw the line.
      '***************************************************************************
      Dim x1, y1, x2, y2 As Double
      x1 = cumper(0)
      x2 = cumper(count - 1)
      y1 = a + b * x1
      y2 = a + b * x2
      Dim j1, j2, l1, l2 As Integer
      j1 = zeroX + (x1 * scaleX)
      j2 = zeroX + (x2 * scaleX)
      l1 = zeroY - (y1 * scaleY)
      l2 = zeroY - (y2 * scaleY)
      g.DrawLine(Pens.Black, j1, l1, j2, l2)
End Sub

Function get_Y_value(ByVal X As Double) As Double
      Dim Y As Double
      Y = a + b * X
      Return Y
End Function

Private Sub draw_graph()
      Dim count As Integer = DataGridView1.Rows.Count - 1
      Dim bmp As Bitmap = New Bitmap(PictureBox1.Width, PictureBox1.Height)
      g = Graphics.FromImage(bmp)
      g.Clear(Color.White)
      Dim w As Integer = PictureBox1.Width - 4
      Dim h As Integer = PictureBox1.Height - 4
      Dim max_Y As Double = get_max(Partsz, count)
      Dim max_X As Double = get_max(cumper, count)
      Dim scaleY As Double = h / max_Y
      Dim scaleX As Double = w / max_X
      Dim countX As Integer = 5
```

```vbnet
        Dim countY As Integer = 5
        Dim zeroX As Integer = 2
        Dim zeroY As Integer = h + 2
        Dim i, j, k As Integer
        Dim f As Font = New Font("Arial", 8)
        'Draw X axis
        g.DrawLine(Pens.Black, zeroX, zeroY, zeroX + w, zeroY)
        Dim x As Double = max_X / 5
        For i = 1 To countX
          j = zeroX + (i * x * scaleX)
          g.DrawLine(Pens.Black, j, zeroY, j, zeroY - 2)
          g.DrawString(FormatNumber(i * x, 0), f, Brushes.Black, j, zeroY - 12)
        Next
        'Draw Y axis
        g.DrawLine(Pens.Black, zeroX, zeroY, zeroX, 2)
        Dim y As Double = max_Y / 5
        For i = 1 To countY
          j = zeroY - (i * y * scaleY)
          g.DrawLine(Pens.Black, zeroX, j, zeroX + 2, j)
          g.DrawString(FormatNumber(i * y, 0), f, Brushes.Black, 4, j)
        Next
        'Draw major points
        For i = 1 To count
          j = zeroX + (cumper(i - 1) * scaleX)
          k = zeroY - (Partsz(i - 1) * scaleY)
          g.DrawEllipse(Pens.Black, j - 2, k - 2, 4, 4)
        Next
        draw_straight_line()
        PictureBox1.Image = Image.FromHbitmap(bmp.GetHbitmap)
        g.Dispose()
        bmp.Dispose()
    End Sub

    Private Sub Form1_Load(ByVal sender As System.Object, ByVal e As System.EventArgs)
Handles MyBase.Load
        Me.Text = "Exercise 7.17: Calculates mole fractions"
        Me.MaximizeBox = False
        Me.FormBorderStyle = Windows.Forms.FormBorderStyle.FixedSingle
        Label1.Text = "Determine if particles size distribution data is log normal"
        Label1.Text += vbCrLf + "and compute the geometric SD."
        Label1.Text += vbCrLf + "Procedure:"
        Label1.Text += vbCrLf + "First determine the cumulative total for the particles and
        the cumulative percentages of each."
        Label1.Text += vbCrLf + "Then the geometric SD is computed."
        Label2.Text = "Type in the particle sizes (in um) and weights:"
        DataGridView1.Columns.Clear()
        DataGridView1.Columns.Add("sizeCol", "Particle size")
        DataGridView1.Columns.Add("wgtCol", "Particle Wgt")
        TextBox1.Multiline = True
        TextBox1.Height = 135
        TextBox1.ScrollBars = ScrollBars.Vertical
        TextBox1.Font = New Font(FontFamily.GenericMonospace, 9)
        Button1.Text = "&Calculate"
    End Sub

    Sub calculateResults()
        If DataGridView1.Rows.Count <= 1 Then
          MsgBox("Type in at least one entry!", vbOKOnly + vbExclamation, "Prompt")
          Exit Sub
        End If
        Dim count = DataGridView1.Rows.Count - 1
        Dim cumtota, cumpera As Double
```

```
    ReDim Partsz(count), Wgt(count), cumtot(count), per(count), cumper(count)
    cumtota = 0
    cumpera = 0
    For i = 0 To count - 1
      Partsz(i) = Val(DataGridView1.Rows(i).Cells("sizeCol").Value)
      Wgt(i) = Val(DataGridView1.Rows(i).Cells("wgtCol").Value)
      cumtota += Wgt(i)
      cumtot(i) = cumtota
    Next
    For i = 0 To count - 1
      per(i) = (Wgt(i) / cumtota) * 100
      cumpera = cumpera + per(i)
      cumper(i) = cumpera
    Next

    Dim s As String
    s = "Particle size Wgt of Cumulative Cumulative"
    s += vbCrLf + "d in um particle total paercent"
    For i = 0 To count - 1
      s += vbCrLf + Format(Partsz(i), "00.000 ").ToString
      s += Format(Wgt(i), "000.000 ").ToString
      s += Format(cumtot(i), "0000.000 ").ToString
      s += Format(cumper(i), "00.000").ToString
    Next
    draw_graph()
    Dim d84 As Double = get_Y_value(84.13)
    Dim d50 As Double = get_Y_value(50.0)
    Dim d15 As Double = get_Y_value(15.87)
    Dim SD1, SD2 As Double
    SD1 = d84 / d50
    s += vbCrLf + "d[84.13] = " + FormatNumber(d84, 2)
    s += vbCrLf + "d[50.00] = " + FormatNumber(d50, 2)
    s += vbCrLf + "Geometric SD = d[84.13]/d[50.00] = " + FormatNumber(SD1, 2)
    s += vbCrLf + "d[50.00] = " + FormatNumber(d50, 2)
    s += vbCrLf + "d[15.87] = " + FormatNumber(d15, 2)
    SD2 = d50 / d15
    s += vbCrLf + "Geometric SD = d[50.00]/d[15.87] = " + FormatNumber(SD2, 2)
    TextBox1.Text = s
  End Sub

  Private Sub Button1_Click(ByVal sender As System.Object, ByVal e As System.EventArgs)
Handles Button1.Click
    calculateResults()
  End Sub
End Class
```

If we assume that Stoke's law applies, we can derive a formula that can be used to calculate the minimum diameter of particles to be collected at 100% theoretical efficiency in a chamber of given length $L$.

The following equation can be used to predict the largest sized particle that can be removed with 100% efficiency in a settling chamber with given dimensions:

$$d_p = \sqrt{\left(\frac{18\mu v_h H}{gL\rho_p}\right)} \qquad (7.31)$$

where:

$\mu$ is the gas viscosity, g/cm*s
$v_h$ is the horizontal gas velocity, cm/s

$H$ is the height of chamber, cm
$g$ is the gravitational constant 981, cm/s$^2$
$L$ is the length of chamber, cm
$\rho_p$ is the particle density, g/cm$^3$

### Example 7.18

Estimate the minimum size of particle that will be removed with 100% efficiency from a settling chamber, given the following data:

Dimensions of gravity settling chamber: 8 m wide, 6 m high, and 15 m long
Volumetric gas flow of 1.3 m$^3$/s
Particle density of 2.0 g/cm$^3$

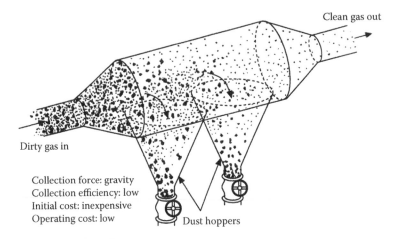

**FIGURE 7.1**   Settling chamber.

Viscosity of the air in the gas stream of $1.85 \times 10^{-4}$ g/cm*s
Gravitational constant of 981 cm/s²

**Solution**

The velocity $v_h$ can be determined from the continuity equation:

$$Q = VA$$

$$1.3 \ m^3/s = V(6 \times 8 \ cm)$$

$$V = 0.027 \ m/s = 2.7 \ cm/s$$

Substituting in Equation 7.31:

$$d_p = \sqrt{\left[\frac{18 \times (1.85 \times 10^{-4} \ cm/s) \times 2.7 \ cm/s \times 600 \ cm}{981 \ cm/s^2 \times 2.0 \ g/cm^3 \times 1500 \ cm}\right]}$$

$$= 0.000135 \ cm - 1.35 \ \mu m$$

**Example 7.19**

1. Prepare a computer program that can be used to predict the largest sized particle that can be removed with 100% efficiency in a settling chamber (Equation 7.31).
2. Using the computer program developed in (1), estimate the largest sized particle that can be removed with 100% efficiency in a settling chamber, given the following data:

Gravity settling chamber dimensions: 4 m wide, 1.2 m high, and 8 m long
Volumetric gas flow of 1.2 m³/s
Particle density of 2.5 g/cm³
Viscosity of the air at 27°C of $1.86 \times 10^{-4}$ g*cm/s

**Solution**

1. For solution to Example 7.19 (1), see the listing of Program 7.19.
2. A solution to Example 7.19 (2):
   a. Calculate $v_h$ from $Q = VA$:

$$1.2 \ m^3/s = V(1.2 \times 4)$$

$$V = 0.25 \ ms = 25 \ cm/s$$

b. Substituting in Equation 7.31:

$$d_p = \sqrt{\left(\frac{18 \times 1.85 \times 10^{-4} \times 25 \times 120}{981 \times 2.5 \times 800}\right)}$$

$$= 0.00 \ 23 \ cm = 23 \ \mu m$$

### 7.3.2   Cyclones

Cyclones are the most commonly used air pollution devices for the removal of particulate from an air stream. They are inertial separators depending on the centrifugal force as the separating mechanism. The separation of particulate in a cyclone is similar to that of gravitational settling, except that centrifugal forces instead of gravitational forces are used.

Cyclones are inexpensive to operate and have no moving parts. The cyclone operates with two vortexes, one spiraling downward at the outside of the cone and the other spiraling upward on the inside of the cone. The efficiency of a cyclone depends upon its size; a single, high-efficiency cyclone is most efficient at the removal of particles in the 40–50 μm size range, whereas small cyclones 25 cm in diameter are most efficient in the 15–20 μm size range. Figure 7.2 shows a conventional single, high-efficiency cyclone.

An equation that can be used to calculate the size of particles that can be collected with a 50% efficiency (cut diameter) is as shown in the following equation:

$$d_{50} = \sqrt{\left(\frac{9 \mu B_c}{2 \pi N_e v_i \rho_p}\right)} \qquad (7.32)$$

where:
  $d_{50}$ is the diameter of the particle that is collected with 50% efficiency (cut diameter)
  $\mu$ is the gas viscosity, kg/m*s
  $B_c$ is the width of cyclone inlet, m
  $N_e$ is the number of effective turns within the cyclone
  $v_i$ is the inlet gas velocity, m/s
  $\rho_p$ is the density of the PM, kg/m³

## LISTING OF PROGRAM 7.19   (CHAP7.19\FORM1.VB): SETTLING CHAMBER

```vb
'********************************************************************************
'Example 7.19: predict the largest sized particle that can be
'removed with 100% efficiency in a settling chamber.
'********************************************************************************
Public Class Form1
    Dim measq, meash1, measw1, visc, hgt, lnth, dens As Double
    Dim comp1, comp2, comp3 As Double
    Dim comp6, comp7, comp8 As Double

    Private Sub Form1_Load(ByVal sender As System.Object, ByVal e As System.EventArgs)
Handles MyBase.Load
        Me.Text = "Exercise 7.19"
        Me.MaximizeBox = False
        Me.FormBorderStyle = Windows.Forms.FormBorderStyle.FixedSingle
        Me.Height = 252
        Label1.Text = "Predict the largest sized particle that can be"
        Label1.Text += vbCrLf + "removed with 100% efficiency in a settling chamber."
        Label1.Text += vbCrLf + "The formula:"
        Label1.Text += vbCrLf + "d[p] = (18 um v[h] H / g L r[p]){1/2}"
        Label1.Text += vbCrLf + "The value for v[h] is computed first."
        Label1.Text += vbCrLf + "This formula is Q = VA"
        Label2.AutoSize = False
        Label2.Height = 120
        Label2.Width = 433
        Label2.Text = "Enter the value of the volumetric gas flow in m{3}/s"
        Label3.Text = "Enter the height of chamber (m):"
        Label4.Text = "Enter the width of chamber (m):"
        Label5.Text = ""
        Label6.Text = "Enter the gas viscosity (g/cm s)"
        Label6.Text += vbCrLf + "Enter as scientific -ex; 1.86E-5"
        Label7.Text = "Enter the length of chamber (cm)"
        Label8.Text = "Enter the particle density (g/cm{3})"
        Button1.Text = "&Calculate V"
        Button2.Text = "C&alculate d[p]"
        Label10.Text = ""
    End Sub

    Sub calculateResults()
        measq = Val(TextBox1.Text)
        meash1 = Val(TextBox2.Text)
        measw1 = Val(TextBox3.Text)
        comp1 = meash1 * measw1
        Comp2 = measq / comp1
        Comp3 = comp2 * 100
        Label5.Text = FormatNumber(measq, 2) + " = V (" + FormatNumber(meash1, 2) + " x " +
        FormatNumber(measw1, 2) + ")"
        Label5.Text += vbCrLf + "V = " + FormatNumber(measq, 2) + " / " +
        FormatNumber(comp2, 2)
        Label5.Text += vbCrLf + "V = " + FormatNumber(comp2, 2) + " m/sec"
        Label5.Text += vbCrLf + "V = " + FormatNumber(comp3, 2) + " cm/sec"
        Me.Height = 480
    End Sub

    Private Sub Button1_Click(ByVal sender As System.Object, ByVal e As System.EventArgs)
Handles Button1.Click
        calculateResults()
    End Sub
```

```
    Private Sub Button2_Click(ByVal sender As System.Object, ByVal e As System.EventArgs)
Handles Button2.Click
      visc = Val(TextBox4.Text)
      hgt = Val(TextBox2.Text)
      lnth = Val(TextBox5.Text)
      dens = Val(TextBox6.Text)
      comp6 = (18 * visc * comp3 * hgt) / (981 * dens * lnth)
      comp7 = comp6 ^ (0.5)
      comp8 = comp7 * 10000
      Label10.Text = "d[p] = ((18 x " + FormatNumber(visc, 2) + " x " +
      FormatNumber(comp3, 2) + _
        " x " + FormatNumber(hgt, 2) + ") / (981 x " + FormatNumber(dens, 2) + " x " _
        + FormatNumber(lnth, 2) + ")){1/2}"
      Label10.Text += vbCrLf + "d[p] = " + FormatNumber(comp7, 4) + " cm = " +
FormatNumber(comp8, 2) + " um"
    End Sub
End Class
```

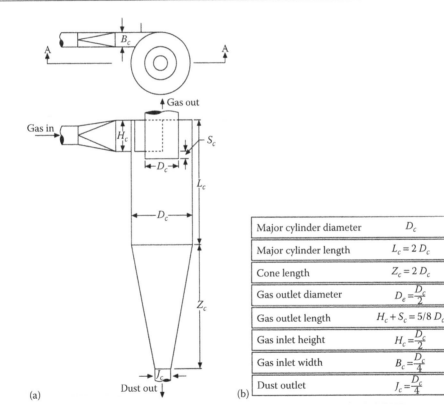

| | |
|---|---|
| Major cylinder diameter | $D_c$ |
| Major cylinder length | $L_c = 2\,D_c$ |
| Cone length | $Z_c = 2\,D_c$ |
| Gas outlet diameter | $D_e = \dfrac{D_c}{2}$ |
| Gas outlet length | $H_c + S_c = 5/8\,D_c$ |
| Gas inlet height | $H_c = \dfrac{D_c}{2}$ |
| Gas inlet width | $B_c = \dfrac{D_c}{4}$ |
| Dust outlet | $J_c = \dfrac{D_c}{4}$ |

(a)                                                                    (b)

**FIGURE 7.2**    (a) Typical dimension for single, high-efficiency cyclone. (b) Nomenclature for a single, high-efficiency cyclone.

The number of effective turns ($N$) within the cyclone can be approximated by the following equation 7.29 (Cooper and Alley 2010):

$$N_e = \frac{1}{H_c}\left(\frac{L_c + Z_c}{2}\right) \tag{7.33}$$

where:

$H_c$ is the height of inlet duct, m
$L_c$ is the length of vertical cylinder, m
$Z_c$ is the length of cone section, m

The collection efficiency of particles both larger and smaller than the $d_{50}$ (cut diameter) can be estimated by using the following equation or from Figure 7.3 (Danielson 1973, Theodore and DePaola 1980):

$$E = \frac{1.0}{\left[1 + \left(d_{50}/d\right)^2\right]} \times 100 \tag{7.34}$$

where:

$E$ is the collection efficiency
$d_{50}$ is the particle diameter with 50% collection efficiency (cut diameter), μm
$d$ is the particle diameter of a specified size, μm

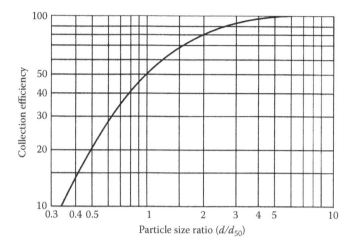

**FIGURE 7.3**  Cyclone efficiency versus particle size ratio.

The pressure drop through a cyclone can be estimated by using the following empirical equation (Vesilind, Morgan and Heine 2009):

$$\Delta P = \frac{3950KQ^2 P \rho}{T} \qquad (7.35)$$

where:

$\Delta P$ is the pressure drop through cyclone, m of water
$Q$ is the flow rate, m³/s
$P$ is the absolute pressure, atm
$\rho$ is the gas density, kg/m³
$T$ is the temperature, $K$
$K$ is the factor, function of cyclone diameter (approximate values are shown in Table 7.7)

Example 7.20 demonstrates the use of Equations 7.32 through 7.35.

### Example 7.20

Determine the size of the particles that can be collected with 50% efficiency (cut diameter) for a cyclone with the following dimensions and operating under the following conditions at a cement kiln plant. Also estimate the pressure drop through the cyclone as well as the removal efficiency for particles with a 20 μm diameter.

**TABLE 7.7**

**Values of $K$ for Calculating Pressure Drop through Cyclone**

| Cyclone Diameter (cm) | K |
| --- | --- |
| 74 | $10^{-4}$ |
| 40 | $10^{-3}$ |
| 20 | $10^{-2}$ |
| 10 | $10^{-1}$ |

*Source:* Vesilind, P.A. et al. *Environmental Engineering.* 2nd ed., Butterworth-Heinemann, Stoneham, MA, 1988. With permission.

Flow rate, $Q = 1.6$ m³/s
Length of vertical cyclone cylinder, $L_e = 1$ m
Length of cyclone cone section, $Z_c = 1$ m
Cyclone diameter, $D_c = 0.50$ m
Cyclone inlet width, $B_c = 0.15$ m
Cyclone inlet height, $H_c = 0.25$ m
Specific gravity of particles, $\rho_p = 2000$ kg/m³
Gas viscosity (air) $\mu = 0.0756$ kg/m*h
Density of air, $\rho_a = 0.965$ kg/m³
Pressure $P = 1$ atm
Temperature $= 93°C$

**Solution**

From Equation 7.33, the estimated number of effective turns ($N_e$) within the cyclone would be

$$N_e = \frac{1}{0.25}\left(1\,m + \frac{1\,m}{2}\right) = \frac{1.50}{0.25} = 6$$

The gas inlet velocity $v_i = Q/A$:

$$v_i = \frac{1.6\ m^3/s}{0.25\ m \times 0.15\ m} = \frac{42.67\ m}{s} = 153\ mh$$

From Equation 4.32, the $d_{50}$ or cut diameter can be calculated as follows:

$$d_{50} = \sqrt{\left(\frac{9 \times 0.0756\ kg/m*h \times 0.15\ m}{2 \times 3.14 \times 6 \times 153{,}600\ m/h \times 2000\ kg/m^3}\right)}$$

$$= 2.96 \times 10^{-6} = 3\ \mu$$

Use Equation 7.35 to estimate the pressure drop through the cyclone:

$Q = 1.6$ m³/s
$P = 1$ atm
$\rho_a = 0.965$ kg/m³
$T = 93°C$
$K =$ factor for cyclone, interpolation for a 50 cm diameter from Table 7.7 gives $K = 0.00074$ or $7.4 \times 10^{-4}$

$$\Delta P = \frac{3950(7.4 \times 10^{-4})(1.6\ m^3/s)^2 (1\ atm)\left(0.965\ kg/m^3\right)}{\left(273 + 93\ K\right)}$$

$$= 0.0197\ m = 1.97\ cm$$

For a 20 μm diameter particle, the removal efficiency can be estimated from Equation 7.34 as follows:

$$E = \frac{1}{\left[1 + \left(3/20\right)^2\right]} \times 100 = 98\%$$

From Figure 7.3, removal efficiency for a 20 μm diameter particle would be as follows:

$$\text{Relative size} = \frac{d_{20}}{d_{50}} = \frac{20}{2.6} = 7.7$$

From the graph, the efficiency would approach 100%.

## Example 7.21

1. Prepare a computer program that can be used to estimate the size of particles that can be collected with 50% (cut diameter) efficiency with a cyclone of given dimensions. Include in the program a system for the determination of the number of effective turns ($N_e$) within the cyclone, the collection efficiency for particles both smaller and larger than $d_{50}$ size, as well as an estimate of the pressure drop ($\Delta P$) through the cyclone.
2. For a single high-efficiency cyclone with the following dimensions that is operating under the given conditions, use the computer program developed in (1) to estimate the size of particles that can be collected with 50% efficiency, $N_e$, $\Delta P$, and the collection efficiency for a 15 μm sized particle.

Flow rate $Q = 6$ m³/s
Length of vertical cyclone cylinder $L_c = 3$ m
Length of cyclone section $Z_c = 3$ m
Cyclone diameter $D_c = 1.5$ m
Cyclone inlet width $B_c = 0.375$ m
Cyclone inlet height $H_c = 0.75$ m
Specific gravity of particle $\rho_p = 1700$ kg/m³
Gas viscosity at 80°C $\mu = 0.0745$ kg/m*h
Density of air at 80°C $\rho_a = 1.001$ kg/m³
Pressure = 1 atm
$K$ = factor for cyclone ($D = 1.5$ m), extrapolation from Table 4.7 gives $5 \times 10^{-5}$

## Solution

1. For a solution to Example 7.21 (1), see the listing of Program 7.21 for the determination of $d_{50}$, Ne, $\Delta P$, and collection efficiency for particles smaller and larger than the $d_{50}$ size.

2. A solution to Example 7.21 (2):
   a. From Equation 7.33:

   $$N_e = \frac{1}{0.75}\left(3\,m + \frac{3\,m}{2}\right) = 6$$

   b. The gas inlet velocity $v_i = Q/A$:

   $$v_i = \frac{6\,m^3/s}{0.375\,m \times 0.75\,m} = \frac{21.33\,m}{s} = 76{,}000\,m/h$$

   c. From Equation 7.32:

   $$d_{50} = \left(\frac{9 \times \dfrac{0.0745\,kg}{m}/h \times 0.375}{2 \times 3.14 \times 6 \times 76{,}800\,m/h \times 1700\,kg/cm^3}\right)^{1/2}$$

   $$= 0.00007\,m = 7\,m$$

   d. From Equation 7.35:

   $$\Delta P = \frac{3950\left(5 \times 10^{-5}\right)(6\,m^3/s)^2(1\,atm)\left(1.001\,kg/m^3\right)}{(273 + 80\,K)}$$

   $$= 0.020\,m = 2.0\,cm$$

   e. From Equation 7.34, the efficiency for a 15 μm particle is calculated as follows:

   $$E = \frac{1.0}{\left[1 + (7/15)^2\right]} \times 100 = \frac{1.0}{1.218} \times 100 = 82\%$$

   f. From Figure 7.3, $d/d_{50} = 15/7 = 2.14$. The collection efficiency = 82%.

---

### LISTING OF PROGRAM 7.21　(CHAP7.21\FORM1.VB): EFFICIENCY OF CYCLONE

```
'****************************************************************************
'Example 7.21: Estimate the size of particles that can be collected
'with a 50% (cut diameter) efficiency with a cyclone of given dimensions
'****************************************************************************
Public Class Form1
    Dim screen As Integer = 1
    Dim hgt1, vert1, length1, bq, bwidth, dens, visc As Double
    Dim factor, pres, gasdens, temp, diam As Double
    Dim acomp1, acomp2, acomp3, aansw As Double
    Dim bcomp1, bansw, banswb As Double
    Dim ccomp1, ccomp2, ccomp3, cansw, canswb As Double
    Dim tcomp, qs, dcomp1, dansw, danswb As Double
    Dim ecomp1, ecomp2, ecomp3, ecomp4, eansw As Double

    Private Sub Form1_Load(ByVal sender As System.Object, ByVal e As System.EventArgs)
Handles MyBase.Load
        Me.Text = "Example 7.21: Cyclone Efficiency"
        Me.FormBorderStyle = Windows.Forms.FormBorderStyle.FixedSingle
        Me.MaximizeBox = False
```

```
        Label1.Text = "Estimate the size of particles that can be collected"
        Label1.Text += vbCrLf + "with a 50% (cut diameter) efficiency with a cyclone of
        given dimensions"
        Label1.Text += vbCrLf + "The formulas:"
        Label1.Text += vbCrLf + "Ne = (1/h[c]) * (L[c] + Z[c]/2))"
        Label1.Text += vbCrLf + "V[i] = Q/A"
        Label1.Text += vbCrLf + "d[50] = (9uB[c] / 2piN[e]V[i]r[p]1/2"
        Label1.Text += vbCrLf + "Delta P = 3950 K Q2 P r / T"
        Label1.Text += vbCrLf + "E = 1 / (1 + (d[50]/d){2}) * 100"
        GroupBox1.Text = ""
        changeScreen()
End Sub
'********************************************************************************
'This sub prepares the screen every time by resetting
'the labels' text, the button text, and empties the
'textboxes. It shows/hides the controls as needed.
'********************************************************************************
    Sub changeScreen()
            TextBox1.Text = ""
            TextBox2.Text = ""
            TextBox3.Text = ""
            TextBox4.Text = ""
    Select Case screen
        Case 1
            Label2.Text = "Ne = (1/h[c]) * (L[c] + Z[c]/2))"
            Label3.Text = "Enter the height of the inlet duct - H[c] (in m)"
            Label4.Text = "Enter the length of vertical cylinder - L[c] (in m)"
            Label5.Text = "Enter the length of cone section - Z[c] (in m)"
            Label6.Text = ""
            Button1.Text = "&Calculate 'Ne'"
        Case 2
            Label2.Text = "v[i] = Q/A"
            Label3.Text = "Enter the Flow Rate - Q (in m{3}/s)"
            Label4.Text = "Enter the cyclone inlet width - B[c] (in m)"
            Label5.Text = "Enter specific gravity of patricles - Rho[p] (in kg/m{3})"
            Button1.Text = "&Calculate 'v[i]'"
        Case 3
            Label2.Text = "d[50] = (9uB[c] / 2piN[e] V[i] Rho[p])1/2"
            Label3.Text = "Enter the gas viscosity - u(in kg/m*h)"
            Button1.Text = "&Calculate 'd[50]'"
        Case 4
            Label2.Text = "Delta P = 3950 K Q{2} P Rho[a] / T"
            Label3.Text = "Enter k factor from Table 7.7 - use scientific notation x.xxE+/-x"
            Label4.Text = "Enter the absolute pressure (in atm)"
            Label5.Text = "Enter the gas density - Rho[a] (in kg/m{3})"
            Label7.Text = "Enter the temperature - T (in C)"
            Button1.Text = "&Calculate 'delta[P]'"
        Case 5
            Label2.Text = "E = 1 / (1 + (d[50]/d){2})* 100"
            Label3.Text = "Enter particle diameter (in um)"
            Button1.Text = "&Calculate 'E'"
    End Select

    Select Case screen
        Case 1, 2
            Label4.Visible = True
            Label5.Visible = True
            Label7.Visible = False
            TextBox2.Visible = True
            TextBox3.Visible = True
            TextBox4.Visible = False
```

```
        Case 3, 5
            Label4.Visible = False
            Label5.Visible = False
            Label7.Visible = False
            TextBox2.Visible = False
            TextBox3.Visible = False
            TextBox4.Visible = False
        Case 4
            Label4.Visible = True
            Label5.Visible = True
            Label7.Visible = True
            TextBox2.Visible = True
            TextBox3.Visible = True
            TextBox4.Visible = True
    End Select
    End Sub

Sub calculateResults()
    Select Case screen
        Case 1
            hgt1 = Val(TextBox1.Text)
            vert1 = Val(TextBox2.Text)
            length1 = Val(TextBox3.Text)
            Acomp1 = 1 / hgt1
            Acomp2 = length1 / 2
            Acomp3 = vert1 + acomp2
            Aansw = acomp1 * acomp3
            Label6.Text = "Ne = " + FormatNumber(aansw, 2)
            screen = 2
            changeScreen()
        Case 2
            bq = Val(TextBox1.Text)
            bwidth = Val(TextBox2.Text)
            dens = Val(TextBox3.Text)
            bcomp1 = bwidth * hgt1
            bansw = bq / bcomp1
            banswb = bansw * 3600
            Label6.Text = "v[i] = " + FormatNumber(bansw, 2) + " m/s = " +
            FormatNumber(banswb, 2) + " m/hr"
            screen = 3
            changeScreen()
        Case 3
            visc = Val(TextBox1.Text)
            Ccomp1 = 9 * visc * bwidth
            Ccomp2 = 2 * 3.14 * aansw * banswb * dens
            Ccomp3 = ccomp1 / ccomp2
            Cansw = ccomp3 ^ (0.5)
            Canswb = cansw * 1000000
            Label6.Text = "d[50] = " + FormatNumber(cansw, 6) + " m = " +
            FormatNumber(canswb, 2) + " um"
            screen = 4
            changeScreen()
        Case 4
            factor = Val(TextBox1.Text)
            pres = Val(TextBox2.Text)
            gasdens = Val(TextBox3.Text)
            temp = Val(TextBox4.Text)
            Tcomp = 273 + temp
            Qs = bq * bq
            Dcomp1 = 3950 * factor * qs * pres * gasdens
            Dansw = dcomp1 / tcomp
```

```
                    Danswb = dansw * 100
                    Label6.Text = "Delta P = " + FormatNumber(dansw, 2) + " m = " +
                    FormatNumber(danswb, 2) + " cm"
                    screen = 5
                    changeScreen()
                Case 5
                    diam = Val(TextBox1.Text)
                    ecomp1 = canswb / diam
                    ecomp2 = ecomp1 ^ 2
                    ecomp3 = 1 + ecomp2
                    ecomp4 = 1 / ecomp3
                    eansw = ecomp4 * 100
                    Label6.Text = "E = " + FormatNumber(eansw, 2) + " percent"
                    screen = 6
                    Button1.Text = "Start over"
                Case 6
                    screen = 1
                    changeScreen()
            End Select
        End Sub


    Private Sub Button1_Click(ByVal sender As System.Object, ByVal e As System.EventArgs)
Handles Button1.Click
            calculateResults()
        End Sub
End Class
```

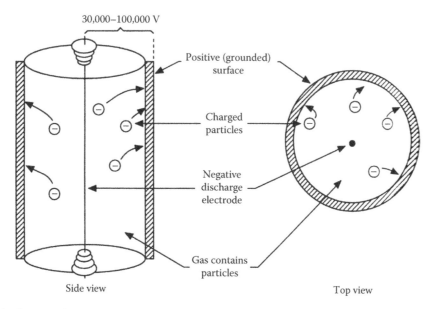

30,000–100,000 V

Positive (grounded) surface

Charged particles

Negative discharge electrode

Gas contains particles

Side view

Top view

**FIGURE 7.4**  Schematic diagram of an ESP.

### 7.3.3 ELECTROSTATIC PRECIPITATORS

ESPs operate by ionizing air molecules into positive and negative ions. This ionization takes place when the air is passed through a very high direct current voltage, anywhere from 30,000 to 100,000 V. As the negative ions move to the positive collecting electrode, they attach themselves to the particles in the gas stream. When the particles become charged by the negative ions, they move to the positively charged collection surface (see Figure 7.4). The particles are removed from the collection surface either by rapping it or by washing it.

There are two types of precipitators: plate type and tube type (see Figure 7.5). The plate type tends to be used most often. The size of the collection plates used ranges from 1 to 3 m long and 15 m high. The distance between the plates ranges from 15 to 40 cm. ESPs can handle huge volumes of gas and are widely used at power plants where electrical power is readily available.

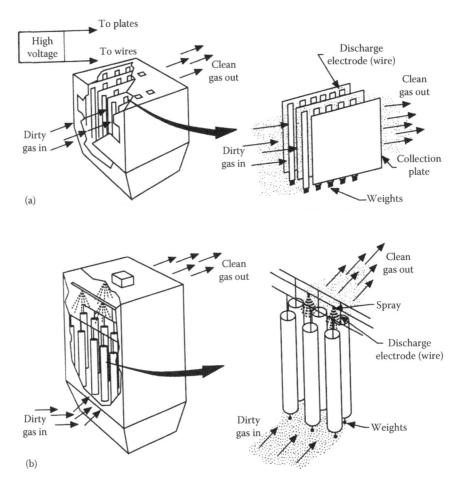

**FIGURE 7.5** High-voltage electrostatic precipitators: a) plate type precipitator and b) Tube type precipitator.

ESPs have many advantages, including the following (U.S. Department of Health and Human Services 1973):

- High efficiency, especially for submicron particles
- Dry collection of dusts
- Low pressure drop
- Ability to collect mists and corrosive acids
- Low maintenance costs
- Low operating costs
- No moving parts
- Collection efficiency can be adjusted by unit size
- Ability to handle gases up to 815°C

The disadvantages of ESPs include the following:

- High initial cost
- Frequently needs a precleaner
- Large space requirements
- Not suitable for collecting combustile particles such as grain or wood dust

The efficiency of an ESP can be estimated by the following empirical equation (Deutsch-Anderson equation) (Air Pollution Training Institute 1984):

$$\eta = 1 - \exp\left(-\frac{Aw}{Q}\right) \qquad (7.36)$$

where:

$\eta$ is the fractional collection efficiency
$A$ is the area of the collection plates, m²
$w$ is the drift velocity of the charged particles, m/s
$Q$ is the flow rate of the gas stream, m³/s

The drift velocity $w$ is the velocity at which a particle approaches the collection surface and can be estimated by the following equation (Air Pollution Training Institute 1984):

$$w = ad_p \qquad (7.37)$$

where:

$w$ is the drift velocity, m/s
$a$ is the constant, a function of the charging field, the carrier gas properties, and the ability of the particle to accept an electrical charge, s⁻¹
$d_p$ is the particle size, μm

Drift velocities commonly range from 0.02 to 0.2 m/s and can be determined experimentally. The following example demonstrates the calculations that can be used to evaluate the operation or design of an ESP.

## Example 7.22

An ESP is to remove fly-ash particles of 0.5 μm diameter from a gas stream flowing at 0.7 m³/s. Determine the plate area required to collect the 0.5 μm particles at 98% efficiency. The value for the constant $a$ has been found to be $0.24 \times 10^6$/s.

### Solution

Using Equation 7.37, the drift velocity will be

$$w = 0.24 \times 106 / s \times 0.5 \times 10^{-6} m = 0.12 m/s$$

Using Equation 7.36 and 98% removal efficiency:

$$0.98 = 1 - \exp\left(-\left(\frac{0.12 m/s}{0.7 m^3/s} \times A\right)\right)$$

which yields $A = 23$ m².

Considering that both sides of the plate are available for collecting particles, the plate area required is 23/2 = 11.5 m². Use a square plate 3.4 by 3.4 m³.

## Example 7.23

1. Prepare a computer program that can be used to estimate the drift velocity for a plate-type ESP, given the constant $a$ in Equation 7.37. Also include in the program the elements in Equation 7.36.

Design the program so that any of the other elements can be calculated, provided any of the other two elements are given.
2. Using the computer program developed in (1), determine the number of 6 by 6 m plates needed to collect 0.4 μm particles with a drift velocity constant $a$ of $2.5 \times 10^5$/s. The total air flow is 30 m³/s, and the collection efficiency is to be 90%.

### Solution

1. For a computer program for Example 7.23 (1), see the listing of program 7.23, which includes the elements in Equations 7.36 and 7.37.
2. A solution to Example 7.23 (2):
   a. Using Equation 7.37:

   $$w = 2.5 \times 105/s \times 0.4 \times 10^{-6} m = 0.10 \ m/s$$

   b. Using Equation 7.36:

   $$0.90 = 1 - \exp\left(-\left(\frac{0.10 \ m/s}{30 \ m^3/s} \times A\right)\right)$$

which yields $A = 691$ m². Each plate surface has $6 \times 6 \times 2 = 72$ m² surface area. Use 10 plates.

---

### LISTING OF PROGRAM 7.23   (CHAP7.23\FORM1.VB): ELECTROSTATIC PRECIPITATOR

```
'******************************************************************************
'Example 7.23: Estimate the drift velocity for a plate-type
'electrostatic precipitator given the constant in eqn. 7.32.
'The program can calculate any of the elements - given the other two.
'******************************************************************************
Imports System.Math

Public Class Form1
      Dim aconst, partsize, wcompa, compvel As Double
      Dim area, rate, perc As Double
   Private Sub Form1_Load(ByVal sender As System.Object, ByVal e As System.EventArgs)
Handles MyBase.Load
      Me.Text = "Example 7.23:"
      Me.MaximizeBox = False
      Me.FormBorderStyle = Windows.Forms.FormBorderStyle.FixedSingle
      Label1.Text = "Estimate the drift velocity for a plate-type"
      Label1.Text += vbCrLf + "electrostatic precipitator given the constant in eqn. 7.32."
      Label1.Text += vbCrLf + "The program can calculate any of the elements - given the
      other two."
      Label1.Text += vbCrLf + "The formulas:"
      Label1.Text += vbCrLf + "w = a*d[p]"
      Label1.Text += vbCrLf + "n = 1 -exp(-(A*w/Q))"
      Label8.Text = "Enter value for constant a - Use scientific notation (x.xxE+/-x)"
      Label9.Text = "Enter particle size - d[p] (in um)"
      Label6.Text = ""
      Label10.Text = ""
      Label2.Text = "n = 1 -exp(-(A*w/Q))"
      Label2.Text += vbCrLf + "In the above formula, given w, this program solves for:"
      RadioButton1.Text = "n, the fractional collection efficiency"
      RadioButton2.Text = "A, the area of the collection plates"
```

```
                RadioButton3.Text = "Q, the flow rate of the gas stream"
                RadioButton4.Text = "Start a new problem (recompute w)"
                RadioButton1.Checked = True
                Button1.Text = "Calculate 'w'"
                Button2.Text = "Calculate 'n'"
                GroupBox1.Text = ""
                Me.Height = 223
        End Sub
        '*********************************************************************************
        'calculates the equation, solving for:
        ' n if Radiobutton1 is selected
        ' A if Radiobutton2 is selected
        ' Q if Radiobutton3 is selected
        ' or,
        ' reset the textboxes and start over
        ' if Radiobutton4 is selected.
        '*********************************************************************************
        Sub calculateResults()
                Dim comp1, comp2, comp3, comp4, comp5, pcomp, acomp, qcomp As Double
                If RadioButton1.Checked Then
                        area = Val(TextBox1.Text)
                        rate = Val(TextBox2.Text)
                        comp1 = (area * compvel) / rate
                        comp2 = (Exp(-2.32))
                        comp3 = 1 / (comp2)
                        comp4 = 1 / comp3comp5 = 1 - comp4
                        pcomp = comp5 * 100
                        Label6.Text = "The fractional collection efficiency, n = " +
                        FormatNumber(pcomp, 2) + "%"
                ElseIf RadioButton2.Checked Then
                        Dim platewidth, plateheight, platearea, numplates As Double
                        perc = Val(TextBox1.Text)
                        If (perc > 1) Then
                            perc = perc * 0.01
                        End If
                        rate = Val(TextBox2.Text)
                        comp1 = 1 - perc
                        comp3 = Log(comp1)
                        comp4 = rate / compvel
                        acomp = (comp4 * comp3) * -1
                        Label12.Text = "The area of the collection plates is " +
                        FormatNumber(acomp, 2) + " m{2}"
                        Label12.Text += vbCrLf + "Determining the number of plates of a given size"
                        platewidth = Val(TextBox4.Text)
                        plateheight = Val(TextBox7.Text)
                        platearea = platewidth * plateheight * 2
                        numplates = acomp / platearea
                        Label6.Text = "The plate area is computed as width*height*2 (plates have
                        two sides)"
                        Label6.Text += vbCrLf + Format(numplates, "##")
                        Label6.Text += " plates will be required for this collection"
                        ElseIf RadioButton3.Checked Then
                        perc = Val(TextBox1.Text)
                        If (perc > 1) Then
                            perc = perc * 0.01
                        End If
                        area = Val(TextBox2.Text)
                        comp1 = 1 - perc
                        comp2 = 1 / comp1
                        comp3 = Log(comp2)
                        comp4 = area * compvel
```

```
                qcomp = comp4 / comp3
                Label6.Text = "The flow rate Q is " + FormatNumber(qcomp, 2) + " m{3}/s"
        End If
End Sub
'*******************************************************************************************
'This sub prepares the screen every time by resetting
'the labels' text, the button text, and empties the
'textboxes. It shows/hides the controls as needed.
'*******************************************************************************************
Sub changeDisplay(ByVal n As Integer)
                Select Case n
            Case 1
                Label3.Text = "Solving for n, the fractional collection efficiency"
                Label3.Text += vbCrLf + "n = 1 -exp(-(A*w/Q))"
                Label4.Text = "Enter the plate area - A (in m{2})"
                Label5.Text = "Enter the flow rate of the stream - Q (in m{3}/s)"
                Label7.Visible = False
                Label11.Visible = False
                Label12.Visible = False
                TextBox4.Visible = False
                TextBox7.Visible = False
                Button2.Text = "Calculate 'n'"
            Case 2
                Label3.Text = "Solving for A, area of the collection plates"
                Label3.Text += vbCrLf + "A = (Q(ln(1/1-n))/w"
                Label4.Text = "Enter the efficiency - n (in %)"
                Label5.Text = "Enter the flow rate of the stream - Q (in m{3}/s)"
                Label7.Text = "Enter the width of the collection plate (in m)"
                Label11.Text = "Enter the height of the collection plate (in m)"
                Label12.Visible = True
                Label12.Text = ""
                Label7.Visible = True
                Label11.Visible = True
                TextBox4.Visible = True
                TextBox7.Visible = True
                Button2.Text = "Calculate 'A'"
            Case 3
                Label3.Text = "Solving for Q, the flow rate of the gas stream"
                Label3.Text += vbCrLf + "Q = (A w)/(ln(1/1 - n)"
                Label4.Text = "Enter the efficiency - n (in %)"
                Label5.Text = "Enter the plate area - A (in m{2})"
                Label7.Visible = False
                Label12.Visible = False
                Label11.Visible = False
                TextBox4.Visible = False
                TextBox7.Visible = False
                Button2.Text = "Calculate 'Q'"
            Case 4
                Me.Height = 223
                RadioButton1.Checked = True
                Label10.Text = ""
                TextBox1.Text = ""
                TextBox2.Text = ""
                TextBox4.Text = ""
                TextBox7.Text = ""
                TextBox5.Text = ""
                TextBox6.Text = ""
                TextBox5.Focus()
        End Select
        Label6.Text = ""
End Sub
```

```
    Private Sub Button1_Click(ByVal sender As System.Object, ByVal e As System.EventArgs)
Handles Button1.Click
        aconst = Val(TextBox5.Text)
        partsize = Val(TextBox6.Text)
        Wcompa = aconst * partsize
        Compvel = wcompa * 0.000001
        Label10.Text = "The Drift velocity - w = " + FormatNumber(compvel, 2) + " m/s"
        Me.Height = 562
    End Sub

    Private Sub Button2_Click(ByVal sender As System.Object, ByVal e As System.EventArgs)
Handles Button2.Click
        calculateResults()
    End Sub

    Private Sub RadioButton1_CheckedChanged(ByVal sender As System.Object, ByVal e As
System.EventArgs) Handles RadioButton1.CheckedChanged
        If RadioButton1.Checked Then changeDisplay(1)
    End Sub

    Private Sub RadioButton2_CheckedChanged(ByVal sender As System.Object, ByVal e As
System.EventArgs) Handles RadioButton2.CheckedChanged
        If RadioButton2.Checked Then changeDisplay(2)
    End Sub

    Private Sub RadioButton3_CheckedChanged(ByVal sender As System.Object, ByVal e As
System.EventArgs) Handles RadioButton3.CheckedChanged
        If RadioButton3.Checked Then changeDisplay(3)
    End Sub

    Private Sub RadioButton4_CheckedChanged(ByVal sender As System.Object, ByVal e As
System.EventArgs) Handles RadioButton4.CheckedChanged
        If RadioButton4.Checked Then changeDisplay(4)
    End Sub
End Class
```

### 7.3.4 Venturi Scrubbers

A venturi scrubber passes a contaminated gas stream through a duct that has a converging venturi-shaped throat, followed by a diverging section (see Figure 7.6). The scrubbing liquid is injected at right angles to the incoming gas stream, which breaks the liquid into small droplets that can then combine with the gaseous or particulate contaminants.

A venturi scrubber will remove from 92% to 99% of particulate in the size range from 0.02 to 0.5 μm, which makes them very effective for removal of submicron particles in smoke and fumes (Danielson 1973). The advantages of using a venturi scrubber include the following:

- Collecting both particulate and gaseous contaminants
- Cooling high-temperature gas streams
- Elimination of problems with fires or explosions
- Variable collection efficiency
- Recovery of a valuable by-product is made possible

Its disadvantages include the following:

- High power costs
- Collecting large volumes of liquid that require disposal
- Elimination of buoyancy of gas at the stack exit by reduced gas temperature
- White vapor cloud at the stack exit that may concern the surrounding community
- Problems with corrosion

Johnstone's equation can be used to evaluate the operation and design of venturi scrubbers (Air Pollution Training Institute 1984):

$$E = 1 - \exp\left[-k\left(Q_L / Q_G\right)\sqrt{\Psi}\right] \tag{7.38}$$

where:

$E$ is the fractional collection efficiency
$k$ is the correlation coefficient depending upon system geometry and operating conditions

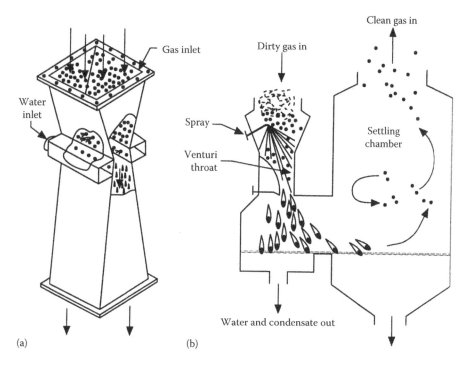

**FIGURE 7.6**    Venturi wet collector: a) isometric view b) sectional view.

$Q_L/Q_G$ is the liquid-to-gas ratio ($Q_L$, m³/s; $Q_G$ m³/s)
$\Psi$ is the interial impaction parameter

$$\Psi = \frac{C\rho_p v d_p^2}{18 d_o \mu} \qquad (7.39)$$

where:
  $C$ is the Cunningham correction factor, dimensionless
  $\rho_p$ is the particle density, kg/m³
  v is the gas velocity at venturi throat, m/s
  $d_p$ is the particle diameter, m
  $d_o$ is the average droplet diameter, m
  $\mu$ is the gas viscosity, kg/m*s

The Cunningham correction factor may be approximated by using the following equation (Air & Waste Management Association and Davis 2000):

$$C = 1 + \frac{\left(6.21 \times 10^{-4}\right)T}{d_p} \qquad (7.40)$$

where:
  $T$ is the absolute gas temperature, K
  $d_p$ is the diameter of particle, µm

Under normal conditions, the Cunningham factor is taken as 1 for particles larger than 1 µm.

The average liquid droplet $d_o$ may be determined by using the following equation 7.41 (Air Pollution Training Institute 1984):

$$d_o = \frac{5000}{v} + 29.67\left(1000 \times \frac{Q_L}{Q_G}\right)^{1.5} \qquad (7.41)$$

where $d_o$ is the average droplet diameter, µm
  The other terms in Equation 7.41 are the same as presented earlier.

The pressure drop through a venturi scrubber can be estimated by the following empirical equation (Vesilind, Morgan and Heine 2009):

$$\Delta P = v^2 L \times 10^{-6} \qquad (7.42)$$

where:
  $\Delta P$ is the pressure drop across the venturi scrubber, cm of water
  v is the gas velocity through throat, cm/s
  $L$ is the water-to-gas volume ratio, L/m³

Example 7.24 demonstrates the calculations that can be made to evaluate the operation and design of a venturi scrubber.

**Example 7.24**

Determine the collection efficiency of a venturi scrubber for particles 0.7 µm in diameter from a gas stream containing fly ash. The liquid-to-gas ratio is 1.12 × 10⁻³ m³ of liquid per m³ of gas. Also determine the pressure drop across the venturi scrubber, given the scrubber described below:

Throat velocity = 90 m/s
Density of particles = 900 kg/m³
Gas viscosity = 2.2 × 10⁻⁵ kg/m/s
Gas temperature = 100°C
Correlation coefficient K = 2000

**Solution**

- Using Equation 7.40, estimate the Cunningham correction factor:

$$C = 1 + \frac{\left(6.21 \times 10^{-4}\right)T}{d_p} = 1.33$$

- Using Equation 7.41, estimate the average droplet diameter $d_o$:

$$d_o = \frac{5000}{90} + 29.67\left(1000 \times 1.12 \times 10^{-3}\right)^{1.5} = 90.7\,\mu m$$

- Using Equation 7.39, calculate the interial impaction parameter:

$$\Psi = \frac{\left(1.33 * 900 \text{ kg/m}^3\right)\left(90 \text{ m/s}\right)\left(0.7 \times 10^{-6} \text{ m}\right)^2}{18\left(90.7 \times 10^{-6} \text{ m}\right)\left(2.20 \times \dfrac{10^{-5}\text{kg}}{\text{m}} / s\right)} = 1.47$$

$$\sqrt{\Psi} = 1.21$$

- Using Equation 7.39, determine the venturi scrubber's collection efficiency: $E = 1 - \exp[-2000 \times 1.12 \times 10^{-3} \times 1.12] = 1 - 0.07 = 93\%$ for removal of 0.7 sized particles
- The pressure drop across the venturi scrubber can be calculated from Equation 7.42: $\Delta P = (9000 \text{ cm/s})^2(1.12 \text{ L/m}^3)10^{-6} = 90.72 \text{ cm} = 35.7$ in $H_2O$

---

**LISTING OF PROGRAM 7.25   (CHAP7.25\FORM1.VB): VENTURI SCRUBBER**

```
'********************************************************************************
'Example 7.25: Estimate the efficiency of a Venturi scrubber to remove
'particles of a specific size as well as the pressure drop through
'the scrubber.
'********************************************************************************
Public Class Form1
  Dim screen As Integer = 1
  Dim temp, dia, qg, ql, vel, visc, dens, cc As Double
  Dim ratio, tcomp, acomp1, acomp2, acomp3, acomp4, acomp5, acomp6, d, c As Double
  Dim dcomp, diacomp, diasquar, nom, denom, IIP, para As Double
  Dim ecomp1, ecomp2, eff As Double
  Dim lfr, pcomp1, pcomp2, pcomp3, expon, drop, dropb As Double

  Private Sub Form1_Load(ByVal sender As System.Object, ByVal e As System.EventArgs)
Handles MyBase.Load
      Me.Text = "Example 7.25:"
      Me.FormBorderStyle = Windows.Forms.FormBorderStyle.FixedSingle
      Me.MaximizeBox = False
      Label1.Text = "Example 7.25: Estimate the efficiency of a venturi scrubber to
      remove"
      Label1.Text += vbCrLf + "particles of a specific size as well as the pressure drop
      through"
      Label1.Text += vbCrLf + "the scrubber."
      Label1.Text += vbCrLf + "The formulas:"
      Label1.Text += vbCrLf + "E = 1 - exp [-k (Q[L]/Q[G]) SQR-IIP]"
      Label1.Text += vbCrLf + "(SQR-IIP = Square root of interial impaction parameter)"
      Label1.Text += vbCrLf + "IIP = C r[p] v d[p]{2} / 18 d[o] u"
      Label1.Text += vbCrLf + "C = 1 + ((6.21 x 10{04})T) / d[p]"
      Label1.Text += vbCrLf + "d[o] = (5000/v) + 29.67 (1000 (Q[L]/Q[G]){1.5}"
      Label1.Text += vbCrLf + "Delta P = v{2} L x 10{-6}"
      GroupBox1.Text = ""
      changeScreen()
  End Sub

'********************************************************************************
'This sub prepares the screen every time by resetting
'the labels' text, the button text, and empties the
'textboxes. It shows/hides the controls as needed.
'********************************************************************************
  Sub changeScreen()
      TextBox1.Text = ""
      TextBox2.Text = ""
      TextBox3.Text = ""
```

```
        TextBox4.Text = ""
        TextBox5.Text = ""

    Select Case screen
        Case 1
           Label7.Text = "C = 1 + ((6.21 x 10{04})T) / d[p]"
           Label7.Text += vbCrLf + "d[o] = (5000/v) + 29.67 (1000 (Q[L]/Q[G]){1.5}"
           Label2.Text = "Enter the gas temperature - T (in C)"
           Label3.Text = "Enter the diameter of the particle - d[p] (in um)"
           Label4.Text = "Enter the gas flow rate - (Q[G]) (in m{3}/s)"
           Label5.Text = "Enter the liquid flow rate - (Q[L]) (in m{3}/s)"
           Label6.Text = "Enter the throat velocity - v (in m/s)"
           Label8.Text = ""
           Button1.Text = "&Calculate 'C' and 'd[o]'"
        Case 2
           Label7.Text = "IIP = C r[p] v d[p]{2} / 18 d[o] u"
           Label7.Text += vbCrLf + "Enter the gas viscosity - u (in kg/m/s)"
           Label2.Text = "- use scientific notation (x.xxE+/-x)"
           Label3.Text = "Enter the particle density - r[p] (in kg/m{3})"
           Button1.Text = "&Calculate 'IIP'"
        Case 3
           Label7.Text = "E = 1 -exp[-k (Q[L]/Q[G]) SQR-IIP]"
           Label7.Text += vbCrLf + "(SQR-IIP = Square root of interial impaction
    parameter)"
           Label2.Text = "Enter the correlation coefficient"
        Case 4
           Button1.Text = "Start over"
    End Select
    Select Case screen
        Case 1
           Label3.Visible = True
           Label4.Visible = True
           Label5.Visible = True
           Label6.Visible = True
           TextBox1.Visible = True
           TextBox2.Visible = True
           TextBox3.Visible = True
           TextBox4.Visible = True
           TextBox5.Visible = True
        Case 2
           Label3.Visible = True
           Label4.Visible = False
           Label5.Visible = False
           Label6.Visible = False
           TextBox1.Visible = True
           TextBox2.Visible = True
           TextBox3.Visible = False
           TextBox4.Visible = False
           TextBox5.Visible = False
        Case 3
           Label3.Visible = False
           Label4.Visible = False
           Label5.Visible = False
           Label6.Visible = False
           TextBox1.Visible = True
           TextBox2.Visible = False
           TextBox3.Visible = False
           TextBox4.Visible = False
           TextBox5.Visible = False
        End Select
End Sub
```

```
Sub calculateResults()
    Select Case screen
        Case 1
            temp = Val(TextBox1.Text)
            dia = Val(TextBox2.Text)
            qg = Val(TextBox3.Text)
            ql = Val(TextBox4.Text)
            vel = Val(TextBox5.Text)
            tcomp = 273 + temp
            acomp1 = 0.000621 * tcomp
            acomp2 = acomp1 / dia
            C = 1 + acomp2
            acomp3 = 5000 / vel
            Ratio = ql / qg
            Acomp4 = 1000 * ratio
            Acomp5 = acomp4 ^ (1.5)
            Acomp6 = acomp5 * 29.67
            D = acomp3 + acomp6
            Label8.Text = "C = " + FormatNumber(c, 2)
            Label8.Text += vbCrLf + "d[o] = " + FormatNumber(d, 2) + " um"
            screen = 2
            changeScreen()
        Case 2
            visc = Val(TextBox1.Text)
            dens = Val(TextBox2.Text)
            Dcomp = d * 0.000001
            Diacomp = dia * 0.000001
            Diasquar = diacomp ^ 2
            Nom = c * dens * vel * diasquar
            denom = 18 * dcomp * visc
            IIP = nom / denom
            para = Math.Sqrt(IIP)
            Label8.Text = "IIP = " + FormatNumber(IIP, 2) + " Square Root of IIP = " +
        FormatNumber(para, 2)
            screen = 3
            changeScreen()
        Case 3
            cc = Val(TextBox1.Text)
            ratio = ql / qg
            Ecomp1 = cc * ratio * para
            ecomp2 = Math.Exp(-ecomp1)
            Eff = (1 - ecomp2) * 100
            Label6.Text = "The efficiency is " + FormatNumber(eff, 2) + "%"

            Label8.Text = "Delta P = v{2} L x 10{-6}"
            Lfr = ql * 1000
            ratio = lfr / qg
            Pcomp1 = vel * 100
            pcomp2 = pcomp1 ^ 2
            Pcomp3 = pcomp2 * ratio
            Expon = 1 / 1000000
            Drop = pcomp3 * expon
            Dropb = drop * 0.3937
            Label8.Text += vbCrLf + "The pressure drop is " + FormatNumber(drop, 2) + _
            " cm or " + FormatNumber(dropb, 2) + " inches of H[2]O"
            screen = 4
            changeScreen()
        Case 4
            screen = 1
            changeScreen()
    End Select
```

```
   End Sub
   Private Sub Button1_Click(ByVal sender As System.Object, ByVal e As System.EventArgs)
Handles Button1.Click
        calculateResults()
   End Sub
End Class
```

## Example 7.25

1. Prepare a computer program that will estimate the efficiency of a venturi scrubber to remove particles of a specific size as well as the pressure drop through the venturi scrubber. Include in the program all the elements contained in Equations 7.38 through 7.40 and 7.42.
2. Using the program developed in (1), estimate the collection efficiency of a venturi scrubber at removing 3.2 μm sized particles. The particles have a density of 3000 kg/m$^3$. Also determine the pressure drop across the venturi scrubber. The venturi scrubber characteristics are as follows:

Gas flow rate = 310 m$^3$/min = 5.2 m$^3$/s
Liquid flow rate = 0.00139 m$^3$/s
Gas flow velocity at throat = 100 m/s
Correlation coefficient = 1500
Temperature = 90°C
Gas viscosity = 2.1 × 10$^{-5}$ kg/m/s

## Solution

1. For a solution to Example 7.25 (1), see the listing of Program 7.25, which includes elements in Equations 7.38 through 7.42.
2. A solution to Example 7.25 (2):
   a. Using Equation 7.40, estimate the Cunningham correction factor:

$$C = 1 + \frac{\left(6.21 \times 10^{-4}\right)(273 + 90)}{3.2} = 1.07$$

   b. Using Equation 7.41, estimate the average droplet diameter:

$$d_o = \frac{5000}{100 \text{ m/s}} + 29.67\left(\frac{1000 \times 0.00139 \text{ m}^3/\text{s}}{5.2 \text{ m}^3/\text{s}}\right)^{1.5}$$

$$= 50 + 4 = 54 \text{ μm}$$

   c. Using Equation 7.39, calculate the interial impaction parameter:

$$\Psi = \frac{1.07\left(3000 \text{ kg/m}^3\right)(100 \text{ m/s})\left(3.2 \times 10^{-6} \text{ m}\right)^2}{18\left(54 \times 10^{-6}\right)\left(2.1 \times 10^{-5} \text{ kg/m*s}\right)} = 161$$

$$\sqrt{\Psi} = 12.69$$

   d. Using Equation 7.38, determine the venturi scrubber's collection efficiency for collecting 3.2 μm sized particles: $E = 1 - \exp(-1500 \times 0.00139/5.2 \times 12.69) = .994 = 99.4\%$

   e. Using Equation 7.42, estimate the pressure drop across the throat of the venturi scrubber: $\Delta P = (10{,}000 \text{ cm/s})^2(1.39 \text{ L/s liquid}/5.2 \text{ m}^3/\text{s gas})10^{-6} = 26.73 \text{ cm} = 10.52$ inches of H$_2$O

## 7.3.5   BAGHOUSE OR FABRIC FILTERS

Baghouse or fabric filters are air pollution control devices that are used to separate solid particles from a gas stream. The particulate-laden gas stream is passed through woven or felted fabric; the fabric used for the filter varies from cotton to glass. The type of fabric used depends upon operational conditions such as temperature, pressure drop, chemical or physical degradation, cleaning methods, and, of course, cost and the life of the fabric. Filter bags usually are tubular or envelope shaped and are capable of removing > 99% of particles down to 0.3 μm, as well as substantial quantities of particles as small as 0.1 μm. Filter bags range from 1.8 to 12 m in length and 0.1 to 0.4 m in diameter. A typical baghouse filter is shown in Figure 7.7.

The particles are captured and retained on the fibers of the fabric by means of interception, inertial impaction, Brownian diffusion, thermal precipitation, gravitational settling, and electrostatic attraction. Once a mat or cake of particulate has been collected on the fabric surface, further collection of particles is accomplished by sieving. The fabric then serves mainly as a supporting structure for the dust mat responsible for the high collection efficiency. Periodically, the accumulated particulates are removed by mechanical shaking, reverse air cleaning, or pulse jet cleaning (Colls and Tiwary 2009). The advantages of fabric filters include the following (U.S. Department of Health and Human Services 1973):

- High efficiency for collection of particles as small as 0.1 μm
- Moderate power requirements
- Reasonably low pressure drops
- Dry disposal of recovered particulate
- Operation over a wide range of particulates
- Modular design for simple add-ons

Their disadvantages include the following:

- High capital cost.
- Large space required
- High maintenance and replacement costs
- Cooling required for high-temperature gas streams
- Fabric harmed by corrosive chemicals
- Control of moisture in particulate required
- Potential for fire or explosions

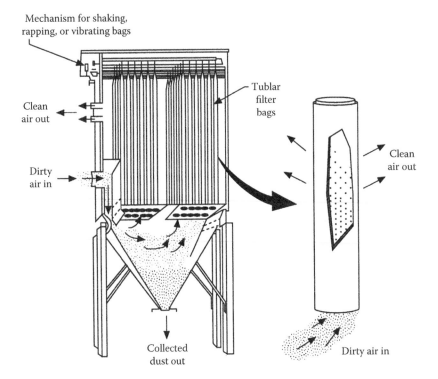

**FIGURE 7.7**  Baghouse filter.

Equation that can be used to evaluate and aid in the design of fabric filters is as follows:

$$Af = \frac{Q}{vf} \tag{7.43}$$

where:

$A_f$ is the total filtering area, m$^2$
$Q$ is the volume of gas stream, m$^3$/min
$v_f$ is the filtering velocity (air-to-cloth ratio), m/min

For cylindrical bags:

$$A_b = \pi dh \tag{7.44}$$

where:

$A_b$ is the filtering area for each bag, m$^2$
$\pi = 3.1416$
$d$ is the diameter of the bag, m
$h$ is the length of the bag, m

The number of bags required then is

$$N = \frac{A_f}{A_b} \tag{7.45}$$

where $N$ is the number of bags required.

The cleaning frequency for a baghouse filter can be estimated from Equation 7.46, provided the desirable pressure drop is known, or, if the cleaning frequency is known, then the pressure drop can be estimated (Williams, Hatch and Greenburg 1940, Air & Waste Management Association and Davis 2000):

$$\Delta P = S_e v_f + k_2 c v_f^2 t \tag{7.46}$$

where:

$\Delta P$ is the pressure drop
$S_e$ is the effective residual drag
$v_f$ is the gas velocity at filter surface
$k_2$ is the specific resistance coefficient of collect particulate
$c$ is the concentration of particulate in the gas stream
$t$ is the filtration time

Values for $S_e$ and $k_2$ can be found in publications or handbooks such as Reference 17. Equation 7.47 includes the values for $S_e$ and $k_2$. This equation can be used under specified conditions to estimate the pressure drop through a fabric filter, provided the required time between cleanings is known or vice versa (Air Pollution Training Institute 1984):

$$\Delta P = 1.6 v_f + 8.52 c v_f^2 t \tag{7.47}$$

where:

$\Delta P$ is the pressure drop, cm
$v_f$ is the filtering velocity, m/min
$c$ is the particulate concentration in the air stream, kg/m$^3$
$t$ is the time required between cleanings, min

Example 7.26 demonstrates the calculations that can be made in order to evaluate the operation or design of a fabric filter.

**Example 7.26**

Determine the filter cloth area and the required frequency of cleaning, given the following data:
Volume of contaminated air stream $Q = 400$ m$^3$/min

Filtering velocity $v_f$ = 0.8 m/min

Particulate concentration in the air stream $c$ = 0.03 kg/m³

Diameter of the filter bag = 0.2 m

Length of the filter bag = 5.0 m

The system is designed for cleaning of filter when pressure drop reaches 9 cm of water.

**Solution**

- The total filter area required according to Equation 7.43:

$$A_f = \frac{400 \text{ m}^3/\text{min}}{0.8 \text{ m/min}} = 500 \text{ m}^2$$

- Using Equation 7.44, the filtering bag area, which is 0.2 m in diameter and 5.0 m long, can be calculated as follows: $A_b$ = 3.14(0.2 m)(5.0 m) = 3.14 m² per bag
- The number of bas required can be determined from Equation 7.45:

$$N = \frac{50 \text{ m}^2}{3.14 \text{ m}^2} = 159 \text{ or } 160 \text{ bags}$$

- The frequency for cleaning these bags can be estimated from Equation 7.47:
9 cm = 1.6(0.8 m/min) + 8.52(0.03 kg/m³)(0.80 m/min)²$t$ which yields $t$ = 47 min.

## Example 7.27

1. Prepare a computer program that will aid in the design and operation of a baghouse (fabric filter unit). Include the elements in Equations 7.43 through 7.45 and 7.47 in the program.

2. Using the program developed in (1), determine the number of filtering bags required and the cleaning frequency of a plant employing a baghouse filter, given the following design data:

Volume of contaminated gas stream $Q$ = 12 m³/s

Filtering velocity $v_f$ = 2 m/min

Particulate concentration in the air stream $c$ = 0.025 kg/m³

Diameter of filter bag = 0.3 m

Length of filter bag = 8 m

The baghouse filter system is to be cleaned when pressure drop reaches 15 cm.

**Solution**

1. For a solution to Example 7.27 (1), see the listing of Program 7.27, which includes the elements in Equations 7.43 through 7.45 and 7.47.
2. A solution to Example 7.27 (2):
   a. From Equation 7.43, the total filter area required is

   $$A_f = \frac{12 \text{ m}^3/\text{min} \times 60 \text{ s/min}}{2 \text{ m/min}} = 360 \text{ m}^2$$

   b. From Equation 7.44, the surface area for each bag is as follows:
   Ab = 3.14(0.3 m)(8 m) = 7.54 m²
   c. From Equation 7.45, the number of bas required is as follows:

   $$N = \frac{360 \text{ m}^2}{7.54 \text{ m}^2} = 47.7 \text{ or } 48 \text{ bags}$$

   d. From Equation 7.47, the estimated frequency of cleaning is as follows: 15 cm = 1.6(2 m/min) + 8.52(0.025 kg/m³)(2 m/min)²$t$ which yields $t$ = 13.8 min.

---

**LISTING OF PROGRAM 7.27   (CHAP7.27\FORM1.VB): DESIGN AND OPERATION OF A BAGHOUSE**

```
'********************************************************************************
'Example 7.27: A program to aid in the design and operation
' of a baghouse (Fabric filter unit)
'********************************************************************************


Public Class Form1
  Dim screen As Integer = 1
  Dim temp, dia, qg, ql, vel, visc, dens, cc As Double
  Dim vol, volcomp, af, ab, n, length, drop, conc As Double
  Dim comp1, comp2, comp4, compt, compvel As Double

  Private Sub Form1_Load(ByVal sender As System.Object, ByVal e As System.EventArgs)
Handles MyBase.Load
    Me.Text = "Example 7.27: Baghouse operation"
    Me.FormBorderStyle = Windows.Forms.FormBorderStyle.FixedSingle
    Me.MaximizeBox = False
    Label1.Text = "Example 7.27: A program to aid in the design and operation of a
    baghouse"
    Label1.Text += vbCrLf + " (Fabric filter unit)"
    Label1.Text += vbCrLf + "The formulas:"
    Label1.Text += vbCrLf + "A[f] = Q / v[f]"
```

```
      Label1.Text += vbCrLf + "A[b] = pi d h"
      Label1.Text += vbCrLf + "N = A[f] / A[b]"
      Label1.Text += vbCrLf + "Delta P = 1.6 v[f] + 8.52 c v[f]{2}t"
      Label4.Font = New Font(FontFamily.GenericSerif, 9, FontStyle.Bold)
      GroupBox1.Text = ""
      changeScreen()
  End Sub

  '********************************************************************************
  'This sub prepares the screen every time by resetting
  'the labels' text, the button text, and empties the
  'textboxes. It shows/hides the controls as needed.
  '********************************************************************************
  Sub changeScreen()
    TextBox1.Text = ""
    TextBox2.Text = ""

    Select Case screen
      Case 1
        Label7.Text = "A[f] = Q / v[f]"
        Label2.Text = "Enter the volume of contaminated gas stream Q (in m{3}/s)"
        Label3.Text = "Enter the filtering velocity v[f] (in m/min)"
        Label4.Text = ""
        Button1.Text = "&Calculate 'A[f]'"
      Case 2
        Label7.Text = "A[b] = pi d h"
        Label2.Text = "Enter the diameter of the bag - d (in m)"
        Label3.Text = "Enter the length of the bag - h (in m);length"
        Button1.Text = "&Calculate 'A[b]'"
      Case 3
        Label7.Text = "N = A[f] / A[b]"
        Label7.Text += vbCrLf + "All the factors for this equation are from previous
        equations"
        N = af / ab
        Label7.Text += vbCrLf + "N = "
        Label7.Text += vbCrLf + Format(N, "#####") + " bags"
        Label7.Text += vbCrLf + "(Rounded to the nearest whole number)"
        Label2.Text = ""
        Label3.Text = ""
        Button1.Text = "Next"
      Case 4
        Label7.Text = "Delta P = 1.6 v[f] + 8.52 c v[f]{2}t"
        Label2.Text = "Enter the pressure drop - Delta P (in cm)"
        Label3.Text = "Enter the particulate concentration in the air stream (kg/m{3})"
        Button1.Text = "&Calculate 't'"
      Case 5
        Button1.Text = "Start over"
    End Select

    Select Case screen
      Case 3
        TextBox1.Visible = False
        TextBox2.Visible = False
        Case 1, 2, 4
        TextBox1.Visible = True
        TextBox2.Visible = True
    End Select
  End Sub

  Sub calculateResults()
    Select Case screen
```

```
                Case 1
                  vol = Val(TextBox1.Text)
                  vel = Val(TextBox2.Text)
                  volcomp = vol * 60
                  af = volcomp / vel
                  Label4.Text = "A[f] = " + FormatNumber(af, 2) + " m{2}"
                  screen = 2
                  changeScreen()
                Case 2
                  dia = Val(TextBox1.Text)
                  length = Val(TextBox2.Text)
                  ab = 3.1416 * dia * length
                  Label4.Text = "A[b] = " + FormatNumber(ab, 2) + " m{2}"
                  screen = 3
                  changeScreen()
                Case 3
                  screen = 4
                  changeScreen()
                Case 4
                  drop = Val(TextBox1.Text)
                  conc = Val(TextBox2.Text)
                  comp1 = 1.6 * vel
                  compvel = vel ^ 2
                  comp2 = 8.52 * conc * compvel
                  comp4 = drop - comp1
                  compt = comp4 / comp2
                  Label4.Text = "t = " + FormatNumber(compt, 2) + " minutes"
                  screen = 5
                  changeScreen()
                Case 5
                  screen = 1
                  changeScreen()
            End Select
        End Sub

      Private Sub Button1_Click(ByVal sender As System.Object, ByVal e As System.EventArgs)
    Handles Button1.Click
          calculateResults()
        End Sub
    End Class
```

### 7.3.6 COMBUSTION PROCESSES

Control of gaseous air contaminants can be categorized as adsorption, absorption, condensation, or combustion. Gaseous pollutants can be controlled by a wide variety of devices applying one or more of these basic principles. Although combustion is one of the major sources of air pollution, it is also an important air pollution control method.

Combustion is defined as the burning or rapid oxidation of organic (fuel) compounds accompanied by the release of energy in the form of heat and light. Scientifically, the terms combustion and incineration have the same definition and often are used interchangeably (Lee 2005). Combustion or incineration can be further categorized as direct-flame combustion, thermal combustion, or catalytic combustion.

Only one catalytic combustion system will be considered here. In 1971, a homogeneous catalytic system for the oxidation of carbon monoxide was developed and was later patented in 1974 (Lloyd and Rowe 1971, 1974). This catalytic system was further developed, and by 1985, 10 patents were issued in the United States, Canada, England, and Japan (Zackay and Rowe 1984/1985). The first homogeneous and heterogeneous catalytic system (1974) contained palladium (II) salts and copper (II) salts, along with a specified balance of copper (II) halide and a nonhalide copper (II) salt. This catalyst was effective at oxidizing CO to $CO_2$ and $SO_2$ and $SO_3$.

After modifications and further development, this oxidizing catalyst in 1985 was composed of palladium, copper, and nickel on an alumina substrate. The catalyst was produced by impregnating the alumina substrate with a halide salt solution of palladium chloride, nickel chloride, copper chloride, and copper sulfate. The catalyst could remove by oxidation, adsorption, or decomposition gases such as carbon monoxide,

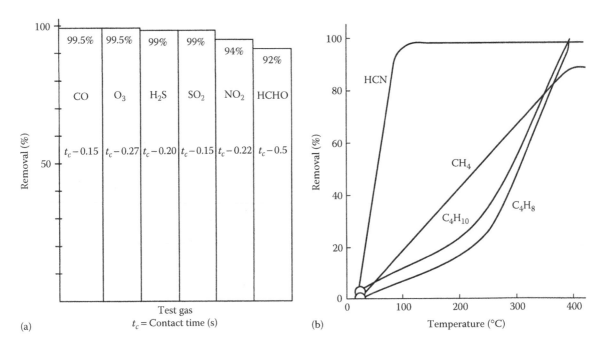

**FIGURE 7.8** The gaseous contaminants that can be removed by catalytic system at both room temperatures and elevated temperatures: (a) Teledyne's room-temperature catalyst at room temperature effectively removes carbon monoxide (CO), ozone ($O_3$), hydrogen sulfide ($H_2S$), sulfur dioxide ($SO_2$), formaldehyde (HCHO), and nitrogen dioxide ($NO_2$). (b) Teledyne's room-temperature catalyst at elevated temperature effectively removes hydrogen cyanide (HCN), propane ($C_3H_8$), methane ($CH_4$), and isobutane ($C_4H_{10}$). (From *Teledyne Technology Introduces a New Room Temperature Catalyst,* From 1007-F-33, Teledyne Water Pik Research & Development Laboratory, Ft. Collins, CO, 1983. With permission.)

hydrogen sulfide, hydrogen cyanide, sulfur dioxide, and ozone present in dilute concentrations in air (Zackay and Rowe 1984/1985). See Figure 7.8 for the gaseous contaminants that can be removed by this catalytic system at both room temperatures and elevated temperatures (Air Pollution Training Institute 1992).

This catalytic system has many potential applications, some of which are listed as follows (Rowe and Lloyd 1978, Teledyne Technology 1983):

- Home appliances (kerosene heaters, stove hoods, etc.)
- Indoor air filtration (home and office)
- Industrial air filtration
- Automotive emissions and passenger compartment (air-conditioning unit)
- Home or public parking garages
- Warehouse and dock areas
- Airline industry
- Mine safety equipment
- First safety equipment
- Carbon monoxide analytical equipment and monitors
- Cigarette filters

One application of this catalyst was marketed in 1985 by Teledyne Water Pik. One stage of the four-stage "Teledyne Water Pik Instapure Filtration System" used this catalyst for the removal of indoor gaseous contaminants (Teledyne Instapure Filtration System 1985).

Not only is the percentage of reduction of the air contaminant important but also parameters such as contact time and the reaction rate coefficient are important. To calculate the percentage of reduction of the air contaminant by the catalyst, the following equation 7.48 can be used:

$$\text{Removal (\%)} = \frac{C_0 - C_e}{C_0} \times 100 \qquad (7.48)$$

where:

$C_0$ is the initial air contaminant concentration, $\mu g/m^3$
$C_e$ is the effluent air contaminant concentration, $\mu g/m^3$

The contact time can be determined by using the following equation 7.49:

$$t = \frac{S_c}{f_r} \qquad (7.49)$$

where:

$t$ is the contact time, s
$S_c$ is the space around the catalyst, mL
$f_r$ is the contaminated gas flow rate, mL/s

A mathematical model (first-order reaction) proved valuable in evaluating the various catalysts. This mathematical model was verified by testing various catalytic systems using a fixed quantity of catalyst and a constant initial air contaminant concentration ($C_0$), but the gas flow was varied, which in turn varied the contact time ($t$). The data from these tests yielded a straight line when $C_e/C_0$ versus $t$ was plotted on a semilog paper. This indicated that the air contaminant removal by the catalyst was a first-order reaction.

Equation 7.50 is the mathematical model describing this process:

$$\frac{C_e}{C_0} = e^{-kt} \tag{7.50}$$

where:

$e$ is the base of natural logarithm, 2.71828
$k$ is the reaction rate, coefficient, $s^{-1}$
$t$ is the catalytic contact time s
$C_0$ and $C_e$ are the same as previously indicated

In general, the longer the contact time with the catalyst, the greater the reduction in the air contaminant levels. For most catalytic systems, the contact time needs to be on the order of a few hundredths of a second.

The reaction rate coefficient $k$ is a single number that measures the overall activity of a catalyst. With equal contact time, the higher the $k$ value, the greater the ability of the catalyst to remove air contaminants. It was found that for this catalytic system, $k$ values > 10 $s^{-1}$ indicated a catalytic system worthy of further development. Example demonstrates how Equations 7.48 through 7.50 can be used to evaluate a catalyst.

## Example 7.28

1. A 2.5 g catalytic sample was tested for its removal efficiency of carbon monoxide (CO) from an air stream. The initial CO concentration was 23,300 $\mu g/m^3$ ($C_0$). The CO concentration in the air stream after passing through the catalyst was 1770 $\mu g/m^3$ ($C_e$). The airstream flow rate was 700 mL/min. The space around the catalytic support system was 0.6 mL. Determine the percentage of CO removal, the contact time in seconds, and the reaction rate coefficient.

2. Prepare a short computer program to solve this problem.

**Solution**

1. A solution to Example 7.28 (1):
   a. Percentage of CO removal can be determined by using Equation 7.48:

$$\text{Removal (\%)} = \frac{23,300\ \frac{\mu g}{m^3} - 1,770\ \mu g/m^3}{23,300\ \mu g/m^3} \times 100 = 92.4\%$$

   b. Contact time can be calculated by using Equation 4.49:

$$\text{contact time} = \frac{0.6\ \text{mL}}{700\ \text{mL/min}\,/\,60\ \text{s/min}} = 0.051\,\text{s}$$

   c. Reaction rate coefficient can be determined by using Equation 7.50:

$$\frac{C_e}{C_0} = e^{-kt}$$

$$\frac{1,770\ \mu g/m^3}{23,300\ \mu g/m^3} = e^{-k(0.051\,\text{s})}$$

   which yields K = 50.6 $s^{-1}$.

   d. This would indicate that with 92.4% removal of CO at a contact time ($t$) of 0.051 s and a reaction rate coefficient ($k$) of 50.6 $s^{-1}$, further development of this catalyst should be considered.

2. For a solution to Example 7.28 (2), see the listing of Program 7.28.

---

**LISTING OF PROGRAM 7.28    (CHAP7.28\FORM1.VB): CATALYTIC SYSTEM**

```
'******************************************************************************
'Example 7.28: A program that can be used to evaluate
' a catalytic system
'******************************************************************************
Public Class Form1
    Dim screen As Integer = 1
    Dim co, ce, perc, space, rate, time, acomp, k As Double

    Private Sub Form1_Load(ByVal sender As System.Object, ByVal e As System.EventArgs)
Handles MyBase.Load
        Me.Text = "Example 7.28: Catalytic System evaluation"
        Me.FormBorderStyle = Windows.Forms.FormBorderStyle.FixedSingle
        Me.MaximizeBox = False
        Label1.Text = "Example 7.28: A program that can be used to evaluate a catalytic
        system"
        Label1.Text += vbCrLf + "The formulas:"
        Label1.Text += vbCrLf + "Removal (%) = ((C[o] - C[e])/C[o]) * 100"
        Label1.Text += vbCrLf + "t = s[c]/f[r]"
        Label1.Text += vbCrLf + "C[e]/C[o] = e{-kt}"
        Label4.Font = New Font(FontFamily.GenericSerif, 9, FontStyle.Bold)
```

```
        Label5.Text = "Decimals:"
        GroupBox1.Text = ""
        NumericUpDown1.Value = 2
        NumericUpDown1.Maximum = 10
        NumericUpDown1.Minimum = 0
        changeScreen()
End Sub

'*******************************************************************************
'This sub prepares the screen every time by resetting
'the labels' text, the button text, and empties the
'textboxes. It shows/hides the controls as needed.
'*******************************************************************************
Sub changeScreen()
    TextBox1.Text = ""
    TextBox2.Text = ""

    Select Case screen
        Case 1
            Label7.Text = "Removal(%) = ((C[o] - C[e])/C[o]) * 100"
            Label2.Text = "Enter the initial concentration C[o] (in um/m{3})"
            Label3.Text = "Enter the initial concentration after the catalyst C[e]
            (in um/m{3})"
            Label4.Text = ""
            Button1.Text = "&Calculate 'percentage'"
        Case 2
            Label7.Text = "t = s[c]/f[r]"
            Label2.Text = "Enter the space around the catalyst s[c] (in mL)"
            Label3.Text = "Enter the air stream flow rate f[r] (in mL/min)"
            Button1.Text = "&Calculate 'time'"
        Case 3
            Label7.Text = "C[e]/C[o] = e{-kt}"
            Label7.Text += vbCrLf + "All factors for this equation are from the previous
            equations"
            acomp = Math.Log(ce / co)
            K = acomp / time * -1
            Label7.Text += vbCrLf + "k = " + FormatNumber(k, NumericUpDown1.Value) + "
            sec{-1}"
            Label2.Text = ""
            Label3.Text = ""
            Button1.Text = "Start over"
    End Select

    Select Case screen
        Case 3
            TextBox1.Visible = False
            TextBox2.Visible = False
        Case 1, 2
            TextBox1.Visible = True
            TextBox2.Visible = True
    End Select
End Sub

Sub calculateResults()
    Dim decimals As Integer = NumericUpDown1.Value
    If decimals < 0 Or decimals > 10 Then decimals = 2
    Select Case screen
        Case 1
            co = Val(TextBox1.Text)
            ce = Val(TextBox2.Text)
            perc = ((co - ce) / co) * 100
```

```
                Label4.Text = "The percentage of removal = " + FormatNumber(perc, decimals)
                screen = 2
                changeScreen()
            Case 2
                space = Val(TextBox1.Text)
                rate = Val(TextBox2.Text)
                time = space / (rate / 60)
                Label4.Text = "The contact time is = " + FormatNumber(time, decimals) + " sec"
                screen = 3
                changeScreen()
            Case 3
                screen = 1
                changeScreen()
        End Select
    End Sub

    Private Sub Button1_Click(ByVal sender As System.Object, ByVal e As System.EventArgs)
Handles Button1.Click
        calculateResults()
    End Sub
End Class
```

## Example 7.29

1. Prepare a computer program that can be used to evaluate a catalytic system including the percentage of reduction of an air contaminant, contact time, and reaction rate coefficient. This program should incorporate all the elements in Equations 7.48 through 7.50.
2. Using the computer program developed in (1), indicate which of the following catalysts should be considered for further development. The air space around the catalyst was found to be 0.24 mL/g of catalyst.

| Catalyst | Weight of Catalyst | Flow Rate (mL/s) | CO Initial ($CO_0$) ($\mu m/m^3$) | CO Final ($CO_e$) ($\mu m/m^3$) |
|---|---|---|---|---|
| 1 | 0.5 | 20 | 33,700 | 26.600 |
| 2 | 1.0 | 13 | 30,490 | 14,670 |
| 3 | 1.5 | 4 | 13,210 | 3100 |
| 4 | 2.0 | 12 | 19,720 | 1610 |

## Solution

1. For solution to Example 7.29 (1), see the listing of Program 7.29, which includes all the elements in Equations 7.48 through 7.50.
2. Solution to Example 7.29 (2):

| Catalyst | Removal (%) | Contact Time (s) | Reaction Rate Constant $k$ ($s^{-1}$) |
|---|---|---|---|
| 1 | 21 | 0.006 | 39.4 |
| 2 | 52 | 0.018 | 39.6 |
| 3 | 77 | 0.09 | 16.1 |
| 4 | 92 | 0.04 | 62.6 |

Of these four catalysts, number 4 shows the greatest promise for further development.

### LISTING OF PROGRAM 7.29   (CHAP7.29\FORM1.VB): MULTIPLE CATALYSTS

```
'**************************************************************************************
'Example 7.29: A program that can be used to evaluate a
' catalytic system.
' Evaluates multiple catalysts.
'**************************************************************************************
Public Class Form1
    Dim cat(), wgt(), flow(), initial(), final() As Double
    Dim aspace(), perc(), ctime(), comp1(), rrate() As Double
    Dim space As Double
```

```
   Private Sub Form1_Load(ByVal sender As System.Object, ByVal e As System.EventArgs)
Handles MyBase.Load
      Me.Text = "Exercise 7.29:"
      Me.MaximizeBox = False
      Me.FormBorderStyle = Windows.Forms.FormBorderStyle.FixedSingle
      Label1.Text = "Example 7.29: a program that can be used to evaluate a catalytic
      system."
      Label1.Text += vbCrLf + "Evaluates multiple catalysts."
      Label1.Text += vbCrLf + "The formulas:"
      Label1.Text += vbCrLf + "Removal (%) = ((C[o] - C[e])/C[o]) * 100"
      Label1.Text += vbCrLf + "t = s[c]/f[r]"
      Label1.Text += vbCrLf + "C[e]/C[o] = e{-kt}"
      Label2.Text = "Type in the catalyst no., weight, flow rate, and initial "
      Label2.Text += "&& final gas concentrations:"
      DataGridView1.Columns.Clear()
      DataGridView1.Columns.Add("NCol", "No.")
      DataGridView1.Columns.Add("wgtCol", "Catalyst Wgt")
      DataGridView1.Columns.Add("FCol", "Flow rate (mL/s)")
      DataGridView1.Columns.Add("ICCol", "Gas initial conc. (um/m3)")
      DataGridView1.Columns.Add("FCCol", "Gas final conc. (um/m3)")
      DataGridView1.AutoResizeColumns()
      TextBox1.Multiline = True
      TextBox1.Height = 140
      TextBox1.ScrollBars = ScrollBars.Vertical
      TextBox1.Font = New Font(FontFamily.GenericMonospace, 9)
      Button1.Text = "&Calculate"
   End Sub

   Sub calculateResults()
      If DataGridView1.Rows.Count <= 1 Then
        MsgBox("Type in at least one entry!", vbOKOnly + vbExclamation, "Prompt")
        Exit Sub
      End If

      Dim count = DataGridView1.Rows.Count - 1

      ReDim cat(count), wgt(count), flow(count), initial(count), final(count)
      ReDim aspace(count), perc(count), ctime(count), comp1(count), rrate(count)

      For i = 0 To count - 1
        cat(i) = DataGridView1.Rows(i).Cells("NCol").Value
        Wgt(i) = Val(DataGridView1.Rows(i).Cells("wgtCol").Value)
        flow(i) = Val(DataGridView1.Rows(i).Cells("FCol").Value)
        initial(i) = Val(DataGridView1.Rows(i).Cells("ICCol").Value)
        final(i) = Val(DataGridView1.Rows(i).Cells("FCCol").Value)
      Next

      space = InputBox("Enter the air space around the catalyst (ml/g)", "Enter space", 0)

      For i = 0 To count - 1
        Aspace(i) = space * wgt(i)
        perc(i) = ((initial(i) - final(i)) / initial(i)) * 100
        ctime(i) = aspace(i) / flow(i)
        comp1(i) = Math.Log(final(i) / initial(i))
        rrate(i) = (comp1(i) / ctime(i)) * -1
      Next

      Dim s As String
      s = " Space Initial Final"
      s += vbCrLf + " Wt of around Flow conc. conc."
      s += vbCrLf + "Catalyst Catalyst Catalyst Rate um/m{3} um/m{3}"
```

```
      For i = 0 To count - 1
        s += vbCrLf + Format(cat(i), "000 ").ToString
        s += Format(wgt(i), "00.000 ").ToString
        s += Format(aspace(i), "00.000 ").ToString
        s += Format(flow(i), "00.000 ").ToString
        s += Format(initial(i), "00.000 ").ToString
        s += Format(final(i), "00.000").ToString
      Next
      TextBox1.Text = s

      s = "----------------------------------------------------------"
      s += vbCrLf + " Removal Contact time Reaction rate"
      s += vbCrLf + "Catalyst % (sec) Constant k (s{-1})"
      For i = 0 To count - 1
        s += vbCrLf + Format(cat(i), "000 ").ToString
        s += Format(perc(i), "00.000 ").ToString
        s += Format(ctime(i), "00.000 ").ToString
        s += Format(rrate(i), "00.000").ToString
      Next
      TextBox1.Text += vbCrLf + s
    End Sub

    Private Sub Button1_Click(ByVal sender As System.Object, ByVal e As System.EventArgs)
  Handles Button1.Click
      calculateResults()
    End Sub
End Class
```

### 7.3.7 Packed Column Absorption Towers

Absorption is a process by which a liquid (absorbent) is used to remove one or more soluble gases (absorbate) from a contaminated air stream. The liquid absorbent may react chemically or physically with the absorbate to remove it from the gas stream. Typical absorbents are water or dilute basic or acidic solutions.

Chemical absorption involves a liquid absorbent reacting with a pollutant to yield a nonvolatile product. A typical process here is the reaction of $SO_2$ and aqueous $H_2O_2$ to produce sulfuric acid. Physical absorption involves the physical dissolving of the pollutant in a liquid. This process is generally reversible. The ideal absorbent would be relatively nonvolatile, inexpensive, noncorrosive, stable, nonviscous, nonflammable, and nontoxic. In many cases, distilled water fulfills many of these characteristics (Air Pollution Training Institute 1983). The types of absorption equipment used for air pollution control include the following:

- Spray towers
- Spray chambers
- Venturi scrubbers
- Packed column (towers)

The amount of absorbate (gas) that dissolves in the absorbent (liquid) depends upon the properties of the absorbent and the absorbate, the absorbate (gas) concentration, the partial pressure of the pollutant, the temperature of the system, the turbulence, the flow rate, and the type of air pollution control equipment used. For dilute solutions, as is the usual case in air pollution control, the relationship between the partial pressure and the concentration of the absorbate (gas) in the absorbent (liquid) is given by Henry's law (see Section 7.2.6).

Only the packed column absorption type of air pollution control equipment will be considered here. One of the most efficient ways to contact the absorbent with absorbate is the countercurrent flow system shown in Figure 7.9a (Corman and Black 1974). In this system, the gas and liquid run in opposite directions. The absorbate (gas) enters at the bottom and the absorbent (liquid) at the top of the column. With this arrangement, the high-concentration pollutant (gas) is absorbed into the absorbent (liquid) with a high-pollutant concentration, and the lower concentration pollutant (gas) is absorbed into the absorbent (liquid) that has no contaminants present. As indicated by Henry's law, the concentration of a gas dissolved in a solvent under equilibrium conditions is proportional to the partial pressure of the gas above the solution at constant temperature.

Applying this principle from the bottom to the top of the packed tower makes it possible to determine the mole fraction of the pollutant (absorbate) in the gas phase in equilibrium with the mole fraction of the pollutant (absorbate) in the liquid phase. When these data are plotted with the gas mole fraction on the ordinate (y-axis) and the liquid mole

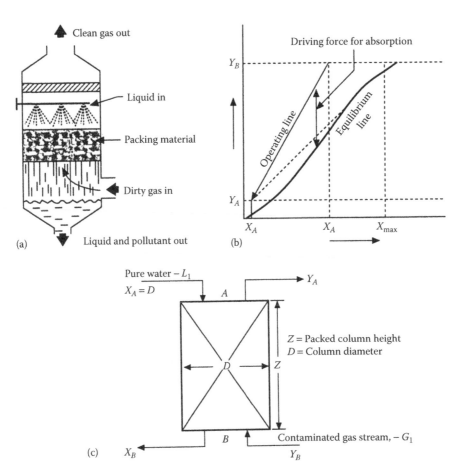

**FIGURE 7.9** (a) Countercurrent-packed column: a typical countercurrent-packed column absorber system. (b) Countercurrent-packed column. (c) Countercurrent-packed column: graphical presentation of the packed column.

fraction on the abscissa (*x*-axis), an equilibrium line is produced (see Figure 7.9b).

For absorption to take place, the operating line must be above the equilibrium line. The vertical distance between the equilibrium line and the operating line indicates the degree of saturation of the liquid with the pollutant (see Figure 7.9b). The operating line in this case is considered to be a straight line, and an equation depicting this line follows:

$$L_1\left(\frac{X_A}{1-X_A}-\frac{X_B}{1-X_B}\right)=G_1\left(\frac{Y_A}{1-Y_A}-\frac{Y_B}{1-Y_B}\right) \quad (7.51)$$

where:

$L_1$ is the liquid flow, kg*mol/h

$G_1$ is the gas flow, kg*mol/h

$X_A$ is the mole fraction of the pollutant in the liquid phase into the column

$X_B$ is the mole fraction of the pollutant in the liquid phase out of the column

$Y_A$ is the mole fraction of the pollutant in the gas phase out of the column

$Y_B$ is the mole fraction of the pollutant in the gas phase into the column

See Figure 7.9c for a graphical representation of these terms.

Example 7.30 demonstrates how Equation 7.51 can be used to evaluate and aid in the design of packed columns used for air pollution control. While the following example illustrates some of the principle used for packed column evaluation and design, there are many other factors that must be taken into account, such as the number of transfer units, the height of an overall gas transfer unit, the type of packing, and the required packed column height and diameter. However, it is not possible to go into all these functions and formula here.

## Example 7.30

A 500 kg/h air stream contains a mixture of acetone in air ($CH_3-CO-CH_3$; molecular weight of acetone = 58.08; molecular weight of air at 25°C = 28.970) to be treated in a countercurrent-packed column absorber, which has a diameter of 0.45 m and is packed with Raschig rings. The equilibrium equation was found to be $Y_B = 2.53X_B$, where $Y_B$ is the mole fraction of acetone in air, and $X_B$ is the mole fraction of acetone in water.

The operating temperature is 25°C at 1 atm pressure. The acetone in the air stream entering at the bottom of the column is 50 mg/L, and it is required that the allowable level of acetone leaving the column to be reduced to 2.5 mg/L. What is the required flow rate of water in kg/h (acetone free, $X_A = 0$) to lower the acetone concentration to the desired level?

**Solution**

- Inlet acetone concentration = 50 mg/L = 0.050 kg/m³
- Air at 25°C has a density of 1.185 kg/m³:

$$Y_B = \frac{\dfrac{0.050 \ kg/m^3}{58 \ MW\text{-}acetone}}{\dfrac{1.185 \ kg/m^3}{28.970 \ MW \ of \ air}} = \frac{0.000862}{0.0409} = 0.021$$

- Outlet acetone concentration = 2.5 mg/L = 0.0025 kg/m:

$$Y_A = \frac{\dfrac{0.025 \ kg/m^3}{58 \ MW\text{-}acetone}}{\dfrac{1.185 \ kg/m^3}{28.970 \ MW \ of \ air}} = \frac{0.000431}{0.0409} = 0.001$$

- The incoming distilled water has no acetone present, so

$$X_A = 0 \ Y_B = 2.53 X_B \ X_B = \frac{0.021}{2.53} = 0.0083$$

- Substituting these values in Equation 7.51 gives

$$L_1\left(\frac{0}{1-0} - \frac{0.0083}{1-0.0083}\right) = G_1\left(\frac{0.001}{1-0.001} - \frac{0.021}{1-0.021}\right)$$

- which yields

$$\frac{L_1}{G_1} = \frac{-0.0204}{-0.00837} = 2.437$$

- The number of moles of inlet air based on the following air–acetone mixture is calculated as follows:

$$\begin{aligned} Acetone\,(0.021 \times 58.08) &= 1.22 \\ Air\,(0.979 \times 28.970) &= 28.36 \\ \hline Combined \ mixtures \ MW &= \overline{29.58} \end{aligned}$$

- The kg/mol of air = 0.979 × 500 kg/h/29.58 = 16.55 kg*mol/h
- $L_1 = 2.437G$
- $G_1 = 16.55$ kg*mol/h
- $L_1 = 2.437 \ (16.55)$ kg*mol/h of $H_2O$ = 40.33 kg*mol/h of $H_2O$
- 1 kg*mol of $H_2O$ = 18 kg
- Water flow rate = 40.33 × 18 = 726 kg/h

**Example 7.31**

1. Prepare a computer program that incorporates all the elements in Equation 7.51 and thus provides a means to evaluate or aid in the design of packed columns used for air pollution control.
2. Using the computer program prepared in (1) and given the following data, determine the liquid flow required to remove 97% of $SO_2$ from the stack gas of a coal-fired furnace.

Column diameter = 3.00 m
Operating temperature = 25°C
Operating pressure = 1 atm
Gas flow rate = 30,000 m³/h
$SO_2$ inlet concentration = 20,000 ppm
Incoming liquid is pure water, $X_A = 0$

**Solution**

1. For solution to Example 7.31 (1), see the listing of Program 7.31, which includes all the elements in Equation 7.51.
2. Convert 20,000 ppm $SO_2$ to µg/m³ using Equation 7.1 at STP:

$$\mu g/m^3 = \frac{20,000 \ ppm \times 64 \times 10^3}{24.46 \ L/mol}$$

$$= 52,330,335 \ \mu g/m^3$$

$$= 52.3 \ g/m^3 = 0.0523 \ kg/m^3$$

Inlet concentration $SO_2$ = 0.0523 kg/m³
a. Air at 25°C has a density of 1.185 kg/m³:

$$Y_B = \frac{\dfrac{0.0523 \ kg/m^3}{64 \ MW \ of \ SO_2}}{\dfrac{1.185 \ kg/m^3}{28.970 \ MW \ of \ air}} = \frac{0.0008171}{0.04090} = 0.02$$

b. Outlet $SO_2$ concentration = 0.03 × 0.0523 kg/m³ = 0.001569.

$$Y_A = \frac{\dfrac{0.001569 \ kg/m^3}{64 \ MW \ of \ SO_2}}{\dfrac{1.185 \ kg/m^3}{28.970 \ MW \ of \ air}} = \frac{0.0000245}{0.04090} = 0.0006$$

c. $X_A = 0$ for the incoming distilled water.
d. Given: $Y_B = 30 X_B$,

$$X_B = \frac{0.02}{30} = 0.00067$$

e. Substituting these values in Equation 7.51 gives

$$L_1\left(\frac{0}{1-0} - \frac{0.00067}{1-0.00067}\right) = G_1\left(\frac{0.0006}{1-0.0006} - \frac{0.02}{1-0.02}\right)$$

which yields

$$\frac{L_1}{G_1} = \frac{-0.0198}{-0.00067} = 29.69$$

f.  The kg*mol of air is 30,000 m³/h × 1.185 kg/m³ = 35,550 kg/h.

g.  Inlet gas composition:
    0.98 = percentage of air

0.02 = percentage of $SO_2$

h.  The kg*mol of air is 0.98 × 35,550 kg/h/28.970 MW of air = 1202.6 kg*mol/h
    $L_1 = 29.69G_1$
    $L_1$ = 29.69(1202.6) = 35,707 kg*mol/h of $H_2O$
    1 kg*mol of $H_2O$ = 18 kg

i.  Water flow rate = 35,707 × 18 = 642,729 kg/h, or 178 kg/s.

---

**LISTING OF PROGRAM 7.31   (CHAP7.31\FORM1.VB): PACKED COLUMNS ABSORPTION TOWERS**

```
'*********************************************************************************
'Example 7.31: A program that can be used to evaluate
'      or aid in the design of packed columns
'      absorption towers used for air pollution
'      control.
'*********************************************************************************

Public Class Form1
  Dim inconc, gasmolwgt, airdens, airmolwgt, outconc As Double
  Dim inconck, outconck, yb1, yb2, yb, ya1, ya, xa As Double
  Dim factor, strea As Double

  Private Sub Form1_Load(ByVal sender As System.Object, ByVal e As System.EventArgs)
Handles MyBase.Load
    Me.Text = "Exercise 7.31:"
    Me.MaximizeBox = False
    Me.FormBorderStyle = Windows.Forms.FormBorderStyle.FixedSingle
    Label2.Text = "A program that can be used to evaluate or aid in the"
    Label2.Text += vbCrLf + "design of packed columns absorption towers used for air
pollution control."
    Label2.Text += vbCrLf + "The formula:"
    Label2.Text += vbCrLf + "L[1] * ((X[a]/(1-X[a]) - (X[b]/1-X[b]))) = "
    Label2.Text += vbCrLf + "G[1] * ((Y[a]/(1-Y[a]) - (Y[b]/1-Y[b]))) "
    Label3.Text = "Enter the inlet concentration of the gas (ppm)"
    Label4.Text = "Enter the molecular weight of the gas"
    Label5.Text = "Enter the air density (kg/m{3})"
    Label6.Text = "Enter the molecular weight of the air"
    Label7.Text = "Enter the percentage of removal of the gas (%)"
    Label1.Text = "The incoming distilled water has none of the gas; therefore X[A] = 0"
    Label8.Text = "Enter the value for the # in the following given equilibrium
equation"
    Label8.Text += vbCrLf + "- Y[B] = #X[B]"
    Label9.Text = "Enter the gas flow rate (m3/hr)"
    Button1.Text = "&Calculate"
  End Sub

  Sub calculateResults()
    inconc = Val(TextBox1.Text)
    gasmolwgt = Val(TextBox2.Text)
    airdens = Val(TextBox3.Text)
    airmolwgt = Val(TextBox4.Text)
    outconc = Val(TextBox5.Text)
    If outconc > 1 Then outconc /= 100
    outconc = 1 - outconc
    factor = Val(TextBox6.Text)
    strea = Val(TextBox7.Text)

    inconck = inconc * gasmolwgt * 10 ^ -6 / 24.46
    outconck = inconck * outconc
```

```
        yb1 = inconck / gasmolwgt
        yb2 = airdens / airmolwgt
        yb = yb1 / yb2
        ya1 = outconck / gasmolwgt
        ya = ya1 / yb2
        xa = 0

        Dim xb, compl1, compg1, compd1, gas, compare, air, combmolwgt As Double
        Dim stream, g1, l1, flowrate As Double
        xb = yb / factor
        compl1 = (xa / (1 - xa)) - (xb / (1 - xb))
        compg1 = (ya / (1 - ya)) - (yb / (1 - yb))
        compd1 = compg1 / compl1
        gas = yb * gasmolwgt
        compare = 1 - yb
        air = compare * airmolwgt
        combmolwgt = gas + air
        stream = strea * airdens
        g1 = compare * (stream / airmolwgt)

        Dim s As String
        s = "A kg*mole of water = 18 kg"
        s += vbCrLf + "L[1] = " + FormatNumber(compd1, 2) + "G"
        s += vbCrLf + "G[1] = " + FormatNumber(g1, 2) + " kg.mole/hr of H[2]O"
        L1 = compd1 * g1
        Flowrate = l1 * 18
        s += vbCrLf + "L[1] = " + FormatNumber(l1, 2) + " kg.mole of water"
        s += vbCrLf + "Water flow rate = " + FormatNumber(flowrate, 2) + " kg/hr"
        MsgBox(s, vbOKOnly, "Results")
    End Sub

    Private Sub Button1_Click(ByVal sender As System.Object, ByVal e As System.EventArgs)
    Handles Button1.Click
        calculateResults()
    End Sub
End Class
```

## 7.3.8 Adsorption

Adsorption is a phenomenon by which gases, liquids, and solutes within liquids are attracted, concentrated, and retained at a boundary surface (Air Pollution Training Institute 1983). In air pollution control, this process involves the retention of molecules from the gas phase onto a solid surface. A typical process consists of passing a contaminated gas stream through a container filled with an absorbent such as activated carbon, activated alumina, silica gel, acid-treated clay, molecular sieve, Fuller's earth, or magnesia.

In the air pollution field, gas adsorption can be used to remove VOCs, $H_2S$, $SO_2$, and $NO_2$ from contaminated air streams. In industry, adsorption is used to recover valuable volatile solvents such as benzene, ethanol, trichloroethylene, and Freon.

Adsorption involves either physical adsorption (physiosorption) or chemical adsorption (chemisorption). With physical adsorption, the attractive forces consist of Van der Waal's interactions, dipole–dipole interactions, and/or electrostatic interactions (Air Pollution Training Institute 1983). An example of physical adsorption is when a contaminant in a gas stream is adsorbed onto activated carbon. In physical adsorption, the adsorbed layer generally is considered to be several molecules thick (multilayer adsorption).

In chemical adsorption, the contaminant gas molecules form a chemical bond with the adsorbent, and the gas (adsorbate) is held strongly to the solid surface (adsorbent) by valence forces. An example of chemical adsorption is the adsorption of oxygen by activated charcoal to form carbon monoxide (CO) and carbon dioxide ($CO_2$). The CO and $CO_2$ are not easily removed from the activated carbon, and this can only be accomplished at elevated temperatures. Chemical adsorption is usually limited to the formation of a single layer of molecules on the surface of the adsorbent (monolayer adsorption) (Air Pollution Training Institute 1983). The variables affecting gas adsorption are as follows:

- Concentration of the contaminant in the gas stream
- Surface area of the adsorbent
- Temperature of the system
- Presence of other competing contaminant gases
- Characteristics of the adsorbent, such as weight, electrical polarity, and chemical reactivity

The relationship between the amount of pollutant adsorbed by the adsorbent under equilibrium conditions and at a constant temperature is called an adsorption isotherm. An isotherm is a graphical representation of the adsorbent's capacity versus the partial pressure of the absorbate (contaminant) at a particular temperature (Lee 2005). Adsorption isotherms are useful in providing a means of evaluating the following:

• Quantity of gas adsorbed at various concentrations
• Adsorption capacity of the adsorbent at various gas concentrations
• Adsorption capacity of the adsorbent for various gases
• Surface areas for a given amount of adsorbent

A graphical plot of the amount of the gas absorbed per gram of adsorbent at various gas concentrations at equilibrium and under conditions of constant temperature produces isotherms with many shapes. Specific isotherms have been developed by Freundlich, Langmuir, and Brunauer, Emmett, and Teller (BET).

Only the Freundlich isotherm will be considered here. The Freundlich isotherm is a purely empirical equation valid for monomolecular physical and chemical adsorptions. The equation is

$$\frac{X}{M} = KC^{\frac{1}{n}} \qquad (7.52)$$

where:

$X/M$ is the mass ($X$) of the element or contaminant adsorbed from solution per unit mass of adsorbent ($M$)
$K$ and $n$ are the constants fitted from the experimental data
$C$ is the concentration of the contaminant in the gas phase at equilibrium

By taking the logarithm of both sides, this equation is converted to a linear form:

$$\log\left(\frac{X}{M}\right) = \log K + \frac{1}{n}\log C \qquad (7.53)$$

If the experimental data fit the Freundlich adsorption isotherm, a plot of log $X/M$ versus log $C$ gives a straight line as shown in Figure 7.10. If a vertical line is erected from a point

on the horizontal scale corresponding to the initial contaminant concentration ($C_0$) and the isotherm extrapolated to intersect that line, the $X/M$ value at this point of intersection can then be read from the vertical scale. The value of $(X/M)_{C_0}$ represents the amount of contaminant adsorbed per unit weight of adsorbent when that adsorbent is in equilibrium with the initial contaminant concentration. This represents the ultimate sorption capacity of the adsorbent for that contaminant (Culp et al. 1978).

The Freundlich equation is most useful for dilute solutions over small concentration ranges. The $1/n$ value represents the slope or change in rate of effectiveness in uptake with varying amounts of adsorbent, and $K$, the ordinate intercept, the fundamental effectiveness of the adsorbent. High $K$ and high $n$ values indicate high adsorption capacities; low $K$ and high $n$ values indicate low adsorption capacities; low $n$ values or a steep slope indicates high adsorption at high contaminant levels and low adsorption at low contaminant levels (Ford 1975). Both graphical and computer analyses can be made for the Freundlich adsorption isotherm (Chansler et al.1995).

In the air pollution control field, adsorption systems can be either nonregenerative or regenerative. They also can be fixed, moving, or fluidized beds. Nonregenerative adsorption systems have thin beds and are economical only when the contaminant level in the gas stream is in the $\mu g/m^3$ or ppm range. Regenerative beds are generally thick and are designed to handle considerable contaminant loadings with the additional advantage of the recovery of a valuable solvent (see Figure 7.11a,b).

Adsorption of a contaminant from an air stream continues until the bed capacity has been reached; at this point, the adsorbent is saturated with the adsorbate. The concentration of the contaminant in the exit gas stream begins to rise rapidly, and the adsorber must be regenerated or disposed of. This point in the adsorption system is called the breakthrough capacity of the bed.

An equation that can be used to calculate the breakthrough capacity of an adsorption bed is as follows (Air Pollution Training Institute 1984):

$$C_B = \frac{0.5C_s(\text{MTZ}) + C_s(D - \text{MTZ})}{D} \qquad (7.54)$$

where:

$C_B$ is the breakthrough capacity, fractional
$C_s$ is the saturation capacity, fractional
MTZ is the mass transfer zone, cm
$D$ is the adsorption bed depth, cm

The degree of saturation of an adsorption bed is defined as follows (Turk 1977):

$$C_s = \frac{\text{WAE}}{\text{WAT}} \qquad (7.55)$$

where:

$C_s$ is the saturation capacity, fractional
WAE is the weight of adsorbate, g or kg
WAT is the weight of adsorbent, g or kg

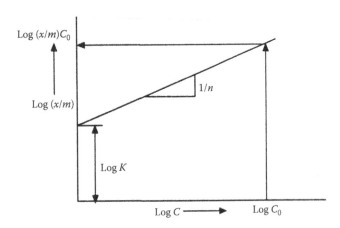

**FIGURE 7.10** Straight-line form of the Freundlich adsorption isotherm.

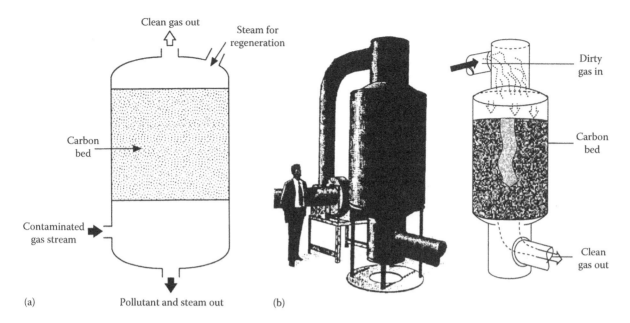

**FIGURE 7.11** (a) Schematic of a regenerative adsorption-type system. (From Lloyd, W.G. and Rowe, D.R., *Environ. Sci. Technol.*, 5, 11, 1971. With permission.) (b) Schematic of a nonregenerative adsorption-type system. (From Corman, R.D., and Black, D., *Controlling Air Pollution,* American Lung Association, Washington, DC, 1974. With permission.)

When an adsorbent bed is regenerated, it is not economical to remove absolutely all the adsorbate (contaminant). The residual contaminant in the adsorbent after regeneration is called the "heel." The practical capacity or working capacity of an adsorbent can be computed as follows (Air Pollution Training Institute 1984):

$$W_c = C_B - H - PF \qquad (7.56)$$

where:

$W_c$ is the working capacity of adsorbent, fractional
$H$ is the heel, fractional
PF is the packing factor, fractional

The working capacity is lower than the breakthrough or saturation capacity. The packing factor is determined experimentally for each adsorbent. The time required before an adsorbent needs to be regenerated is an important parameter in the operation of an adsorption system. An estimate of the time an adsorbent can be left in operation can be computed by using the following equation (Turk 1977):

$$t = \frac{2.41 \times 10^7 \, SW}{EQMC} \qquad (7.57)$$

where:

$t$ is the duration of adsorbent service before saturation, h
$S$ is the proportionate saturation of sorbent, fractional
$W$ is the weight of adsorbent, kg
$E$ is the sorption efficiency, fractional
$Q$ is the airflow rate through sorbent bed, m³/h
$M$ is the average molecular weight of sorbed vapor, g/mol
$C$ is the entering vapor concentration, ppm by volume

Examples 7.32 and 7.33 demonstrates the use of Equations 7.52 through 7.57.

### Example 7.32

Given the following data for the uptake or adsorption of benzene by activated carbon from a gas stream, determine if these data fit the Freundlich isotherm; if so, then determine the $K$ and $n$ values as well as the ultimate adsorption capacity of the activated carbon $(X/M)_{C_0}$ for benzene at its initial concentration.

| Activated Carbon (g)($M$) | Residual Benzene Concentration (g/m³) ($C_0 = 30$) | Benzene Adsorbed (g)($X$) | ($X/M$) |
|---|---|---|---|
| 0 | – | – | – |
| 200 | 20 | 11 | 0.055 |
| 300 | 15 | 15 | 0.050 |
| 500 | 11 | 19 | 0.038 |
| 650 | 8 | 22 | 0.034 |
| 900 | 5 | 25 | 0.028 |
| 1600 | 2 | 28 | 0.018 |

### Solution

The data plot a straight line on log–log paper, indicating the data fit the Freundlich isotherm (see the following graph). The $K$ and $1/n$ constants are 0.012 and 0.52, respectively. The ultimate capacity of the activated carbon for benzene adsorption at 30 g/m³ is 0.07 g/g, or 7 g of benzene can be adsorbed by 100 g of this activated carbon.

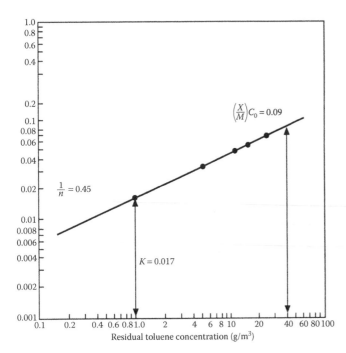

### Example 7.33

1. Prepare a computer program incorporating all the elements in Equation 7.52 (Freundlich isotherm) that makes it possible to determine the constants $K$ and $n$ as well as the adsorbent's ultimate capacity $(X/M)_{C_0}$.
2. Using the computer program developed in (1), determine the $K$ and $n$ values as well as the adsorbent's ultimate capacity $(X/M)_{C_0}$ for toluene adsorption.

### Solution

1. For solution to Example 7.33 (1), see the listing of Program 7.33.
2. These data plot a straight line on log–log paper indicating it follows the Freundlich isotherm (see the following graph). The $K$ and $1/n$ constants are 0.017 and 0.45, respectively. The ultimate capacity of this activated carbon for toluene adsorption at 40 g/m³ was 0.09 g/g, or 9 g of toluene can be adsorbed on 100 g of activated carbon.

| Activated Carbon (g)(M) | Residual Toluene Concentration (g/m³)($C_0$ = 40 g/m³) | Toluene Adsorbed (g) (X) | (X/M) |
|---|---|---|---|
| 100 | 32 | 8 | 0.08 |
| 240 | 24 | 16 | 0.067 |
| 430 | 15 | 25 | 0.058 |
| 600 | 11 | 29 | 0.048 |
| 1000 | 5 | 35 | 0.035 |
| 2500 | 1 | 39 | 0.016 |

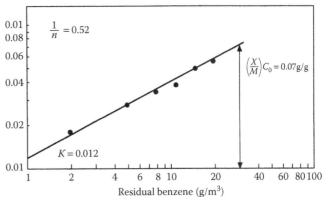

---

### LISTING OF PROGRAM 7.33   (CHAP7.33\FORM1.VB): FREUNDLICH ADSORPTION ISOTHERM

```
'*********************************************************************************
'Example 7.33: Freundlich adsorption isotherm:
' a retyped version from older program.
'*********************************************************************************

Imports System.Math

Public Class Form1
    Dim M(), C(), X(), Q() As Double
    Dim N, Cont As String
    Dim T, Co As Double

    Dim a, b As Double
    Dim sum1X, sum1Y, sum2X, sum2Y As Double
'*********************************************************************************
    'Calculates the straight line equation from the
    'scattered point data. The algorithm used is:
    '(1) Divide data into two sets
```

```
'(2) Find a mid-point in each set
'(3) Find line equation from these two points
'**************************************************************************
Private Sub get_line_equ()
  Dim count As Integer = DataGridView1.Rows.Count - 1
  Dim mid_count As Integer = count / 2
  Dim i As Integer
  sum1X = 0 : sum2X = 0
  sum1Y = 0 : sum2Y = 0
  For i = 0 To mid_count - 1
    sum1X += C(i)
    sum1Y += Q(i)
  Next
  sum1X /= mid_count
  sum1Y /= mid_count
  For i = mid_count To count - 1
    sum2X += C(i)
    sum2Y += Q(i)
  Next
  sum2X /= (count - mid_count)
  sum2Y /= (count - mid_count)
  b = (sum2Y - sum1Y) / (sum2X - sum1X)
  'find straight line equation:
  'y = a + bx
  'y2 = mx2 - mx1 + y1
  a = sum1Y - (b * sum1X)
End Sub


Function get_Y_value(ByVal X As Double) As Double
  Dim Y As Double
  Y = a + b * X
  Return Y
End Function


Private Sub Form1_Load(ByVal sender As System.Object, ByVal e As System.EventArgs)
Handles MyBase.Load
  Me.Text = "Example 7.33: Freundlich adsorption isotherm"
  Me.MaximizeBox = False
  Me.FormBorderStyle = Windows.Forms.FormBorderStyle.FixedSingle
  Label1.Text = "Name of adsorption material (optional)"
  Label2.Text = "Name of contaminant material (optional)"
  Label3.Text = "Initial concentration (g/m{3})"

  Label4.Text = "Enter measurements and g/m{3} of contaminant remaining:"
  DataGridView1.Columns.Clear()
  DataGridView1.Columns.Add("MCol", "Measurement (gm adsorbent)")
  DataGridView1.Columns.Add("CCol", "Contaminant remaining (g/m{3})")
  DataGridView1.AutoResizeColumns()
  TextBox1.Multiline = True
  TextBox1.Height = 140
  TextBox1.ScrollBars = ScrollBars.Vertical
  TextBox1.Font = New Font(FontFamily.GenericMonospace, 9)
  Button1.Text = "&Calculate"
End Sub


Sub calculateResults()
  If DataGridView1.Rows.Count <= 1 Then
    MsgBox("Type in at least one entry!", vbOKOnly + vbExclamation, "Prompt")
    Exit Sub
  End If
```

```
        N = TextBox2.Text
        Cont = TextBox3.Text
        Co = Val(TextBox4.Text)

        Dim count = DataGridView1.Rows.Count - 1
        ReDim M(count), C(count), X(count), Q(count)
        For i = 0 To count - 1
          M(i) = DataGridView1.Rows(i).Cells("MCol").Value
          C(i) = Val(DataGridView1.Rows(i).Cells("CCol").Value)
          X(i) = CO - C(i)
          Q(i) = X(i) / M(i)
        Next

        'PRINT DATA
        Dim s As String
        s = "[M] [C] [X] [X/M]"
        s += vbCrLf + " Concentration Amount "
        s += vbCrLf + " remaining absorbed "
        s += vbCrLf + "grams g/m{3} g/m{3} mg/g"
        s += vbCrLf + "----- ------------- -------- -----"
        For i = 0 To count - 1
          s += vbCrLf + Format(M(i), "00.00 ").ToString
          s += Format(C(i), "00.000 ").ToString
          s += Format(X(i), "00.000 ").ToString
          s += Format(Q(i), "00.000").ToString
        Next
        s += vbCrLf + "Initial concentration was " + Co.ToString + " g/m{3}"
        get_line_equ()
        s += vbCrLf + "Slope = " + FormatNumber(b, 4)
        s += vbCrLf + "Intercept = " + FormatNumber(a, 4)
        s += vbCrLf + "Ultimate sorption capacity of absorbent = " + FormatNumber(get_Y_
        value(Co), 4) + " g/g"
        TextBox1.Text = s
    End Sub

    Private Sub Button1_Click(ByVal sender As System.Object, ByVal e As System.EventArgs)
Handles Button1.Click
        calculateResults()
    End Sub
End Class
```

## Example 7.34

Calculate the breakthrough capacity and the working capacity of an adsorption bed using Equations 7.54 and 7.56, given the following data (neglect the packing factor, PF):

Depth of adsorption bed = 1 m (100 cm)
Saturation capacity = 42%
MTZ (mass transfer zone) = 10 cm
Heel = 2%

### Solution

- $C_B$ = Breakthrough capacity:

$$C_B = \frac{0.5(.42)(10) + 0.42(100-10)}{100}$$

$$= \frac{2.10 + 37.8}{100} = 0.399 = 39.9\%$$

(7.58)

- $W_c$ = Working capacity (neglect PF):

$$Wc = C_B - H - PF = 39.9 - 2.0 = 37.9\% \quad (7.59)$$

## Example 7.35

1. Prepare a computer program that incorporates all the elements in Equations 7.50 and 7.52 so that the breakthrough and working capacities of an absorbent can be determined, given the depth of bed, the saturation capacity, mass transfer zone, and heel.
2. Using the computer program developed in (1) and given the following data, calculate the breakthrough capacity and the working capacity for this absorbent bed.
Depth of adsorption bed = 50 cm
Saturation capacity = 0.75

MTZ = 3 cm

Heel = 2.5%

**Solution**

1. For a solution to Example 7.35 (1), see the listing of Program 7.35.

2. A solution to Example 7.35 (2):

Breakthrough capacity, $C_B = 0.5(.75)(3) + .75(50 - 3)/50 = 0.7275 = 72.8\%$

Working capacity, $W_c = 72.8 - 2.5 = 70.3\%$

---

### LISTING OF PROGRAM 7.35 (CHAP7.35\FORM1.VB): BREAKTHROUGH WORKING CAPACITY OF AN ADSORPTION BED

```
'*********************************************************************************
' Example 7.35: A program to calculate the breakthrough
' capacity and the working capacity of an
' adsorption bed.
'*********************************************************************************

  Public Class Form1
     Dim cap, MTZ, depth, heel As Double
     Dim cb, Wc As Double

  Private Sub Form1_Load(ByVal sender As System.Object, ByVal e As System.EventArgs)
Handles MyBase.Load
     Me.Text = "Example 7.35: Adsorption bed capacity"
     Me.FormBorderStyle = Windows.Forms.FormBorderStyle.FixedSingle
     Me.MaximizeBox = False
     Label1.Text = "Example 7.35: A program to calculate the breakthrough capacity and"
     Label1.Text += vbCrLf + "the working capacity of an adsorption bed"
     Label1.Text += vbCrLf + "The formulas:"
     Label1.Text += vbCrLf + "C[B] = (0.5 * C[s] * (MTZ) + C[s] * (D - MTZ))/D"
     Label1.Text += vbCrLf + "W[c] = C[B] - H - PF"
     Label2.Text = "Enter the saturation capacity (C[s])"
     Label3.Text = "Enter the value of MTZ (in cm)"
     Label4.Text = "Enter the adsorption bed depth (D) in cm"
     Label5.Text = "Enter the heel value (H) in %"
     Label6.Text = ""
     Label7.Text = "Decimals:"
     Button1.Text = "&Calculate"
     GroupBox1.Text = ""
     NumericUpDown1.Value = 2
     NumericUpDown1.Maximum = 10
     NumericUpDown1.Minimum = 0
  End Sub

  Sub calculateResults()
     cap = Val(TextBox1.Text)
     MTZ = Val(TextBox2.Text)
     depth = Val(TextBox3.Text)
     heel = Val(TextBox4.Text)
     Dim comp1, comp2, comp3, comp4, cb1 As Double
     Comp1 = 0.5 * cap * MTZ
     Comp2 = depth - MTZ
     Comp3 = cap * comp2
     Comp4 = comp1 + comp3
     Cb1 = comp4 / depth
     Cb = cb1 * 100
     Wc = cb - heel
  End Sub

  Sub showResults()
     Label6.Text = "Breakthrough capacity (C[B]) = " + FormatNumber(cb, NumericUpDown1.
     Value) + "%"
```

```
    Label6.Text += vbCrLf + "Working capacity (W[c]) = " + FormatNumber(Wc,
    NumericUpDown1.Value) + "%"
  End Sub

  Private Sub Button1_Click(ByVal sender As System.Object, ByVal e As System.EventArgs)
Handles Button1.Click
    calculateResults()
    showResults()
  End Sub

  Private Sub NumericUpDown1_ValueChanged(ByVal sender As System.Object, ByVal e As
System.EventArgs) Handles NumericUpDown1.ValueChanged
    showResults()
  End Sub
  End Class
```

## Example 7.36

An activated carbon adsorption bed is used to remove trichloroethylene (TCE) from an air stream at 21°C and 1 atm. The actual flow rate is 12,730 m³/h, and the inlet TCE concentration is 2000 ppm by volume (MW of TCE = 131.5). Estimate the service time of the adsorbent before saturation. The proportionate saturation of the adsorbent is taken to be 0.2. The activated carbon bed is in a cylinder-type configuration, 1.36 m in radius and 1 m deep ($V = \pi r^2 h = 3.14 \times 1.85 \times 1$ m = 5.8 m³). The density of the adsorbent is 480 kg/m³. The sorption efficiency is 99.5%.

### Solution

Using Equation 7.57, the estimated service time of the adsorbent would be

$$t = \frac{2.41 \times 10^7 \times 0.2 \times 5.8 \text{ m}^3 \times 480 \text{ kg/m}^3}{0.995 \times 12,730 \text{ m}^3/\text{h} \times 2000 \text{ ppm} \times 131.5} = 4.03 \text{ h}$$

## Example 7.37

1. Prepare a computer program that incorporates all the elements in Equation 7.57, which makes it

possible to estimate the adsorbent's service time before regeneration.
2. An adsorption bed of activated carbon is to be used to remove acetone (MW = 58.08) from an air stream. The bed is 4 m thick and has a surface area of 6 m². The gas flow rate is 5,000 m³/h. The density of the activated carbon is 400 kg/m³. The inlet acetone concentration is 110,000 pp. The proportionate saturation of the adsorbent is taken to be 0.2. The sorption efficiency is 99.5%. Using Equation 7.57, estimate the service time of the adsorbent before regeneration is required.

### Solution

1. For a solution to Example 7.37 (1), see the listing of Program 7.37, which incorporates all the elements in Equations 7.52, and 7.57.
2. A solution to Example 7.37 (2):

$$t = \frac{2.41 \times 10^7 \times 0.2 \times 4 \text{ m} \times 6 \text{ m}^2 \times 400 \text{ kg/m}^3}{0.995 \times 5000 \text{ m}^3/\text{h} \times 58.08 \times 110,000 \text{ ppm}} = 1.46 \text{ h}$$

---

**LISTING OF PROGRAM 7.37   (CHAP7.37\FORM1.VB): ADSORBENT'S SERVICE TIME BEFORE REGENERATION**

```
'****************************************************************************
'Example 7.37: A program to estimate the adsorbent's
' service time before regeneration.
'****************************************************************************
Public Class Form1
Dim V, thick, surface As Double
  Dim S, dens, e, q, m, C As Double
  Dim comp1, comp2, t As Double

  Private Sub Form1_Load(ByVal sender As System.Object, ByVal e As System.EventArgs)
Handles MyBase.Load
    Me.Text = "Example 7.37:"
```

```
    Me.MaximizeBox = False
    Me.FormBorderStyle = Windows.Forms.FormBorderStyle.FixedSingle
    Label1.Text = "Example 7.37: A program to estimate the adsorbent's service time"
    Label1.Text += vbCrLf + "'before regeneration."
    Label1.Text += vbCrLf + "The formula:"
    Label1.Text += vbCrLf + "t = (2.41E7 * S * W)/(E * Q * M * C)"
    Label2.Text = "Is the adsorption bed square (1) or cylindrical (2)?"
    RadioButton1.Text = "Square"
    RadioButton2.Text = "Cylindrical"
    RadioButton1.Checked = True
    Label5.Text = "Enter the proportionate saturation of adsorbent (fractional) (x.x)"
    Label6.Text = "Enter the density of the adsorbent (kg/m{3})"
    Label7.Text = "Enter the adsorbent efficiency (fractional .xxx)"
    Label8.Text = "Enter the gas flow rate through the bed (m{3}/h)"
    Label9.Text = "Enter the molecular weight of the sorbed vapor (gm/mol)"
    Label10.Text = "Enter the entering vapor concentration (ppm by volume)"
    Label11.Text = ""
    Label12.Text = "Decimals:"
    Button1.Text = "&Calculate"
    NumericUpDown1.Value = 2
    NumericUpDown1.Maximum = 10
    NumericUpDown1.Minimum = 0
    changeDisp()
End Sub


'Changes the the first textboxes' labels according to the radiobutton selected.
Sub changeDisp()
    If RadioButton1.Checked Then
        Label3.Text = "Enter the depth of the bed of adsorbent (m)"
        Label4.Text = "Enter the diameter of the cylinder (m)"
    Else
    Label3.Text = "Enter the thickness of the bed of adsorbent (m)"
    Label4.Text = "Enter the surface area of the adsorbent (m{2})"
    End If
End Sub


Sub calculateResults()
    thick = Val(TextBox1.Text)
    surface = Val(TextBox2.Text)
    If RadioButton1.Checked Then
        V = 3.14 * thick * surface
    Else
        V = thick * surface
    End If

    S = Val(TextBox3.Text)
    dens = Val(TextBox4.Text)
    e = Val(TextBox5.Text)
    q = Val(TextBox6.Text)
    m = Val(TextBox7.Text)
    C = Val(TextBox8.Text)
    comp1 = 24100000.0 * S * V * dens
    comp2 = e * q * m * C
    t = comp1 / comp2
End Sub


Sub showResults()
    Label11.Text = "The time, t = " + FormatNumber(t, NumericUpDown1.Value) + " hours"
End Sub
```

```
    Private Sub Button1_Click(ByVal sender As System.Object, ByVal e As System.EventArgs)
Handles Button1.Click
        calculateResults()
        showResults()
    End Sub

    Private Sub RadioButton1_CheckedChanged(ByVal sender As System.Object, ByVal e As
System.EventArgs) Handles RadioButton1.CheckedChanged
        changeDisp()
    End Sub

    Private Sub RadioButton2_CheckedChanged(ByVal sender As System.Object, ByVal e As
System.EventArgs) Handles RadioButton2.CheckedChanged
        changeDisp()
    End Sub

    Private Sub NumericUpDown1_ValueChanged(ByVal sender As System.Object, ByVal e As
System.EventArgs) Handles NumericUpDown1.ValueChanged
        showResults()
    End Sub
End Class
```

## 7.4  AIR QUALITY MODELING

### 7.4.1  INTRODUCTION

Dispersion modeling is a procedure used to estimate the ambient air pollutant concentrations at various locations (receptors) downwind of a source, or any array of sources, based on emission rates, release specifications, and meteorological factors such as wind speed, wind direction, atmospheric stability, mixing height, and ambient temperature (Office of Air Quality Planning and Standards 1989). Air quality models can be categorized into four generic classes: Gaussian, numerical, statistical or empirical, and physical. Gaussian models generally are considered to be state-of-the-art techniques used for estimating the environmental impact of nonreactive pollutants. Numerical models are more appropriate than Gaussian models for multisource applications that involve reactive pollutants. Statistical or empirical models frequently are used in situations where incomplete scientific understanding of the physical and chemical processes makes use of a Gaussian or numerical model impractical. Physical modeling involves the use of wind tunnels or other fluid modeling facilities. Physical modeling is a complex process and applicable to a limited geographic area of only a few square kilometers (Zackay and Rowe 1984/1985).

In the United States, the 1977 Clean Air Act Amendments required that air quality models be used to identify potential violations of the NAAQS and to determine emission limits (Office of Air Quality Planning and Standards 1978).

A wide variety of air quality models are available to evaluate and simulate atmospheric dispersion processes. Two modeling systems used by the EPA are the Industrial Source Complex (ISC2) model and the SCREEN2 model. The SCREEN2 model is a scoping model that can be used to evaluate the air quality and estimate whether or not a given source is likely to pose a threat and cause the NAAQS to be exceeded. This type of dispersion model is used first, before going to a regulatory model in order to evaluate air quality conditions. The Industrial Source Complex model is more precise that a screening model uses local data to predict the levels of pollutants at a specific place (Air Pollution Training Institutre 1992). The EPA ISC2 model is a Gaussian dispersion model and has both a long-term module (ISCLT) and a short-term module (ISCST) (Lee 2005).

The Office of Air Quality Planning and Standards (OAQPS) of the U.S. EPA, through the Technology Transfer Network (TTN), has established an electronic bulletin board system (BBS), which allows remote users with either terminals or microcomputers to dial up and have access to numerous air quality models (U.S. EPA 2013). Only two air quality models will be considered here: the basic Gaussian model and the SCREEN2 model. First, however, the equations, formulas, models, or procedures used to estimate the effective stack height will be presented.

### 7.4.2  EFFECTIVE STACK HEIGHT

To apply the various Gaussian dispersion models, one important element in the models is the height at which the pollutant is emitted. Most stack emissions have an initial upward momentum and buoyancy. The buoyancy is due to the hot gases being ejected. The temperature differential between the hot stack gases (less dense) and the surrounding cooler ambient air causes the plume to rise into the atmosphere. The distance above the stack that a plume will rise into the atmosphere before leveling off is called the plume rise, $\Delta h$. The plume rise is affected not only by temperature but also by wind speed, the molecular weight of the gases being emitted, the physical stack height, its inside diameter, and the ambient pressure.

The emission height used in the dispersion models is the effective stack height, which is not only the physical stack height ($h$) but also the plume rise (see Figure 7.12a) (Office of Air Quality Planning and Standards 1983).

$$H = h + \Delta h \qquad (7.60)$$

where:
$H$ is the effective stack height, m
$h$ is the physical stack height, m
$\Delta h$ is the rise of the plume above the stack, m

The plume rise which is part of the effective stack height is an important element in estimating the maximum downwind ground-level concentration of a pollutant emitted from a stack. The maximum downwind ground-level concentration of a pollutant is reduced approximately by the inverse square of the effective stack height. For example, if the effective stack height is doubled, the maximum downwind ground-level

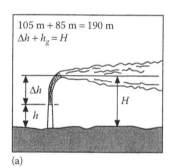

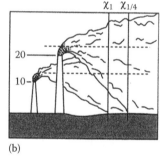

$$105\ m + 85\ m = 190\ m$$
$$\Delta h + h_g = H$$

**FIGURE 7.12** (a) An example of plume rise calculations. (b) An example of the reduced ground-level contaminant concentration due to doubling of the effective stack height.

concentration on the centerline of the plume will be reduced by a factor of 4 (see Figure 7.12b).

Numerous systems are available to estimate the effective stack height. The names of a few of these equations, models, or procedures are as follows.

- Bryant-Davidson
- Bosanquet
- Concawer
- Briggs
- Moses and Carson
- Holland

Only the well-known Holland equation will be presented here (Holland 1953):

$$\Delta h = \frac{d_s d}{u}\left[1.5 + 2.68(10)^{-3}\,p\left(\frac{T_s - T_a}{T_s}\right)d\right] \qquad (7.61)$$

where:
$v_s$ is the stack gas velocity, m/s
$u$ is the mean wind speed at stack height, m/s
$d$ is the stack inner diameter, m
$p$ is the atmospheric pressure, mb
$T_s$ is the stack gas temperature, K
$T_a$ is the atmospheric temperature, K

Equation 7.61 is valid for neutral stability conditions; however, Holland suggests that the plume rise be adjusted by a factor of from 1.1 to 1.2 for unstable conditions such as stability types A and B and from 0.8 to 0.9 for stability conditions E and F (see Table 7.8) (Turner 1969, Air Pollution Training Institute 1982).

To estimate the mean wind speed ($u$) at the top of the stack, an empirical formula can be used. This formula generally is considered appropriate for estimating wind speed at various heights up to 700–1000 m. This simple formula, or power law, is as follows:

**TABLE 7.8**
**Pasquill Stability Types**

| Surface Wind Speed (m/s) | Day | | | Night | |
|---|---|---|---|---|---|
| | Incoming Solar Radiation (Sunshine) | | | Thinly Overcast or ≥ 4/8 Low Cloud | ≤3/8 Low Cloud |
| | Strong | Moderate | Slight | | |
| <2 | A | A–B | B | | |
| 2 | A–B | B | C | E | F |
| 4 | B | B–C | C | D | E |
| 6 | C | C–D | D | D | D |
| >6 | C | D | D | D | D |

*Sources:* Turner, D.B., *Workbook of Atmospheric Dispersion Estimates*, Cincinnati, OH, National Air Pollution Control Administration, 31, 1969; Air Pollution Training Institute, *Basic Air Pollution Meteorology*, (SI:409), Research Triangle Park, NC, U.S. Environmental Protection Agency, 1982.

*Note:* A, extremely unstable; B, moderately unstable; C, slightly unstable; D, neutral; E, slightly stable; F, moderately stable. Neutral class D should be assumed for overcast conditions during day or night.

**TABLE 7.9**

**Average Values of Wind Profile Power Law Exponents (k) by Stability Class**

| Pasquill Stability Class | Average Value of Exponent |
|---|---|
| A | 0.141 |
| B | 0.176 |
| C | 0.174 |
| D | 0.209 |
| E | 0.277 |
| F | 0.414 |
| G[a] | 0.435 |

*Source:* From Touma, J.S., *J. Air Pollut. Control Assoc.*, 27, 863, 1977. With permission.

[a] Pasquill stability class G is considered to be very stable.

$$\frac{v}{v_0} = \left(\frac{z}{z_0}\right)^k \qquad (7.62)$$

where:

$v$ is the wind speed at height $z$, m/s

$v_0$ is the wind speed at anemometer level $z_0$, m/s

$k$ is the exponent or coefficient, dimensionless

The exponent $k$ in the past has been taken generally as 1/7; however, recent research has provided values for $k$ depending upon the stability class. Table 7.9 presents $k$ values for each stability class.[39] Examples 7.38 and 7.39 demonstrate the use of Equations 7.60 through 7.62.

## Example 7.38

Determine the effective stack height, given the following data:

Physical stack height, $h$ = 183 m inside diameter, with $d$ = 6 m

Wind velocity at anemometer level (2 m above ground) = 5 m/s

Air temperature = 10°C

Atmospheric pressure = 1000 mbar

Stack gas velocity = 16 m/s

Stack gas temperature = 135°C

Class B Pasquill stability type

### Solution

- Determine the mean wind speed at the top of the stack; use Equation 7.62.

$$u = (5 \text{ m/s})\left(\frac{183 \text{ m}}{2 \text{ m}}\right)^{0.176} = 11 \text{ m/s}$$

For Class B stability, $k$ = 0.176 (see Table 7.9).

- Convert temperature to K:

$T_a$ = 273 + 10 = 283 K

$T_s$ = 273 + 135 = 408 K

- Substitute the given and calculated values in Holland's equation (7.61) to determine the plume rise, $\Delta h$:

$$\Delta h = \frac{(16 \text{ m/s})(6\text{m})}{11\text{m/s}}$$

$$\left[1.5 + 2.68(10)^{-3}(1000 \text{ mbar})\left(\frac{408 \text{ K} - 283 \text{ K}}{408 \text{ K}}\right)6 \text{ m}\right] = 56 \text{ m}$$

- Use Equation 7.60 to determine the effective stack height:

$H$ = 183 m + 56 M = 239 m

## Example 7.39

1. Prepare a computer program that will determine the effective stack height, given all the required elements contained in Equations 7.60 through 7.62.
2. Using the computer program developed in (1), compute the effective stack height, given the following data:

Physical stack height $h$ = 50 m with inside diameter $d$ = 3.5 m

Wind velocity at anemometer level (4 m above ground level) = 10 m/s

Air temperature = 0°C

Atmospheric pressure = 1000 mbar

Stack gas velocity = 20 m/s

Stack gas temperature = 150°C

Class C Pasquill stability type

### Solution

1. For a solution to Example 7.39 (1), see the listing of Program 7.39, which includes the elements in Equations 7.60 through 7.62. This computer program can be used to estimate the effective stack height.
2. A solution to Example 7.39 (2):
   a. Determine the mean wind speed at the top of the stack (50 m); use Equation 7.62:

$$u = 10 \text{ m/s}\left(\frac{50 \text{ m}}{4 \text{ m}}\right)^{0.174} = 15.5 \text{ m/s}$$

For Class C stability, $k$ = 0.174 (see Table 7.9).

   b. Convert temperature to K:

$T_a$ = 0 + 273 = 273 K

$T_s$ = 150 + 273 = 423 K

   c. Substitute values in Holland's equation (7.61) to determine the plume rise, $\Delta h$:

$$\Delta h = \frac{20 \text{ m/s}(3.5 \text{ m})}{15.5 \text{ m/s}}$$

$$\left[1.5 + 2.68(10)^{-3}(1000 \text{ mbar})\left(\frac{423\text{K} - 273\text{K}}{423\text{K}}\right)3.5\text{m}\right] = 21.8\text{m}$$

   d. Use Equation 7.60 to determine the effective stack:

$H$ = 50 m + 21.8 m = 71.8 m

## LISTING OF PROGRAM 7.39    (CHAP7.39\FORM1.VB): EFFECTIVE STACK HEIGHT

```vb
'******************************************************************************
'Example 7.39: A program to determine the effective stack height.
'******************************************************************************
Public Class Form1
  Dim hT, h, dh, k, z0 As Double
  Dim v, v0, vs, d, p, Ts, Ta As Double
  Dim k_table() As Double =
    {0.141, 0.176, 0.174, 0.209, 0.277, 0.414, 0.435}

  Private Sub Form1_Load(ByVal sender As System.Object, ByVal e As System.EventArgs)
Handles MyBase.Load
    Me.Text = "Example 7.39: Effective stack height"
    Me.FormBorderStyle = Windows.Forms.FormBorderStyle.FixedSingle
    Me.MaximizeBox = False
    Label1.Text = "Example 7.39: A program to determine the effective stack height."
    Label1.Text += vbCrLf + "The formulas:"
    Label1.Text += vbCrLf + "H = h + (delta)h"
    Label1.Text += vbCrLf + "delta(h) = ((v[s] * d) / u) * (1.5 + 2.68(10){-3} r((T[s]
    - T[a]) / T[s])d)"
    Label1.Text += vbCrLf + "v/v[o] = (z / z[o]){k}"
    Label2.Text = "Enter the physical stack height, h (m)"
    Label3.Text = "Enter the inside diameter of the stack, d (m)"
    Label4.Text = "Enter the wind velocity at anemometer level, v0 (m/s)"
    Label5.Text = "Enter the anemometer height above ground level, z0 (m)"
    Label6.Text = "Enter the air temperature, Ta (C)"
    Label7.Text = "Enter the atmospheric pressure, p (mbar)"
    Label8.Text = "Enter the stack gas velocity, vs (m/s)"
    Label9.Text = "Enter the stack gas temperature, Ts (C)"
    Label10.Text = "Pasquill stability class"
    Label11.Text = ""
    Label12.Text = "Decimals:"
    NumericUpDown1.Value = 2
    NumericUpDown1.Maximum = 10
    NumericUpDown1.Minimum = 0
    GroupBox1.Text = ""
    ComboBox1.Items.Clear()
    ComboBox1.Items.Add("A")
    ComboBox1.Items.Add("B")
    ComboBox1.Items.Add("C")
    ComboBox1.Items.Add("D")
    ComboBox1.Items.Add("E")
    ComboBox1.Items.Add("F")
    ComboBox1.Items.Add("G")
    ComboBox1.SelectedIndex = 0
    Button1.Text = "&Calculate"
  End Sub

  Sub calculateResults()
    Dim index As Integer = ComboBox1.SelectedIndex
    If index = -1 Then
      MsgBox("Please select Pasquill stability class.",
        vbOKOnly Or vbExclamation)
      Exit Sub
    End If
    k = k_table(index)
    h = Val(TextBox1.Text)
    d = Val(TextBox2.Text)
    v0 = Val(TextBox3.Text)
    z0 = Val(TextBox4.Text)
```

```
      Ta = Val(TextBox5.Text) + 273
      p = Val(TextBox6.Text)
      vs = Val(TextBox7.Text)
      Ts = Val(TextBox8.Text) + 273
      v = v0 * ((h / z0) ^ k)
      dh = ((vs * d) / v) * (1.5 + (0.00268 * p * ((Ts - Ta) / Ts) * d))
      hT = h + dh
   End Sub


   Sub showResults()
      Dim decimals As Integer = NumericUpDown1.Value
      If decimals < 0 Or decimals > 10 Then decimals = 2
      Label11.Text = "Wind speed at the top of the stack = " + FormatNumber(v, decimals) +
      " m/s"
      Label11.Text += vbCrLf + "The plume rise dh = " + FormatNumber(dh, decimals) + " m"
      Label11.Text += vbCrLf + "The effective stack height = " + FormatNumber(hT,
      decimals) + " m"
   End Sub


   Private Sub Button1_Click(ByVal sender As System.Object, ByVal e As System.EventArgs)
Handles Button1.Click
      calculateResults()
      showResults()
   End Sub


   Private Sub NumericUpDown1_ValueChanged(ByVal sender As System.Object, ByVal e As
System.EventArgs) Handles NumericUpDown1.ValueChanged
      showResults()
   End Sub
End Class
```

### 7.4.3 Dispersion Models

#### 7.4.3.1 Basic Gaussian Dispersion Model

Many dispersion models have been developed; most of the models in use today are based on the work of Pasquill and modified by Gifford. The binormal Gaussian plume equation (7.57) relates dispersion in the $x$ (downwind) direction as a function of variables in all directions of a three-dimensional space.

The concentration ($\chi$) of a gas or aerosol ($<20$ μm) calculated at the ground level for a distance downwind ($x$) is expressed as follows:

$$\chi(x, y) = \frac{Q}{\pi \sigma_y \sigma_z \bar{u}} \exp\left[-\frac{1}{2}\left(\frac{y}{\sigma_y}\right)^2\right] \exp\left[-\frac{1}{2}\left(\frac{H}{\sigma_z}\right)^2\right] \quad (7.63)$$

where:

($x$, $y$) are the receptor coordinates, m
$\chi$ is the ground-level concentration, g/m$^3$
$Q$ is the emission rate, g/s
$H$ is the effective stack height, m
$\bar{u}$ is the mean wind speed, m/s
$\sigma_y$ and $\sigma_z$ are the dispersion coefficients, m
$\pi = 3.14159$
exp is the base of natural log, 2.7182818

Figure 7.13a depicts a graphical presentation of this equation (Turner 1969).

The Gaussian distribution equation expresses downwind ground-level contaminant concentration when the terrain is approximately flat (see Figure 7.13b). Uneven terrain such as valleys, hills, and mountains makes it necessary to modify the Gaussian plume distribution. These modifications are exponential and vary depending upon the specific air quality model used (Air Pollution Training Institute 1982).

Values for the dispersion coefficients $\sigma_y$ and $\sigma_z$ depend not only on the downwind distances but also on the atmospheric stability (lapse rates). Values for $\sigma_y$ and $\sigma_z$ for various distances downwind ($x$) with various stability categories can be determined by using Figure 7.14a,b (Turner 1969). The Pasquill stability types or categories were previously presented in Table 7.8.

Equation 7.63 can be simplified if only the ground-level downwind concentrations along the centerline of the plume are needed. In this case, $y = 0$, and the equation then becomes

$$\chi_{x,0} = \frac{Q}{\pi \bar{u} \sigma_y \sigma_z} \exp\left[-\frac{1}{2}\left(\frac{H}{\sigma_z}\right)^2\right] \quad (7.64)$$

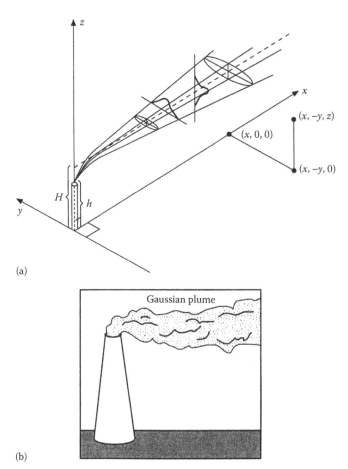

(a)

(b)

**FIGURE 7.13** (a) Coordinate system showing Gaussian distribution in the horizontal and vertical directions. (b) Flat terrain for Gaussian distribution.

The units for this equation are the same as for Equation 7.63. See Figure 7.15a for a schematic presentation of this equation (Turner 1969, Air Pollution Training Institute 1982).

Equation 7.64 may be further simplified if the effective stack height is $H = 0$. In this case, the source, such as a burning dump, is at ground level. Equation 7.59 can then be used to calculate the ground-level downwind concentrations (Turner 1969, Air Pollution Training Institute 1982):

$$\chi = \frac{Q}{\pi \bar{u} \sigma_y \sigma_z} \qquad (7.65)$$

The units for this equation are the same as for Equations 7.63 and 7.64. Figure 7.15b is a graphical schematic presentation of this equation (Turner 1969, Air Pollution Training Institute 1982).

The maximum downwind ground-level concentration of a contaminant occurs on the centerline of the plume. The maximum downwind ground-level concentration can be estimated by taking the differential of a modified version of Equation 7.63, resulting in the following expression (Colls and Tiwary 2009):

$$\sigma_z = 0.707H \qquad (7.66)$$

This equation holds true, provided $\sigma_z/\sigma_y$ are constants with downwind distance $x$. This gives a rough approximation of the maximum downwind ground-level concentration and is a much better approximation for unstable conditions than for stable conditions (Turner 1969).

The distance at which the maximum downwind ground-level concentration occurs can then be estimated by using the vertical dispersion coefficient $\sigma_z$ (Figure 7.14b) and the

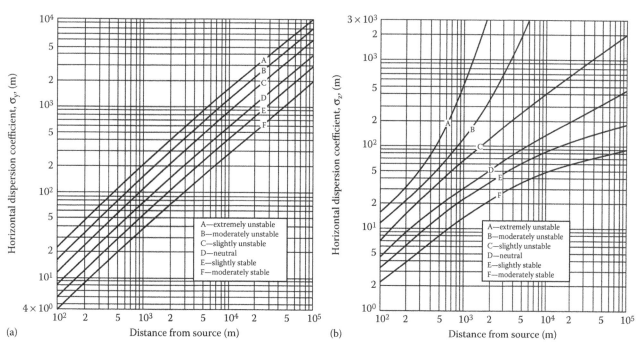

(a)

(b)

**FIGURE 7.14** (a) Horizontal dispersion coefficient, $\sigma_y$, as a function of downwind distance from the source. (b) Vertical dispersion coefficient, $\sigma_z$, as a function of downward distance from the source.

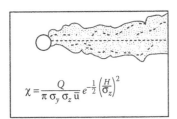

$$\chi = \frac{Q}{\pi\,\sigma_y\,\sigma_z\,\bar{u}}\,e^{-\frac{1}{2}\left(\frac{H}{\sigma_z}\right)^2}$$

(a)

$$\chi = \frac{Q}{\pi\,\sigma_y\,\sigma_z\,\bar{u}}$$

(b)

**FIGURE 7.15**   (a) Schematic of ground-level concentration underneath the center line of the plume. (b) Schematic of contaminants generated under ground-level conditions.

appropriate stability type. Examples 7.40 and 7.41 demonstrate the use of the dispersion modeling equations.

## Example 7.40

A modern 700-MW coal-fired power plant operates under the following conditions:

Stack height = 203 m
Inside stack diameter = 7 m
Stack gas exit velocity = 16 m/s
Stack gas temperature = 160°C
Ambient air temperature = 24°C
The average wind speed at anemometer level (2 m above ground level) = 4 m/s
Atmospheric pressure = 1000 mbar
Atmospheric conditions are neutral stability (type D)
Emission rate for $SO_2$ from the stack = 4000 g/s

Determine the effective stack height using Holland's equation (7.61). What is the maximum downwind ground-level concentration and how far is this from the plant? What would the expected ground-level concentration be at 300 m crosswind from the point on the plume centerline where the maximum downwind ground-level concentration occurred? Do the emissions from this plant cause the U.S. NAAQS, 24-h standards for $SO_2$ at ground-level downwind from the plant to be exceeded?

### Solution

• Determine the mean wind speed at the top of the stack; use Equation 7.62:

$$u = 4\ \text{m/s}\left(\frac{203\ \text{m}}{2\ \text{m}}\right)^{0.209} = 10.5\ \text{m/s}$$

For Class D stability, $k = 0.209$ (see Table 7.9).
• Convert temperature to K:
   $T_a = 24 + 273 = 297$ K
   $T_s = 160 + 273 = 433$ K
• Use the Holland's equation (7.61) to determine the plume rise, $\Delta h$:

$$\Delta h = \frac{16\ \text{m/s}\ (7\ \text{m})}{10.5\ \text{m/s}}$$

$$\left[1.5 + 2.68(10)^{-3}(1000\,\text{mbar})\left(\frac{433\text{K} - 297\text{K}}{433\text{K}}\right)7\,\text{m}\right] = 78.9\,\text{m}$$

• Use Equation 7.60 to determine the effective stack height: $H$ = 203 m + 78.9 m = 281.9 m (usse 282 m).
• Determine the location for the maximum ground-level concentration; use Equation 7.66:
   $\sigma_z = 0.707H = 0.707(282\ \text{m}) = 199$ m.
   From Figure 7.14b and under neutral-type D stability, $\sigma_z$ reaches a value of 199 m at a distance of 20 km from the stack.
• Thus, $\chi_{max}$ occurs ~20 km downwind from the plant. Determine the maximum ground-level concentration at 20 km; use Equation 7.64. From Figure 7.14a and b:

$$\sigma_y = 1000\ \text{m}$$

$$\sigma_z = 199\ \text{m}$$

$$\chi_{max} = \frac{4000\ \text{g/s}}{\pi \times 10.5\ \text{m/s} \times 199\ \text{m} \times 1000\ \text{m}}\exp-\frac{1}{2}\left(\frac{282\ \text{m}}{199\ \text{m}}\right)^2$$

$$= \frac{0.0002233\ \text{g}}{\text{m}^3} = 223\,\mu\text{g/m}^3$$

• Determine the expected ground-level concentration at 300 m crosswind from the point of maximum downwind ground-level concentration; use Equation 7.63.

$$\chi = \frac{4000}{\pi \times 10.5 \times 199 \times 1000}\exp-\frac{1}{2}\left(\frac{300}{1000}\right)^2$$

$$\exp-\frac{1}{2}\left(\frac{282}{199}\right)^2 = 213\,\mu\text{g/m}^3$$

• From Table 7.1, the primary U.S. NAAQS standard for the 24-h $SO_2$ level is 365 $\mu\text{g/m}^3$, whereas the 3-h secondary $SO_2$ standard is 1300 $\mu\text{g/m}^3$. In this case, the emissions of $SO_2$ from this plant do not cause the downwind $SO_2$ ambient air concentrations to exceed the standards. However, field testing would be necessary to confirm this.

## Example 7.41

1. Prepare a computer program that incorporates all the elements in Equations 4.63 through 4.66.
2. Using the computer program developed in (1), compute the maximum downwind ground-level concentration, given the following data:

   a. A manufacturing plant emits nitrogen dioxide ($NO_2$) on a continuous 24-h/day basis. What is the concentration of $NO_2$ at a receptor 8 km downwind of the plant? Does this plant cause concentrations at the receptor to be in excess of the primary and secondary U.S. NAAQS?
   b. The effective stack height for this plant is 60 m.
   c. The $NO_2$ emission rate is 100 g/s.
   d. The wind speed at the top of the stack is 15.5 m/s.
   e. Pasquill stability type is moderately stable, type F.

## Solution

1. For a solution to Example 7.41 (1), see the listing of Program 7.41.
2. A solution to Example 7.41 (2):

   a. Use Equation 7.64. From Figure 7.14a and b at 8 km and F stability:

   $$\sigma_y = 220 \, m$$

   $$\sigma_z = 42 \, m$$

   $$\chi = \frac{100 \, g/s}{3.14 \times 15.5 \, m/s \times 220 \, m \times 42 \, m} \exp -\frac{1}{2} \left( \frac{60 \, m}{42 \, m} \right)^2$$

   $$= 0.000080 \, g/m^3 = 80 \, mg/m^3$$

   b. The U.S. NAAQS for $NO_2$ (Table 7.1) for the primary and secondary standards is 100 $\mu g/m^3$, with an averaging time of 1 year. The receptor in this case would not be exposed to $NO_2$ concentrations above the standard.

---

### LISTING OF PROGRAM 7.41   (CHAP7.41\FORM1.VB): DOWNWIND GROUND-LEVEL CONCENTRATION

```
'*******************************************************************************
'Example 7.41: A program to compute the max. downwind
' groundlevel concentration
'*******************************************************************************
Public Class Form1
  Dim u, wind, sheight, aheight, stabil As Double
  Dim vel, dia, pres, airtemp, stemp As Double
  Dim q, sy As Double
  Dim k_table() As Double =
  {0.141, 0.176, 0.174, 0.209, 0.277, 0.414, 0.435}

  Private Sub Form1_Load(ByVal sender As System.Object, ByVal e As System.EventArgs)
  Handles MyBase.Load
    Me.Text = "Example 7.41: Max. downwind groundlevel concentration"
    Me.MaximizeBox = False
    Label1.Text = "Dispersion coefficients (sigma)[y] and (sigma)[z] will be represented
    by the letter S."
    GroupBox1.Text = "Is the mean wind speed at the top of the stack is provided?"
    RadioButton1.Text = "Yes"
    RadioButton2.Text = "No"
    RadioButton1.Checked = True
    Label2.Text = "Enter wind speed at the top of the stack u (m/s)"
    Label3.Text = "Enter the wind velocity at anemometer level u (m/s)"
    Label4.Text = "Enter the stack height (m)"
    Label5.Text = "Enter the anemometer height above ground level (m)"
    Label6.Text = "Select stability type class"
    Label7.Text = ""

    GroupBox2.Text = "Is the effective stack height for this plant is provided?"
    RadioButton4.Text = "Yes"
    RadioButton3.Text = "No"
    RadioButton4.Checked = True
    Label13.Text = "Enter the effective stack height for this plant (m)"
    Label12.Text = "Enter the emission rate (g/s)"
    Label11.Text = "The Pasquill stability type data is provided. S[y] is read from the
    tables"
```

```
         Label11.Text += vbCrLf + "Enter the value of S[y] from the tables"
         Label10.Text = "The value of S[z] is = .707 * effective stack height"
         Label9.Text = "Enter the stack gas velocity (m/s)"
         Label8.Text = "enter the inside diameter of the stack (m)"
         Label15.Text = "Enter the atmospheric pressure (mbar)"
         Label14.Text = "Enter the air temperature (C)"
         Label16.Text = "Enter the stack gas temperature (C)"
         Label17.Text = ""
         Label18.Text = ""
         Button1.Text = "&Calculate"
         Label19.Text = "Decimal places:"
         NumericUpDown1.Value = 2
         NumericUpDown1.Maximum = 10
         NumericUpDown1.Minimum = 0
         ComboBox1.Items.Clear()
         ComboBox1.Items.Add("A")
         ComboBox1.Items.Add("B")
         ComboBox1.Items.Add("C")
         ComboBox1.Items.Add("D")
         ComboBox1.Items.Add("E")
         ComboBox1.Items.Add("F")
         ComboBox1.Items.Add("G")
         ComboBox1.SelectedIndex = 0
         changeDisp()
End Sub
'********************************************************************************
'According to the selected Radiobutton,
'enables/disables textboxes and labels
'on the form, to modify user's inputs.
'********************************************************************************
Sub changeDisp()
  If RadioButton1.Checked Then
     TextBox1.Enabled = True
     TextBox2.Enabled = False
     TextBox3.Enabled = False
     TextBox4.Enabled = False
     Label2.Enabled = True
     Label3.Enabled = False
     Label4.Enabled = False
     Label5.Enabled = False
     Label6.Enabled = False
     Label7.Enabled = False
     ComboBox1.Enabled = False
  Else
     TextBox1.Enabled = False
     TextBox2.Enabled = True
     TextBox3.Enabled = True
     TextBox4.Enabled = True
     Label2.Enabled = False
     Label3.Enabled = True
     Label4.Enabled = True
     Label5.Enabled = True
     Label6.Enabled = True
     Label7.Enabled = True
     ComboBox1.Enabled = True
  End If
  If RadioButton4.Checked Then
     TextBox6.Enabled = False
     TextBox7.Enabled = False
     TextBox8.Enabled = True
     TextBox10.Enabled = True
```

```
      TextBox11.Enabled = False
      TextBox12.Enabled = False
      TextBox13.Enabled = False
      Label9.Enabled = False
      Label8.Enabled = False
      Label10.Enabled = True
      Label11.Enabled = True
      Label13.Enabled = True
      Label14.Enabled = False
      Label15.Enabled = False
      Label16.Enabled = False
      Label17.Enabled = False
   Else
      TextBox6.Enabled = True
      TextBox7.Enabled = True
      TextBox8.Enabled = False
      TextBox10.Enabled = False
      TextBox11.Enabled = True
      TextBox12.Enabled = True
      TextBox13.Enabled = True
      Label8.Enabled = True
      Label9.Enabled = True
      Label10.Enabled = False
      Label11.Enabled = False
      Label13.Enabled = False
      Label14.Enabled = True
      Label15.Enabled = True
      Label16.Enabled = True
      Label17.Enabled = True
   End If
End Sub

   Sub calculateResults()
      Dim COMP1, COMP2, COMP3, COMP4, COMP5, COMP6, COMPT As Double
      Dim ta, ts, dh, h As Double
      Dim decimals As Integer = NumericUpDown1.Value
      If decimals < 0 Or decimals > 10 Then decimals = 2
      If RadioButton1.Checked Then
        u = Val(TextBox1.Text)
      Else
        Dim k_index As Integer = ComboBox1.SelectedIndex
        If k_index = -1 Then
        MsgBox("Please select Pasquill stability class.", vbOKOnly Or vbExclamation)
        Exit Sub
      End If
      stabil = k_table(k_index)
      wind = Val(TextBox2.Text)
      sheight = Val(TextBox3.Text)
      aheight = Val(TextBox4.Text)
      COMP1 = sheight / aheight
      COMP2 = COMP1 ^ stabil
      u = wind * COMP2
      Label7.Text = "Wind speed at the top of the stack = " + FormatNumber(u, decimals) + " m/s"

      End If
      q = Val(TextBox9.Text)
      If RadioButton4.Checked Then
        h = Val(TextBox10.Text)
        sy = Val(TextBox8.Text)
      Else
        vel = Val(TextBox7.Text)
```

```
            dia = Val(TextBox6.Text)
            pres = Val(TextBox12.Text)
            airtemp = Val(TextBox11.Text)
            stemp = Val(TextBox13.Text)
            sy = 1000
            ta = airtemp + 273
            ts = stemp + 273
            COMP3 = (vel * dia) / u
            COMPT = (ts - ta) / ts
            COMP4 = pres * COMPT * 0.001
            COMP5 = 2.68 * COMP4 * dia
            COMP6 = 1.5 + COMP5
            dh = COMP3 * COMP6
            Label17.Text = "The plume rise, (delta)h = " + FormatNumber(dh, decimals) + " m"
            h = sheight + dh
            Label17.Text += vbCrLf + "The effective stack height = " + FormatNumber(h,
            decimals) + " m"
        End If
        Dim sz, den1, half1, den2, half2, xg, xu As Double
        sz = h * 0.707
        den1 = Math.PI * u * sy * sz
        half1 = q / den1
        den2 = -0.5 * ((h / sz) * (h / sz))
        half2 = Math.Exp(den2)
        xg = half1 * half2
        xu = xg * 1000000
        Label18.Text = "The maximum downwind groundlevel concentration "
        Label18.Text += vbCrLf + "is " + FormatNumber(xu, decimals) + " ug/m{3}"
    End Sub


    Private Sub RadioButton1_CheckedChanged(ByVal sender As System.Object, ByVal e As
System.EventArgs) Handles RadioButton1.CheckedChanged
        changeDisp()
End Sub


    Private Sub RadioButton2_CheckedChanged(ByVal sender As System.Object, ByVal e As
System.EventArgs) Handles RadioButton2.CheckedChanged
        changeDisp()
End Sub

    Private Sub Button1_Click(ByVal sender As System.Object, ByVal e As System.EventArgs)
Handles Button1.Click
        calculateResults()
End Sub

    Private Sub RadioButton4_CheckedChanged(ByVal sender As System.Object, ByVal e As
System.EventArgs) Handles RadioButton4.CheckedChanged
        changeDisp()
End Sub

    Private Sub RadioButton3_CheckedChanged(ByVal sender As System.Object, ByVal e As
System.EventArgs) Handles RadioButton3.CheckedChanged
        changeDisp()
    End Sub
End Class
```

### 7.4.3.2 U.S. EPA Dispersion Models

As indicated at the beginning of this section, the U.S. EPA OAQPS, through the TTN, has established an electronic bulletin board system that allows anyone in the world to dial up and exchange information about air pollution (Office of Air Quality Planning and Standards 1994). EPA personnel, state and local agencies, the private sector, and foreign countries all have access to this network (Office of Air Quality Planning and Standards 1994).

To access the 18 bulletin boards currently available on the network, a computer, modem, and communications software set with the following parameters are needed:

Data bits: 8
Parity: *N*
Stop bits: 1
Terminal emulation: VT100 or VT/ANSI
Duplex: Full

A short explanation of the communications software parameters is as follows: Each 0 or 1 in a binary system is a bit; 0 and 1 are binary numbers. A byte is a string of binary digits and usually contains eight. "Data bits" indicates that a packet of information is being processed, whereas "Stop bits" indicates that no information is being transmitted. "Parity" indicates that an extra bit has been added to a computer word to detect errors. "Terminal emulation" designates a communications program that makes a personal computer act like a terminal for the purpose of interacting with a remote computer. The terms VT100 or VS/ANSI are options that are used in order to access particular computer programs. ANSI stands for the American National Standards Institute. "Full duplex" indicates a method of transmitting data that allow for the simultaneous sending and receiving of data. Using the communications software, the number to call is (919) 541–5742 for modems up to 14,400 (bps), or a voice help line is available at (919) 541–5384. Of the bulletin boards currently available, the SCRAM (Support Center for Regulatory Air Models) provides regulatory air quality model computer codes, meteorological data, and documentation, as well as modeling guidance (Office of Air Quality Planning and Standards 1994).

### 7.4.3.3 SCREEN2 Dispersion Model

The SCREEN2 dispersion model uses the Gaussian plume model that incorporates source-related factors and meteorological factors to estimate pollutant concentration from continuous sources. It is assumed that the pollutant does not undergo any chemical reactions and that no other removal processes, such as wet or dry deposition, act on the plume during its transport from the source (Office of Air Quality Planning and Standards 1992). The Gaussian model equations and the interactions of the source-related and meteorological factors are presented in Section 7.4.3.1, and more details are presented in the work of Turner (1969).

The basic equation used in the SCREEN2 model for determining ground-level concentrations under the plume center-line is as follows (Office of Air Quality Planning and Standards 1992):

$$
\begin{aligned}
\chi = Q/(2\pi u_s \sigma_y \sigma_z) \Bigg[ &\exp\left\{-\frac{1}{2}\left[(z_r - h_e)/\sigma_z\right]^2\right\} \\
&+\exp\left\{-\frac{1}{2}\left[(z_r + h_e)/\sigma_z\right]^2\right\} \\
&+\sum_{N=1}^{k}\left(\exp\left\{-\frac{1}{2}\left[(z_r - h_e - 2Nz_i)/\sigma_z\right]^2\right\}\right. \\
&+\exp\left\{-\frac{1}{2}\left[(z_r + h_e - 2Nz_i)/\sigma_z\right]^2\right\} \\
&+\exp\left\{-\frac{1}{2}\left[(z_r - h_e + 2Nz_i)/\sigma_z\right]^2\right\} \\
&\left.+\exp\left\{-\frac{1}{2}\left[(z_r + h_e + 2Nz_i)/\sigma_z\right]^2\right\}\right)\Bigg]
\end{aligned}
\tag{7.61}
$$

where:

$\chi$ is the concentration, $g/m^3$
$Q$ is the emission rate, $g/s$
$\pi = 3.141593$
$u_s$ is the stack height wind speed, m/s
$\sigma_y$ is the lateral dispersion parameter, m
$\sigma_z$ is the vertical dispersion parameter, m
$z_r$ is the receptor height above ground, m
$h_e$ is the plume center line height, m
$z_i$ is the mixing height, m
$k$ is the summation limit for multiple reflections of plume off of ground and elevated inversion, usually $\leq 4$

This SCREEN2 model can be used for point sources or area sources (Office of Air Quality Planning and Standards 1992).

For point sources, the inputs needed are as follows:

- Emission rate, g/s
- Stack height, m
- Stack inside diameter, m
- Stack gas exit velocity, m/s, or flow rate (ACFM or $m^3/s$)
- Stack gas temperature, K
- Ambient temperature, K (use default of 293 K if not known)
- Receptor height above ground (may be used to define flagpole receptors)
- Urban/rural option (U = urban, R = rural)

For area sources, the inputs needed are as follows:

- Emission rate, $g/s/m^2$
- Source release height, m
- Length of side of the square area, m

- Receptor height above ground, m
- Urban/rural option (U = urban, R = rural)

Other information that must be considered in the SCREEN2 model is as follows:

- Mixing height
- Worst-case meteorological conditions
- Plume rise for point source
- Dispersion parameters
- Buoyancy-induced dispersion
- Building downwash
- Fumigation
- Shoreline fumigation
- Complex terrain 24-h screen

For stable conditions and/or mixing heights $\geq$ 10,000 m, unlimited mixing is assumed, and the summation term in Equation 7.67 is assumed to be zero. The mixing height used in Equation 7.67 is based on Randerson calculations (see Randerson 1984). The worst-case meteorological conditions use the various stability classes (see Table 7.8) and wind speeds to identify the "worst-case" scenario.

The plume rise equations developed by Briggs are used in the SCREEN2 model to estimate the plume rise. The plume rise equations come from References Briggs 1969, 1973, 1975. The dispersion coefficients $\sigma_z$, vertical, and $\sigma_y$, lateral, were designed to be used in the Gaussian plume models and were first published by Pasquill in 1959 (see Figure 7.14a,b). The dispersion coefficients used for rural and urban sites in the SCREEN2 model come from Reference U.S. EPA 1992. The dispersion coefficients ($\sigma_y$ and $\sigma_z$) are adjusted in most cases to account for the effects of buoyancy-induced dispersion using the following equations:

$$\sigma_{ye} = \sqrt{\left[ \sigma_y^2 + \left( \frac{\Delta h}{3.5} \right)^2 \right]}$$

$$\sigma_{ze} = \sqrt{\left[ \sigma_z^2 + \left( \frac{\Delta h}{3.5} \right)^2 \right]}$$

where:
  $\sigma_{ye}$ is the adjusted lateral dispersion coefficient, m
  $\sigma_{ze}$ is the adjusted vertical dispersion coefficient, m
  $\Delta h$ is the distance-dependent plume rise, m

The building downwash, cavity, and wake-screening problems encountered in dispersion modeling are dealt with in References Hosker 1984 and U.S. EPA 1992.

When an inversion layer occurs a short distance above a plume source and superdiabatic conditions prevail below the stack, the plume is said to be fumigating. In this case, the air contaminants are suddenly brought to ground level. In the SCREEN2 model, the inversion breakup calculations are based on procedures described by Turner (1969).

A special-purpose model has been developed to consider the worst-case impact for complex terrain, 24-h screen. This dispersion model is called VALLEY and is a modified Gaussian model (Burt 1977). This model assumes F stability conditions (E for urban) and a stack height wind speed of 2.5 m/s. The width of the sector of uniform air contaminant concentration is assumed to be 22.5°. The following information is taken directly from the Reference (Office of Air Quality Planning and Standards 1992).

If the plume height is at or below the terrain height for the distance entered, then SCREEN2 will make a 24-h average concentration estimate using the VALLEY screening technique. If the terrain is above the stack height but below the plume center-line height, then SCREEN2 will make a VALLEY 24-h estimate (assuming F or E and 2.5 m/s) and also will estimate the maximum concentration across a full range of meteorological conditions using simple terrain procedures with terrain "chopped off" at physical stack height, and select the higher estimate. Calculations continue until a terrain height of zero is entered. For the VALLEY model concentration, SCREEN2 will calculate a sector-averaged ground-level concentration with the plume center-line height ($h_e$) as the larger of 10.0 m or the difference between the plume height and the terrain height.

## 7.5 HOMEWORK PROBLEMS IN COMPUTER MODELING APPLICATIONS FOR AIR POLLUTION CONTROL TECHNOLOGY

### 7.5.1 DISCUSSION PROBLEMS

1. Define:
   a. Charles' law
   b. Boyle's law
   c. Ideal gas law
   d. Henry's law
   e. Dalton's law
   f. Reynold's law
   g. Stoke's law
   h. Air pollution
2. What are the four major components that make up the troposphere of the earth and what is the percentage of each?
3. What are the two basic physical forms of air pollution?
4. What are the six criteria pollutants designated by the U.S. EPA?
5. In the United States, what are the four principal sources (categories) of air pollution, and what are the major pollutants emitted by these sources?
6. If you were to read an air pollution control report, in what units would you expect to see the following concentrations reported?
   a. $SO_2$
   b. CO

c. Pb

d. $PM_{10}$

7. Indicate the basic principles involved in the control of gaseous air pollution emissions.

8. Indicate the commonly used devices for control of particulate air pollution emissions.

9. In the following equations, what does each term represent? Give the units for each, as well as the name associated with each equation.

$$\left(P + \frac{n^2 a}{V^2}\right)(V - nb) = nRT$$

Name: _____

$$\ln\left(\frac{C_2}{C_1}\right) = \frac{\Delta H}{R}\left(\frac{1}{T_1} - \frac{1}{T_2}\right)$$

Name: _____

$$\frac{X}{M} = KC^{\frac{1}{n}}$$

Name: _____

$$v = \frac{gd_p^2(\rho_p - \rho_a)}{18\mu}$$

Name: _____

10. What industrial plant-operating conditions would dictate the use of an electrostatic air-cleaning device instead of a fabric filter?

11. What control technology do we have available for control of particulate airborne contaminants? Indicate the size range for which each type of unit is most efficient for removal.

12. What is meant by effective stack height and what are the controlling factors?

13. What are the four general categories of air quality models?

14. Multiple choice and matching questions:

a. Dispersion models estimate the ambient air concentrations as a function of
   i. Source location
   ii. Emission strengths
   iii. Terrain features
   iv. Meteorological conditions
   v. All of the above

b. The primary mobile source of air pollution in the United States is
   i. Diesel trucks
   ii. Automobiles
   iii. Airplanes
   iv. Lawn mowers

c. Which air pollution control device generates a large volume of wastewater?
   i. Venturi scrubbers
   ii. Cyclone
   iii. Settling chamber
   iv. Electrostatic precipitator

d. A substance that alters the rate of a chemical reaction without the substance itself being changed or consumed by the reaction is a(n):
   i. Additive
   ii. Catalyst
   iii. Add-on
   iv. Stimulus

e. Match the type of atmospheric stability to the appropriate Pasquill–Gifford stability categories:
   ____ Unstable (i) E–F
   ____ Neutral (ii) A–B–C
   ____ Stable (iii) D

f. The $\sigma_y$ and $\sigma_z$ used in the Gaussian dispersion formulas are defined as
   i. Atmospheric pressure at points $y$ and $z$
   ii. Standard deviations of pollutant concentration in the horizontal and vertical directions
   iii. Temperature variations in the $y$ and $z$ directions
   iv. None of the above

g. Match the category of air quality model with its definition.
   ___Empirical i. Investigate pollutant dispersion for complicated situations
   ___Numerical ii. Derived from an analysis of source data, meteorological data, and air quality
   ___Physical iii. Use complex equations to simulate the effects of turbulence, chemical transformations, deposition, and so on on pollutant transport and dispersion
   ___Gaussian iv. Techniques for estimating the impact of nonreactive pollutants; use simple algebraic expressions for reactive pollutants

h. Plume rise from a stack is due to
   i. Heat and type of pollutant
   ii. Momentum and buoyancy
   iii. Composition of the stack
   iv. None of the above

i. Settling chambers use _____ to remove solid particles from a gas stream.
   i. Electrostatic pressure
   ii. Gravity
   iii. Hydraulic simulation
   iv. Filtration

15. Draw a schematic diagram for each of the following air pollution control devices and briefly describe the mechanisms, principles, or processes involved in their operation.

a. Settling chambers

b. Electrostatic precipitators

c. Cyclone clean

d.  Venturi scrubber
e.  Packed tower

16. Indicate the equipment needed to access the EPA, OAQPS, BBS, and SCRAM regulatory dispersion models on the Internet; also indicate the parameters required on the communications software. Define or explain each of these parameters.

17. What are the three categories for the SCRAM BBS models?

18. How are the regulatory dispersion models and the screening models used by the EPA?

19. What data or information is needed in order to use the SCREEN2 model?

## 7.5.2  Specific Mathematical Problems

1a. Write a computer program to determine the partial pressure exerted by a gas, given the temperature of the gas mixture, the prevailing pressure, and the gas concentration.

1b. Use the program developed in (a) to find the partial pressure exerted by carbon monoxide in a gas mixture. The gas mixture which is at 25°C and 1 atm pressure contains 150 mg/L of carbon monoxide (CO) gas.

2a. Write a computer program to find the percentage of the margin of safety lost if the temperature of a gas in a tank rises to $T_2$°C from a temperature of $T_1$, given the gas pressure and the pressure the tank is able to withstand.

2b. A tank is tested to be able to withstand 17 atm of pressure. The tank is filled with gas at 25°C and 10 atm of pressure. What percentage of the margin of safety is lost if the temperature goes to 100°C? Use the program developed in (a) to check your answer.

3a. Write a short computer program that allows the determination of the volume of a certain gas, given its weight, temperature, and pressure.

3b. Use the computer program developed in (a) to calculate the volume of 36 g of $SO_2$ at 25°C and an absolute pressure of 1 atm.

4a. A cylinder containing 100 g of carbon monoxide (CO) fell off a laboratory table and broke the release valve, permitting the gas to escape into the closed laboratory room that was 20 m long, 10 m wide, and 3 m high. What would the average CO concentration be in this room in $\mu g/m^3$ and ppm?

4b. Write a short computer program to solve the problem outlined in (a).

5a. Write a computer program to find the new volume of a gas, the temperature of which has been increased from a certain value to another, given its initial pressure.

5b. Use the program developed in (a) to solve the following problem: When the temperature of 25 mL of dry carbon dioxide ($CO_2$) is increased from 10°C to 35°C at a pressure of 1 atm, what is the volume of the gas?

6a. Write a short computer program to compute the number of moles present of each gas in a gas mixture composed of three different gaseous substances enclosed in a container of a known volume, given the weight of each gas. Let the program determine the total pressure exerted by the gas mixture. The program should also estimate the percentage of each volume of gas present in the container.

6b. Use the program developed in (a) to solve the following problem: a 2.5 L volume cylinder at 25°C contains 1 g of methane ($CH_4$), 2 g of nitrogen ($N_2$), and 12 g of oxygen ($O_2$). What is the number of moles of each gas present? What is the partial pressure exerted by each gas? What is the total pressure exerted by the gas mixture, and what is the percentage by volume of each gas present in the cylinder?

7. Write a simple computer program to change the concentration of a certain pollutant from $\mu g/m^3$ to ppm by volume, given the relevant atomic weights. Use the program to fill in the blanks in the following table:

| Pollutant | Concentration ($\mu g/m^3$) | Concentration (ppm by Volume) |
|-----------|-----------------------------|-------------------------------|
| $SO_2$    | 60                          |                               |
| $NO_2$    |                             | 0.70                          |
| CO        | 55                          |                               |

*Note:*  S = 32; O = 16; N = 14; C = 12.

8. Use the computer program developed for Example 7.15 to solve the following problem: According to Stoke's law, what would be the terminal settling velocity of a 1.0 μm radius particle of quartz that has a specific gravity of 2.75, and the temperature is 27°C?

9a. It is desired that all particles with a size of 50 μm and larger be removed with 100% efficiency in a settling chamber, which has a height of 0.8 m. The air moves at 1.6 ft/s, and its temperature is 80°F. The particle density is 2.2 g/cm³. What minimum length of chamber should be provided?

9b. Write a short computer program to solve the problem of (a).

10a. Write a computer program to determine the particle size collected in a cyclone cleaner, given the outer diameter of the cyclone, the number of effective turns for the cyclone, air flow, the temperature and pressure of gas, the density of particles, the gas velocity at the entrance to the cyclone, and the efficiency of the system.

10b. Use the program developed in (a) to solve the following problem: A cyclone cleaner has an outer circumference of 251 mm and handles 3.7 m3/s of contaminated air at a temperature of 30°C and 1 atm of pressure. The density of the particles is 2.0 g/cm³. The gas velocity at the entrance to the cyclone is 15 m/s. Determine the particle size in microns collected with 50% efficiency if the number of effective turns for the cyclone is 4.6.

11a. Calculate the fabric area, in square meters, required in a baghouse to treat 6500 m3/min of particulate-laden gas stream at a removal efficiency of 99.82%.

The baghouse unit operates at an air-to-cloth ratio of 0.72 m/min.

| | |
|---|---|
| 1. 6,000 | 3. 9,000 |
| 2. 5,500 | 4. 15,000 |

11b. Calculate the area $a$, in square meters, for an ESP required to process 4000 m³/min of a particulate-laden gas stream with a removal efficiency of 99.5%. The drift velocity $w$ has been determined to be 0.07 m/s.

| | |
|---|---|
| 1. 6,500 | 3. 55,000 |
| 2. 10,000 | 4. 60,500 |

11c. A plant has an inlet loading to a baghouse of 23,000 mg/m³. The average filtration velocity is 3 m/min, and the gas flow rate is 700 m³/min. What is the total filtering area for this operation?

| | |
|---|---|
| 1. 76 m/min | 3. 233 m²/min |
| 2. 3 m³/min/m² | 4. 2 m/min |

11d. At the point of maximum concentration downwind from a source with an effective stack height of 70 m, $\sigma_z$ is closest to

| | |
|---|---|
| 1. 15 m | 3. 50 m |
| 2. 58 m | 4. 80 m |

11e. The concentration at 1000 m downwind of a 60 m (effective stack height) source emitting 100 g/s under type B stability conditions with a 6 m/s mean wind speed is

| | |
|---|---|
| 1. $7.0 \times 10^{-5}$ g/m³ | 3. $2.6 \times 10^{-4}$ g/m³ |
| 2. $1.5 \times 10^{-3}$ g/m³ | 4. $5.3 \times 10^{-3}$ g/m³ |

11f. Write short computer programs to solve each of the aforementioned multiple choice questions.

12a. Write a short computer program to estimate the collection efficiency of a venturi scrubber at removing particles of given densities. Let the program determine the pressure drop across the venturi scrubber, given the relevant characteristics of the venturi scrubber.

12b. Use the program developed in (a) to solve the following problem: Estimate the collection efficiency of a

venturi scrubber at removing 4 μm sized particles. The particles have a density of 2750 kg/m³. Also, determine the pressure drop across the venturi scrubber. The venturi scrubber has the following characteristics:

Gas flow rate = 7 m³/s
Liquid flow rate = 0.002 m³/s
Gas flow rate at the throat = 80 m/s
Correlation coefficient = 1700
Temperature = 100°C

13a. Write a computer program to find the geometric mean diameter and the geometric standard deviation for a certain aerosol, given the results for the mass distribution of the aerosol.

13b. Use the program developed in (a) to solve the following problem: The results for the mass distribution of an aerosol are given in the following table. From this, determine the geometric mean diameter and the geometric standard deviation for this aerosol.

| Average Size (μm) | Percentage by Weight |
|---|---|
| 0.25 | 0.1 |
| 1.0 | 0.4 |
| 2.0 | 9.5 |
| 3.0 | 20.0 |
| 4.0 | 20.0 |
| 5.0 | 15.0 |
| 6.0 | 11.0 |
| 7.0 | 8.0 |
| 8.0 | 5.5 |
| 10.0 | 5.5 |
| 14.0 | 4.0 |
| 20.0 | 0.8 |
| >20.0 | 0.2 |

14a. Write a computer program to compute the collection efficiency of an ESP for a certain particle size having a certain drift velocity. Let the program determine the efficiency for other particle diameters, given the ESP specifications.

14b. Use the program developed in (a) to determine the collection efficiency of the ESP described below for a particle size of 0.70 μm diameter having a drift velocity of 0.20 m/s. What would be the efficiency for 1.0 μm particles? ESP specifications are given as follows:

Plate height = 8 m
Plate length = 6 m
Number of passages = 6
Plate spacing = 0.3 m
Gas flow rate = 20 m³/s

15a. A 2.5 g sample of catalyst was tested for its efficiency for removal of CO from an air stream. The initial CO concentration entering the catalyst was 160 ppm; the CO concentration after passing through the catalyst was found to be 2.5 ppm. Determine the percent removal of CO for this catalyst, the contact time or space velocity, and the reaction rate coefficient for the catalyst. The space around the catalyst was found to be 1.3 cm$^3$, and the air stream in which the CO is present had a flow rate of 800 cm$^3$/min.

15b. Write a computer program to solve the problem indicated in (a).

16a. Develop a computer program to find the effective stack height for a power plant, given the physical stack height, stack inside diameter, stack gas effluent velocity, effluent gas temperature, ambient air temperature, and mean wind speed.

16b. Use the computer program developed in (a) to determine the effective stack height for a power plant operating under the following conditions:
Physical stack height = 45 m
Stack inside diameter = 3 m
Stack gas effluent velocity = 16 m/s
Effluent gas temperature = 160°C
Ambient air temperature = 22°C
Mean wind speed = 5 m/s

17a. A manufacturing plant emits nitrogen oxides under the conditions set forth below on a continuous 24-h/day basis. What is the concentration of $NO_2$ at a receptor 760 m downwind of the plant? Does this plant by itself cause concentrations at the receptor to exceed U.S. primary and secondary ambient annual air quality standards for $NO_2$?
Physical stack height = 55 m (situated on level terrain)
Stack inside diameter = 1 m
Stack exit velocity = 4.6 m/s
Stack temperature = 120°C
$NO_2$ emission rate = 450 kg/h
Mean wind speed $u$ = 3.8 m/s
$t_{air}$ = 10°C
$p$ = 1000 mbar
Pasquill stability category D

17b. Develop a computer program to solve the problem mentioned in (a).

18a. Improve the computer program for Example 7.33 to solve the following problem given in 18b, given the following data for the adsorption of vinyl chloride by activated carbon:

| Activated Carbon (g) | Concentration of Vinyl Chloride Remaining (g/m$^3$) |
|---|---|
| 2.0 | 80 |
| 3.5 | 50 |
| 5.0 | 31 |
| 16.0 | 2.7 |
| 25.0 | 1.0 |

Note: Control or initial concentration at the start of the test = 155 g/m$^3$.

Determine if the data follow the Freundlich isotherm. If the data do follow the Freundlich isotherm,
1. Determine the $k$ and $n$ values.
2. Determine the ultimate capacity of this activated carbon $(X/M)_{C_0}$ for vinyl chloride adsorption at the control or initial concentration.

19a. An effluent air stream contains 5.0% $SO_2$ by volume. It is desired to reduce this concentration to 25000 mg/m$^3$ by passing the gas stream through a packed column adsorption tower before releasing the air stream to the atmosphere. The inlet gas flow rate is 30 m$^3$/min measured at 1 atm and 22°C, and the absorber is the counterflow type. The equilibrium of $SO_2$ in water is assumed to be reasonably represented by the following relationship: $Y_B = 30X_B$ ($Y_B$ = mole fraction of $SO_2$ in the air, and $X_B$ = mole fraction of $SO_2$ in the water). Estimate the minimum amount of water ($SO_2$-free) required as a solvent, in cubic meters per second, to meet the desired requirements.

19b. Develop a suitable computer program to solve the problem outlined in (a).

# REFERENCES

Air & Waste Management Association and Davis, W. R. 2000. *Air Pollution Engineering Manual*, 2nd ed. Wiley-Interscience.

Air Pollution Training Institute. 1981. *Control Techniques for Gaseous and Particulate Pollutants*, (S1:422). Research Triangle Park, NC: U.S. Environmental Protection Agency.

Air Pollution Training Institute. 1982. *Basic Air Pollution Meteorology*, (SI:409). Research Triangle Park, NC: U.S. Environmental Protection Agency.

Air Pollution Training Institute. 1983. *Atmosphere Sampling*, (S1:435). Research Triangle Park, NC: U.S. Environmental Protection Agency.

Air Pollution Training Institute. 1984. *Control of Gaseous and Particular Emissions*, (S1:412D). Research Triangle Park, NC: U.S. Environmental Protection Agency.

Air Pollution Training Institute. 1992. *Air Pollution Control Orientation Course*, (S1:422). Research Triangle Park, NC: U.S. Environmental Protection Agency.

Briggs, G.A. 1969. *Plume Rise*, USAEC Critical Review Series, TID-25075. Springfield, VA: National Technical Information Service.

Briggs, G.A. 1973. *Diffusion Estimation for Small Emissions*, NOAA ATDL, Contribution File No. 79 (draft), Oak Ridge, TN.

Briggs, G.A. 1975. Plume Rise Predictions, in *Lectures on Air Pollution and Environmental Impact Analysis*, Haugen, D.A., Ed. Boston, MA: American Meteorological Society, 59.

Burt, E.W. 1977. *VALLEY Model User's Guide*, 450/2-77-018. Research Triangle Park, NC: U.S. Environmental Protection Agency.

Chansler, J.M., Lloyd, W.G., and Rowe, D.R. 1995. Soil sorption of zinc according to the Freundlich isotherm. *Florida Water Resources Journal* 47(9):32–34.

Colls, J. and Tiwary, A. 2009. *Air Pollution: Measurement, Modelling and Mitigation*, 3rd ed. CRC Press.

Cooper, C.D., and Alley, F.C. 2010. *Air Pollution Control: A Design Approach*, 4th ed. Waveland Press Inc.

Corman, R. D. and Black, D. 1974. *Controlling Air Pollution*. Washington, DC: American Lung Association.

Council on Environmental Quality. 1993. *24th Annual Report of the Council on Environmental Quality.* Washington, DC.

Culp, R.L., Wesner, G.M., and Culp, G.L. 1978. *Handbook of Advanced Wastewater Treatment*, 2nd ed., 180. New York: Van Nostrand Reinhold.

Danielson, J.A., Ed. 1973. *Air Pollution Engineering Manual.* Office of Air and Water Programs, 92, 107. Research Triangle Park, NC: U.S. Environmental Protection Agency.

Environmental Protection Agency (EPA). 2016. National Ambient Air Quality Standards (NAAQS). https://www3.epa.gov/ttn/naaqs/criteria.html.

Ford, D.L. 1975. *Process Design in Water Quality Engineering, New Concepts and Developments. XI: Coagulation and Precipitation and Carbon Adsorption.* Nashville, TN: Sponsored by the Department of Environmental and Water Resources Engineering, Vanderbilt University.

Holland, J.Z. 1953. *A Meteorological Survey of the Oak Ridge Area*, Atomic Energy Commission, Rep. ORO-99, Washington DC, 540.

Hosker, R.P. 1984. Flow and Diffusion Near Obstacles, in *Atmospheric Science and Power Production*, Randerson, D., Ed., DOE/TIC-27601. Washington, DC: U.S. Department of Energy.

Lee, C. C. 2005. *Environmental Engineering Dictionary,* 4th ed. Government Institutes.

Lide, D.R., and Frederiske, H.P.R., Eds. 1995/1996. *CRC Handbook of Chemistry and Physics*, 76th ed., 6–48. Boca Raton, FL: CRC Press.

Lloyd, W. G., and Rowe, D. R. 1971. Homogeneous catalytic oxidation of carbon monoxide. *Environ. Sci. Technol.* 5:11.

Lloyd, W. G., and Rowe, D. R. 1974. Palladium Compositions Suitable as Oxidation Catalysts. U.S. Patent 3,849,336.

Manahan, S.E. 2009. *Environmental Chemistry*, 9th ed. Boca Raton, FL: CRC Press.

Office of Air Quality Planning and Standards. 1978. *Guideline on Air Quality Models*, 450/2-78-027. Research Triangle Park, NC: U.S. Environmental Protection Agency.

Office of Air Quality Planning and Standards. 1983. *Introduction to Dispersion Modeling*, (SI:410). Research Triangle Park, NC: U.S. Environmental Protection Agency, 4–19.

Office of Air Quality Planning and Standards. 1989. *Assessing Multiple Pollutant, Multiple Source Cancer Risks from Urban Air Toxics*, 450/2-89-010. Research Triangle Park, NC: U.S. Environmental Protection Agency, A-xii.

Office of Air Quality Planning and Standards. 1992. *SCREEN2 Model User's Guide*, 450/4-92-006. Research Triangle Park, NC: U.S. Environmental Protection Agency.

Office of Air Quality Planning and Standards. 1994. U.S. Environmental Protection Agency, Technology Transfer Network, Research Triangle Park, NC, 27711.

Peavey, H.S., Rowe, D.R., and Tchobanoglous, G. 1985. *Environmental Engineering.* New York: McGraw-Hill.

Randerson, D. 1984. Atmospheric Boundary Layer, in *Atmospheric Science and Power Production*, Randerson, D., Ed., DOE/TIC-27601. Washington, DC: U.S. Department of Energy.

Rowe, D. R., and Lloyd, W. G. 1978. Catalytic cigarette filter for carbon monoxide reduction. *J. Air Pollut. Control Assoc.* 28(3):253.

Rowe, D.R. et al. 1985. Indoor-outdoor relationship of suspended particulate matter in Riyadh, Saudi Arabia. *APCA J.* 35(1).

Teledyne Technology Introduces a New Room Temperature Catalyst, Form 1007-F-33, 1730 E. Prospect St., Fort Collins, CO, 1983.

Teledyne Instapure Filtration System, Form 1035-F-35, 1730 e. Prospect St., Fort Collins, CO, 1985.

Theodore, L. and DePaola, V. 1980. Predicting cyclone efficiency. *J. Air Pollut. Control Assoc.* 80:1132.

Touma, J.S. 1977. Dependence of the wind profile power law on stability for various locations. *J. Air Pollut. Control Assoc.* 27:863.

Turk, A. 1977. Adsorption, in *Air Pollution*, Vol. IV, 3rd ed., Engineering Control of Air Pollution, Stern, A., Ed., 329.

Turner, D.B. 1969. *Workbook of Atmospheric Dispersion Estimates.* Cincinnati, OH: National Air Pollution Control Administration, 31.

U.S. Code of Federal Regulations. 1994. Definitions, *Fed. Regist.*, 52.741 (40).

U.S. Department of Health and Human Services. 1973. *The Industrial Environment—Its Evaluation and Control.* National Institute for Occupational Safety and Health, Cincinnati, OH, 159, 641.

U.S. Environmental Protection Agency. 1992. *Industrial Source Complex (ISC2) Dispersion Model User's Guide*; 450/4-92 - 008. Research Triangle Park, NC: U.S. Environmental Protection Agency.

U.S. Environmental Protection Agency. 2013. Office of Air Quality Planning and Standards (Oaqps) Technology Transfer Network (TNN), User's ManualJul. BiblioGov.

U.S. Environment Protection Agency. 2014. *National Air Quality Status and Trends Through 2007.* CreateSpace Independent Publishing Platform.

Vesilind, A., Morgan, S. M. and Heine, L. G. 2009. *Introduction to Environmental Engineering*, 3rd ed. Cengage Learning.

Williams, C.E., Hatch, T., and Greenburg, L. 1940. Determination of cloth area for industrial air filters. *Heating Piping Air Cond.* 12:259.

Zackay, V. F. and Rowe, D. R. 1984/1985. Catalyst of Palladium, Copper and Nickel on a Substrate. U.S. Patent 4,459,269, July 10, 1984; U.S. Patent 4,521,530, June 4, 1985.

# 8 Computer Modeling Applications for Noise Pollution and Abatement

## 8.1 SOUND AND NOISE

Sound is oscillatory motion of small amplitude in elastic media (Thumann 1990). The vibrating surface causes a disturbance of the pressure and density in the air as well as particle velocity. Each element of air transfers momentum and energy to the adjacent air element, and the pressure disturbance is propagated through the air in the form of a wave (see Figure 8.1).

Sound, hence noise, results from periodic disturbances of the air, and at room temperature, sound is propagated in air at a speed of 340 m/s. In water and steel, for example, the speed is much greater, being respectively ~1500 m/s and 5000 m/s (see Tables 8.1 and 8.2). As the disturbance spreads geometrically, its effect will decrease with its distance from the sound source, but the diminution in sound intensity will also be affected by the damping of the sound waves by the transmitting medium (Isaac 1985).

Importance of sound detection is reflected on communications with people, reception of useful information from the environment, and provision of warning (alarm) and enjoyment (music).

However, noise is unwanted or extraneous sound (Anderson and Bratos-Anderson 1993), that is, unwanted by-product of civilization such as transportation mechanisms (trucks, airplanes, trains, etc.), industrial sector (machinery and equipment), appliances (construction, air conditioners, etc.), and similar sound producers.

Urban noise sources include (Vesilind and Morgan 2010, Hansen 2005, Bhatia 2007, Harris 2013) transportation, aircraft/airport, motor vehicle, rapid transit, railroads (freight and passenger), industrial, construction or domestic (in the home, e.g., air conditioners, lawn mower, vacuum cleaner), and outdoors, for example, garbage disposal.

Noise is any undesired sound. It is measured on a decibel scale ranging from the threshold of hearing (0 dB) to the threshold of pain (130 dB) or any unwanted disturbance within a useful frequency band in a communication channel (Saenz and Stephens 1986).

Decibel is a unit used to compare two power levels, usually applied to sound or electrical signals. Although the decibel is one-tenth of a Bel, It is the decibel, not the Bel, that is invariably used.

Two power levels $P$ and $P_0$ differ by $n$ decibels as presented in the following equation:

$$n = 10 * \log\left(\frac{P}{P_0}\right) \qquad (8.1)$$

where:

  $n$ is the sound intensity level, dB
  $P$ is the level of sound intensity to be measured, W
  $P_0$ is the reference level, usually the intensity of a note of the same frequency at the threshold of audibility = intensity of the least audible sound, usually given as $P_0 = 10^{-12}$ W (Anderson and Bratos-Anderson 1993).

The logarithmic scale is convenient as human audibility has a range of 1 (just audible) to $10^{12}$ (just causing pain), and one decibel, representing an increase of some 26%, is about the smallest change the ear can detect (Saenz and Stephens 1986).

Threshold level is the number of decibels by which the sound intensity must be increased to be heard, relative to 0 dB hearing level, which is the sound level that can be heard by an ear that presumably has never been affected by any deleterious agent (Isaac 1985).

Sound signifies a vibration in an elastic medium at a frequency and intensity that is capable of being heard by the human ear. The frequency of sound lies in the range 20–20,000 Hz, but the ability to hear sounds in the upper part of the frequency range declines with age. Vibrations that have a lower frequency than sound are called infrasound, and those with a higher frequency are called ultra sounds (Saenz and Stephens 1986) (see Figure 8.2).

Sound is in effect a transfer of energy (Anderson and Bratos-Anderson 1993); it travels at different speeds in different materials, depending on the material's elasticity.

Velocity of sound in a given medium can be found from the following equation:

$$c = \lambda f \qquad (8.2)$$

where:

  $c$ is the velocity of the sound in a given medium, m/s
  $\lambda$ is the wavelength, m
  $f$ is the frequency, cycle/s, Hz

### Example 8.1

1. Write a computer program to evaluate the wavelength of a sound generated by a machine, given the frequency and the sound waves traveling velocity.
2. Find the wavelength of a sound from a machine, given that the frequency is 70 cycle/s, and the sound waves travel at about 2.1 km/s.
3. Use program set in (1) to verify your computations in (2).

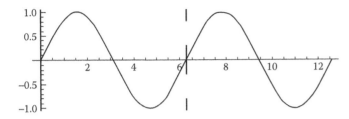

**FIGURE 8.1** Sound wave travel in space.

**TABLE 8.1**

**Speed of Sound in Different Media at 0°C**

| Medium | Velocity of Sound at 0°C (m/s) |
|---|---|
| Air | 331 |
| Oxygen | 460 |
| Alcohol | 1213 |
| Water | 1435 |
| Copper | 3560 |
| Iron | 5130 |

**TABLE 8.2**

**Speed of Sound in Air at Different Temperatures**

| Temperature (°C) | Velocity of Sound (m/s) |
|---|---|
| 0 | 331 |
| 20 | 344 |
| 100 | 386 |
| 500 | 553 |
| 1000 | 700 |

**Solution**

1. For the solution of Example 8.1 (1), see the listing of Program 8.1.
2. Given: $c = 2100/s$ m, $f = 70$ cycle/s.
3. Use Equation 8.2 to determine wavelength as follows: $\lambda = c/f = 2100$ m/s/70 cycle/s = 30 m.

Noise as a sound sensation has its origin in the mechanical vibrations of matter, either in the solid or in the fluid state. The transmission of these vibrations through air is received at the ear to become interpreted as sound by the human sensory system (Isaac 1985).

The intensities of some typical environmental noises

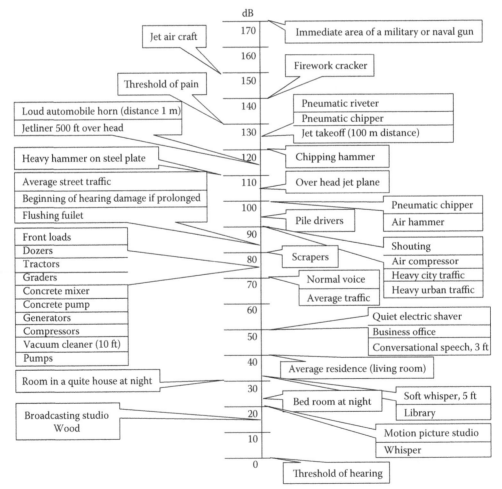

**FIGURE 8.2** Examples of environmental noise on a decibel scale.

### LISTING OF PROGRAM 8.1 (CHAP8.1\FORM1.VB): SOUND VELOCITY AND WAVELENGTH

```vb
'********************************************************************************
'Example 8.1: Sound velocity and wave length
'********************************************************************************

Public Class Form1
    Dim c, d, f As Double
    Dim result As Double

    Private Sub Form1_Load(ByVal sender As System.Object, ByVal e As System.EventArgs)
Handles MyBase.Load
        Me.Text = "Example 8.1: Velocity and Wave length"
        Me.FormBorderStyle = Windows.Forms.FormBorderStyle.FixedSingle
        Me.MaximizeBox = False
        Label1.Text = "Calculate sound velocity and wavelength"
        Label1.Text += vbCrLf + "Using the equation: c = d * f"
        Label2.Text = "Solving for:"
        Label5.Text = "Decimal places:"
        Button1.Text = "&Calculate"
        ComboBox1.Items.Clear()
        ComboBox1.Items.Add("c, sound velocity")
        ComboBox1.Items.Add("d, wave length")
        ComboBox1.Items.Add("f, frequency")
        ComboBox1.SelectedIndex = 0
        NumericUpDown1.Value = 2
        NumericUpDown1.Maximum = 10
        NumericUpDown1.Minimum = 0
    End Sub

    Private Sub Button1_Click(ByVal sender As System.Object, ByVal e As System.
EventArgs) Handles Button1.Click
        Select ComboBox1.SelectedIndex
            Case 0
                'Solving for c
                d = Val(TextBox1.Text)
                f = Val(TextBox2.Text)
                c = d * f
                result = c
            Case 1
                'Solving for d
                c = Val(TextBox1.Text)
                f = Val(TextBox2.Text)
                d = c / f
                result = d
            Case 2
                'Solving for f
                c = Val(TextBox1.Text)
                d = Val(TextBox2.Text)
                f = c / d
                result = f
            Case Else
                MsgBox("Please select an item from the list.",
                    vbOKOnly Or vbInformation)
                Exit Sub
        End Select
        showResult()
    End Sub
```

```
    Sub showResult()
        TextBox3.Text = FormatNumber(result, NumericUpDown1.Value)
    End Sub

    Private Sub NumericUpDown1_ValueChanged(ByVal sender As System.Object, ByVal e As
System.EventArgs) Handles NumericUpDown1.ValueChanged
        showResult()
    End Sub

    Private Sub ComboBox1_SelectedIndexChanged(ByVal sender As System.Object, ByVal e As
System.EventArgs) Handles ComboBox1.SelectedIndexChanged
        Select Case ComboBox1.SelectedIndex
            Case 0
                'Solving for c
                Label3.Text = "Wave length, d, (m):"
                Label4.Text = "Frequency, f (Hz):"
                Label6.Text = "Sound velocity, c (m/s):"
            Case 1
                'Solving for d
                Label3.Text = "Sound velocity, c (m/s):"
                Label4.Text = "Frequency, f (Hz):"
                Label6.Text = "Wave length, d, (m):"
            Case 2
                'Solving for f
                Label3.Text = "Sound velocity, c (m/s):"
                Label4.Text = "Wave length, d, (m):"
                Label6.Text = "Frequency, f (Hz):"
        End Select
    End Sub
End Class
```

Vibrations of matter are controlled by its density and its elasticity (or springiness).

Due to elastic bonds in matter, any local vibration is transmitted to the neighboring elements. The process of vibration transmission through condensed media constitutes the elastic wave propagation, which is responsible for the transport of mechanical (acoustical) energy (Isaac 1985).

The sound wave is commonly associated with elastic waves in the frequency range of the human hearing (i.e., 20 Hz to 20 kHz). Waves below and above this range are, respectively, known as infra- and ultrasounds.

The propagation of vibrational energy takes a finite time, and the distance traveled by the wave front (all points of which are in the same phase of the oscillation), in unit time, is the so-called phase velocity or speed of sound, $c$. The value of $c$ depends on the type of wave and physical characteristics of the propagation medium.

Wave is a periodic disturbance in a medium or in space. In a traveling wave (progressive wave), energy is transferred from one place to another by the vibrations.

Sound waves are longitudinal waves whereby the air is alternately compressed and rarefied by displacements in the direction of propagation (Saenz and Stephens 1986).

The chief characteristics of a wave are its speed of propagation, its frequency, its wave length, and its amplitude (Saenz and Stephens 1986) (see Figure 8.3).

Speed propagation is the distance covered by the wave in unit time.

Frequency is the number of complete disturbances (cycles) in unit time (hertz). It is interrelated to pitch: the lower the frequency, the lower is the sound pitch, and vice versa.

Wave length is the distance (m) between successive points of equal phase in a wave.

Amplitude is the maximum distance of the disturbed quantity from the mean value. It determines how loud is the sound to the receiver: Large amplitude means louder sounds and small amplitude means quieter sounds (see Figure 8.4).

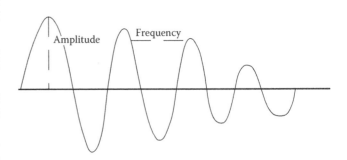

**FIGURE 8.3**   Amplitude and frequency.

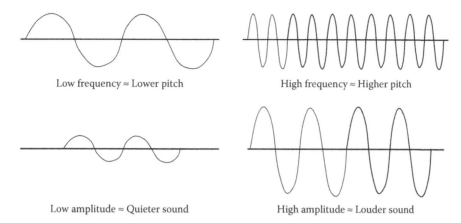

Low frequency ≈ Lower pitch

High frequency ≈ Higher pitch

Low amplitude ≈ Quieter sound

High amplitude ≈ Louder sound

**FIGURE 8.4** Frequency, pitch, amplitude, and sound level.

## 8.2 EFFECTS OF NOISE

Effects of noise may include its irreversible effect on humans physiologically (including temporary or permanent hearing impairment), psychologically, or socially (Barrett 2010, Berglund and Lindvall 1995, Munjal 2013).

Prolonged exposure to noise of a certain frequency pattern (Saenz and Stephens 1986) may cause temporary or permanent hearing loss. Temporary hearing loss, or temporary threshold shift, disappears in a few hours or days (e.g., performers in rock bands, industrial applications).

Permanent loss is also referred to as permanent threshold shift.

Factors affecting hearing (Saenz and Stephens 1986, Thumann 1990, Kamboj 2002, Abdel-Magid 1995, Abdel-Magid 2002, Abdel-Magid and Abdel-Magid 2015) include the following:

- Age: Hearing becomes less acute with age (presbyacusis)
- Environmental noise (motor cycles, rock music)
- Education and illness
- Intensity, duration, and exposure of sound

Disadvantages of noise (Faulkner 1976, Isaac 1985, Saenz and Stephens 1986, Thumann 1990, Vesilind and Morgan 2010, Bies and Hansen 2013) include the following:

- Deterioration of level of hearing
- Introduction or increase of certain diseases such as cardiovascular, rhythm of heartbeat, blood thickening, dilation of blood vessels, headache, irritability, relaxation prevention, and mental anguish like that in a supersonic aircraft
- Sociological problems such as focusing difficulties, annoyance to people during close work, for example, watch makers, and increased convalescence time in noisy hospitals
- Economic difficulties, for example, cost to compensate hearing disorders, cost for sleeping pills, lost time in industry and apartment sound proofing, property damage, and so on

Primary effects of noise (Vesilind and Morgan 2010) could be direct or indirect effects.

Direct effects include sleep disturbance, speech interference, hearing loss, noisiness, task interference, masking, unconditioned physiological response, disturbances in auditive and nonauditive performance, communication, information processing, and rest and relaxation.

Indirect effects address annoyance, social behavior, health and well-being, sleep disturbance, other task interference or fatigue, conditioner, physiological response, impairment of people's well-being (closing windows, reducing speech communication or refraining from it, changing dwelling areas), and annoyance (psychic effects).

Auditory aftereffects of noise signify impairment and handicap (partial deafness), hearing damage (inability to hear weak sounds), tinnitus (ringing in the ears), paracusis (hearing sounds incorrectly), speech misperception, and physiological damage (alterations in the peripheral blood circulation).

## 8.3 DECIBEL SCALE

The decibel scale is a logarithmic scale applied in acoustics to scaling the ratio of sound intensities or the ratio of sound pressures.

The benefits of a decibel are as follows (Thumann 1990, Sound Research Laboratories 1991; Anderson and Bratos-Anderson 1993):

- It describes intensity or energy level of a physical quantity
- It compresses the quantity into convenient numbers for data presentation
- It may be applied in acoustics and electrical power and mechanical energy presentations

The human response to sound may be approximated by the laws of Weber and Fechner. Law of Weber states that "The minimum increase of stimulus which will produce a perceptible increase of sensation is proportional to the pre-existent stimulus." The Fechner's law states that "the intensity of human sensation changes logarithmically with the energy which causes the sensation" (Thumann 1990, Anderson and Bratos-Anderson 1993, Nathanson and Schneider 2014).

These two laws may be used in the definition of a decibel as shown in the following equation:

$$SWL = 10 * \log\left(\frac{W}{W_0}\right) \qquad (8.3)$$

where:

SWL is the sound power level, dB

$W$ is the power of sound wave, W

$W_0$ is the some reference sound power $= 10^{-12}$ W, pW

The basic equation used for expressing the decibel concept in terms of sound intensities from a single sound source is as indicated in the following equation (Thumann 1990, Sound Research Laboratories 1991; Anderson and Bratos-Andreson 1993):

$$SPL = 20 * \log\left(\frac{P}{P_{ref}}\right) dB(A) \qquad (8.4)$$

where:

SPL is the sound pressure level, dB

$P$ is the pressure of sound wave, Pa

$P_{ref}$ is some reference sound pressure $= 2 * 10^{-5}$ Pa

## Example 8.2

1. Write a computer program to determine the sound pressure level, given the pressure of a sound wave.
2. The pressure of a sound wave is 5 Pa. Determine the sound pressure level.
3. Use program set in (1) to verify your computations in (2).

### Solution

1. For the solution of Example 8.2 (1), see the listing of Program 8.2.
2. Given: $P = 5$ Pa.
3. Find the sound pressure level from Equation 8.4 as follows:

$$SPL = 20 * \log\left(\frac{5}{2 * 10^5}\right) = 107.96 \, dB(A)$$

The intensities of some typical environmental noises are shown in Figure 8.2 (Faulkner 1976, Isaac 1985, Anderson and Bratos-Andreson 1993, Kamboj 2002, Vesilind and Morgan 2010).

---

### LISTING OF PROGRAM 8.2    (CHAP8.2\FORM1.VB): SOUND PRESSURE LEVEL

```
'*******************************************************************************
'Example 8.2: Sound Pressure Level
'*******************************************************************************

Public Class Form1
    Dim SPL, P As Double
    Const Pref As Double = 2 / 100000

    Private Sub Form1_Load(ByVal sender As System.Object, ByVal e As System.EventArgs)
Handles MyBase.Load
        Me.Text = "Example 8.2: SPL"
        Me.FormBorderStyle = Windows.Forms.FormBorderStyle.FixedSingle
        Me.MaximizeBox = False
        Label1.Text = "Calculate Sound Pressure Level"
        Label1.Text += vbCrLf + "Using equation: SPL = 20 * Log [P / Pref]"
        Label2.Text = "Pressure of sound wave, P (Pa):"
        Label3.Text = "Decimal places:"
        Label4.Text = "Sound pressure level, SPL (dB):"
        NumericUpDown1.Value = 2
        NumericUpDown1.Maximum = 10
        NumericUpDown1.Minimum = 0
        Button1.Text = "&Calculate"
    End Sub

    Private Sub NumericUpDown1_ValueChanged(ByVal sender As System.Object, ByVal e As
System.EventArgs) Handles NumericUpDown1.ValueChanged
        showResult()
    End Sub

    Sub showResult()
        TextBox2.Text = FormatNumber(SPL, NumericUpDown1.Value)
    End Sub
```

```
        Private Sub Button1_Click(ByVal sender As System.Object, ByVal e As System.
EventArgs) Handles Button1.Click
            P = Val(TextBox1.Text)
            SPL = 20 * Math.Log10(P / Pref)
            showResult()
        End Sub
End Class
```

## 8.4 SOUND FIELDS FROM DIFFERENT SOUND SOURCES

For sound fields generated from different sound sources (e.g., different machines), the $SPL_N$ can be found from the following equation (Faulkner 1976, Thumann 1990, Nathanson and Schneider 2014):

$$SPL_N = 10 * Log \sum_{i=1}^{N} \left( 10^{SPL_i /10} \right) \qquad (8.5)$$

where:

$SPL_N$ is the net sound pressure level, dB
$SPL_i$ is the sound pressure level from source $i$, dB
$N$ is the number of sound sources

### Example 8.3

1. Write a computer program to evaluate SPL sound produced in a workshop by a number of machines when they work independently together, given sound pressure levels produced by each machine.

2. In a workshop, four machines independently produce sound pressure levels of 85, 66, 72, and 103 dB. Determine the SPL sound produced by the four machines when they work together.

3. Use program set in (1) to verify your computations in (2).

**Solution**

1. For the solution of Example 8.3 (1), see the listing of Program 8.3.

2. Given: $SPL_1 = 85$ dB, $SPL_2 = 66$ dB, $SPL_3 = 72$ dB, $SPL_4 = 103$ dB.

3. Use Equation 8.5 to compute the overall noise generated by all machines when they operate together.

$$SPL_N = 10 * \log \sum_{i=1}^{N} \left( 10^{85/10} + 10^{66/10} + 10^{72/10} + 10^{103/10} \right)$$

$$= 10 * \log \sum_{i=1}^{N} \left( 10^{8.5} + 10^{6.6} + 10^{7.2} + 10^{10.3} \right) = 103.1$$

$$SPL_N = 103.1 dB$$

---

**LISTING OF PROGRAM 8.3    (CHAP8.3\FORM1.VB): NET SOUND PRESSURE LEVEL**

```
'*********************************************************************************
'Example 8.3: Net Sound Pressure Level
'*********************************************************************************

Public Class Form1
    Private Sub Form1_Load(ByVal sender As System.Object, ByVal e As System.EventArgs)
Handles MyBase.Load
        Me.Text = "Example 8.3: SPLn"
        Me.FormBorderStyle = Windows.Forms.FormBorderStyle.FixedSingle
        Me.MaximizeBox = False
        Label1.Text = "Enter sound pressure levels (dB):"
        Label2.Text = "Decimal places:"
        Label3.Text = "Net sound pressure level, SPLn (dB):"
        NumericUpDown1.Value = 2
        NumericUpDown1.Maximum = 10
        NumericUpDown1.Minimum = 0
```

```
            Button1.Text = "&Calculate"
            DataGridView1.Rows.Clear()
            DataGridView1.Columns.Clear()
            DataGridView1.Columns.Add("SPLCol", "SPL (dB)")
        End Sub

    Private Sub Button1_Click(ByVal sender As System.Object, ByVal e As System.
EventArgs) Handles Button1.Click
            Dim count As Integer = DataGridView1.RowCount - 1
            If count = 0 Then
                MsgBox("Please enter at least one SPL!",
                        vbOKOnly Or vbInformation)
                Exit Sub
            End If
            Dim sum As Double = 0
            Dim SPLi, SPLn As Double
            For i = 0 To count - 1
                SPLi = Val(DataGridView1.Rows(i).Cells("SPLCol").Value) / 10
                sum += Math.Pow(10, SPLi)
            Next
            SPLn = 10 * Math.Log10(sum)
            TextBox1.Text = FormatNumber(SPLn, NumericUpDown1.Value)
        End Sub
End Class
```

The daily personal noise exposure level ($L_{EP},d$) may be determined from the following equation (Thumann 1990, Nathanson and Schneider 2014):

$$L_{EP},d = 10* \log\left[\frac{1}{T_o}\int_0^{T_e}\frac{P_A(t)}{P_{ref}}\right]^2 \qquad (8.6)$$

where:
$L_{EP},d$ is the daily personal noise exposure level, dB
$T_e$ is the duration of the person's personal exposure to noise
$T_o$ is the length of the working day (usually taken to be 8 hrs)
$P_A(t)$ is the time-varying value of $A$-weighted instantaneous sound pressure, Pa
$P_{ref}$ is some reference sound pressure = $2 * 10^{-5}$, Pa

A weekly average $L_{EP},w$ of the daily values can be found from Equation 8.7 (Thumann 1990):

$$L_{EP},w = 10* \log\left[\frac{1}{5}\sum_{i=1}^{N}10^{0.1(L_{EP},d)_i}\right] \qquad (8.7)$$

where:
$L_{EP},w$ is the weekly average of the daily personal noise exposure, dB(A)
$(L_{EP},d)_i$ is the daily personal noise exposure of day $i$, dB(A)
$N$ is the number of working days taken into account during the week

## Example 8.4

1. Write a computer program to determine the weekly average of the daily personal noise exposures for a factory worker who works for five days during the week, given the daily personal noise exposures for the working days.
2. A factory worker works for 5 days during the week. The daily personal noise exposures for the 5 days are 80, 77, 84, 70, and 93 dB($A$). Determine the weekly average of the daily personal noise exposures.
3. Use program set in (1) to verify your computations in (2).

### Solution

1. For the solution of Example 8.4 (1), see the listing of Program 8.4.
2. Given: $L_{EP1} = 73$ dB, $L_{EP2} = 85$ dB, $L_{EP3} = 68$ dB, $L_{EP4} = 57$ dB, $L_{EP5} = 97$ dB.
3. Determine the weekly average of the daily personal noise exposures from Equation 8.7 as follows:

$$L_{EP},w = 10* \log\left[\frac{1}{5}\sum_{i=1}^{N}\left(10^{0.1*73}+10^{0.1*85}+10^{0.1*68}\atop +10^{0.1*57}+10^{0.1*97}\right)\right]$$

$$L_{EP},w = 90.30 \text{ dB}(A).$$

---

**LISTING OF PROGRAM 8.4   (CHAP8.4\FORM1.VB): WEEKLY AVERAGE OF NOISE EXPOSURES**

```vb
'*****************************************************************************
'Example 8.4: Weekly average of noise exposures
'*****************************************************************************
Public Class Form1
    Private Sub Form1_Load(ByVal sender As System.Object, ByVal e As System.EventArgs)
Handles MyBase.Load
        Me.Text = "Example 8.4: LEP Week"
        Me.FormBorderStyle = Windows.Forms.FormBorderStyle.FixedSingle
        Me.MaximizeBox = False
        Label1.Text = "Enter daily personal noise exposures (dB):"
        Label2.Text = "Decimal places:"
        Label3.Text = "Weekly average of noise exposures, LEPw (dB):"
        NumericUpDown1.Value = 2
        NumericUpDown1.Maximum = 10
        NumericUpDown1.Minimum = 0
        Button1.Text = "&Calculate"
        DataGridView1.Rows.Clear()
        DataGridView1.Columns.Clear()
        DataGridView1.Columns.Add("LEPdCol", "LEP,d (dB)")
    End Sub

    Private Sub Button1_Click(ByVal sender As System.Object, ByVal e As System.
EventArgs) Handles Button1.Click
        Dim count As Integer = DataGridView1.RowCount - 1
        If count = 0 Then
            MsgBox("Please enter at least one (LEP,d)!",
                vbOKOnly Or vbInformation)
            Exit Sub
        End If
        Dim sum As Double = 0
        Dim LEPd, LEPw As Double
        For i = 0 To count - 1
            LEPd = Val(DataGridView1.Rows(i).Cells("LEPdCol").Value) / 10
            sum += Math.Pow(10, LEPd)
        Next
        sum /= 5
        LEPw = 10 * Math.Log10(sum)
        TextBox1.Text = FormatNumber(LEPw, NumericUpDown1.Value)
    End Sub
End Class
```

---

Sound levels drop significantly with increasing distance from the noise source (Kamboj 2002). The relationship between the sound level and the distance from a line source can be written as indicated in the following equation:

$$SPL_B = SPL_A - 10 * \log\left(\frac{D_B}{D_A}\right) \quad (8.8)$$

where:

$SLP_A$ is the sound level at distance $D_A$ from the noise source
$SLP_B$ is the sound level at distance $D_B$ from the noise source

## Example 8.5

1. Write a computer program to find the distance from the sound source at which the sound will fall to a certain level, given the sound level at a particular distance from the sound source.

2. The sound level at a distance of 2 m from a sound source is 89 dB($A$). Find the distance from the sound source at which the sound will fall to 72 dB($A$).

3. Use program set in (1) to verify your computations in (2).

**Solution**

1. For the solution of Example 8.5 (1), see the listing of Program 8.5.

2. Given: $SPL_A = 89$ $D_A = 2$, $SPL_B = 72$, $D_B = ?$

3. Determine the distance from the sound source at which the sound will fall to 86 dB($A$):

$$SPL_B = SPL_A - 10 * \log\left(\frac{D_B}{D_A}\right) = 72 = 89 - 10 * \log\left(\frac{D_B}{2}\right)$$

Thus, $D_B = 100.24$ m.

**LISTING OF PROGRAM 8.5     (CHAP8.5\FORM1.VB): SOUND LEVEL AT A DISTANCE**

```vb
'********************************************************************************
'Example 8.5: Sound level at a distance
'********************************************************************************

Public Class Form1
    Dim result As Double
    Private Sub Form1_Load(ByVal sender As System.Object, ByVal e As System.EventArgs)
Handles MyBase.Load
        Me.Text = "Example 8.5: Sound level"
        Me.FormBorderStyle = Windows.Forms.FormBorderStyle.FixedSingle
        Me.MaximizeBox = False
        Label1.Text = "Equation: SPLB = SPLA - 10*Log(DB/DA)"
        Label2.Text = "Solving for:"
        Label6.Text = "Decimal places:"
        Button1.Text = "&Calculate"
        ComboBox1.Items.Clear()
        ComboBox1.Items.Add("SPLB")
        ComboBox1.Items.Add("SPLA")
        ComboBox1.Items.Add("DB")
        ComboBox1.Items.Add("DA")
        ComboBox1.SelectedIndex = 0
        NumericUpDown1.Value = 2
        NumericUpDown1.Maximum = 10
        NumericUpDown1.Minimum = 0
    End Sub

    Private Sub ComboBox1_SelectedIndexChanged(ByVal sender As System.Object, ByVal e As
System.EventArgs) Handles ComboBox1.SelectedIndexChanged
        Select Case ComboBox1.SelectedIndex
            Case 0
                'Solving for SPLB
                Label3.Text = "Sound level at distance DA, SPLA (dB):"
                Label4.Text = "Distance DB (m):"
                Label5.Text = "Distance DA (m):"
                Label7.Text = "Sound level at distance DB, SPLB (dB):"
            Case 1
                'Solving for SPLA
                Label3.Text = "Sound level at distance DB, SPLB (dB):"
                Label4.Text = "Distance DB (m):"
                Label5.Text = "Distance DA (m):"
                Label7.Text = "Sound level at distance DA, SPLA (dB):"
            Case 2
                'Solving for DB
                Label3.Text = "Sound level at distance DA, SPLA (dB):"
                Label4.Text = "Sound level at distance DB, SPLB (dB):"
                Label5.Text = "Distance DA (m):"
                Label7.Text = "Distance DB (m):"
            Case 3
                'Solving for DA
                Label3.Text = "Sound level at distance DA, SPLA (dB):"
                Label4.Text = "Sound level at distance DB, SPLB (dB):"
                Label5.Text = "Distance DB (m):"
                Label7.Text = "Distance DA (m):"
        End Select
    End Sub

    Private Sub Button1_Click(ByVal sender As System.Object, ByVal e As System.EventArgs)
Handles Button1.Click
```

```
        Dim SPLA, SPLB, DA, DB As Double
        Select Case ComboBox1.SelectedIndex
            Case 0
                'Solving for SPLB
                SPLA = Val(TextBox1.Text)
                DB = Val(TextBox2.Text)
                DA = Val(TextBox3.Text)
                SPLB = SPLA - (10 * Math.Log10(DB / DA))
                result = SPLB
            Case 1
                'Solving for SPLA
                SPLB = Val(TextBox1.Text)
                DB = Val(TextBox2.Text)
                DA = Val(TextBox3.Text)
                SPLA = SPLB + (10 * Math.Log10(DB / DA))
                result = SPLA
            Case 2
                'Solving for DB
                SPLA = Val(TextBox1.Text)
                SPLB = Val(TextBox2.Text)
                DA = Val(TextBox3.Text)
                DB = DA * (10 ^ ((SPLA - SPLB) / 10))
                result = DB
            Case 3
                'Solving for DA
                SPLA = Val(TextBox1.Text)
                SPLB = Val(TextBox2.Text)
                DB = Val(TextBox3.Text)
                DA = DB / (10 ^ ((SPLA - SPLB) / 10))
                result = DA
            Case Else
                MsgBox("Please select an item from the list.",
                        vbOKOnly Or vbInformation)
                Exit Sub
        End Select
        showResult()
    End Sub

    Sub showResult()
        TextBox4.Text = FormatNumber(result, NumericUpDown1.Value)
    End Sub

    Private Sub NumericUpDown1_ValueChanged(ByVal sender As System.Object, ByVal e As
System.EventArgs) Handles NumericUpDown1.ValueChanged
        showResult()
    End Sub
End Class
```

## 8.5 NOISE MEASUREMENT

A number of techniques exist for measuring loudness easily, accurately, and precisely, such as sound level meter. It consists of a microphone, an amplifier, a frequency-weighing circuit (filters), and an output scale. The weighing network filters out specific frequencies to make the response more characteristic of human hearing (see Figure 8.5).

Sound measurement instruments (Crocker and Kessler 1982) include microphones, electronic weighing networks, and sound level meters.

Factors affecting noise control (Vesilind and Morgan 2010) incorporate regulations, economic concerns, and operational and environmental considerations. In controlling noise in the open air, it is very common and important to build an obstacle (e.g., screen, solid fence, buildings, earth berms, terrain that blocks source from observer) between a noise source and observers to reduce noise that is received. This effect is called acoustic shielding (Isaac 1985).

Noise control is possible at three levels (Saenz and Stephens 1986, Singal 2005, Tan and Chan 2006): reducing the sound

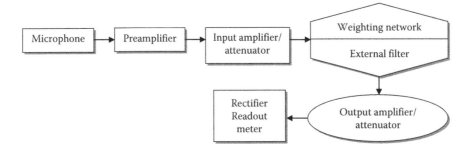

**FIGURE 8.5**   Schematic diagram of a sound level meter.

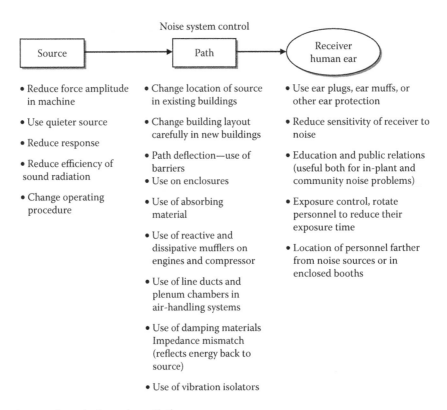

**FIGURE 8.6**   Important means for reducing noise pollution.

produced, interrupting the path of the sound, or protecting the recipient (see Figure 8.6).

For industrial noise control, the following are used: replacement of noise-producing machinery or equipment with quiet alternatives, interruption of path of noise (covering a noisy motor with insulating material), or protecting recipient by distributing earmuffs to the employees.

Community noise control may be done through control by local ordinances; muffling of air compressors, jack hammers, and hand compactors; or diverting noise from populated areas.

Noise control and reduction techniques are listed in terms of source-path receiver system. In source approaches use is made of a quieter source, choosing the quietest machine available in new installations or substituting quieter machine (or part) for machines in existing installations. Reducing force amplitude in a machine addresses reducing imparts or impulses by applying

forces more slowly or using softer materials for contacting surfaces, balancing moving parts, reducing friction by lubrication, or alignment of bearings, or using dynamic absorbers that have compensating force certain degrees out of phase with driving force. In reducing response, one may increase internal energy dissipation (damping) of machine structure, after natural frequencies of machine structure, reduce efficiency of sound radiation, or change operating procedure.

Sound insulation is used to control the transmission path of the sound and can be achieved by inserting a wall or panel between the path of the source and the listeners. It is a technique used for controlling air-borne sound and is particularly applied to problems of sound propagation from one room to another in a building. For effective insulation, heavy walls of brick or concrete are required. Sound absorption is used to reduce the level of reflected sound in a room in which both the noise source and the listener, whose acoustic conditions are intended to improve, are situated, for example, lead. Silencers are used to reduce the transmitted sound through a duct to an observer (e.g., use of silencers in room ventilation) (Thumann 1990).

Silencers are applied for the control of noise due to mechanical equipment, engines (internal combustion, diesel, etc.), and high-velocity air flows (Faulkner 1976).

## 8.6 HOMEWORK PROBLEMS IN COMPUTER MODELING APPLICATIONS FOR NOISE POLLUTION AND ABATEMENT

### 8.6.1 Discussion Problems

1. Define all of the following: sound, noise, acoustic wave, frequency, and ultrasonic wave amplitude.
2. What is the use of sound in human life?
3. What causes noise in your area?
4. How sound travels from one medium to another?
5. Write briefly about each of the following:
   a. Life in a sound isolated area
   b. Risk of noise pollution on humans
   c. Use of decibel unit for measuring sound level
   d. Noise pollution control
6. What are the main factors that affect the ability of hearing and its continuity?
7. How can you use the equation of finding the average daily personal exposure of sound pollution?
8. Which of the following represents a danger to hearing? Why is that?
   a. Jet aircraft
   b. Car horn
   c. Dirt suction home cleaner
9. How do you measure the intensity of noise?
10. What are the objectives of public control of noise pollution?

### 8.6.2 Special Mathematical Problems

1. a. Write a computer program to find the resulting sound level from combining the different decibels.
   b. Find the resulting sound level from combining the following decibels: 91, 66, 78, 95, 109, and 83.
2. a. Write a computer program to determine sound transmission speed from a machine, given the length of a sound wave and its frequency.
   b. Determine sound transmission speed from a machine assuming the length of sound wave of 96 m and frequency of 71 Hz.
3. a. Write a computer program to compute the sound pressure level of a sound wave of a certain pressure.
   b. Compute the sound pressure level of a sound wave of pressure of 5 Pa.
4. a. Write a computer program to find the value of the sound pressure level of a number of machines ($n$)

when working together in a mechanical workshop that includes the given machines, given the sound produced from each one.
   b. A mechanical workshop includes five machines, each of which produces a sound estimated at 113, 72, 52, 90, and 105 dB($A$), respectively. Find the value of the sound pressure level of the five machines when working together.
5. a. Write a computer program to determine the sound level at a certain distance ($x$) from a noise source, given the sound level at another fixed distance ($y$) from the sound source.
   b. Sound level at a distance of 3 m is estimated to be 91 dB. Determine the sound level at a distance of 15 m from noise source.

## REFERENCES

Abdel-Magid, I. M. 1995. *Environmental Engineering.* Amman, Jordan: Dar Al-Mustaqbal Publishers and Distributors (Number at the National Library 464/5/1995) (A reviewed book) (doi:10.13140/RG.2.1.1688.4244).

Abdel-Magid, I. M. 2002. *Pollution: Hazards and Control.* AlGabbadah, Al-Aslia, P.O. Box 1120, Tunis: Arab League Education Culture Science Organization ALECSO. Won ALECSO prize for a book in engineering, 2002, ISBN 9973-15-100-3. (doi: 10.13140/RG.2.1.1364.2722)

Abdel-Magid, I. M. Sr., and Abdel-Magid, M. I. M. Jr. 2015. *Noise Pollution.* CreateSpace Independent Publishing Platform.

Anderson, J. S., and Bratos-Anderson, M. 1993. *Noise and Its Measurement, Analysis, Rating and Control.* Hants: Avebury Technical.

Barrett, K., Brooks, H., Boitano, S., Barman, S. 2010. *Ganong's Review of Medical Physiology,* 23rd ed. New York: McGraw Hill.

Berglund, B., and Lindvall, T. 1995. Community noise. *Archives of the Center for Sensory Research.* 2:1–195. http://www.who.int/docstore/peh/noise/guidelines2.html (accessed November 9, 2015).

Bhatia, S. C. 2007. *Textbook of Noise Pollution and Its Control.* New Delhi, India: Atlantic Publishers & Distributors.

Bies, D. A., and Hansen, C. H. 2013. *Engineering Noise Control: Theory and Practice,* 4th ed. New York: Spon Press.

Crocker, M. J., and Kessler, F. M. 1982. *Noise and Noise Control,* Vol II. Boca Raton, FL: CRC Press.

Faulkner, L. L. (Ed.). 1976. *Handbook of Industrial Noise Control.* New York: Industrial Press, Inc.

Hansen, C. H. 2005. *Noise Control: From Concept to Application.* London: Taylor & Francis.

Harris, D. A. 2013. *Noise Control Manual: Guidelines for Problem-Solving in the Industrial/Commercial Acoustical Environment.* Boston, MA: Springer.

Isaac, A. (Ed.). 1985. *Concise Dictionary of Physics.* Oxford: Oxford Science Publications/Oxford University Press.

Kamboj, N. S. 2002. *Control of Noise Pollution,* 2nd ed. New Delhi, India: Deep & Deep Publications.

Munjal, M. L. 2013. *Noise and Vibration Control* (IIsc Lecture Notes Series - Vol 3), 1st ed. Hackensack, NJ: World Scientific Publishing Company.

Nathanson, J. A., and Schneider, R. A. 2014. *Basic Environmental Technology: Water Supply, Waste Management and Pollution Control,* 6th ed. Upper Saddle River, NJ: Prentice Hall.

Saenz, A. L., and Stephens, R. W. B. (Ed.) 1986. *Noise Pollution: Effects and Control*. Chichester: Published on Behalf of the Scope of the ICSU, by John Wiley & Sons.

Singal, S. P. 2005. *Noise Pollution and Control*, 1st ed. Oxford: Alpha Science International, Ltd.

Sound Research Laboratories. 1991. *Noise Control in Industry*, 3rd ed. London: E.&F.N.Spon, An imprint of Chapman & Hall.

Tan, K. T., and Chan, M.O.Y. (2006) *Industrial Noise Control—The Singapore Experience*. Singapore: Occupational Health Department, Occupational Safety and Health Division, Ministry of Manpower.

Thumann, A. 1990. *Fundamentals of Noise Control Engineering*, 2nd ed. Englewood Cliffs, NJ: Prentice-Hall.

Vesilind, P. A., and Morgan, S. M. 2010. *Introduction to Environmental Engineering—SI Version*, 3rd ed. Stamford, CT: CL Engineering.

# Appendices

## APPENDIX A1

### PHYSICAL PROPERTIES OF WATER

| Temperature (°C) | Temperature (°F) | Density at 1 atm (g/cm³) | Dynamic Viscosity (cP) ($10^{-2}$ dyn-s/cm²) | Kinematic Viscosity (cS) ($10^{-2}$ cm²/s) | Surface Tension against Air (dyn/cm) | Vapor Pressure (mmHg) |
|---|---|---|---|---|---|---|
| 5 | 41.0 | 0.999965 | 1.5188 | 1.5189 | 74.92 | 6.543 |
| 6 | 42.8 | 0.999941 | 1.4726 | 1.4727 | 74.78 | 7.013 |
| 7 | 44.6 | 0.999902 | 1.4288 | 1.4289 | 74.64 | 7.513 |
| 8 | 46.4 | 0.999849 | 1.3872 | 1.3874 | 74.50 | 8.045 |
| 9 | 48.2 | 0.999781 | 1.3476 | 1.3479 | 74.36 | 8.609 |
| 10 | 50.0 | 0.999700 | 1.3097 | 1.3101 | 74.22 | 9.209 |
| 11 | 51.8 | 0.999605 | 1.2735 | 1.2740 | 74.07 | 9.844 |
| 12 | 53.6 | 0.999498 | 1.2390 | 1.2396 | 73.93 | 10.518 |
| 13 | 55.4 | 0.999377 | 1.2061 | 1.2069 | 73.78 | 11.231 |
| 14 | 57.2 | 0.999244 | 1.1748 | 1.1757 | 73.64 | 11.987 |
| 15 | 59.0 | 0.999099 | 1.1447 | 1.1457 | 73.49 | 12.788 |
| 16 | 60.8 | 0.998943 | 1.1156 | 1.1168 | 73.34 | 13.634 |
| 17 | 62.6 | 0.998774 | 1.0875 | 1.0889 | 73.19 | 14.530 |
| 18 | 64.4 | 0.998595 | 1.0603 | 1.0618 | 73.05 | 15.477 |
| 19 | 66.2 | 0.998405 | 1.0340 | 1.0357 | 72.90 | 16.477 |
| 20 | 68.0 | 0.998203 | 1.0087 | 1.0105 | 72.75 | 17.535 |
| 21 | 69.8 | 0.997992 | 0.9843 | 0.9863 | 72.59 | 18.650 |
| 22 | 71.6 | 0.997770 | 0.9608 | 0.9629 | 72.44 | 19.827 |
| 23 | 73.4 | 0.997538 | 0.9380 | 0.9403 | 72.28 | 21.068 |
| 24 | 75.2 | 0.997296 | 0.9161 | 0.9186 | 72.13 | 22.377 |
| 25 | 77.0 | 0.997044 | 0.8949 | 0.8976 | 71.97 | 23.756 |
| 26 | 78.8 | 0.996783 | 0.8746 | 0.8774 | 71.82 | 25.209 |
| 27 | 80.6 | 0.996512 | 0.8551 | 0.8581 | 71.66 | 26.739 |
| 28 | 82.4 | 0.996232 | 0.8363 | 0.8395 | 71.50 | 28.349 |
| 29 | 84.2 | 0.995944 | 0.8181 | 0.8214 | 71.35 | 30.043 |
| 30 | 86.0 | 0.995646 | 0.8004 | 0.8039 | 71.18 | 31.824 |
| 31 | 87.8 | 0.995340 | 0.7834 | 0.7871 | 71.02 | 33.695 |
| 32 | 89.6 | 0.995025 | 0.7670 | 0.7708 | 70.86 | 35.663 |
| 33 | 91.4 | 0.994702 | 0.7511 | 0.7551 | 70.70 | 37.729 |
| 34 | 93.2 | 0.994371 | 0.7357 | 0.7399 | 70.53 | 39.898 |
| 35 | 95.0 | 0.99403 | 0.7208 | 0.7251 | 70.38 | 42.175 |
| 36 | 96.8 | 0.99368 | 0.7064 | 0.7109 | 70.21 | 44.563 |
| 37 | 98.6 | 0.99333 | 0.6925 | 0.6971 | 70.05 | 47.067 |
| 38 | 100.4 | 0.99296 | 0.6791 | 0.6839 | 69.88 | 49.692 |

*Source:* Van der Leeden, F., *The Water Encyclopedia*, 2nd ed, Celsea, MI, CRC Press, 1990. With permission.

## APPENDIX A2

### SATURATION VALUES OF DISSOLVED OXYGEN IN WATER EXPOSED TO WATER-SATURATED AIR CONTAINING 20.9% OXYGEN UNDER A PRESSURE OF 760 MMHG[a]

| Temperature (°C) | Dissolved Oxygen (mg/L) | | | Difference per 100 mg Chloride | Temperature (°C) | Vapor Pressure (mm) |
|---|---|---|---|---|---|---|
| | Chloride Concentration in Water (mg/L) | | | | | |
| | 0 | 5,000 | 10,000 | | | |
| 0 | 14.6 | 13.8 | 13.0 | 0.017 | 0 | 5 |
| 1 | 14.2 | 13.4 | 12.6 | 0.016 | 1 | 5 |
| 2 | 13.8 | 13.1 | 12.3 | 0.015 | 2 | 5 |
| 3 | 13.5 | 12.7 | 12.0 | 0.015 | 3 | 6 |
| 4 | 13.1 | 12.4 | 11.7 | 0.014 | 4 | 6 |
| 5 | 12.8 | 12.1 | 11.4 | 0.014 | 5 | 7 |
| 6 | 12.5 | 11.8 | 11.1 | 0.014 | 6 | 7 |
| 7 | 12.2 | 11.5 | 10.9 | 0.013 | 7 | 8 |
| 8 | 11.9 | 11.2 | 10.6 | 0.013 | 8 | 8 |
| 9 | 11.6 | 11.0 | 10.4 | 0.012 | 9 | 9 |
| 10 | 11.3 | 10.7 | 10.1 | 0.012 | 10 | 9 |
| 11 | 11.1 | 10.5 | 9.9 | 0.011 | 11 | 10 |
| 12 | 10.8 | 10.3 | 9.7 | 0.011 | 12 | 11 |
| 13 | 10.6 | 10.1 | 9.5 | 0.011 | 13 | 11 |
| 14 | 10.4 | 9.9 | 9.3 | 0.010 | 14 | 12 |
| 15 | 10.2 | 9.7 | 9.1 | 0.010 | 15 | 13 |
| 16 | 10.0 | 9.5 | 9.0 | 0.010 | 16 | 14 |
| 17 | 9.7 | 9.3 | 8.8 | 0.010 | 17 | 15 |
| 18 | 9.5 | 9.1 | 8.6 | 0.009 | 18 | 16 |
| 19 | 9.4 | 8.9 | 8.5 | 0.009 | 19 | 17 |
| 20 | 9.2 | 8.7 | 8.3 | 0.009 | 20 | 18 |
| 21 | 9.0 | 8.6 | 8.1 | 0.009 | 21 | 19 |
| 22 | 8.8 | 8.4 | 8.0 | 0.008 | 22 | 20 |
| 23 | 8.7 | 8.3 | 7.9 | 0.008 | 23 | 21 |
| 24 | 8.5 | 8.1 | 7.7 | 0.008 | 24 | 22 |
| 25 | 8.4 | 8.0 | 7.6 | 0.008 | 25 | 24 |
| 26 | 8.2 | 7.8 | 7.4 | 0.008 | 26 | 25 |
| 27 | 8.1 | 7.7 | 7.3 | 0.008 | 27 | 27 |
| 28 | 7.9 | 7.5 | 7.1 | 0.008 | 28 | 28 |
| 29 | 7.8 | 7.4 | 7.0 | 0.008 | 29 | 30 |
| 30 | 7.6 | 7.3 | 6.9 | 0.008 | 30 | 32 |

*Source:* Hammer, M. J., *Water and Wastewater Technology*, 2nd ed, Englewood Cliffs, NJ, Prentice Hall, 1986. With permission.

[a] Saturation at barometric pressures other than 760 mm (29.92 in.), $C_s'$, is related to the corresponding tabulated values, $C$, by the equation:

$$C_s' = C_s \frac{P - p}{760 - p}$$

where:

$C_s'$ is the solubility at barometric pressure $P$ and given temperature, mg/L

$C_s$ is the saturation at given temperature from table, mg/L

$P$ is the barometric pressure, mm

$p$ is the pressure of saturated water vapor at temperature of the water selected from table, mm

# APPENDIX A3

## THE PERIODIC TABLE OF ELEMENTS

**Group headings (three numbering schemes):** New notation / Previous IUPAC form / CAS version

**Key to chart:**
- Atomic number → 50
- Symbol → Sn
- 1993 Atomic weight → 118.71
- Oxidation states → +2 +4
- Electron configuration → 18-18-4

Columns below are given as: Group (New notation, Previous IUPAC, CAS) — Atomic number, Symbol, Atomic weight, Oxidation states, Electron configuration (shell).

| Group (New / IUPAC / CAS) | At. no. | Symbol | Atomic weight | Oxidation states | Electron config | Shell |
|---|---|---|---|---|---|---|
| 1 / IA / IA | 1 | H | 1.00794 | +1, −1 | 1 | K |
| | 3 | Li | 6.941 | +1 | 2-1 | K-L |
| | 11 | Na | 22.989768 | +1 | 2-8-1 | K-L-M |
| | 19 | K | 39.0983 | +1 | -8-8-1 | -L-M-N |
| | 37 | Rb | 85.4678 | +1 | -18-8-1 | -M-N-O |
| | 55 | Cs | 132.90543 | +1 | -18-8-1 | -N-O-P |
| | 87 | Fr | (223) | +1 | -18-8-1 | O-P-Q |
| 2 / IIA / IIA | 4 | Be | 9.012182 | +2 | 2-2 | |
| | 12 | Mg | 24.3050 | +2 | 2-8-2 | |
| | 20 | Ca | 40.078 | +2 | -8-8-2 | |
| | 38 | Sr | 87.62 | +2 | -18-8-2 | |
| | 56 | Ba | 137.327 | +2 | -18-8-2 | |
| | 88 | Ra | 226.025 | +2 | -18-8-2 | |
| 3 / IIIA / IIIB | 21 | Sc | 44.955910 | +3 | -8-9-2 | |
| | 39 | Y | 88.90585 | +3 | -18-9-2 | |
| | 57* | La | 138.9055 | +3 | -18-9-2 | |
| | 89** | Ac | 227.028 | +3 | -18-9-2 | |
| 4 / IVA / IVB | 22 | Ti | 47.867 | +2 +3 +4 | -8-10-2 | |
| | 40 | Zr | 91.224 | +4 | -18-10-2 | |
| | 72 | Hf | 178.49 | +4 | -32-10-2 | |
| | 104 | Unq | (261) | +4 | -32-10-2 | |
| 5 / VA / VB | 23 | V | 50.9415 | +2 +3 +4 +5 | -8-11-2 | |
| | 41 | Nb | 92.90638 | +3 +5 | -18-12-1 | |
| | 73 | Ta | 180.9479 | +5 | -32-11-2 | |
| | 105 | Unp | (262) | | -32-11-2 | |
| 6 / VIA / VIB | 24 | Cr | 51.9961 | +2 +3 +6 | -8-13-1 | |
| | 42 | Mo | 95.94 | +6 | -18-13-1 | |
| | 74 | W | 183.84 | +6 | -32-12-2 | |
| | 106 | Unh | (263) | | -32-12-2 | |
| 7 / VIIA / VIIB | 25 | Mn | 54.93805 | +2 +3 +4 +7 | -8-13-2 | |
| | 43 | Tc | (98) | +4 +6 +7 | -18-13-2 | |
| | 75 | Re | 186.207 | +4 +6 +7 | -32-13-2 | |
| | 107 | Uns | (262) | | -32-13-2 | |
| 8 / VIIIA / VIII | 26 | Fe | 55.845 | +2 +3 | -8-13-2 | |
| | 44 | Ru | 101.07 | +3 | -18-15-1 | |
| | 76 | Os | 190.23 | +3 +4 | -32-14-2 | |
| | 108 | Uno | (265) | | -32-14-2 | |
| 9 / VIIIA / VIII | 27 | Co | 58.93320 | +2 +3 | -8-15-2 | |
| | 45 | Rh | 102.90550 | +3 | -18-16-1 | |
| | 77 | Ir | 192.217 | +3 +4 | -32-15-2 | |
| | 109 | Une | (266) | | -32-15-2 | |
| 10 / VIIIA / VIII | 28 | Ni | 58.6934 | +2 | -8-16-2 | |
| | 46 | Pd | 106.42 | +2 +3 | -18-18-0 | |
| | 78 | Pt | 195.08 | +2 +4 | -32-17-1 | |
| | 110 | Uun | (269) | | -32-18-0 | |
| 11 / IB / IB | 29 | Cu | 63.546 | +1 +2 | -8-18-1 | |
| | 47 | Ag | 107.8682 | +1 | -18-18-1 | |
| | 79 | Au | 196.96654 | +1 +3 | -32-18-1 | |
| 12 / IIB / IIB | 30 | Zn | 65.39 | +2 | -8-18-2 | |
| | 48 | Cd | 112.411 | +2 | -18-18-2 | |
| | 80 | Hg | 200.59 | +1 +2 | -32-18-2 | |
| 13 / IIIB / IIIA | 5 | B | 10.811 | +3 | 2-3 | |
| | 13 | Al | 26.981539 | +3 | 2-8-3 | |
| | 31 | Ga | 69.723 | +3 | -8-18-3 | |
| | 49 | In | 114.818 | +3 | -18-18-3 | |
| | 81 | Tl | 204.3833 | +1 +3 | -32-18-3 | |
| 14 / IVB / IVA | 6 | C | 12.011 | +2 +4 −4 | 2-4 | |
| | 14 | Si | 28.0855 | +2 +4 −4 | 2-8-4 | |
| | 32 | Ge | 72.61 | +4 | -8-18-4 | |
| | 50 | Sn | 118.710 | +2 +4 | -18-18-4 | |
| | 82 | Pb | 207.2 | +2 +4 | -32-18-4 | |
| 15 / VB / VA | 7 | N | 14.00674 | +1 +2 +3 +4 +5 −1 −2 −3 | 2-5 | |
| | 15 | P | 30.973762 | +3 +5 −3 | 2-8-5 | |
| | 33 | As | 74.92159 | +3 +5 −3 | -8-18-5 | |
| | 51 | Sb | 121.760 | +3 +5 −3 | -18-18-5 | |
| | 83 | Bi | 208.98037 | +3 +5 | -32-18-5 | |
| 16 / VIB / VIA | 8 | O | 15.9994 | −2 | 2-6 | |
| | 16 | S | 15.9994 | +4 +6 −2 | 2-8-6 | |
| | 34 | Se | 78.96 | +4 +6 −2 | -8-18-6 | |
| | 52 | Te | 121.760 | +4 +6 −2 | -18-18-6 | |
| | 84 | Po | (209) | +2 +4 | -32-18-6 | |
| 17 / VIIB / VIIA | 9 | F | 18.9984032 | −1 | 2-7 | |
| | 17 | Cl | 35.4527 | +1 +5 +6 +7 −1 −2 | 2-8-7 | |
| | 35 | Br | 79.904 | +1 +5 −1 | -8-18-7 | |
| | 53 | I | 126.90447 | +1 +5 +7 −1 | -18-18-7 | |
| | 85 | At | (209) | | -32-18-6 | |
| 18 / VIIA / VIIIA | 2 | He | 4.0020602 | 0 | 2 | |
| | 10 | Ne | 20.1797 | 0 | 2-8 | |
| | 18 | Ar | 39.948 | 0 | -2-8-8 | |
| | 36 | Kr | 83.80 | 0 | -8-18-8 | |
| | 54 | Xe | 131.29 | 0 | -18-18-8 | |
| | 86 | Rn | (222) | | -32-18-8 | |

**\* Lanthanides**

| At. no. | Symbol | Atomic weight | Oxidation states | Electron config |
|---|---|---|---|---|
| 58 | Ce | 140.115 | +3 +4 | -19-9-2 |
| 59 | Pr | 140.90765 | +3 | -21-8-2 |
| 60 | Nd | 144.24 | +3 | -22-8-2 |
| 61 | Pm | (145) | +3 | -23-8-2 |
| 62 | Sm | 150.36 | +2 +3 | -24-8-2 |
| 63 | Eu | 151.965 | +2 +3 | -25-8-2 |
| 64 | Gd | 157.25 | +3 | -25-9-2 |
| 65 | Tb | 158.92534 | +3 | -27-8-2 |
| 66 | Dy | 162.50 | +3 | -28-8-2 |
| 67 | Ho | 164.93032 | +3 | -29-8-2 |
| 68 | Er | 167.26 | +3 | -30-8-2 |
| 69 | Tm | 168.93421 | +3 | -31-8-2 |
| 70 | Yb | 173.04 | +3 | -32-8-2 |
| 71 | Lu | 174.967 | +3 | -32-9-2 |

Shell (lanthanides): N-O-P

**\*\* Actinides**

| At. no. | Symbol | Atomic weight | Oxidation states | Electron config |
|---|---|---|---|---|
| 90 | Th | 232.0381 | +4 | -18-10-2 |
| 91 | Pa | 231.03588 | +4 +5 | -20-9-2 |
| 92 | U | 238.0289 | +3 +4 +5 +6 | -21-9-2 |
| 93 | Np | 237.048 | +3 +4 +5 +6 | -22-9-2 |
| 94 | Pu | (244) | +3 +4 +5 +6 | -24-8-2 |
| 95 | Am | (243) | +3 +4 +5 +6 | -25-8-2 |
| 96 | Cm | (247) | +3 | -25-9-2 |
| 97 | Bk | (247) | +3 +4 | -27-8-2 |
| 98 | Cf | (251) | +3 | -28-8-2 |
| 99 | Es | (252) | | -29-8-2 |
| 100 | Fm | (257) | | -30-8-2 |
| 101 | Md | (258) | +2 +3 | -31-8-2 |
| 102 | No | (259) | +2 +3 | -32-8-2 |
| 103 | Lr | (206) | +3 | -32-9-2 |

Shell (actinides): O-P-Q

The new IUPAC format numbers the groups from 1 to 18. The previous IUPAC numbering system and the system used by Chemical Abstracts Service (CAS) are also shown. For radioactive elements that do not occur in nature, the mass number of the most stable isotope is given in parentheses.

*Source:* Leigh, G. L. (Ed.), *Nomenclature of Inorganic Chemistry*, Oxford, Blackwell Scientific Publications, 1990; American Chemical Society, Group notation revised in periodic table. *Chemical and Engineering News*, 63(5), 26–27, 1985; International Union of Pure and Applied Chemistry, Atomic weights of the elements, *Pure & Appl. Chem.*, 66, 2423–2444, 1994; Manahan, S. E., *Fundamentals of Environmental Chemistry*, Celsea, MI, Lewis Publishers, 1993. With permission.

*Note:* The larger and smaller labels reflect two different numbering schemes in common usage.

## APPENDIX A4

### HYDRAULIC ELEMENTS GRAPH FOR CIRCULAR SEWERS

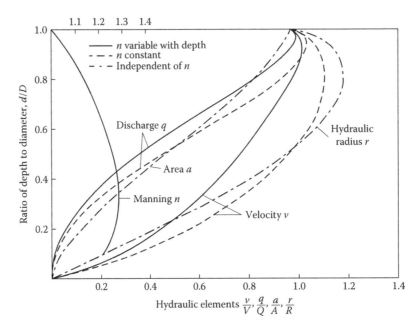

*Source:*   Qasim, S. R., *Waste Water Treatment Plants: Planning, Design and Operation*, Boca Raton, FL, CRC Press, p. 163, Figure 7-6, 1999. With permission.

# APPENDIX A5

## ALIGNMENT CHART FOR FLOW IN PIPES

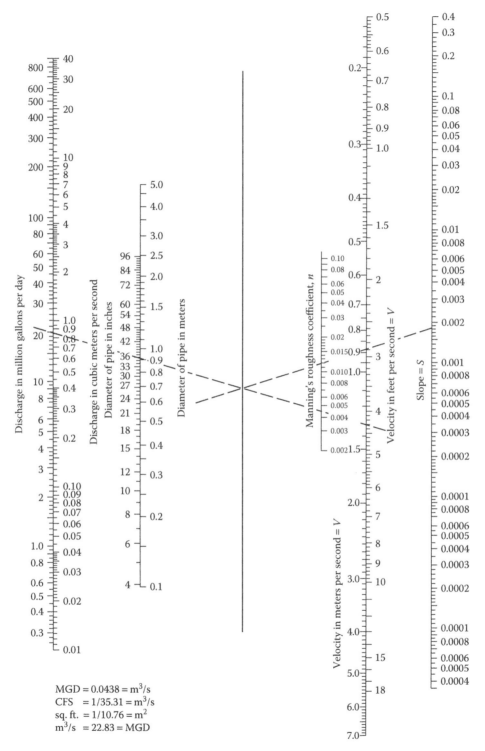

MGD = 0.0438 = m³/s
CFS   = 1/35.31 = m³/s
sq. ft. = 1/10.76 = m²
m³/s   = 22.83 = MGD

*Source:* Joint Taskforce of the American Society of Civil Engineers and the Water Pollution Control Federation. 1982. *Gravity Sanitary Sewer Design and Construction*, ASCE Manuals and Reports on Engineering Practice, No. 60. New York: ASCE, WPCF. Reproduced by permission of the publisher American Society of Civil Engineers.

## APPENDIX A6

### MOODY'S DIAGRAM FOR REYNOLDS NUMBER VERSUS RELATIVE ROUGHNESS AND FRICTION FACTOR

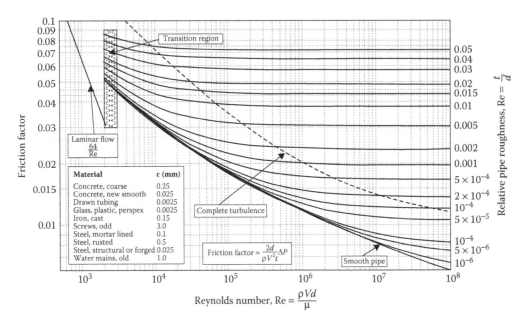

*Source:* Rayaprolu, K., *Boilers: A Practical Reference*, Boca Raton, FL: CRC Press, p. 163, 2012. With permission.

## APPENDIX A7

### TYPICAL SPECIFIC WEIGHT AND MOISTURE CONTENT DATA FOR RESIDENTIAL, COMMERCIAL, INDUSTRIAL, AND AGRICULTURAL WASTES

| Type of Waste | Specific Weight (lb/yd³) | | Moisture Content, Weight (%) | |
|---|---|---|---|---|
| | Range | Typical | Range | Typical |
| **Residential (Compacted)** | | | | |
| Food wastes (mixed) | 220–810 | 490 | 50–80 | 70 |
| Paper | 70–220 | 150 | 4–10 | 6 |
| Cardboard | 70–135 | 85 | 4–8 | 5 |
| Plastics | 70–220 | 110 | 1–4 | 2 |
| Textiles | 70–170 | 110 | 6–15 | 10 |
| Rubber | 170–340 | 220 | 1–4 | 2 |
| Leather | 170–440 | 270 | 8–12 | 10 |
| Yard wastes | 100–380 | 170 | 30–80 | 60 |
| Wood | 220–540 | 400 | 15–40 | 20 |
| Glass | 270–810 | 330 | 1–4 | 2 |
| Tin cans | 85–270 | 150 | 2–4 | 3 |
| Aluminum | 110–405 | 270 | 2–4 | 2 |
| Other metal | 220–1940 | 540 | 2–4 | 3 |
| Dirt, ash, and so on | 540–1685 | 810 | 6–12 | 8 |
| Ashes | 1095–1400 | 1255 | 6–12 | 6 |
| Rubbish | 150–305 | 220 | 5–20 | 15 |
| **Residential Yard Wastes** | | | | |
| Leaves (loose and dry) | 50–250 | 100 | 20–40 | 30 |
| Green grass (loose and moist) | 350–500 | 400 | 40–80 | 60 |
| Green grass (wet and compacted) | 1000–1400 | 1000 | 50–90 | 80 |
| Yard waste (shredded) | 450–600 | 500 | 20–70 | 50 |
| Yard waste (composted) | 450–650 | 550 | 40–60 | 50 |

*(Continued)*

## APPENDIX A7 (*Continued*)

### TYPICAL SPECIFIC WEIGHT AND MOISTURE CONTENT DATA FOR RESIDENTIAL, COMMERCIAL, INDUSTRIAL, AND AGRICULTURAL WASTES

| Type of Waste | Specific Weight (lb/yd³) | | Moisture Content, Weight (%) | |
|---|---|---|---|---|
| | Range | Typical | Range | Typical |
| **Municipal** | | | | |
| In compactor truck | 300–760 | 500 | 15–40 | 20 |
| In landfill | | | | |
|   Normally compacted | 610–840 | 760 | 15–40 | 25 |
|   Well compacted | 995–1250 | 1010 | 15–40 | 25 |
| **Commercial** | | | | |
| Food wastes (wet) | 800–1600 | 910 | 50–80 | 70 |
| Appliances | 250–340 | 305 | 0–2 | 1 |
| Wooden crates | 185–270 | 185 | 10–30 | 20 |
| Tree trimmings | 170–305 | 250 | 20–80 | 5 |
| Rubbish (combustible) | 85–305 | 200 | 10–30 | 15 |
| Rubbish (noncombustible) | 305–610 | 505 | 5–15 | 10 |
| Rubbish (mixed) | 235–305 | 270 | 10–25 | 15 |
| **Construction and Demolition** | | | | |
| Mixed demolition (noncombustible) | 1685–2695 | 2395 | 2–10 | 4 |
| Mixed demolition (combustible) | 505–675 | 605 | 4–15 | 8 |
| Mixed construction (combustible) | 305–605 | 440 | 4–15 | 8 |
| Broken concrete | 2020–3035 | 2595 | 0–5 | – |
| **Industrial** | | | | |
| Chemical sludges (wet) | 1350–1855 | 1685 | 75–99 | 80 |
| Fly ash | 1180–1515 | 1350 | 2–10 | 4 |
| Leather scraps | 170–420 | 270 | 6–15 | 10 |
| Metal scrap (heavy) | 2530–3370 | 3000 | 0–5 | – |
| Metal scrap (light) | 840–1515 | 1245 | 0–5 | – |
| Metal scrap (mixed) | 1180–2530 | 1515 | 0–5 | – |
| Oils, tars, and asphalts | 1350–1685 | 1600 | 0–5 | 2 |
| Sawdust | 170–590 | 490 | 10–40 | 20 |
| Textile wastes | 170–370 | 305 | 6–15 | 10 |
| Wood (mixed) | 675–1140 | 840 | 30–60 | 25 |
| **Agricultural** | | | | |
| Agricultural (mixed) | 675–1265 | 945 | 40–80 | 50 |
| Dead animals | 340–840 | 605 | – | – |
| Fruit wastes (mixed) | 420–1265 | 605 | 60–90 | 75 |
| Manure (wet) | 1515–1770 | 1685 | 75–96 | 94 |
| Vegetable wastes (mixed) | 340–1180 | 605 | 60–90 | 75 |

*Source:* Abdel-Magid, I. M. and Abdel-Magid, M. I. M., *Solid Waste Engineering and Management*, CreateSpace Independent Publishing Platform, North Charleston, SC: (In Arabic), 2015; Blackman, W. C., *Basic Hazardous Waste Management*, 3rd ed, CRC Press, 2001; de Bertoldi, M., *Science of Composting*, Springer, 1996; Cheremisinoff, N. P., *Handbook of Solid Waste Management and Waste Minimization Technologies*, Burlington, MA, Elsevier Science 2003; CEHA, Solid waste management in some countries of the Eastern Mediterranean region. CEHA Document No., Special studies, ss-4. Amman, Jordan, WHO, Eastern Mediterranean Regional Office, Regional Centre for Environmental Health Activities, 1995; McDougall, F. R. et al., *Integrated Solid Waste Management: A Life Cycle Inventory*, Oxford, Blackwell Science, 2009; Ojovan, M. I. (Ed.), *Handbook of Advanced Radioactive Waste Conditioning Technologies*, Cambridge, Woodhead Publishing, 2011; Peavy, H. S. et al., *Environmental Engineering*, New York, McGraw-Hill Book Co, 1985; Walsh, P., and O'Leary, P., *Implementing Municipal Solid Waste to Energy Systems*, Madison, WI, University of Wisconsin—Extension for Great Lakes Regional Biogas Energy Program, 1986.

# APPENDIX A8

## CONVERSION TABLE

| Multiply | By | To Obtain |
|---|---|---|
| **Area** | | |
| acre | 0.4047 | ha |
| acre | 43560 | ft$^2$ |
| acre | 4047 | m$^2$ |
| cm$^2$ | 0.155 | in$^2$ |
| ft$^2$ | 0.0929 | m$^2$ |
| ha | 2.471 | acre |
| ha | 10$^4$ | m$^2$ |
| in$^2$ | 6.452 | cm$^2$ |
| km$^2$ | 0.3861 | mile$^2$ |
| m$^2$ | 10.76 | ft$^2$ |
| mm$^2$ | 0.00155 | in$^2$ |
| **Concentration** | | |
| mg/L | 8.345 | lb/M gal (U.S.) |
| ppm | 1 | mg/L |
| **Density** | | |
| g/cm$^3$ | 1000 | kg/m$^3$ |
| g/cm$^3$ | 1 | kg/L |
| g/cm$^3$ | 62.428 | lb/ft$^3$ |
| g/cm$^3$ | 10.022 | lb/gal (Br.) |
| g/cm$^3$ | 8.345 | lb/gal (U.S.) |
| kg/m$^3$ | 0.001 | g/cm$^3$ |
| kg/m$^3$ | 0.001 | kg/L |
| kg/m$^3$ | 0.06242 | lb/ft$^3$ |
| **Flow Rate** | | |
| ft$^3$/s | 448.83 | gal (U.S.)/min |
| ft$^3$/s | 28.32 | L/s |
| ft$^3$/s | 0.02832 | m$^3$/s |
| ft$^3$/s | 0.6462 | M gal (U.S.)/day |
| gal (U.S.)/min | 0.00223 | ft$^3$/s |
| gal (U.S.)/min | 0.06309 | L/s |
| L/s | 15.85 | gal (U.S.)/min |
| M gal (U.S.)/day | 1.547 | ft$^3$/s |
| m$^3$/h | 4.403 | gal (U.S.)/min |
| m$^3$/s | 35.31 | ft$^3$/s |
| **Length** | | |
| ft | 30.48 | cm |
| in | 2.54 | cm |
| km | 0.6214 | mile |
| km | 3280.8399 | ft |
| m | 3.281 | ft |
| m | 39.37 | in |
| m | 1.094 | yd |
| mile | 5280 | ft |
| mile | 1.60934 | km |
| mm | 0.03937 | in |
| yard | 0.914 | m |
| **Mass** | | |
| g | 2.205 * 10$^{-3}$ | lb |
| kg | 2.20462 | lb |
| lb | 0.453592 | kg |
| lb | 16 | Oz |
| ton | 2240 | lb |
| ton | 1.1023 | ton (2000 lb) |
| **Power** | | |
| Btu | 252.16 | Cal |
| Btu | 778.169 | ft-lb |
| Btu | 3.93 * 10$^4$ | Hp-h |
| Btu | 1055.06 | J |
| Btu | 2.9307 * 10$^{-4}$ | kW·h |
| Hp | 0.7457 | kW |
| **Pressure** | | |
| atm | 33.93 | ft H$_2$O |
| atm | 29.921 | in Hg |
| atm | 1.033 * 10$^4$ | kg/m$^2$ |
| atm | 760 | mmHg |
| atm | 10.33 | m water |
| atm | 1.013 * 10$^5$ | N/m$^2$ |
| Bar | 10$^5$ | N/m$^2$ |
| cm water | 98.096 | N/m$^2$ |
| in H$_2$O | 1.8651 | mmHg |
| in Hg | 0.491154 | lb/in$^2$ (psi) |
| in Hg | 25.4 | mmHg |
| in Hg | 3386.4 | N/m$^2$ |
| kPa | 0.145 | lb/in$^2$ (psi) |
| lb/in$^2$ (psi) | 0.070307 | kg/cm$^2$ |
| lb/in$^2$ (psi) | 6895 | N/m$^2$ |
| mmHg | 13.595 | kg/m$^2$ |
| mmHg | 0.019336 | lb/in$^2$ (psi) |
| mmHg | 133.3 | N/m$^2$ |
| mmHg | 1 | Torr |
| Torr | 133.3 | N/m$^2$ |
| **Temperature** | | |
| Celsius (C) | (9C/5) + 32 | Fahrenheit (F) |
| Fahrenheit (F) | 5(F-32)/9 | Celsius (C) |
| Celsius (C) | C + 237.16 | Kelvin (K) |
| Fahrenheit (F) | F + 459.67 | Rankine (R) |
| **Velocity** | | |
| cm/s | 0.03281 | ft/s |
| cm/s | 0.6 | m/min |
| m/s | 196.85 | ft/min |
| m/s | 3.281 | ft/s |
| ft/min | 0.508 | cm/s |
| ft/s | 30.48 | cm/s |
| ft/s | 1.097 | km/h |
| mile/h | 1.609 | km/h |
| **Viscosity** | | |
| P (g/cm·s) | 0.1 | N·s/m$^2$ |
| cP | 0.01 | g/cm·s |
| cP | 0.01 | Dyne·s/cm$^2$ (Pa·s) |
| Stoke | 10$^{-4}$ | m$^2$/s |
| **Volume** | | |
| ft$^3$ | 6.22884 | gal (Br.) |
| ft$^3$ | 7.4805 | gal (U.S.) |
| ft$^3$ | 28.3168 | L |
| ft$^3$ | 0.02832 | m$^3$ |
| gal (Br.) | 0.16054 | ft$^3$ |
| gal (U.S.) | 0.13368 | ft$^3$ |
| gal (U.S.) | 0.83267 | gal (Br.) |
| gal (U.S.) | 3.7854 | L |
| in$^3$ | 16.3871 | cm$^3$ |
| L | 0.03531 | ft$^3$ |
| L | 0.219969 | gal (Br.) |
| L | 0.26417 | gal (U.S.) |
| L | 0.001 | m$^3$ |
| m$^3$ | 35.3147 | ft$^3$ |
| m$^3$ | 1000 | L |

# REFERENCES

Abdel-Magid, I. M. and Abdel-Magid, M. I. M. 2015. *Solid Waste Engineering and Management*, North Charleston, SC: CreateSpace Independent Publishing Platform (in Arabic).

American Chemical Society. Group notation revised in periodic table. *Chemical and Engineering News*, 63(5), 26–27, 1985.

Blackman, W. C. 2001. *Basic Hazardous Waste Management*, 3rd ed. Boca Raton, FL: CRC Press.

CEHA. 1995. Solid waste management in some countries of the Eastern Mediterranean region. CEHA Document No., Special studies, ss-4. Amman, Jordan: WHO, Eastern Mediterranean Regional Office, Regional Centre for Environmental Health Activities.

Cheremisinoff, N. P. 2003. *Handbook of Solid Waste Management and Waste Minimization Technologies*. Burlington, MA: Elsevier Science.

de Bertoldi, M., (Ed.). 1996. *The Science of Composting*. London, UK: Blackie Academic & Professional.

Hammer, M. J. 1986. *Water and Wastewater Technology*, 2nd ed. Englewood Cliffs, NJ: Prentice Hall.

International Union of Pure and Applied Chemistry. 1994. Atomic weights of the elements, *Pure & Appl. Chem.*, 66, 2423–2444.

Joint Taskforce of the American Society of Civil Engineers and the Water Pollution Control Federation. 1982. *Gravity Sanitary Sewer Design and Construction,* ASCE Manuals and Reports on Engineering Practice, No. 60. New York: ASCE, WPCF.

Leigh, G. L. (Ed.) 1990. *Nomenclature of Inorganic Chemistry*, Oxford: Blackwell Scientific Publications.

Manahan, S. E. 1993. *Fundamentals of Environmental Chemistry*. Celsea, MI: Lewis Publishers.

McDougall, F. R., White, P. R., Franke, M., and Hindle, P. 2009. *Integrated Solid Waste Management: A Life Cycle Inventory*. Oxford: Blackwell Science.

Ojovan, M. I. (Ed.) 2011. *Handbook of Advanced Radioactive Waste Conditioning Technologies*. Cambridge: Woodhead Publishing.

Peavy, H. S., Rowe, D. R., and Tchobanglous, G. 1985. *Environmental Engineering*. New York: McGraw-Hill Book Co.

Qasim, S. R. 1999. *Waste Water Treatment Plants: Planning, Design and Operation*. Boca Raton, FL: CRC Press, p. 163.

Rayaprolu, K. 2012. *Boilers: A Practical Reference*. Boca Raton, FL: CRC Press. p. 163.

van der Leeden, F. 1990. *The Water Encyclopedia*, 2nd ed. Celsea, MI: CRC Press.

Walsh, P. and O'Leary, P. 1986. *Implementing Municipal Solid Waste to Energy Systems*. Madison, WI: University of Wisconsin— Extension for Great Lakes Regional Biogas Energy Program.

# Index

Note: Page numbers followed by f and t refer to figures and tables, respectively.

Printed and bound by CPI Group (UK) Ltd, Croydon, CR0 4YY

01/11/2024

01782600-0018